Developmental biology of higher fungi

Developmental biology of higher fungi

SYMPOSIUM OF
THE BRITISH MYCOLOGICAL SOCIETY
HELD AT THE UNIVERSITY OF MANCHESTER
APRIL 1984

EDITED BY
D. MOORE, L. A. CASSELTON,
D. A. WOOD & J. C. FRANKLAND

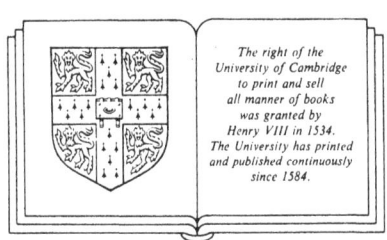

The right of the
University of Cambridge
to print and sell
all manner of books
was granted by
Henry VIII in 1534.
The University has printed
and published continuously
since 1584.

CAMBRIDGE UNIVERSITY PRESS
CAMBRIDGE
LONDON   NEW YORK   NEW ROCHELLE
MELBOURNE   SYDNEY

CAMBRIDGE UNIVERSITY PRESS
Cambridge, New York, Melbourne, Madrid, Cape Town,
Singapore, São Paulo, Delhi, Tokyo, Mexico City

Cambridge University Press
The Edinburgh Building, Cambridge CB2 8RU, UK

Published in the United States of America by Cambridge University Press, New York

www.cambridge.org
Information on this title: www.cambridge.org/9780521106276

First published 1985
First paperback edition 2011

A catalogue record for this publication is available from the British Library

Library of Congress Catalogue Card Number: 84–28486

ISBN 978-0-521-30161-9 Hardback
ISBN 978-0-521-10627-6 Paperback

Contents

List of contributors   vii

Preface   xi

1   Resource relations – an overview   1
    A. D. M. Rayner, R. Watling and Juliet C. Frankland

2   Tropical agarics: resource relations and fruiting periodicity   41
    J. N. Hedger

3   *Armillaria*: resources and hosts   87
    J. Rishbeth

4   The growth of *Crinipellis perniciosa* in living and dead cocoa
    tissue   103
    B. E. J. Wheeler

5   Mycorrhizal dynamics during forest tree development   117
    J. Dighton and P. A. Mason

6   A comparative ultrastructural analysis of the host–fungus
    interface in mycorrhizal and parasitic associations   141
    J. A. Duddridge

7   The challenge of the dolipore/parenthesome septum   175
    Royall T. Moore

8   Dikaryon formation   213
    Lorna A. Casselton and Androulla Economou

9   Cytology of hyphal interactions and reactions in
    Schizophyllum commune   231
    N. K. Todd and R. C. Aylmore

10 Morphogenesis of vegetative organs   249
    A. D. M. Rayner, K. A. Powell, W. Thompson and D. H. Jennings

11 Developmental characters of agarics   281
    Roy Watling

12 Elongation of the stipe of *Coprinus cinereus*   311
    G. W. Gooday

13 Differentiation and pattern formation in the fruit body cap of
    Coprinus cinereus   333
    Isabelle V. Rosin, Jacqueline Horner and David Moore

14 Determination of the initial steps in differentiation in *Coprinus*
    congregatus   353
    Ian K. Ross

15 Production and roles of extracellular enzymes during
    morphogenesis of basidiomycete fungi   375
    D. A. Wood

16 The biochemistry of *Agaricus* fructification   389
    J. B. W. Hammond

17 Growth and development of *Lentinus edodes* on a chemically
    defined medium   403
    Gary F. Leatham

18 Ultrastructural aspects of fruit body differentiation in
    Flammulina velutipes   429
    M. A. J. Williams, A. Beckett and N. D. Read

19 Developmental genetics – from spore to sporophore   451
    T. J. Elliott

20 Nucleic acid studies in *Schizophyllum*   467
    Robert C. Ullrich, Charles P. Novotny and Charles A. Specht

21 Molecular biology of fruit body formation in *Schizophyllum*
    commune   485
    J. G. H. Wessels, J. J. M. Dons and O. M. H. De Vries

22 Meiosis and genetic recombination in *Coprinus cinereus*   499
    Patricia J. Pukkila, David M. Binninger, Jeane R. Cassidy,
    Beverly M. Yashar and Miriam E. Zolan

23 Strategies for mushroom breeding   513
    Carlene A. Raper

24 Biological and technological aspects of commercial mushroom
    growing   529
    P. B. Flegg

25 Composting technology   541
    Lung-Chi Wu

26 Secondary metabolic products of selected agarics   561
    Norman Claydon

27 Developmental biology of agarics – an overview   581
    A. F. M. Reijnders and David Moore

    Species index   597

    Subject index   603

Contributors

R. C. Aylmore, *Department of Biological Sciences, The University, Washington Singer Laboratories, Perry Road, Exeter EX4 4QG, UK.*

A. Beckett, *Department of Botany, University of Bristol, Woodland Road, Bristol BS8 1UG, UK.*

David M. Binninger, *Department of Biology, UNC–Chapel Hill, North Carolina 27514, USA.*

Lorna A. Casselton, *School of Biological Sciences, Queen Mary College, University of London, Mile End Road, London E1 4NS, UK.*

Jeane R. Cassidy, *Department of Biology, UNC–Chapel Hill, North Carolina 27514, USA.*

Norman Claydon, *Glasshouse Crops Research Institute, Worthing Road, Littlehampton, West Sussex BN17 6LP, UK.*

O. M. H. De Vries, *Biologisch Centrum, Rijksuniversiteit Groningen, Vakgroep Plantenfysiologie, Kerklaan 30, 9751 NN Haren, Nederland.*

J. Dighton, *Institute of Terrestrial Ecology, Merlewood Research Station, Grange-over-Sands, Cumbria LA11 6JU, UK.*

J. J. M. Dons, *Biologisch Centrum, Rijksuniversiteit Groningen, Vakgroep Plantenfysiologie, Kerklaan 30, 9751 NN Haren, Nederland.*

J. A. Duddridge, *Department of Agricultural and Forest Sciences, Commonwealth Forestry Institute, South Parks Road, Oxford OX1 3RB, UK.*

Androulla Economou, *School of Biological Sciences, Queen Mary College, University of London, Mile End Road, London E1 4NS, UK.*

T. J. Elliott, *Glasshouse Crops Research Institute, Worthing Road, Littlehampton, West Sussex BN17 6LP, UK.*

P. B. Flegg, *Glasshouse Crops Research Institute, Worthing Road, Littlehampton, West Sussex BN17 6LP, UK.*

Juliet C. Frankland, *Institute of Terrestrial Ecology, Merlewood Research Station, Grange-over-Sands, Cumbria LA11 6JU, UK.*

G. W. Gooday, *Department of Microbiology, Marischal College, University of Aberdeen, Aberdeen AB9 1AS, UK.*

J. B. W. Hammond, *Glasshouse Crops Research Institute, Worthing Road, Littlehampton, West Sussex BN17 6LP, UK.*

J. N. Hedger, *Department of Botany and Microbiology, University College of Wales, Aberystwyth SY23 3DA, UK.*

Jacqueline Horner, *Department of Botany, The University, Manchester M13 9PL, UK.*

D. H. Jennings, *Department of Botany, The University, PO Box 147, Liverpool L69 3BX, UK.*

Gary F. Leatham, *USDA Forest Products Laboratory, PO Box 5130, Madison, Wisconsin 53705, USA.*

P. A. Mason, *Institute of Terrestrial Ecology, Bush Estate, Penicuik, EH26 0QB, UK.*

David Moore, *Department of Botany, The University, Manchester M13 9PL, UK.*

Royall T. Moore, *Biology Department, University of Ulster, Coleraine, Northern Ireland BT52 1SA, UK.*

Charles P. Novotny, *Department of Medical Microbiology, Given Medical Building, University of Vermont, Burlington, Vermont 05405, USA.*

K. A. Powell, *Department of Material Science, University of Bath, Claverton Down, Bath BA2 7AY, UK.*

Patricia J. Pukkila, *Department of Biology, UNC–Chapel Hill, Wilson Hall 046A, Chapel Hill, North Carolina 27514, USA.*

Carlene A. Raper, *Department of Medical Microbiology, University of Vermont, School of Medicine, Burlington, Vermont 05405, USA.*

A. D. M. Rayner, *School of Biological Sciences, University of Bath, Claverton Down, Bath BA2 7AY, UK.*

N. D. Read, *Department of Botany, The University, The King's Buildings, Mayfield Road, Edinburgh EH9 3JH, UK.*

A. F. M. Reijnders, *De Schuilenburght B72, Schuilenburgerplein, 1, 3816TD, Amersfoort, The Netherlands.*

J. Rishbeth, *Botany School, University of Cambridge, Downing Street, Cambridge CB2 3EA, UK.*

Isabelle V. Rosin, *Department of Botany, The University, Manchester M13 9PL, UK.*

Ian K. Ross, *Department of Biological Sciences, University of California at Santa Barbara, Santa Barbara, California 93106, USA.*

Charles A. Specht, *Department of Medical Microbiology, Given Medical Building, University of Vermont, Burlington, Vermont 05405, USA.*

W. Thompson, *Department of Botany, The University, PO Box 147, Liverpool L69 3BX, UK.*

N. K. Todd, *Department of Biological Sciences, Washington Singer Laboratories, The University, Perry Road, Exeter EX4 4QG, UK.*

Robert C. Ullrich, *Department of Botany, Marsh Life Science Building, University of Vermont, Burlington, Vermont 05405, USA.*

Roy Watling, *Royal Botanic Garden, Edinburgh EH3 5LR, UK.*

J. G. H. Wessels, *Biologisch Centrum, Rijksuniversiteit Groningen, Vakgroep Plantenfysiologie, Kerklaan 30, 9751 NN Haren, Nederland.*

B. E. J. Wheeler, *Department of Pure and Applied Biology, Imperial College, Silwood Park, Ascot, Berks. SL5 7PY, UK.*

M. A. J. Williams, *Department of Botany, University of Bristol, Woodland Road, Bristol BS8 1UG, UK.*

D. A. Wood, *Glasshouse Crops Research Institute, Worthing Road, Littlehampton, West Sussex BN17 6LP, UK.*

Lung-Chi Wu, *Campbell Institute for Research and Technology, Napoleon, Ohio 43545, USA.*

Beverly M. Yashar, *Department of Biology, UNC–Chapel Hill, Chapel Hill, North Carolina 27514, USA.*

Miriam E. Zolan, *Department of Biology, UNC–Chapel Hill, Chapel Hill, North Carolina 27514, USA.*

Preface

There are probably as many definitions of developmental biology as there are developmental biologists. In the context of this book we consider that it includes not only structural form – both vegetative and reproductive – but also how that form relates to an organism's ecological niche. We feel that it would be a grave error to deal with development and differentiation as an abstraction, independent of the role that an organism may have to play in the ecological community to which it belongs. Many of the most obvious, and most intellectually challenging, morphogenetic features are a direct contribution to ecological performance, so ignorance of the ecological aspects can limit appreciation of the developmental process under consideration.

Over recent years the British Mycological Society has published a number of Symposium volumes which provide excellent background for this one. We particularly draw attention to: *Fungal Walls and Hyphal Growth* (1979; ed. J. H. Burnett & A. P. J. Trinci); *The Fungal Nucleus* (1981; ed. K. Gull & S. G. Oliver); *Decomposer Basidiomycetes* (1982; ed. J. C. Frankland, J. N. Hedger & M. J. Swift); and *Ecology and Physiology of the Fungal Mycelium* (1984; ed. D. H. Jennings & A. D. M. Rayner). In a very real sense these separate volumes contribute to a common narrative which reaches a climax with the publication of this volume on *Developmental Biology of the Higher Fungi*, uniting form, function and ecology.

The resource relations of agaric and agaricoid fungi with respect to developmental factors are discussed in the first six chapters, including aspects poorly represented in earlier volumes: namely tropical, parasitic and mycorrhizal fungi. Although only a small selection of such articles appears here, the connection between ecological habit and developmental form is clearly demonstrated.

The chapters on 'pure' developmental biology that follow relate cell structure, biochemistry and molecular biology, and extend to consideration of the ways in which such cellular processes are integrated in construction of multicellular fungal structures, including those of commercial importance. The most recent observations and interpretations are stressed throughout and we hope the content of this volume will remain of value for many years to come. Our authors have been imaginative and provocative where necessary, however, so we also hope to generate lively debate.

The idea for this Symposium topic was initiated in early 1981. It culminated in the 2nd General Meeting of the British Mycological Society held at Manchester University in April 1984, and we hope that now it will be disseminated more widely. We have many people to thank for their help with this enterprise, and too little space to name them; our thanks, though, are no less sincere for that. The British Mycological Society records its appreciation to the H. J. Heinz Co. Ltd., the Mushroom Growers Association and the British Council for donations in support of the Symposium.

Finally, we recall with sadness that just a few months before the meeting in Manchester we heard of the death of Dr G. H. Banbury, whose keen interest in basidiomycete physiology was an inspiration to many students of these organisms.

David Moore
Lorna Casselton
David Wood
Juliet Frankland
January 1985

1

Resource relations – an overview

A. D. M. RAYNER, R. WATLING* AND
JULIET C. FRANKLAND†

*School of Biological Sciences, University of Bath, Claverton Down, Bath
BA2 7AY, UK*
* Royal Botanic Garden, Edinburgh EH3 5LR, UK
*† Institute of Terrestrial Ecology, Merlewood Research Station,
Grange-over-Sands, Cumbria LA11 6JU, UK*

Introduction

An important constituent of the developmental cycle of fungi is
the way in which they interact with those living or non-living materials
(resources/substrata[1]) which provide them with a source of organic
nutrients. In this overview, which focuses on Agaricales and agaricoid
Aphyllophorales (e.g. *Schizophyllum* and *Pleurotus* spp.), we hope, by
consideration of these interactions, to provide a framework which will aid
understanding of the diversity of habitats in which these fungi are found.
An attempt will also be made to characterise the widely differing degrees
of selectivity shown by these fungi for particular types of resource
(resource selectivity), and to seek underlying causes. Our discussion will
be based on the occurrence of interactions during four distinctive phases:
arrival at, *establishment* in, *exploitation* of, and *exit* from a resource.

Resource and habitat types

To set the background, we will first attempt to summarise what
is actually known about the diversity of locations in which agarics are
found and their relation to fungal modes of nutrition and type of resource.
Then we will consider how the distribution patterns can be related to
interactions during each of the four phases itemised above. At the outset,
however, we must point out that information regarding the distribution
of agarics is still overwhelmingly based on that of their fruit bodies

---

[1] In this chapter the term 'habitat' is used to describe the place where a fungus
lives, 'substratum' is the medium within the habitat which physically supports
the fungus during development, and 'substrate' is a specific biochemical
component of the substratum. 'Resource' denotes any material which sustains
fungal growth, and a 'resource unit' is a physically determinate unit of such
material.

1

(basidiomata). Production of the latter is dependent on a wide range of variable exogenous and endogenous influences, so that their distribution probably rarely reflects adequately the 'real' distribution of mycelia.

Another source of bias results from the longer tradition of mycology in northern temperate regions, and we must acknowledge that our own understanding is based very much on the mycoflora of western Europe, with only a limited knowledge of the agarics of other parts of the world. The British vascular plant flora in particular is depauperate, associations occurring on the continent of Europe are not found here and *vice versa*; such geographical differences undoubtedly influence the occurrence and roles of agarics, which are distributed from the most extreme tundra regions to the tropics of all continents. In desert regions some agarics have even acquired a specialised xerophytic state, as in *Montagnites*, *Galeropsis* and *Cyttarophyllum*.

Occurrence of biotrophy, necrotrophy and saprotrophy

Three distinctive nutritional modes whereby fungi obtain organic nutrients from living or non-living substrata are now widely recognised (e.g. Lewis, 1973). In biotrophy, readily assimilable, soluble nutrients, such as simple sugars, are absorbed directly from living cells with which intimate contact is made, often via specialised haustoria, and sustained by a lack of cellular rejection responses concomitant with minimal damage to the tissues. In necrotrophy, living tissues are killed by the fungus prior to utilisation of constituent substrates, which often include polymers such as pectin, cellulose, lignin and chitin. In saprotrophy, non-living material, other than that killed by the fungus itself, is utilised.

The distinction between these nutritional modes may not always be clear, however, for two reasons. First, with respect to colonisation of living tissues, essentially saprotrophic behaviour, involving utilisation of diffusates or dead cells (which occur extensively in such tissues as wood, bark and skin), can easily be confused with necrotrophy and even biotrophy. In the latter case intimacy of contact with living cells and possession of mechanisms causing a redirection of nutrient flow towards the fungus may be important criteria enabling distinction from saprotrophy to be made. In the former case, perhaps only those examples where killing of living tissues is a *precondition* for, rather than an adjunct of, successful colonisation are worth distinguishing as 'true' or 'strict' necrotrophy. The problems of delineation are particularly acute in senescing, damaged or otherwise physiologically stressed tissues where the balance between a host and its symbiotic associates may be altered, as is frequently seen in mycorrhizal roots when they age and become decorticated. Furthermore, in such tissues there may be release of nutrients otherwise unavailable to obligate saprotrophs. Subsequent establishment of the latter may *then* result in acceleration of tissue death, giving the appearance of necrotrophy, and indeed fungi behaving in this way are often described as 'weak parasites'. Similarly, a variety of fungi – whose number may be much greater than is currently recognised – cause 'symptomless infections', and some of these only develop extensive mycelium following senescence or injury. It may be best to regard these fungi as essentially saprotrophic but exhibiting a 'latent invasion' or, more graphically, a 'wooden horse' strategy whereby colonisation is effected – via buds, spores, sparse mycelium or mycelial knots or fragments – under conditions of stress militating against full mycelial development. Subsequent alleviation of stress, which in living tissues may be due to lack of nutrients, aeration problems or host-resistance mechanisms, then allows extensive mycelial outgrowth and a territorial advantage in resource capture (Cooke & Rayner, 1984; Rayner *et al.*, 1984).

The second problem arises from the possibility of adoption, under different circumstances, of different nutritional modes by an individual fungus. This is due partly to the mycelial body-form which, almost uniquely, provides a fungus with the potential to be in two places at once, with the different parts fulfilling different roles. Our knowledge of the extent to which this potential is actually realised is limited by the lack of adequately rigorous studies of the vegetative mycelium in natural habitats. However, that the potential exists is clear from the classic associations of *Monotropa* species with ectomycorrhizal fungi and of the orchid *Gastrodia* with *Armillaria*, whereby the fungus provides a bridge across which nutrients can flow from infected tree roots to the herbaceous plant (Hamada, 1939, 1940; Campbell, 1971; Duddridge: Chapter 6). There is also evidence that some normally saprotrophic fungi can, under appropriate circumstances, become necrotrophic or biotrophic, and *vice versa*. Thus *Marasmius androsaceus*, typically a decomposer of conifer needles, can be necrotrophic on *Calluna vulgaris*, causing heather blight (Macdonald, 1949). In some cases, exploitation of one nutritional mode may play a vital role in establishment of a fungus, and precede a second, often dominant, mode. Thus temporary saprotrophy may precede establishment of a wide range of essentially biotrophic and necrotrophic fungi, including lichen- and mycorrhiza-formers as well as parasites. By the same token, necrotrophy and possibly sometimes even biotrophy, by allowing occupation of living tissues, may facilitate saprotrophic survival or colonisation in the presence of non-symbiotic competitors after death of the tissues (see *Crinipellis perniciosa* in Wheeler: Chapter 4). With respect to competition, necrotrophic mycoparasitism represents one important combative (antagonistic) mechanism enabling capture of resources from other fungi ('secondary resource capture'; see Cooke & Rayner, 1984; Rayner & Webber, 1984).

Given these problems, it is clearly difficult to classify *fungi* in terms of their nutritional modes; even the superficially useful prefixes 'obligate' and 'facultative' acquire shades of meaning and cannot be defined precisely. However, it *is* possible to distinguish the three modes as entities in themselves and to identify those fungi which characteristically exhibit them during all or part of their life cycles. This will be the approach in the following treatment of agarics.

Biotrophy. Agaric species and groups in which biotrophy has been detected with reasonable certainty are listed in Table 1.1. It will be evident that whereas, in general, biotrophic fungi form both mutualistic (lichens and mycorrhizas) and parasitic associations (with animals, above-ground parts of plants, and other fungi), amongst agarics biotrophy is overwhelmingly associated with the formation of ectomycorrhizas (sheathing mycorrhizas). Thus, although mycorrhiza formation is known, or reasonably suspected, in over 1975 species belonging to nine families, very few agarics are recognised as biotrophic parasites. One example is *Crinipellis perniciosa*, the cause of witches' broom of cocoa, (Hedger: Chapter 2; Wheeler: Chapter 4). Similarly, lichen formation is unusual amongst agarics, being particularly associated with the genus *Omphalina* which has many lichenised members in many parts of the world, particularly in Arctic–alpine and boreal plant communities.

We should mention here that not all mutualistic associations between agarics and other organisms are based on biotrophy. Thus, termites and attine ants cultivate, respectively, species of *Termitomyces* and *Leuco-agaricus*. Here saprotrophy underlies the relationships, the fungi being grown on comminuted plant materials.

Necrotrophy. Examples of necrotrophy are given in Table 1.2. Although in general commoner than biotrophic parasitism, necrotrophy is apparently unusual in agarics, occurring sporadically amongst the various taxonomic groups. However, necrotrophic mycoparasitism may ultimately prove to be more common than is presently realised, due to the fact that many agarics only become dominant at relatively late stages during decomposition processes, when opportunistic parasitism may facilitate their establishment

Table 1.1. *Examples of biotrophy among agarics*

Ectomycorrhizal	Parastic on angiosperms	Lichenised forms
Russulaceae: Russula, Lactarius Amanitaceae: Amanita, Limacella Gomphidiaceae: Gomphidius, Chroogomphus, Cystogomphus Paxillaceae: Paxillus Boletaceae (including Xerocomaceae & Strobilomycetaceae): Boletus, Suillus, Xerocomus, Strobilomyces, Boletellus, Heimiella, Austroboletus, Fistulinella Cortinariaceae: Cortinarius, Descolea, Stephanopus, Inocybe, Hebeloma, Naucoria[a], Rozites Tricholomataceae: Tricholoma s.s., Laccaria, Leucopaxillus Hygrophoraceae Hygrophorus (excluding Hygrocybe, Camarophyllus)	Tricholomataceae: Crinipellis perniciosa	Tricholomataceae: Omphalina ericetorum group O. grisella O. chrysophylla O. hudsoniana (= luteolilacina) O. flava (= luteovitellina) Myxomphalia maura

[a] An unexplained association with *Alnus* and *Salix*, not generally considered to be ectomycorrhizal.

6

Table 1.2. *Examples of necrotrophy in agarics*

Fungi	Hosts
Fungicolous	
Armillaria spp.	*Entoloma abortivum*
Claudopus parasiticus	*Polyporus squarrosus*
Collybia cirrhata/tuberosa group	Tricholomataceae
Nyctalis spp.	Russulaceae
Omphalina cupulatoides	*Peltigera*
Psathyrella epimyces	*Coprinus comatus*
Voltariella surrecta	*Clitocybe nebularis* and allies
On animals	
Hohenbuehelia spp.	Nematoda
Coprinus cinereus	Human beings (but see text)
Schizophyllum commune	
On plants	
Agrocybe parasitica	*Beilschmiechia tawa* and introduced dicotyledonous trees
Armillaria spp.	Gymnospermous and angiospermous trees and shrubs
Flammulina velutipes[a]	Deciduous trees
Marasmius androsaceus	*Calluna*
Mycena citricolor	*Coffea*
Oudemansiella mucida, O. radicata[a]	*Fagus sylvatica*
Pholiota spp.[a]	Deciduous trees
Pleurotus spp.[a]	
Russula annulata	Shore forest, Madagascar
Xerocomus radicicola	Amazonian rain forest

[a] These wood-inhabitants may not be strict necrotrophs; see text.

and replacement of other fungi. At present, however, much of our knowledge of mycoparasitism amongst agarics is based on those which produce fruit bodies on those of other fungi, notably other agarics, e.g. species of *Nyctalis* on Russulaceae; *Armillaria* inducing gasteroid fruit bodies in *Entoloma abortivum* (Watling, 1974*a*), and *Psathyrella epimyces* on *Coprinus comatus* (McDougall, 1919).

Very few agarics have been reported to be necrotrophic on animals. True necrotrophy occurs in *Hohenbuehelia*, where the anamorphic state, *Nematoctonus*, is nematophagous. *Schizophyllum commune* has been associated with infections of the mouth (Watling & Sweeney, 1971), although more familiar are the records of colonisation of skin and other keratinised tissues, the resulting pathogenesis is not considered to be true necrotrophy. *Coprinus cinereus*, as *C. delicatulus* (see Kemp, 1975), has been isolated as an opportunistic but fatal pathogen of human heart (De Vries, Kemp & Speller, 1971), but little is known of the exact nature of the association.

Amongst agarics recorded as necrotrophic on plants, it is probable that many should be regarded as only secondarily or not strictly necrotrophic since they predominantly exhibit other nutritional modes, or pathogenesis is incidental to, rather than a prerequisite for, colonisation. Examples include various *Armillaria* and *Pholiota* species, *Agrocybe parasitica* in New Zealand, *Russula annulata* in Madagascar and *Marasmius androsaceus* (see above). Species of *Armillaria*, in particular, exhibit a wide spectrum of behaviour ranging from virtually pure saprotrophy, as in *A. bulbosa*, to strongly pathogenic species exhibiting a high degree of necrotrophy, as in *A. mellea sensu stricto*, *A. ostoyae* and *A. hiemii* (see also Rishbeth: Chapter 3). Clear examples of necrotrophy are provided by *Mycena citricolor* which causes a sometimes severe leaf spot on *Coffea* (Buller, 1934, as *Omphalia flavida*), and *Tephrocybe palustris*, which causes distinctive white patches in *Sphagnum* beds, probably utilising proteinaceous material from the moss cells (Redhead, 1981).

Saprotrophy. The occurrence of saprotrophy amongst agarics of western Europe, together with an indication of the range of resource and habitat types, is given in Table 1.3. The widespread occurrence of agarics exhibiting saprotrophy and the diversity of their resource relations will be immediately apparent. However, the exact nature of these relations often remains obscure in the absence of rigorous studies. For example, in the case of those fungi listed as 'terricolous', largely because they are not obviously mycorrhizal or attached to visible particulate debris, it would

Table 1.3. *Examples from western Europe of agarics exhibiting saprotrophy, with a general indication of their typical habitats*

Litter-inhabitants[a]
Agaricus, Agrocybe, Calocybe, Clitocybe, Clitopilus, Collybia, Coprinus, Crepidotus, Entoloma, Hygrocybe, Lepiota, Lepista, Leptonia, Leucopaxillus, Marasmiellus, Marasmius, Melanophyllum, Mycena, Nolanea, Psathyrella, Psilocybe, Rhodocybe spp.

Nitrophiles
   Bolbitius vitellinus group
   Coprinus cinereus group
   Hebeloma sarcophyllum group
   Panaeolus rickenii group
   Psilocybe spp., e.g. *P. semilanceata*
   Stropharia aeruginea group

Peat inhabitants
   Galerina spp.
   Hypholoma ericaeum/udum group
   Pholiota myosotis

Cultivated by insects
   Leucoagaricus spp.
   Termitomyces spp.

Coprophilous
Conocybe coprophila (base-rich dung)
Coprinus (many taxa)
Panaeolus semiovatus (base-rich dung)
P. sphinctrinus group (neutral to slightly acidic dung)
Psathyrella coprobia group
Psilocybe coprophila/merdaria group
Stropharia semiglobata (rather acidic dung)

Terricolous
Agrocybe spp.
Conocybe spp. (nitrophile)
Lacrymaria velutina
Lyophyllum spp.
Melanoleuca
Pluteus spp. (e.g. *P. nanus, P. lutescens*)
Psilocybe (*P. montana* group)

Pyrophilous[b]
   Coprinus angulatus
   Geopetalum carbonarium
   Hebeloma anthracophilum
   Pholiota highlandensis
   Psathyrella pennata
   Tephrocybe atrata group (= *Lyophyllum*)

Table 1.3 *cont.*

Wood-inhabitants
Agrocybe cylindrica
Clitocybula spp.
Crepidotus mollis
Hypholoma capnoides/fasciculare group
Mycena spp. (e.g. *M. galericulata, M. inclinata*)
Panellus spp.
Pholiota spp.
Pleurotus spp.
Pluteus spp. (e.g. *P. cervinus, P. salicinus, P. boudieri*)
Psathyrella spp. (e.g. *P. hydrophila, P. caput-medusae,*
   P. multipedata)
Tricholomopsis spp.

[a] Those attached to visible, small, woody fragments and non-woody debris in litter accumulations and soil.
[b] A whole series of fungi, not yet catalogued, is associated with naturally occurring burns in Australia.

be valuable to know with more precision the extent to which they grow on humic materials. The processes of comminution and depletion of readily assimilable nutrients which characterise humus formation create an inhospitable habitat for mycelial fungi, and if truly humicolous species exist it would be of interest to know how they manage to do so.

Unclassified symbiosis. Several agarics characteristically grow in association with particular living plants, but the underlying nutritional relations are uncertain. Thus many *Galerina* species produce fruit bodies in moss-cushions, some being associated with particular species of *Sphagnum*, and there is an unexplained association between *Naucoria* and members of the Salicales (See Table 1.4). Many wood-inhabiting species, including those listed as 'necrotrophs' in Table 1.2, grow on particular trees or shrubs, but since even functional sapwood rarely contains more than 10% by volume of living cells, and heartwood no or negligible amounts of living tissue, they must be regarded as predominantly saprotrophic although initial access to woody tissues may be via necrotrophy or pathogenesis. An unusual example is provided by the agaricoid members of the Polyporaceae *Pleurotus eryngii* and *P. fossulatus* which associate with certain umbellifers.

When considering associations based on the occurrence of fruit bodies it is worth remembering that these may not always have a direct nutritional basis. It may sometimes be that the associated organism provides a suitable base, perhaps sometimes related to specific physical or chemical stimulation, upon which fruiting is facilitated. Thus *Boletellus ananiceps* will ascend stringy-bark eucalypts, thereby producing fruit bodies in the turbulent air above the trash-covered forest floor. Again, *Boletus parasiticus* and *Xerocomus astraeicola* are *not* parasitic on Sclerodermatales as is often thought, but simply require the presence of the latter to stimulate fruiting. Similarly, it appears that *Gomphidius roseus* and *Suillus bovinus* mutually stimulate fruiting.

Selectivity and ubiquity

A noticeable feature of the distribution of agarics, as well as of other fungi, is that, regardless of nutritional mode, they exhibit a spectrum of behaviour ranging from selectivity towards a particular resource type to occurrence on a wide range of substrata. In the following discussion we describe preference for a particular resource type as 'selectivity' rather than 'specificity', since the latter term implies an absoluteness of association which is rarely found in practice. Indeed Harley & Smith (1983) have argued strongly against the concept of specificity in mycorrhizal fungi, because exceptions – however rare – can usually be found. Three distinctive but overlapping forms or levels of selectivity are identified below: taxon selectivity, resource-unit restriction, and habitat selectivity.

Taxon selectivity. This involves preferential occurrence on living or dead material derived from particular or related groups of organisms. It can be subdivided hierarchically according to the taxa concerned, for example, into selectivity of family, genus, species and variety. Examples of taxon selectivity shown by agarics with primarily biotrophic, necrotrophic and saprotrophic nutrition are listed in Tables 1.4, 1.2 and 1.5, respectively.

Taxon selectivity has often been regarded as a characteristic feature of biotrophy, consistent with the notion that here the association with a living host is most intimate and hence most specialised. Accordingly, some mycorrhizal agarics show marked selectivity, such as *Suillus* species and the Gomphidiaceae with the Coniferae, a classic example being the association between *Suillus grevillei* and larch (*Larix*). However it is also clear that some putatively mycorrhizal agarics such as *Russula ochroleuca* occur with a wide range of coniferous and angiospermous hosts, whereas others occur with a few, sometimes distantly related, hosts. Thus *Amanita muscaria* occurs most consistently with species of birch (*Betula*), but may occur with other frondose trees, e.g. hornbeam (*Carpinus betula*), and conifers, e.g. pine (*Pinus*); *Boletus spadiceus* occurs with oak (*Quercus*), birch, and dwarf willows (*Salix repens* and *S. herbacea*; Watling, 1981); and *Rozites caperata* occurs with pine in Britain, oak in central Europe

Table 1.4. *Examples of taxon selectivity in biotrophic agarics*

Fungi	Hosts
Crinipellis perniciosa	*Cacao*
Lactarius glyciosmus, L. torminosus,	*Betula*
L. vietus, Russula betularum,	
R. claroflava	
Russula alpina	dwarf Arctic/alpine
	Salix spp.
Lactarius deliciosus, L. hepaticus	*Pinus*
Russula caerulea, R. decolorans	
Lactarius blennius, Russula fellea	*Fagus*
Lactarius cyathula, L. obscuratus	*Alnus*
Cortinarius violaceus, C. armillatus,	*Betula*
C. crocolitus	
C. mucosus, C. pinicola	*Pinus*
Suillus spp.	Coniferae; in
	Britain with
	Pinus, except
	S. grevillei
	with *Larix*
Gomphidius/Chroogomphus	Coniferae; in
	Britain with
	Pinus
Paxillus rubicundulus, Naucoria	*Alnus*
escharoides/scolecina group[a]	
Uloporus (Gyrodon) lividus	
Naucoria salicis[a]	*Salix*
Tricholoma fulvum	*Betula*
T. flavovirens	*Pinus*
T. acerbum	*Quercus*
T. psammopus	*Larix*

[a] An unexplained association with *Alnus* and *Salix*, not generally considered to be ectomycorrhizal.

and the USA, and with dwarf birch (*Betula nana*) in Finnmark (Watling, 1974*b*).

Man's activities in planting exotic trees or otherwise modifying the environment may often, as with genetic barriers, break down ecological barriers, hence disrupting previous close associations. Thus in coniferous plantations on former hardwood sites it is common for mycorrhizal associates of the latter to continue fruiting for many years after felling. Similarly, in coastal hazel (*Corylus*) scrub in northern and western Scotland, several agarics formerly associated with oak, e.g. *Tricholoma sulphureum*, continue to fruit, and, in *Salix repens* scrub, agarics typical of the former dominant vegetation of the locality can be seen in many northern areas of the British Isles (Watling, 1981).

Such apparent non-selectivity within species of mycorrhizal fungi may be because, as in other fungal groups, they are genuinely non-exacting in their requirements and competitive on a range of hosts. Alternatively, there may be ecotypic or strain differences within species with respect to their ability to associate with different hosts, such as occur between *formae speciales* of certain biotrophic and necrotrophic plant pathogens. Such differences might or might not be associated with reproductive isolation and discontinuous variation. In general, information on strain differences within mycorrhizal fungi is limited, as is that on the ecological preferences of virtually all fungi, by the absence of relevant population studies. However, Mason (1975) found some variation between isolates of *Amanita muscaria* from birch and pine in their ability to infect birch seedlings, this being partially consistent with origin. In some cases strain differences in ability to tolerate distinctive environmental conditions associated with different hosts may be more important than the type of host *per se*. Thus, certain strains of *Suillus* which grow better at high altitudes and low temperatures in comparison with other strains are used for inoculation of nursery stock destined for recolonisation of areas adjacent to ski runs (Horak, personal communication).

Within agarics exhibiting necrotrophy, although there are few examples, a range of behaviour is found again. However, since this is well exhibited by *Armillaria* and is discussed by Rishbeth (Chapter 3), we will not pursue this further.

Although widely thought of as versatile in their ecological requirements and capacity for production of extracellular enzymes, many saprotrophic fungi, including agarics, show very marked taxon selectivity. However, just as apparent non-selectivity in mycorrhizal fungi could sometimes be a taxonomic artifact resulting from ecotypic differences, the obverse also applies due to the tendency, particularly with saprotrophs, to use habitat as a criterion in descriptions of fungal species. This probably applies more to microfungi – especially Ascomycotina – than to agarics, but the possibility remains that forms relatively similar morphologically but occurring on different resource types – and hence described as different species – could nonetheless have a common gene pool.

Examples of taxon selectivity in saprotrophic agarics are given in Table 1.5. Many of these also exhibit resource-unit (component) restriction (see below), e.g. *Baeospora myosura*, *Flammulaster carpophila* and *Tubaria autochthona*. Amongst non-component-restricted litter fungi, *Clitocybe*

Table 1.5. *Examples from western Europe of taxon selectivity in saprotrophic agarics*

Fungi	Resources/habitats
Mycena pterigena	*Pteridium, Dryopteris* petioles
Baeospora myosura	*Pinus* and *Picea* cones (also *Cedrus?*)
Mycena strobilina	*Pinus* cones
Strobilurus spp.	Cones of Coniferae and 'cones' of Magnoliaceae in North America
Marasmius buxi	*Buxus* leaves
M. hudsonii	*Ilex* leaves
M. epiphylloides	*Hedera* leaves
Galerina hypnorum/mycenopsis group	Moss-cushions
Leptoglossum spp.	Moss-cushions, usually Mniaceae
Galerina tibiicystis/sphagnorum group, *G. paludosa*, *Hypholoma elongatum*, *Omphalina sphagnorum/ philonotis* group	*Sphagnum* bogs
Conocybe utriformis, Psathyrella typhae, Melanotus phillipsii	Marsh plants
Flammulaster carpophila, F. carpophiloides	*Fagus* cupules
Tubaria autochthona	*Crataegus* berries
Flammulaster erinaceus	*Salix* branches
Galerina ampullaeocystis/ camerina group	Conifer branches
Paxillus atrotomentosus	*Pinus* stumps
Lyophyllum ulmarium, Pleurotus cornucopiae, Rhodotus palmatus	*Ulmus* trunks
Pleurotus eryngii, P. fossulatus	Umbelliferae stocks
Micromphale perforans, Mycena capillaripes, Fayodia bisphaerigera	*Picea* needles
Clitocybe langei	Conifer litter
C. odora	Angiosperm litter
Mycena crocata, M. pelianthina	*Fagus* litter

langei, C. odora, Mycena crocata and *M. pelianthina* show varying degrees of selectivity, whereas *Clitocybe flaccida, C. nebularis, Collybia butyracea, Mycena galopus* and *M. pura* can grow on both angiospermous and coniferous litter. Amongst wood-inhabiting agarics, *Lyophyllum ulmarium* and *Rhodotus palmatus* grow selectively on elm (*Ulmus*) whereas *Hypholoma fasciculare* occurs on wood from a wide range of both coniferous and angiospermous trees. In some cases, differences in selectivity occur with geographical distribution, thus *Agrocybe cylindrica* is widespread on poplar (*Populus*) in France, but in Britain it is commoner on elm and elder (*Sambucus*), and in South America it is recorded from a wide range of native and exotic plants including one Monocotyledon.

Resource-unit restriction. This occurs when a fungus is restricted within the boundary of a resource unit (Swift, 1976), be this a mycorrhizal root, fruit, petiole, twig or branch. Thus, amongst agarics inhabiting accumulations of plant litter, a distinction can be made between those occurring only within distinct components of the litter (component restricted) and others whose mycelia can grow freely within litter, colonising all its components (non-component restricted) (Cooke & Rayner, 1984). Resource-unit restriction is often related on the one hand to taxon selectivity, for instance *Tubaria autochthona* on hawthorn (*Crataegus*) berries, and on the other to habitat selectivity, as shown by many coprophilous fungi. However, non-component-restricted fungi can be taxon selective, such as *Mycena pelianthina* on beech, and habitat selective, e.g. many non-component-restricted litter species. Equally, many unit-restricted agarics occur on similar substrata from a wide range of genera, such as *Marasmiellus ramealis* on dead stems of nettles (*Urtica* spp.), brambles (*Rubus*), etc., and *Marasmius rotula* on small dead twigs.

Habitat selectivity. This broad category of selectivity embraces both taxon selectivity and resource-unit restriction but includes all cases of selectivity for a particular habitat type, such as wood, faeces, leaf-litter, and animal remains. Even here, a spectrum of behaviour from selectivity to non-selectivity exists, with some agarics able to occupy more than one distinctive habitat type, perhaps related in some cases to different nutritional modes. For agarics, it has often been implied that wood and leaf-litter represent distinct habitats with exclusive fungal floras (e.g. Hintikka, 1982). However, it is clear that although the vast majority of wood- and litter-inhabiting agarics are exclusive to these habitats, the distinction is again one of degree. Thus several non-component-restricted litter fungi may embrace small woody fragments within their activities. Also, species which are predominantly wood-inhabiting, such as the cord-formers *Hypholoma fasciculare* and *Tricholomopsis platyphylla*, frequently colonise and decompose leaf-litter, particularly where this underlies bulky woody substrata. However, if a contender for the most ubiquitous agaricoid fungus is to be found, this must surely be *Schizophyllum commune* which, especially in the tropics, can occur on almost any lignocellulosic resource including coconut matting, palm fronds, sapwood of all types including drift and processed woods from a variety of species, and vegetable extracts. It has even been reported as a dermatophyte, isolated from spinal fluid and sputum, and, as mentioned above, as a necrotroph causing a palate ulcer in a young child. When tested, intercompatibility between strains of such diverse origin indicates that they are indeed all referable to a single biological species.

Interactions and modes of arrival

Arrival at a resource is largely dictated by patterns of exit (see below), being accomplished either by vegetative mycelium or via propagules which in turn may be either genetically homogeneous or heterogeneous. The greater the dependence on propagules, the greater is the tendency for resource-unit restriction. Patterns of colonisation resulting from these modes of arrival are also likely to be distinctive in relation to the different opportunities they provide for interactions both within the resource and between separate colonisation units (Rayner *et al.*, 1984).

Arrival by propagules

Characteristic features of arrival via propagules are that colonisation is effected from localised foci where germination has occurred; opportunities for input of water and nutrients are limited, and buffering against hostile environmental influences at the resource surface is absent. Since agaric propagules are non-motile – after initial ejection from the basidium in the case of basidiospores – they play an essentially passive role in their own dispersal which is probably effected principally by air currents, although rain splash, water flow and animal vectors can all be important. Actual arrival on the resource surface will depend largely on random selection operating on spore dispersal, and the relative proportions of spores in the air above the resource will influence the subsequent pattern of colonisation (Frankland, 1981). Animal vectors, by effective inoculation below the resource surface, may obviate hostile surface effects but, with the exception of the associations of *Termitomyces* and *Leucoagaricus* with termites and attine ants, there are few cases where this is known to occur. However, viable spores are carried by beetles either in the integument or in the gut, as in *Cartodere filum* (Gordon, 1938), and Collembola, mites and nematodes can often be seen coated with basidiospores. The *Nematoctonus* state of *Hohenbuehelia* is notable for the production of adhesive knobs on conidia or mycelium, by which it attaches itself to nematodes (Barron & Dierkes, 1977).

If the propagules are genetically homogeneous, there is the possibility of coalescence between separate colonisation foci into a single unit, as originally envisaged by Buller (1931) for *Coprinus sterquilinus*. However, if they are genetically different, the resulting mycelial individuals may be expected to occupy discrete domains demarcated by somatic incompatibility reactions (Rayner *et al.*, 1984). Since agaric species appear to be predominantly heteromictic (heterothallic), basidiospores are typically heterogeneous and thus may give rise to populations of antagonistic dikaryotic individuals via an initial homokaryotic phase. Therefore, where colonisation by basidiospores is favoured, it is probable that numerous antagonistic individuals will occur in close proximity and in extreme cases this can actually inhibit their activity, as has been indicated in *Hypholoma fasciculare* (Coates, 1984; Rayner *et al.*, 1984). It is interesting that several coprophilous fungi are homomictic or homoheteromictic (primary or secondary homothallic), including, amongst agarics, *Coprinus sterquilinus* and *C. bisporus*, respectively. This could be related to one or more of several factors: obviation of the requirement for mating for reproduction in a small ephemeral resource; facilitation of synergistic resource capture (see above); and avoidance of the deleterious effects of antagonism.

Asexually produced propagules are relevant to the discussion here, since when they are produced by the same original mycelium they will be genetically homogeneous (or, at most, of two types in the case of monokaryotic conidia produced by dikaryons). Although the range is nowhere near as great as in Ascomycotina, agarics do produce a wide variety of asexual propagules (Kendrick & Watling, 1979; Watling, 1979). Thallic arthroconidia, sometimes referred to as oidia, are the predominant conidial type (see discussion of evolutionary aspects below). In homokaryons they can play a role in sexual conjugation via homing reactions with mycelia as in *Coprinus* (Kemp, 1977), as can basidiospores in *Leccinum* (Fries, 1978). In *Hypholoma fasciculare*, production of arthroconidia is associated, in both monokaryotic and dikaryotic cultures, with a characteristic dense, slow-growing, mycelial phase which can revert spontaneously to a fast effuse phase (Rayner *et al.*: Chapter 10).

Chlamydospores are produced by a range of agarics and are sometimes a particular feature of those exposed to fluctuating environmental conditions, particularly moisture and temperature. They are produced *in situ* and as such may sometimes best be regarded as 'continuation' rather than exit or arrival structures. Thus *Clitocybe flaccida*, a non-component-restricted decomposer of upper litter layers, which is intolerant of temperatures above 25 °C and which is presumably subject to alternating moisture conditions, produces abundant spiny chlamydospores, even in young, actively growing cultures (Mitchell, unpublished). *Rhodotus palmatus*, which is characteristic of the decorticated upper portions of dead elm trunks, is even more extreme, mycelia in culture and on wood rapidly becoming powdery masses of chlamydospores with little sign of interconnecting hyphae (Gibbs & Gulliver, 1977; Rayner & Hedges, 1982). Similarly, in *Nyctalis* the teleomorphic condition may be totally replaced by the production of chlamydospores, and it may be hypothesised that insects, attracted to putrefying agarics colonised by the fungus, are active in dispersal, the thick-walled chlamydospores being better suited to passage through the gut than the thin-walled tricholomataceous basidiospores. Interestingly and remarkably, chlamydospores of *N. asterophora* have recently been shown to be capable of germination into a yeast-like phase (Jahrmann & Prillinger, 1983), but how this is related to the resource is not yet clear.

Sclerotia, microsclerotia and pseudosclerotia represent an alternative, perhaps longer-term, solution to the problem of survival of intermittent environmental stresses, and their arrival has some of the features characteristic of mycelial arrival, due to their origin as mycelial aggregations and their consequently greater stored energy reserves. Furthermore, many are 'myceliogenic', germinating directly into vegetative mycelium (Coley-Smith & Cooke, 1971). Other sclerotial structures, germinate directly into reproductive structures (e.g. in *Agrocybe arvalis; Coprinus sclerotiger*; Watling, 1972), and where the latter are used uncritically as an indicator of distribution a mistaken impression of resource type can easily be obtained. Knowledge of sclerotial structures in agarics is limited and whereas they are known in a few cases, as in the *Collybia tuberosa* group, *Leucocoprinus* (Warcup & Talbot, 1962) and *Paxillus* (W. Bigg, personal communication), many more species probably produce sclerotia than we currently realise.

Finally, it should be remembered that small fragments of mycelium can themselves, under appropriate circumstances, act as propagules. Thus, velar cells of *Coprinus* are capable of germination and hence of acting as dikaryotic propagules. In the case of fungi cultivated by ants and termites (not to mention man) new cultures are initiated by mycelial inocula, and this can be expected to lead to a clonal population structure.

Arrival by mycelium

Arrival by mycelium is a particular feature of those agarics which can loosely be described as soil- or litter-inhabiting, whence mycelial extension between resource units is possible under non-desiccating conditions. Characteristic features of arrival by mycelium can be contrasted with those of arrival by propagules. Colonisation is not necessarily localised in separate foci and, even where it is, coalescence and synergism are possible because of the lack of antagonism resulting from genotypic differences. There are also opportunities for import of nutrients and water, and hence much greater energy of invasion or inoculum potential (see Garrett, 1970). Then there is the potential, via formation of aggregated systems, for buffering against hostile abiotic conditions and effective combat of actual or potential competitors. In consequence, single genotypes may be capable of occupying considerable volumes of individual resource units as well as considerable areas of ground. Thus, in a study of a population of the cord-forming wood-decomposer *Tricholomopsis platyphylla*, 22 somatically incompatible, dikaryotic, mycelial types were identified from a sample of 113 isolates from different locations. However, pieces of wood colonised by more than one type were not found, and isolates of the same mycelial type often occurred in fairly contiguous groups and as much as 160 m apart (Thompson & Rayner, 1982*a*). In the rhizomorphic *Armillaria*, isolates of the same type have been located as much as 450 m apart (Anderson *et al.*, 1979; Kile, 1983), whereas the distance between mycelium which produced intermingling isolates of the non-rhizomorphic, non-cord forming, *Mycena galopus* in a coniferous woodland was only 2–3 m or less. The number of individual mycelia per unit area of this species was also far greater than that described for *T. platyphylla* (Frankland, 1984). Such a strategy would be suited to a coniferous litter resource which changes constantly as new supplies fall from the canopy.

As already intimated, arrival via mycelium may be effected either by diffuse mycelium or via specialised migratory organs such as cords and rhizomorphs (see also Rayner *et al.*: Chapter 10). In the first case the possibility arises for the exploitation of different nutritional modes by internal and interstitial mycelium. The type of interstitial mycelium is likely to be related to the distribution of resources (Cooke & Rayner, 1984; Thompson, 1984). Thus non-component-restricted agarics inhabiting accumulations of mostly non-woody litter tend to form diffuse mycelium, related to the fact that individual resource units occur in a more or less continuous layer. Where resource units are in discontinuous and/or uneven supply, for example roots, fruits, wood pieces and faecal pellets, there will be an increased tendency to form cords and rhizomorphs. Fungi producing such structures are somewhat intermediate between resource-unit restricted and unrestricted species, being able to extend between units, but being restricted to a particular type.

It is understandable that some of the largest and/or most highly developed migratory structures are produced by agarics inhabiting bulky woody substrata, such as *Armillaria* species and *T. platyphylla* which, respectively, produce rhizomorphs and cords up to several millimetres in diameter. Smaller but still well-defined structures are produced by certain *Marasmius* species and by some mycorrhizal fungi such as *Suillus bovinus* (Duddridge, Malibari & Read, 1980; Read, 1984). The first genus is common in tropical forests and, as described by Hedger (Chapter 2), the rhizomorphs exhibit a remarkable capacity for extension through aerial or sub-aerial environments and for developing either as agents of colon-isation or as fruit bodies. This raises an important point in that, although cords and rhizomorphs can and often do act as exploratory structures *per se*, they are often preceded by diffuse mycelium and then function as connectives crossing domains lacking available nutrients or linking with developing fruit bodies. The many cord-forming species in the Agaricaceae and Lepiotaceae probably fall into this category.

Arrival, selectivity and recognition

Patterns of arrival may be dictated in two ways, resulting in resource selectivity. First, environmental stresses operative outside the resource surface may act selectively, inhibiting mycelial growth, propagule dissemination or germination of some fungi but not of others. Alternatively, and of more relevance here, there may be more or less specific recognition mechanisms resulting on the one hand in directed growth of mycelium and, on the other, in stimulation of propagule germination in the vicinity or on the surface of an appropriate resource.

Stimulation of spore germination by factors associated with a resource is widely recognised as being of ecological importance. In some cases it may be due to relatively simple physical or nutritional factors, such as elevation of temperature following passage through the gut of a homeo-thermic animal, soil heating after fire, or alleviation of fungistasis by supply of a carbon source. In other cases there may be more specific stimulation (Watling, 1971). Such chemical stimulants may be produced by the mycelium of the same or different species (Fries, 1977, 1978) and in the case of *Agaricus campestris* has been identified as a 7-C-olefin (Lösel, 1964;

see also Elliott: Chapter 19, p. 453). A classic example amongst agarics was provided for certain mycorrhizal fungi by Melin (1959, 1963) who described stimulation of mycelial growth and spore germination by a substance or substances (the 'M-factor') in living roots. Interestingly, responsiveness ranged from non-specific, as in *Russula xerampelina* which responded to diffusates from a wide range of species including tomato (*Lycopersicon*), to specific, as in *Russula adusta* which responded only to pine. Since then it has been suggested that the stimulation may be due partly to reduction of auxins rather than the presence of a factor (Harley & Smith, 1983).

Directed growth would be a valuable adaptive feature in fungi growing as mycelium between discontinuously distributed resource units, avoiding wastage of mycelial biomass by growth in the wrong direction (Thompson & Rayner, 1983; Thompson, 1984). There are two ways in which this could be brought about, correlated with the degree to which specialised vegetative organs act as agents of colonisation (see above). First, randomly orientated and sparse exploratory mycelium could be succeeded by extensive development only *after* contact with an appropriate resource (or diffusates from it) were established. This appears to be the case in *Suillus bovinus* where exploratory fans of mycelium are consolidated into cord systems only after contact with host roots has been made (Read, 1984). Alternatively, hyphae and more or less apically growing organs such as rhizomorphs and cords could grow directly towards suitable resource units, and there is some evidence that this can occur (Mowe, King & Senn, 1983; Thompson & Rayner, 1983; Thompson, 1984). Such behaviour would depend on sensitivity to a diffusible, possibly volatile, chemical stimulus, sometimes operating over distances of several centimetres or more, and it has been suggested that this might be facilitated by amplification of responsiveness by a multi-hyphal system (Thompson & Rayner, 1983; Thompson, 1984). However, there is no information on the nature of any such directional chemical stimulus, although a wide range of substances has been reported to induce rhizomorph formation in *Armillaria* (see Chapter 10), and nonanal, for example, induces aerial mycelial development in some other Basidiomycotina (Fries, 1961; Sortkjaer & Alderman, 1972).

Interactions during establishment and exploitation

Following arrival and, in the case of propagules, germination, further development is dependent on interactions leading to successful establishment of mycelium within the resource and exploitation of the nutrients it contains. In understanding these processes it is important to appreciate the dynamic nature both of fungal thalli and of the resources themselves. Thus changes in the physiological properties and functioning of the thalli are associated with changing properties of the resources as nutrients are extracted, the relation between cause and consequence often being inextricably involved. The nature of the balance between these dynamic systems will determine the type of resource relation and will be an important factor underlying selectivity. Unfortunately, critical work on the fundamental nature of resource interactions during establishment and exploitation by agarics, particularly non-mycorrhizal species, has been very sparse and mainly restricted to a very few species actually or potentially of commercial importance, such as *Agaricus bisporus* and *Lentinus edodes* (cf. Wood: Chapter 15; Leatham: Chapter 17). Our approach is to highlight factors which we believe are important determinants of the interactions and then relate them to what is known, or should be known, about actual patterns of establishment and spread.

Availability and uptake of nutrients

An obviously crucial determinant of resource relations is the availability and type of organic and mineral nutrients present, and the corresponding ability of fungi to utilise them. With respect to organic nutrients, these range from easily assimilable substances such as simple sugars (including hexoses and pentoses) which can be absorbed directly, to more refractory compounds such as cellulose and lignin which are utilised only after breakdown into assimilable units by extracellular enzyme action. Probably all agarics, in common with other fungi, are capable of utilising simple sugars. However, not all can utilise the more refractory substrates, possession of the requisite enzymes being an important determinant of relations with resources containing significant quantities of these materials. Related to the habitats of saprotrophic agarics (Table 1.5), ligninocellulolytic ability is widespread, contrasting with the general lack (or ignorance) of keratinolytic and chitinolytic ability.

Various aspects of the relation between extracellular enzyme production and agaric development are discussed by Wood in Chapter 15 and we shall raise here only some general points. First, since many extracellular enzymes, including the cellulase complex, are inducible and subject to catabolite repression, the stage at which they become important during colonisation is of interest; early establishment, for example, may often be dependent on supplies of assimilable compounds. Even more extreme is the situation with ligninolytic systems since these appear to be brought into action as a result of secondary metabolic activity resulting from nitrogen starvation (Kirk & Fenn, 1982). Relevant to such events are possible changes in extracellular enzyme production resulting from switching between nutritional modes. Biotrophy involves absorption of simple substrates, any production of extracellular enzymes having deleterious effects on structural and physiologically important components of host tissues with the further possibility of elicitation of phytoalexins (Hahn, Darvill & Albersheim, 1981). It is therefore not surprising that most mycorrhizal agarics have been found to be capable of utilising only simple sugars and amino acids as carbon sources. However, several, such as *Tricholoma fulvum*, appear to have a ligninocellulolytic capacity, possibly associated with facultative saprotrophy in litter. How is the transition between such irreconcilable roles effected? Here it may be noted that even apparently obligate mycorrhiza-formers, such as species of *Russula*, possess polyphenoloxidase systems in their fruit bodies and these are specifically tested for using guiac resin or *o*-toluidine to aid taxonomic diagnosis.

Besides carbon, nitrogen is widely recognised as being of vital ecological importance. The majority of fungi, presumably including most if not all agarics, can utilise inorganic nitrogen, but there is an important distinction between ammonium, which is generally utilisable, and nitrate, which many species cannot utilise. The concentration of nitrogen in relation to carbon (C:N ratio) is often considered highly relevant to nitrogen availability (Waksman & Tenney, 1928; Swift, Heal & Anderson, 1979), although this view has been criticised in cases where the carbon is present in insoluble form and hence can act as an inert matrix (Park, 1976). Generally, the C:N ratio decreases during decomposition due to removal of carbon as carbon dioxide, and this may allow colonisation by more nitrogen-demanding fungi. Nitrogen availability can also often be greatly increased locally by animal urine or remains and this can favour certain ammonogenous species, including certain *Tephrocybe* species and the *Hebeloma sarcophyllum* group. The association of *H. vinosophyllum* with buried carcasses led to the suggestion that it could aid forensic detection of buried murder victims (Sagara, 1976), whilst *H. radicosum* is associated with animal nests and burrows (Watling, 1978).

Microenvironmental factors

Different resource types inevitably contain different microenvironments, important components of which are light (supply of this to the host plant, incidentally, indirectly affects mycorrhizal fungi), temperature, water potential, aeration and gaseous composition, pH, and solute concentration. The role of the microenvironment in the distribution of Basidiomycotina in temperate woodland has recently been reviewed by Boddy (1984). Effects of many of the factors listed above are, as will be evident from subsequent chapters, also an important consideration in morphogenetic studies. Obviously variation in microenvironmental factors will affect fungi (including different individuals or strains of the same species) differentially, according to their differing requirements and optima, and hence is likely to cause resource selectivity (see discussion of *Suillus* strains above). Nonetheless, the fact remains that critical studies on the effects of microenvironmental parameters on agaric morphogenesis (especially vegetative development) have been limited to very few species, so that the requirements of the vast majority are virtually unknown and their influence on resource relations has been almost entirely neglected. In discussions of fungi exhibiting extreme requirements, such as xerophiles, osmophiles and thermophiles, agarics rarely feature. Nevertheless many do occur under extreme conditions, for example *Montagnites* in deserts (see above), *Psathyrella ammophila* in sand dunes and *Conocybe halophila* in salt-pan soils in desert regions. Yet the nature of their adaptation to such conditions, other than in connection with their fruit bodies (see below), is not clear. One exception to the general dearth of critical studies, and one which provides a signpost for future work, has been provided by Hintikka (reviewed in Hintikka, 1982). He pointed out that Basidiomycotina inhabiting bulky woody substrata were likely to encounter higher carbon dioxide concentrations than, for example, those inhabiting leaf-litter, and he showed that the tolerance of the former (including the agaric species *Mycena haematopus*, *Hypholoma capnoides*, *H. sublateritium* and *Pholiota squarrosa*) to high carbon dioxide concentrations was generally far greater than that of the latter (including *Clitocybe infundibuliformis*, *Clitopilus prunulus*, *Collybia confluens*, *Lepiota clypeolaria* and *Mycena epipterygia*). Significantly, the heart-rot species *Pholiota squarrosa* grew well at 70 kPa ($= 70\%$) partial pressure of carbon dioxide.

Stimulation and inhibition by chemical factors; 'recognition'

Besides major carbon sources, a wide variety of 'incidental' chemical substances occur in natural substrata and affect fungi physiologically during establishment and exploitation. These may be inhibitory, causing selection of tolerant species, or selectively stimulatory (see Dighton & Mason: Chapter 5). In living substrata, such substances may be produced dynamically as part of a recognition response, mediating successful or unsuccessful symbiosis.

In non-living substrata, although the probable role of chemical factors as determinants of selectivity is widely appreciated, again they have been surprisingly neglected experimentally. Most widely recognised as inhibitors are a range of phenolic substances in wood and leaf-litter, including the flavonoids, terpenoids, tropolones, stilbenes and tannins which can be obtained as extractives from heartwood. The latter provide a good example because, although their inhibitory properties are well documented (e.g. Scheffer & Cowling, 1966), they have not been related adequately to actual occurrence and concentrations (Hart & Shrimpton, 1979), and the possibility of differential effects on a range of fungi seems rarely to have been tested, with the exception of a few studies such as that of Harrison (1971) and others reviewed by Hintikka (1982). These gave largely expected results on the basis of known preferences of the fungi tested.

Specific stimulatory effects have been reported for a variety of fungi (Glasare, 1970; Rice, 1970; Flodin & Fries, 1978) but rarely, except for *Armillaria*, involving agarics. However, certain flavonoids present in conifer needles, notably taxifolin glucoside and taxifolin, selectively stimulate growth of fungi, including *Marasmius androsaceus* and *Micromphale perforans* which colonise these substrata (Lindeberg *et al.*, 1980). Similarly, mycelial growth of the agarics *Lyophyllum ulmarium*, *Pleurotus cornucopiae* and *Rhodotus palmatus*, all of which are strongly selective towards wood of elm, is markedly stimulated by addition of elm sap to malt agar (Rayner & Hedges, 1982).

With respect to chemically mediated recognition responses during colonisation of living resources, little appears to be known about agarics which are necrotrophs or biotrophic parasites, perhaps because of the relative scarcity of these fungi. However, lectin–carbohydrate recognition mediates attachment of certain nematophagous fungi to their prey (Nordbring-Hertz, 1984) and this may well apply to the agaric *Hohenbuehelia*. Phytoalexins, which probably mediate many plant–pathogen interactions, have not, so far as we know, been implicated yet in resistance to, or selectivity for, pathogenic agarics.

The whole question of recognition during mycorrhiza formation has recently been reviewed by Harley & Smith (1983), and the nature of the host–fungus interface is discussed by Duddridge in Chapter 6. A few general points will be mentioned here. Obviously, the ability to form a well-developed mycorrhiza rapidly will be an important competitive factor, underlying selectivity except under circumstances where competitors are absent. Thus *Suillus luteus*, which is regularly associated with pine in the field, forms a well-developed mycorrhiza with this genus in pure culture, whereas *S. grevillei*, which rarely occurs with pine and is typically associated with larch, forms only a poorly developed mycorrhiza in pure culture with pine (Duddridge: Chapter 6). That factors derived from the host affect the structure of a mycorrhiza formed by a particular fungus is indicated by the different types of mycorrhiza formed by certain fungi with different hosts. Thus *Cortinarius zakii* forms arbutoid mycorrhizas with *Arbutus* and ectomycorrhizas with *Pseudotsuga* (Zak, 1974).

With regard to structure, two facets of ectomycorrhiza formation should be considered: formation of the sheath and formation of the Hartig net. There are several indications that the former is not influenced specifically by the host *per se*, not least being the fact that a sheath can form on silicone tubes (Read & Armstrong, 1972). Rather, it has been suggested, as it has been for mycelial cord formation (Rayner *et al.*: Chapter 10), that aggregation of hyphae to form the sheath is due to exudation of nutrients from the main hyphae attached to a food base, which attract other hyphal branches under low nutrient conditions. However, the lack of differentiation of hyphae seen in mycelial cords may indicate a hormonal influence in sheath formation.

By contrast with sheath formation, specific recognition processes are likely to be important in the development of the Hartig net, and Harley & Smith favour the hypothesis of Vanderplank (1976, 1978) that this involves protein copolymerisation (see Duddridge: Chapter 6). This hypothesis suggests that the fungi possess enzymes or proteins on their hyphal walls which deactivate those enzymes of the host involved in building wall polymers. Wall formation would thus be inhibited or altered and soluble carbohydrates made available to the fungus. It introduces the question of how mycorrhizal fungi and host roots interact at the molecular level during recognition and whether, for example, lectins, which are involved in other symbioses, have any role. In studies of ectomycorrhizal formation in *Eucalyptus* (Malajczuk, Molina & Trappe, 1984), little difference was found in the morphology of ectomycorrhizas formed by compatible, host-selective fungi and broad-host-range species. However, interaction between eucalypt roots and incompatible, conifer-selective fungi induced deposition of tannins in the roots, or lysis of hyphae and host cells (see Duddridge: Chapter 6).

Patterns of establishment and spread

A common view, perpetuated in elementary texts, is that development following arrival is a simple matter of mycelial extension through a resource, culminating in reproduction. That this is indeed too simple we hope has been established by the foregoing discussion, since it is evident that a fungal thallus will be subject, in both space and time, to a wide variety of often conflicting demands during exploitation of a resource. A crucial issue, barely understood for fungi in general, let alone agarics, is the extent to which such conflicting demands are met by morphologically and/or physiologically different or specialised forms. Here a useful concept, recently established by Gregory (1984), is that of mycelial 'modes' or operationally distinctive phases, transitions between which characterise processes of colonisation and exploitation. Examples of such transitions include those between unicellular and mycelial thalli, homokaryons and heterokaryons, trophophase and idiophase, diffuse and aggregated systems, and vegetative and reproductive development.

The nature of these transitions, as they occur under laboratory conditions, provides the basis for many subsequent chapters, but the question as to how they apply to natural colonisation processes following arrival is substantially unresolved. A few pointers related to wood-inhabiting agarics will be given here. The importance of cords and rhizomorphs as migratory organs has already been emphasised. Following arrival, effective resource capture is dependent on dedifferentiation into effuse mycelium, and a competitive advantage is often gained by subcortical mycelial extension in the plane of weakness provided by the cambium, allowing rapid occupation of peripheral regions and exclusion of competitors invading radially from outside (Rayner & Todd, 1979). In non-living tissues, this transition to effuse growth is more readily achieved by the arguably less-specialised cord-formers than by rhizomorphic *Armillaria* species, and this often results in substantial competitive exclusion of the latter by the former (Thompson & Boddy, 1983; Cooke & Rayner, 1984). The possibility of latent invasion as a mechanism of establishment is another example of the type of information needed. In general, such a mechanism is suspected when development of an extensive mycelium occurs more rapidly than expected following normal patterns of arrival, establishment and spread from a colonisation court, one mechanism being the dissemination of propagules within fluids (Boddy & Rayner, 1983; Cooke & Rayner, 1984). Possible examples among agaricoid fungi are provided by *Pleurotus ostreatus* and *Flammulina velutipes*, which develop remarkably rapidly and on a broad front in sapwood of recently dead or stressed deciduous trees such as elm and sycamore (*Acer pseudoplatanus*). Both these agarics produce abundant conidia, although their potential role during establishment in sapwood requires verification.

Exit patterns

The nature of resources can determine patterns of exit from substrata and consequent colonisation of new domains. This will be discussed here in relation to the two distinctive modes of exit via reproduction or mycelium.

Exit via reproduction

As discussed by Rayner *et al.* (1984), there is a complex three-way inter-relation between a mycelial domain (space within the immediate sphere of influence of the thallus), resource pool (nutrient and water supply available to the thallus) and reproductive commitment (energy diverted from vegetative to reproductive development). The larger the resource pool on which a mycelium can draw, the greater, ultimately, can be its reproductive commitment. In turn, the size of the resource pool will often be directly related to that of the mycelial domain, so that agarics forming extensive mycelia due to their patterns of arrival and establishment, resource preferences, or tolerance of selective stresses (see above) will be capable ultimately of relatively greater reproductive commitment. The distinction in size between the minute fruit bodies of agarics restricted to small resource units, such as *Marasmius* and *Mycena* species on twigs, fruits, leaf laminae and petioles, and the much larger, grouped, fruit bodies of species such as *Pholiota squarrosa* causing heart-rot is easily understood. However, the relation between mycelial domain and resource pool need not be absolute or direct. The resource pool may depend, for example, on ability to utilise recalcitrant substrates and, if nutrients are mobilised towards the mycelium, as in mycorrhizas, considerable commitment to reproduction may be possible without development of extensive mycelium.

An important factor in these considerations is time, particularly in relation to the longevity of mycelia. The latter is dependent on the longevity of resources although the reverse relation (that durable resources always contain long-lived mycelia) does not necessarily apply, due to effects of competition. In non-durable systems there will be pressure for early and complete commitment to reproduction, whereas in durable systems such commitment may be either slow or partial and intermittent over an extended time period. In general, rapid commitment to reproduction may be effected either by production of asexual propagules, which in terms of energy are least expensive, or of fruit bodies containing a minimum amount of material in relation to spore output. Agarics exploiting ephemeral resources would therefore be expected to exhibit such characteristics; species of *Coprinus*, *Conocybe* and *Bolbitius* provide good examples, at least with respect to their fruit bodies. By contrast, species of *Tricholoma* take longer to mature and they withhold spore production. Boletes, although bulky, develop quite quickly and the stipe tissue often forms a spore-producing hymenial palisade both before the pileus has expanded and well into maturity. In *Leccinum*, for instance, the scabrosities on the stipe produce spores (Watling: Chapter 11).

Longevity of resources may result from physical inaccessibility, allelopathic chemicals, refractory substrates or continuity of supply in space and/or time. Continuity of supply occurs, for example, in litter accumulations which, besides forming a continuous layer over the soil surface, are replenished seasonally or throughout the year by litter fall. Similarly, mycorrhizal agarics are, subject to seasonal cycles, in contact with a replenishable resource pool. For agarics exploiting spatiotemporally discontinuous resources, a cue for reproductive differentiation may come with nutrient exhaustion or attainment of the physical boundary of the substratum. No such cue exists for those with continuous supplies, and it has been suggested that here rhythmic growth of mycelia between alternating phases of extension and sporulation, mediated by external or endogenous factors, could help fulfil the requirement for reproduction. Such rhythms, associated with seasonal factors, could well explain much of the behaviour of agarics forming fairy rings (Lysek, 1984).

Although rhythms could explain some aspects, seasonal fruiting of agarics and other fungi is still not fully understood. A major influence must certainly be that of climate and microclimate and, as a whole, agaric fruit bodies are extremely sensitive to frost and desiccation. Only those buffered in some way against these extremes will be independent of season.

Buffering against extremes may be due to features either of the resource or of the fruit bodies. In Arctic/alpine conditions, saprotrophic agarics tend to produce small fruit bodies obtaining protection from the surrounding vegetation whereas fruit bodies of mycorrhizal species are a more normal size, water reserves originating from the host. Agarics inhabiting bulky woody substrata are also likely to be buffered against effects of desiccation, relative to those occupying small or exposed substrata. This is supported by the frequency of coarsely hairy surfaces which must inhibit water loss, as in *Schizophyllum*.

Adaptations of fruit bodies are important in allowing completion of the life cycle in resources exposed to environmental stresses. These include viscidness of the pileus and gills, as seen in *Flammulina velutipes* (see Williams *et al.*: Chapter 18); this absorbs water and maintains water balance when the pellicle dries out, sealing the water in (Moser, 1982) and playing a similar role to hairy surfaces. More extreme adaptations result in a tendency towards gasteroid forms, as in *Montagnites* and *Cytarrophyllum* in arid regions, and in the hypogeous forms which develop within the 'protected' environment of soil and occur in many families in various parts of the world, for example, the astrogastraceous species which have affinities with *Russula* and *Lactarius* (Watling: Chapter 11).

Here it is worth reiterating that the occurrence of reproductive bodies may sometimes provide a misleading picture of resource relations. One way in which this can happen is where reproduction occurs only on a particular resource type, giving the impression of ecological specificity whereas colonisation of other resource types is still possible. Thus there are examples of mycorrhizal species which fruit only with particular hosts but form mycorrhizas with a wider range (Molina & Trappe, 1982). Since ability to reproduce is an important fitness character, such fungi must nevertheless be on the road to selectivity.

Chlamydospores and sclerotial structures, discussed above in relation to survival of environmental stresses and patterns of arrival, are not treated here as *exit* structures.

Exit via mycelium

Exit can occur by outgrowth of diffuse or aggregated mycelium, and many aspects have been discussed in relation to arrival and establishment. An important consideration is the timing of exit, which may vary between diffuse and aggregated systems and, in turn, may be related to whether nutrients are available to interstitial mycelium, that is, to the extent to which growth occurs between nutrient depots across nutrient deserts. In the latter situation, for example, it might be of adaptive value for there to be a delay between arrival and exit, allowing build-up of energy reserves in the intervening period. This possibility appears to have received little experimental consideration, but some data obtained by Thompson & Rayner (1982*b*, 1983) may provide some clues. They examined outgrowth of a range of cord-forming Basidiomycotina from inoculated wood blocks into tubes containing non-sterile or sterile (γ-irradiated or autoclaved) soil. In sterile soil the mycelia were not aggregated into cords to the extent seen in non-sterile soil and, excepting *Tricholomopsis platyphylla*, had slower extension rates. Again with the exception of *T. platyphylla*, the inoculum blocks (including those occupied by the agaric *Hypholoma fasciculare*) underwent significantly greater losses in dry weight in sterile soil than did blocks in non-sterile soil. This suggests that in generally high-nutrient conditions (sterile soil) there may be a tendency for an established mycelium to dwell. Where resources are more limited and also localised (non-sterile soil), channelling of nutrients occurs, so facilitating increased exploration of the environment through rapid mycelial extension (Cooke & Rayner, 1984).

Evolutionary considerations: ecological strategies and taxonomic relationships

It should now be apparent that resource relations constitute an important, complex and often neglected component of the biology of agarics. A major problem is how to rationalise the bewildering array of information within an evolutionary framework. Here we shall briefly attempt to provide some clues.

One approach could be through the application of the concepts of ecological strategies which have proved helpful in animal and plant ecology and have recently attracted the attention of mycologists (Swift, 1976; Pugh, 1980; Andrews & Rouse, 1982; Cooke & Rayner, 1984). A starting point is the recognition of two basic types of selection, *r*- and *K*-selection, by which organisms are fitted for an ephemeral or sustained existence, characterised, respectively, by rapid and by slow or intermittent commitment to reproduction (MacArthur & Wilson, 1967; Harper & Ogden, 1970). Within the *r–K* spectrum it has subsequently been recognised that three distinctive types of selection are involved (Grime, 1979), *R*-, *S*- and *C*-selection, which result, respectively, in three primary ecological strategies: ruderal, stress-tolerant and combative (the latter term is preferred to competitive, for reasons discussed by Cooke & Rayner (1984) and by Rayner & Webber (1984)). Ruderal organisms are *r*-selected, whereas stress-tolerant and combative ones are *K*-selected. The three primary strategies provide extreme points in a spectrum of behaviour and many organisms, including fungi, combine characteristics of two or all three, resulting in secondary and tertiary strategies, respectively. For reasons similar to those relating to nutritional modes (see above), strategies are best regarded as entities in themselves, useful for classifying fungal *behaviour*, but not necessarily fungi *per se*.

Very generally, ruderal fungi are ephemeral and are characteristic of disturbed habitats where competitors are absent and nutrients readily available and assimilable. Stress-tolerant fungi are specialised towards habitats in which other species are disadvantaged by abiotic stress factors such as unavailability or intractability of nutrients, microenvironmental extremes, and presence of allelopathic chemicals. Combative fungi occur in undisturbed habitats where stress factors are relatively absent but where competitors must be overcome by antagonistic mechanisms.

Relative to other groups of fungi, agarics tend to exhibit stress-tolerant or combative strategies. However, relative to each other, it is easier to distinguish types exhibiting all three primary strategies. Thus, ruderal species tend to be relatively non-selective in their resource relations and to occur in ephemeral substrata, as do many Coprinaceae, or only early during community development in durable substrata, for example *Schizophyllum commune* and *Flammulina velutipes*. Stress-tolerant species, because they are specialised towards habitats where others are disadvantaged, often exhibit strong resource selectivity, for example many wood-decay species attacking standing trees. Combative species tend to be non-selective and to be dominant at later stages of community development in durable substrata; *Hypholoma fasciculare*, which occurs on wood of a wide range of both coniferous and angiospermous trees, is one example.

Turning to another aspect of eco-evolutionary considerations, one of the most important benefits of natural classification systems is that, by reflecting relationships, they are more likely to be of predictive value. It is therefore of interest to consider the degree to which current classifications of agarics are in accord with the range of resource relations which we have described.

The classification of the agarics has only recently stabilised and even now some areas are in a state of flux. However, the modern classification, based on anatomical features and morphology of hymenial structures, is more in keeping with the findings of ecologists than the traditional approach, persisting in certain areas and based on the end products of developmental patterns, although these same patterns fit with the current approach (Watling: Chapter 11). A good correlation has already been indicated in general terms for major groups of basidiomycetes (Watling, 1982) and this generally holds good for agarics.

The object of this account is to offer briefly some idea of how resource relations dovetail into the scheme if, as we believe, one exists. This can be done most successfully by reference to two traditional genera such as *Boletus* and *Tricholoma* as conceived by Rea (1922), each containing a spectrum of life-forms.

Pearson (1946) produced the first account of British boleti, comprising six genera, the largest being *Boletus*, a scheme which was retained almost unchanged even as late as 1960 (Dennis, Orton & Hora, 1960). Traditionally, *Boletus* included nearly all soft-pored fungi, but taking Pearson's circumscription one would now distinguish seven distinct genera, excluding *Boletus* and *Strobilomyces*: *Xerocomus*, known to have a worldwide distribution; *Leccinum*, a predominantly northern temperate genus with only a few subtropical/tropical South-East Asian taxa; *Pulveroboletus*, an essentially tropical or subtropical genus; *Tylopilus*, a cosmopolitan genus apparently more concentrated in South-East Asia and common in Australasia; *Chalciporus*, a genus with three distinct centres of distribution – Central Africa, the Caribbean and northern temperate forests; *Suillus*, confined to the boreal forests; and *Boletus sensu stricto*, found in temperate and mediterranean areas of the southern and northern hemispheres. Of these genera, *Xerocomus*, which is one of the few boletoid genera to occur in hygrophytic forests, has been placed in its own family (Pegler & Young, 1981). *Boletus*, *Leccinum* and *Pulveroboletus* are placed in the Boletaceae, and the last group, comprising *Tylopilus*, *Chalciporus* and *Suillus*, is linked with *Strobilomyces*, a mostly tropical–subtropical genus abundant in Australasia. Finally, *Fuscoboletinus*, a genus not universally accepted, has been related to *Strobilomyces* and *Suillus*; it is a small specialised genus confined to larch, especially in the New World.

The distribution of *Xerocomus* reflects the varied nutritional and ecological requirements of its members: a few species are obligately ectomycorrhizal and tend to possess compressed fusiform basidiospores (boletoid) as do members of the genera *Boletus* and *Leccinum*, whereas other species are facultatively mycorrhizal or non-mycorrhizal and show a range of spore morphology; *X. radicicola* is considered to be parasitic (Singer, 1978). This is paralleled in the nutritionally similar *Pulveroboletus* and in *Chalciporus* as was demonstrated by Pegler & Young (1981). *X. parasiticus* is always associated with *Scleroderma* species; as mentioned above, however, there is no evidence of necrotrophy but rather of stimulation of bolete fruiting, a similar phenomenon being found with *Suillus grevillei* and *Gomphidius maculatus*.

In the strobilomycetoid group, *Suillus* species show a high degree of selective biotrophism, some associating with single genera of the Coniferae and others with even a single species or group of closely related species. Except where introduced, as in Australia, they are found only in northern boreal forests. In parallel, *Fuscoboletinus* is also very selective. By contrast, *Chalciporus* associations are known with Fagales and Salicales, indeed some may associate with either Fagales (*Betula*) or Coniferae (*Pinus*), and some of the tropical species are not mycorrhizal at all. The genus *Tylopilus* and the related *Porphyrellus* have been badly confused by the 'dumping' of many unrelated taxa in the genera. As delimited today, both can be defined as obligately ectomycorrhizal genera associating with conifers and broadleaved trees. Those related species with sticky/viscid pilei are recognised as a distinct genus *Fistulinella* (= *Mucilopilus*), exclusively tropical and non-ectomycorrhizal except for an arm of distribution into New Zealand where members may be mycorrhizal. Indeed it seems possible that the stresses of the Mediterranean/temperate climate select for the mycorrhizal condition. *Austroboletus* is closely related and exhibits similar patterns.

Boletus itself is obligately ectomycorrhizal, usually with Fagales and Coniferae (Pinaceae), whereas in *Pulveroboletus* one sees the only examples within the boletoid fungi of a lignicolous saprotrophic lifestyle; the terrestrial species are always associated with the same range of hosts as *Boletus* and are probably ectomycorrhizal. Finally, *Leccinum*, parallel to *Suillus* in the Strobilomycetaceae, is associated rather strictly with Fagales and Salicales and less frequently with the Pinaceae. Indeed extensive speciation seems to have taken place amongst the taxa associated with arborescent weed species such as birch (*Betula*) and poplar (*Populus*).

The gasteroid *formae speciales* of the boletes fit in admirably with these decisions; thus within the Boletales parallel linear series can be demonstrated. These linear series are seen in the Gomphidiaceae, a family which is similar to *Suillus* in distribution and host taxon, and in the Paxillaceae, also related to the boletes, which includes active wood-rotters as well as facultative mycorrhiza-formers.

The traditional genus *Tricholoma* includes in the British Isles six genera in addition to *Tricholoma* itself; these provide a second example of the correlation between generic segregates and ecological strategy. *Tricholoma*, in its modern restricted sense based on anatomical data, contains about 42 species in the British Isles, all of which are obligate ectomycorrhiza-formers with Fagales and Pinaceae, often showing considerable selectivity. *Tricholomopsis*, on the other hand, with only a small number of species, is saprotrophic, the constituent members growing on woody debris, particularly of conifers, and causing rapid decomposition; *T. platyphylla*, however, occurs in deciduous woodland and, because of differences in morphology correlated with this habitat preference, has been recently transferred to the genus *Megacollybia* by Kotlaba & Pouzar (1972). *Melanoleuca* appears to be a temperate genus occurring in both pastures and woodlands and is undoubtedly litter-decomposing as are the members of *Lepista*, a genus including the familiar 'Wood Blewit' and 'Common Blewit'. *Squamanita odorata* and its relatives are almost totally boreal and their ecology is little understood; they are characterised by the production of protocarpic tubers covered in chlamydospores (Watling, 1980), first demonstrated by Bas (1965), but the part played by these structures is still unknown. Remaining are two closely related taxa, *Calocybe* and *Lyophyllum*, which are rather remote from *Tricholoma sensu stricto*. They demonstrate a linear series, as seen in the boletes, and are more closely related to some fungi formerly placed in the genus *Collybia* than they are to *Tricholoma* or indeed to *Collybia*. Many *Lyophyllum* species produce clusters of fruit bodies, often of quite gigantic size, and have an advantage in producing mycelial fans which permeate out into mineral soil from the leafy debris from which they take their nutrients. *Calocybe*, however, is a genus of litter-decomposers, particularly of Gramineae, although some species are woodland fungi. Both genera are related to *Tephrocybe* which contains many species extending the preference for fruiting on mineral soil substrata to bonfire sites, and species which utilise nitrogenous (ammonogenous) material.

Thus using only two examples, both in Britain and on a world scale, the pattern appears to hold true and gives hope for unification of our ideas. Bondarstev (1963) was of the opinion that in the polypores ecology must be correlated with anatomy, in that environmental pressures selected those very microstructures used in delimiting taxa. This is true, but by calling on a whole series of disciplines a picture, if only a glimpse, is becoming clearer. Taxonomy is based on correlations and such correlations are demonstrated throughout this chapter and in the tables offered. However, one must expect similar levels of evolution or exploitation of similar natural phenomena in unrelated groups. As an example, association of an agaric genus with insects is found in two quite unrelated groups: in the New World *Leucoagaricus* (Lepiotaceae, a group related to the true mushrooms) with attine ants, and in South-East Asia and Africa *Termitomyces* (Tricholomataceae) with termites. The latter genus is confined to termite gardens whereas species of *Leucoagaricus* are worldwide litter-decomposers. In addition, the different *Termitomyces*–termite relationships can be related to theories of continental drift.

Finally, an understanding of the part played by anamorphs is paramount. Specialised, complex anamorphs have been described in *Mycena citricolor* (Tricholomataceae-cephalosus) and *Coprinus clastophyllus* (Coprinaceae-caterva, Watling, 1979). These are, however, exceptional, as the main anamorphic condition in the agarics is thalloarthrosporic although special chlamydospores have also been described (Watling, 1979). Thallic arthroconidia are found in culture, but again one can only speculate as to how they contribute to the fungal strategy in nature. *Mycena inclinata*, common on hardwood, produces arthroconidia; *Collybia peronata*, an important litter-fungus, does not. *Lepista nuda* and *Tricholomopsis platyphylla* are also litter-decomposers, the former utilising leafy debris, the latter twiggy debris; both produce cord-like structures but neither readily produces conidia. However, *Agrocybe praecox* and its relatives, also litter-decomposers but within cultivated or disturbed soils (glacier retreat areas, laval soils and volcanic ash), produce both cords and conidia. The rule therefore is not infallible, but we are at an early stage in understanding the principles involved.

We end with the hope that this overview will have provided an orientation, allowing the more specific chapters on resource relations and agaric morphogenesis that follow to be viewed within a general ecological context.

References

Anderson, J. B., Ullrich, R. C., Roth, L. F. & Filip, G. M. (1979). Genetic identification of clones of *Armillaria mellea* in coniferous forests in Washington. *Phytopathology*, **69**, 1109–11.

Andrews, J. H. & Rouse, D. I. (1982). Plant pathogens and the theory of r- and k-selection. *The American Naturalist*, **120**, 283–96.

Barron, G. L. & Dierkes, Y. (1977). Nematophagous fungi: *Hohenbuehelia*, the perfect stage of *Nematoctonus*. *Canadian Journal of Botany*, **55**, 3054–62.

Bas, C. (1965). The genus *Squamanita*. *Persoonia*, 3, 331–59.

Boddy, L. (1984). The micro-environment of basidiomycete mycelia in temperate deciduous woodlands. In *The Ecology and Physiology of the Fungal Mycelium*, ed. D. H. Jennings & A. D. M. Rayner, pp. 261–89. Cambridge University Press.

Boddy, L. & Rayner, A. D. M. (1983). Origins of decay in living deciduous trees: the role of moisture content and a re-appraisal of the expanded concept of tree decay. *New Phytologist*, **94**, 623–41.

Bondarstev, M. A. (1963). On the anatomical criterion in the taxonomy of Aphyllophorales. *Botanicheskiĭ Zhurnal SSSR*, **48**, 362–72.

Buller, A. H. R. (1931). *Researches on Fungi*, vol. 4, part 2, pp. 139–86. London: Longmans, Green & Co.

Buller, A. H. R. (1934). *Researches on Fungi*, vol. 6, pp. 397–443. London: Longmans, Green & Co.

Campbell, E. O. (1971). Notes on the fungal association of two *Monotropa* species in Michigan. *The Michigan Botanist*, **10**, 63–7.

Coates, D. (1984). *The biological consequences of somatic incompatibility in wood decaying basidiomycetes and other fungi.* Ph.D. thesis, University of Bath.

Coley-Smith, J. R. & Cooke, R. C. (1971). Survival and germination of fungal sclerotia. *Annual Review of Phytopathology*, **9**, 65–92.

Cooke, R. C. & Rayner, A. D. M. (1984). *Ecology of Saprotrophic Fungi.* London & New York: Longman.

De Vries, G. A., Kemp, R. F. O. & Speller, D. C. E. (1971). Endocarditis caused by *Coprinus delicatulus*. In *Comptes Rendus 5e Congrès de la Société Internationale de Mycologie Humaine et Animale*, ed. J. G. O'Sullivan, pp. 185–6. Paris: Institut Pasteur.

Dennis, R. W. G., Orton, P. D. & Hora, F. B. (1960). New check list of British agarics and boleti. *Supplement to the Transactions of the British Mycological Society*, **43**, 1–225.

Duddrige, J. A., Malibari, A. & Read, D. J. (1980). Structure and function of mycorrhizal rhizomorphs with special reference to their role in water transport. *Nature, London*, **287**, 834–6.

Flodin, K. & Fries, N. (1978). Studies on volatile compounds from *Pinus sylvestris* and their effect on wood-decomposing fungi: II. Effects of some volatile compounds on fungal growth. *European Journal of Forest Pathology*, **8**, 300–10.

Frankland, J. C. (1981). Mechanisms in fungal successions. In *The Fungal Community. Its Organization and Role in the Ecosystem*, ed. D. T. Wicklow & G. C. Carroll, pp. 403–26. New York & Basel: Marcel Dekker.

Frankland, J. C. (1984). Autecology and the mycelium of a woodland litter decomposer. In *The Ecology and Physiology of the Fungal Mycelium*, ed. D. H. Jennings & A. D. M. Rayner, pp. 241–60. Cambridge University Press.

Fries, N. (1961). The growth-promoting activity of some aliphatic aldehydes on fungi. *Svensk Botanisk Tidskrift*, **55**, 1–16.

Fries, N. (1977). Germination of *Laccaria laccata* spores *in vitro*. *Mycologia*, **69**, 848–50.

Fries, N. (1978). Basidiospore germination in some mycorrhiza-forming Hymenomycetes. *Transactions of the British Mycological Society*, **70**, 319–24.

Garrett, S. D. (1970). *Pathogenic Root-infecting Fungi*. Cambridge University Press.

Gibbs, J. N. & Gulliver, C. C. (1977). Fungal decay of dead elms. *European Journal of Forest Pathology*, **7**, 193–200.

Glasare, P. (1970). Volatile compounds from *Pinus sylvestris* stimulating the growth of wood-rotting fungi. *Archiv für Mikrobiologie*, **72**, 333–43.

Gordon, H. D. (1938). Note on a rare beetle, *Cartodere filum* Aubé, eating fungus spores. *Transactions of the British Mycological Society*, **21**, 193–7.

Gregory, P. H. (1984). The first Benefactor's Lecture. The fungal mycelium: an historical perspective. *Transactions of the British Mycological Society*, **82**, 1–11.

Grime, J. P. (1979). *Plant Strategies and Vegetation Processes*. Chichester: John Wiley.

Hahn, M. G., Darvill, A. G. & Albersheim, P. (1981). Host–pathogen interactions. XIX. The endogenous elicitor, a fragment of plant cell wall polysaccharide that elicits phytoalexin accumulation in soybean. *Plant Physiology*, **68**, 1161–9.

Hamada, M. (1939). Studien über die Mycorrhiza von *Galeola septentrionalis* Reichb.f. Ein neuer Fall der Mykorrhiza – Bildung durch intraradicale Rhizomorpha. *Japanese Journal of Botany*, **10**, 151–211.

Hamada, M. (1940). Physiologisch–morphologische Studien über *Armillaria mellea* (Vahl) Quél., mit besonderer Rücksicht auf die Oxalsäure-bildung. Ein Nachtrag zur Mycorrhiza von *Galeola septentrionalis* Reichb.f. *Japanese Journal of Botany*, **10**, 387–463.

Harley, J. L. & Smith, S. E. (1983). *Mycorrhizal Symbiosis*. London & New York: Academic Press.

Harper, J. L. & Ogden, J. (1970). The reproductive strategy of higher plants. I. The concept of strategy with special reference to *Senecio vulgaris* L. *Journal of Ecology*, **58**, 681–98.

Harrison, A. F. (1971). The inhibitory effect of oak leaf litter tannins on the growth of fungi in relation to litter decomposition. *Soil Biology and Biochemistry*, **3**, 167–72.

Hart, J. H. & Shrimpton, D. M. (1979). Role of stilbenes in resistance of wood to decay. *Phytopathology*, **69**, 1138–43.

Hintikka, V. (1982). The colonization of litter and wood by basidiomycetes in Finnish forests. In *Decomposer Basidiomycetes: their Biology and Ecology*, ed. J. C. Frankland, J. N. Hedger & M. J. Swift, pp. 227–39. Cambridge University Press.

Jahrmann, H. J. & Prillinger, H. (1983). Das Vorkommen eines 'Hefe'-Stadiums bei dem Homobasidiomyceten *Asterophora* (*Nyctalis*) *lycoperdoides* (Bull.) Ditm. ex S. F. Gray und seine Bedeutung für die Phylogenese der Basidiomyceten. *Zeitschrift für Mykologie*, **49**, 195–235.

Kemp, R. F. O. (1975). Breeding biology of *Coprinus* species in the Section Lanatuli. *Transactions of the British Mycological Society*, **65**, 375–88.

Kemp, R. F. O. (1977). Oidial homing and the taxonomy and speciation of basidiomycetes with special reference to the genus *Coprinus*. In *Herbette Symposium on the Species Concept in Hymenomycetes*, 1976, ed. H. Clémençon, pp. 259–76. Vaduz: Cramer.

Kendrick, B. & Watling, R. (1979). Mitospores in basidiomycetes. In *The Whole Fungus*, vol. 2, ed. B. Kendrick, pp. 473–545. Ottawa: National Museum of Natural Sciences, National Museums of Canada & the Kananaskis Foundation.

Kile, G. A. (1983). Identification of genotypes and the clonal development of *Armillaria luteobubalina* Watling & Kile in eucalypt forests. *Australian Journal of Botany*, **31**, 657–71.

Kirk, T. K. & Fenn, P. (1982). Formation and action of the ligninolytic system in basidiomycetes. In *Decomposer Basidiomycetes: Their Biology and Ecology*, ed. J. C. Frankland, J. N. Hedger & M. J. Swift, pp. 67–90. Cambridge University Press.

Kotlaba, F. & Pouzar, Z. (1972). Taxonomic and nomenclatural notes on some Macromycetes. *Ceská Mykologie*, **26**, 217–22.

Lewis, D. H. (1973). Concepts in fungal nutrition and the origin of biotrophy. *Biological Reviews*, **48**, 261–78.

Lindeberg, G., Lindeberg, M., Lundgren, L., Popoff, T. & Theander, O. (1980). Stimulation of litter-decomposing basidiomycetes by flavonoids. *Transactions of the British Mycological Society*, **75**, 455–9.

Lösel, D. (1964). The stimulation of spore germination of *Agaricus bisporus* by living mycelium. *Annals of Botany, NS*, **28**, 541–54.

Lysek, G. (1984). Physiology and ecology of rhythmic growth and sporulation in fungi. In *The Ecology and Physiology of the Fungal Mycelium*, ed. D. H. Jennings & A. D. M. Rayner, pp. 323–42. Cambridge University Press.

MacArthur, R. H. & Wilson, E. O. (1967). *The Theory of Island Biogeography*. New Jersey: Princeton University Press.

Macdonald, J. A. (1949). The heather rhizomorph fungus, '*Marasmius androsaceus*' Fries. *Proceedings of the Royal Society of Edinburgh, Section B*, **63**, 230–41.

McDougall, W. B. (1919). Development of *Stropharia epimyces*. *Botanical Gazette*, **67**, 258–63.

Malajzuk, N., Molina, R. & Trappe, J. M. (1984). Ectomycorrhiza formation in *Eucalyptus*. II. The ultrastructure of compatible mycorrhizal fungi and associated roots. *New Phytologist*, **96**, 43–53.

Mason, P. (1975). The genetics of mycorrhizal associations between *Amanita muscaria* and *Betula verrucosa*. In *The Development and Function of Roots*, ed. J. G. Torrey & D. T. Clarkson, pp. 567–74. London: Academic Press.

Melin, E. (1959). *Mycorrhiza. Encyclopedia of Plant Physiology*, vol. 11, ed. W. Ruhland, pp. 605–38. Berlin: Springer-Verlag.

Melin, E. (1963). Some effects of forest tree roots on mycorrhizal Basidiomycetes. In *Symbiotic Associations*, ed. B. Mosse & P. S. Nutman, pp. 125–45. Cambridge University Press.

Molina, R. J. & Trappe, J. M. (1982). Patterns of ectomycorrhizal host specificity and potential amongst Pacific Northwest conifers and fungi. *Forest Science*, **28**, 423–57.

Moser, M. (1982). Mycoflora of the transitional zone from subalpine forests to alpine tundra. In *Arctic and Alpine Mycology*, ed. G. A. Laursen & J. F. Ammirati, pp. 371–89. Seattle & London: University of Washington Press.

Mowe, G., King, B. & Senn, S. J. (1983). Tropic responses of fungi to wood volatiles. *Journal of General Microbiology*, **129**, 779–84.

Nordbring-Hertz, B. (1984). Mycelial development and lectin-carbohydrate interactions in nematode-trapping fungi. In *The Ecology and Physiology of the Fungal Mycelium*, ed. D. H. Jennings & A. D. M. Rayner, pp. 419–32. Cambridge University Press.

Park, D. (1976). Carbon and nitrogen levels as factors influencing fungal decomposers. In *The Role of Terrestrial and Aquatic Organisms in Decomposition Processes*, ed. J. M. Anderson & A. Macfadyen, pp. 41–59. Oxford: Blackwell Scientific Publications.

Pearson, A. A. (1946). Notes on the Boleti, with a short monograph and key. *Naturalist*, 85–99.

Pegler, D. N. & Young, T. W. K. (1981). A natural arrangement of the Boletales, with reference to spore morphology. *Transactions of the British Mycological Society*, **76**, 103–46.

Pugh, G. W. F. (1980). Strategies in fungal ecology. *Transactions of the British Mycological Society*, **75**, 1–14.

Rayner, A. D. M., Coates, D., Ainsworth, A. M., Adams, T. J. H., Williams, E. N. D. & Todd, N. K. (1984). The biological consequences of the individualistic mycelium. In *The Ecology and Physiology of the Fungal Mycelium*, ed. D. H. Jennings & A. D. M. Rayner, pp. 509–40. Cambridge University Press.

Rayner, A. D. M. & Hedges, M. J. (1982). Observations on the specificity and ecological role of basidiomycetes colonizing dead elm wood. *Transactions of the British Mycological Society*, **78**, 370–3.

Rayner, A. D. M. & Todd, N. K. (1979). Population and community structure and dynamics of fungi in decaying wood. *Advances in Botanical Research*, **7**, 333–420.

Rayner, A. D. M. & Webber, J. F. (1984). Interspecific mycelial interactions – an overview. In *The Ecology and Physiology of the Fungal Mycelium*, ed. D. H. Jennings & A. D. M. Rayner, pp. 383–417. Cambridge University Press.

Rea, C. (1922). *British Basidiomycetae*. Cambridge University Press.

Read, D. J. (1984). The structure and function of the vegetative mycelium of mycorrhizal roots. In *The Ecology and Physiology of the Fungal Mycelium*, ed. D. H. Jennings & A. D. M. Rayner, pp. 215–40. Cambridge University Press.

Read, D. J. & Armstrong, W. (1972). A relationship between oxygen transport and the formation of the ectotrophic mycorrhizal sheath in conifer seedlings. *New Phytologist*, **71**, 49–53.

Redhead, S. A. (1981). Parasitism of bryophytes by agarics. *Canadian Journal of Botany*, **59**, 63–7.

Rice, P. F. (1970). Some biological effects of volatiles emanating from wood. *Canadian Journal of Botany*, **48**, 719–35.

Sagara, N. (1976). Presence of a buried mammalian carcass indicated by fungal fruiting bodies. *Nature*, **262**, 816.

Scheffer, T. C. & Cowling, E. B. (1966). Natural resistance of wood to microbial deterioration. *Annual Review of Phytopathology*, **4**, 147–70.

Singer, R. (1978). Notes on bolete taxonomy – II. *Persoonia*, **9**, 421–38.

Sortkjaer, O. & Alderman, K. (1972). Rhizomorph formation in fungi: I. Stimulation by ethanol and acetate and inhibition by disulfiram of growth and rhizomorph formation in *Armillaria mellea*. *Physiologia Plantarum*, **26**, 376–80.

Swift, M. J. (1976). Species diversity and the structure of microbial communities in terrestrial habitats. In *The Role of Terrestrial and Aquatic Organisms in Decomposition Processes*, ed. J. M. Anderson & A. Macfadyen, pp. 185–222. Oxford: Blackwell Scientific Publications.

Swift, M. J., Heal, O. W. & Anderson, J. M. (1979). *Decomposition in Terrestrial Ecosystems*. Oxford: Blackwell Scientific Publications.

Thompson, W. (1984). Distribution, development and functioning of mycelial cord systems of decomposer basidiomycetes of the deciduous woodland floor. In *The Ecology and Physiology of the Fungal Mycelium*, ed. D. H. Jennings & A. D. M. Rayner, pp. 185–214. Cambridge University Press.

Thompson, W. & Boddy, L. (1983). Decomposition of suppressed oak trees in even-aged plantations. II. Colonization of tree roots by cord- and rhizomorph-producing basidiomycetes. *New Phytologist*, **93**, 277–91.

Thompson, W. & Rayner, A. D. M. (1982a). Spatial structure of a population of *Tricholomopsis platyphylla* in a woodland site. *New Phytologist*, **92**, 103–14.

Thompson, W. & Rayner, A. D. M. (1982b). Structure and development of mycelial cord systems of *Phanerochaete laevis* in soil. *Transactions of the British Mycological Society*, **78**, 193–200.

Thompson, W. & Rayner, A. D. M. (1983). Extent, development and function of mycelial cord systems in soil. *Transactions of the British Mycological Society*, **81**, 333–45.

Vanderplank, J. E. (1976). Four essays. *Annual Review of Phytopathology*, **14**, 1–10.

Vanderplank, J. E. (1978). *Genetic and Molecular Basis of Plant Pathogenesis*. Berlin & New York: Springer-Verlag.

Waksman, S. A. & Tenney, F. G. (1928). Composition of natural organic materials and their decomposition in the soil. III. The influence of nature of plant upon the rapidity of its decomposition. *Soil Science*, **26**, 155–71.

Warcup, J. H. & Talbot, P. H. B. (1962). Ecology and identity of mycelia isolated from soil. *Transactions of the British Mycological Society*, **45**, 495–518.

Watling, R. (1971). Basidiomycetes: Homobasidiomycetidae. In *Methods in Microbiology*, vol. 4, ed. C. Booth, pp. 219–36. London & New York: Academic Press.

Watling, R. (1972). Notes on British agarics: III. *Notes, Royal Botanic Garden, Edinburgh*, **32**, 127–33.

Watling, R. (1974a). Dimorphism in *Entoloma abortivum*. *Bulletin de la Société Linnéene de Lyon*. Numéro special, **43**, 449–70.

Watling, R. (1974b). Macrofungi in the oak woods of Britain. In *The British Oak*, ed. M. G. Morris & F. H. Perring, pp. 221–34. Faringdon: Classey.

Watling, R. (1978). The distribution of larger fungi in Yorkshire. *Naturalist*, **103**, 39–57.

Watling, R. (1979). The morphology, variation and ecological significance of anamorphs in the Agaricales. In *The Whole Fungus*, vol. 2, ed. B. Kendrick, pp. 453–72. Ottawa: National Museum of Natural Sciences, National Museums of Canada & the Kananaskis Foundation.

Watling, R. (1980). Notes on some British agarics: IV. *Notes, Royal Botanic Garden, Edinburgh*, **33**, 325–31.

Watling, R. (1981). Relationships between macromycetes and the development of higher plant communities. In *The Fungal Community. Its Organization and Role in the*

Ecosystem, ed. D. T. Wicklow & G. C. Carroll, pp. 427–58. New York & Basel: Marcel Dekker.

Watling, R. (1982). Taxonomic status and ecological identity in the basidiomycetes. In *Decomposer Basidiomycetes: Their Biology and Ecology*, ed. J. C. Frankland, J. N. Hedger & M. J. Swift, pp. 1–32. Cambridge University Press.

Watling, R. & Sweeney, J. (1971). Observations on *Schizophyllum commune* Fries. *Sabouraudia*, **12**, 214–26.

Zak, B. (1974). Ectendomycorrhiza of Pacific madrone (*Arbutus menziesii*). *Transactions of the British Mycological Society*, **62**, 202–4.

2

Tropical agarics: resource relations and fruiting periodicity

J.N.HEDGER

Department of Botany and Microbiology, University College of Wales, Aberystwyth, SY23 3DA, UK

Introduction

In Chapter 1, Rayner, Watling & Frankland have considered what general principles may underlie relations between agarics and the resources and habitat types in which they are encountered. Similar exercises have been carried out on broader fronts in recent years, for example, considerations of the whole of the Basidiomycotina by Swift (1982), or indeed the whole of the terrestrial decomposition subsystem (Lynch & Poole, 1979; Swift, Heal & Anderson, 1979). All are characterised by a 'temperate' tone, reflecting the knowledge of the microbial ecology of the northern boreal and temperate climatic zones, and few references to work carried out in other major ecosystems of the world can be found in these reviews.

This paucity of evidence is particularly acute in the case of the agarics. As pointed out in Chapter 1, much information is limited to Western Europe, North America and Japan, and we know little of the ecology of agarics which occur in equivalent areas of the less-developed southern hemisphere, such as in the *Nothofagus* forests and boreal zones of South America (Singer & Moser, 1965) or in the cool, high-altitude equatorial zones of East Africa (Pegler, 1977) and South America (Dennis, 1970; Boekhout, 1983). The ideas put forward by Rayner *et al.* (Chapter 1) have yet to be tested in these environmentally parallel ecosystems and also in the very different conditions of the lowland tropical zones. The paucity of evidence as to even the agaric mycofloras of these tropical areas makes an analysis of general principles of resource relations in tropical agarics a very difficult task. Yet such is the aim of the first section of this chapter, which is amplified in the second half by some data obtained during a 2-year study of agarics in the lowland tropical zone of Ecuador. These two threads are then joined together to offer a simplified model of resource relations in the saprotrophic agaric.

Ecosystems and agarics

The tropics are by no means an ecological continuum and include desert, grassland and forest as major formations, each of which can be further sub-divided into many different sub-formations on local and continental scales (Holdridge, 1971). This account will be largely restricted to the lowland tropical forests, or, indeed, given the alarming rate at which they are disappearing, the climatic zones in which these formations would naturally predominate. In passing, it should be noted that in terms of resource availability and environmental factors, considerable differences exist between different tropical forest formations and between the grassland and desert ecosystems into which they merge. A very general comparison will be made.

In intact tropical forest the environment is generally considered to be uniform, with temperature near the ground in the diurnal range of 18–25 °C and high relative humidity (over 80%). Such a description is over simplistic and considerable variation occurs. Annual rainfall at 1500–2000 mm or more can, for example, have an uneven distribution in seasonal (monsoon) formations. Such seasonal formations may exhibit some synchrony in leaf fall from the trees, but in moist tropical rain forests litter deposition is continuous throughout the year (Longman & Jenik, 1974). New resources are constantly presented to the saprotrophic agaric community in such forests, albeit separated throughout the vegetation stand in time and space. Growth is also continuous, or at least in a coordinated series of root and shoot flushes (Alvim, 1964), so that new infection sites are always available to necrotrophic and biotrophic agarics. However, the very high species diversity in these formations (Richards, 1952) means that resources for taxon-selective and component-restricted agarics (see Rayner *et al.*: Chapter 1) will be highly disjunct in distribution. The multistoried structure of the forest, within which exist a number of synusiae of different life-forms (for instance, epiphytes and climbers – see Richards (1952)), further increases the heterogeneity of resources and habitats available to agarics. As an example, an area of tropical moist lowland forest in Ecuador, the agaric flora of which will be discussed later in this chapter, may be quoted. So far, 800 species of vascular plants have been recorded from an 87-ha reserve of this forest (Dodson & Gentry, 1978). For the whole of Amazonia, Gentry (1977) estimated that the species diversity of plants may be as high as 20000.

In such forests, most of the biomass is held above ground (perhaps 75–85%), mostly as wood (Longman & Jenik, 1974), so that the ligno-cellulose habitat is dominant. Leaves may, however, constitute a higher proportion of litter input to the decomposition subsystem and hence to the agaric saprotrophs. Swift *et al.* (1979) estimated the average litter input in tropical forests to be 30 t ha^{-1} yr^{-1} in comparison to 7.5–11.5 t ha^{-1} yr^{-1} for boreal and temperate forests. Golley (1983) put these figures at about 5 t ha^{-1} yr^{-1} and 1 t ha^{-1} yr^{-1}, respectively. Decomposition is continuous and may equal the annual litter fall. Saprotrophic agarics exist as part of a very active community with a high fungal biomass (Swift *et al.*, 1979), of which agaric mycelium makes up a considerable fraction.

In contrast to the relative uniformity of conditions and heterogeneity of resources in moist tropical forests, the more extreme climatic variations of the other tropical formations place constraints on the productivity of the plant community, its species diversity, and subsequently the resources and habitats available to agarics. In dry formations, such as the scattered-tree grassland of Africa, resource availability may be restricted to a very short wet season, when soil moisture levels may be high enough to permit mycelial growth to take place (perhaps in a cycle of 3 years in East Africa). Burning also reduces the amount of litter available to decomposers (Hopkins, 1965). It seems likely that the relatively slow-growing mycelium of agarics may be a disadvantage under such conditions and might be even more so in desert areas. The strategies of resource capture and retention described by Rayner *et al.* in Chapter 1 may not be appropriate to the brief periods of intense microbial and faunal activity which take place following the rains in these areas (Singh, 1977). However, agarics do occur and their fruit bodies may show xeromorphic characters (Rayner *et al.*: Chapter 1). *Podaxis*, which although a member of the Secotiaceae is probably of agaric origin (Watling: Chapter 11), represents such an evolutionary pathway to xerophily. We know little of the mycelium of these agarics. The onset of rains in dry tropical areas may result in an almost simultaneous flush of fruit bodies of terricolous agarics. This flush must be supported by nutrient reserves within the mycelium. A slow, but significant, growth of a stress-tolerant mycelium may be possible at low moisture tensions in the previous dry season. Alternatively, nutrients derived from resource exploitation during the previous wet season may be stored in the dormant mycelium. Parallels exist between these stressed environments and the long periods of physiological dormancy imposed on agarics of the boreal zone by removal of free water by freezing in the winter periods (Laursen & Chmielewski, 1982).

Another factor which reduces the availability of resources to saprotrophic agarics in the tropics, particularly in grassland systems, is competition from other decomposer organisms, especially arthropods, for lignocellulose as both wood and leaf-litter. African tropics are dominated by the Macro-termitinae (Lee & Wood, 1971) which consume much of the lignocellulose production of the plant community, up to 50% in some cases (Maldague, 1964). However, since the *de facto* decomposer in this system is the termite mycobiont *Termitomyces*, agarics turn out to be the dominant saprotrophs after all! Neotropical and Malaysian termites do not have agaric symbionts and therefore may exclude a proportion of the lignocellulose from the agaric orbit. However, they may have less importance within the decomposition system than termites in Africa (Matsumoto, 1976). Attine ants in the neotropics do have an agaric mycobiont, *Leucoagaricus* (see Rayner *et al.*: Chapter 1), but in this case the deflection of the nutrient flow is from the leaf component of the litter, which is cropped as living material by the ants (Stradling, 1977). The agaric mycobiont is thus presented with material of a higher resource quality than the fallen litter available to free-living saprotrophic agarics in these neotropical ecosystems.

The co-evolution of these two types of agaric mycobiont and their arthropod partners from saprotrophic litter-inhabiting progenitors is an example of a type of very effective resource capture which is, however, maintained by the animal symbiont rather than by the agaric itself. Once this protection is removed other agarics may colonise, and the degraded bed material provides a concentrated resource, which is exploited by a cord-forming species of *Lyophyllum* in the South American attine symbiosis (Haines, 1975; Hedger & Watling, unpublished) and by species of *Lepiota*, e.g. *Lepiota termitophila*, in termite nests (Heim, 1977).

Habitat considerations

If selectivity occurs in tropical agarics, the sheer diversity of life-forms and taxa of higher plants in tropical rain forest should logically result in a high diversity of agarics, both ubiquitous and resource-selected. These can be envisaged as distributed amongst many habitats, including the primary agaric habitats of other ecosystems, that is the soil, plant litter, or even standing dead trees and branches, and also in the synusiae unique to the tropical forest, for example, in epiphytic communities. The high relative humidity and rainfall in moist tropical rain forests allow the development of saprotrophic agarics on tree boles, branches and on litter trapped within understorey trees, stranglers, climbers and epiphytes (Longman & Jenik, 1974). Such aerial agaric communities, usually dominated by small agarics such as *Marasmius*, *Psathyrella*, *Gerronema*, *Crinipellis*, *Marasmiellus* and *Mycena*, may have a distinct species composition compared to those developed on equivalent resources which have reached the ground litter layer. They may be interconnected by an aerial network of cords and rhizomorphs, paralleling in complexity those within the soil litter system.

In this aerial environment, sites for biotrophic infection are also available, for example, mycorrhizas between epiphytic orchid species and agarics. Rishbeth (Chapter 3) notes the association between *Gastrodia* and *Armillaria mellea*, and other similar associations occur, e.g. between *Gymnopilus aculeatus* and Mexican orchids (Singer, 1962). Johnsson & Nylund (1979) considered *Favolaschia dybowskyana* (which has an agaricoid form but is a member of the Aphyllophorales) to be a common mycobiont in epiphytic orchid communities in West African forests. The distribution of this fungus between different tree species, i.e. its taxon selectivity, may determine the development of orchid communities upon them (Johanssen, 1974): an example of an agaric determining habitat selection in higher plants.

Retaining the focus on biotrophism but returning to ground level, ectomycorrhizal relationships are to be expected with root systems. The levels of taxon selectivity noted by Rayner *et al.* (Chapter 1) for temperate mycorrhizal agarics, combined with the expected totals of saprotrophs in aerial and soil litter communities, should result in a very large number of agaric species for tropical forest types.

Saprotrophs, biotrophs and necrotrophs – nutritional modes

Direct evidence as to the nutritional modes of most British agarics is not available (Hering, 1982), and data for tropical agarics are even sparser. However, some indirect evidence of the relative distribution of the nutritional modes proposed by Rayner *et al.* (Chapter 1) can be gained from considerations of mycofloras of tropical areas, such as those of East Africa (Pegler, 1977), the Lesser Antilles (Pegler, 1983) and the Venezuelan region (Dennis, 1970). Dennis makes a revealing comparison between the Venezuelan region and the British Isles. The agaric data are reproduced in Table 2.1.

Venezuela has a lower total of agaric species but this must be viewed as a result of undercollection, which Dennis considered in 1970 to be 2.5–3.5 times in favour of the British Isles for the whole mycoflora. Dennis also calculated the ratios of different agaric families, Venezuela: British Isles (see Table 2.1). A further comparison can be made by grouping the families according to their predominant modes of nutrition (following Watling, 1982), effectively biotrophic (ectomycorrhizal) and saprotrophic. Some of the biotrophs show Venezuela: British Isles ratios much higher

Table 2.1. *Numbers of species in families of Agaricales recorded for the Venezuelan region and the British Isles*

Predominant nutritional mode of family	Family	Number of species recorded		Ratio (Venezuela to British Isles)
		Venezuela	British Isles	
Biotrophic	Amanitaceae	26	68	1:2.6
Biotrophic	Russulaceae	15	143	1:9.5
Biotrophic	Cortinariaceae	20	408	1:20
Biotrophic/ saprotrophic	Paxillaceae	2	3	1:1.5
Biotrophic	Boletaceae	7	53	1:7.6
Biotrophic	Gomphidiaceae	—	4	—
	'Biotrophs' group total	70	679	
Saprotrophic	Tricholomataceae	287	423	1:1.5
Saprotrophic/ biotrophic	Hygrophoraceae	21	87	1:4.1
Saprotrophic	Crepidotaceae	7	25	1:3.6
Saprotrophic	Strophariaceae	29	66	1:2.3
Saprotrophic	Bolbitiaceae	14	49	1:3.5
Saprotrophic	Rhodophyllaceae	42	125	1:3.0
Saprotrophic	Agaricaceae	78	111	1:1.4
Saprotrophic	Coprinaceae	53	132	1:2.5
	'Saprotrophs' group total	531	1018	
	Total number of species	601	1697	1:2.8

Data derived from Dennis (1970) and Watling (1982).

than the 'undercollection' ratio, e.g. Cortinariaceae 1:20, whilst the saprotrophs show values which may be below this ratio, e.g. 1:1.5 for the Tricholomataceae. Of course this could be explainable on mycogeographical grounds, but it is more likely the figures show that, in the tropical area, the saprotrophic mode of nutrition is better developed and the biotrophic mode less well represented than in the temperate area. This difference can be emphasised by comparing the biotrophic and saprotrophic species totals from the two regions – the biotrophs:saprotrophs ratio is 1:1.5 for the British Isles but 1:7.6 for Venezuela. The fact that the British Isles higher plant communities are species-poor, in comparison with Venezuela, further underlines the paradoxical species poverty in agaric biotroph families in Venezuela. Calculations can also be made for a southern temperate area – the *Nothofagus* forests of Chile, utilising data from Singer (1971). In this case the ratio between mycorrhizal and saprotrophic agaric species is 1:1.4, very similar to the figures for the British Isles.

Ectomycorrhizal biotrophs

The under-representation of agaric families in which the mycorrhizal nutritional mode is well developed (at least when viewed in temperate terms) applies not only to the Venezuelan region. Most mycologists visiting the tropics are struck by the absence of the larger 'mycorrhizal' toadstools such as *Russula*, *Lactarius* and *Cortinarius*. Watling (1982) felt that this apparent difference was really due to the lack of a sharp distinction in the tropics between the obligate ectomycorrhizal (biotrophic) and saprotrophic modes of nutrition which are combined in (apparently) saprotrophic agaric species, rather than being separated in different saprotrophic and obligately ectomycorrhizal species, as is classically considered to be the case in the boreal and temperate zones. Such agarics, termed facultative by Singer (1971), would provide a route for the direct cycling of nutrients from plant litter to plant roots, which Went & Stark (1968) proposed took place in the poor soils of the areas of Amazonia which they studied. Unfortunately, no actual evidence exists for such a nutritional mode in tropical agarics, although the concept is apparently accepted by a number of authors (Frankland, 1982; Golley, 1983). Experimental studies are obviously desirable, and there are indications of such duality in a number of temperate agarics such as *Laccaria* (Singer, 1971; see Dighton & Mason: Chapter 5).

Alternatively, we may accept the views of Black (1980) and others – simply that the vesicular-arbuscular (VA) mycorrhizas are dominant in lowland tropical forests. Surveys in all the major tropical zones, for example, in Brazil (Thomazini, 1974), Nigeria (Redhead, 1968a) and West Java (Janse, 1896), confirmed that only a very small proportion of the trees had sheathing mycorrhizas, whilst the majority had VA endophytes. The absence of obligately mycorrhizal agarics in the tropics is difficult to explain. That it is solely due to the higher temperature is unlikely; although 'typical' biotrophic agarics such as *Russula* begin to appear with increasing altitude (2500–3000 m, 10–14 °C) in the Ecuadorean Andes (Hedger, unpublished), so do members of the Ericaceae which are absent lower down (Acosta-Solis, 1968). This leads to the argument that tropical plant families are non-selective for agaric or basidiomycete mycobionts. Evidence is difficult to assemble on this point but in Table 2.2 a number of tropical

Table 2.2. *Examples of ectomycorrhizas associated with agarics and boletes in tropical areas*

Higher plant family	Approximate numbers of mycorrhizal species	Geographical location	Family of agaric mycobiont
Caesalpin-aceae	Many	West Africa South America	Cortinariaceae (Redhead, 1968a, b) Russulaceae (Pegler & Fiard, 1979)
Dipterocar-paceae	Many	Malaysia	Boletaceae (Corner, 1972)
Euphorbi-aceae	Many	Madagascar	Russulaceae (Heim, 1937)
Bignoni-aceae	Few	Madagascar	Russulaceae (Heim, 1937)
Fagaceae	Many	Malaysia	Boletaceae (Corner, 1972)
Myrtaceae (*Eucalyptus*)	Many	Australia	Boletaceae (Chilvers & Prior, 1965)
Sapotaceae	?	South America	Amanitaceae (Singer & Araujo, 1979)
Sapindaceae	Few	—	—
Pinaceae	Many	Malaysia	Boletaceae (Ivory, 1975) Amanitaceae (Ivory, 1975)

Most evidence is based on the association of agaric fruit bodies with trees bearing sheathing mycorrhizas. In a few cases mycelial connections or inoculation experiments have been carried out (for example, in Pinaceae, Redhead (1980)).

examples of families in the Angiospermae and Coniferae are listed in which proven ectomycorrhizas occur or there is strong circumstantial evidence linking agaric fruit bodies with sheathing mycorrhizas. Of these, only the Fagaceae and Pinaceae are largely temperate families, whilst others such as the Dipterocarpaceae and Caesalpinaceae are tropical in their distributions (Hutchinson, 1959).

There thus seems to be no clear climatic or taxonomic reason why agaric ectomycorrhizas should not form on tropical trees, although ectotroph families represent a very small percentage of the tropical tree flora. The sporadic, but widespread, presence of fruit bodies of 'typical' ectomycorrhizal agarics such as *Lactarius* in lowland forests of a predominantly endotrophic type (Singer & Araujo, 1979; Pegler, 1983) lends weight to this argument. In addition, Singer & Araujo (1979), in an investigation of the lower Rio Negro area of Amazonia, distinguished areas of predominantly ectotrophic forest (for example the *campina* type). In these areas they found 30 species of basidiomycete, mostly agarics, which they presumed to be ectomycorrhizal. They claimed definite links for ten (species of

Amanita, Cantharellus, Lactarius, Russula, Tylopilus and *Xerocomus*) with members of the Sapotaceae and Leguminosae. However, the fruit bodies of these fungi only accounted for 1 % of the total for these ectotroph areas and no data are supplied as to the relative levels of endotroph and ectotroph infections of the root systems.

We are forced to conclude, in the absence of evidence to the contrary, that the biotrophic (ectomycorrhizal) mode of nutrition is a poor option for the tropical agaric. Paradoxically, this may be related to the high level of resource diversity which we have seen characterises many areas of tropical forest. The plant communities of this ecosystem are associated with relatively few non-taxon-selective mycobionts (the VA fungi). In contrast, the species-poor northern temperate and boreal areas contain a high diversity of basidiomycete mycobionts, including many taxon-selective agarics (Rayner *et al.*: Chapter 1). Malloch, Pirozynski & Raven (1980) made a similar point in their discussion of the evolution of the mycorrhizal symbiosis. It would seem possible that the selectivity of the agarics means that they are at a disadvantage in an ecosystem where many families and species of plant occur per unit area, with individuals of the same species sometimes spaced at kilometre intervals (Richards, 1952). It is also possible that resource capture by cords and mycelium of agarics conveys an advantage in species-poor forests but is a disadvantage in highly diverse communities. In this respect, it is perhaps significant that areas of predominantly ectotrophic tropical forests are often characterised by a relatively low species diversity, usually imposed by extreme conditions, such as poor soil (Pegler & Fiard, 1979) or annual flooding (Singer & Araujo, 1979). We may also note that a predominantly ectomycorrhizal family, the Dipterocarpaceae, is often the dominant element in the Malaysian forests in which it occurs (Singh, 1966; Whitmore, 1975).

The exploitation of the biotrophic nutritional mode by the agarics of tropical forest remains something of an enigma and must await studies of their resource relations before it can begin to be resolved.

Necrotrophy

Tropical examples of agarics with a predominantly necrotrophic mode of nutrition are difficult to delineate from the mass of saprotrophs, some of which may exhibit varying degrees of necrotrophy. This is no better illustrated than by the *Armillaria* species complex (Rishbeth: Chapter 3), members of which have been recorded as root pathogens in a variety of tropical tree crops, for example *A. mellea* and *A. tabescens* on cocoa (*Theobroma*) (Thorold, 1975), but which has saprotrophic species

Table 2.3. *Examples of tropical necrotrophic agarics*

Agaric	Host	Disease	Description
Armillaria species	Tree crops	Root rots	Thorold (1975)
Mycena citricolor	Coffee (*Coffea*)	Leaf spot	Buller (1934)
Marasmius pulcher	Tea (*Thea*)	Thread blight	Petch (1915)
M. cyphella	Rubber (*Hevea*)	Thread blight	Dennis & Reid (1957)
M. scandens	Cocoa		
	(*Theobroma*)	Thread blight	Bunting & Dade (1924)
M. equicrinis	Cocoa	Horsehair blight	Bunting & Dade (1924)
Crinipellis stupparia	Cocoa	Horsehair blight	Dennis (1951)
Marasmiellus semiustus	Banana (*Musa*)	Stem rot	Stover (1972)
M. cocophilus	Coconut (*Cocos*)	Bole rot	Pegler (1969)

in tropical forests (Swift, 1968). In fact, our knowledge of necrotrophy stems largely from the impact of this nutritional grouping on tropical crops. Nothing is known of their role in undisturbed ecosystems, but because of their economic importance, necrotrophs are better understood than most other tropical agarics.

Table 2.3 lists some of these necrotrophs. One of them, *Mycena citricolor*, has already been mentioned by Rayner *et al.* in Chapter 1. This agaric, which has a Caribbean and South and Central American distribution, shows little resource selectivity and occurs on the leaves and fruits of many angiosperms, including timber trees (de Segura, 1969), but especially on coffee (Kranz, Schmutterer & Koch, 1978) where its presence may cause severe losses in heavily shaded plantations in high-rainfall areas. Typically small spreading lesions occur on leaves and developing fruits. These become coloured yellow by the development of one to four mainly pedicillate stilboids (Singer, 1962) or *cabecitas*. Heavy infection may cause partial defoliation, apparently due to auxin inactivation (Kranz *et al.*, 1978). Buller (1934) carried out studies on these stilboids (gemmifers), which he showed were easily detached from the pedicel and are splash dispersed by rain in the field (Kranz *et al.*, 1978). He concluded that these stilboids represented modified sterile pilei, and the pedicels, stipes. These are illustrated in Fig. 2.1 and represent a unique type of exit structure. Buller was also able to obtain fertile fruit bodies of *M. citricolor* (see Fig. 2.1). These represent two different exit strategies from the resource: the stilboids are an efficient means of local dispersal and impaction within the coffee crop (Buller, 1934; Kranz *et al.*, 1978), whilst the fruit bodies, which naturally occur on rotted leaves under the bushes (Buller, 1934), provide long-distance dispersal and genetic exchange. It would be of interest to know if resource quality and environment regulate the production of these two different but developmentally linked reproductive structures.

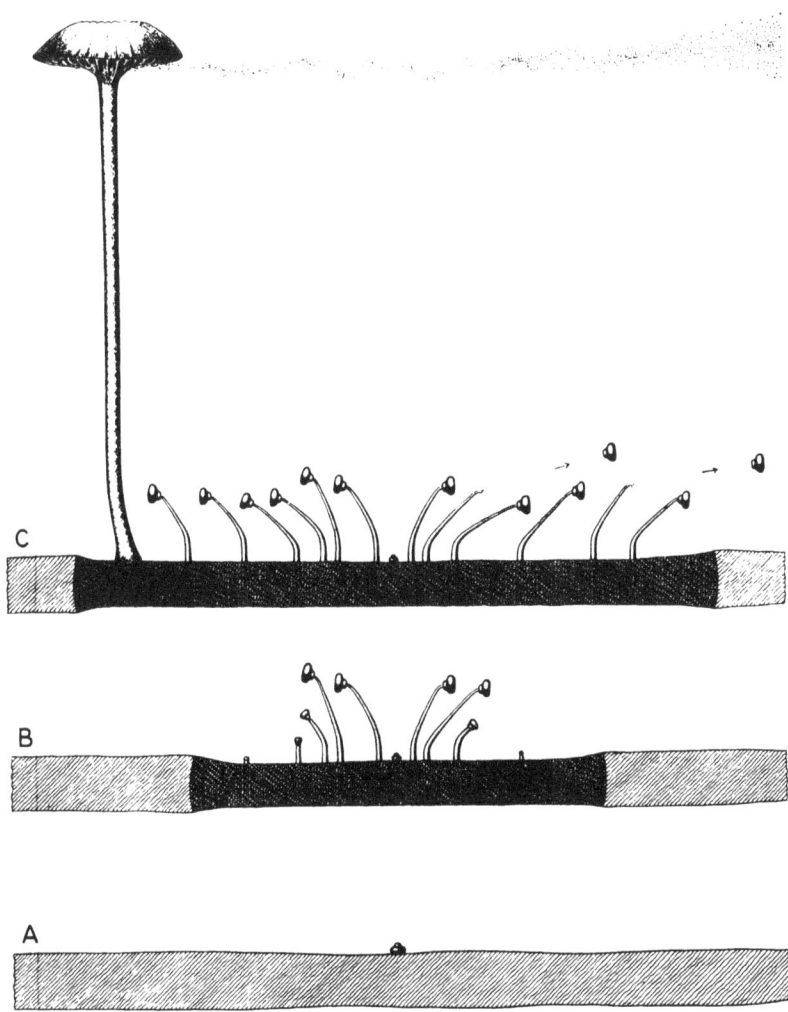

Fig. 2.1. *Mycena citricolor*. Illustrations made by Buller (1934) of a colony developing from a stilboid on a leaf of *Bryophyllum calycinum*, showing stilboids and fruit bodies. Note the different modes of liberation and dispersal of the basidiospores and stilboids. A, B and C represent development of the colony over a 4-week period.

The 'thread blight' and 'horsehair blight' disease complexes of a number of tropical crops are caused by agarics of varying degrees of necrotrophy; most represent saprotrophs with a very limited necrotrophic phase which can cause problems in crop monocultures. Examples are listed in Table 2.3. The thread blight complex has the distinction of being first reported by Berkeley in 1873 from Indian tea plantations, and is characterised by ramifications of white or brown cords over stems and leaves of the host, possibly leading to death of shoots and internal colonisation. 'Horsehair blights' differ somewhat in that colonisation is by looser masses of black rhizomorphs, usually about 1–2 mm in diameter. Both types of disease are usually linked to fruit bodies of *Crinipellis*, *Marasmius* and *Marasmiellus*, which occur either directly on the rhizomorphs or on dead plant tissues (see Table 2.3). The identities of the taxa and their status as pathogens rather than epiphytes are confused. Some degree of taxon selectivity seems to occur, for example, *Marasmius pulcher* on tea (Petch, 1915, 1923; Dennis & Reid, 1957), *M. cyphella* on rubber (Dennis & Reid, 1957) and *Marasmiellus cocophilus* on coconuts in East Africa (Pegler, 1969). Other species have a wider host range and can also be found as saprotrophs in forests; for example, *Marasmius equicrinis* (*M. crinesqui*), a pantropical species, has been recorded on tea, cocoa, coffee and other crops (Thorold, 1975) and as a foliicolous saprotroph (Dennis, 1970), whilst *Marasmiellus semiustus* causes a stem rot of banana and is also a common lignicolous saprotroph in South America (Stover, 1972; Pegler, 1983).

This group of agarics is characterised by a common saprotroph colonisation strategy, with necrotrophy as a secondary nutritional mode in some species, given the opportunity of crop monocultures. The primary selective factor in their evolution must have been epiphytic colonisation, via cords and rhizomorphs, and these fungi are part of the saprotrophic aerial agaric communities of tropical forests described above. The degree of necrotrophy in undisturbed forest is unknown, but it is probably confined to stressed plants – in cocoa, for example, *M. equicrinis* causes death of leaves only in very heavily shaded, poorly maintained stands (Hedger, unpublished), and *M. pulcher* is a problem on underfertilised overcropped tea (Hadfield, personal communication).

All the examples of thread and hair blights in Table 2.3 belong to the genera *Marasmius*, *Marasmiellus* and *Crinipellis*. Species of these genera are often characterised by the desiccation tolerance of their fruit bodies (Kramer, 1982). These fruit bodies, cords and rhizomorphs are well adapted to the extremities of the aerial environment where, even in tropical, moist forest, vegetation dries off within hours of rainfall (Whitmore, 1975). The epiphytic habit confers the advantages of ease of resource capture, either of senescent *in situ* plant material or of trapped material which has fallen from above.

Crinipellis perniciosa

The nutrition of the tropical agaric *Crinipellis perniciosa* is mentioned separately in order to emphasise its uniqueness. For a fuller discussion of this important pathogen of cocoa see Wheeler (Chapter 4). Although clearly related to the saprotrophic/necrotrophic *Crinipellis* and *Marasmius* species described in the previous section, its nutritional modes include a biotrophic phase, which actually provides the only example of physiological biotrophy, *sensu* Ingram, Sargent & Tommerup (1976), within the agarics.

In the context of this chapter, two aspects of the nutrition of this fungus will be very briefly considered: the significance of the biotrophic and saprotrophic modes in its life cycle, and the relationship of *C. perniciosa* to other *Crinipellis* species in the neotropics.

As described by Wheeler (Chapter 4), infection of developing cocoa tissues by basidiospores of *C. perniciosa* may result in abnormal vegetative growths (brooms) and distorted pods (*chirimoyas*), within which a mono-karyotic biotrophic mycelium grows. Eventually such tissues die, possibly aided by a switch to necrotrophy by the fungus mycelium, and accompanied by a change to a culturable dikaryon. This mycelium may continue to occupy dead brooms attached to trees as a saprotroph and to produce fruit bodies for several years, in spite of competition from invading fungi including other agarics such as *Crepidotus* (Hedger, unpublished). In nutritional terms the biotrophic phase acts as a resource capture mechanism; the host-adapted monokaryon is able to expand and fill the developing broom and pod tissues in the absence of competition from other fungi. This 'captured' resource is then exploited by the saprotrophic and apparently combative dikaryon for many months, during which a flow of nutrients is maintained to repeated crops of fruit bodies. This capture mechanism differs from those outlined by Rayner *et al.* (Chapter 1) in the mode of initial occupation, and indeed generation of the resource, which is unique to *C. perniciosa*.

The origins of this strategy must be sought in other *Crinipellis* species of the neotropics, possibly in the aerial agaric communities mentioned earlier. At present we know that *C. perniciosa*, readily identified on cultivated and wild cocoa, as well as other Sterculiaceae such as *Herrania*

(Evans, 1981*a*), also appears to occur as non-selective saprotrophic strains apparently without a biotrophic phase (see Wheeler: Chapter 4). Nevertheless, some of these strains caused some infection in cocoa seedling tests, as did a brown *Crinipellis* species isolated from dead *Calathea marantifolia* (Marantaceae) in Ecuador (Hedger & Aragundi, unpublished). Since Bastos & Evans (1984) have reported a type of *C. perniciosa* which actually forms brooms in Solanaceae in the Amazon, it is possible that biotrophic infections of a range of host plants will eventually be discovered.

It seems probable that 'non-pathogenic' species of *Crinipellis*, including strains of *C. perniciosa*, *C. eggersii*, *C. trinitatis* and others as yet undescribed, exist as a saprotroph population in the neotropical forest. These saprotrophs show little component restriction but should be regarded as the pool from which component restriction has evolved, via biotrophic monokaryotic infections with subsequent efficiency in resource capture. In this respect it is of interest to note that Evans (1981*b*) has speculated that *Moniliophthora roreri* (Deuteromycotina), another cocoa pathogen with a similar distribution to *C. perniciosa*, may have evolutionary affinities with *C. perniciosa*, since its mycelium has dolipore septa (see R. T. Moore: Chapter 7).

Saprotrophy

We have already noted that saprotrophy is the probable nutritional mode of many agarics in tropical forests and it seems likely that the total number of saprotrophic species will eventually exceed that known for other ecosystems. Since we know very little about tropical saprotrophs, the presence of resource and habitat selectivity and subsequent species diversity can be conjectured but not proved.

A few general characteristics of habitat selection by saprotrophic agarics in tropical forests can be pointed out. Habitat affinities are usually delimited in descriptions of tropical agarics by the distribution of fruit bodies. Such criteria may not accurately reflect mycelial distribution (see Rayner *et al.*: Chapter 1) but are in most cases the only ones available for tropical areas.

Some information emerges from ecological studies such as that of Singer & Araujo (1979) on forest plots in the Rio Negro region of the Amazon. They classified the agarics into lignicolous, foliicolous and terricolous (humicolous) populations, on the basis of fruit body distribution. Only a few species were observed to be both lignicolous and foliicolous, although the same genera (especially *Mycena*, *Marasmius*, *Hemimycena* and *Hydropus*) were represented in both habitats. The environmental differen-

ces between these two habitats have been discussed by Rayner *et al.*
(Chapter 1) and Hintikka (1982). It is significant that, of those species
which appeared non-selective of habitat in their study area, some were
actually component-selective – *Mycena polyadelpha* was primarily foliicolous but was also able to colonise small pieces of wood and twig in the
leaf-litter, a woody habitat of a similar component size to leaves.

Another point which emerges from their study is that the majority of
the agarics (some 25–50%) were recorded from wood, rather than soil and
litter. Watling (1977) makes the observation that a higher proportion of
species of fungi occur on wood in the tropics than in temperate regions.
Obviously this might relate to the abundance of lignocellulose in the forest
biomass. It must be borne in mind that the buffered environment of large
wood masses may be a contributory factor, enabling more prolonged
fruiting of agarics to occur, in contrast to the wide fluctuations in leaf-litter
water content which can occur in the tropical forest climate. The pitfalls
of using fruit body distribution as an indication of mycelial activity are
again emphasised.

The ecology of tropical agarics – a case study

Some general principles underlying the resource relations and
habitat distribution of tropical agarics have been reviewed. These principles
will now be considered in relation to specific information obtained by a
2-year study (1981–83) of the agarics associated with stands of cocoa
(*Theobroma cacao*) in Pichilingue, which is sited in the tropical lowland
coastal area of Ecuador (1°06′S 79°29′W). The stands of cocoa represented
a very simplified plant community in which it proved possible to recognise
patterns of resource selectivity and habitat relations.

Climate and plant community structure

Pichilingue has a seasonal tropical climate. A dry season usually
extends from June to November, with little precipitation and relatively low
insolation (see Fig. 2.2). Rains usually begin in December, are accompanied
by a rise in mean maximum temperature to 26–30 °C, and extend to late
May or early June (see Wood, 1975, for climatic data). The effect of this
seasonality on agaric activity will be examined later.

The natural vegetation of the area would have been a transition between
a moist tropical forest, which lies to the north, and a seasonal (monsoon)
type, which lies further south (Dodson, personal communication). In
general terms it falls within the neotropical moist forest life zone of
Holdridge (1971). Disturbed forests of this type still existed within 1 km

of the cocoa and were used, together with a forest reserve at Palenque, some 50 km to the north, as reference areas for the study.

The investigation was carried out within a number of adjacent stands of 10-year-old clonal cocoa (*Theobroma cacao*) about 5–10 m high, irregularly shaded by trees of *Inga edulis* (Leguminosae). The ground flora was poor and consisted of species of *Thelypteris* (Filicales), *Piper* and *Sida acuta*. Weeding kept lianes to a minimum, but the orchid *Cryptarrhena lunata* was a common epiphyte.

Primary production and decomposition

Measurements were made of the primary production of the two tree components and the rate of detritus accumulation/decomposition in the litter layer beneath them, with a view to establishing the availability of resources to the agaric communities (Table 2.4).

Inputs to the decomposition system on the soil surface, as might be

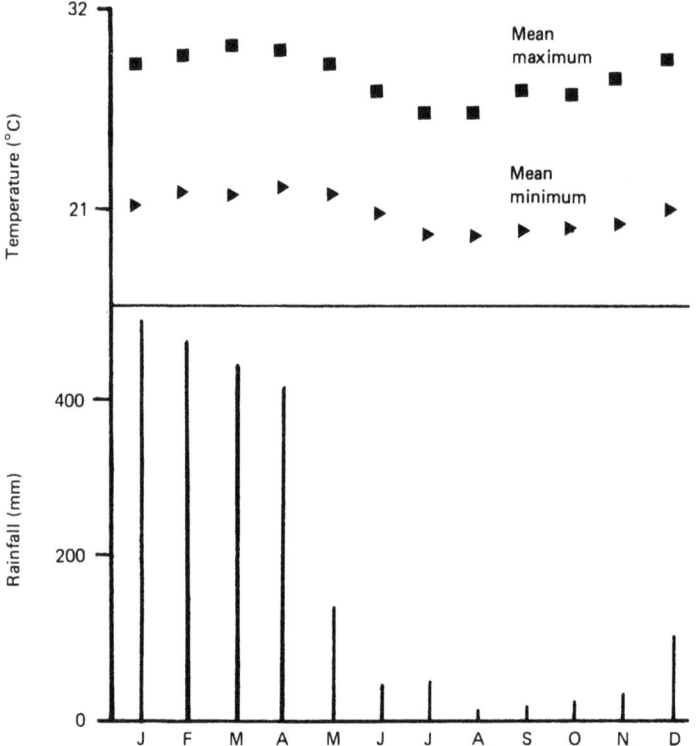

Fig. 2.2. Annual rainfall and temperature patterns at Pichilingue, Ecuador: 14-yr average. Data from Wood (1975).

Table 2.4. *Annual production and decomposition of litter within stands of*
Theobroma cacao *at Pichilingue*

Annual litter fall ($t\ ha^{-1}\ yr^{-1}$)	Standing crop of litter			Per cent loss in dry weight of litter per year
	Average ($t\ ha^{-1}$)		After 1 year ($t\ ha^{-1}$)[a]	
8.03	Twigs + brooms	1.88	Twigs + brooms 0.37	88.5^a (65.6^b)
	Leaves	7.09	Leaves 0.76	
	Total	8.97	Total 1.13	

Standing crops were measured by the use of 1-m² quadrats (Swift *et al.*, 1979) within which all litter was removed. Rate of annual loss of standing crop (^{a}) was measured by excluding 1-m² areas from further litter fall by 1-mm² nylon mesh. Annual loss (^{b}) was also measured for cocoa leaves only, by use of the litter bag method (10-mm² and 1-mm² mesh).

expected, were dominated by the leaves, branches and fruits of the tree species, but the added dimension in cocoa is the heavy deposition of abscissed brooms and pods resulting from infection of the trees by *Crinipellis perniciosa*. At times these made up over 50% of the litter fall by weight. The estimations in Table 2.4 do not take into account large trunks and branches, but indicate that k values (Olson, 1963) for the smaller litter components were in the order of 3.3–3.5, i.e. almost total degradation within 1–2 years of deposition. These values may be compared to the figure of 6.0 given by Swift, Heal & Anderson (1979) for moist tropical forest, and indicate that the litter may have degraded somewhat more slowly than in a natural forest. However, use of tagged leaves in the rainy season showed that there was a component difference in decomposition rates: *Inga* leaves were almost totally destroyed in 5 months whereas cocoa leaves were still partly intact after the same period. The proportion of cocoa material in the litter standing crop increased in proportion to its residence time on the soil surface, as did the proportion of cocoa twigs and brooms as compared to leaves (see Table 2.4).

Resources and habitats

Woody resources available to saprotrophs consisted of standing dead and fallen trunks and branches of the two tree species, both fresh and in an advanced state of decay. A proportion of material (leaves, etc.) remained attached or trapped within the canopy, perhaps for several months prior to deposition onto the litter layer beneath the trees. The depth of this layer varied at different times of the year, between 40–80 mm and 10–20 mm. Like many tropical trees, cocoa grows by a series of vegetative flushes, which are often followed by a heavy fall of leaves (Alvim & Alvim, 1978). A major fall occurred after the first flush of the wet season (January and February). Litter depth also increased during the dry season (May–October), possibly reflecting a decrease in decomposition activity, and water-stress-related leaf abscission during September and October (Orchard, personal communication). The loss in weight of litter held within bags confirmed the low decomposer activity during the dry periods. A 12% loss in dry weight of leaves occurred in the bags between May and December 1982 (dry season) and 40% between December 1981 and May 1982 (wet season).

Very similar litter structures could be found in nearby areas of forest, although the thickness varied according to the adjacent tree species. Members of the Moraceae developed a deep litter, but many Leguminosae often had no litter layer at all. Presumably this difference was related to the differences in the qualities of the resources presented to the decomposers. In the study area, the rate of decomposition of the *Inga* leaves (Leguminosae) has already been contrasted with that of the cocoa (Sterculiaceae).

Lignicolous agaric communities

In Table 2.5 a selection of the higher fungi recorded on wood of *Inga* or *Theobroma* during the study period has been grouped according to the types of material on which they were found. Pioneer colonisers of freshly dead bark and sapwood included *Coprinus cubensis*, *Psathyrella* species and *Crepidotus cuneiformis*. These occurred on large branches, boles and twigs. The main phase of decomposition of boles and branches, which was largely by white-rot attack by *Hexagona*, *Pogonomyces* and *Amauroderma* (Aphyllophorales), was characterised by a different community of agarics (*sensu lato*) principally *Pleurotus eugrammus* and *Gymnopilus chrysopellis*. This phase seemed to occupy about 2–3 years in the case of *Inga* and 1–2 years for cocoa. At the end of this period the very soft, much decayed wood seemed to be colonised principally by species of *Gerronema* and *Pluteus*. The larger fruit bodies of *Volvariella bombycina* var. *microspora* and *Leucocoprinus cepaestipes* occurred at the bases of well-rotted trees.

Small branches that fell onto the litter eventually became incorporated into the A_1 horizon of the soil, and usually became covered by white mycelial cords. Fruit bodies of *Trogia* species and *Collybia fibrosipes* grew

Table 2.5. *Resource groupings of higher fungi on woody substrata* (Inga edulis *and* Theobroma cacao)

1. Freshly dead boles and branches	
Coprinus cubensis	*Pleurotus subhaedinus*
Crepidotus cuneiformis	*Psathyrella coprinocepes*
Favolaschia pygmaea	*P. araguana*
Marasmiellus cubensis	*Schizophyllum commune*
Omphalia flavella	
2. Boles and branches in intermediate states of decay	
Gymnopilus chrysopellis	
Oudemansiella canarii	
Pleurotus eugrammus	
3. Boles and branches at advanced stages of decay	
Gerronema icterinum	
Leucocoprinus cepaestipes	
Pluteus laetifrons	
Volvariella bombycina	
4. Branches buried in soil	
Collybia fibrosipes	
Trogia buccinalis	
T. cantharelloides	
5. Roots	
Oudemansiella steffendii	
6. Abandoned termite nests	
Tricholoma pachymeres	

Groupings are based on observations of the distribution of fruit bodies in areas of cocoa at Pichilingue (Ecuador).

upwards from these buried resources and fruited at the litter surface. *Oudemansiella steffendii* was another apparently litter-inhabiting agaric which, on investigation, proved to be utilising lignocellulose resources deeper in the soil. The fruit bodies of *O. steffendii* resemble the temperate *O. radicata* and, like those of this species, they possess a rooting base. This base was connected via mycelial cords about 5 mm in diameter to soft, white-rotted tree roots at depths of up to 0.1–0.3 m.

The final habitat included in Table 2.5 is that of abandoned termite nests. These nests were built on trunks and branches of the trees and normally had no agarics upon them. However, in nests in which the colony had died out, flushes of large fruit bodies of *Tricholoma pachymeres* appeared, which were connected to white mycelial cords throughout the nest and which were presumably exploiting the nest material.

Surveys of natural forest indicated that similar, though more species-diverse, groupings of agarics occurred on lignocellulose substrata. The *Gerronema icterinum/Pluteus laetifrons* association on much-decayed standing tree boles was particularly easy to identify. All the agaric species recorded on *Inga* and *Theobroma* were found in these areas, indicating a low level of taxon selectivity. However, as in the cocoa, the lignicolous species were strongly habitat-selective and, even in leaf-litter, were usually associated with lignocellulose components such as twigs.

The physiological ecology of these lignicolous agarics is conjectural in the absence of experimental data. It might be argued that the resource utilised by the *Psathyrella/Coprinus* grouping is the outer sapwood/phloem of fresh material, that the *Gymnopilus/Pleurotus* group utilises the primary lignocellulose source, whilst the *Gerronema/Pluteus* grouping is able to exploit the secondary resources present in much decayed wood (Hedger, 1982). The position of *Oudemansiella steffendii* is of particular interest: presumably it is able to exploit lignocellulose under the conditions of low oxygen concentrations and high water content found at such depths in the soil.

Litter-inhabiting (foliicolous) agaric communities

The internal structure of the litter (A_0) layer was usually well defined and is summarised in Fig. 2.3. Freshly deposited leaves, twigs and brooms could usually be distinguished as a surface L layer. The cocoa leaves, some 60–100 mm in length, made up the bulk of the material and when freshly deposited were reddish brown in colour. Tagging of leaves showed that in the wet season bleached patches often developed on these leaves within one month of deposition.

L material became integrated into the F_1 layer by further deposition of leaves, usually within 1–2 months, depending on rates of litter fall. F_1 cocoa leaves were characterised by a paler brown colour, presumably related to leaching, and by a further extension of the bleached areas to as much as 100% of the lamina in some cases. Loss of strength occurred in these areas

---

Fig. 2.3. Examples of cocoa litter material. (*a*) Partly fragmented cocoa leaf from the F_1 fermentation layer. Note the bleached area at the axial end of the leaf caused by colonisation by a species of *Marasmius*. (*b*) Totally bleached cocoa leaf from the F_1 fermentation layer. Note the development of black zone lines between two confronting *Marasmius* mycelial systems. (*c*) Partly fragmented and bleached cocoa leaves from the lower F_1 layer. Note the presence of a rhizomorph (*Marasmius* sp.) (*d*) Fragmented material from the F_2 fermentation layer. Masses of white hyphae, which formed part of a *Lepiota* mycelium, can be seen. The thread-like structures are mostly mycorrhizal roots of cocoa.

but the leaves usually remained intact, although completely bleached areas were often removed from tagged leaves by animal feeding. Little mycelium was seen in the F_1 layer (Fig. 2.3.*a–c*).

After 3–4 months, leaf materials had begun to be comminuted into smaller fragments 10–30 mm in length, probably by the action of Diplopoda, and formed a heterogeneous F_2 layer in which broken fragments of brooms and twigs were mixed with faecal pellets of Lepidoptera. Dense white or cream mycelia were observed advancing through this material, binding it into amorphous bleached masses in which the structure of the leaves was entirely lost. These masses were usually penetrated by fine mycorrhizal roots of cocoa, which were abundant in this layer (Fig. 2.3*d*). Brooms and twigs were by this time fragmented and, although the periderm was intact, the interior had often been completely destroyed.

Beneath this material, the soil surface had a poorly defined A_1 horizon into which material was mixed by earthworm action; leaf material, twigs and brooms could be extracted from a depth of 50 mm by wet sieving.

Agaric mycelial systems were found to have species-related distribution patterns in the litter, which were related both to the vertical structure of the litter and also to the different components within it, notably to the leaves and to woody resources, especially witches' brooms.

L-layer agarics

The rapid bleaching of the L-layer leaves was found to be caused by an assemblage of *Marasmius* species of which the commonest were *M. haematocephalus*, *M. foliicola* and *M. ferrugineus* (see Table 2.6 & Fig. 2.4). These species were also found in a similar position in litter in nearby forest. Colonisation seemed to take place following litter fall and these species were never observed on trapped brooms and leaves within the tree canopy, although *M. equicrinis* was recorded from this habitat in abandoned cocoa and in natural forest. It is possible that a degree of colonisation via basidiospores occurred prior to leaf fall, as on occasions happens in the temperate agaric *Mycena galopus* (Frankland, 1984), but observations indicated that the usual mode of arrival was by rhizomorphs and adhesion zones.

Adhesion zones were characteristic of all the species of *Marasmius*. Bleached areas of F_1 leaves, most particularly margins and veins, were found to adhere to recently fallen leaves by cream or white mycelial pads of irregular shape, usually about 1–2 mm in diameter. From these points of adhesion, bleaching extended into the laminas. Bleached zones 10–20 mm in diameter were produced within 1–2 months in the rainy season and

Table 2.6. *A summary of the distribution of the mycelia of agaric species and other Basidiomycotina in cocoa litter*

Upper litter	Lower litter	Soil
Hydropus paraensis	*Clitocybe jamaicensis*	*Chlorophyllum molybdites*
Marasmiellus nigripes	*C. microspora*	*Cystoderma luteohemisphericum*
Marasmius aripoensis	*Collybia hemileuca*	*Cystolepiota eriophora*
M. cohortalis	*Geastrum minimum*	*Lepiota abruptibulba*
M. confertus	*G. velutinum*	*L. erythrosticta*
M. ferrugineus	*Hemimycena gigaspora*	*L. lactea*
M. foliicola	*Lepiota guatapoensis*	*Leptonia howellii*
M. haematocephalus	*L. lilacea*	
M. orinocensis	*L. pseudoroseola*	
M. pallescens	*L. roseolamellata*	
M. personatus	*Lepiota* spp.	
Mycena osmundicola	*Leucocoprinus birnbaumii*	
Mycena spp.	*Omphalia flavella*	
Omphalia flavella	*Pterula plumosa*	
Pterula plumosa		

The distribution of mycelia was assessed by dissection of the litter in the field and their identity established by observations of fruit body production. The categories upper and lower litter and soil refer to the zones in which the mycelia of these species were observed: 'upper litter' includes the L and F_1 fermentation layers (pH = 7.3–7.5); 'lower litter' is the F_2 fermentation layer (pH = 5.8–5.9); and 'soil' represents the upper 20–40 mm of the A_1 horizon of the soil (pH 6.1–6.3).

during this period primordia formed on the leaf margins or veins. Many of the resultant fruit bodies were very small – *M. foliicola*, for example, had a cap diameter of only 2–4 mm and a stipe length of 10–20 mm. One bleached leaf patch about 40 mm² in area produced 17 fruit bodies of this species within a period of one month.

A second mode of entry was undoubtedly the system of rhizomorphs which ramified through the F_1 and upper F_2 litter layers. Two types of rhizomorph were observed. The first type was produced by a number of species. They were 0.5–1.0 mm in diameter, black glabrous rhizomorphs similar to those of *Marasmius androsaceus*, and were principally associated with areas colonised by *M. ferrugineus* and *M. foliicola*. A second widespread type was brown, about 2.0 mm in diameter and strongly pubescent. This type (45) is as yet unidentified. On occasions it terminated in a cone-shaped structure which was covered in broom cells, indicating a *Marasmius* affinity.

In both types of rhizomorph, colonisation of L-layer leaves took place following contact with the tip or, more usually, part of the rhizomorph length. Observations made in February 1982 of individual rhizomorphs tagged with pieces of cotton showed that within one week of lateral contact with a leaf, the sections of rhizomorph, usually about 2 mm in length, seemed to swell to about twice the usual diameter. Beneath these swollen sections, dense white mycelium formed an adhesion zone similar to those already described. Repeated contact by a rhizomorph crossing a leaf resulted in a series of coalescing bleached patches (Fig. 2.4b). On occasion, under conditions of higher water content, usually in the F_1 rather than the L layer, rhizomorph contact resulted in zones of lateral, fine fan-shaped branches on the leaf surface, about 0.1 mm or less in diameter and about 2–3 mm long, from which bleached patches developed. These represented miniature mycelial fans (see Rayner *et al.*: Chapter 10).

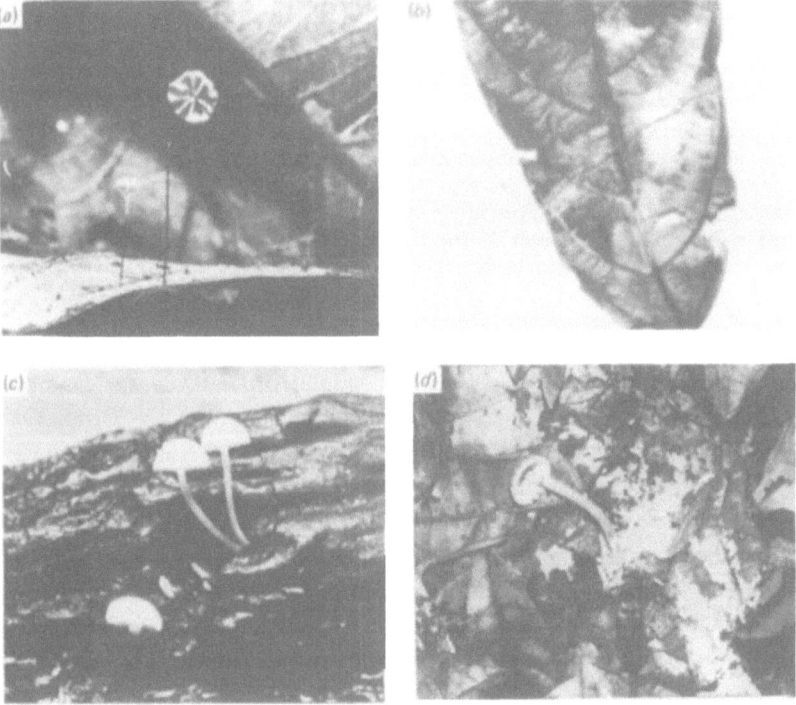

Fig. 2.4. Examples of agaric colonisers of cocoa litter. (*a*) *Marasmius foliicola*. Fruit bodies on L-layer cocoa leaves. (*b*) Rhizomorph type 45 (*Marasmius* sp.) crossing an L-layer leaf. Note the bleached areas extending from contact points between the rhizomorph and the leaf. (*c*) *Omphalia flavella*. Fruit bodies growing from a decayed cocoa pod. (*d*) *Lepiota pseudoroseola*. Fruit bodies associated with white masses of mycelium in the lower F_2 fermentation layer.

An attempt was made to determine the distribution of the rhizomorph system within the litter layers. Rhizomorphs were found to be connected vertically down through the litter layers, although those present in the lower F_1 and A_1 horizons were fragmented and brittle. They did not grow on potato dextrose agar. Viable rhizomorphs were largely restricted to the upper F_2 and the F_1 litter layers. Dissection of several 0.5 m² areas of litter showed that they contained 5.5–6 m of rhizomorph type 45, and the total length of the finer black rhizomorph type was probably in excess of this figure.

In some areas of litter, rhizomorphs were observed projecting well beyond the L layer vertically into the air. Type 45 proved easier to study and the growth of the vertical rhizomorphs was followed by tagging with cotton thread. Initiation took place from rounded primordia on leaves within the F_2 layer, similar to the primordia of fruit bodies of *Marasmius* species. The rhizomorphs, which had a pointed apex, grew vertically for up to 100 mm, and then looped back towards the litter layer, where tip and lateral contact with freshly fallen leaves was followed by adhesion zones and subsequent bleaching. The average length of some 30 of these loops was calculated to be 411 ± 105 mm. The rate of growth varied considerably according to the onset of rainfall, but some rhizomorphs were observed to extend at the rate of 8 mm day^{-1}.

These aerial rhizomorphs provided a means of rapid interception and capture of the freshly fallen cocoa leaves by the *Marasmius* species. Further rhizomorph growth and formation of contact zones continued the colonisation of new areas as the material moved down into the F_1 layer.

F_1-layer agarics

Reference to Table 2.6 shows that agarics in the F_1 layer included the *Marasmius* species which had colonised the leaves in the L layer. The amount of bleaching of these leaves varied considerably, as did the numbers of agarics which had colonised them. Some leaves were entirely bleached and apparently occupied by one species. In others, bleached patches occurred, each of which could sometimes be associated by fruit body formation with different species of *Marasmius* – usually *M. foliicola* and *M. haematocephalus*. In many bleached leaves, networks of fine black lines could be seen which presumably marked zones of intra- and inter-specific confrontation between different *Marasmius* mycelial systems (see Fig. 2.3b).

In the wet season, fruit bodies of the agarics *Hydropus paraensis*, *Omphalia flavella*, *Mycena osmundicola* and other *Mycena* and

Hemimycena species, together with *Pterula plumosa*, were observed originating from aggregates of mycelia on leaves and brooms within the F_1 zone (see Table 2.6). A diffuse, randomly orientated and non-aggregated weft of hyphae occupied large areas of the litter, with a low-density mycelium closely applied to the surfaces, of bleached and unbleached leaves. No further bleaching seemed to be associated with its presence.

This grouping of agarics represents a quite different mode of resource capture to that shown by the *Marasmius* species. Unlike the internal colonisation of rhizomorphs of *Marasmius*, the external mycelium systems of this group appeared to be vulnerable to desiccation and some species, especially *Mycena osmundicola*, were only abundant during the wettest periods of the rainy season.

Reference to Table 2.8 will show that *Marasmiellus nigripes* is included as a coloniser of the F_1 litter. It appeared to be component specific, and the fruit bodies nearly always formed on leaves of *Inga*.

F_2-layer agarics

Bleached masses of material were frequent but irregular in distribution in the F_2 layer and the size of the associated mycelial systems varied from a few centimetres to as much as 0.5 m in diameter. They grew actively in the wet season. Some measurements were made of some large advancing fronts, and rates of growth of 0.5–1.0 m month^{-1} were recorded. Growth of these mycelial systems continued well beyond the end of the rainy season for a period of 2–4 weeks, although it ceased once the A_1 horizon of the soil had dried out. Subsequently, many of these mycelial patches were disrupted by armadillos and other animals, but a number were observed to recommence growth in the following wet season from the fragmented areas which were left.

Production of fruit bodies was infrequent, so that the identity of these mycelia usually remained uncertain. However, flushes did occur and the genera *Lepiota*, *Leucocoprinus* and *Clitocybe* predominated. Although several species were relatively common, especially *Lepiota pseudoroseola* (see Fig. 2.4*d*), many other species were recorded only once or twice during the study period (see Table 2.7). Mycelial systems of different species appeared to be scattered at a low population density throughout the area.

A_1 soil-layer agarics

Mycelial continuity existed between the dense masses of mycelium in the F_2 litter and mycelial cords in the soil. However, for a number of agarics it was clear that their primordia were initiated from well within the soil, and the associated mycelial cords seemed to be primarily within the soil and not associated with specific resources such as twigs or brooms. These agarics were felt to be truly soil-inhabiting (terricolous or humicolous) and included species of *Leucocoprinus*, *Cystoderma luteohemisphericum*, *Leptonia howellii* and *Chlorophyllum molybdites* (see Table 2.6).

The resources utilised by this group of agarics are not clear but some are likely to be secondary. Their nutrition will be discussed later, in conjunction with that of the other groups of litter agarics.

Higher fungi associated with witches' brooms

These twig-like structures are precolonised internally by the mycelium of *Crinipellis perniciosa* (Wheeler: Chapter 4), and fruit bodies of this species appeared on them during the wet season whilst they were still attached to the tree. However, other fungi were observed fruiting on these brooms, notably *Crepidotus cuneiformis*, *Omphalia flavella* and, on very decayed brooms, *Favolaschia cinnabarina*. These three species also colonised dead cocoa pods. Sectioning of the brooms colonised by these fungi revealed the presence of black zone lines about 1–2 mm below the fruit bodies, indicating that a confrontation had occurred between their mycelia and that of *C. perniciosa*. Normally *C. perniciosa* did not seem to be replaced to any great extent by such competitors and fruit bodies were produced on some brooms for 2 years (Hedger, unpublished).

Once the brooms entered the litter system, however, there was a rapid displacement of the mycelium of *C. perniciosa* within the resource. Brooms on the litter layers only continued to produce fruit bodies whilst still exposed and ceased to do so once they had been incorporated into the F_1 layer. This has a partly physiological explanation, since fruit body initiation required alternate wetting and drying conditions (Wheeler: Chapter 4); but use of tagged brooms showed that after 6 months in the litter during the wet season, recovery of *C. perniciosa* mycelium from the brooms by use of culture media proved impossible, and they did not produce fruit bodies on exposure.

The part played by other agarics in this process was not clear. However, *Omphalia flavella* was commonly observed on the brooms in the F_1 layer. In the F_2 layers the brooms were incorporated into the colonising mycelial fronts and destroyed together with the surrounding leaf material.

Resource utilisation by agarics

In the previous section the association was made between different agaric species and particular types of decomposition of cocoa leaf-litter.

An attempt was made to confirm these observations by autecological studies on agarics in pure culture. Isolations were made from fruit bodies of a selection of litter agarics and the non-agaric *Pterula plumosa* (Clavariales). Pure cultures were then inoculated onto approximately 15 g of sterilised cocoa leaves, in plugged conical flasks which were incubated at 25–32 °C. The leaves had been collected from the L layer in the study areas, had been dried and weighed prior to autoclaving in the flasks.

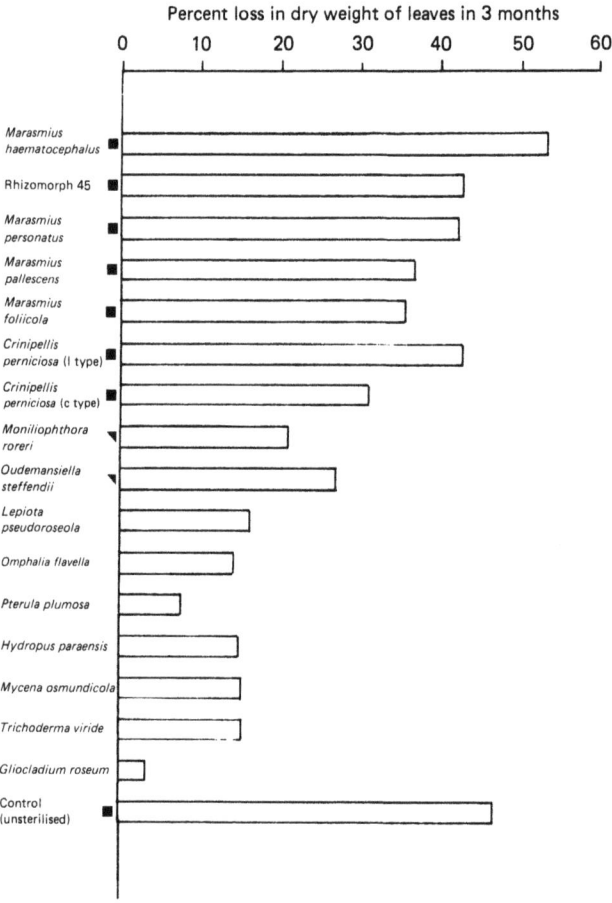

Fig. 2.5. Decomposition of cocoa leaves by pure cultures of fungi. The bar graphs represent loss in dry weight of cocoa leaves previously sterilised and inoculated with pure cultures of fungi. The symbols represent total bleaching (■) and partial bleaching (◥) of the leaves. For a full explanation see the text; l type and c type refer to strains of *C. perniciosa* derived from fruit bodies on liana and cocoa, respectively.

Moisture content was adjusted to 70–80% with distilled water. Unsterilised control flasks were also prepared and, for comparative purposes, other flasks were inoculated with the hyphomycetes *Trichoderma viride* and *Gliocladium roseum*, and with the cocoa pathogen *Moniliophthora roreri*.

Loss in weight of the leaves was determined after 3 months' growth and is shown in Fig. 2.5 together with an indication of the degree of bleaching of the laminae. All the *Marasmius* and *Crinipellis perniciosa* isolates caused a high rate of weight loss, and this was accompanied by the growth of a dense, closely applied cream or white mycelium over the leaves and by complete bleaching and partial destruction. Lignocellulose degradation of this type occurred in the L/F_1 layers colonised by these *Marasmius* species in the field, although here colonisation was internal. This reflects the saturated conditions which existed in the flasks as compared to the drier conditions in the field. The similarity of the decomposition by *C. perniciosa* of the leaves to that of the saprotrophic *Marasmius* species and indeed that of another cocoa pathogen, *Moniliophthora roreri*, raises again the question of the connection between these pathogens and saprotrophic populations.

In contrast, a number of other isolates colonised the leaf material but caused no bleaching and comparatively little decomposition, although the interstices between the leaves were filled with mycelium. These taxa included *Hydropus paraensis*, *Omphalia flavella*, *Mycena osmundicola* and *Pterula plumosa*. Levels of resource utilisation by this group are comparable to the hyphomycete controls, implying that they exploit simple carbohydrates and possibly cell wall polymers without much modification of the lignocellulose content. Again this confirms the ideas of resource utilisation by this group in the F_1/F_2 litter zone deduced from mycelium and fruit body distribution in the field.

The behaviour of the *Lepiota* species was unexpected (see Fig. 2.5). Only one of the six species which were studied grew away from the inoculum on to the sterilised leaves. This species, *L. pseudoroseola*, grew slowly to form a small colony 10–20 mm in diameter which was associated with local bleaching of the laminae. Repetition of the inoculations with all six species failed to establish any growth. However, when *L. pseudoroseola* was inoculated into flasks previously colonised by *Marasmius pallescens* and *M. foliicola* (after 3 months' growth by these fungi) a dense white mass of *L. pseudoroseola* mycelium occupied the whole of the leaves within 2 months.

These results imply that, at least for these six *Lepiota* species, colonisation of freshly fallen cocoa leaves by the mycelium is not possible. Preconditioning of the resources in the upper litter layers may be required by these species and possibly other F_2/soil decomposer agarics, a concept which will be returned to below.

Fruit body production and fruiting periodicity

The relation between fruit body production and mycelial growth in litter-inhabiting agarics is relatively unknown in natural environments. Detailed data now exist for cultivated species such as *Agaricus bisporus* (see Wood: Chapter 15) but not for wild species of this genus or other agarics. At present some information exists on seasonal rates of fruit body production of both mycorrhizal and saprotrophic species for areas of temperate woodland (see for example, Hintikka, 1963; Richardson, 1979; Hering, 1982) and for some species annual production has been related to mycelial systems in the litter, for example the data provided by Frankland (1984) for *Mycena galopus* in conifer woodland. In the tropics few studies of this kind have been attempted, although Singer & Araujo (1979) did carry out a quantitative study of the agarics in areas of Amazonas.

As pointed out by Hering (1982), production of fruit bodies represents only a proportion of the carbon assimilated by the mycelium of saprotrophic litter agarics, and seasonal distribution of fruit bodies does not reflect changes in mycelial activity. It rather represents a response by the mycelium to environmental stimuli. These 'cues' may coordinate the appearance of the fruit bodies at an appropriate time for successful release of propagules and colonisation of newly available substrate (in temperate climates usually the later summer/early autumn for most species), as well as ensuring that genetic exchange can take place between different mycelial systems. Examples of seasonal coordination of fruit body production can be quoted for many temperate species (Hering, 1982) and the winter fruiting of *Flammulina velutipes* (see Williams, Beckett & Read: Chapter 18) provides a further excellent example.

In the humid tropics, temperature changes cannot provide a stimulus to primordium induction by mycelial systems, but daily and seasonal variations in hydration of the substrata may well do so, as pointed out in the introduction to this chapter. In order to establish if such a coordinated seasonal distribution of fruit body production existed in the litter-inhabiting agarics of cocoa, records were made during weekly visits to the study area: first by a qualitative count of species present in two 300-m² plots and, secondly, by quantitative records of the fruit bodies present within ten 1.0-m² permanent quadrats randomly distributed in the two study areas. Fruit bodies were removed after counting, a procedure which undoubtedly

Table 2.7. *Occurrence of fruit bodies of higher fungi within fixed quadrats in cocoa litter*

	Total number of fruit bodies in 10 m²	Number of occasions seen	Substratum
Collybia fibrosipes	3	1	W/S
Clitocybe jalapensis	8	3	F_2
Coprinus cubensis	3	1	W
Cystoderma luteohemisphericum	1	1	F_2/S
Cystolepiota eriophora	3	2	S
Crepidotus cuneiformis	10	3	B
Crinipellis perniciosa	128	25	B
Favolaschia cinnabarina	3	2	P
Gymnopilus bakeri	3	2	W
G. chrysopellis	5	1	W
Hydropus paraensis	95	15	F_1/F_2
Hemimycena gigaspora	1	1	F_2/S
Lepiota pseudoignicolor	3	2	F_2
L. pseudoroseola	6	4	F_2
Lepiota sp.	10	6	F_2
Marasmiellus nigripes	195	20	F_1
Marasmius ferrugineus	15	3	L/F_1
M. foliicola	25	6	L/F_1
M. haematocephalus	31	6	L/F_1
M. orinocensis	1	1	L/F_1
M. personatus	13	1	L/F_1
Marasmius 45 (rhizomorphs only)	—	73	L/F_1
Mycena osmundicola	1005	10	F_1
Mycena sp.	18	7	F_1
Mycena sp.	3	1	F_1
Omphalia flavella	1340	42	F_1/F_2
Oudemansiella steffendii	1	1	S
Pleurotus subhaedinus	13	3	W
Psathyrella coprinoceps	3	1	W
Pterula plumosa	130	13	F_1/F_2

The period of recording extended from September 1981 to June 1983, and records were made on a weekly basis (total 88 weeks) within ten fixed 1-m² quadrats laid out in cocoa litter. Probable substrata of the mycelia of these agaric species are indicated by the following code: W = fallen branches; B = fallen witches' brooms; W/S = branches incorporated into the upper soil horizons; P = fallen cocoa pods; S = soil; L_1, F_1, F_2 = litter fermentation layers.

Table 2.8. *Frequency of fruit bodies of selected higher fungi in cocoa litter*

	Total fruit bodies m^{-2} at 88 weeks	Group total
Group I (L/F$_1$ litter)		
Hydropus paraensis	9.5	
Marasmius spp.	8.5	
Marasmiellus nigripes	19.5	
Crinipellis perniciosa	12.8	
		50.3
Group II (F$_1$/F$_2$ litter)		
Omphalia flavella	134.0	
Pterula plumosa	13.0	
Mycena spp.	2.1	
Mycena osmundicola	100.5	
		249.6
Group III (F$_2$ litter and soil)		
Collybia and *Clitocybe* spp.	1.1	
Lepiota spp.	2.3	
Oudemansiella steffendii	0.1	
		3.5

Species are grouped according to the usual vertical distribution of their mycelial systems in cocoa litter and the soil beneath. L/F$_1$ (Group I) represents species with mycelia which were observed to be inactive or absent in lower litter layers. F$_1$/F$_2$ (Group II) were species whose mycelia extended upwards to the F$_1$ layer, but never the L layer, and downwards into the F$_2$ layer. F$_2$/soil (Group III) represents species whose mycelia did not penetrate the F$_1$ layer and were restricted to the F$_2$ and soil horizons.

disturbed some of the litter structure and may have influenced the mycelial systems (see Frankland, 1984).

Fruit body production

A consideration of Table 2.7 reveals that out of the total of 88 weekly recordings of the quadrats, fruit bodies of 10 taxa were seen on one occasion only, 13 on two to nine occasions, and six taxa on ten or more occasions (the agarics *Crinipellis perniciosa*, *Hydropus paraensis*, *Omphalia flavella*, *Marasmiellus nigripes* and *Mycena osmundicola*, together with *Pterula plumosa* (Aphyllophorales)).

All these taxa have comparatively small fruit bodies (usually 1–10 mm cap diameter) so that in biomass terms the productivity of their mycelial systems is overemphasised by fruit body counts, whilst that of the larger agarics, such as the *Lepiota* species, which had pileus diameters of 30–40 mm, is underestimated. There seems to be a correlation between fruit body numbers, the size of fruit bodies and the position of the mycelial system in the litter (see Table 2.8). The species from the lower litter layers were characterised by relatively large, infrequent fruit bodies, whilst those of the upper layers produced smaller fruit bodies with a greater frequency. Strict comparisons with other systems are not easy, but *Mycena galopus*, a temperate species of a similar size to the smaller agarics, had a productivity of 20–40 fruit bodies m^{-2} yr^{-1} in a conifer woodland (Frankland, 1984).

Seasonal periodicity

The seasonal production of fruit bodies during the study period is illustrated in Fig. 2.6. Presence and absence of fruit bodies of different species and actual numbers of fruit bodies of selected genera and species are presented in relation to rainfall, the latter being assessed in terms of hydration of the litter layers rather than actual precipitation (for an explanation see Fig. 2.6). Typical precipitation and temperature data have been presented in Fig. 2.2 for previous years at Pichilingue.

Recording was started in October 1981 and December 1982 during the dry seasons. Rains began in the first week of December and, within 2 weeks fruit bodies of *Chlorophyllum molybdites*, *Oudemansiella steffendii* and *Lepiota pseudoroseola* had appeared. The fruit bodies of *L. pseudoroseola* were found to be attached to mycelial systems in the F$_2$ litter layers, which were partially destroyed by animal activity and were assumed to have been established in the previous wet season (which ended in mid May 1981). The fruit bodies of *C. molybdites* were found to be connected to mycelial cords in the upper 50–100 mm of the soil. The connections of *O. steffendii* to resources buried in the soil have been discussed above.

These three species produce medium-sized to large, fleshy fruit bodies. It was noticeable that the general appearance in the litter of fruit bodies of the smaller species such as *Marasmius* and *Hydropus* did not take place until 2–4 weeks after the start of the rains (see Fig. 2.6). An exception was *Crinipellis perniciosa* which fruited on brooms exposed in the L layer of the litter within a week of the first rains.

74

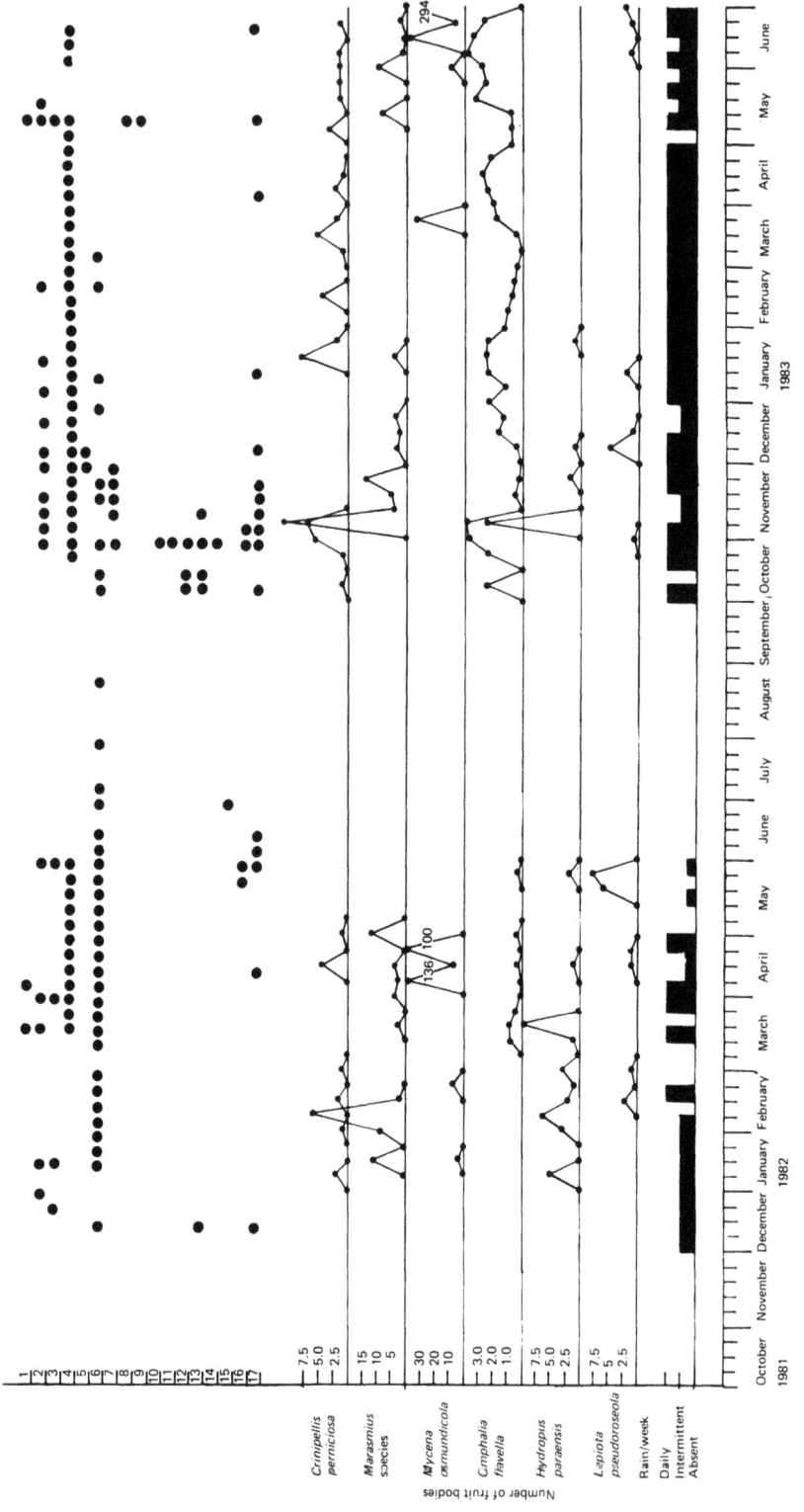

The main wet season period from January onwards in both 1982 and 1983 was characterised by irregular flushes of agarics rather than scattered continuous fruiting. The environmental factors which triggered these flushes were not easy to ascertain but on occasions they occurred in the first few days of rain following dry periods. This was particularly clear in the case of *Marasmius* species. However, observations indicated that the fruiting by *Marasmius* species was reduced during periods when the litter was continuously wet. The interruption of wet weather by only 24 h without rain at the beginning of May 1983 was sufficient to dry out the upper litter layers and was followed by abundant fruiting of *Marasmius* species, especially *M. haematocephalus* and *M. foliicola*, in the following week. In these agarics, brief interruption of the growth of the mycelia by a dry period might, therefore, be the basis of the irregular flushing seen in the study area.

The other small agarics showed less clear association of flushing with alternating wet and dry periods. *Hydropus paraensis* showed the most similar behaviour, and fruiting periodicity in the 1983 wet season can be compared to that of the *Marasmius* species (see Fig. 2.6). In contrast, *Omphalia flavella* fruited continuously, even during very wet weather. The tiny fruit bodies of *Mycena osmundicola* in F_1 material were associated with conditions of continuously wet litter, but appeared in discrete seasonal flushes (see Fig. 2.6). No obvious environmental control of flushing can be postulated in this species but its appearance late in both wet seasons leads to speculation that internal factors in the litter, perhaps the build-up

---

Fig. 2.6. Production of fruit bodies in litter of cocoa (*Theobroma cacao*) in Pichilingue, Ecuador, 1981–83. Rainfall is expressed qualitatively in terms of relative hydration of the litter layers during a one-week period. Daily = daily precipitation; upper litter layers partially dry for a few hours per day. Intermittent = at least one 24-h period with no rain in the week; upper litter layers dry for at least 12 h. Absent = one week without significant rain; upper litter and upper F_2 fermentation layers dry for several days but the lower F_2 and soil horizon remaining moist. Numbers of fruit bodies are expressed as the total in 10 m^2 of cocoa litter (total in ten 1-m^2 quadrats) per week. Presence (●) of fruit bodies represents data derived from weekly visits to two 300-m^2 plots of cocoa. The numerical code for the species is as follows: 1. *Marasmiellus nigripes*; 2. *Marasmius foliicola*; 3. *M. haematocephalus*; 4. *Omphalia flavella*; 5. *Geastrum velutinum*; 6. *Oudemansiella steffendii*; 7. *Collybia fibrosipes*; 8. *Clitocybe jamaicensis*; 9. *Lepiota phaeosticta*; 10. *Cystoderma luteohemisphericum*; 11. *Leucocoprinus cepaestipes*; 12. *L. birnbaumii*; 13. *Chlorophyllum molybdites*; 14. *Lepiota lactea*; 15. *L. lilacea*; 16. *Leucocoprinus* sp.; 17. *Lepiota pseudoroseola*.

of correct conditions for its mycelial growth during the wet season, were likely to be responsible for the control of flushing.

The influence of wetting and drying cycles extended to the mycelial systems growing in the lower litter layers and soil. The initial flush of fruit bodies of *Lepiota* and other genera at the beginning of the first wet season (1981–82) has already been noted and was repeated at the somewhat earlier start of the second wet season (in October 1982; Fig. 2.6). During both rainy seasons, flushes of litter F_2-/soil-inhabiting agarics continued. The commonest was *Lepiota pseudoroseola*.

Oudemansiella steffendii provided the only example of an agaric which continued to produce fruit bodies in the dry season, albeit at a lower frequency than in the wet season. Its mycelium grew at a depth in the soil where moisture content only fell slowly in the dry season, but the rapid penetration of the rains at the beginning of the wet season seems to have resulted in an increase in fruiting (Fig. 2.6).

These fruit body records only apply to limited areas of cocoa plantations. An attempt was made to compare the fruiting of agarics in adjacent forests. Although the species diversity of agarics was higher, a similar fruiting pattern emerged. Although it was noticeable that the microclimate and therefore the litter layers were better buffered than in the plantation, drying out of the upper litter occurred during rainless intervals in the wet season. This was often followed by a flush of *Marasmius*, especially *M. ferrugineus*, and other small agarics in the genera *Hemimycena*, *Marasmiellus*, *Hydropus*, *Gerronema* and *Psathyrella*. No clear pattern was established for the agarics of the lower litter layers and soil, such as *Cystoderma luteohemis-phericum*, although the early wet season and late wet season flushes were observed to occur.

Little comparative data can be quoted, with the exception of Singer & Araujo (1979). In a detailed study of the permanent 5×5-m quadrats in forest areas near Manaus, these workers recorded the numbers of agaric fruit bodies in litter, soil and wood on a daily basis. These observations were combined with detailed climatic records, and Singer & Araujo were able to analyse the results for correlations between fruit body numbers and rainfall. The numbers of fruit bodies of species of agarics characteristic of the litter material (foliicolous types) were strongly influenced by precipitation; those of the lignicolous and terricolous species were influenced less strongly. This parallels the behaviour of agarics in cocoa litter. A few fruit bodies occurred even during rainless periods but an immediate response to rain was clear. The numbers of fruit bodies of foliicolous types were often not proportional to the amount of rain but rather to the absence of rainfall for a few days prior to the rain episode.

The structure of agaric communities in tropical forest

The ecological position of any agaric in tropical forest will be the result of the interactions between its genome and the physical, organic and biotic environment within which its mycelium grows. Many responses are possible and these have been explored in general terms by Rayner *et al.* in Chapter 1. Development of nutritional modes, exit and entry strategies, resource specificity, and component capture have determined in evolutionary time the role of an agaric species within the tropical microbial community, be it as a generalist saprotroph such as *Schizophyllum commune*, or a specialist biotroph such as *Crinipellis perniciosa*.

These types of attribute were reviewed in a general way earlier in this chapter. The object of this section is to use the agarics of cocoa and forest in coastal Ecuador to explore in more detail some aspects of interactions between agaric mycelia and the environment and resources present in tropical forests.

At the habitat level a sharp dichotomy of preference existed between wood- and litter- or soil-inhabiting agaric species. Rayner *et al.* (Chapter 1) also referred to the division between foliicolous and lignicolous agarics in temperate areas. We await studies to determine the physicochemical basis of this distribution pattern, but it may be the same in both biomes (Hintikka, 1982).

Within the foliicolous and terricolous agaric communities saprotrophy appears to be the predominant nutritional mode. Future investigations may yield more proven cases of ectomycorrhizal biotrophs or of dual modes. However, the data gained from the studies on cocoa indicate that 'direct cycling' of nutrients (Went & Stark, 1968) from saprotrophs, such as *Lepiota*, whose mycelial masses were freely penetrated by cocoa mycorrhizal roots, was via an intermediary, the vesicular-arbuscular mycobiont *Acaulospora scrobiculata*, rather than truly direct (Hedger & Walker, unpublished).

The saprotrophic litter agarics might be expected to exhibit a degree of resource specialisation. It must be concluded that little evidence of taxon selectivity could be found. The spatial heterogeneity of tropical forest ecosystems may make taxon selectivity a disadvantage for both biotrophs, as noted earlier, and for saprotrophs. For saprotrophs it may be that only gross differences in physicochemical composition or plant form may be of importance. In this respect it is significant that the only clear cases of taxon selectivity in litter-inhabiting agarics noted in tropical forest in Ecuador during the study were associated with such gross differences – for example between palms and tree species. A particularly clear case of resource selectivity and component restriction was afforded by bamboo (*Bambusa guadua*), where *Marasmius cladophyllus* was restricted to the leaf litter and *M. griseoviolaceus* to standing dead twigs. These species were absent on

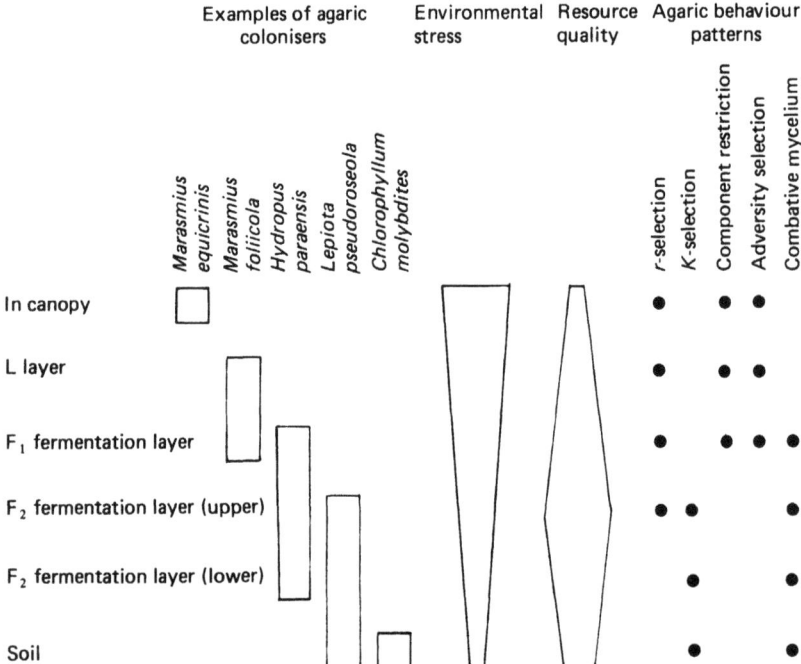

Fig. 2.7. A general scheme of resource relations and behavioural patterns in litter-inhabiting tropical agarics. The diagrams represent *relative* development of mycelia, environmental stress or resource quality in tropical plant litter. The agaric examples are chosen from species associated with the litter of cocoa (*Theobroma cacao*) in Ecuador. *Environmental stress* is principally that imposed by water stress. *Resource quality* assumes temporary increase in quality of leaf litter with time due to (1) leaching and (2) accumulation of secondary resources (Swift, Heal & Anderson, 1979). *Component restriction* implies restriction of agaric mycelium to particular litter units (Swift, 1976). K-*selection* and r-*selection* refer to the relation between mycelial biomass, numbers and sizes of fruit bodies (Southwood, 1976); *adversity selection* refers to tolerance of negative environmental and resource characteristics; *combative mycelium* refers to the ability of the mycelium to replace existing mycelial systems (Hedger, 1982; Cooke & Rayner, 1984).

similar components of the litter produced by surrounding tree species. Here the species assemblage of litter agarics was similar for many different taxa, in fact almost identical to that described in this chapter for cocoa litter.

Cocoa litter may therefore be taken as a very simplified example of tropical rain forest litter. The results of such a general analysis are presented in Fig. 2.7. Early colonisers, whether of litter trapped in the tree canopy (aerial colonisers) or of L-layer material, were characterised by a degree of component restriction (for example, *Marasmius* species on leaves, *Crepidotus* and *Crinipellis perniciosa* on brooms). As decomposition proceeded, there was a population shift, in the lower litter layers, from these component-restricted mycelia, which usually occupied small volumes of the primary resources, to non-component-restricted mycelia, which occupied large areas of partly decomposed (secondary) resources.

The agarics of the aerial communities and upper litter layers, typified by *Crinipellis* and *Marasmius*, existed in an environment in which extreme daily fluctuations of water content took place in the wet season. However, the resources were not occupied by existing mycelia, and resource capture in foliicolous species was possible from contact *points*, either derived from rhizomorphs or adhesion zones formed by existing colonies – the role of basidiospores remains uncertain. The resulting mycelium expanded rapidly within the leaves, exploiting lignocellulose substrata. Confrontation took place with mycelia of the same species and of other agarics, sometimes resulting in a mosaic of small colonies of several species on one leaf. Exit from these colonies was achieved by desiccation-resistant fruit bodies, rhizomorphs and adhesion zones.

These strategies allowed these agarics to cope with the considerable, fluctuating stress of the upper litter/aerial environment in the humid tropics. Additional stress may have been imposed by the internal chemical environment of primary resources, such as leaves, which may, in the case of cocoa, have high levels of polyphenols and tannins (Orchard, personal communication). These materials must be tolerated and ameliorated by degradation if the unoccupied primary resource is to be captured by the mycelia of these agarics.

The position of these colonisers can be related to the behavioural concepts discussed in Chapter 1 by Rayner *et al.* These are based on general principles developed in recent years by Southwood (1976) and Grime (1979), and more recently applied to fungal ecological strategies by, for example, Pugh (1980), Hedger (1982) and Cooke & Rayner (1984). The adaptation of the primary colonising group of agarics to somewhat adverse physical and chemical environments implied a degree of adversity selection.

They also appeared to have a comparatively short period of rapid growth, followed by conversion of biomass to exit structures, such as fruit bodies which are produced in frequent flushes in response to short wetting and drying cycles in the litter. Such a strategy can be described as ruderal or *r*-selected (Southwood, 1976) and the mycelia were ephemeral, relying on continual recolonisation to maintain their position in the upper litter layers.

Unlike true ruderals, the mycelia of these agarics appeared competitive. However, there was a deadlocked confrontation between mycelial fronts and retention of resources, rather than replacement of existing mycelia from a resource, i.e. combative in the sense of Cooke & Rayner (1984).

As the cocoa litter material reached the lower layers, soil conditions and agaric communities changed (see Fig. 2.7). A *K*-selection strategy can be ascribed to the behaviour of these agarics, as typified by the large, slow-growing mycelia of *Lepiota* and *Chlorophyllum*. Their species diversity was high, when compared to that of the pioneer *Marasmius* community, but the number of individual mycelia was much lower. Each mycelium was comparatively long lived and could survive the dry season intact. It had the capacity to extend almost continuously through the litter and soil, without any component restriction. Much biomass appeared to be concentrated in these mycelial systems, which could combatively occupy substrata by mycelial fronts, fans and cords.

Investment in competitive ability, comparatively low numbers but large biomass, are all characteristic of *K*-selection, as is the channelling of reproductive effort into widely spaced production of large reproductive structures (Southwood, 1976). The periodic, sometimes seasonally related, mycelial biomass-dependent, flushing of this group of litter agarics and their comparatively large fruit body size but low numbers, reinforce this point. In some cases, as with *Chlorophyllum*, flushes seemed to correlate with the availability of new resources, i.e. at the beginning of the wet season, a feature noted by Corner (1935) in his study of the larger agarics in Malaysia and Singapore.

K-selected species are not usually adversity-selected (Southwood, 1976) and this was also true of these lower litter or soil agarics. Their mycelium remained in the uniform environment of the lower litter layers and did not extend upwards beyond the F_2 fermentation layer. Environmental fluctuations in the upper litter may have caused this colonisation boundary, but resource quality probably also played a part. Conditioning of the litter material by the upper-litter colonisers created a suitable resource for the lower-litter and soil agarics (or at least the *Lepiota* species). Such conditioning of the primary resources, such as cocoa leaves, probably included removal of phenolics and tannins, and partial degradation of lignocellulose, principally by agarics such as *Marasmius*. In addition, the biomass of the previous decomposers, plus material of animal origin such as faeces and arthropod exoskeletons, probably provided a series of secondary resources (Paul & Voroney, 1980).

The mycelial systems of the lower-litter and soil agarics must compete directly with microtrophic arthropods and annelids for these secondary resources. The existence of this nutritional mode in tropical agarics is at present conjectural, although Fermor & Wood (1981) and Atkey & Wood (1983) have shown that *Agaricus bisporus*, a *K*-selected temperate agaric, may obtain up to 10% of its nutritional requirements from microbial biomass in wheat straw compost (see also Wood: Chapter 15; Hedger, 1982).

These types of behavioural pattern represent the extremes of a continuum and intermediate examples are to be expected. Thus some agaric species of the middle cocoa litter layers, such as *Hydropus paraensis*, occupied less clear positions, being neither *r*-selected nor *K*-selected, but they combined some adversity selection with a degree of combative replacement of the previous upper-litter *Marasmius* colonisers, by virtue of external colonisation. Their mycelia, however, appeared comparatively ruderal in nature, and were replaced by the dense combative mycelial fronts in the lower litter layers. Fluctuations in water content limited their degree of penetration of the upper litter, so that an intermediate behaviour pattern resulted in an intermediate vertical position in the litter (Fig. 2.7).

Conclusions

The sequence of agaric colonisers of tropical plant-litter is capable of categorisation, by use of a series of criteria to define the type of resources within which the mycelia are likely to be successful colonists. These sequences must also exist in temperate and boreal litter systems but are less easy to define. The early aerial and L-layer communities dominated by *Marasmius*, in particular, seem to be partially absent in temperate systems, although Newell (1984*a*,*b*) has found this sequence in *Picea* litter (consisting of *Marasmius androsaceus* in the L layer, and *Mycena galopus* and *Cystoderma amianthinum* in the F layers).

Newell's study emphasised the difficulty in interpreting the type of observational data which are the basis of this chapter. She considered that selective grazing by Collembola was responsible for the restriction of *Marasmius* mycelium to the upper litter, where the Collembola population was lower. Biotic factors of this type must also influence the agaric mycelium in tropical plant-litter. At present, these and other interactions between the mycelium of the tropical agaric and its environment remain largely unexplored. We await the appearance of agaric ecology as a significant component of the research at present being directed towards endangered tropical ecosystems. This lack of research effort (when compared to that on decomposers such as termites) is especially sad in view of the primary role which these Basidiomycotina play in the decomposition cycle in these ecosystems, a role which has been amply demonstrated in this chapter.

Acknowledgements. I wish to thank Juliet Frankland, Roy Watling and Alan Rayner for much useful discussion and helpful criticism during the preparation of this manuscript. In Ecuador, I should like to thank particularly Carmen Suarez and Jaime Aragundi of INIAP, Pichilingue, and also fellow members of the British ODA team at that Station, and Calway Dodson of the Palenque Field Station, for much assistance rendered. I should also like to thank David Pegler and Roy Watling for advice on the taxonomic identity of the agarics. These identifications are provisional and will be subject to revision in future.

References

Acosta-Solis, M. (1968). *Divisiones Fitogeograficas y Formaciones Geobotanicas del Ecuador.* Quito: Casa de la Cultura Ecuatoriana.

Alvim, P. de T. (1964). Tree growth periodicity in tropical climates. In *Formation of Wood in Forest Trees*, ed. M. H. Zimmerman, pp. 479–95. New York: Academic Press.

Alvim, P. de T. & Alvim, R. (1978). Relation of climate to growth periodicity in tropical trees. In *Tropical Trees as Living Systems*, ed. P. A. Tomlinson & M. H. Zimmerman, pp. 445–64. Cambridge University Press.

Atkey, P. T. & Wood, D. A. (1983). An electron microscope study of wheat straw composted as a substrate for the cultivation of the edible mushroom (*Agaricus bisporus*). *Journal of Applied Bacteriology*, **55**, 293–304.

Bastos, C. N. & Evans, H. C. (1985). A new pathotype of *Crinipellis perniciosa* (witches' broom disease) on solanaceous hosts. *Plant Pathology*, **34** (in press).

Berkeley, M. J. (1873). *Gardener's Chronicle*, **24**, 810–11.

Black, R. (1980). The role of mycorrhizal symbiosis in the nutrition of tropical plants. In *Tropical Mycorrhiza Research*, ed. P. Mikola, pp. 191–203. Oxford: Oxford University Press.

Boekhout, T. (1983). Distribution and ecology of Andean macroscopic fungi. In *Studies on Tropical Andean Ecosystems*, vol. 1. *La Cordillera Central Colombiana*,

Transecto Parque los Nevados, ed. T. van der Hammen, A. P. Precido & P. Pinto, pp. 210–15. Weinheim: J. Cramer.

Buller, R. (1934). *Researches on Fungi*, vol. 4. London: Longman, Green & Co.

Bunting, R. H. & Dade, H. A. (1924). *Gold Coast Plant Diseases*. London: Waterlow & Sons Ltd.

Chilvers, G. A. & Prior, C. D. (1965). The structure of eucalypt mycorrhizas. *Australian Journal of Botany*, **13**, 245–59.

Cooke, R. C. & Rayner, A. D. M. (1984). *Ecology of Saprotrophic Fungi*. London & New York: Longman.

Corner, E. J. H. (1935). The seasonal fruiting of agarics in Malaya. *Gardens' Bulletin, Straits Settlement*, **9**, 79–88.

Corner, E. J. H. (1972). Boletus *in Malaysia*. Singapore: Government Printing Office.

de Segura, C. B. (1969). Ojo de gallo (*Mycena citricolor*) en Kadam, *Anthocephalus cadamba*. *Turrialba*, **19**, 553–4.

Dennis, R. W. G. (1951). Some tropical American Agaricaceae referred by Berkeley and Montagne to *Marasmius*, *Collybia* and *Heliomyces*. *Kew Bulletin*, **3**, 386–410.

Dennis, R. W. G. (1970). Fungus flora of Venezuela and adjacent countries. (*Kew Bulletin*, additional series, vol. 3.) London: HMSO.

Dennis, R. W. G. & Reid, D. A. (1957). Some marasmioid fungi allegedly parasitic on leaves and twigs in the tropics. *Kew Bulletin*, **2**, 287–92.

Dodson, C. H. & Gentry, A. H. (1978). Flora of the Rio Palenque Science Center, Los Rios Province, Ecuador. *Selbyana*, **4**, i–xxx, 1–628.

Evans, H. C. (1981*a*). Witches' broom disease – a case study. *Cocoa Growers' Bulletin*, **32**, 5–19.

Evans, H. C. (1981*b*). Pod rot of cacao caused by *Moniliophthora* (*Monilia*) *roreri*. *Phytopathological Papers*, 24.

Fermor, T. R & Wood, D. A. (1981). Degradation of bacteria by *Agaricus bisporus* and other fungi. *Journal of General Microbiology*, **126**, 377–87.

Frankland, J. C. (1982). Biomass and nutrient cycling by decomposer basidiomycetes. In *Decomposer Basidiomycetes: their Biology and Ecology*, ed. J. C. Frankland, J. N. Hedger & M. J. Swift, pp. 241–61. Cambridge University Press.

Frankland, J. C. (1984). Autecology and the mycelium of a woodland litter decomposer. In *The Physiology and Ecology of the Fungal Mycelium*, ed. D. H. Jennings & A. D. M. Rayner, pp. 241–60. Cambridge University Press.

Gentry, A. H. (1977). Endangered plant species and habitats of Ecuador and Amazonian Peru. In *Extinction is Forever*, ed. G. Prance & T. Elias, pp. 136–49. New York: New York Botanical Garden.

Golley, F. B. (1983). Decomposition. In *Tropical Rain Forest Ecosystems: Structure and Function*, ed. F. B. Golley, pp. 157–66. Amsterdam, Oxford & New York: Elsevier Scientific Publishing Company.

Grime, J. P. (1979). *Plant Strategies and Vegetation Processes*. Chichester: John Wiley.

Haines, B. (1975). Impact of leaf cutting ants on vegetation development at Barro Colorado Island. In *Tropical Ecological Systems*, ed. F. B. Golley & F. Medina, pp. 99–111. Berlin: Springer-Verlag.

Hedger, J. N. (1982). The role of basidiomycetes in composts: a model system for decomposition studies. In *Decomposer Basidiomycetes, their Biology and Ecology*, ed. J. C. Frankland, J. N. Hedger & M. J. Swift, pp. 263–306. Cambridge University Press.

Heim, R. (1937). Les Lactario-Russules du domaine oriental de Madagascar. *Flore Mycologie de Madagascar*, **1**, 1–196.

Heim, R. (1977). *Termites et Champignons. Les Champignons Termitophiles d'Afrique Noire et d'Asie Meridionale*. Paris: Editions Boubée.

Hering, T. F. (1982). Decomposing activity of basidiomycetes in forest litter. In
    Decomposer Basidiomycetes: their Biology and Ecology, ed. J. C. Frankland,
    J. N. Hedger & M. J. Swift, pp. 213–25. Cambridge University Press.
Hintikka, V. (1963). Studies on the genus *Mycena* in Finland. *Karstenia*, **6/7**, 77–87.
Hintikka, V. (1982). The colonisation of litter and wood by basidiomycetes in Finnish
    forests. In *Decomposer Basidiomycetes: their Biology and Ecology*, ed.
    J. C. Frankland, J. N. Hedger & M. J. Swift, pp. 227–39. Cambridge University
    Press.
Holdridge, L. R. (1971). *Life Zone Ecology*. San Jose, Costa Rica: Tropical Science
    Center.
Hopkins, B. (1965). *Forest and Savanna: an Introduction to Tropical Plant Ecology with
    Special Reference to West Africa*. Ibadan & London: Heinemann.
Hutchinson, J. (1959). *The Families of Flowering Plants: Arranged According to a New
    System Based on their Probable Phylogeny*, vols 1 & 2. Oxford: Clarendon
    Press.
Ingram, D. S., Sargent, J. A. & Tommerup, I. C. (1976). Structural aspects of infection by
    biotrophic fungi. In *Biochemical Aspects of Plant Parasite Relationships*, ed.
    J. Friend & D. R. Threlfall, pp. 43–78. London: Academic Press.
Ivory, M. H. (1975). Mycorrhizal studies on exotic conifers. *Malaysian Forester*, **38**,
    149–52.
Janse, J. M. (1896). Les endophytes radicaux de quelques plantes javanaises. *Annales du
    Jardin Botanique de Buitenzorg*, **14**, 53–212.
Johanssen, D. (1974). Ecology of vascular epiphytes in West African rain forest. *Acta
    Phytogeographica Suecica*, **59**, 1–129.
Johnsson, L. & Nylund, J. E. (1979). *Favolaschia dybowskyana* (Singer) Singer, a new
    orchid mycorrhizal fungus from tropical Africa. *New Phytologist*, **83**, 121–8.
Kramer, C. L. (1982). Production, release and dispersal of basidiospores. In *Decomposer
    Basidiomycetes: their Biology and Ecology*, ed. J. C. Frankland, J. N. Hedger &
    M. J. Swift, pp. 33–49. Cambridge University Press.
Kranz, J., Schmutterer, H. & Koch, W. (1978). *Diseases, Pests and Weeds in Tropical
    Crops*. New York: Wiley.
Laursen, G. A. & Chmielewski, M. A. (1982). The ecological significance of soil fungi in
    Arctic Tundra. In *Arctic and Alpine Mycology, the First International
    Symposium on Arcto-Alpine Mycology*, ed. G. A. Laursen & J. F. Ammirati,
    pp. 432–92. Seattle & London: University of Washington Press.
Lee, K. E. & Wood, T. G. (1971). *Termites and Soils*. London & New York: Academic
    Press.
Longman, K. A. & Jenik, J. (1974). *Tropical Forest and its Environment*. London:
    Longman.
Lynch, J. M. & Poole, M. J. (1979). *Microbial Ecology: a Conceptual Approach*. Oxford:
    Blackwell Scientific Publications.
Maldague, M. E. (1964). Importance des populations de termites dans les sols
    equatoriaux. *Transactions of the 8th International Congress of Soil Science,
    Bucharest*, **3**, 743–51.
Malloch, D. W., Pirozynski, K. A. & Raven, R. H. (1980). Ecological and evolutionary
    significance of mycorrhizal symbioses in vascular plants. *Proceedings of the
    National Academy of Science, USA*, **77**, 2113–18.
Matsumoto, T. (1976). The rôle of termites in an equatorial rain forest ecosystem in West
    Malaysia. 1. Population density, biomass, carbon, nitrogen and calorific
    content and respiration rate. *Oecologia*, **22**, 153–78.
Newell, K. (1984*a*). Interaction between two decomposer basidiomycetes and a

collembolan under Sitka spruce: distribution, abundance and selective grazing. *Soil Biology and Biochemistry*, **16**, 227–33.

Newell, K. (1984*b*). Interaction between two decomposer basidiomycetes and a collembolan under Sitka spruce: grazing and its potential effects on fungal distribution and litter decomposition. *Soil Biology and Biochemistry*, **16**, 235–9.

Olson, J. S. (1963). Energy storage and the balance of producers and decomposers in ecological systems. *Ecology*, **44**, 322–31.

Paul, E. A. & Voroney, R. P. (1980). Nutrient and energy flows through soil microbial biomass. In *Contemporary Microbial Ecology*, ed. D. C. Ellwood, J. N. Hedger, M. J. Latham, J. M. Lynch & J. H. Slater, pp. 215–37. London: Academic Press.

Pegler, D. N. (1969). A new pathogenic species of *Marasmiellus* Murr. (Tricholomataceae). *Kew Bulletin*, **23**, 523–5.

Pegler, D. N. (1977). A preliminary agaric flora of East Africa. (*Kew Bulletin*, additional series, vol. 6.) London: HMSO.

Pegler, D. N. (1983). Agaric Flora of the Lesser Antilles. *Kew Bulletin Additional Series*, **9**. London: HMSO.

Pegler, D. N. & Fiard, J. P. (1979). Taxonomy and ecology of *Lactarius* (Agaricales) in the Lesser Antilles. *Kew Bulletin*, **3**, 601–28.

Petch, T. (1915). Horse-hair blights. *Annals of the Royal Botanic Gardens, Peradeniya*, **6**, 1–26.

Petch, T. (1923). *The Diseases of the Tea Bush*. London: Macmillan.

Pugh, G. J. F. (1980). Presidential address: strategies in fungal ecology. *Transactions of the British Mycological Society*, **75**, 1–14.

Redhead, J. F. (1968*a*). Mycorrhizal associations in some Nigerian forest trees. *Transactions of the British Mycological Society*, **51**, 377–87.

Redhead, J. F. (1968*b*). *Inocybe* sp. associated with ectotrophic mycorrhiza on *Afzelia bella* in Nigeria. *Commonwealth Forestry Review*, **47**, 63–5.

Redhead, J. F. (1980). Mycorrhiza in natural tropical forests. In *Tropical Mycorrhiza Research*, ed. P. Mikola, pp. 127–42. Oxford: Oxford University Press.

Richards, P. W. (1952). *The Tropical Rain Forest, an Ecological Study*. Cambridge University Press.

Richardson, M. J. (1970). Studies on *Russula emetica* and other agarics in a Scots Pine plantation. *Transactions of the British Mycological Society*, **55**, 217–29.

Singer, R. (1962). *The Agaricales in Modern Taxonomy*, 2nd edn. Weinheim: J. Cramer.

Singer, R. (1971). Forest mycology and forest communities in South America. II. Mycorrhiza sociology and fungus succession in the *Nothofagus dombeyi–Austrocedrus chilensis* woods of Patagonia. In *Mycorrhizae*, ed. E. Hacskaylo, pp. 204–15. Washington: US Government Printing Office.

Singer, R. & Araujo, I. J. da Silva (1979). Litter decomposition and ectomycorrhiza in Amazonian forests. 1. A comparison of litter decomposing and ectomycorrhizal Basidiomycetes in latosol–terra firma forest and white podzol *campinarana*. *Acta Amazonica*, **9**, 25–41.

Singer, R. & Moser, M. (1965). Forest mycology and forest communities in South America. *Mycopathologia et Mycologia Applicata*, **26**, 129–91.

Singh, K. G. (1966). Ectotrophic mycorrhiza in equatorial rain forest. *Malaysian Forester*, **39**, 13–19.

Singh, U. R. (1977). Relationships between the population density of soil microarthropods and mycoflora associated with litter and total litter respiration on the floor of a Sal forest in Varanasi, India. In *Soil Organisms as Components of Ecosystems*, ed. U. Lohm & T. Persson, pp. 463–72. Stockholm: Swedish Natural Science Research Council.

Southwood, T. R. E. (1976). Binomic strategies and population parameters. In *Theoretical Ecology, Principles and Applications*, ed. R. M. May, pp. 26–48. Oxford: Blackwell Scientific Publications.

Stover, R. H. (1972). *Banana, Plantain and Abaca Diseases*. Kew: Commonwealth Mycological Institute.

Stradling, D. J. (1977). Food and feeding habits of ants. In *Production Ecology of Ants and Termites*, ed. M. V. Brian, pp. 81–106. Cambridge University Press.

Swift, M. J. (1968). Inhibition of rhizomorph development by *Armillaria mellea* in Rhodesian forest soils. *Transactions of the British Mycological Society*, **51**, 241–7.

Swift, M. J. (1976). Species diversity and the structure of microbial communities in terrestrial habitats. In *The Role of Terrestrial and Aquatic Organisms in Decomposition Processes*, ed. J. M. Anderson & A. Macfadyen, pp. 185–222. Oxford: Blackwell Scientific Publications.

Swift, M. J. (1982). Basidiomycetes as components of forest ecosystems. In *Decomposer Basidiomycetes: their Biology and Ecology*, ed. J. C. Frankland, J. N. Hedger & M. J. Swift, pp. 307–37. Cambridge University Press.

Swift, M. J., Heal, O. W. & Anderson, M. J. (1979). *Decomposition in Terrestrial Ecosystems*. (Studies in Ecology, vol. 5.) Oxford: Blackwell Scientific Publications.

Thomazini, L. I. (1974). Mycorrhiza in plants of the Cerrado. *Plant and Soil*, **41**, 707–11.

Thorold, C. A. (1975). *Diseases of Cocoa*. Oxford: Oxford University Press.

Watling, R. (1977). An analysis of the taxonomic characters used in defining the species of the Bolbitiaceae. In *The Species Concept in Hymenomycetes*, ed. H. Clemençon, pp. 11–53. Weinheim: J. Cramer.

Watling, R. (1982). Taxonomic status and ecological identity in the basidiomycetes. In *Decomposer Basidiomycetes: their Biology and Ecology*, ed. J. C. Frankland, J. N. Hedger & M. J. Swift, pp. 1–32. Cambridge University Press.

Went, F. W. & Stark, N. (1968). Mycorrhiza. *Bioscience*, **18**, 1035–9.

Whitmore, T. (1975). *Tropical Rain Forests of the Far East*. London: Oxford University Press.

Wood, G. A. R. (1975). *Cocoa*, 3rd edn. London: Longman.

3

Armillaria: resources and hosts

J.RISHBETH

Botany School, University of Cambridge, Cambridge CB2 3EA, UK

Introduction

Although it has been apparent for a long time that *Armillaria* can utilise a wide variety of woody resources, more detailed information is required now that several distinct species are known to occur. Korhonen (1978) was the first to identify species by using incompatibility factors as genetic markers. Of the three species occurring in Finland, he found that two are predominantly saprotrophs and occur to some extent in butt-rott of Norway spruce (*Picea abies*), whilst the third is a pathogen of Scots pine (*Pinus sylvestris*) and apparently uses stumps as a food base. This demonstration that *Armillaria* species may differ in ecological status has provided a stimulus for investigation elsewhere. Preliminary observations have been made on the occurrence of *Armillaria* in southern England and its role in woodlands (Rishbeth, 1982). The species commonly encountered were *Armillaria mellea*, *A. ostoyae* and *A. bulbosa*, whilst *A. tabescens* was locally abundant in some woodlands. This chapter reports the results of further surveys in this area, laboratory experiments to elucidate the ability of the different species to colonise natural resources, and more tests of their pathogenicity. Wherever possible several isolates of each species were used.

Host preferences

Since trees and shrubs attacked by *Armillaria* constitute important resources for the fungus, it seems appropriate to consider first on what hosts the different species are found. In plantations of broadleaved trees on sandy soil in East Anglia, *Armillaria* kills well-established birch (*Betula pendula*) and occasionally ash (*Fraxinus excelsior*). In gardens of the Cambridge area, *Armillaria* kills a considerable variety of broadleaved trees; when records of these were added to records from plantations, the distribution of species occurring in 44 samples, restricted to one per infection focus, was *A. mellea* 39, *A. ostoyae* 2 and *A. bulbosa* 3. *A. mellea* was therefore much the commonest species involved. Of the two records of *A. ostoyae*, one was in ash and the other in young gean (*Prunus avium*) that had been weakened by other agencies. Both were in mixed plantings with conifers, which according to observations made in France is characteristic of the relatively rare attacks made by *A. ostoyae* on broadleaved trees. If all records of killing by *Armillaria* in Cambridge gardens are considered, *A. mellea* was found in 96 out of 103 broadleaved trees or shrubs. The other seven were infected by *A. bulbosa*, and there was generally evidence that a predisposing factor such as considerable moisture stress had been involved. This species may also cause a partial die-back of garden shrubs. The commonest genera to be attacked by *A. mellea* and the number of cases that occurred were as follows: *Prunus* 12, *Malus* 7, *Betula*, 7, *Salix* and *Ligustrum* 6, *Acer* and *Viburnum* 5. A further 24 genera had fewer than five cases. Of all deaths recorded, 32% occurred in members of the Rosaceae and 16% in those of the Oleaceae. Conifer genera killed by *A. mellea* in gardens included *Cupressus*, *Cupressocyparis*, *Libocedrus* and *Sequoiadendron*. The very broad host range of this species is evident. *A. ostoyae* was not found in these gardens, which had neutral or alkaline soil, but may well occur on conifers in other districts where the soil is acidic.

Conifer plantations are often vulnerable, and the results of isolations from well-established trees of various species that had been killed are given in Table 3.1. The age of trees ranged from about 10 to 50 years; once again samples were limited to one per focus. *A. ostoyae* regularly occurred on Scots pine and Corsican pine (*P. nigra* var. *maritima*), whereas *A. mellea* was found only once. Sample numbers for other conifers were low, but it seems likely that *A. ostoyae* also predominates in Norway spruce, European larch (*Larix decidua*) and Douglas fir (*Pseudotsuga menziesii*). By contrast, *A. mellea* was commoner in Western hemlock (*Tsuga heterophylla*) and Western red cedar (*Thuja plicata*). Of the 78 samples collected from all conifers, 62 contained *A. ostoyae* and 16 *A. mellea*, in marked contrast to the distribution of species occurring in broadleaved trees.

Since pathogenicity tests on 2-year-old pines had shown that these are very susceptible to attack by *A. mellea*, it was decided to survey a variety of plantations in the hope of discovering to what extent it occurs naturally on pines. Pines established on former woodland are the most suitable because infection sources in the form of stumps are usually present. The younger plantings were predominantly of Corsican pine, though a few were

Table 3.1. Armillaria *species isolated from well-established conifers that had been killed*

Species	Total number of samples	Number of samples from which the following species were obtained	
		A. mellea	A. ostoyae
Pinus sylvestris P. nigra var. maritima	50	1	49
Picea abies	4	0	4
Larix decidua	6	1	5
Pseudotsuga menziesii	3	0	3
Tsuga heterophylla	7	6	1
Thuja plicata	8	8	0

Table 3.2. Armillaria *species isolated from various categories of pine planted on former woodland*

Type of former woodland	Age (years) and state of trees	Number of trees sampled[a]	Number of trees from which the following species were obtained		
			A. mellea	A. ostoyae	A. bulbosa
Broadleaved	2–5 vigorous	31 (9)	11	16	4
Conifer	2–5 vigorous	45 (6)	0	43	2
Mixed broadleaved and conifer	> 10 vigorous	50 (12)	1	49	0
	> 10 suppressed	51 (6)	8	40	3

[a] The number of forest compartments surveyed is given in parentheses.

of Scots pine. The main findings are recorded in Table 3.2. At sites previously occupied by broadleaved woodland, *A. mellea* and *A. ostoyae* were both common in young dead pines, whereas *A. bulbosa* was relatively infrequent and occurred in trees growing at badly drained sites which were probably subject to frequent waterlogging. The susceptibility of young pines to *A. mellea* is thus confirmed. The situation differed, however, where conifers, such as pines or spruces, comprised the former crop. Here *A. ostoyae* predominated, no *A. mellea* being detected, whilst *A. bulbosa* again appeared occasionally in pines growing in very wet situations. At sites where the previous woodland had been broadleaved, of total area 35 ha,

Table 3.3. Armillaria *species isolated from butt-rotted trees and from stumps*

	Total number of samples	Number of samples from which the following species were obtained			
		A. mellea	A. ostoyae	A. bulbosa	A. tabescens
Butt-rotted trees					
Broadleaved	15	0	0	15	0
Conifer	7	0	7	0	0
Stumps					
Broadleaved	77	35	2	32	8
Conifer	20	3	14	3	0

the percentage of deaths caused by *Armillaria* ranged from 0 to 2% (mean 0.3%). The site having 2% mortality was unusual in having several large foci of *A. ostoyae*. By contrast, at sites originally bearing conifers, covering 20 ha, the percentage of deaths varied from 3 to 10% (mean 6.5%). The possible reasons for these differences will be discussed later. Results obtained from older plantations, established on sites previously bearing a mixture of conifers and broadleaved trees, provided a contrast between vigorous trees and those heavily suppressed by competition. The latter were more likely to be infected by *A. mellea* or occasionally *A. bulbosa* than were vigorous trees; nevertheless *A. ostoyae* was still the dominant species. Killing of pines by *A. mellea* is therefore generally restricted to very young trees or those seriously weakened in some way.

The incidence of *Armillaria* species in butt-rotted trees and in stumps is shown in Table 3.3. All broadleaved trees so far sampled that had large cavities were infected by *A. bulbosa*, whereas Norway spruce, the only conifer investigated, was infected by *A. ostoyae*. *A. mellea* and *A. bulbosa* were almost equally common in stumps of broadleaved trees, whilst *A. tabescens* sometimes occurred but was restricted to those in heavier soils; *A. ostoyae* was rare. This last species predominated in conifer stumps, whilst *A. mellea* and *A. bulbosa* were less common.

Pathogenicity tests

Most of the work on pathogenicity has been carried out with seedlings of Scots pine 2 years old, growing in field plots on a medium loam of pH 7.2 (Rishbeth, 1982). A woody inoculum containing *Armillaria* was placed in contact with the main root 50 mm below soil level, ten replicates being set up for each isolate. Any infection that developed was scored as

Table 3.4. *Results of inoculating 2-year-old Scots pines with* Armillaria

Species	Number of isolates tested	Mean score per isolate $\pm$ S.E.[a]	% Trees killed or infected	Rhizomorph abundance[b]
A. mellea	22	13.3 ± 1.4	63	0.3
A. ostoyae	13	8.5 ± 1.3	55	0.5
A. bulbosa	22	0.5 ± 0.2	0.3	1.7
A. tabescens	7	0	0	0.03

[a] The following scores were allocated: 3 for trees killed during the first growing season, 2 for trees killed during the second season; and at the end of the second season: 1 for moderate infection, 0.5 for slight infection. The maximum possible score for ten replicates was 30.
[b] Based on scores of 0 to 3: mean for all inocula.

indicated in Table 3.4, which summarises results obtained over a period of several years. *A. mellea* and *A. ostoyae* were very pathogenic, the former species tending to kill pines more rapidly than the latter. *A. bulbosa* was only slightly pathogenic, whilst *A. tabescens* caused no infection. Analysis of results, restricted to isolates that were tested at the same time and within one year of first being cultured, showed that within *A. mellea* and *A. ostoyae* there were significant differences between isolates in ability to cause infection. Inocula of *A. bulbosa* generally produced an abundant growth of rhizomorphs, whereas rhizomorph production by *A. mellea* and *A. ostoyae* was less and varied considerably between isolates; with *A. tabescens* rhizomorph production was minimal. The density of inocula was usually determined at the end of the experiment, which lasted about 18 months. For 16 isolates of *A. mellea*, inoculum density ranged from 0.36 to 0.43 g cm^{-3} (mean 0.38 ± 0.06 g cm^{-3}) and for 18 of *A. bulbosa* it varied from 0.20 to 0.36 g cm^{-3} (mean 0.29 ± 0.05 g cm^{-3}). These means differ significantly ($P < 0.01$) and represent density losses of 27% and 44%, respectively, during the period of burial. Fewer estimates were made for inocula of *A. ostoyae* and *A. tabescens*, but their densities were similar to those of *A. mellea*.

In view of the wide variety of hosts infected by *Armillaria* it is desirable also to have information about broadleaved species. Preliminary tests have been carried out on seedlings of birch and ash, and rooted suckers of wild black currant (*Ribes nigrum*). The few results so far obtained are given in Table 3.5, together with comparable results from Scots pine. Birch and ash were generally more resistant to infection than was pine.

Table 3.5. *Results of inoculating young broadleaved species with* Armillaria *and comparisons with Scots pine*

	Mean score for infection[b]			
Species[a]	Birch	Ash	Currant	Pine
A. mellea	0.5	1	–	6
	3.5	0	–	6.5
	–	2	–	17.5
	–	3	9	17.5
A. ostoyae	0.5	–	–	7
	–	0	–	9.5
	–	0	–	8
A. bulbosa	0	–	–	0
	–	–	0.5	0

[a] Results obtained from four, three and two isolates, respectively.
[b] As in Table 3.4.

Colonisation of natural substrates
Growth in fresh excised roots and stems

Since roots, and to a lesser extent stems, provide natural resource units for *Armillaria*, it is useful to discover how rapidly they are colonised and whether species differ in this respect. Discs of hazel stem (*Corylus avellana*) 20 mm in diameter and 10 mm thick, which had been rasped on one surface to produce a channel 3 mm deep, were used as inocula after complete colonisation by *Armillaria*. Roots of pine or oak (*Quercus robur*) were washed, surface sterilised, cut into 100 mm lengths and inoculated centrally by attaching a disc to the surface with a taut rubber band. The ends of each length were painted in order to prevent entry of rhizomorphs. Inoculated lengths were incubated in moist sterile sand held in 180-mm plastic pots. They were inspected after 6 weeks and thereafter at intervals of 4 weeks, when the extent of any colonisation by *Armillaria* was determined. This was often evident by swelling, cracking or discoloration of the bark but, if not, mycelial sheets could be detected by making small incisions in the bark. At the end of the experiment the bark was removed and the extent of any growth measured. The proportion of lengths in which mycelial sheets were 50 mm or more in length is recorded in Table 3.6. Since this was essentially a pilot experiment to detect any major differences between species, results from isolates of each species have been combined. Caution is required in interpreting the results of experiments on excised material, and conclusions are therefore tentative.

Table 3.6. *Growth of* Armillaria *in excised roots inoculated over intact bark*

Species	Mean relative extent of growth at 20 °C[a]	
	Pine	Oak
A. mellea	0.72	0.82
A. ostoyae	0.06	0.06
A. bulbosa	0.27	0.29
A. tabescens	0.06	0

[a] The proportion of lengths in which the fungus had grown 50 mm or more within 18 weeks: means derived from four isolates of each species and four replicates of each isolate.

Under these conditions, *A. mellea* colonised a much greater proportion of root lengths than the other species. Although at the end of the experiment the mean relative extents of growth of *A. mellea* in pine and oak roots were similar, pine roots tended to be colonised more rapidly: thus after 10 weeks the proportion of pine lengths fully colonised was 0.63 and that of oak lengths only 0.11. *A. bulbosa* showed a greater ability than *A. ostoyae* to penetrate intact bark. By far the commonest mode of infection was directly through the bark under the inoculum. Penetration generally occurred through an area of bark only 1–3 mm², although the total area of contact with the inoculum often exceeded 70 mm². On a vastly different scale, this provides an interesting parallel with hyphae of wood-rotting fungi when penetrating cell walls. Especially with *A. bulbosa*, but more rarely with *A. mellea*, rhizomorphs produced from the inoculum grew over the root surface and formed small knots locally, beneath which penetration sometimes occurred. There was no entry through cut ends.

In some circumstances *Armillaria*, having gained entry, has the opportunity to invade considerable lengths of root or stem which have little or no resistance, as for example in stump roots and recently killed trees. A further series of experiments was therefore set up in which *Armillaria* could enter the cut end of excised material prepared as before. The inoculum was similar, but with no channelled surface, and was firmly attached to one end of each length by a rubber band. The same method of incubation was used; lengths were examined after 2 weeks and then at intervals of 4–10 days, depending on the growth rate of *Armillaria*. The final examination was made after 4 or 6 weeks. Results obtained with oak roots, stems of sycamore (*Acer pseudoplatanus*) and coppice shoots of willow (*Salix*

Table 3.7. *Growth of* Armillaria *in excised roots or stems inoculated at one end*

Species	Mean relative extent of growth at 20 °C[a] (number of isolates tested)		
	Oak roots	Sycamore stems	Willow shoots
A. mellea	0.79 (4)	0.92 (8)	0.60 (12)
A. ostoyae	0.19 (6)	0.66 (6)	0.20 (10)
A. bulbosa	0.36 (5)	1.00 (6)	0.35 (12)
A. tabescens	0.39 (5)	0.29 (6)	0.05 (6)

[a] The proportion of lengths in which the fungus had grown 50 mm or more within 4 weeks (oak and sycamore) or 6 weeks (willow): means derived from four replicates of each isolate.

caprea or *S. alba*) are summarised in Table 3.7; once again, a single mean is given for all isolates of a species. If results for oak roots are compared with those shown in Table 3.6, it will be seen that growth of *Armillaria* through cut ends was much more rapid, as would be expected, the proportion of lengths colonised by the first three species listed being roughly comparable after only 4 weeks. However, whilst *A. tabescens* was unable to enter oak lengths through intact bark, its ability to colonise them through cut ends was similar to that of *A. bulbosa*. Sycamore and willow stems provided a contrast since the former were generally soon colonised by *A. mellea*, *A. bulbosa* and *A. ostoyae*, whereas in the latter their growth was slower and more erratic. Particularly with willow collected in winter, this was associated with the production of callus at the cut ends and the appearance of shoots and sometimes roots. In oak roots, *A. mellea* often grew at about 4 mm day^{-1} and the other species at about 2.5 mm day^{-1}, whereas in sycamore stems growth was more rapid, often at a rate of 8 mm day^{-1}. All species grew faster in willow stems collected in summer than in those cut during winter. Considerable differences also existed between material obtained from different species at the same time of year. Thus roots of Scots pine collected in summer were rapidly colonised by all *Armillaria* species, whereas roots of the hybrid poplar Androscoggin were colonised very little.

The results of these experiments may be summarised by stating that *A. mellea* showed a greater ability than other species to colonise roots or stems with appreciable resistance to entry, as in the case of intact bark. However in material with little resistance, such as pine roots or sycamore stems, *A. bulbosa* grew as fast as *A. mellea*. *A. bulbosa* generally colonised tissues more rapidly than the two remaining species. *A. ostoyae* grew more extensively than *A. tabescens* except after entering through cut ends of oak roots, which is noteworthy in view of the fact that *A. tabescens* often occurs in roots of oak stumps.

Growth of tissues partially colonised by microorganisms

Under natural conditions, roots of stumps and dying trees are often colonised by a variety of microorganisms. It is therefore desirable to have information about the ability of *Armillaria* to grow in such tissues. Whereas in the experiments just described it was aimed to keep competition to a minimum, in the following one it was promoted by first incubating the material in unsterile soil. The principle is similar to that employed by Garrett (1960), although the method differed considerably in detail. Oak stems, rather than roots, were used for ease of standardisation. Lengths 20 mm in diameter and 40 mm long were buried in moist acidic soil held in plastic boxes and incubated at 20 °C for 5 weeks. The soil had been found suitable for production of rhizomorphs. One set of lengths was incubated whilst still fresh and another after autoclaving for 30 min; it was hoped that these would be colonised more extensively by soil microorganisms. After incubation, lengths were placed in glass jars of capacity 350 ml, containing the same type of soil, at a distance of 20 mm from a woody inoculum fully colonised by *Armillaria*. Jars were capped with polythene and incubated at 20 °C for 20 weeks. Ten replicates were set up for each combination of isolate and type of oak length. At the end of the experiment the bark was removed and the proportion colonised by *Armillaria* estimated visually. The wood was then cut serially into five equal discs which were incubated on moist sterile sand at 8 °C for 2 weeks; the proportion of cross-sectional area occupied by *Armillaria* was then determined.

In most cases rhizomorphs had reached the lengths, many of which had been colonised by *Armillaria*, apparently through the cut ends. Growth was generally more extensive in the bark than in the wood. The results were so variable that no conclusions could be drawn about the effect of autoclaving, but it was clear from the appearance of the discs that autoclaved lengths had been more extensively occupied by various wood-staining fungi than had non-autoclaved ones. Nonetheless all isolates colonised autoclaved wood to some extent, one isolate of *A. bulbosa* being particularly effective. Garrett used an isolate that had killed a garden shrub and was probably *A. mellea sensu stricto*. His assessment, although based on different criteria, showed that it possessed a considerable degree of

Table 3.8. *Growth in length (cm) of* Armillaria *rhizomorphs produced from inocula buried for 2 years at a depth of 150 mm at the sites shown*

Soil texture...	Sand		Sand		Sand	
Soil pH...	4.0		5.5		7.0	
Tree cover...	Oak		Pine		Pine	
	M[a]	T[b]	M	T	M	T
A. mellea	9	4±2	15	29±4	6	5±2
A. ostoyae	37	127±25	40	114±36	66	112±27
A. bulbosa	130	158±38	150	248±110	230	344±116

Soil texture...	Sand		Clay-loam	
Soil pH...	7.0		7.1	
Tree cover...	None		None	
	M	T	M	T
A. mellea	4	8±1	4	14±2
A. ostoyae	50	64±13	43	102±41
A. bulbosa	94	200±21	170	354±55

[a] Maximum length recorded in five replicates.
[b] Total length per inoculum: mean of five replicates ± standard error.

competitive saprophytic ability. The current observations confirmed this and provided no evidence that *A. mellea* differs very much in this respect from the less pathogenic *A. bulbosa*.

Ability of rhizomorphs to grow through soil

The availability of new resources to *Armillaria* is likely to depend partly upon the distance to which rhizomorphs grow through soil. That rhizomorphs can reach potential resources at a distance from the food base is shown by the situation at edges of expanding foci in woodland, where roots in various stages of colonisation can be seen. The ability of rhizomorphs to spread from standard inocula in soil was investigated at sites that had been selected to provide differences in soil texture, soil reaction and tree cover. One isolate each of *A. mellea, A. ostoyae* and *A. bulbosa*, known to produce rhizomorphs in soil, was used, fully colonised lengths of oak stem 40 mm in diameter and 120 mm long being buried at a depth of 150 mm. After 2 years the soil around inocula was examined and any rhizomorphs found were collected and measured. Results obtained from five sites are given in Table 3.8.

At each site *A. mellea* produced the shortest rhizomorphs and *A. bulbosa* the longest. There were no consistent differences between sites, and if the data from all of them are combined, the mean maximum lengths produced by *A. mellea*, *A. ostoyae* and *A. bulbosa* were 8, 47 and 155 cm, respectively, whilst the mean total lengths per inoculum were 12, 104 and 261 cm, respectively. In the woodland sites, rhizomorphs of *A. bulbosa* often grew towards the soil surface and then ran horizontally in the humus layer, branching extensively. By contrast, rhizomorphs of *A. ostoyae* tended to grow horizontally or vertically downwards, and where pines were present some of the latter rhizomorphs had infected roots up to 30 mm in diameter. A fresh crop of rhizomorphs had often been produced from resinous areas on these roots, as is characteristic of natural infections by *A. ostoyae*. No root infections were seen near inocula of *A. mellea* or *A. bulbosa*. Thus of the three species, *A. bulbosa* had the greatest ability to produce rhizomorphs. This was also indicated by the results of sampling rhizomorphs in woodland: of 84 samples from which different mycelial types were obtained, 82 were of *A. bulbosa* and 2 were of *A. ostoyae*. Furthermore *A. bulbosa* created the largest focus so far discovered in woodland, having a diameter of about 330 m. The extensive growth of *A. bulbosa* rhizomorphs in some gardens leads to their appearance in a variety of potential resources, such as composted leaves, sawdust and pulverised bark. Infection of pine roots by *A. ostoyae* some 200–300 mm from inocula demonstrates the potential advantage of producing rhizomorphs even over a relatively short range. It seems probable that most infections by *A. mellea* occur at an even shorter distance or as a result of root contacts. Over a comparable period, rhizomorphs are likely to grow further from natural food sources than from the relatively small inocula used in this experiment; measurements made around the edges of foci showed that for *A. ostoyae* and *A. bulbosa* the mean rate of rhizomorph extension was 0.9–1.7 m yr^{-1}. Since isolates of all three species vary in ability to form rhizomorphs, and many isolates of *A. ostoyae* produce few or none, further experiments of this type are needed to give a more representative picture.

Discussion

Probably the most useful way to evaluate the situation is to consider each species in turn. *A. mellea* has a wide host range amongst broadleaved trees and shrubs and also attacks certain conifers, although more resinous types such as pines are seldom killed unless very young or much weakened. It is uncommon to find young naturally regenerated birch or ash killed by *A. mellea*, which bears out the results of inoculating young seedlings of these trees. *A. mellea* has a much greater ability than the other species to penetrate intact bark, and this may well compensate for its limited capacity to spread by means of rhizomorphs. The results of inoculating roots or stems at one end suggest that this species can readily exploit weakened tissues once it has gained entry, and can invade tissues with some residual resistance more effectively than can the other species tested. Although there is evidence that *A. mellea* can infect freshly cut stumps by means of basidiospores (Rishbeth, 1983), it is probable that most infection sources in recently felled woodland represent trees infected whilst still standing. This would account for the observation that foci of *A. mellea*, as indicated by the presence of fruit bodies, are generally small and often confined to single stumps. Groups of young pines killed by this species after such sites have been replanted are correspondingly small. However, much larger foci may develop in closely planted trees of susceptible species, as happens in gardens and with birch on light-textured soils.

A. ostoyae has a different pattern of behaviour. It often kills well-established conifers, especially the more resinous types, but rarely kills broadleaved trees. Inoculations confirmed that young pines are very susceptible and young broadleaved trees moderately resistant. There is some evidence from a field experiment (Rishbeth, 1982), that *A. ostoyae*, unlike *A. mellea*, can infect considerable areas of pine bark without inducing much resin flow. Moreover, even when much resin has been produced *A. ostoyae* still grows very freely under the bark and may produce abundant rhizomorphs. This ability was overlooked in earlier work when the relative values of conifer and broadleaved tissues as a food base for *Armillaria* were investigated. Observations on excised roots suggest that *A. ostoyae* has a limited capacity to penetrate intact bark: this may be true of mycelial sheets, but not necessarily of rhizomorphs, few of which were produced in the experiment. It would be useful to repeat this under conditions favouring rhizomorph production. The results of inoculating roots or stems at one end suggest that *A. ostoyae* is less able to exploit tissues having some residual resistance than either *A. mellea* or *A. bulbosa*. This may partly account for its relative scarcity in stumps of broadleaved trees. However, *A. ostoyae* sometimes forms large foci in broadleaved woodlands, perhaps more particularly where soil conditions favour rhizomorph growth, and it colonises suppressed trees in even-aged oak plantations (Thompson & Boddy, 1983). It seems possible that *A. ostoyae* only enters tissues when their resistance has fallen to a very low level, as is suggested by its comparatively rapid growth in sycamore stems. *A. ostoyae* probably also infects stumps by means of spores. In mixtures of broadleaved trees and conifers its ability to spread on conifer roots may create

substantial food bases, thus accounting for its frequency in pines subsequently replanted at such sites. *A. ostoyae* was almost the sole species
found killing young pines established after conifers, even when these had
been growing for only a few years after the replanting of previously broadleaved woodland. This again is probably due to the rapidity with which
A. ostoyae invades conifer roots from infected stumps.

Unlike the two previous species, which kill trees in the absence of any
obvious predisposing factor, *A. bulbosa* is generally confined to severely
weakened trees, either broadleaved or conifer. Small pines are seldom
killed or even infected after inoculation. *A. bulbosa* was less able to
penetrate intact bark than *A. mellea*, although rhizomorphs were generally
abundant. By contrast it was at least as effective as *A. mellea* in colonising
tissues having little resistance, through the cut end. It is frequently found
in stumps of broadleaved trees, but less often in those of conifers.
Rhizomorphs of *A. bulbosa* are remarkably common, and, confusingly,
may occur on roots of trees killed by other species of *Armillaria*. In
addition, they often form a network over the surface of living roots and
stem bases and are therefore well placed to invade them when resistance
declines sufficiently. Under certain conditions, *A. bulbosa* can decay roots
in apparently healthy trees, judging by the extensive basal trunk rots
sometimes present in elms and poplars, for example. Invasion by *A.
bulbosa* probably explains the early disappearance of suppressed trees in
some stands of first-rotation oak in Thetford Forest (Rishbeth, 1978). The
foci in such stands are often centred on stumps created by thinning, which
were almost certainly infected by air-borne spores of *A. bulbosa*. This
species had also colonised a very high proportion of suppressed oaks
sampled in the Forest of Dean (Thompson & Boddy, 1983). Possession
of a considerable degree of competitive saprophytic ability is likely to be
advantageous in colonising stumps and heavily suppressed trees. Although
A. bulbosa probably decays some types of wood more rapidly than the
other species, it often survives for a long time: for instance it was present
in an oak stump 53 years after felling and still producing rhizomorphs from
a very small amount of infected tissue. Growth of rhizomorphs is greatly
affected by soil organic matter (Redfern, 1973; Morrison, 1982), and
nutrients derived from the humus layer may well be of particular importance
to *A. bulbosa*. This is the species upon which the colourless orchids
Gastrodia elata and *Galeola septentrionalis* are parasitic; neither will
flower unless the fungus is present.

A. tabescens did not kill any young trees during pathogenicity tests, nor
is there any record of it killing standing trees in woodland. In experiments

with excised oak roots it was unable to penetrate intact bark; however it colonised them through the cut end as effectively as *A. bulbosa*. In south-eastern England *A. tabescens* occurs in broadleaved woodland on heavy soils, where it sometimes decays roots sufficiently to cause windthrow of shrubs such as blackthorn (*Prunus spinosa*) and trees such as oak. It is present in stumps of a wide variety of species. Rhizomorphs of *A. tabescens* have not been detected in woodland but are so small as easily to be overlooked. Even if they are present, it seems doubtful whether they could play a significant role in colonising new substrates or obtaining nutrients from organic matter in the soil. The existence of a focus at least 100 m in diameter in a very wet part of Hayley Wood, Cambridgeshire, shows that in certain circumstances it can spread extensively. Possibly the association of *A. tabescens* with heavy soils is related to periodic waterlogging and its weakening effect on roots, enabling this species to become established.

The four species exhibit a broad range of resource relations. They provide a series of increasing pathogenicity in the order: *A. tabescens*, *A. bulbosa*, *A. ostoyae* and *A. mellea*. Of the last two species, *A. mellea* can perhaps be considered more pathogenic because it kills small pines more rapidly; it also has an extensive host range. By contrast *A. ostoyae* is specialised in that it chiefly attacks resinous conifers. Although these species have retained a considerable saprotrophic ability, as indicated by their growth in tissues partially colonised by other microorganisms and their long survival in roots of dead trees, *A. bulbosa* probably ranks as the most effective saprotroph. Since it commonly produces an extensive system of rhizomorphs, by means of which woody tissues are penetrated when their resistance declines, a wide variety of resources may become available. Thus, not only root systems but also logs lying on the soil surface are commonly invaded. A further feature that may well contribute to the success of *A. bulbosa* is its ability to continue producing rhizomorphs until the food base is virtually exhausted. It is difficult to assess *A. tabescens* adequately on the present evidence. It should be remembered that this species is characteristic of warmer countries and in England is probably at or near the northern limit of its distribution. *A. tabescens* certainly kills trees in Portugal and the southern USA, for example, but perhaps only when they are weakened by some other agency. Although *A. tabescens* is known to survive in roots for many years, more information is required about its saprotrophic ability.

References

Garrett, S. D. (1960). Rhizomorph behaviour in *Armillaria mellea* (Fr.) Quél. III. Saprophytic colonisation of woody substrates in soil. *Annals of Botany*, **24**, 275–85.

Korhonen, K. (1978). Infertility and clonal size in the *Armillariella mellea* complex. *Karstenia*, **18**, 31–42.

Morrison, D. J. (1982). Effects of soil organic matter on rhizomorph growth by *Armillaria mellea*. *Transactions of the British Mycological Society*, **78**, 201–7.

Redfern, D. B. (1973). Growth and behaviour of *Armillaria mellea* rhizomorphs in soil. *Transactions of the British Mycological Society*, **61**, 569–81.

Rishbeth, J. (1978). Infection foci of *Armillaria mellea* in first-rotation hardwoods. *Annals of Botany*, **42**, 1131–9.

Rishbeth, J. (1982). Species of *Armillaria* in southern England. *Plant Pathology*, **31**, 9–17.

Rishbeth, J. (1983). The importance of honey fungus (*Armillaria*) in urban forestry. *Arboricultural Journal*, **7**, 217–25.

Thompson, W. & Boddy, L. (1983). Decomposition of suppressed oak trees in even-aged plantations. II. Colonization of tree roots by cord- and rhizomorph-producing basidiomycetes. *New Phytologist*, **93**, 277–91.

4

The growth of *Crinipellis perniciosa* in living and dead cocoa tissue

B.E.J. WHEELER

Department of Pure and Applied Biology, Imperial College, Silwood Park, Ascot, Berks. SL5 7PY, UK

Introduction

The basidiomycete fungus *Crinipellis perniciosa* is endemic to the forests of the Amazon basin where it occurs on wild cocoa (*Theobroma cacao*), related species of *Theobroma* and *Herrania*, lianes and also on wood from the understorey and litter (Thorold, 1975; Evans, 1978; Hedger: Chapter 2). It is a highly destructive pathogen in cocoa plantations in the northern regions of South America and in the Caribbean islands of Grenada, Tobago and Trinidad, causing the disease known as witches' broom (Baker & Holliday, 1957). Since 1895 when the disease was first reported in Surinam it has severely curtailed cocoa production in Surinam, Guyana, Ecuador, Trinidad and Tobago. It now threatenes new plantings of cocoa in the Amazon basin of Brazil, particularly in the state of Rondonia, where cocoa is one of the more ecologically acceptable crops in the agricultural development of the rain forests and its devastation by witches' broom would have serious effects beyond those of a solely commercial nature.

The fungus was first described by Stahel (1915) as *Marasmius perniciosus* and then transferred to the genus *Crinipellis* by Singer (1942). Three varieties have been described by Pegler (1978) and at least three different pathotypes identified on cocoa, lianes and solanaceous hosts, respectively (Evans, 1978; Bastos & Evans, 1985). On cocoa, basidiospores of *C. perniciosa* infect actively growing (meristematic) tissue and induce a range of symptoms on vegetative shoots, flower cushions, flowers and pods. The diversity of these symptoms is indicated by Baker & Holliday (1957), Thorold (1975) and Evans (1981), and reflects to a great extent the type of tissue involved and its stage of development (Cronshaw & Evans, 1978).

Green brooms which develop from buds on vegetative shoots and sometimes from flower cushions are the most spectacular expressions of

103

infection (see Hedger: Chapter 2). The mycelium of the fungus is found within the parenchymatous tissues of these brooms (Calle, 1978; Calle, Cook & Fernando, 1982). It is composed of relatively wide (5–20 μm), intercellular hyphae which are often swollen and flexuous, but without clamp connections, and is generally considered to be monokaryotic (Pegus, 1972; Evans, 1980), though this has been disputed by Delgado & Cook (1976). This form, the parasitic mycelium (Stahel, 1915), has never been grown on artificial media though similar mycelium was obtained when basidiospores germinated on actively growing cocoa callus tissues (Evans, 1980). As the broom dies, the wide, intercellular mycelium gives rise to thinner (1.5–3 μm) hyphae with clamp connections and binucleate cells (Pegus, 1972). These hyphae grow through all tissues both inter- and intracellularly and eventually aggregate beneath the bark of the dead broom and form fruit bodies. This form, the saprotrophic mycelium (Stahel, 1915), can be isolated readily on agar media, and its ability to produce degenerative enzymes *in vitro* has been established (Lindeberg & Molin, 1949; Krupasagar & Sequeira, 1969), but all attempts to infect cocoa with it have failed.

These two distinct mycelial phases in the life cycle pose different, challenging problems of both fundamental and practical interest which relate correspondingly to host–pathogen interactions and environment–saprotroph relationships. This chapter deals with recent work on particular aspects within these broad topics.

The parasitic phase: host–pathogen interactions

Two techniques have proved particularly valuable for studying interactions between isolates of *C. perniciosa* and different types of *T. cacao*.

In one technique, cocoa seed is germinated and after 4–5 days is inoculated, either by dipping in a suspension of basidiospores (Holliday, 1955) or by applying to the hypocotyl a disc cut from a basidiospore print made by suspending an excised pileus of *C. perniciosa* over water agar (Evans & Bastos, 1980). The inoculated seedlings are kept in a damp chamber at 25 °C for 2 days before planting. These hypocotyl inoculations result in varying amounts of swelling at the base of the seedling stem and the maximum diameter of this swelling can be used as an indicator of host reaction. Particular types of cocoa react in more specific ways; on some, brooms are initiated at the cotyledon node and on others inoculations result in seedling death after some 12 weeks. These features can be recorded quantitatively (Wheeler & Mepsted, 1982).

Table 4.1. *Reactions of Amelonado cocoa seedlings to hypocotyl inoculations with isolates of* Crinipellis perniciosa

Origin of isolate		Diameter(mm)	
Country	Locality and code	stem base at 8 weeks	Mean days to death
Colombia	Chigorodo C	16.3 a	91 c
	Chigorodo D	15.5 a	103 c
Ecuador	Pichilingue B	13.0 b	168 a
	Esmeraldas A	11.1 c	131 b
Brazil	Ouro Preto C	9.0 d	109 c
	Benevides A	11.6 bc	89 c
	Manaus B	11.0 c	109 c
Uninoculated control		0.43 e	—

Values are means based on 23 to 25 replicates for each isolate; those in the same column with no letter in common are significantly different at $P < 0.05$.
Data from Wheeler & Mepsted (1982).

In the second technique, healthy seed is planted and a young seedling obtained with expanded cotyledons and a leading shoot. The shoot is then cut off to induce development of the cotyledon buds. When growth of these is just visible one bud is removed, the other is inoculated with basidiospores on an agar disc and the seedling is enclosed in a polyethylene bag for 2 days at 25 °C. These cotyledon bud inoculations result in small, vegetative brooms, the main features of which, such as maximum diameter of the main shoot and the number of side branches over 10 mm long, are measured (Wheeler & Mepsted, 1984).

Hypocotyl inoculations have been used for many years to screen cocoa seedlings for resistance to *C. perniciosa* (Holliday, 1955; Bartley, 1959). More recently, Wheeler & Mepsted (1982, 1984) used both hypocotyl and cotyledon bud inoculations to determine the reactions of different types of cocoa to isolates of *C. perniciosa* collected from various localities within Bolivia, Brazil, Colombia, Ecuador, Trinidad and Venezuela. Their data indicate that different isolates of the fungus induce different degrees of host reaction in the same cocoa. Thus, an isolate from Colombia (Chigorodo) caused most stem-base swelling on Amelonado seedlings inoculated on the hypocotyl, but seedlings inoculated with isolates from Ecuador (Pichilingue and Esmeraldas) took longer to die (Table 4.1). Similar inoculations of Scavina 6 seedlings with these isolates from Colombia and Ecuador resulted in significantly more proliferating brooms at the cotyledonary

Table 4.2. *Reactions of Scavina 6 cocoa seedlings to hypocotyl
inoculations with isolates of* Crinipellis perniciosa

| Origin of isolate | | Diameter (mm) of stem base at 8 weeks[a] | Number of plants with brooms | | Number of plants dead at 37 weeks |
Country	Locality and code		'Thin'[b]	'Pro- liferating'[b]	
Colombia	Chigorodo E	6.8 bc	5	7	10
	F	6.2 c	10	6	8
Ecuador	Pichilingue B	9.7 a	2	16	14
	Esmeraldas B	8.0 b	8	11	9
Brazil	Benevides B	5.0 cd	12	1	1
	Castanhal B	5.2 cd	14	0	2
Trinidad	A	5.2 cd	15	1	1
Uninoculated control		4.6 d	0	0	0

[a] Values for stem base diameters are means based on 23 to 25 replicates for each
isolate; those with no letter in common are significantly different at $P < 0.05$.
[b] 'Thin' brooms = a few thin shoots at the cotyledon node which soon withered
and died. 'Proliferating' brooms = many swollen shoots arising from a considerably
thickened cotyledon node.
Data from Wheeler & Mepsted (1982).

nodes than with isolates from Brazil (Benevides and Castanhal) and
Trinidad (Table 4.2). Inoculations of cotyledon buds indicated further
distinctive features. The isolate from Chigorodo in Colombia caused
significantly more swelling of shoots on EET 400 seedlings than did other
isolates from Colombia (Palmira and Caldas) or those from Brazil (Ouro
Preto), Venezuela (Apure) or Trinidad (Table 4.3). On all cocoas tested,
particularly on Scavina 6, isolates from Bolivia induced considerable
branching of the inoculated shoot and abnormal development of stipules.

In an appraisal of the results of 30 experiments in which, overall, 14
different types of cocoa were inoculated with selected isolates from 16
localities, Wheeler & Mepsted (1984) concluded that two populations of
C. perniciosa existed on cultivated cocoa. These were most readily
distinguished on seedlings with Scavina 6 as one parent. On this type of
cocoa, one population (A), comprising isolates from Bolivia, Colombia
and Ecuador, caused extensive swelling of the stem base and brooming at
the cotyledon node when the hypocotyl was inoculated, and marked
swelling and branching of the shoot which developed from an inoculated
cotyledon bud. The other population (B), comprising isolates from Brazil,
Trinidad and Venezuela, induced only limited reactions from similar
inoculations. To some extent these populations could be separated on other

Table 4.3. *Reactions of EET 400 cocoa seedlings to cotyledon bud inoculations with isolates of* Crinipellis perniciosa

Origin of isolate		Maximum width (mm) of inoculated shoot at 9 weeks	Mean number of side shoots > 10 mm at 9 weeks
Country	Locality and code		
Colombia	Chigorodo J	8.9 a	2.4 a
	Palmira A	7.5 b	4.1 b
	Caldas D	7.2 b	4.4 b
Brazil	Ouro Preto G	7.1 b	5.0 b
Venezuela	Apure C	6.9 b	4.2 b
Trinidad	I	6.9 b	5.3 b

Values are means based on 19 or 20 replicates for each isolate; those in the same column with no letter in common are significantly different at $P < 0.05$.
Data from Wheeler & Mepsted (1984).

types of cocoa using the same criteria, but the reactions on these cocoas were generally less distinct. Within the two populations, but especially within population A, variants of *C. perniciosa* could be further distinguished by the reactions they evoked on different cocoas.

Clearly data of this type are directly relevant to programmes for screening cocoa lines for resistance to witches' broom because they indicate that selections made in one area against one isolate of *C. perniciosa* will not necessarily be suitable for other areas where a different isolate of the fungus exists.

In this connection, the differential reaction of cocoa seedlings with Scavina 6 parentage is of particular interest. The Scavina 'family' was derived from pods collected by Pound (1938) in the upper Amazon valley. One clone, designated Scavina 6, proved highly resistant to witches' broom in Trinidad and this resistance was incorporated into a breeding programme. However, when similar material was planted in Ecuador it gradually became infected and now is devastated by witches' broom (Bartley, 1977). The apparent diversity of isolates of *C. perniciosa* within Ecuador, which is further indicated by the studies of Wheeler & Mepsted (1984), suggests that isolates with ability to infect cocoa with Scavina 6 resistance may have already existed there.

The results also pose questions of a fundamental nature regarding the mechanism of broom formation. It is inferred that the hyperplasia and hypertrophy associated with the development of a broom result from hormonal changes within the infected cocoa shoot (Pegus, 1972) but, as

Table 4.4. *Growth of isolates of* Crinipellis perniciosa *from Ecuador* (*Pichilingue*) *and Trinidad in Scavina 6 cocoa seedlings inoculated at the hypocotyl*

		Number of hyphal fragments 10 mm^{-2} host tissue at week					
Site	Isolate	7	8	9	10	11	12
60 mm above inoculation point	Pichilingue	9	9	25	35	119	574
	Trinidad	8	1	69	15	0	9

Each value is based on transverse sections taken from three seedlings.
Data from Danquah (1985).

Evans (1980) pointed out, the only physiological work attempted has used culture filtrates of the saprophytic mycelium (Lindeberg & Molin, 1949; Dudman & Nichols, 1959; Krupasagar & Sequeira, 1969) and thus is hardly relevant to the biotrophic phase. The specific interactions between a single isolate of *C. perniciosa* and a particular cocoa provide several systems in which to examine changes in plant hormone levels. For example, why does the interaction of the Chigorodo isolate with EET 400 result in a grossly swollen stem with suppressed side shoots whilst other isolates cause excessive branching on this cocoa?

Relationships between *C. perniciosa* and Scavina 6 cocoa are of special interest because some isolates of the fungus induce marked host responses and others do not. Studies of these systems could yield valuable information not only on mechanisms of resistance but also on those features necessary for successful parasitism. Only one limited study has been attempted so far. Danquah (1985) examined the growth of *C. perniciosa* in seedlings with Scavina 6 as one parent which were inoculated on the hypocotyl with isolates from Trinidad and from Pichilingue in Ecuador. The Pichilingue isolate extensively colonised the parenchymatous tissues of the hypocotyl and grew rapidly within the developing shoot causing buds to proliferate at the cotyledon node some 6–8 weeks after inoculation. The Trinidad isolate grew more slowly and sparsely (Table 4.4). Within 6–8 weeks, cells surrounding the intercellular hyphae of the Trinidad isolate became brown and the number of cells within these areas appeared to increase with time. Browning occurred around relatively few hyphae of the Pichilingue isolate and, when it did, fewer cells seemed to be involved (Table 4.5).

There is now a need to examine in more detail these two features, slow growth and induction of cell browning, which apparently confer some

Table 4.5. *Cell browning in tissues of Scavina 6 cocoa seedlings inoculated at the hypocotyl with isolates of* Crinipellis perniciosa *from Ecuador* (*Pichilingue*) *and Trinidad*

Parameter	Isolate	Weeks after inoculation			
		9	10	11	12
Number of sites 10 mm^{-2}	Pichilingue	0.6	1	1	1
with brown cells	Trinidad	10	8	10	6
Number of brown cells	Pichilingue	22	60	105	6
10 mm^{-2}	Trinidad	835	1403	994	1416

Each value is based on transverse sections taken at the hypocotyl from three seedlings.
Data from Danquah (1985).

resistance to the Trinidad isolate. The experiments of Evans & Bastos (1980) suggested one possible way of exploring further the different growth rates of the two isolates in Scavina 6. They found that basidiospores of a Brazilian isolate germinated differently in extracts of cocoa clones resistant to *C. perniciosa* and in those of susceptible clones. The difference was primarily in the morphology of the germ-tube. In a preliminary experiment, Danquah (1985) prepared extracts of flush tissue from seedlings of Scavina 6 and germinated basidiospores of the Trinidad and Pichilingue isolates in these. Germ-tube growth of both isolates was much reduced in the extracts compared with the corresponding water controls, but in each extract germ-tube growth of the Trinidad isolate was less than that of the Pichilingue isolate. This is of particular interest because previous work (Delgado, 1974; Delgado & Cook, 1976) has indicated that for a short period the primary mycelium arising from the germinating basidiospore is monokaryotic like the parasitic mycelium.

The ease with which the dikaryotic state arises from this mycelium on a wide range of synthetic media suggests that the living host itself inhibits dikaryotisation. Evans & Bastos (1980) concluded from their experiments that young tissues of susceptible cocoa contain excess of a heat-stable metabolite or modifier, possibly a reducing sugar, which inhibits dikaryotisation. They suggested further that resistant cultivars contain either insufficient amounts of this substance or excess of an inhibitor, possibly a bound monophenol, which causes basidiospore plasmolysis. Many aspects remain to be tested but it is a useful working hypothesis.

The saprotrophic phase: environment–saprotroph relationships

There are problems no less intriguing related to the saprotrophic phase of *C. perniciosa*. Much attention has been given to the induction of fruit bodies in axenic culture. However, despite the many and diverse experiments, particularly of Delgado (1974) and Suarez-Capello (1977), there has been only limited success and in some instances pure culture conditions were not maintained. Stahel (1919) obtained fruit bodies by exposing mycelium from an agar culture to field conditions in a porcelain vessel and Pegus (1972) obtained one small fruit body, which soon aborted, 8 days after transferring a portion of an almost dry culture to fresh medium. Merchan (1979) reported that fruit bodies formed on cocoa stems about 4 months after the stems were sterilised and inoculated with mycelium of *C. perniciosa*. Attempts at Imperial College, Silwood Park, to repeat this procedure with autoclaved cocoa stems failed, but fruit bodies developed when mycelial *C. perniciosa* was allowed to grow for about 3 months on surface-sterilised cocoa stems which were then hung in cabinets with intermittent wetting and drying. Purdy, Trese & Aragundi (1983) also obtained fruit bodies after 2 weeks when mycelial mats from 30-day-old cultures on Trione's medium were hung in a chamber with an intermittent water spray and 10–13 months after autoclaved brooms or brooms sterilised with propylene oxide were dual inoculated with isolates of *C. perniciosa* from Brazil and Ecuador in flasks of potato dextrose agar.

In many respects all these experiments pose as many problems as they attempted to solve. The detailed studies of Suarez-Capello (1977) indicated that a prolonged period of vegetative growth on various synthetic media and depletion of nutrients are not themselves generally sufficient to initiate fruit bodies. Only in one experiment did growth on staled media apparently induce the formation of mycelial aggregates which resembled fruit body primordia. It could be that during mycelial growth, substances accumulate either around or within hyphae which, if present in certain amounts, inhibit fruit body production. It may also be a feature of sporulation on brooms that such substances are removed by leaching due to rain. Certainly the relative periods of wetting and drying of brooms constitute an important factor in fruit body production. In experiments reported by Rocha & Wheeler (1982), the number of fruit bodies produced over 5 months on brooms subjected to a daily regime of 8 h wet/16 h dry was more than double, 10 times and 100 times that on comparable brooms in daily regimes of 16 h wet/8 h dry, 23 h wet/1 h dry and 1 h wet/23 h dry, respectively (Table 4.6). The production of fruit bodies appeared to be linked with the water content of the brooms which in the 8 h wet/16 h dry regime was about 50% (w/w) during the wet period and 15% in the dry period. It was

Table 4.6. *Production of fruit bodies of* Crinipellis perniciosa *on brooms kept in four different regimes*

Daily regime (h wet/h dry)	Fruit body primordia	Mature fruit bodies
1/23	9	0
8/16	1036	653
16/8	401	302
23/1	104	65

Each figure is the total production on 60 brooms over 15 weeks.
Data from Rocha & Wheeler (1982).

an additional feature of brooms in this regime that successive flushes of fruit bodies were produced with little evidence of other microorganisms whereas in wetter regimes, especially 23 h wet/1 h dry, many other fungi developed on the brooms which then rotted (Rocha, 1983). One aspect of the prolific fruiting in the daily regime of 8 h wet/16 h dry could thus be the extent to which the growth of *C. perniciosa* is favoured relative to other microorganisms. In this respect, colonisation of the cortical tissues is likely to be a key factor. The production of fruit bodies on dead brooms is preceded by a massing of hyphae in the cortex and it is this tissue which is most readily available to saprotrophs invading via the surface of the dying brooms.

Given a favourable regime of wetting and drying, the production of fruit bodies is influenced further by host factors, which possibly reflect the extent to which *C. perniciosa* colonises the green broom, and by temperature and light. Thus, other experiments by Rocha (1983) showed that significantly more fruit bodies formed at nodes than at internodes, and the productivity on brooms of comparable size from different clones of cocoa also differed significantly. More fruit bodies formed and matured at 20–25 °C than at 25–30 °C, none were produced at 30–35 °C and only one fruit body formed on 20 brooms kept in the dark compared with 113 fruit bodies on a similar sample in the light (Table 4.7).

There is also some evidence that isolates of *C. perniciosa* vary in their ability to produce fruit bodies. Wheeler & Mepsted (1984) inoculated seedlings of Amelonado cocoa at the hypocotyl and subsequently, when the seedlings died, counted the fruit bodies produced over 12 months in a daily regime of 8 h wet/16 h dry. Considerably more fruit bodies formed on seedlings infected with isolates from Brazil than on those infected with isolates from Colombia and Ecuador and this difference was not related to broom size (Table 4.8). This result supported the conclusion from the

Table 4.7. *Effect of various factors on the production of fruit bodies of* Crinipellis perniciosa *on brooms kept in a daily regime of 8 h wet/ 16 h dry*

	Factor examined	Comparisons	Total fruit bodies
1.	Broom tissue	Node	102
		Internode	8
2.	Cocoa clone	Scavina 6	210
		Scavina 12	124
		IMC 67	100
		Catongo	86
		EET 400	80
		ICS 39	72
3.	Temperature	20–25 °C	75
		25–30 °C	23
		30–35 °C	0
4.	Light	Light	113
		Dark	1

Values are totals for: 1, seven brooms over 36 weeks; 2, ten brooms over 30 weeks; 3, ten brooms over 24 weeks; 4, 20 brooms over 24 weeks. Brooms in 1 and 2 came from Pichilingue, Ecuador; those in 3 and 4 came from Castanhal, Brazil. Data from Rocha (1983).

Table 4.8. *Production of fruit bodies on Amelonado cocoa seedlings infected with different isolates of* Crinipellis perniciosa

Origin of isolate		Mean weight (g) dead seedling	Mean number of fruit bodies per broom (52 weeks)
Country	Locality and code		
Colombia	Chigorodo C	0.96	5.6
	D	1.08	2.7
Ecuador	Pichilingue C	1.94	7.5
	Esmeraldas A	1.09	8.3
Brazil	Ouro Preto C	1.50	42.7
	Benevides A	1.04	25.1
	Manaus B	1.10	28.9

Values are based on ten replicates per isolate.
Data from Wheeler & Mepsted (1984).

Table 4.9. *Compatible* (+) *and non-compatible* (−) *reactions between paired isolates of* Crinipellis perniciosa[a]

Origin of isolates		Isolate number											
		1	2	3	4	5	6	7	8	9	10	11	12
1. Trinidad		+											
2. Venezuela		−	+										
3. Brazil	− Benevides	−	−	+									
4.	− Castanhal	−	−	+	+								
5.	− Belem	−	−	+	+	+							
6.	− Manaus	−	−	+	+	+	+						
7.	− Ouro Preto	−	−	−	−	−	−	+					
8.	− Rio Branco	−	−	−	−	−	−	−	+				
9. Colombia	− Caldas	−	−	−	−	−	−	−	−	+			
10.	− Chigorodo	−	−	−	−	−	−	−	−	(+)[b]	+		
11. Ecuador	− Esmeraldas	−	−	−	−	−	−	−	−	(+)[b]	+	+	
12.	− Pichilingue	−	−	−	−	−	−	−	−	(+)[b]	+	+	+

[a] Combined results of two experiments: experiment 1 with isolates 3–12; experiment 2 with isolates 1–12.
[b] Compatible with these isolates in experiment 1 when first obtained; not compatible in experiment 2 after maintaining on V8 agar.
Data from Wheeler & Mepsted (1982).

pathogenicity tests that two populations of *C. perniciosa* existed on cultivated cocoa.

In this latter connection, the results of experiments in which mycelial cultures of *C. perniciosa* were paired on agar plates are of interest. Delgado (1974) made 412 pairings on various media of isolates from Brazil, Ecuador, Surinam and Trinidad derived from basidiospores, gill tissue and brooms. Although he recognised three pairing reactions – complete intermingling of mycelia, a marked line of demarcation between mycelia, and mutual aversion – he presents only an overall summary table from which it is difficult to discern any consistent features. Wheeler & Mepsted (1982) defined only two pairing reactions, compatible (+) where the mycelia intermingled freely and incompatible (−) where they did not. Their tests demonstrated differences between isolates from various localities which did not conflict with the general conclusion that two populations of *C. perniciosa* exist on cultivated cocoa but indicated a greater diversity amongst isolates than could be observed by infecting cocoa seedlings from basidiospore inocula (Table 4.9).

Although the problem of linking information from compatibility tests with differences in pathogenicity remains, such tests might determine how many isolates to use initially in screening cocoa for resistance in one locality.

Epilogue

Many of the problems outlined in this chapter, for example those related to biotrophy or the induction of fruit bodies, are not individually unique to *C. perniciosa*. There are similar problems with many other fungi which are either plant pathogens or saprotrophs. What is perhaps unique is that so many problems in resource relations are posed by the life cycle of one fungus.

References

Baker, R. E. D. & Holliday, P. (1957). Witches' broom disease of cocoa (*Marasmius perniciosus* Stahel). *Phytopathological Papers*, 2. Kew: Commonwealth Mycological Institute.

Bartley, B. G. D. (1959). The efficiency of a test of the resistance of cocoa seedlings to *Marasmius perniciosus* Stahel. *Report of Cocoa Research*, 1957–1958, pp. 49–52. St. Augustine, Trinidad: Imperial College of Tropical Agriculture.

Bartley, B. G. D. (1977). The status of genetic resistance in cacao to *Crinipellis perniciosa* (Stahel) Singer. *Proceedings of the Sixth International Cocoa Research Conference*, 1977, *Caracas, Venezuela*. pp. 1–18. Lagos: Cocoa Producers Alliance.

Bastos, C. N. & Evans, H. C. (1985). A new pathotype of *Crinipellis perniciosa* (witches' broom disease) on solanaceous hosts. *Plant Pathology* (in press).

Calle, H. (1978). Relation of internal infection by *Crinipellis perniciosa* to witches' broom formation in cacao. M.Sc. thesis, University of Florida, Gainesville.

Calle, H., Cook, A. A. & Fernando, S. Y. (1982). Histology of witches' broom caused in cacao by *Crinipellis perniciosa*. *Phytopathology*, **72**, 1479–81.

Cronshaw, D. K. & Evans, H. C. (1978). Witches' broom disease of cocoa (*Crinipellis perniciosa*) in Ecuador. II. Methods of infection. *Annals of Applied Biology*, **89**, 193–200.

Danquah, O.-A. (1985). Growth of *Crinipellis perniciosa* in cocoa resistant and susceptible to witches' broom. Ph.D. thesis, University of London.

Delgado, J. C. (1974). Studies of *Marasmius perniciosus*. Ph.D. thesis, University of Florida, Gainesville.

Delgado, J. C. & Cook, A. A. (1976). Nuclear condition of the basidia, basidiospores, and mycelium of *Marasmius perniciosus*. *Canadian Journal of Botany*, **54**, 66–72.

Dudman, W. F. & Nichols, R. (1959). Absence of gibberellin-like substances in filtrates of *Marasmius perniciosus* Stahel (witch broom disease of cacao). *Nature, London*, **183**, 899–900.

Evans, H. C. (1978). Witches' broom disease of cocoa (*Crinipellis perniciosa*) in Ecuador. I. The fungus. *Annals of Applied Biology*, **89**, 185–92.

Evans, H. C. (1980). Pleomorphism in *Crinipellis perniciosa*, causal agent of witches' broom disease of cocoa. *Transactions of the British Mycological Society*, **74**, 515–23.

Evans, H. C. (1981). Witches' broom disease – a case study. *Cocoa Growers' Bulletin*, **32**, 5–19.

Evans, H. C. & Bastos, C. N. (1980). Basidiospore germination as a means of assessing resistance to *Crinipellis perniciosa* (witches' broom disease) in cocoa cultivars. *Transactions of the British Mycological Society*, **74**, 525–36.

Holliday, P. (1955). A test for resistance to *Marasmius perniciosus* Stahel. *Report of Cacao Research*, 1954, 50–5. St. Augustine, Trinidad: Imperial College of Tropical Agriculture.

Krupasagar, V. & Sequeira, L. (1969). Auxin destruction by *Marasmius perniciosus*. *American Journal of Botany*, **56**, 390–7.

Lindeberg, G. & Molin, K. (1949). Notes on the physiology of the cocoa parasite, *Marasmius perniciosus*. *Physiologia Plantarum*, **2**, 138–44.

Merchan, V. M. (1979). Formación en medios de cultivo de basidiocarpos de *Crinipellis perniciosa*. *Boletin de la Associacion Colombiana de Fitopatologia*, **5**, 54–6.

Pegler, D. N. (1978). *Crinipellis perniciosa* (Agaricales). *Kew Bulletin*, **32**, 731–6.

Pegus, J. E. (1972). Aspects of the host–parasite relationships in the *Theobroma cacao* L./*Marasmius perniciosus* Stahel disease complex. M.Sc. thesis, University of the West Indies, Trinidad.

Pound, F. J. (1938). Cacao and witch broom disease (*Marasmius perniciosus*) of South America. With notes on other species of *Theobroma*. *Report on a visit to Ecuador, the Amazon Valley and Colombia, April 1937–April 1938*. Port of Spain, Trinidad: Yuille's Printerie. (Reprinted in *Archives of Cocoa Research*, **1**, 20–72.)

Purdy, L. H., Trese, A. T. & Aragundi, J. A. (1983). Proof of pathogenicity of *Crinipellis perniciosa* to *Theobroma cacao* by using basidiospores produced in *in vitro* cultures. *Revista Theobroma*, **13**, 157–63.

Rocha, H. M. (1983). The ecology of *Crinipellis perniciosa* (Stahel) Singer in witches' brooms on cocoa (*Theobroma cacao* L.). Ph.D. thesis, University of London.

Rocha, H. M. & Wheeler, B. E. J. (1982). The water balance as an important factor in basidiocarp production by *Crinipellis perniciosa*, the causal fungus of cocoa witches' broom. *Proceeding of the Eighth International Cocoa Research Conference*, 1981, *Cartagena, Colombia*, 381–6. Lagos: Cocoa Producers Alliance.

Singer, R. (1942). A monographic study of the genera *Crinipellis* and *Chaetocalathus*. *Lilloa*, **8**, 441–534.

Stahel, G. (1915). *Marasmius perniciosus* nov. spec. *Bulletin Departement van den Landbouw in Suriname*, **33**, 1–26.

Stahel, G. (1919). Bijdrage tot de Kennis der Krullotenziekte. Bulletin no. 39 of the Department of Agriculture in Surinam. (Translation by B. C. Montserin (1932) in *Tropical Agriculture, Trinidad*, **9**, 167–76.)

Suarez-Capello, C. (1977). Growth of *Crinipellis perniciosa* (Stahel) Singer *in vivo* and *in vitro*. Ph.D. thesis, University of London.

Thorold, C. A. (1975). *Diseases of Cocoa*. Oxford: Clarendon Press. (See especially pp. 11–26.)

Wheeler, B. E. J. & Mepsted, R. (1982). Pathogenic races of *Crinipellis perniciosa*. *Proceedings of the Eighth International Cocoa Research Conference, Cartagena, Colombia*, 1981, 365–70. Lagos: Cocoa Producers Alliance.

Wheeler, B. E. J. & Mepsted, R. (1984). Pathogenic races of *Crinipellis perniciosa* (Stahel) Singer, the causal fungus of witches' broom of cocoa (*Theobroma cacao* L.). *Final Report to the Cocoa, Chocolate and Confectionery Alliance*. London: Imperial College.

5

Mycorrhizal dynamics during forest tree development

J. DIGHTON AND P. A. MASON*

*Institute of Terrestrial Ecology, Merlewood Research Station,
Grange-over-Sands, Cumbria, LA11 6JU, UK and *Institute of Terrestrial
Ecology, Bush Estate, Penicuik, EH26 0QB, UK*

Introduction

The wide variety of necrotrophic, saprotrophic and biotrophic fungi living within the forest habitat form a dynamic interacting community in which niches are both lost and gained and the composition of the community regularly changes. We know little, however, of the structure of these changing communities or the actual mechanisms whereby one species replaces another (Frankland, 1981).

One of the most important groups of fungi within the forest flora are the mycorrhizal fungi. They are important in nutrient cycling as, in conjunction with their host plant, they act as a sink for mineral ions derived from within the soil. They also produce vitamins and hormones which appear to increase root size and longevity (Slankis, 1973). Mycorrhizal fungi, in turn, depend largely on their hosts for their energy and carbon requirements. Thus mycorrhizas broadly function as a mutualistic symbiotic biotrophy between a fungus and a higher plant (Lewis, 1973).

Although mycorrhizas are a key link in nutrient and energy cycling within the forest ecosystem, we know little about the changes in mycorrhizal species composition or function as forests develop and reach a climax, nor of the processes which may control or affect such changes.

During the development and growth of natural and planted forests, a number of changes in vegetation occur which, in turn, affect the physico-chemical attributes of the soil ecosystem in which the mycorrhizas are active. Thus, as forest trees age, the resources available to the mycorrhizas and their host plants will alter (Fig. 5.1). Plant, animal and probably fungal species diversities increase as trees age, and then decline following canopy closure and the development of a climax community (Usher & Parr, 1977). The resource quality available to the decomposer community declines with the decline of understorey herbaceous vegetation (Ford & Newbould, 1977; Cromack, 1981) as more lignin and polyphenols become incorporated into the litter. As a consequence, litter residence time increases, organic matter accumulates and organisms tend to adopt '*K*-' rather than '*r*-' strategies (Southwood, 1977; Gerson & Chet, 1981; Heal & Ineson, 1984; Heal & Dighton, 1985; see also Rayner, Watling & Frankland: Chapter 1). As trees become larger it is hypothesised that their demand for nutrients from soil decreases as internal recycling of nutrient within the tree becomes more efficient (Miller, 1979; Miller *et al.* 1979). These changes are important in terms of the rate of release of labile nutrients available to the mycorrhizal roots and the tree's nutrient demand. It is also envisaged that the amount of carbohydrate available to support a mycorrhizal fungal symbiont might differ with tree size/age, with small trees having limited supplies and supporting fungi with limited carbohydrate demands, while larger trees support carbohydrate-demanding fungi. It is probable, therefore, that the stage of development of the ecosystem will strongly influence the resources available to mycorrhizal fungi.

Thus it is expected that the species of fungi forming mycorrhizas with

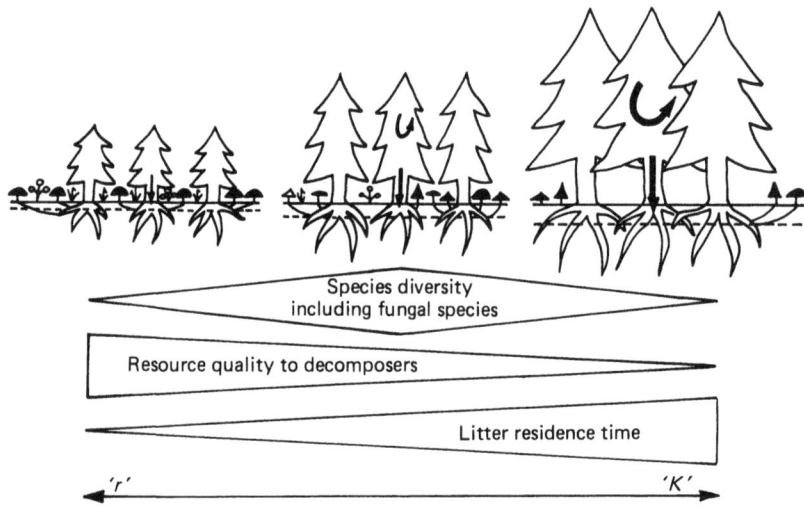

Fig. 5.1. Diagrammatic representation of forest succession indicating growth of trees, shading of ground flora and changes in ectomycorrhizal composition with stand development. Changes in resources are indicated; arrows within trees represent the hypothesised increase in internal recycling of nutrients in the tree and possible changes in carbohydrate supply to the root system with stand age. These concepts and the change from '*r*' to '*K*' attributes of organisms are discussed in the text.

a tree 40 years old will differ from those associated with saplings. Nevertheless, the concept of 'mycorrhizal succession' is only now being unravelled as a result of several pieces of circumstantial evidence. In theory, fungal successions are not uncommon (Frankland, 1981) but until recently there has been scant information concerning changes in a mycorrhizal flora with stand development. As a result, the purpose of this chapter is to document the current information regarding changes in the ectomycorrhizal flora of temperate forests, which is chiefly agaricoid, and to highlight some of the processes which might be responsible for this mycorrhizal succession.

Changes in mycorrhizal dynamics

Evidence from fruit body observations

Terrestrial plant communities tend to be dominated by either endomycorrhizal or ectomycorrhizal plants (Moser, 1967). During ecosystem succession to climax forest in northern temperate regions the fungal associates of higher plants change from endo- or vesicular-arbuscular mycorrhizas in herbaceous and scrub communities to ectomycorrhizas associated with trees (Rose, 1980). The ectomycorrhizal fungi associated with most coniferous species and many angiosperm trees belong to the Basidiomycotina, Zygomycotina and Ascomycotina. It is estimated that over 2000 fungal species are potential ectomycorrhizal symbionts (Trappe, 1962, 1977). Most of these species have been identified by fruit body characteristics e.g. toadstools and earthballs, which they produce each autumn. The host–fungus specificity is varied, birch (*Betula* spp.), for example, being associated with a number of fungal genera including *Amanita, Boletus, Cantharellus, Cortinarius, Hebeloma, Inocybe, Laccaria, Leccinum, Paxillus, Russula, Scleroderma* and *Tricholoma* (Trappe, 1962; Watling, 1973; Pegler, 1981). Many of these fungi, however, are not host-specific and are also associated with coniferous and other deciduous trees. Some fungi, e.g. *Suillus* and *Rhizopogon*, associate almost exclusively with the Pinaceae, and some tree species such as alder (*Alnus* spp.) have a restricted mycorrhizal flora (Molina, 1979). Mixed forests tend to be more diverse in fungal composition than pure stands (Mosse, Stribley & Le Tacon, 1981).

Most lists showing the association of fruit bodies of sheathing ectomycorrhizas have paid little attention to soil conditions, climate, and age and size of tree. We can see, however, that superimposed upon the tree/fungus heterogeneity there is a temporal shift in the species composition of the mycorrhizal flora of a monospecific tree stand as the trees age. In a stand

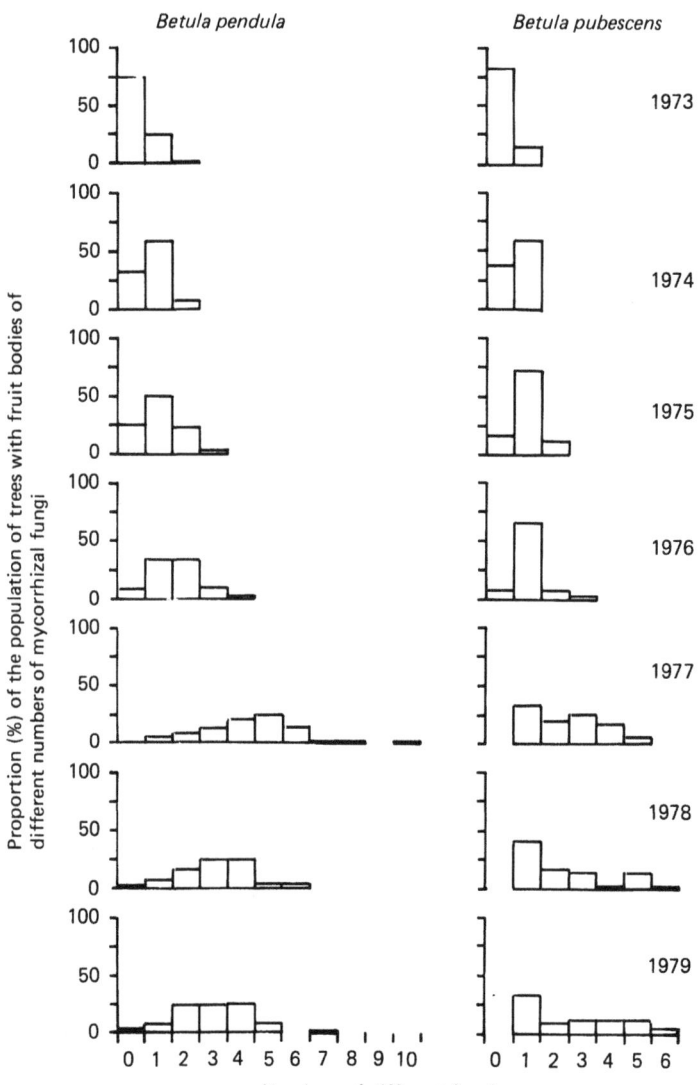

Fig. 5.2. Number of trees (expressed as a proportion of the total trees)
under which different numbers of mycorrhizal fruit body species appear
in successive years after planting saplings of *Betula pendula* and *B.
pubescens* in 1971. (After Mason *et al.* 1982.)

of birches of *Betula pendula* and *B. pubescens*, clear evidence has been obtained of a sequence of ectomycorrhizal fruit bodies in both time and space (Mason *et al.*, 1983). *Hebeloma crustuliniforme* and *Laccaria tortilis* were observed within 2 years of planting; *Inocybe lanuginella* and *Lactarius pubescens* appeared in year 4, while species of *Cortinarius* and *Leccinum* were recorded in year 6 and *Russula* spp. in year 10 (Last *et al.*, 1983). Similar data are present in the literature for coniferous species where *Thelephora terrestris*, *Hebeloma* spp., *Laccaria* spp. and *Inocybe* spp. are associated with young trees (Trappe & Strand, 1969; Chu-Chou, 1979; Chu-Chou & Grace, 1981, 1983). Fungi often associated with nursery soils, e.g. *Thelephora terrestris* and *Hebeloma crustuliniforme*, may be lost at outplanting (Mosse *et al.*, 1981) but others, including *Laccaria* spp., may be found in both young and mature stands. Chu-Chou (1979) associated fruit bodies of *Suillus* and *Inocybe* species with stands of *Pinus radiata* 5 or more years old, while *Amanita muscaria* toadstools were only found in plantations at least 10 years old. Distinct successions have also been reported from ageing stands of Norway spruce in the French Jura (Mosse *et al.*, 1981).

Not only do the dominant species of mycorrhizal fungi alter with tree age, but the species diversity also changes. Last *et al.* (1983) indicated that on a brown earth the number of fungal species associated with *Betula pendula* and *B. pubescens* increased from four species in year 3 to nearly 30 species in year 10, with a consistently greater variety of fungi associated with *B. pendula* (Mason *et al.*, 1982) (Fig. 5.2). In a survey of fruit body production of fungi under *Pinus contorta* of different ages planted on peat, clear changes in fungal diversity with stand development could be seen with *Lactarius rufus* being dominant under smaller trees, a diverse group of fungi including *Laccaria*, *Inocybe longicystis* and *Cortinarius* species occurring at canopy closure, and a reduced diversity of mycorrhizal fungi dominated by *Russula emetica* in the well-developed sites, after canopy closure (Fig. 5.3). The changes in species diversity of the most commonly occurring fruit bodies with stand age is indicated in Fig. 5.3, the reduction of diversity in older stands agrees with data from root assessment in 250-year-old Douglas fir/larch (*Pseudotsuga menziesii*/*Larix*) forests (Harvey, Larsen & Jurgensen, 1976) indicating that fir and larch were each associated with a single dominant mycorrhizal fungus, *Russula brevipes* and *Suillus cavipes*, respectively.

In summary it is observed that some mycorrhizal fungi (e.g. *Hebeloma*, *Laccaria* and *Inocybe*) are characteristic of young stands, whereas others (e.g. *Cortinarius*, *Russula* and *Amanita*) are associated with older stands.

The concept of mycorrhizal succession based on fruit body observation is given credence by the close relation between fruit bodies and their own mycorrhizal type occurring on roots, observed by Warcup (see Mason *et al.*, 1982). The sequence of fungi can, however, be modified by soil type and environmental factors; *Paxillus involutus* was dominant on young birch growing in coal spoil (Last, unpublished) although it was not observed on young trees growing in a brown earth soil (Mason *et al.*, 1982), indicating possible selection for adverse environmental conditions, i.e.

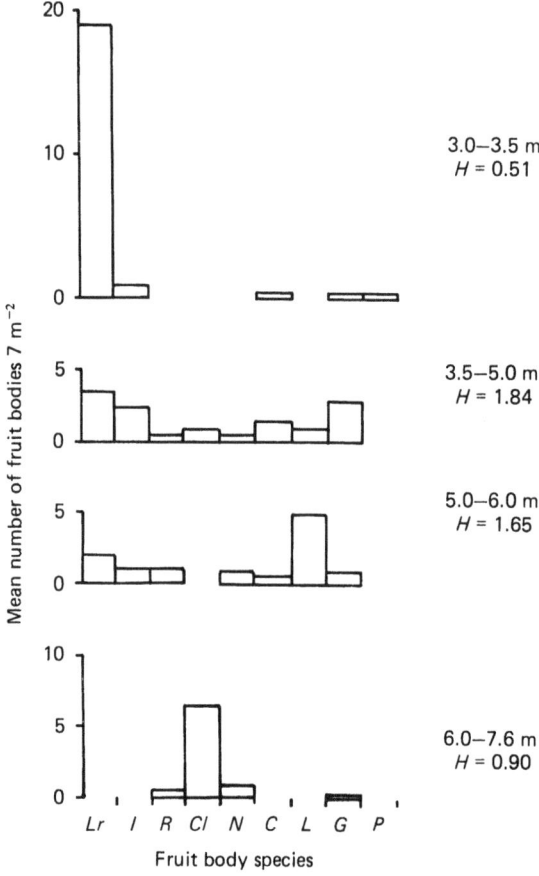

Fig. 5.3. Changes in the populations of fungal fruit bodies occurring under *Pinus contorta* stands at different stages in stand development (height classes) on peat in northern England. The *H*-value is the Shannon Weaver species diversity index. (*Lr = Lactarius rufus*; *I = Inocybe longicystis*; *R = Russula emetica*; *Cl = Clitocybe* sp.; *N = Nolanea cetrata*; *C = Cortinarius* sp.; *L = Laccaria* sp.; *G = Galerina* sp.; *P = Paxillus involutus*.)

adversity strategy (Southwood, 1977) or stress-tolerant (*S*) strategy (Grime, 1979).

Evidence from mycorrhizas

More definitive evidence for an age-related succession of mycorrhizal fungi has been presented by Deacon, Donaldson & Last (1983) who examined the distribution of mycorrhizas in soil cores taken at 0.25-m intervals from the base of an 8-year-old *Betula pubescens*. *Hebeloma*-type mycorrhizas were most frequent in the outer sampling positions on the newest part of the root system, while *Leccinum*-type mycorrhizas were found mostly in the inner sampling positions. *Leccinum* species are associated with later stages of the succession in contrast to *Hebeloma*, an 'early-stage' fungus. *Lactarius*-type mycorrhizas were found in all positions, although their frequency 'peaked' in the middle samples (Fig. 5.4). Evidence of this nature strongly supports the concept of mycorrhizal succession as observed by the pattern of fruit bodies with changes in tree age. On roots from trees of uniform age in a stand of *Pinus radiata*, Marks & Foster (1967) observed replacement of one mycorrhizal fungal associate by another. They believed their extensive anatomical observations indicated that mycorrhizas of a forest stand probably alter considerably during the lifetime of the stand.

Within 5 years of outplanting, Lamb (1979) noticed that *Pinus elliottii* plantations yielded up to 40 ectomycorrhizal species, while 20 different

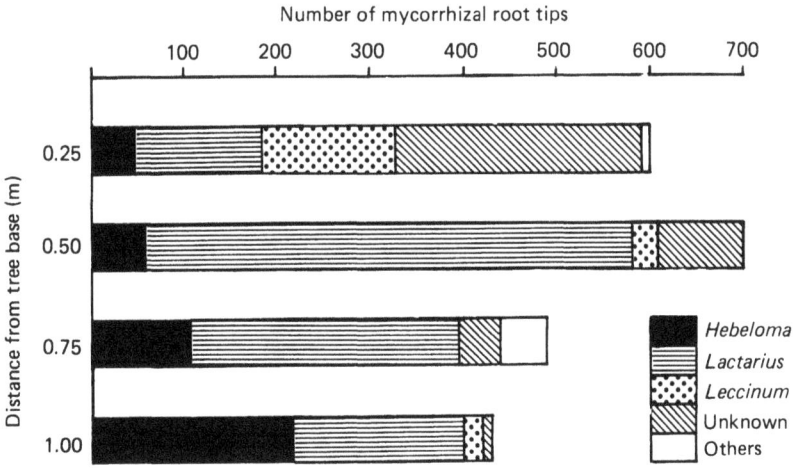

Fig. 5.4. Number of mycorrhizas of different types in 15 soil cores taken at increasing distances from the base of a *Betula pubescens* tree. (After Deacon *et al.*, 1983.)

Table 5.1. *Mycorrhizal formation on birch seedlings planted into unsterile soil supplemented with spores of 'early-' and 'late-stage' fungi. (After Fox, 1983a)*

	Per cent of mycorrhizas formed by inoculant fungus	Number of contaminant fungi
'Early-stage'		
Hebeloma sacchariolens	44	1
H. leucosarx	59	1
Inocybe geophylla	23	0
I. lacera	52	1
Laccaria proxima	67	1
L. tortilis	11	0
'Late-stage'		
Cortinarius delibutus	0	3
Lactarius pubescens	0	3
L. vietus	0	3
Leccinum roseofracta	0	3
Russula grisea	0	2

species were isolated from *Pinus radiata* stands of equivalent age. Similarly, Thomas, Rogers & Jackson (1983) revealed that in 2–4 years after outplanting Sitka spruce (*Picea sitchensis*) only E-strain mycorrhizas persisted from the nursery and the number of ectomycorrhizal fungi increased, such that in mature stands 24 different fungi were found.

An increased understanding of mycorrhizal changes in ageing stands is being achieved by the use of baiting techniques. Deacon *et al.* (1983) grew birch seedlings in soil cores taken from beneath fruit bodies of a range of ectomycorrhizal fungi. After 8–10 weeks, seedlings grown in cores from beneath fruit bodies of the 'early-stage' fungi *Laccaria* and *Inocybe* always developed mycorrhizas attributable to these fungi. In contrast, seedlings seldom or never developed mycorrhizas of the late-stage fungi *Lactarius* or *Leccinum*, although cores contained abundant mycorrhizas of both types. This demonstrates differences in the abilities of fungi which occur early and late in the sequence to establish mycorrhizas in field conditions, whereas in monoxenic culture both 'early-' and 'late-stage' fungi readily form mycorrhizas. Additional evidence for this distinction between early and late fungi is provided by Deacon *et al.* (1983) and Mason *et al.* (1983) using mycelial inocula, and by Fox (1983*b*) using basidiospores. Fox found that birch seedlings planted into basidiospore-amended, unsterile soil readily developed mycorrhizas with species of *Hebeloma, Inocybe*

Table 5.2. *Mean number of root tips of different mycorrhizal types on birch seedlings grown in non-isolated and cored positions around a mature birch tree. (After Fleming, 1984.)*

Mycorrhizal type	Non-isolated	Cored
Lactarius	34[a]	3[a]
Inocybe, Hebeloma and *Laccaria*	20[a]	50[a]
Total mycorrhizal tips	54	53
Total root tips	61	82

Superscript refers to pairs in each row differing from each other at $P = 0.05$.

and *Laccaria* but not with *Cortinarius, Lactarius, Leccinum* or *Russula* (Table 5.1).

Clearly the observations have important implications for the selection of mycorrhizal fungi for inoculating into nurseries. The choice is dependent upon the position of the fungus in the succession, its ability to colonise roots in unsterile soil and its tolerance of the nutrient levels in the nursery soil. Further knowledge regarding the behaviour of mycorrhizal fungi in a variety of soil and environmental conditions is necessary in order to explain phenomena such as the replacement of *Hebeloma crustuliniforme* and *Laccaria laccata* inoculated onto Douglas fir seedlings by native mycorrhizal fungi within 5 months after outplanting onto dry, burned-over sites in the USA (Bledsoe, Tennyson & Lopushinsky, 1982).

So far we have mainly highlighted the events occurring in first-rotation forest plantations. The phenomena occurring under unmanaged forests or woodlands, where natural regeneration occurs in the understorey, may be different. Here we expect a stable and possibly declining microflora of mainly 'late-stage' fungi (Harvey *et al.*, 1976; Malajczuk, Molina & Trappe, 1982). In contrast to outplanted forest seedlings, the majority of the roots of regenerating tree seedlings appear to be infected with late-stage fungi such as *Lactarius, Cortinarius, Russula* and *Amanita*. Fleming (1983) suggested that a common feature of the late-stage fungi is the ability to form strands. When he planted axenically raised birch seedlings into soil in contact with the root system of a parent birch tree, late-stage *Lactarius pubescens* mycorrhizas formed on the seedling roots. However, where soil was isolated, by coring, from the parent tree roots and the characteristic mycorrhizal strands of late-stage fungi, the seedlings developed mycorrhizas of early-stage fungi and generally not those associated with *L. pubescens* (Table 5.2). Separation of these 'late-stage' fungi, like other root-infecting

Table 5.3. *Possible factors influencing the succession of mycorrhizal fungi on tree roots based on expected soil, fungal and host tree characteristics as viewed from* r–K *strategies*

	Stage of succession	
	Early	Late
Carbohydrate demand (supply by host tree)	Low	High
Available nutrients	In inorganic pool	Mainly in organic pool
Competitive ability of fungus	Due to rapid mycelial growth	Due to production of mycelial strands
Associations of mycorrhizal fungi with saprotrophic fungi	Competitive	Synergistic if not able to break down organic complexes, or competitive if mycorrhizal fungi are also decomposers

fungi, from a well-established food base appears to reduce their infection potential (Garrett, 1951; Fleming, 1983). These findings clearly raise interesting questions with respect to the roles of early- and late-stage fungi in forest plantations and in naturally regenerating woodlands, some of which will be explored in the next section.

Considerable evidence is thus now accumulating to support the concept of mycorrhizal successions on ageing trees, but what forces dictate changes in mycorrhizal associations? What physiological differences are there between 'early-' and 'late-stage' mycorrhizas?

Resources – their role in mycorrhizal succession

Some of the main changes occurring during the development of a forest stand are summarised in Fig. 5.1. A number of factors thought to be involved in determining mycorrhizal successions were proposed by Dighton, Harrison & Mason (1981), some of which are discussed here in the light of evidence in the literature (Table 5.3).

Carbohydrate requirement of ectomycorrhizas

The supply of carbohydrates (photosynthates) from the tree host to the fungal symbiont is of paramount importance to the development and functioning of mycorrhizas. This supply not only satisfies the energy and carbon demand for fungal growth but is intimately connected with nutrient uptake by mycorrhizal roots (France & Reid, 1983).

The requirement of light (and hence photosynthates) for mycorrhizal

Table 5.4. *Growth of mycorrhizal fungi at three glucose levels (number of crosses is arbitrary scale of growth from 1 to 4 based on colony diameter on agar)*

	Glucose level		
	0.1 g l^{-1}	1.0 g l^{-1}	10 g l^{-1}
Hebeloma spp.	+ + + +	+ +	+
Leccinum spp.	+	+ +	+ + +
Amanita muscaria	+	+ +	+ + +(+)

development was demonstrated by Björkman (1949) in pines where no mycorrhizal development was observed at 6% daylight but occurred only on nutrient-poor soils at 12% daylight and on all soils at 23% and 49% daylight. Hacskaylo (1973) discusses this and other evidence from trench plots, tree shading and phloem destruction in trees, to demonstrate the carbohydrate demand of the fungus from its tree host. The carbohydrate demand by mycorrhizal root systems can be very large. St John & Coleman (1983) cite figures for vesicular-arbuscular mycorrhizal plants equivalent to 1% of the total plant weight, or 6% of the weight of infected root can be accounted for by the carbon allocation to the fungus. This can be equivalent to 40–60% of the net photosynthesis. Both the increased fungal component (i.e. the sheath) of ectomycorrhizas and the production of large fruit bodies may be an even greater carbon drain on the tree. In one study Trappe & Fogel (1977) demonstrated that the fruit bodies of mycorrhizal fungi may require the carbohydrate equivalent of 1m^3 of spruce timber or as much as 25% of a tree's assimilate (Newman, 1978). Cultural experiments on solid media have shown that early-stage fungi are less glucose-demanding than late-stage fungi (Table 5.4), and that early-stage fungi produce small fruit bodies (*Laccaria*, 0.03 g dry weight/fruit body) compared with large late-stage fruit bodies (*Leccinum*, 5–10.5 g dry weight/fruit body) (Deacon *et al.*, 1983) and adds support to the possibility that early-stage fungi have a lower demand for, or reduced access to, host-derived carbohydrate.

An additional important feature of carbohydrate nutrition was demonstrated by Björkman (1960). He demonstrated interplant carbohydrate transfer when ^{14}C-labelled glucose injected into spruce and pine trees was detected in the achlorophyllous plant *Monotropa hypopitys* but not in the other sub-canopy vegetation. A common mycorrhizal fungal symbiont between the trees and *Monotropa* was implicated in this preferential carbon transfer. Interplant carbohydrate transfer between green plants has been shown by Reid & Woods (1969) using *Pinus taeda* seedlings of similar age in association with *Pisolithus tinctorius* and *Thelephora terrestris*. Where seedling trees were allowed to become connected by mycorrhizal hyphal bridges between root systems, ^{14}C applied as $^{14}CO_2$ to one of the pair was detected in the recipient plant from which $^{14}CO_2$ was excluded. No transfer was observed where hyphal bridges were not allowed to form.

With the potential for interplant transfer of carbohydrates and the probable increased carbohydrate demand by 'late-stage' fungi compared to 'early-stage' fungi, we can propose that one factor controlling the succession of mycorrhizas on trees is the ability of the photosynthetic capacity of the tree at different ages to satisfy the carbon demands of the array of fungi at any one time. Young trees may not have enough 'excess' photosynthate to support 'late-stage' fungi. In the experiments of Fleming (1983), seedlings planted into soil separated from the mother tree became mycorrhizal only with 'early-stage' fungi. Where the seedling birches were planted into soil not isolated from the mother tree's root system, late-stage mycorrhizal fungi were formed on the seedlings. This, we suggest, is possibly due to the fact that the late-stage fungi forming mycorrhizas on the roots of seedlings are also associated with a mother tree, thus they obtain their carbohydrates from the mother tree rather than from the seedling alone. This facility is an important consideration in comparing seedling survival under afforestation and natural regeneration. In afforestation programmes, the mycorrhizal fungi associating with the seedling trees will tend to be early-stage fungi (Chu-Chou, 1979; Lamb, 1979) derived from spore populations in the soil. In natural regeneration, however, it is possible that the dominant late-stage fungi may form mycorrhizal associations between existing mature trees and young seedlings by mycorrhizal strands (Fleming, 1983) and the carbohydrate flow from mature donor tree to recipient seedling may enhance seedling growth and mycorrhizal establishment even under severe shading conditions, a facility observed in vesicular-arbuscular mycorrhizas of *Plantago* sp. and *Festuca* sp. (Francis, 1983). Similar data exist for carbohydrate transfer from donor mother pine trees to recipient seedlings in naturally regenerating woodland (Read, 1985) and for inter-tree transfer of phosphorus and calcium (Woods & Brock, 1964) in mixed-aged mixed deciduous forests.

Evidence of changes in carbohydrate content or supply to roots with tree age is scant. Seasonal variation in sugar and starch contents of fine roots is expected to change according to photosynthetic rate and partitioning of end products to above- and below-ground biomass (Vogt *et al.*, 1980). Vogt *et al.* (1983*b*) suggested that fine root and mycorrhizal biomass

Table 5.5. *The effect of fertiliser on mycorrhizal development on* Pinus contorta *in Japanese Paper Pot culture*

	Percent mycorrhizal root tips		
	0.13 kg m^{-3} E^a 11 ppm P	0.75 kg m^{-3} E 66 ppm P	1.5 kg m^{-3} E 132 ppm P
Non-mycorrhizal	0.18	0.1	0
Paxillus involutus	19.3	50.3	0.2
Hebeloma sacchariolens	55.0	16.4	0.4

^{a} E = Enmag commercial fertiliser.

fluctuate with stand development with maximum biomass at canopy closure and that nutrient-poor sites maintain a significantly higher mycorrhizal biomass than nutrient-rich sites, with trees on poor sites probably allocating a greater proportion of photosynthate to mycorrhizal production thereby satisfying mineral nutrient demands. In this manner the ageing stand would be able to sustain a most diverse mycorrhizal fungal flora around the time of canopy closure.

In addition to direct energy supply from the tree host for fungal growth, there are mutually dependent links between the supply of carbohydrates available for mycorrhizal development on root systems, the nutritional status of soil and the feedback to photosynthetic rates which have only recently been investigated (Marx, Hatch & Mendicino, 1977; France & Reid, 1983; Reid, Kidd & Ekwebelam, 1983). High levels of nitrogen and phosphorus in soil suppress the development of mycorrhizas on roots (Table 5.5). Marx *et al.* (1977) showed that the suppressed development of *Pisolithus tinctorius* on *Pinus taeda* was a result of decreased root sugar concentration, where sucrose concentration accounted for 85% of the variation in susceptibility of short roots to infection by *P. tinctorius* but fructose concentration was not correlated with infection. France & Reid (1983) produced a conceptual model of carbon and nitrogen interaction in ectomycorrhizal roots, where sucrose is immobilised in the fungus into storage forms, and nitrogen, primarily ammonium, is absorbed by the fungus and assimilated into amino acids in the sheath. When the host plant acts as a sink for nitrogen, amino acids are transported to the root tissue as glutamic acid and glutamine. Reid *et al.* (1983) demonstrated in experimental conditions with *P. tinctorius*-infected *Pinus taeda* that the percentage mycorrhizal infection and net photosynthesis increased with seedling age and that photosynthesis of mycorrhizal seedlings was greater than that of non-mycorrhizal seedlings. The addition of *P. tinctorius* or *Suillus granulatus* to *Pinus contorta* showed similar effects of increased net photosynthesis and biomass. Foliar analysis, however, suggested that phosphorus rather than nitrogen was responsible for the mycorrhizal seedling response. Increased photosynthesis as a result of mycorrhization is consistent with the ideas of source–sink relations in that the mycorrhizal fungus creates a carbohydrate sink and that both nitrogen and phosphorus concentrations, being enhanced by the presence of the mycorrhiza, could account for increased photosynthesis by, for example, increased chlorophyll concentration, decreased resistance to carbon dioxide diffusion, increased activity of carboxylating enzymes, and improved utilisation of assimilated carbon by sinks.

In terms of mycorrhizal succession on ageing trees, the carbohydrate supply may be an important determinant in the selection of mycorrhizal symbionts. An increased interdependence of the tree and symbiont for carbohydrate and nutrient dynamics with increased tree age would be consistent with the '*r*'–'*K*' concept. 'Early-stage' fungi exhibit rapid growth on simple media containing low levels of sugars (Table 5.4) and produce small fruit bodies. Late-stage fungi are more difficult to culture, requiring larger amounts of sugars and probably complex mixtures of vitamins which in nature they obtain from their tree host. This close dependence on the host tree and greater investment of energy resources into large fruit bodies and more persistent mycelial structures (strands) are consistent with '*K*' attributes.

Resource quality – a determinant of nutrient availability

As the forest stand ages, the resource quality of materials supplying nutrients to the trees decreases due to the increased tree-litter component containing a higher proportion of complex molecules (lignin and polyphenols). At canopy closure, temperature and moisture conditions become less favourable for decomposition (Swift, Heal & Anderson, 1979) and litter breakdown and mobilisation of mineral elements decline (Vogt *et al.*, 1983*a*) (Fig. 5.1). Rates of mineralisation become increasingly dependent on close-linked, nutrient-conservative cycling processes in the soil (Heal & Dighton, 1985). It is not surprising, therefore, to find the presence of mycorrhizas on seedlings to be favoured by mineral rather than organic soils (Alvarez, Rowney & Cobb, 1979) and mycorrhizas of mature forest associated with organic fractions (Harvey *et al.*, 1976). Under conditions of forest maturation the production of extracellular enzymes by mycorrhizal fungi would be an advantageous strategy in order to derive mineral nutrients and supplementary carbohydrates directly from organic sources. The development of close interrelations between decomposition and mycorrhizal roots could be hypothesised from the principles underlying '*r–K*' strategies (MacArthur & Wilson, 1967) as suggested by Went & Stark (1968) and Malloch, Pirozynski & Raven (1980). The 'direct cycling' hypothesis, generated by Went & Stark (1968) and based on observation from tropical forest floors, proposed that mycorrhizal fungi, in direct contact with dead plant remains, were able to act as decomposers. In this way it was possible for trees to extract mineral nutrients directly from decomposing litter and thus to circumvent the nutrient mineralisation pathway via saprotrophs. The arguments presented by Went & Stark were discussed in relation to further evidence by St John (personal communication) and Janos (1983). They suggested that although certain enzymatic capabilities are shown by mycorrhizal fungi which would enable them to enhance decomposition, the spatial distribution of mycorrhizal roots and hyphae in relation to decomposing litter and saprotrophic organisms may be of great significance in that they are in juxtaposition to localised sites of nutrient mineralisation.

Data supporting the direct cycling hypothesis, or indeed the idea that mycorrhizal fungi are involved in decomposition, are scarce and often conflicting. In experiments using trenched, dug and undisturbed forest soil plots under *Pinus radiata*, and in experimental systems, Gadgil & Gadgil (1971, 1975) demonstrated that there was competition between mycorrhizal and saprotrophic fungi which decreased the rate of decomposition of *P. radiata* litter. Where living mycorrhizal roots were excluded, 60–80% more weight loss of litter was recorded after 12 months (Gadgil & Gadgil, 1971). When this experiment was repeated under Scots pine (*Pinus sylvestris*) in Sweden (Berg & Lindberg, 1980), however, litter weight loss increases of 1–7% per 12 months could be attributed to removal of mycorrhizal roots from the system. Berg & Lindberg suggested that the magnitude of the effect may not be as great as that expressed by Gadgil & Gadgil and put forward a number of alternative suggestions to explain the results. Digging and sieving soil may present new energy resources to microorganisms, resulting in faster mineralisation. Removal of a nutrient sink (plant roots) may allow more nutrients to become available for the microorganisms, allowing higher turnover rates (similar competitive effects between roots and microorganisms were explored by Bosatta (1981)) and the formation of inhibiting substances in the soil. These are all possible alternative explanations for the observed effect.

The possession of enzyme capacity compatible with the hypothesis that

Table 5.6. *Acid phosphatase production by fungal mycelia in Hagem's medium with 10 ppm orthophosphate-P or inositol hexaphosphate-P after 42 days growth at 20 °C.* (After Dighton, 1983.)

	Phosphatase production (μg phenol mg^{-1} mycelium)	
Fungus	orthophosphate-P	inositol hexaphosphate-P
Hebeloma crustuliniforme	29.8	23.9
Lactarius rufus	26.3	22.2
Paxillus involutus	17.1	10.2
Lactarius pubescens	12.8	7.6
Amanita muscaria	5.8	8.8
Suillus luteus	5.0	9.2
Marasmius androsaceus (s)	20.0	1.7
Mycena galopus (s)	0.9	0.2

(s) = saprotrophic fungus, the remainder are mycorrhizal.

Table 5.7. *Release of PO$_4$-P from 10 ppm inositol hexaphosphate (IHP) in Hagem's medium by fungi grown in liquid culture at 20 °C.* (After Dighton, 1983.)

Fungus	P released (μg P mg^{-1} fungus)	Percent P release from IHP and retained in fungus
Lactarius rufus	86.1	9.2
Paxillus involutus	62.4	16.7
Suillus luteus	7.1	92.5
Mycena galopus (s)	13.4	76.7
Marasmius androsaceus (s)	2.7	34.0

(s) = saprotrophic fungus, the remainder are mycorrhizal.

mycorrhizal fungi express *K*-strategist attributes is documented in the literature. Bartlett & Lewis (1973) demonstrated phosphatase and phytase activity of mycorrhizal beech (*Fagus* spp.) roots, although surface contamination by rhizosphere microflora may have complicated the interpretation. Similar information on phosphatase production by excised roots is available from spruce (Alexander & Hardy, 1981). The phosphatase capacity is a facility of the fungal partner, as demonstrated in pure cultures (Ho & Zak, 1979; Dighton, 1983) and the production may be greater than that of recognised saprotrophic basidiomycetes (Table 5.6). Dighton (1983) also demonstrated a greater orthophosphate production as a result of hydrolysis of inositol hexaphosphate by pure cultures of mycorrhizal fungi than could be utilised by fungal growth (Table 5.7). This phenomenon was also observed by Bartlett & Lewis (1973) and may be important in terms of supplying phosphorus to the tree host from organic sources.

A number of mycorrhizal fungi are shown to produce polyphenoloxidases (Giltrap, 1982), particularly species of *Lactarius*, which tend to be unspecialised facultative mycorrhizal fungi with a broad host range. Cellulase activity by mycorrhizal fungi is also documented; Oelbe (1982), following the work of Norkrans (1950), demonstrated cellulolytic activity of mycorrhizal varieties of *Tricholoma aurantium*. Similarly high cellulase activity of mycorrhizal *Salix rotundifolia* roots was reported by Linkins & Antibus (1981) and by St John (personal communication). It is interesting here that the mycorrhizal strands have a higher exocellulase activity than the mantle (60 and 140 times greater activity of exocellulase and β-glucosidase, respectively). *S. rotundifolia*, growing in Arctic tundra, may well rely on its mycorrhizal fungi for litter decomposition and nutrient release due to the short active season for decomposition. If this is so it would be in agreement with adaptive strategies for stressed environments (adversity selection) according to the '*r*'–'*K*'–'*A*' matrix of Southwood (1977) (Heal & Dighton, 1985). In semi-aseptic culture, *Suillus luteus* in association with *Pinus contorta* seedlings significantly increased the decomposition of cellulose (cotton) and chitin (Latter & Dighton, personal communication).

Work in culturing mycorrhizal fungi to examine their enzyme capabilities and their acquisition of nutrients and carbohydrates from organic sources (Lundeberg, 1970; Hacskaylo, 1973) may provide some information on the physiology of the mycorrhizal fungi. The demonstration of these physiological capabilities in a wide variety of fungi from early, mid, and late successional stages, in conjunction with the host root tissue, is, however, much needed in order to understand fully the physiological basis of succession.

Discussion

Changes in the structure of above-ground plant communities during ecosystem succession or the development of forest stands can be seen to influence the mycorrhizal flora. The mechanisms of change in mycorrhizal dynamics is far from obvious, changes in carbohydrate supply from the host tree and increasing recalcitrance of forest floor organic matter are two major factors which might influence successional changes of mycorrhizal flora. Evidence that the change in mycorrhizal fungi is of a successional nature is strong, with similar patterns of fruit bodies occurring sequentially with age in a number of tree species (Lamb, 1979; Chu-Chou & Grace, 1981, 1983; Deacon *et al.*, 1983; Last *et al.* 1983). Of the little that is known regarding the differences in the physiology of the fungi during succession, a number of tentative hypotheses show them to have characteristics predicted by the '*r*'–'*K*'–'*A*' model of Southwood (1977). 'Early-stage' fungi have a low carbohydrate demand for rapid growth and, as a result, many 'nursery' species (e.g. *Thelephora terrestris*) are not competitive after outplanting into nutrient-poor soils but are highly so in the very fertile soils of the nursery which may often be fumigated before seeding. The formation of mycelial strands by many of the 'late-stage' fungi may be seen as evidence of long-lived structures with a longer turnover time (*K*-characteristics *sensu* MacArthur & Wilson, 1967), along with the potential for litter decomposition by exoenzymes supplementing carbohydrate and mineral nutrient demand in later stages of forest development (Giltrap, 1982; Oelbe, 1982; Dighton, 1983), particularly where unfavourable environmental conditions exist, tending to select for '*A*'-strategies (Southwood, 1977; Linkins & Antibus, 1981).

What benefit is there in looking for the physiological factors behind mycorrhizal successions? Where man wishes to intervene and attempt to enhance tree growth by the artificial inoculation of forest seedlings with mycorrhizal fungi, a correct choice of fungus is necessary to optimise growth and survival of the seedlings and persistence of the introduced fungal strain. In first-rotation forestry the choice of effective strains of early-stage fungi for the appropriate soil type and climatic zone appears to be essential (Mason *et al.*, 1983). We have seen that late-stage fungi cannot successfully compete with early-stage fungi under such conditions, while other mycorrhizal fungi may have a limited effective climatic tolerance. *Pisolithus tinctorius*, for example, has proved to be a very successful mycorrhizal fungus on a number of sites in the southern USA (Marx, 1980), but appears to be ill adapted to cooler conditions (Momoh & Gbadegesin, 1980; Grossnickle & Reid, 1983) where indigenous, and therefore more ecologically adapted, fungi appear to perform better. In second-rotation forests the indigenous mycorrhizal flora on roots of cut stumps will consist mainly of late-stage fungi. Their influence on the mycorrhizal flora of newly planted seedlings has yet to be evaluated, but second-rotation decline in some parts of the world may, in part, be due to the availability of a restricted and rather host/age-specific mycorrhizal flora (Robinson, 1973; Malajczuk *et al.*, 1982). In the UK, however, the small scale of the landscape and size of plantations would allow the maintenance of a broad spectrum spore population at all times, thus second-rotation decline due to a depauperate mycorrhizal flora would not be expected. In naturally regenerating woodlands, on the other hand, the predominance of late-stage fungi might be more beneficial to seedling trees than early-stage fungi. In such situations the possibility of carbohydrate transfer from mother trees to seedlings via mycorrhizal bridges may be of benefit to seedlings growing in sub-optimal light conditions on the forest floor. Can any similar advantage occur in clear-cut areas? If mycorrhizal fungi are still viable on roots of cut stumps they may act in a similar manner to benefit second-rotation trees by the transfer of carbohydrates and mineral ions as a result of saprotrophic activity of the fungus at the stump end.

A number of questions have been raised by this discussion of the physiological basis for the succession of mycorrhizal agarics in relation to resources available to them. Their implications are broad ranging in the selection of mycorrhizas for optimisation of forest tree growth. These questions, we feel, are the challenge for future mycorrhizal research.

References

Alexander, I. J. & Hardy, K. (1981). Surface phosphatase activity of Sitka spruce mycorrhizas from a Serpentine site. *Soil Biology and Biochemistry*, **13**, 301–5.

Alvarez, I. F., Rowney, D. L. & Cobb, F. W., Jr (1979). Mycorrhizae and growth of white fir seedlings in mineral soil with and without organic layers in a California forest. *Canadian Journal of Forest Research*, **9**, 311–15.

Bartlett, E. M. & Lewis, D. H. (1973). Surface phosphatase activity of mycorrhizal roots of beech. *Soil Biology and Biochemistry*, **5**, 249–57.

Berg, B. & Lindberg, T. (1980). Is litter decomposition retarded in the presence of mycorrhizal roots in forest soil? Uppsala: *Swedish Coniferous Forest Project Internal Report*, No. 95.

Björkman, E. (1949). The ecological significance of the ectotrophic mycorrhizal association in forest trees. *Svensk Botanisk Tidskrift*, **43**, 223–62.

Björkman, E. (1960). *Monotropa hypopitys* L. – an epiparasite on tree roots. *Physiologia Plantarum*, **13**, 308–27.

Bledsoe, C. S., Tennyson, K. & Lopushinsky, W. (1982). Survival and growth of outplanted Douglas fir seedlings inoculated with mycorrhizal fungi. *Canadian Journal of Forest Research*, **12**, 720–3.

Bosatta, E. (1981). A qualitative analysis of the root–microorganism soil system. II. Combined effect of several factors. *Ecological Modelling*, **13**, 237–45.

Chu-Chou, M. (1979). Mycorrhizal fungi of *Pinus radiata* in New Zealand. *Soil Biology and Biochemistry*, **11**, 557–62.

Chu-Chou, M. & Grace, L. J. (1981). Mycorrhizal fungi of *Pseudotsuga menziesii* in the north island of New Zealand. *Soil Biology and Biochemistry*, **13**, 247–9.

Chu-Chou, M. & Grace, L. (1983). Characterization and identification of mycorrhizas of Douglas fir in New Zealand. *European Journal of Forest Pathology*, **13**, 251–60.

Cromack, K., Jr (1981). Below-ground processes in forest succession. In *Forest Succession: Concepts and Application*, ed. D. C. West, H. H. Shugart & D. B. Botkin, pp. 361–73. New York: Springer Verlag.

Deacon, J. W., Donaldson, S. J. & Last, F. T. (1983). Sequences and interactions of mycorrhizal fungi on birch. *Plant and Soil*, **71**, 257–62.

Dighton, J. (1983). Phosphatase production by mycorrhizal fungi. *Plant and Soil*, **71**, 455–62.

Dighton, J., Harrison, A. F. & Mason, P. A. (1981). Is the mycorrhizal succession on trees related to nutrient uptake? *Journal of Science Food and Agriculture*, **32**, 629–30.

Fleming, L. V. (1983). Succession of mycorrhizal fungi on birch: infection of seedlings planted around mature trees. *Plant and Soil*, **71**, 263–7.

Fleming, L. V. (1984). Effects of soil trenching and coring on the formation of ectomycorrhizas on birch seedlings grown around mature trees. *New Phytologist*, **98**, 143–53.

Ford, E. D. & Newbould, P. J. (1977). The biomass and production of ground vegetation and its relation to tree cover through a deciduous woodland cycle. *Journal of Ecology*, **65**, 201–10.

Fox, F. M. (1983*a*). *Sources of inoculum of sheathing mycorrhizal fungi of Birch* (Betula *spp.*). Ph.D. thesis, University of Edinburgh.

Fox, F. M. (1983*b*). Role of basidiospores as inocula of mycorrhizal fungi of birch. *Plant and Soil*, **71**, 269–73.

France, R. C. & Reid, C. P. P. (1983). Interactions of nitrogen and carbon in the physiology of ectomycorrhizae. *Canadian Journal of Botany*, **61**, 964–84.

Francis, R. (1983). Inter- and intra-specific transfer of carbon between mycorrhizal plants in light and shade. In *Current Mycorrhizal Research: Abstracts of Communications Presented at the Mycorrhiza Group Meeting, Lancaster University, UK, March* 28–30, 1983, ed. J. Dighton, Merlewood Research and Development Paper No. 95. Grange-over-Sands, Cumbria: Institute of Terrestrial Ecology.

Frankland, J. C. (1981). Mechanisms in fungal successions. In *The Fungal Community: Its Organisation and Role in the Ecosystem*, ed. D. T. Wicklow & G. C. Carroll, pp. 403–26. New York: Marcel Dekker.

Gadgil, R. L. & Gadgil, P. D. (1971). Mycorrhiza and litter decomposition. *Nature (London)*, **233**, 133.

Gadgil, R. L. & Gadgil, P. D. (1975). Suppression of litter decomposition by mycorrhizal roots of *Pinus radiata*. *New Zealand Journal of Forest Science*, **5**, 33–41.

Garrett, S. D. (1951). Ecological groups of soil fungi: a survey of substrate relationships. *New Phytologist*, **50**, 149–66.

Gerson, U. & Chet, I. (1981). Are allochthonous and autochthonous soil microorganisms r- and K- selected? *Revue d'Ecologie et de Biologie du Sol*, **18**, 285–9.

Giltrap, N. J. (1982). Production of polyphenol oxidases by ectomycorrhizal fungi with special reference to *Lactarius* spp. *Transactions of the British Mycological Society*, **78**, 75–81.

Grime, J. P. (1979). *Plant Strategies and Vegetation Processes*. Chichester & New York: Wiley.

Grossnickle, S. C. & Reid, C. P. P. (1983). Ectomycorrhiza formation and root development pattern of conifer seedlings on a high-elevation mine site. *Canadian Journal of Forest Research*, **13**, 1145–58.

Hacskaylo, E. (1973). Carbohydrate physiology of ectomycorrhizae. In *Ectomycorrhizae: Their Ecology and Physiology*, ed. G. C. Marks & T. T. Kozlowski, pp. 207–30. New York: Academic Press.

Harvey, A. E., Larsen, M. J. & Jurgensen, M. F. (1976). Distribution of ectomycorrhizae in a mature Douglas fir/Larch forest soil in western Montana. *Forest Science*, **22**, 393–8.

Heal, O. W. & Dighton, J. (1985). Nutrient cycling and decomposition in natural terrestrial ecosystems. In *Soil Microflora–Microfauna Interactions*, ed. M. J. Mitchell & J. P. Nakas. The Hague: Martinus Nijhoff. (in press)

Heal, O. W. & Ineson, P. (1984). Carbon and energy flow in terrestrial ecosystems – relevance to microflora. In *Current Perspectives in Microbial Ecology*, ed. M. J. Klug & C. A. Reddy, pp. 394–404. Washington: American Society for Microbiology.

Ho, I. & Zak, B. (1979). Acid phosphatase activity of six ectomycorrhizal fungi. *Canadian Journal of Botany*, **57**, 1203–5.

Janos, D. P. (1983). Tropical mycorrhizas, nutrient cycles and plant growth. In *Tropical Rain Forest: Ecology and Management*, ed. S. L. Sutton, T. C. Whitmore & A. C. Chadwick, pp. 327–45. Oxford: Blackwell.

Lamb, R. J. (1979). Factors responsible for the distribution of mycorrhizal fungi of *Pinus* in eastern Australia. *Australian Forest Research*, **9**, 25–34.

Last, F. T., Mason, P. A., Wilson, J. & Deacon, J. W. (1983). Fine roots and sheathing mycorrhizas: their formation, function and dynamics. *Plant and Soil*, **71**, 9–21.

Lewis, D. H. (1973). Concepts in fungal nutrition and the origin of biotrophy. *Biological Reviews*, **48**, 261–78.

Linkins, A. E. & Antibus, R. K. (1981). Mycorrhizae of *Salix rotundifolia* in coastal arctic tundra. In *Arctic and Alpine Mycology*, ed. G. A. Laursen & J. F. Ammirati, pp. 509–31. Washington: University of Washington Press.

Lundeberg, G. (1970). Utilisation of various nitrogen sources, in particular bound nitrogen, by mycorrhizal fungi. *Studia forestalia Suecica*, **79**, 1–95.

MacArthur, R. H. & Wilson, E. D. (1967). *The Theory of Island Biogeography*. Princeton: Princeton University Press.

Malajczuk, N., Molina, R. & Trappe, J. M. (1982). Ectomycorrhiza formation in *Eucalyptus* I. Pure culture synthesis, host specificity and mycorrhizal compatability in *Pinus radiata*. *New Phytologist*, **91**, 467–82.

Malloch, D. W., Pirozynski, K. A. & Raven, P. H. (1980). Ecological and evolutionary significance of mycorrhizal symbioses in vascular plants (a review). *Proceedings of the National Academy of Sciences, USA*, **77**, 2113–8.

Marks, G. C. & Foster, R. C. (1967). Succession of mycorrhizal associations on individual roots of Radiata pine. *Australian Forestry*, **31**, 193–201.

Marx, D. H. (1980). Ectomycorrhizal fungus inoculation: a tool for improving forestation practices. In *Tropical Mycorrhiza Research*, ed. P. Mikola, pp. 13–71. Oxford: Clarendon Press.

Marx, D. H., Hatch, A. B. & Mendicino, J. F. (1977). High soil fertility decreases sucrose content and susceptibility of loblolly pine roots to ectomycorrhizal infection by *Pisolithus tinctorius*. *Canadian Journal of Botany*, **55**, 1569–74.

Mason, P. A., Last, F. T., Pelham, J. & Ingleby, K. (1982). Ecology of some fungi associated with an ageing stand of birches (*Betula pendula* and *Betula pubescens*). *Forest Ecology and Management*, **4**, 19–37.

Mason, P. A., Wilson, J., Last, F. T. & Walker, C. (1983). The concept of succession in relation to the spread of sheathing mycorrhizal fungi on inoculated tree seedlings growing in unsterile soils. *Plant and Soil*, **71**, 247–56.

Miller, H. G. (1979). The nutrient budgets of even-aged forests. In *The Ecology of Even-Aged Forest Plantations*, ed. E. D. Ford, D. C. Malcolm & J. Atterson, pp. 221–56. Cambridge: Institute of Terrestrial Ecology.

Miller, H. G., Cooper, J. M., Miller, J. D. & Pauline, O. J. L. (1979). Nutrient cycles in pine and their adaptation to poor soils. *Canadian Journal of Forest Research*, **9**, 19–26.

Molina, R. (1979). Ectomycorrhizal inoculation of containerized Douglas fir and lodgepole pine seedlings with six isolates of *Pisolothus tinctorius*. *Forest Science*, **25**, 585–90.

Momoh, Z. O. & Gbadegesin, R. A. (1980). Field performance of *Pisolithus tinctorius* as a mycorrhizal fungus of pines in Nigeria. In *Tropical Mycorrhiza Research*, ed. P. Mikola, pp. 72–9. Oxford: Clarendon Press.

Moser, M. (1967). Die Ektotrophe Ernahrungsweke an der Waldgrenze. *Mitteilungen Forstlichen Bundesversuchsanstalt Wien*, **75**, 357–80.

Mosse, B., Stribley, D. P. & Tacon, Le F. (1981). Ecology of mycorrhizae and mycorrhizal fungi. *Advances in Microbial Ecology*, **5**, 137–210.

Newman, E. I. (1978). Root microorganisms: their significance in the ecosystem. *Biological Reviews of the Cambridge Philosophical Society*, **53**, 511–54.

Norkrans, B. (1950). Studies in growth and cellulolytic enzymes of *Tricholoma*. *Symbolae Botanical Upsalienses*, **11**, 1–126.

Oelbe, M. (1982). Untersuchungen über einige Kohlenhydratabbauende Enzyme des Mykorrhizapilzes *Tricholoma aurantium*. M.Sc. thesis, Georg-August-University, Gottingen.

Pegler, D. N. (1981). *The Mitchell Beazley Pocket Guide to Mushrooms and Toadstools*. London: Mitchell Beazley Publishers.

Read, D. J. (1985). Mycorrhizas and nutrient cycling under field conditions. In *Ecological Interactions in Soil: Plants, Microbes and Animals*, ed. A. H. Fitter, D. Atkinson, D. J. Read & M. B. Usher. British Ecological Society Special Publication 4. Oxford: Blackwell. (in press)

Reid, C. P. P., Kidd, F. A. & Ekwebelam, S. A. (1983). Nitrogen nutrition, photosynthesis and carbon allocation in ectomycorrhizal pine. *Plant and Soil*, **71**, 415–32.

Reid, C. P. P. & Woods, F. W. (1969). Translocation of C^{14} labelled compounds in mycorrhizae and its implications in interplant nutrient cycling. *Ecology*, **50**, 179–87.

Robinson, R. K. (1973). Mycorrhizas and 'Second rotation decline' of *Pinus patula* in Swaziland. *South African Forestry Journal*, **84**, 16–19.

Rose, S. L. (1980). Mycorrhizal associations of some actinomycete nodulated nitrogen-fixing plants. *Canadian Journal of Botany*, **58**, 1449–54.

St John, T. V. & Coleman, D. C. (1983). The role of mycorrhizae in plant ecology. *Canadian Journal of Botany*, **61**, 1005–14.

Slankis, V. (1973). Hormonal relationships in mycorrhizal development. In *Ectomycorrhizae: Their Ecology and Physiology*, ed. G. C. Marks & T. T. Kozlowski, pp. 232–98. New York: Academic Press.

Southwood, T. R. E. (1977). Habitat, the templet for ecological strategies? *Journal of Animal Ecology*, **46**, 337–65.

Swift, M. J., Heal, O. W. & Anderson, J. M. (1979). *Decomposition in Terrestrial Ecosystems*. Oxford: Blackwell Scientific Publications.

Thomas, G. W., Rogers, D. & Jackson, R. M. (1983). Changes in the mycorrhizal status of Sitka spruce following outplanting. *Plant and Soil*, **71**, 319–23.

Trappe, J. M. (1962). Fungus associates of ectotrophic mycorrhizae. *Botanical Reviews*, **28**, 538–606.

Trappe, J. M. (1977). Selection of fungi for ectomycorrhizal inoculation in nurseries. *Annual Review of Phytopathology*, **15**, 203–22.

Trappe, J. M. & Fogel, R. D. (1977). Ecosystematic functions of mycorrhizae. In *The Belowground Ecosystem: A Synthesis of Plant-Associated Processes*, ed. J. K. Marshall, pp. 205–13. Colorado: Range Soil Department, Science Series No. 26, Colorado State University.

Trappe, J. M. & Strand, R. F. (1969). Mycorrhizal deficiency in a Douglas fir region nursery. *Forest Science*, **15**, 381–9.

Usher, M. B. & Parr, T. W. (1977). Are there successional changes in arthropod decomposer communities? *Journal of Environmental Management*, **5**, 151–60.

Vogt, K. A., Edmonds, R. L., Grier, C. C. & Piper, S. R. (1980). Seasonal changes in mycorrhizal and fibrous-textured root biomass in 23- and 180-year old Pacific silver fir stands in western Washington. *Canadian Journal of Forest Research*, **10**, 523–9.

Vogt, K. A., Grier, C. C., Meier, C. E. & Keyes, M. R. (1983a). Organic matter and nutrient dynamics in forest floor of young and mature *Abies amabilis* stands in western Washington, as affected by fine-root input. *Ecological Monographs*, **53**, 139–57.

Vogt, K. A., Moore, E. E., Vogt, D. J., Redlin, M. J. & Edmonds, R. L. (1983b). Conifer fine root and mycorrhizal root biomass within the forest floors of Douglas fir stands of different ages and site productivities. *Canadian Journal of Forest Research*, **13**, 429–37.

Watling, R. (1973). *Identification of the Larger Fungi*. Amersham: Hulton Educational Publications Ltd.

Went, F. W. & Stark, N. (1968). The biological and mechanical role of soil fungi. *Proceedings of the National Academy of Sciences, USA*, **60**, 497–504.

Woods, F. W. & Brock, K. (1964). Interspecific transfer of ^{45}Ca and ^{32}P by root systems. *Ecology*, **45**, 887–9.

6

A comparative ultrastructural analysis of the host–fungus interface in mycorrhizal and parasitic associations

J.A.DUDDRIDGE

Department of Agricultural and Forest Sciences, Commonwealth Forestry Institute, South Parks Road, Oxford OX1 3RB, UK

Introduction

This chapter is not intended as a comprehensive review of the current literature on host–symbiont interfaces, for which reference should be made to other sources (Bracker & Littlefield, 1973; Scannerini & Bonfante–Fasolo, 1983). Its scope is restricted to those mycorrhizal associations in which the fungal partner is a member of the Agaricales. These include many ectomycorrhizas, monotropoid and arbutoid mycorrhizas, and a few orchidaceous associations. Thus agaric symbionts are represented in a large number of mycorrhizal groups, the main exceptions being vesicular-arbuscular mycorrhizas, which have phycomycetous symbionts, and ericoid mycorrhizas, where the endophytes have been identified as either ascomycetes (Read, 1974) or basidiomycetes (Seviour, Willing & Chilvers, 1973; Englander & Hull, 1980), the latter being members of the Aphyllophorales and not the Agaricales.

Among the agarics there are relatively few active parasites, especially biotrophic ones, most being either saprotrophs or mycorrhizal symbionts. One exception to this is *Crinipellis perniciosa*, a biotrophic parasite on cocoa (Hedger: Chapter 2; Wheeler: Chapter 4). *Armillaria mellea* agg. is the most widely known necrotrophic parasite in the agarics (Rishbeth: Chapter 3) and probably some species of *Marasmius, Pholiota, Mycena* and *Oudemansiella* may be parasitic to varying degrees (Rayner, Watling & Frankland: Chapter 1). For the purpose of this comparative ultrastructural study, we are particularly interested in the host–agaric interface in both mutualistic and parasitic biotrophs. This creates two problems; the first is that there are very few known biotrophic parasites among the agarics;

and the second is that there are no ultrastructural investigations on the few parasites that there are. For this reason, many of the comparisons used in this chapter have had to be with non-agaric biotrophic parasites.

Ectomycorrhizas

Compatible interactions

In ectomycorrhizas, the fungal symbionts do not penetrate the host cells but form a sheath on the root surface and an intercellular network between the cortical cells, the Hartig net. There are two separate interfaces between the fungus and host cells, that with the inner sheath hyphae and that with the Hartig net. In the early stages of sheath formation, the hyphae become attached to the root surface by polysaccharide mucigel (Fig. 6.1a) (Marks & Foster, 1973; Ling-Lee, Ashford & Chilvers, 1977b; Duddridge, 1980; Duddridge & Read, 1984a), in which the whole of the sheath eventually becomes embedded. In a fully developed ectomycorrhiza the structure of the interface between the sheath and host tissue changes as one moves proximally from the apex of a mycorrhizal lateral to that area in which Hartig net development occurs. The remains of host cells are incorporated into the inner sheath, including root cap cells at the apex and cortical cells and root hairs (Fig. 6.1b,c) further back (Foster & Marks, 1966, 1967; Chilvers, 1968; Marks & Foster, 1973; Strullu & Gourret, 1973; Kiffer, 1974; Strullu, 1974; Atkinson, 1976; Nylund, 1981; Duddridge & Read, 1984a). The electron-dense matrix in which the mature inner sheath is often embedded (Fig. 6.1c), is probably the remains of the phenolic contents of these cells (Ling-Lee, Chilvers & Ashford, 1977a; Duddridge & Read, 1984a).

In the pre-Hartig net zone, that is the part of the root between the apical meristem and the position at which Hartig net formation commences (Atkinson, 1976), occasional examples are found of host cell wall lysis by inner sheath hyphae (Fig. 6.1d) (Duddridge & Read, 1984a). It is not known how often this occurs but in all the instances observed the fungus never breaches the wall. The presence of a small electron-lucent zone, which contains disorganised cell wall microfibrils, adjacent to the invading hypha, is characteristic of enzymic penetration by a fungal pathogen (e.g. McKeen, 1977a,b; Duddridge & Sargent, 1978; Faull & Campbell, 1979) and could represent an area of degraded pectin (Maeda, 1970). However, the disorganisation of the microfibrils could also indicate cellulase activity. A cuticular layer, probably of suberin (Ling-Lee et al., 1977a; Foster, 1981, 1982), on the outer cortical cell walls is also broken down by the fungus.

It has been widely demonstrated that biotrophic fungi have a limited

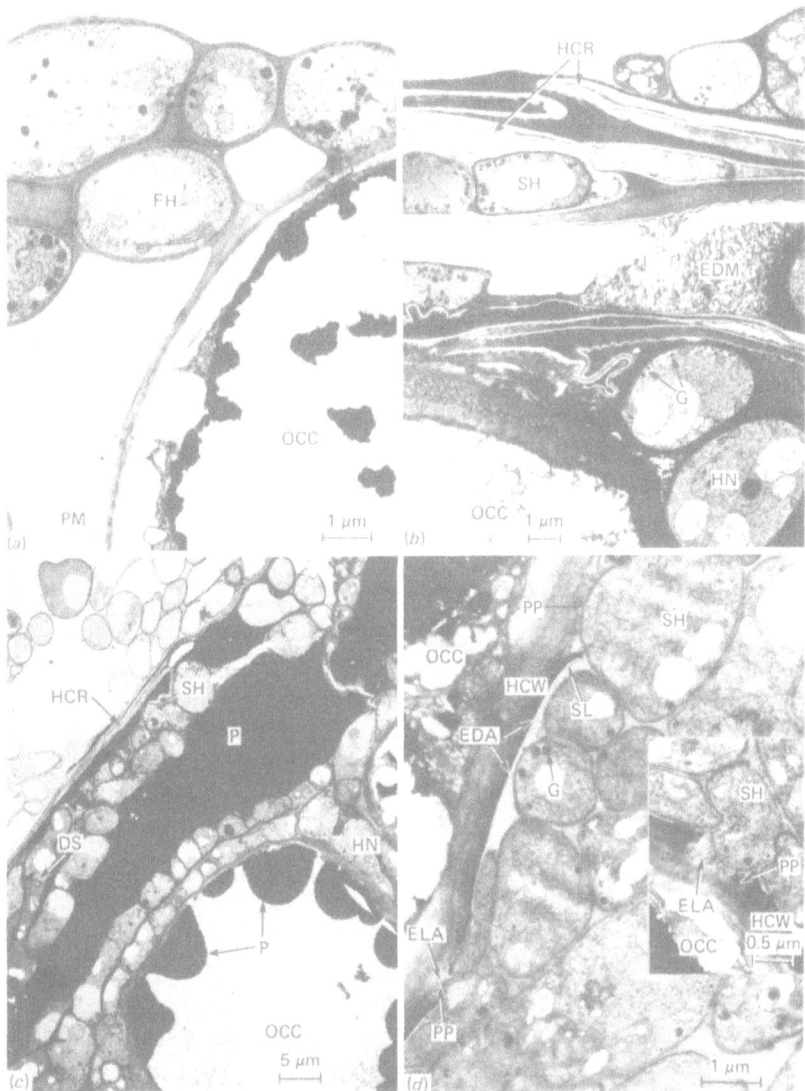

Fig. 6.1. Ultrastructure of the host–agaric interface in ectomycorrhizas. (a) Initial contact between mycorrhizal symbiont and outer cortical cells of host. (b, c) The inner region of the sheath and outer cortical cells of the host showing the incorporation of host cells and their electron-dense contents into the sheath. The sheath hyphae eventually grow through the dead cells. (d) The alteration of the outer cortical cell walls by sheath hyphae in the pre-Hartig net zone of field mycorrhizas. The electron-lucent covering on the surface of these outer cortical cells could be suberin. There is an increased electron density in the cortical cell wall in the area of penetration and an electron-lucent zone directly in advance of penetration. DS = dolipore septum; EDA = electron-dense area; EDM = electron-dense matrix; ELA = electron-lucent area; FH = fungal hypha; G = glycogen; HCR = host cell remains; HCW = host cell wall; HN = Hartig net; OCC = outer cortical cell; P = phenolic material; PM = polysaccharide mucigel; PP = penetration point; SH = sheath; SL = suberin-like layer.

ability to produce extracellular enzymes in culture, their production usually being associated with saprotrophs or with necrotrophic parasites which cause extensive cellular damage. For example, the majority of agarics that form mycorrhizas cannot degrade lignin or cellulose in culture (see Harley & Smith, 1983) although there is variation in the degree to which pectin can (Palmer & Hacskaylo, 1970; Lamb, 1974) or cannot (Lindeberg & Lindeberg, 1977) be used as a carbon source by ectomycorrhizal fungi. Scougal (unpublished) has shown that a number of isolates of *Xerocomus subtomentosus* differ in their ability either to form mycorrhizas or to degrade lignin. Those isolates able to degrade lignin have proved unsuccessful in aseptic syntheses and their hyphae can often be observed penetrating the walls of moribund cortical cells (Fig. 6.2*a*). However, the ability to utilise substrates as carbon sources in culture does not preclude the possibility of enzyme mediated penetration involving high-molecular-weight non-diffusible molecules or molecules bound to the wall of the fungus (Bateman, 1976; Ingram, Sargent & Tommerup, 1976). Stahmann (1973) suggests that in those cases where little damage is incurred during the penetration of the cell wall, the hydrolytic enzymes are bound to the fungal peg and only become active when they are in contact with the host wall. This would represent an extremely economic method of penetration, requiring minimal quantities of enzymes, and such a phenomenon could explain why diffusible hydrolytic enzymes are not detected in most mycorrhizal agarics in culture (Lundeberg, 1970; Giltrap & Lewis, 1982; Ramstedt & Söderhall, 1983). Ectomycorrhizal agarics are therefore able to breach host cell walls, although intracellular penetration of healthy outer cortical cells in ectomycorrhizas is very rare unless there is an upset in the physiological balance of the symbiosis (see next section). However, in a later section on arbutoid mycorrhizas it will be shown that the same ectomycorrhizal agarics can produce intracellular mycorrhizal associations in arbutoid hosts. This ability to penetrate host cell walls may be the result of either the production of enzymes by the hyphal apex or, possibly, the activation of enzymes within the host cell wall.

Two distinct types of interface are found between the Hartig net and host cortical cells of *Pinus* spp. (Duddridge & Read, 1984*a,b*). The first involves direct contact between the host cell wall and the wall of the Hartig net, both walls remaining distinct (Fig. 6.2*b,d*). In the second type of interface the outside of the host cell wall adjacent to the Hartig net, and to a lesser extent the wall of the Hartig net itself, loses its integrity and becomes indistinguishable from the electron-dense matrix in which the Hartig net is embedded (Fig. 6.2*c*). A similar type of interface has been

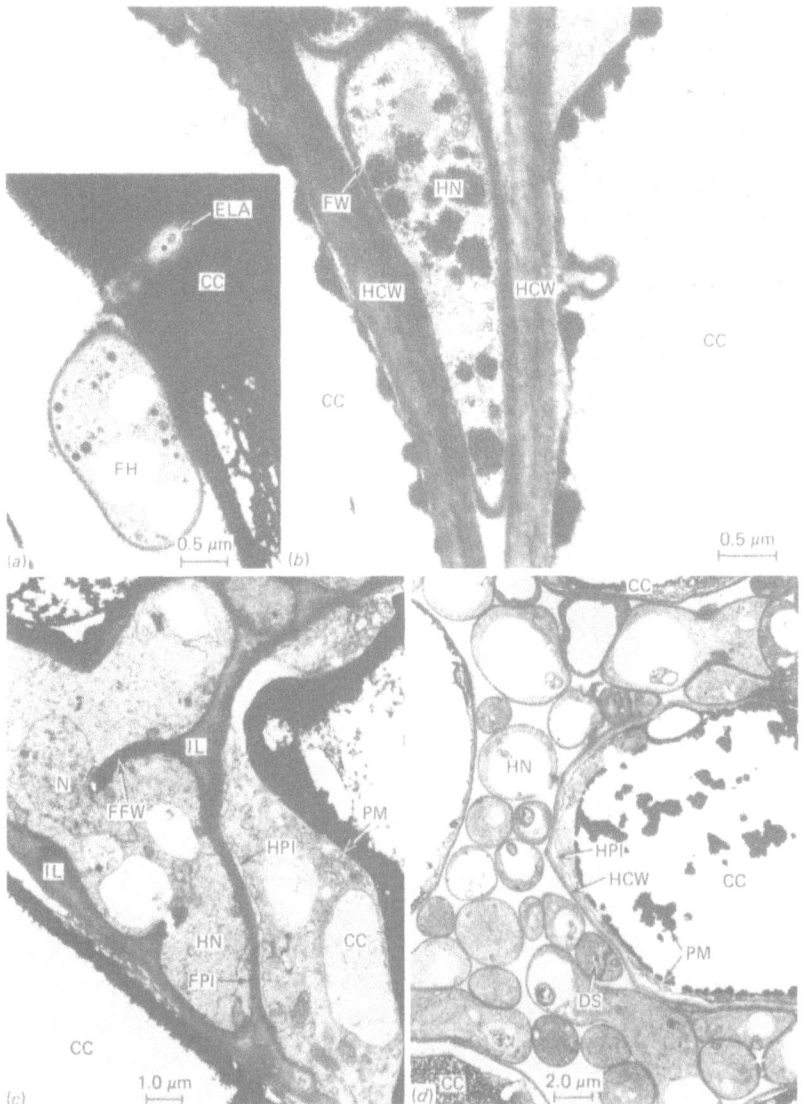

Fig. 6.2. (*a*) Penetration by hyphae of *Xerocomus subtomentosus* through moribund cortical cells of *Pinus sylvestris*. (*b–d*) Host–agaric interface between outer cortical cells and Hartig net. (*b*) Simple wall-to-wall contact in which fungal and host walls remain distinct. Furthermost point of Hartig net penetration. (*c*) Involving layer formed between fungus and host cell walls, both losing their integrity. (*d*) Hartig net formed in a mycorrhiza synthesised between *Larix kaempferi* and *Suillus grevillei*. No sugar was used in the synthesis medium. CC = cortical cell; DS = dolipore septum; ELA = electron-lucent area; FH = fungal hypha; FPl = fungal plasmalemma; FFW = folded fungal wall; FW = fungal wall; HCW = host cell wall; HN = Hartig net; HPl = host plasmalemma; IL = involving layer; N = nucleus; PM = phenolic material.

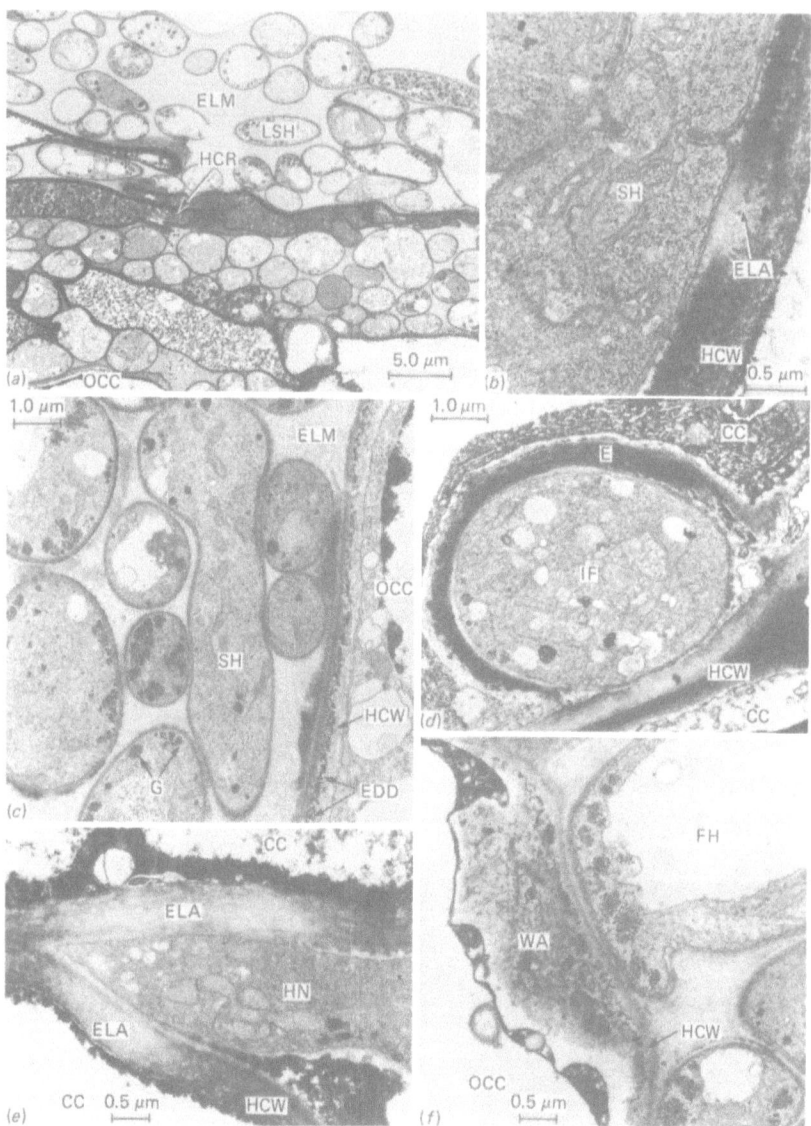

Fig. 6.3. Ectomycorrhizas aseptically synthesised with *Suillus grevillei*. In *b–f*, 10 g l⁻¹ glucose was included in the synthesis medium. (*a*) Loose unstructured sheath of an ectomycorrhiza synthesised between *S. grevillei* and *Pinus sylvestris*. No sugar was used in the synthesis medium. (*b*) An alteration in the host cell wall adjacent to a sheath hypha in a mycorrhiza synthesised between *S. grevillei* and *P. sylvestris*. (*c*) The interface between the sheath and outer cortical cells of a mycorrhiza synthesised between *S. grevillei* and *L. kaempferi*. The host cell wall has become thickened and there are electron-dense deposits in it. (*d*) An intracellular

observed in ectomycorrhizas of *Pseudotsuga menziesii* (Strullu, 1974) and *Betula* (Strullu & Gerrault, 1977) formed with ascomycetes, and in agaricoid ectomycorrhizas of *Betula* (Atkinson, 1976) and *Dryas* (Debaud, Pepin & Bruchet, 1981).

This electron-dense matrix is equivalent to the 'involving layer' observed by Scannerini (1968), and the '*zone d'apposition dense aux electrons*' of Strullu & Gerrault (1977). Strullu (1974) believed that this type of modified interface was only diagnostic of ascomycete associations. He suggested that in the latter the host cell wall became electron-dense but that in basidiomycete (agaric) associations there was little modification. However, recent studies have shown that both modified and unmodified types of interface are present in agaric associations (Duddridge, 1980; Duddridge & Read, 1984*a,b*). It is probable that the modified interface is a feature of mature mycorrhizas and the unmodified type found in developing mycorrhizas.

In compatible ectomycorrhizal associations, where there is no physiological imbalance, the development of the interface described for an agaric symbiont is remarkably constant, regardless of the species of host or fungus. For instance, such diverse examples as ectomycorrhizas formed between *Dryas octopetala* (Rosaceae) and *Hebeloma* spp. (basidiomycete) (Debaud *et al.*, 1981), a *Tilia* sp. (Tiliaceae) and *Cenococcum graniforme* (ascomycete) (Pigott, 1982), and *Pinus strobus* (Pinaceae) and *Endogone flammicorona* (phycomycete) (Bonfante-Fasolo & Scannerini, 1977*b*), all show host–fungus interfaces with a well-developed involving layer.

Incompatible interactions

The host–agaric interface in ectomycorrhizas is not a static structure but undergoes development, maturation and senescence. Its formation may be altered by the compatibility and physiological state of both symbionts and also by prevailing environmental conditions. There

---

hypha of *S. grevillei* surrounded by a callose-type encasement in an outer cortical cell of *P. sylvestris*. (*e*) Alterations in the host cell wall adjacent to the Hartig net in a mycorrhiza synthesised between *S. grevillei* and *P. sylvestris*. (*f*) A wall apposition produced on the inside of a cortical cell wall of *Betula pubescens* adjacent to the point of contact with a hypha of *S. grevillei*. CC = cortical cell; E = encasement; EDD = electron-dense deposits; ELA = electron-lucent area; ELM = electron-lucent matrix; FH = fungal hypha; G = glycogen; HCR = host cell remains; HCW = host cell wall; HN = Hartig net; IF = intracellular fungus; LSH = base of fungal sheath; OCC = outer cortical cell; SH = sheath; WA = wall apposition.

have been no published ultrastructural studies on the effects of environmental conditions such as light intensity, pH and water availability on the host–agaric interface. This is an area of mycorrhizal research where there is a need for much more information. In the artificial conditions of aseptic synthesis, the inclusion of concentrations of sugar (10 g l⁻¹) in the medium (Marx & Zak, 1965) at levels not normally found in the field, does cause changes in the host–fungus interface (Duddridge & Read, 1984c). This was found to be the case during studies on the host specificity of *Suillus grevillei*, a fungus usually thought to be specific for *Larix* spp. (Duddridge, unpublished). For instance, when glucose is omitted from the synthesis medium, a well-developed sheath and Hartig net is formed in interactions between *S. grevillei* and *Larix kaempferi* (Fig. 6.2d), a loose sheath and limited Hartig net form with *Pinus sylvestris* (Fig. 6.3a), and there is no mycorrhizal development at all with *Betula pubescens*. However, with the addition of 10 g l⁻¹ glucose to the growth medium, not only are a well-developed sheath and Hartig net formed between *S. grevillei* and *P. sylvestris* but intracellular penetration can be observed in some of the outer cortical cells (Fig. 6.3d). These intracellular hyphae are surrounded by a callose-like encasement material similar to that found around haustoria of biotrophic parasites that have infected non-susceptible or immune hosts. Papilla formation has also been observed in ectomycorrhizas synthesised with *Picea abies* and *P. sylvestris* (Nylund, Kasimir & Arveby, 1982). Areas of possible enzymic degradation can be observed adjacent to some sheath hyphae (Fig. 6.3b) and also changes in the host cell wall adjacent to the Hartig net (Fig. 6.3e). Even in interactions with *L. kaempferi*, the host–fungus interface is altered when glucose is added to the synthesis medium. The host wall appears thicker and there are electron-dense deposits in the wall (Fig. 6.3c). In the presence of glucose, *S. grevillei* forms a rudimentary sheath around the roots of *B. pubescens* and wall thickening and wall appositions (Fig. 6.3f) are produced on the inside of the outer cortical cell wall where it comes in contact with the fungus. The cortical cell appears dead. This is very similar to the situation found when, for example, cow pea rust infects a non-susceptible host (Heath, 1972). The deposition of tannins in host tissue and the death of cortical cells have been observed in incompatible reactions between *Suillus* and *Rhizopogon* spp. and eucalypts (Malajczuk, Molina & Trappe, 1984), an *Alnus* sp. and *Paxillus involutus* (Molina, 1981), and conifers and both *Fuscoboletinus aeruginascens* and *Alpova diplopholeus* (Molina & Trappe, 1982). However, in all of these syntheses high levels of sugar have been used in the growth medium, making difficult the interpretation of the interactions which

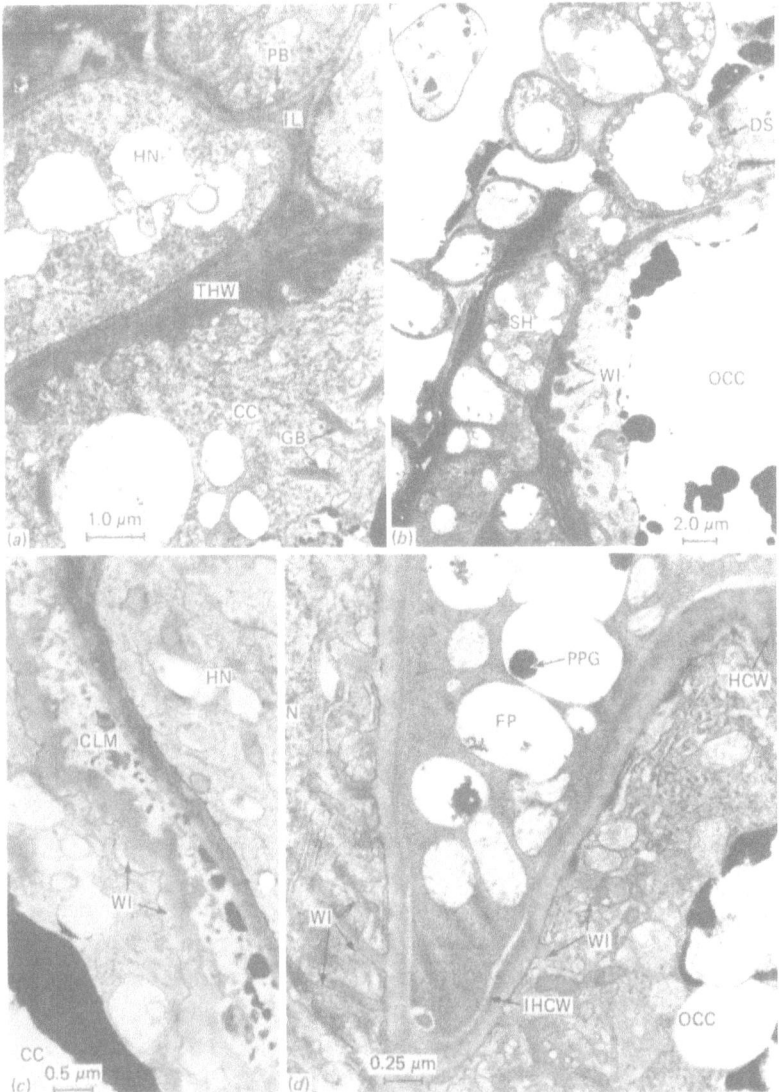

Fig. 6.4. (*a–c*) Ectomycorrhizas synthesised between *Suillus bovinus* and *Pinus sylvestris* where 10 g l⁻¹ glucose has been included in the medium. (*a*) General thickening of the host cell wall adjacent to the Hartig net. (*b*) Wall ingrowths produced from the host cell wall adjacent to the sheath and Hartig net. (*c*) Host–fungus interface showing a band of callose-like material between the host wall ingrowths and the Hartig net. (*d*) A fungal peg in an outer cortical cell of *Monotropa hypopitys* showing the invaginated host cell wall and associated wall ingrowths. CC = cortical cell; CLM = callose-like material; DS = dolipore septum; FP = fungal peg; GB = Golgi body; HN = Hartig net; IHCW = invaginated host cell wall; IL = involving layer; N = nucleus; OCC = outer cortical cell; PB = paramural body; PPG = polyphosphate granule; SH = sheath; THW = thickened host wall; WI = wall ingrowth.

occurred. It is therefore important in specificity studies to omit sugar from the synthesis medium altogether. There have also been previous observations of abnormal reactions produced during syntheses as a result of high glucose levels in the medium (Giltrap, 1979; Malibari, 1979; Duddridge & Read, 1984c). Fig. 6.4a shows a thickening of the host cell wall adjacent to the sheath and Hartig net in mycorrhizas synthesised between *Suillus bovinus* and *P. sylvestris*. This generalised thickening eventually develops into discrete wall ingrowths (Fig. 6.4b), reminiscent of those produced in transfer cells described by Gunning & Pate (1969a,b). In older associations, a layer of 'callose-like' material forms between the wall ingrowths and the host cell wall (Fig. 6.4c). Using Thiéry's method for the demonstration of periodate-sensitive polysaccharides (Thiéry, 1967; Courtoy & Simar, 1974) the wall ingrowths stain similarly to the host cell wall and the 'callose-like' material remains electron-lucent although it contains densely staining areas (Fig. 6.4c). This is the characteristic appearance of callose and may represent a defence mechanism on the part of the host to an increasingly dominant fungal partner (Duddridge & Read, 1984c). 'Callose-type' encasements are often found around the haustoria of many plant pathogens (e.g. Berlin & Bowen, 1964; Heath & Heath, 1971; Littlefield & Bracker, 1972; Coffey, 1975; Hohl & Stössel, 1976; Higgins & Lazarovits, 1978; Littlefield & Heath, 1979; Steinkamp, Martin, Hoefert & Ruppel, 1979; Beakes, Singh & Dickinson, 1982). During the re-differentiation of transfer cells in the nectaries of *Aloe* and *Gasteria* (Schnepf & Pross, 1976) a 'callose-like' material has also been observed. This new material is deposited on a membrane which isolates the wall protruberances and its associated cytoplasm from the rest of the cell.

A similar host–fungus interface is found consistently under field conditions in mycorrhizas of *Pisonia grandis* (Nyctaginaceae) (Ashford & Allaway, 1982). The regular occurrence of agaric fruiting bodies, particularly of *Lactarius* and *Russula*, with *Pisonia fragrans* in the Caribbean led to the suggestion that it may be ectomycorrhizal (Pegler & Fiard, 1979). It was not until the recent study on *P. grandis* that the presence of a sheath and poorly developed Hartig net was confirmed. In addition to these, transfer cells develop in the outer cortex of the root adjacent to the fungus. The host–fungus interface with its labyrinthine wall ingrowths resembles that previously described in ectomycorrhizas synthesised between *S. bovinus* and *P. sylvestris* growing in high glucose levels. *P. grandis* is subjected to large fluctuations in nutrient availability and it is suggested that the development of transfer cells would allow rapid absorption and storage of nutrients that would otherwise be lost by leaching (Ashford &

Allaway, 1982). The role of transfer cells in this and other mycorrhizal associations will be discussed at the end of this chapter.

Monotropoid mycorrhizas

The term 'monotropoid' describes mycorrhizas formed by the achlorophyllous members of the Ericaceae in the sub-family Monotropoideae (Duddridge & Read, 1982*b*; Robertson & Robertson, 1982) because of their unique structure. Members of this group are always found growing in association with autotrophic higher plants, and fungal rhizomorphs can be found connecting the sheaths of both (Kamienski, 1881, 1882; Romell, 1939; Björkman, 1960; Campbell, 1971; Duddridge & Read, 1982*b*). The roots of *Monotropa hypopitys* are found in association with ectomycorrhizal roots of *Pinus* and *Salix* (Duddridge & Read, 1982*b*) and have the same fungal symbiont as their ectomycorrhizal associates. In many cases this will be an agaric. *Boletus* spp. have previously been linked with the mycorrhizas of *M. hypopitys* (Francke, 1934; Björkman, 1960; Khan, 1972). The identities of the fungal symbionts of *Pterospora andromedea* (Fig. 6.5*a*) and *Sarcodes sanguinea* (Fig. 6.5*c*) (Robertson & Robertson, 1982) are unknown, but both possess dolipore septa, indicative of basidiomycetes.

Monotropoid mycorrhizas superficially resemble ectomycorrhizas, possessing both a sheath and limited Hartig net. However, they have an unusual feature which distinguishes them from ectomycorrhizas: fungal pegs (Lutz & Sjolund, 1973; Duddridge & Read, 1982*b*; Robertson & Robertson, 1982) intrude from the sheath into the outer cortical cells, causing both the host cell wall and host plasmalemma to invaginate (Fig. 6.4*d*; 6.5*a,c*). The invagination, rather than penetration of the host cell wall by the agaric symbiont, is reminiscent of the situation seen in some ascolichens where the algal wall is also merely invaginated by the intruding fungus (Jacobs & Ahmadjian, 1969; Galun, Paran & Ben-Shaul, 1970). Unfortunately there are relatively few ultrastructural studies of basidiolichens (e.g. Slocum & Floyd, 1977) and none where an agaric is the fungal symbiont. However, modifications of the monotropoid host–agaric interface distinguish it from that observed in lichens. In *Monotropa* (Fig. 6.4*d*) (Duddridge & Read, 1982*b*), *Sarcodes* (Fig. 6.5*c*) and *Pterospora* (Fig. 6.5*a*) (Robertson & Robertson, 1982) the interface takes the form of intricate convoluted wall ingrowths produced from the invaginated host cell wall surrounding the fungal peg. These ingrowths give the infected host cell the appearance of a transfer cell (Gunning & Pate, 1969*a,b*). As seen earlier, comparable elaborations of the cell wall can be found at the

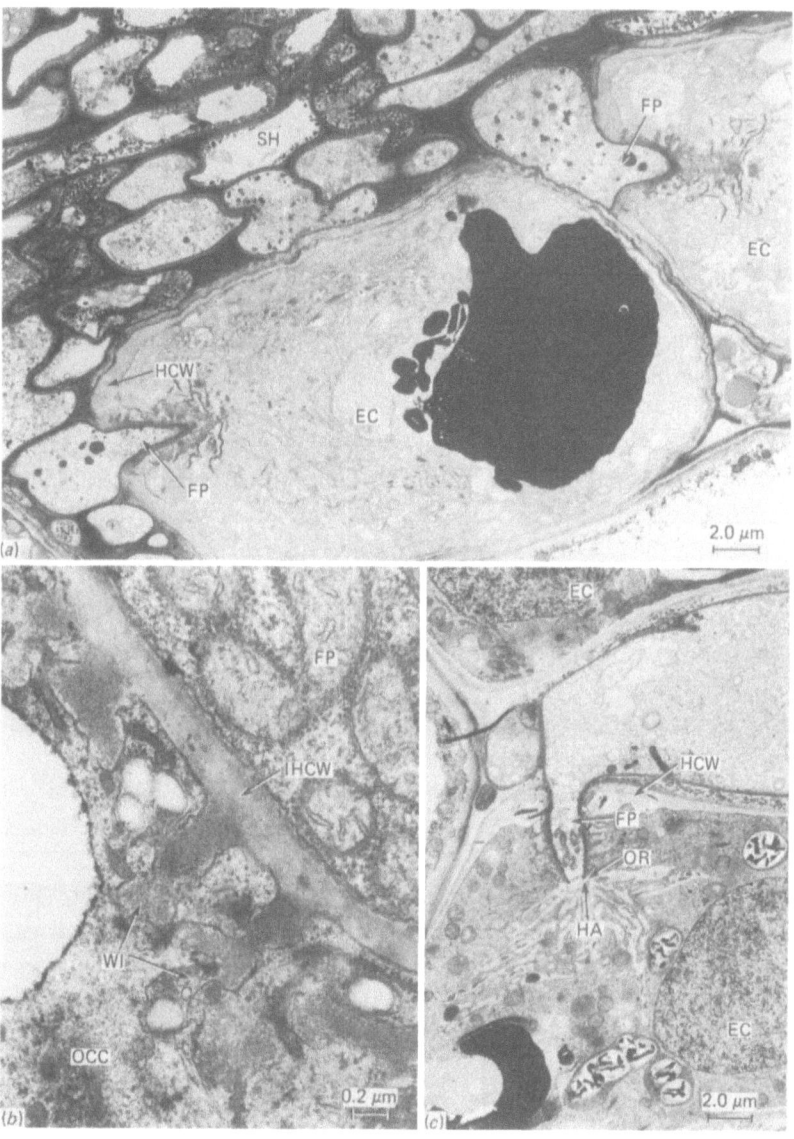

Fig. 6.5. The host–agaric interface in monotropoid mycorrhizas. (*a*) A non-median section through the epidermal cells of *Pterospora andromedea* showing the intrusion of fungal pegs into the host cells. (*b*) A higher magnification of the invaginated host cell wall and associated ingrowths surrounding a fungal peg in a mycorrhiza of *Monotropa hypopitys*. The fungal wall is indistinguishable from the invaginated host wall. (*c*) A longitudinal section through an epidermal cell of *Sarcodes sanguinea*, penetrated by a fungal peg. EC = epidermal cell; FP = fungal peg; HA = hemispherical area of moderate electron density; HCW = host cell wall; IHCW = invaginated host cell wall; OCC = outer cortical cell; OR = osmiophilic ring; SH = sheath; WI = wall ingrowth. (*a*) and (*c*) from Robertson & Robertson (1982) with permission of *The New Phytologist* and the authors.

host–agaric interface of other mycorrhizal associations. The ingrowths stain with Thiéry's method similarly to those described previously and also to primary host wall material. The fungal wall loses its integrity and becomes indistinguishable from the invaginated host cell wall.

In *Monotropa* the development of the greatest number of fungal pegs coincides with the period of rapid shoot extension (Duddridge & Read, 1982*b*). In the terminal stage of the association, the tip of the invaginated host cell wall surrounding the fungal peg appears to 'burst' (Fig. 6.6*a*). At the point where the wall has apparently ruptured, a ring of osmiophilic material can be observed. This is reminiscent of the osmiophilic 'neck rings' found round haustorial necks of rusts (Van Dyke & Hooker, 1969; Ehrlich & Ehrlich, 1971*a*; Hardwick, Greenwood & Wood, 1971; Heath & Heath, 1971; Littlefield & Bracker, 1972; Coffey, 1976; Heath, 1976; Harder, 1978; Harder *et al.*, 1978; Allen, Coffey & Heath, 1979; Chong & Harder, 1980), downy mildew (Woods & Gay, 1983) and powdery mildew (Gil & Gay, 1977) fungi. The neckband in biotrophic parasites is thought to represent a transition region between the wall of the penetration peg and the wall of the haustorium (Littlefield & Bracker, 1972) and is the point where the modified invaginated membrane of the haustorium and the non-invaginated host membrane meet. In biotrophic parasites it has been suggested that the neckband may 'form a stabilising structure which maintains two distinct domains of host plasmamembrane' (Woods & Gay, 1983) and may form at this point because 'antigens' of the modified and unmodified wall meet and become deposited (Gil & Gay, 1977). It has also been compared with the Casparian strip found round the endodermal cells of plants (Bonnett, 1968) and has been shown to prevent apoplastic flow of solutes along the haustorial neck of rusts (Heath, 1976). This is unlikely to be the case in monotropoid mycorrhizas because the ring is formed at a terminal stage in the association when there is no barrier to the movement of nutrients between the associates. However the function of this ring in monotropoid mycorrhizas is unknown. In all the monotropoid mycorrhizas studied so far, a large membranous sac extrudes from the tip of the ruptured peg where the osmiophilic neckband is situated. This membrane appears to be continuous with the host plasma membrane. The integrity of the plasmalemma of the fungal peg, at this stage, remains in doubt (Duddridge & Read, 1982*b*; Robertson & Robertson, 1982). In some cases it can be seen intact after the membranous sac has formed. The contents of the latter do not appear to be fungal cytoplasm derived from the peg but rather the result of breakdown of the host cell wall ingrowths (Duddridge & Read, 1982*b*). An electron-dense hemispherical area can be seen radiating from the point at which the dissolution of the peg wall occurs. This has been observed in *Monotropa*, (Fig. 6.6*a*), *Sarcodes* (Fig. 6.5*c*) and *Pterospora*, but no explanation as to its nature has been found. The function of this stage of the association will be discussed in a later section.

The degenerating peg and associated wall ingrowths eventually become surrounded by a 'callose-like' encasement. Some of these degenerating transfer cells are invaded at this stage by fungal hyphae from the sheath, which may fill the cell. This late invasion probably enables the fungus to absorb products of cytoplasmic degradation from the host which would otherwise leak into the environment. A similar absorption into the fungus may occur in the late Hartig net region (Atkinson, 1976) of ectomycorrhizas. Nutrients are thus tightly cycled and the substrate acquired may help to maintain the fungus during the winter months when autotrophic partners produce little or no assimilate.

Arbutoid mycorrhizas

Hosts forming arbutoid mycorrhizas are found in the Ericaceae and Pyrolaceae, e.g. *Arbutus* and *Arctostaphylos* (Ericaceae) and *Pyrola* (Pyrolaceae). They possess both a sheath and Hartig net characteristic of ectomycorrhizas and an intracellular infection structurally similar to ericoid mycorrhizas. Current research has shown that a wide range of agarics isolated from ectomycorrhizas can form ectendomycorrhizas on arbutoid hosts (Zak, 1976*a,b*; Molina & Trappe, 1982). Largent, Sugihara & Brinitzer (1980) traced rhizomorphs from fruiting bodies of *Amanita gemmata* and *Cortinarius zakii* to mycorrhizas of *Arctostaphylos manzanita* and *Arbutus menziesii*, respectively.

There has been only one detailed ultrastructural study published on the development of arbutoid mycorrhizas (Fusconi & Bonfante-Fasolo, 1984), others being preliminary investigations (Duddridge, 1980; Read, 1983; Scannerini & Bonfante-Fasolo, 1983). This study however dealt with an ascomycete association of *Arbutus unedo*. Although a Hartig net and intracellular penetration are regular features of arbutoid mycorrhizas, the presence of a well-developed sheath is not consistent. An excellent ultrastructural study on the development of mycorrhizas in *Pyrola* has been carried out by Robertson & Robertson (unpublished) in which basidiomycete and ascomycete associations were found. In the former association, a well-developed sheath and Hartig net occurred. Duddridge (1980) observed two kinds of mycorrhizas on *Pyrola rotundifolia* ssp. *maritima*, a more common black type and an unpigmented one. Clamp

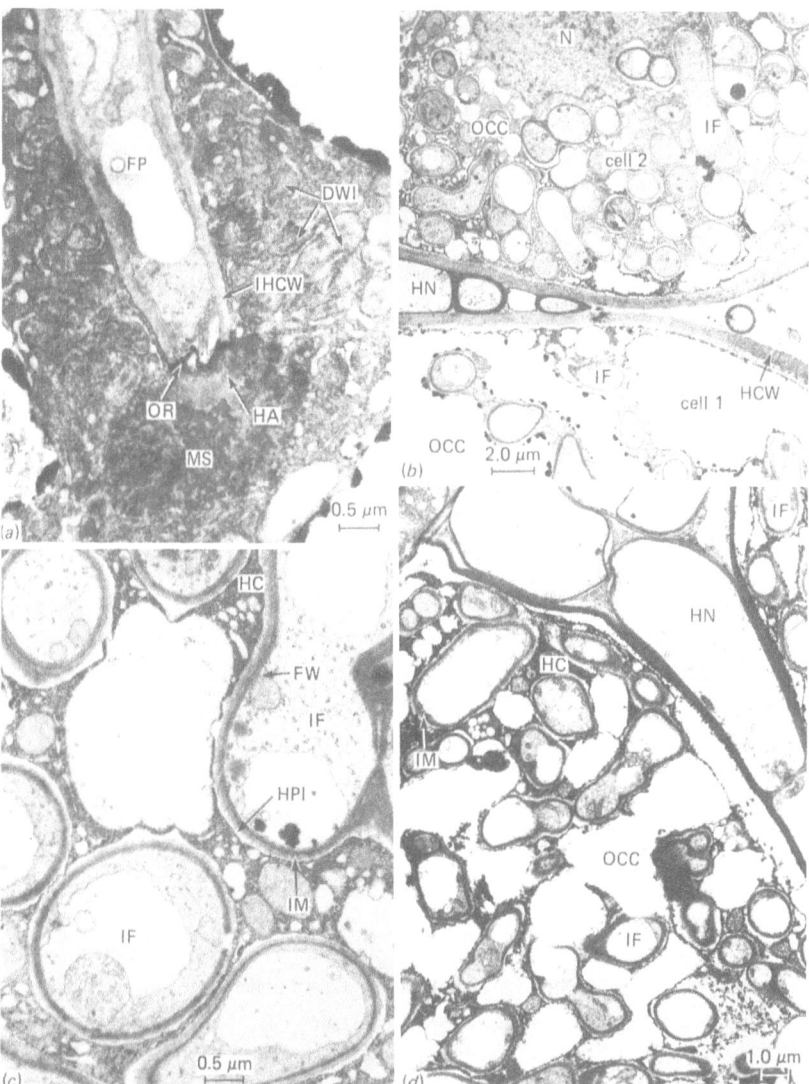

Fig. 6.6. (*a*) In a mycorrhizal root of *Monotropa*, a large membranous sac extrudes from the tip of a ruptured peg where an osmiophilic neckband is situated. (*b–d*) The host–agaric interface in arbutoid mycorrhizas. (*b*) Outer cortical cells of *Pyrola rotundifolia* with intracellular infections. A Hartig net separates the two cells. Cell 1 is newly infected and contains relatively few intracellular hyphae. Cell 2 has been infected for a longer period of time and contains a massive fungal invasion. (*c*) A higher magnification of the interface between the intracellular hyphae and host cell of *Pyrola*. An interfacial matrix separates the fungal wall from the host plasmalemma. (*d*) Intracellular infections in the outer cells of *Arctostaphylos uva-ursi*; the infections are separated by a Hartig net. DWI = degenerating wall ingrowth; FP = fungal peg; FW = fungal wall; HA = hemispherical area of moderate electron density; HC = host cell; HCW = host cell wall; HN = Hartig net; HPl = host plasmalemma; IF = intracellular fungus; IHCW = invaginated host cell wall; IM = interfacial matrix; MS = membranous sac; N = nucleus; OCC = outer cortical cell; OR = osmiophilic ring.

connections and dolipore septa confirm that the fungal symbiont of the former is a basidiomycete. In the black mycorrhizas of *Pyrola* the Hartig net develops first, forming a characteristic black reticulate pattern where it grows along the outer radial walls of the cortical cells (Duddridge, 1980). In *Arctostaphylos uva-ursi* a thick sheath develops, the inner hyphae of which are embedded in an electron-lucent matrix (Duddridge, unpublished; Scannerini & Bonfante-Fasolo, 1983). In both *Pyrola* (Fig. 6.6*b*) and *Arctostaphylos* (Fig. 6.6*d*) a limited Hartig net is formed between the outer layer of cortical cells. In some cases the cell walls of the host and fungus still appear distinct, but an involving layer, similar to that seen in ectomycorrhizas, eventually develops. The ultrastructure of the intracellular fungus suggests that it is a simple coil similar to that of an ericoid infection. It is of interest to note here that agarics have so far not been identified as endophytes of ericoid mycorrhizas nor have syntheses with a number of agarics been successful (Read, personal communication) although they are able to form a similar intracellular infection in related arbutoid hosts.

In a balanced association the intracellular fungus never breaches the host plasmalemma and is separated from the host by an electron-dense interfacial matrix (Fig. 6.6*c*) similar to that observed in vesicular-arbuscular mycorrhizas (Cox & Sanders, 1974; Bonfante-Fasolo & Scannerini, 1977*a*; Carling, White & Brown, 1977; Dexheimer, Gianinazzi & Gianinazzi-Pearson, 1979), ericoid mycorrhizas (Peterson, Mueller & Englander, 1980; Bonfante-Fasolo *et al.*, 1981; Bonfante-Fasolo & Gianinazzi-Pearson, 1982; Duddridge & Read, 1982*a*) and orchidaceous mycorrhizas (Hadley, Johnson & John, 1971; Nieuwdorp, 1972; Hadley, 1975; Strullu & Gourret, 1975).

At some stage in the association a breakdown occurs between the symbionts. In arbutoid mycorrhizas it is not clear whether the host or fungus begins to degenerate first. In *Pyrola* mycorrhizas, Robertson & Robertson (unpublished) have shown that the breakdown of the fungus precedes that of the infected host cells. However, in the ascomycete association of *Arbutus unedo*, the endophyte usually outlives the host epidermal cells (Duddridge, 1980; Fusconi & Bonfante-Fasolo, 1984). The colonisation of moribund host cells of *A. uva-ursi* by the symbiont has been observed (Scannerini & Bonfante-Fasolo, 1983) and in *Pyrola* the formation of a thick black pseudoparenchymatous sheath ('rind'), following the penetration of the outer cortical cells, is indicative that the symbiosis is senescent (Duddridge, 1980). The outer cortical cells of mycorrhizas surrounded by this thick black 'rind' are dead or dying and contain degraded fungal and host remains. Contrary to earlier light microscope investigations of arbutoid mycorrhizas (Kramar, 1901; Henderson, 1919), the outer cortical cells are not killed by the intracellular penetration of the fungus but a balanced symbiosis is set up, in which the translocation of materials occurs across a live interface (Duddridge, 1980; Read, 1983; Scannerini & Bonfante-Fasolo, 1983). Once the symbiosis has broken down, agaric hyphae can be observed in the moribund host cells, just as they are in degenerating host cells of ectomycorrhizas and monotropoid mycorrhizas.

Orchidaceous mycorrhizas

The majority of the fungal symbionts of green orchids belong to the form genus *Rhizoctonia* (Harley, 1969), of which most of the known perfect stages are found in the Tremellaceae and Tulasnellaceae (Warcup, 1981). Clamp-forming basidiomycetes have been isolated from colourless saprotrophic orchids (Harley, 1969). The most frequently occurring agaric symbiont of orchids is *Armillaria mellea* agg. which has been found as an endophyte of *Gastrodia elata* (Kusano, 1911; Sagara & Takayama, 1978; Zhuang *et al.*, 1983) and *Galeola septentrionalis* (Hamada, 1940). Kusano showed that it formed a mycorrhizal symbiosis whilst simultaneously parasitising neighbouring trees. The only other agaric reported as an orchid endophyte is *Marasmius coniatus* var. *didymoplexis* (Burgeff, 1932) but this has never been confirmed. There have been no published ultra-structural studies on a host–agaric interface in orchidaceous mycor-rhizas, although interfaces with other basidiomycetes have been studied (Strullu & Gourret, 1975; Strullu, 1976*a*).

Role of the host–fungus interface in recognition and the translocation of materials between symbionts

Ultracytological studies have shown that a major component of the host–agaric interface in ectomycorrhizal associations is a zone of modified host cell wall material with a high pectin content (Scannerini, 1968; Duddridge, 1980; Nylund, 1981). Intracellular mycorrhizal symbionts (in the case of agaric symbionts, arbutoid or orchidaceous associations) are separated from the host plasmalemma by a similar pectin–hemicellulose layer. This zone is found in a wide range of different symbiotic associations including the matrix around the intracellular symbiont of *Alnus* (Lalonde & Knowles, 1975) and *Comptonia* (Newcomb *et al.*, 1978) nodules, around the haustoria of *Erysiphe pisi* (Gil & Gay, 1977) and around the intracellular endophyte in ericoid and vesicular-arbuscular mycorrhizas (see section on arbutoid mycorrhizas for references). It is called variously the interfacial zone (Bracker & Littlefield, 1973), haustorial sheath (Bracker & Littlefield, 1973), zone of apposition (Peyton & Bowen, 1963; Chou, 1970), encapsulation (Erhlich & Erhlich, 1963, 1971a,b; Berlin & Bowen, 1964; Heath & Heath, 1971; Heath, 1972) and extrahaustorial matrix (Bushnell, 1972). In vesicular-arbuscular and ericoid mycorrhizas there is evidence that the interfacial matrix is of host origin, and although there are no ultracytochemical studies of arbutoid mycorrhizas, this matrix structurally resembles that found in other intracellular mycorrhizal associations. In fungal parasites the haustorial sheath is often believed to be produced as a combination of fungal and host material (Hickey & Coffey, 1977). This sheath has always been considered quite separately from that around intracellular mutualistic symbionts, partly because it is thought to play a role in host resistance. However, this material, whether it surrounds a fungal haustorium or an inter- or intracellular symbiont, must be important in the translocation of nutrients between the host and fungus.

By varying both the packing of different matrix polysaccharides (particularly pectin) and the amount and orientation of the cellulose microfibrils, the plant can control the physicochemical state of its cell walls (Northcote, 1972). It is therefore possible that the mycorrhizal fungus could use enzymic means to increase the ratio of pectic to cellulosic material, thus modifying the host cell wall (Duddridge & Read, 1984b). This would result in the formation of the involving layer in ectomycorrhizas and the interfacial matrix in intracellular mycorrhizal associations. Conversely, the work of Nylund (1981) shows that the zone of maturing cortical cells where ectomycorrhizal infection takes place (the mycorrhizal infection zone of Marks & Foster, 1973) is high in pectic substances before Hartig net development occurs. He suggests that this high pectin content makes the walls more susceptible to fungal penetration. As mentioned by Rayner, Watling & Frankland (Chapter 1), Harley & Smith (1983) favour the theory of Vanderplank (1978) whereby proteins in the fungal membrane copolymerise with wall-building enzymes of the host, thus interfering with the normal development of the cell wall, allowing the fungus to utilise host cell wall precursors. Cell wall precursors are just one of the possible sources of host carbohydrate available to the ectomycorrhizal symbiont.

Another source could be host photosynthate diverted to the fungus, possibly under the control of plant hormones produced by the fungal symbiont (see Harley & Smith, 1983). This phenomenon is well known in infections caused by biotrophic foliar pathogens such as rusts, producing characteristic 'green islands' on the infected leaves (Holligan, Chen, McGee & Lewis, 1974), but there is no evidence for the redirection of large quantities of host photosynthate in vesicular-arbuscular mycorrhizas (Harley & Smith, 1983). The use of radiotracer (^{14}C) techniques to follow the translocation of photosynthate from host to fungus has produced contradictory results. Some workers have shown an eight-fold (w/w) increase in photosynthate translocated to ectomycorrhizas compared with uninfected roots (Bevege, Bowen & Skinner, 1975), while others (Ahrens & Reid, 1973) have found no such difference.

In ectomycorrhizal associations one of the most striking features of the host–agaric interface is the apparent absence of cytoplasmic reaction to the presence of the intercellular fungus (Duddridge & Read, 1984a,b), such as that found in response to intracellular infection by mycorrhizal symbionts (Dörr & Kollman, 1969; Strullu, 1976a; Duddridge, 1980) and biotrophic fungal parasites (Ehrlich & Ehrlich, 1966; Littlefield & Bracker, 1972). There is no obvious increase in the type or number of organelles in the cells adjacent to the Hartig net, although paramural bodies (Marchant & Robards, 1968) are frequently found (Atkinson, 1976; Duddridge, 1980). In higher plants, paramural bodies are often found in cells where active wall synthesis is taking place, suggesting that they are involved in wall formation (Cox & Juniper, 1973). They have also been associated with the production of wall ingrowths in transfer cells (Gunning & Pate, 1969a,b; Kelley, 1969). Previous workers have observed that Hartig net development only occurs between cortical cells which have 'reached or almost reached their full dimensions' (Clowes, 1951; Foster & Marks, 1966; Chilvers & Pryor, 1965; Harley & Smith, 1983). The presence of paramural bodies in cells that are not actively growing suggests that they may be implicated in the formation of the involving layer and the transport of cell wall precursors to the fungus.

Other sources of host carbohydrate for ectomycorrhizas are root exudates (Foster, 1981) and the contents of senescent host cells, the walls of which are eventually incorporated into the sheath. The latter include root cap cells which are still produced in ectomycorrhizal apices (Clowes, 1981) and outer cortical cells. There also appears to be evidence that in the post-Hartig net zone of Atkinson (1976), where the cortex begins to senesce, the agaric symbiont invades the moribund host cells (Atkinson, 1976; Duddridge, 1980; Nylund & Unestam, 1982). The latter workers suggest that 'the fungus starts to behave as a saprophyte where the cortical tissue becomes senescent'. This is just another example of tight nutrient cycling seen in many mycorrhizal associations when the symbiosis breaks down.

There has been some confusion in the literature concerning the viability of the cortical cells adjacent to the Hartig net, some workers reporting that adjacent host cells contain either no cytoplasm or show signs of cytoplasmic disorganisation (Foster & Marks, 1966; Strullu & Gourret, 1973; Strullu & Gerrault, 1977). Others believed that the majority of the cells adjacent to the Hartig net were dead (Scannerini, 1968; Hofsten, 1969). It is likely that these workers were looking at senescent cells either in the post-Hartig net zone (Atkinson, 1976; Harley & Smith, 1983) or in moribund mycorrhizas. Recent ultrastructural studies have shown that in healthy ectomycorrhizal associations, the cortical cells adjacent to the Hartig net are viable, confirming that translocation of substances between the symbionts occurs across a live interface (Atkinson, 1976; Duddridge & Read, 1984*a*,*b*). Nylund (1980) also made the observation that symplastic continuity is maintained during the development of the Hartig net in *Picea abies* although this was not found to be the case in *Pinus strobus* (Piché *et al.*, 1983).

The presence of substances in the immediate proximity of the root, such as sloughed off cells and root exudates, is known to attract a population of rhizosphere organisms, including a wide range of fungal saprotrophs. This has led some workers to suggest that ectomycorrhizal agarics may have evolved from root-surface saprotrophs (Harley, 1948; Cooke & Whipps, 1980), becoming more and more intimate with the outer root cells until they occupied an intercellular position in which they obtained not only root exudates and substances 'leaked' from cells but actively obtained soluble carbohydrate host wall precursors. An opposing view is put forward by Garrett (1970) who proposed that they could have evolved from specialised root parasites, adopting the intercellular habit to avoid encountering host defences.

There is now overwhelming ultrastructural evidence to show that the major translocation of substances between agaric symbionts and their ectomycorrhizal and arbutoid hosts occurs across a live interface. There have been as yet no published ultrastructural studies on the localisation of enzymes at the host–agaric interface in these mycorrhizal associations, as there has in vesicular-arbuscular mycorrhizas for enzymes probably involved in either active transport (ATPase) (Marx *et al.*, 1982) or the metabolism of phosphate (alkaline and acid phosphatases) (Gianinazzi-Pearson & Gianinazzi, 1978; Gianinazzi, Gianinazzi-Pearson & Dexheimer, 1979). In host cells infected by either vesicular-arbuscular mycorrhizas (Marx *et al.*, 1982) or biotrophic parasites (Spencer-Phillips & Gay, 1981), plasmalemma-bound ATPase (which in non-infected cells is concentrated on the peripheral plasmalemma) becomes localised on the invaginated membrane surrounding either the finer arbuscular branches or haustoria. Although the invaginated membrane round the haustoria of biotrophic parasites has been shown to be morphogenetically and cytochemically different (i.e. changes in thickness, stability and composition and staining properties) from the non-invaginated membrane (Littlefield & Bracker, 1972; Gil & Gay, 1977; Bracker & Littlefield, 1973), this is not found to be the case in vesicular-arbuscular mycorrhizas (Bonfante-Fasolo *et al.*, 1981).

There are few biotrophic parasites (and certainly no agarics) that exhibit purely intercellular infection of the kind found in ectomycorrhizas. *Peronospora trifoliorum* is an example of such a fungus and is discussed later with reference to transfer cells. Other intercellular biotrophic parasites are usually included in a group called hemi-biotrophs (Luttrel, 1974; Cooke & Whipps, 1980). In these associations (e.g. *Venturia inaequalis* on apple), the intercellular habit, during which the fungus remains biotrophic, is transient and necrotrophy arises with the onset of asexual reproduction.

In mycorrhizas of *Monotropa* many early workers observing the dissolution of the 'haustorial' tip (Francke, 1934; Björkman, 1960; Boullard, 1968) believed that the major exchange of nutrients between the fungus and host occurred when the fungal peg discharged its contents into the cell. This so-called digestion or lysis of the fungus in the host cell was termed 'ptophagy'. It was thought that the apex of the hypha burst and that fungal cytoplasm became mixed with that of the higher plant in the form of 'ptysomes' or digestion bodies. However, more recent work (Duddridge & Read, 1982*b*; Robertson & Robertson, 1982) has shown that the membranous sac is not the swollen fungal protoplast but an extension of the invaginated host plasmalemma and that the fungal cytoplasm is not released directly into the host cell. These last stages of peg breakdown are unlikely to be of great importance in the nutrition of the host, since by this time the shoot is fully differentiated. This phase is thus seen as one of degeneration of the transfer cells rather than significant nutrient transfer (Duddridge & Read, 1982*b*). The major transfer of materials from fungus to host is thought to occur in specialised cells in the outer cortex which have the characteristics of transfer cells and which develop during the phase of maximum shoot growth. Thus, a mechanism has evolved in *Monotropa* which enables it rapidly to obtain carbohydrates via its mycorrhizal partner during the few weeks in the summer when its flowering scape is being produced.

Transfer cells are formed in a wide variety of plants, in situations where rapid and intensive short-distance solute transport has been observed or assumed (Gunning & Pate, 1969a,b; Pate & Gunning, 1972), although it is still not clear exactly how they function. Wall ingrowths develop on the wall adjacent to which the translocation of solutes will occur. The extent of the wall protruberances is proportional to the intensity of the transmembrane flux (Maier & Maier, 1972). In healthy plants they are normally associated with the shoot vascular system and secretory glands (Fahn & Rachmilevitz, 1970; Browning & Gunning, 1977). However, transfer cells can be induced by a variety of stimuli, both biotic and abiotic. Abiotic factors include nutrient deficiency (MacKenzie, 1983), particularly iron deficiency (Kramer *et al.*, 1980), and high salt concentrations (Kramer *et al.*, 1977; Kramer, 1978; Kramer, Anderson & Preston, 1978). Biotic stimuli include the infection of roots by rhizobia (Pate, Gunning & Briarty, 1969; Briarty, 1978) or root-knot nematodes (Christie, 1936; Knisberg & Nielson, 1958; Huang & Maggenti, 1969; Paulson & Webster, 1970; Jones & Northcote, 1972a,b) and the infection of clover leaves by *Peronospora trifoliorum*, an intercellular biotrophic parasite. However, in the latter association the development of transfer cells is unusual in that wall ingrowths develop in the intercellular fungal wall and not on the adjacent host wall (Martin & Stuteville, 1975). Until recently no transfer cells have been observed to develop in response to mycorrhizal infection. However, there is now evidence for the induction of transfer cells by mycorrhizal agarics in both monotropoid mycorrhizas (Lutz & Sjolund, 1973; Duddridge & Read, 1982b; Robertson & Robertson, 1982) and ectomycorrhizas (Ashford & Allaway, 1982). It is no doubt only a matter of time before transfer cells are observed in other mycorrhizal associations.

One question that has received little attention in mycorrhizal research is how mycorrhizal symbionts recognise each other. The attraction of fungi to roots is not limited to mycorrhizal symbionts. Many fungal saprotrophs live in the rhizosphere on nutrients leaked from the roots, root exudates and sloughed-off cells. Cell junctions are thought to be sites of exudation and this may explain the trend for hyphae of many root-inhabiting fungi, both mycorrhizal (Duddridge, 1980; Brown & Sinclair, 1981) and parasitic (Garrett, 1956), to grow along the indentations. Thus the arrival of a fungus in the rhizosphere and the initial spread on the root surface is not a unique character of mycorrhizal symbionts. However, within the root population is a group of fungi which is selectively stimulated to form mycorrhizas. Harley & Smith (1983) point out that recognition between mycorrhizal symbionts is unlikely to be a 'once and for all' process but the result of a sequence of events.

One of the earliest events in this sequence is the initial contact between the fungus and host surface. There is evidence that this has to be 'tight' to ensure the successful development of the association in both biotrophic pathogens (Staples & Macko, 1980) and mycorrhizal symbionts (Theodorou & Bowen, 1971). On non-hosts hyphae only adhere loosely. Electron-dense fibrillar material has been observed around the hyphae and appressoria of pathogenic (Murray & Maxwell, 1975; Evans, Stempen & Stewart, 1981) and mycorrhizal (Duddridge, 1980; Bonfante-Fasolo, Berta & Gianinazzi-Pearson, 1981) fungi and is thought to be involved in attachment and possibly recognition between the associates. Initial contact between the mycorrhizal associates and the period immediately preceding this stage are vitally important in the recognition process and yet they have received little attention. It is essential to determine the actual morphological location of primary penetration and the constitution and reactions of the cell walls of both host and fungus in this position. No evidence has been obtained on the production of carbohydrases by mycorrhizal hyphae during these early stages. It is therefore important that studies should be carried out on the chemical, ultrastructural and enzymic changes that occur at the hyphal apex.

In ectomycorrhizas the initial spread of the fungus on the root surface is followed by intercellular penetration, which results in the alteration of both the mode of growth of the fungus and the development of the adjacent host cell wall (Melin, 1923; Strullu, 1976*b*, 1979; Atkinson, 1976; Giltrap, 1979; Nylund & Unestam, 1982; Duddridge & Read, 1984*b*) and which is itself a process of recognition (Harley & Smith, 1983). In intracellular mycorrhizal associations, an equivalent process is the development of the interfacial matrix between the fungus and the host plasmalemma. These morphological changes which occur at the host–fungus interface are obvious manifestations of molecular interactions that are occurring between the symbionts. This is another area of mycorrhizal research where more work needs to be done to elucidate the processes that are involved in the development of the host–fungus interface. Whatever mechanism exists it must involve the interactions of surface molecules, enzymes and possibly lectins at this interface. Lectins are plant proteins or glycoproteins, located on or in the host cell wall or membrane, which bind to specific oligosaccharide determinants on the cell surface of the other symbiont, causing conformational changes in the latter which result in a specific response. They have been increasingly implicated in the process of recognition in plant–microbe interactions (Albersheim & Anderson-Prouty, 1975; Callow, 1977; Clarke & Knox, 1978; Sequeira, 1978; Schmidt, 1979). However, concrete evidence of their involvement in recognition is lacking

(Reisert, 1981) except in a few cases (Hamblin & Kent, 1973; Rosen *et al.*, 1973, 1974; Bohlool & Schmidt, 1974; Ray, Shinnick & Lerner, 1979).

The search for specific molecules on the cell surface that may play a part in recognition between agarics and their mycorrhizal hosts may be helped by the fact that a single species of agaric may form mycorrhizas with a large number of diverse hosts, while another may be restricted to a particular genus (Harley & Smith, 1983). Many ectomycorrhizal agaric symbionts can also form ectendomycorrhizas on arbutoid hosts. Future research must therefore be directed towards a comprehensive characterisation of the host–agaric interface from the earliest stage of contact, including ultracytological studies and the localisation of wall enzymes, proteins and other surface molecules that may play a part in the recognition process.

Acknowledgement. Previously published illustrations are reproduced with the permission of *The New Phytologist.*

References

Ahrens, J. R. & Reid, C. P. P. (1973). Distribution of ^{14}C-labelled metabolites in mycorrhizal and non-mycorrhizal lodgepole pine seedlings. *Canadian Journal of Botany*, **51**, 1029–35.

Albersheim, P. & Anderson-Prouty, A. J. (1975). Carbohydrates, proteins, cell surfaces and the biochemistry of pathogenesis. *Annual Review of Plant Physiology*, **26**, 31–52.

Allen, F. H., Coffey, M. D. & Heath, M. (1979). Plasmolysis of rusted flax: a fine structural study of the host–pathogen interface. *Canadian Journal of Botany*, **57**, 1528–33.

Ashford, A. E. & Allaway, W. G. (1982). A sheathing mycorrhiza on *Pisonia grandis* R. Br. (Nyctaginaceae) with development of transfer cells rather than a Hartig net. *New Phytologist*, **90**, 511–19.

Atkinson, M. A. (1976). *The fine structure of mycorrhizas.* D.Phil. thesis, Oxford University, UK.

Bateman, D. F. (1976). Plant cell wall hydrolysis. In *Biochemical Aspects of Plant–Parasite Relationships*, ed. J. Friend & D. R. Threlfall, pp. 79–106. London and New York: Academic Press.

Beakes, G. W., Singh, H. & Dickinson, C. H. (1982). Ultrastructure of the host–pathogen interface of *Peronospora viciae* in cultivars of pea which show different susceptibilities. *Plant Pathology*, **31**, 343–54.

Berlin, J. D. & Bowen, C. C. (1964). The host–parasite interface of *Albugo candida* on *Raphanus sativus. American Journal of Botany*, **51**, 445–52.

Bevege, D. I., Bowen, G. D. & Skinner, M. F. (1975). Comparative carbohydrate physiology of ecto- and endo-mycorrhizas. In *Endomycorrhizas*, ed. F. E. Sanders, B. Mosse & P. B. Tinker, pp. 149–74. London and New York: Academic Press.

Björkman, E. (1960). *Monotropa hypopitys* L. an epiparasite on tree roots. *Physiologia Plantarum,* **13**, 308–29.

Bohlool, B. B. & Schmidt, E. L. (1974). Lectins as rhizobial recognition signals. *Science,* **185**, 269–71.

Bonfante-Fasolo, P., Berta, G. & Gianinazzi-Pearson, V. (1981). Ultrastructural aspects of endomycorrhizas in the Ericaceae. II. Host–endophyte relationships in *Vaccinium myrtillus* L. *New Phytologist,* **89**, 219–24.

Bonfante-Fasolo, P., Dexheimer, J., Gianinazzi, S., Gianinazzi-Pearson, V. & Scannerini, S. (1981). Cytochemical modifications in the host–fungus interface during intracellular interactions in vesicular-arbuscular mycorrhiza. *Plant Science Letters,* **22**, 13–21.

Bonfante-Fasolo, P. & Gianinazzi-Pearson, V. (1982). Ultrastructural aspects of endomycorrhiza in the Ericaceae. III. Morphology of the dissociated symbionts and modifications occurring during their reassociation in axenic culture. *New Phytologist,* **91**, 691–704.

Bonfante-Fasolo, P. & Scannerini, S. (1977*a*). A cytological study of the vesicular-arbuscular mycorrhiza in '*Ornithogalum umbellatum*' L. *Allionia,* **22**, 5–21.

Bonfante-Fasolo, P. & Scannerini, S. (1977*b*). Cytological observations on the mycorrhiza *Endogone flammicorona* × *Pinus strobus. Allionia,* **22**, 23–34.

Bonnett, H. T. (1968). The root epidermis: fine structure and function. *Journal of Cell Biology,* **37**, 199–205.

Boullard, B. (1968). *Les Mycorrhizes (Monographie* 2). Paris: Masson et Cie.

Bracker, C. E. & Littlefield, L. G. (1973). Structural concepts of the host–pathogen interface. In *Fungal Pathogenicity and the Plant's Response,* ed. R. J. W. Byrde & C. V. Cutting, pp. 159–313. London and New York: Academic Press.

Briarty, L. G. (1978). The development of root nodule xylem transfer cells in *Trifolium repens. Journal of Experimental Botany,* **29**, 735–47.

Brown, A. C. & Sinclair, W. A. (1981). Colonisation and infection of primary roots of Douglas fir seedlings by the ectomycorrhizal fungus *Laccaria laccata. Forest Science,* **207**, 111–24.

Browning, A. J. & Gunning, B. E. S. (1977). An ultrastructural and cytochemical study of the wall-membrane apparatus using freeze-substitution. *Protoplasma,* **93**, 7–26.

Burgeff, H. (1932). *Saprophytismus und Symbiose.* Jena: Gustav Fisher.

Bushnell, W. R. (1972). Physiology of fungal haustoria. *Annual Review of Phytopathology,* **10**, 151–76.

Callow, J. A. (1977). Recognition, resistance and the role of plant lectins in host–parasite interactions. *Advances in Botanical Research,* **4**, 1–49.

Campbell, E. (1971). Notes on the fungal association of two *Monotropa* species in Michigan. *Michigan Botanist,* **10**, 63–7.

Carling, D. E., White, J. A. & Brown, M. F. (1977). The influence of fixation procedure on the ultrastructure of the host–endophyte interface of vesicular-arbuscular mycorrhizae. *Canadian Journal of Botany,* **55**, 48–51.

Chilvers, G. A. (1968). Low power electron microscopy of the root cap region of eucalypt mycorrhizas. *New Phytologist,* **67**, 663–5.

Chilvers, G. A. & Pryor, L. D. (1965). The structure of eucalypt mycorrhizas. *Australian Journal of Botany,* **13**, 245–9.

Chong, J. & Harder, D. E. (1980). Ultrastructure of haustorium development in *Puccinia coronata avenae.* I. Cytochemistry and electron probe X-ray analysis of the haustorial ring. *Canadian Journal of Botany,* **58**, 2496–505.

Chou, C. R. (1970). An electron-microscope study of host penetration and early stages of haustorium formation of *Peronospora parasitica* (Fr.) Tul. on cabbage cotyledons. *Annals of Botany*, **34**, 189–204.

Christie, J. R. (1936). The development of root-knot nematode infections. *Phytopathology*, **26**, 1–22.

Clarke, A. E. & Knox, R. B. (1978). Cell recognition in flowering plants. *Quarterly Review of Biology*, **53**, 3–28.

Clowes, F. A. L. (1951). The structure of mycorrhizal roots of *Fagus sylvatica*. *New Phytologist*, **50**, 1–16.

Clowes, F. A. L. (1981). Cell proliferation in ectotrophic mycorrhizas of *Fagus sylvatica* L. *New Phytologist*, **87**, 457–555.

Coffey, M. D. (1975). Ultrastructural features of the haustorial apparatus of the white blister fungus *Albugo candida*. *Canadian Journal of Botany*, **53**, 1285–99.

Coffey, M. D. (1976). Flax rust resistance involving the K gene: an ultrastructural survey. *Canadian Journal of Botany*, **54**, 1443–57.

Cooke, R. C. & Whipps, J. M. (1980). The evolution of modes of nutrition in fungi parasitic on terrestrial plants. *Biological Review of the Cambridge Philosophical Society*, **55**, 341–62.

Courtoy, R. & Simar, L. J. (1974). Importance of controls for the demonstration of carbohydrates in electron microscopy with the silver methenamine or the thiocarbohydrazide – silver proteinate method. *Journal of Microscopy*, **100**, 199–211.

Cox, G. C. & Juniper, B. E. (1973). Autoradiographic evidence for paramural-body function. *Nature*, **243**, 116–17.

Cox, G. C. & Sanders, F. E. (1974). Ultrastructure of the host–fungus interface in a vesicular-arbuscular mycorrhiza. *New Phytologist*, **73**, 901–12.

Debaud, J. C., Pepin, R. & Bruchet, G. (1981). Ultrastructure des ectomycorrhizes synthétiques à *Hebeloma alpinum* et *Hebeloma marginatalum* de *Dryas octopetala*. *Canadian Journal of Botany*, **59**, 2160–6.

Dexheimer, J., Gianinazzi, S. & Gianinazzi-Pearson, V. (1979). Ultrastructural cytochemistry of the host–fungus interface in the endomycorrhizal association *Glomus mosseae/Allium cepa*. *Zeitschrift für Pflanzenphysiologie*, **86**, 189–201.

Dörr, I. & Kollman, R. (1969). Fine structure of mycorrhiza in *Neottia nidus – avis* (L) L. C. Rich (Orchidaceae). *Planta*, **89**, 372–5.

Duddridge, J. A. (1980). *A comparative ultrastructural analysis of a range of mycorrhizal associations*. Ph.D. thesis, University of Sheffield.

Duddridge, J. A. & Read, D. J. (1982a). An ultrastructural analysis of the development of mycorrhizas in *Rhododendron ponticum*. *Canadian Journal of Botany*, **60**, 2345–56.

Duddridge, J. A. & Read, D. J. (1982b). An ultrastructural analysis of the development of mycorrhizas in *Monotropa hypopitys* L. *New Phytologist*, **92**, 203–14.

Duddridge, J. A. & Read, D. J. (1984a). The development and ultrastructure of ectomycorrhizas. I. Ectomycorrhizal development on pine in the field. *New Phytologist*, **96**, 565–73.

Duddridge, J. A. & Read, D. J. (1984b). The development and ultrastructure of ectomycorrhizas. II. Ectomycorrhizal development on pine *in vitro*. *New Phytologist*, **96**, 575–82.

Duddridge, J. A. & Read, D. J. (1984c). Modification of the host–fungus interface in mycorrhizas synthesised between *Suillus bovinus* (Fr.) O. Kuntz. and *Pinus sylvestris* L. *New Phytologist*, **96**, 583–8.

Duddridge, J. A. & Sargent, J. A. (1978). A cytochemical study of lipolytic activity in

Bremia lactucae Regel. during germination of the conidium and penetration of the host. *Physiological Plant Pathology*, **12**, 289–98.

Ehrlich, H. G. & Ehrlich, M. A. (1963). Electron microscopy of the sheath surrounding the haustorium of *Erysiphe graminis*. *Phytopathology*, **53**, 1378–80.

Ehrlich, M. A. & Ehrlich, H. G. (1966). Ultrastructure of the hyphae and haustoria of *Phytophthora infestans* and hyphae of *P. parasitica*. *Canadian Journal of Botany*, **44**, 1495–503.

Ehrlich, M. A. & Ehrlich, H. G. (1971a). Fine structure of the host–parasite interfaces in mycoparasitism. *Annual Review of Phytopathology*, **9**, 155–84.

Ehrlich, M. A. & Ehrlich, H. G. (1971b). Fine structure of *Puccinia graminis* and the transfer of ¹⁴C from uredospores to *Triticum vulgare*. In *Morphological and Biochemical Events in Plant–Parasite Interaction*, ed. S. Akai & S. Ouchi, pp. 279–307. Tokyo, Japan: Mochizuki Publishing Co.

Englander, L. & Hull, R. J. (1980). Reciprocal transfer of nutrients between ericaceous plants and a *Clavaria* sp. *New Phytologist*, **84**, 661–7.

Evans, R. C., Stempen, H. & Stewart, S. J. (1981). Development of hyphal sheaths in *Bipolaris maydis* race T. *Canadian Journal of Botany*, **59**, 453–9.

Fahn, A. & Rachmilevitz, T. (1970). Ultrastructure and nectar secretion in *Lonicera japonica*. *Botanical Journal of the Linnean Society*, **63**, 51–6.

Faull, J. L. & Campbell, R. (1979). Ultrastructure of the interaction between take-all fungus and antagonistic bacteria. *Canadian Journal of Botany*, **57**, 1800–8.

Foster, R. C. (1981). The ultrastructure and histochemistry of the rhizosphere. *New Phytologist*, **89**, 263–73.

Foster, R. C. (1982). The fine structure of epidermal cell mucilages of roots. *New Phytologist*, **91**, 727–40.

Foster, R. C. & Marks, G. C. (1966). Observations on the mycorrhiza of forest trees. I. The fine structure of the mycorrhizas of *Pinus radiata* D. Don. *Australian Journal of Biological Sciences*, **19**, 1027–38.

Foster, R. C. & Marks, G. C. (1967). Observations on the mycorrhiza of forest trees. II. The rhizosphere of *Pinus radiata* D. Don. *Australian Journal of Biological Sciences*, **20**, 915–26.

Francke, H. L. (1934). Beiträge zur Kenntnis der Mykorrhiza von *Monotropa hypopitys* L. Analyse und Synthese der Symbiose. *Flora (Jena)*, **129**, 1–52.

Fusconi, A. & Bonfante-Fasolo, P. (1984). Ultrastructural aspects of host–endophyte relationships in *Arbutus unedo* L. mycorrhizas. *New Phytologist*, **96**, 397–410.

Galun, M., Paran, N. & Ben-Shaul, Y. (1970). The fungus–alga association in the Lecanoraceae: an ultrastructural study. *New Phytologist*, **69**, 599–603.

Garrett, S. D. (1956). *The Biology of Root Infecting Fungi*. Cambridge University Press.

Garrett, S. D. (1970). *Pathogenic Root Infecting Fungi*. Cambridge University Press.

Gianinazzi, S., Gianinazzi-Pearson, V. & Dexheimer, J. (1979). Enzymic studies on the metabolism of vesicular arbuscular mycorrhizas. III. Ultrastructural localisation of acid and alkaline phosphatase in onion roots infected by *Glomus mosseae* (Nicol. & Gerd.). *New Phytologist*, **82**, 127–32.

Gianinazzi-Pearson, V. Gianinazzi, S. (1978). Enzymatic studies on the metabolism of vesicular arbuscular mycorrhizas.II. Soluble alkaline phosphatase specific to mycorrhizal infection in onion roots. *Physiological Plant Pathology*, **12**, 45–53.

Gil, F. & Gay, J. L. (1977). Ultrastructural and physiological properties of the host interfacial components of haustoria of *Erysiphe pisi* in vivo and in vitro. *Physiological Plant Pathology*, **10**, 1–12.

Giltrap, N. J. (1979). *Experimental studies on the establishment and stability of ectomycorrhizas*. Ph.D. thesis, University of Sheffield, UK.

Giltrap, N. J. & Lewis, D. H. (1982). Catabolic repression of the synthesis of pectin degrading enzymes of *Suillus luteus* (L. ex Fr) S. F. Gray and *Hebeloma oculatum* Bruchet. *New Phytologist*, **90**, 485–93.

Gunning, B. E. S. & Pate, J. S. (1969a). 'Transfer cells' – plant cells with wall ingrowths specialised in relation to short distance transport of solutes – their occurrence, structure and development. *Protoplasma*, **68**, 107–33.

Gunning, B. E. S. & Pate, J. S. (1969b). Cells with wall ingrowths (transfer cells) in the placenta of ferns. *Planta*, **87**, 271–4.

Hadley, G. (1975). Organization and fine structure of orchid mycorrhiza. In *Endomycorrhizas*, ed. F. E. Sanders, B. Mosse & P. B. Tinker, pp. 335–51. London and New York: Academic Press.

Hadley, G., Johnson, R. P. C. & John, D. A. (1971). Fine structure of the host–fungus interface in orchid mycorrhiza. *Planta*, **100**, 191–9.

Hamada, M. (1940). Studien über die Mykorrhiza von *Galeola septentrionalis* Reichb. f. Ein neuer Fall der Mykorrhiza-bildung durch intra-radicale Rhizomorpha. *Japanese Journal of Botany*, **10**, 151–212.

Hamblin, J. & Kent, S. P. (1973). Possible role of phytohaemagglutinin in *Phaseolus vulgaris* L. *Nature*, **245**, 28–30.

Harder, D. E. (1978). Comparative ultrastructure of the haustoria in uredial and pycnial infections of *Puccinia coronata avenae*. *Canadian Journal of Botany*, **56**, 214–24.

Harder, D. E., Rohringer, R., Samborski, D. J., Kim, W. K. & Chong, J. (1978). Electron microscopy of susceptible and resistant near isogenic (Sr6/sr6) lines of wheat infected by *Puccinia graminis tritici*. I. The host–pathogen interface in the compatible (Sr6/sr6) interaction. *Canadian Journal of Botany*, **56**, 2955–66.

Hardwick, N. V., Greenwood, A. D. & Wood, R. K. S. (1971). The fine structure of the haustorium of *Uromyces appendiculatus* in *Phaseolus vulgaris*. *Canadian Journal of Botany*, **49**, 383–90.

Harley, J. L. (1948). Mycorrhiza and soil ecology. *Biological Reviews*, **23**, 127–58.

Harley, J. L. (1969). *The Biology of Mycorrhiza*. 2nd edn. London: Leonard Hill.

Harley, J. L. & Smith, S. E. (1983). *Mycorrhizal Symbiosis*. London and New York: Academic Press.

Heath, M. C. (1972). Ultrastructure of host and non-host reactions to cowpea rust. *Phytopathology*, **62**, 27–38.

Heath, M. C. (1976). Ultrastructural and functional similarity of the haustorial neckband of rust fungi and the casparian strip of vascular plants. *Canadian Journal of Botany*, **54**, 2484–9.

Heath, M. C. & Heath, I. B. (1971). Ultrastructure of an immune and susceptible reaction of cowpea leaves to rust infection. *Physiological Plant Pathology*, **1**, 277–87.

Henderson, M. W. (1919). A comparative study of the structure and saprophytism of the Pyrolaceae and Monotropaceae with reference to their derivation from the Ericaceae. *Contributions from the Botanical Laboratory and the Morris Laboratory, University of Pennsylvania*, **5**, 42–51.

Hickey, E. L. & Coffey, M. O. (1977). A fine structural study of pea downy mildew fungus *Peronospora pisi* in its host *Pisum sativum*. *Canadian Journal of Botany*, **55**, 2845–58.

Higgins, V. J. & Lazarovits, G. L. (1978). Histopathogical and ultrastructural comparison of *Stemphylium sarcinaeforme* and *S. botryosum* on red clover foliage. *Canadian Journal of Botany*, **56**, 2079–108.

Hofsten, A. von (1969). The ultrastructure of mycorrhiza. I. Ectotrophic and ectendotrophic mycorrhiza of *Pinus sylvestris*. *Svensk Botanisk Tidskrift*, **63**, 455–63.

Hohl, H. R. & Stössel, P. (1976). Host–parasite interfaces in a resistant and susceptible cultivar of *Solanum tuberosum* inoculated with *Phytophthora infestans*. *Canadian Journal of Botany*, **54**, 900–12.

Holligan, P. M., Chen, C., McGee, E. E. M. & Lewis, D. H. (1974). Carbohydrate metabolism in healthy and rusted leaves of coltsfoot. *New Phytologist*, **73**, 881–8.

Huang, C. S. & Maggenti, A. R. (1969). Wall modification in developing giant cells of *Vicia faba* and *Cucumis sativus* induced by root knot nematode, *Meloidogyne javanica*. *Phytopathology*, **59**, 931–7.

Ingram, D. S., Sargent, J. A. & Tommerup, I. C. (1976). Structural aspects of infection by biotrophic fungi. In *Biochemical Aspects of Plant–Parasite Relationships*, ed. J. Friend & D. R. Threfall, pp. 43–78. London and New York: Academic Press.

Jacobs, J. B. & Ahmadjian, V. (1969). The ultrastructure of lichens. I. A general survey. *Journal of Phycology*, **5**, 227–40.

Jones, M. G. K. & Northcote, D. H. (1972a). Nematode induced syncytium – a multinucleate transfer cell. *Journal of Cell Science*, **10**, 789–809.

Jones, M. G. K. & Northcote, D. H. (1972b). Multinucleate transfer cells induced in *Coleus* roots by the root-knot nematode, *Meloidogyne arenaria*. *Protoplasma*, **75**, 381–95.

Kamienski, F. (1881). Die Vegetationsorganen der *Monotropa hypopitys* L. *Botanische Zeitung*, **39**, 458–61.

Kamienski, F. (1882). Les organes vegetatifs du *Monotropa hypopitys*. L. *Mémoires de la societé nationale des sciences naturelles et mathématiques de Cherbourg*, **24**, 5–40.

Kelley, C. (1969). Wall projections in the sporophyte and gametophyte of *Sphaerocarpos*. *Journal of Cell Biology*, **41**, 910–14.

Khan, A. H. (1972). Mycorrhizae in the Pakistan Ericales. *Pakistan Journal of Botany*, **4**, 183–94.

Kiffer, E. (1974). *Etude des champignons mycorrhiziens et de quelques souches associées à l'Epicea en Lorraine*. Ph.D. thesis, University of Nancy, France.

Knisberg, L. R. & Nielson, L. W. (1958). Pathogenesis of root-knot nematodes to the Porto Rico variety of sweet potato. *Phytopathology*, **48**, 30–9.

Kramar, U. (1901). Studie über die Mikorrhiza von *Pyrola rotundifolia*. *Bulletin of the International Academy of Sciences, Prague*, VI, 1901–15.

Kramer, D. (1978). Transfer cells in the epidermis of roots of the halophyte *Atriplex hastata* L. *Naturwissenschaften*, **65**, 339–474.

Kramer, D., Anderson, W. P. & Preston, J. (1978). Transfer cells in the root epidermis of *Atriplex hastata* L. as a response to salinity. A comparative cytological and X-ray microprobe investigation. *Australian Journal of Plant Physiology*, **5**, 739–47.

Kramer, D., Läuchli, A., Yeo, A. R. & Gullasch, J. (1977). Transfer cells in roots of *Phaseolus coccineus*: ultrastructure and possible function in exclusion of sodium from the shoot. *Annals of Botany*, **41**, 1031–41.

Kramer, D., Römheld, V., Landsberg, E. & Marschner, M. (1980). Induction of transfer cell formation by iron deficiency in the root epidermis of *Helianthus annuus* L. *Planta*, **147**, 335–9.

Kusano, S. (1911). *Gastrodia elata* and its symbiotic association with *Armillaria mellea*. *Journal of the College of Agriculture, University of Tokyo*, **4**, 1–66.

Lalonde, M. & Knowles, R. (1975). Ultrastructure, composition and biogenesis of the encapsulation material surrounding the endophyte in *Alnus crispa* var. *mollis* root nodules. *Canadian Journal of Botany*, **53**, 1951–71.

Lamb, R. J. (1974). Effect of D-glucose on utilization of simple carbon sources by ectomycorrhizal fungi. *Transactions of the British Mycological Society*, **63**, 295–306.

Largent, D. L., Sugihara, N. and Brinitzer, A. (1980). *Amanita gemmata*, a non-host specific mycorrhizal fungus of *Arctostaphylos manzanita*. *Mycologia*, **72**, 435–9.

Lindeberg, G. & Lindeberg, M. (1977). Pectinolytic ability of some mycorrhizal and saprophytic hymenomycetes. *Archives of Microbiology*, **115**, 9–12.

Ling-Lee, M., Chilvers, G. A. & Ashford, A. E. (1977a). A histochemical study of phenolic materials in mycorrhizal and uninfected roots of *Eucalyptus fastigata* Deane & Maiden. *New Phytologist*, **78**, 313–28.

Ling-Lee, M., Ashford, A. E. & Chilvers, G. A. (1977b). A histochemical study of polysaccharide distribution in Eucalypt mycorrhizas. *New Phytologist*, **78**, 329–35.

Littlefield, L. J. & Bracker, C. E. (1972). Ultrastructural specialisation at the host–pathogen interface in rust-infected flax. *Protoplasma*, **74**, 271–305.

Littlefield, L. J. & Heath, M. C. (1979). *Ultrastructure of Rust Fungi*. New York: Academic Press.

Lundeberg, G. (1970). Utilization of various nitrogen sources, in particular bound soil nitrogen by mycorrhizal fungi. *Studia Forestalia Suecica*, **79**, 1–95.

Luttrel, E. S. (1974). Parasitism of fungi on vascular plants. *Mycologia*, **66**, 1–15.

Lutz, R. W. & Sjolund, R. D. (1973). *Monotropa uniflora*: ultrastructural details of its mycorrhizal habit. *American Journal of Botany*, **60**, 339–45.

MacKenzie, K. A. D. (1983). Some aspects of the development of the endodermis and cortex of *Tilia cordata* and *Picea sitchensis*. In *Tree Root Systems and their Mycorrhizas*, ed. D. Atkinson, K. K. S. Bhat, M. P. Coutts, P. A. Mason & D. J. Read, pp. 147–53. The Hague: Martinus Nijhoff/Dr W. Junk.

Maeda, K. M. (1970). *An ultrastructural study of* Venturia inaequalis *(Cke) Wint. infection of* Malus *hosts*. M.Sc. thesis, Purdue University, Lafayette, Indiana.

Maier, K. & Maier, U. (1972). Localisation of beta-glycerophosphatase and Mg++ activated adenosine triphosphatase in a moss haustorium and the relation of these enzymes to the cell wall labyrinth. *Protoplasma*, **75**, 91–112.

Malajczuk, N., Molina, R. & Trappe, J. M. (1984). Ectomycorrhiza formation in *Eucalyptus*. II. The ultrastructure of compatible mycorrhizal fungi and associated roots. *New Phytologist*, **96**, 43–53.

Malibari, A. A. (1979). *Biology of ectomycorrhizas with special reference to their possible role in plant water relations*. Ph.D. thesis, University of Sheffield, UK.

Marchant, R. & Robards, A. W. (1968). Membrane systems associated with the plasmalemma of plant cells. *Annals of Botany*, **32**, 457–71.

Marks, G. C. & Foster, R. C. (1973). Structure, morphogenesis and ultrastructure of ectomycorrhizas. In *Ectomycorrhizae: Their Ecology and Physiology*, ed. G. C. Marks & T. T. Kozlowski, pp. 1–41. New York and London: Academic Press.

Martin, T. J. & Stuteville, D. L. (1975). Cell wall ingrowths of non-haustorial hyphae of *Peronospora trifoliorum*. *Phytopathology*, **65**, 638–9.

Marx, C., Dexheimer, J., Gianinazzi-Pearson, V. & Gianinazzi, S. (1982). Enzymatic studies on the metabolism of vesicular-arbuscular mycorrhizas. IV. Ultracytoenzymological evidence (ATP ase) for active transfer processes in the host–arbuscule interface. *New Phytologist*, **90**, 37–43.

Marx, D. H. & Zak, B. (1965). Effect of pH on mycorrhizal formation of slash pine in aseptic culture. *Forest Science*, **11**, 66–75.

McKeen, W. E. (1977a). *Fusarium* in barley and corn roots. *Canadian Journal of Botany*, **55**, 12–16.

McKeen, W. E. (1977*b*). Growth of *Pythium graminicola* in barley roots. *Canadian Journal of Botany*, **55**, 44–7.

Melin, E. (1923). Experimentelle Untersuchungen über die Konstitution und Ökologie der Mykorrhizen von *Pinus sylvestris* und *Picea abies*. *Mykologische Untersuchungen, Berlin*, **2**, 72–331.

Molina, R. (1981). Ectomycorrhizal specificity in the genus *Alnus*. *Canadian Journal of Botany*, **59**, 325–34.

Molina, R. & Trappe, J. M. (1982). Patterns of ectomycorrhizal specificity and potential among Pacific Northwest conifers and fungi. *Forest Science*, **28**, 423–58.

Murray, G. M. & Maxwell, D. P. (1975). Penetration of *Zea mays* by *Helminthosporium carbonum*. *Canadian Journal of Botany*, **53**, 2872–83.

Newcomb, W., Peterson, R. L., Callaham, D. & Torrey, J. (1978). Structure and host–actinomycete interactions in developing root nodules of *Comptonia peregrina*. *Canadian Journal of Botany*, **56**, 502–31.

Nieuwdorp, P. J. (1972). Some observations with light and electron microscope on the endotrophic mycorrhiza of orchids. *Acta Botanica Neerlandica*, **21**, 128–44.

Northcote, D. H. (1972). Chemistry of the plant cell wall. *Annual Review of Plant Physiology*, **23**, 113–32.

Nylund, J.-E. (1980). Symplastic continuity during Hartig net formation in Norway Spruce ectomycorrhizae. *New Phytologist*, **86**, 373–8.

Nylund, J.-E. (1981). *The formation of ectomycorrhiza in conifers: structural and physiological studies with special reference to the mycobiont*, Piloderma croceum Erikss and Hjortst. Ph.D. thesis, University of Uppsala, Sweden.

Nylund, J.-E., Kasimir, A. & Arveby, A. S. (1982). Cell wall penetration and papilla formation in senescent cortical cells during ectomycorrhiza synthesis in vitro. *Physiological Plant Pathology*, **21**, 71–3.

Nylund, J.-E. & Unestam, T. (1982). Structure and physiology of ectomycorrhizae. I. The process of mycorrhiza formation in Norway spruce in vitro. *New Phytologist*, **91**, 63–79.

Palmer, J. G. & Hacskaylo, E. (1970). Ectomycorrhizal fungi in pure culture. I. Growth on single carbon sources. *Physiologia Plantarum*, **23**, 1187–97.

Pate, J. S. & Gunning, B. E. S. (1972). Transfer cells. *Annual Review of Plant Physiology*, **23**, 173–96.

Pate, J. S., Gunning, B. E. S. & Briarty, L. G. (1969). Ultrastructure and functioning of the transport system of the leguminous root nodule. *Planta*, **85**, 11–34.

Paulson, R. E. & Webster, J. M. (1970). Giant cell formation in tomato roots caused by *Meloidogyne incognita* and *Meloidogyne hapla* (Nematoda) infection. A light and electron microscope study. *Canadian Journal of Botany*, **48**, 271–6.

Pegler, D. N. & Fiard, J. P. (1979). Taxonomy and ecology of *Lactarius* (Agaricales) in the Lesser Antilles. *Kew Bulletin*, **33**, 601–28.

Peterson, T. A., Mueller, W. C. & Englander, L. (1980). Anatomy and ultrastructure of a *Rhododendron* root–fungus association. *Canadian Journal of Botany*, **58**, 2421–33.

Peyton, G. A. & Bowen, C. C. (1963). The host–parasite interface of *Peronospora manshurica* on *Glycine max*. *American Journal of Botany*, **50**, 787–97.

Piché, Y., Peterson, R. L., Howarth, M. J. & Fortin, J.-A. (1983). A structural study of the interaction between the ectomycorrhizal fungus *Pisolithus tinctorius* and *Pinus strobus* roots. *Canadian Journal of Botany*, **61**, 1185–93.

Pigott, C. D. (1982). Survival of mycorrhiza formed by *Cenococcum graniforme* Fr. in dry soils. *New Phytologist*, **92**, 513–17.

Ramstedt, M. & Söderhäll, K. (1983). Protease, phenoloxidase and pectinase activities in mycorrhizal fungi. *Transactions of the British Mycological Society*, **81**, 157–61.

Ray, J., Shinnick, T. & Lerner, R. (1979). A mutation altering the function of a carbohydrate binding protein blocks cell–cell cohesion in developing *Dictyostelium discoideum*. *Nature*, **279**, 215–21.

Read, D. J. (1974). *Pezizella ericae* sp. nov. the perfect state of a typical mycorrhizal endophyte of Ericaceae. *Transactions of the British Mycological Society*, **63**, 381–3.

Read, D. J. (1983). The biology of mycorrhiza in the *Ericales*. *Canadian Journal of Botany*, **61**, 985–1004.

Reisert, P. (1981). Plant cell surface structure and recognition phenomena with reference to symbioses. *International Review of Cytology*, **12**, 71–112.

Robertson, D. C. & Robertson, J. A. (1982). Ultrastructure of *Pterospora andromedea* Nuttall and *Sarcodes saguinea* Torrey mycorrhizas. *New Phytologist*, **92**, 539–51.

Romell, L. G. (1939). The ecological problem of mycotrophy. *Ecology*, **20**, 163–7.

Rosen, S. D., Kafka, J. A., Simpson, D. L. & Barondes, S..H. (1973). Developmentally regulated carbohydrate-binding protein in *Dictyostelium discoideum*. *Proceedings of the National Academy of Science, USA*, **70**, 2554–7.

Rosen, S. D., Simpson, D. L., Rose, J. E. & Barondes, S. H. (1974). Carbohydrate-binding protein from *Polysphondylium pallidum* implicated in intercellular adhesion. *Nature*, **252**, 128 and 149–51.

Sagara, N. & Takayama, S. (1978). An example of the root system of *Galeola septentrionalis*. *Transactions of the Mycological Society of Japan*, **19**, 338–40.

Scannerini, S. (1968). Sull' ultrastruttura delle ectomicorrize. II. Ultrastruttura di una micorriza di ascomicete: *Tuber albidum* × *Pinus strobus* L. *Allionia*, **14**, 77–95.

Scannerini, S. & Bonfante-Fasolo, P. (1983). Comparative ultrastructural analysis of mycorrhizal associations. (Proceedings of the 5th North American Conference on Mycorrhizas.) *Canadian Journal of Botany*, **61**, 917–43.

Schmidt, E. L. (1979). Initiation of plant root–microbe interactions. *Annual Review of Microbiology*, **33**, 355–76.

Schnepf, E. & Pross, E. (1976). Differentiation and redifferentiation of a transfer cell: development of septal nectaries of *Aloe* and *Gasteria*. *Protoplasma*, **89**, 105–15.

Sequeira, L. (1978). Lectins and their role in host–pathogen specificity. *Annual Review of Phytopathology*, **16**, 453–81.

Seviour, R. J., Willing, R. R. & Chilvers, G. A. (1973). Basidiocarps associated with ericoid mycorrhizas. *New Phytologist*, **72**, 381–5.

Slocum, R. D. & Floyd, G. L. (1977). Light and electron microscopic investigations in the Dictyonemataceae (Basidiolichens). *Canadian Journal of Botany*, **55**, 2565–73.

Spencer-Phillips, P. T. N. & Gay, J. L. (1981). Domains of ATPase in plasmamembranes and transport through infected plant cells. *New Phytologist*, **89**, 393–400.

Stahmann, M. A. (1973). In discussion on paper by Williams, Aist & Bhattacharya on 'Host–parasite relations in cabbage clubroot'. In *Fungal Pathogenicity and the Plant's Response*, ed. R. J. W. Byrde & C. V. Cutting, pp. 156–7. London and New York: Academic Press.

Staples, R. & Macko, V. (1980). Formation of infection structures as recognition response in fungi. *Experimental Mycology*, **4**, 2–16.

Steinkamp, M. P., Martin, S. S., Hoefert, L. L. & Ruppel, E. G. (1979). Ultrastructure of lesions produced by *Cercospora beticola* in leaves of *Beta vulgaris*. *Physiological Plant Pathology*, **15**, 13–26.

Strullu, D. G. (1974). Etude ultrastructurale du réseau de Hartig d'une ectomycorrhize à Ascomycète de *Pseudotsuga menziesii* (Mirb.). *Comptes Rendus Hebdomadaire des Séances de l'Académie des Sciences, Paris, Série D*, **278**, 2139–42.

Strullu, D. G. (1976a). *Recherches de biologie et de microbiologie forestières.* Ph.D. thesis, University of Rennes, France.

Strullu, D. G. (1976b). Contribution à l'étude ultrastructurale des ectomycorrhizes à basidiomycètes de *Pseudotsuga menziesii* (Mirb). *Bulletin de la société botanique de France, Paris,* **123**, 5–16.

Strullu, D. G. (1979). The ultrastructure and spatial representation of the fungal mantle of ectomycorrhizae. *Canadian Journal of Botany,* **57**, 2319–24.

Strullu, D. G. & Gerrault, A. (1977). Etude des ectomycorhizes à Basidiomycètes et à Ascomycètes du *Betula pubescens* (Ehrh) en microscopie électronique. *Comptes Rendus Hebdomadaire des Séances de l'Académie des Sciences, Paris, Série D,* **284**, 2243–4.

Strullu, D. G. & Gourret, J. P. (1973). Etude des mycorhizes ectotrophes de *Pinus brutia* Ten. en microscopie électronique à balayage et à transmission. *Comptes Rendus Hebdomadaire des Séances de l'Académie des Sciences, Paris, Série D,* **277**, 1757–60.

Strullu, D. G. & Gourret, J. P. (1975). Ultrastructure et evolution du champignon symbotique des racines de *Dactylorchis maculata* (L.) Verm. *Journal de Microscopie,* **20**, 285–94.

Theodorou, G. & Bowen, G. D. (1971). Effect of non-host plants on the growth of mycorrhizal fungi of radiata pine. *Australian Forestry,* **35**, 17–22.

Thiéry, J. P. (1967). Mise en évidence des polysaccharides sur coupes fines en microscopie électronique. *Journal de Microscopie,* **6**, 987–1017.

Van Dyke, C. G. & Hooker, A. L. (1969). Ultrastructure of host and parasite in interactions of *Zea mays* with *Puccinia sorghi. Phytopathology,* **59**, 1934–46.

Vanderplank, J. E. (1978). *Genetic and Molecular Basis of Plant Pathogenesis.* Berlin, Heidelberg, New York: Springer Verlag.

Warcup, J. H. (1981). The mycorrhizal relationships of Australian orchids. *New Phytologist,* **87**, 371–81.

Woods, A. M. & Gay, J. L. (1983). Evidence for a neckband delimiting structural and physiological regions of the host plasmalemma associated with the haustoria of *Albugo candida. Physiological Plant Pathology,* **23**, 73–88.

Zak, B. (1976a). Pure culture synthesis of bearberry mycorrhizae. *Canadian Journal of Botany,* **12**, 1297–305.

Zak, B. (1976b). Pure culture synthesis of Pacific madrone ectendomycorrhizae. *Mycologia,* **68**, 362–9.

Zhuang, Y., Wang, Y., Zhang, W., Qin, M., Xie, Z., Chen, Y., Chen, X., Shao, Y. & Rulan, Z. (1983). A study on the source of secondary nutrients for *Gastrodia elata. Acta Botanica Yunnanica,* **5**, 83–90.

7

The challenge of the dolipore/parenthesome septum

ROYALL T. MOORE

Biology Department, University of Ulster, Coleraine, Northern Ireland BT52 1SA, UK

The challenge of its origin

Speculations on the origin of the dolipore/parenthesome (d/p) septum must be made in the context of the descent of the septomycete sporothallus. Although the general presence or absence of septa is the salient distinction between the phycomycetes and the septomycetes, little attention has been given to the possible phylogenetic significance of the fact that the septa are, respectively, imperforate and perforate. The phycomycetes are somatically coenocytic, there is little, if any, nuclear differentiation, and when septa are formed they are abscissional. In the septomycetes, however, the septa are not abscissional and they occur at regular intervals. Cytologically, the process of septation has become linked with the growth of the hyphal tip so that plasma membrane invagination is initiated at a constant rate (see Trinci, 1979) and, in some yet to be resolved way, the wall material being deposited in the infolded, ingrowing, tapered membrane mould does not become deposited across the centre. How such a centripetally invaginated cytokinetic membrane can result in a perforate septum is difficult to visualise (see Moore, 1965, 1975). Be that as it may, regular perforate septation underlies the advance of the higher fungi. Compartmentalisation facilitates heterokaryosis and eventually its discipline, in association with other evolutionary changes, restricts the nuclear number to a contrasting diad.

In the haplo-dikaryotic life cycle of these species, fertilisation is, as usual, initiated by plasmogamy but, uniquely, is only completed an indefinite number of cell cycles later when a block to karyogamy is removed in the meiocyte (ascus, ustospore, or basidium). The sporothallus is thus dikaryotic and exists between the two events that constitute syngamy. This strategy of being both haploid and functionally diploid at the same time necessitates

175

the development of means for sustaining the binucleate cell state. The development of the crozier would be a premier event. Initially it may have had a crochet-like repetition in which there was a chain of croziers formed as a consequence of the repetitive fusion (serial plasmogamy) of ultimate and antepenultimate cells. The result would be a number of randomly distributed asci such as occur in *Neosartorya* (*Aspergillus*) *fischeri* (Fig. 7.1). It is unknown what triggers and guides the concomitant formation of the fruiting body produced by the gametothallus but the primitive cleistothecium has no provision for the release of its ascospores. In the transition to the perithecium a number of basic changes occur. The penultimate cell of the crozier becomes paramount, polarising growth to form linear, ascogenous hyphae; its reversion to the meiocyte role occurs only when a number of such cells come to form a hymenium. The ultimate cell continues to fuse with the antepenultimate cell and functions as a shunt to allow reunion of the single nucleus of these cells (see Swart, 1969). We should also expect to find changes in the septa. The gametothallus, though regularly septate, is functionally coenocytic; the nuclei can migrate from cell to cell (Hunsley & Gooday, 1974) and the number of nuclei per cell is indeterminate. The septa sometimes have a ring or other configuration (Brenner & Carroll, 1968) of dense material around the pore but not sufficient to block it and stop nuclear migration (Burnett, 1968). For the limited dikaryotic sporothallus, however, a number of reports show

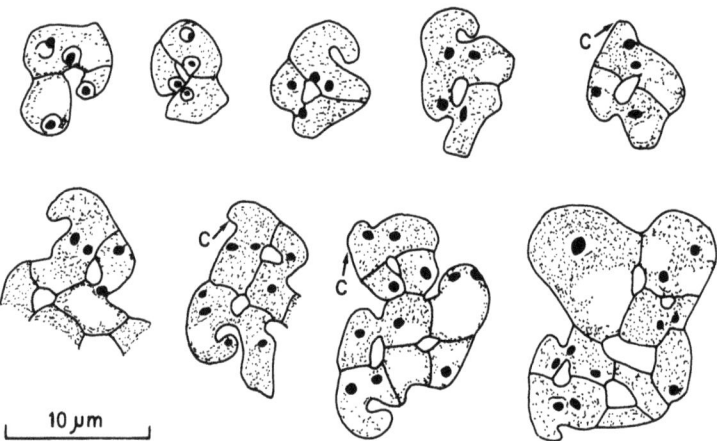

Fig. 7.1. Crozier (c) formation and serial plasmogamy in *Neosartorya fischeri*. A probable early stage in sporothallus evolution in which the dikaryotic condition is limited essentially to single cells. (Redrawn from Olive, 1944.)

micrographs of pore-occluding structures (Maret, 1972; Beckett, 1981; Rosing, 1981; Curry & Kimbrough, 1983; Steffens & Jones, 1983). Beckett (1981) critically studied the ascogenous hyphae of *Sordaria humana* and found a range of pore morphologies. At the base of the ascogenous hyphae the pores had relatively simple pore caps while variously complex pores with swollen, electropaque rims and fans of associated membrane cisternae were found at the apex, in the croziers, and in very young sporulating asci. Also evident in these micrographs are long, parallel profiles of endoplasmic reticulum (ER) similar in appearance to comparable ER arrays that appear to give rise to basidiomycete parenthesomes (Moore, 1975). Although there is little morphological similarity between the pore complexes of ascogenous hyphae and those of basidiomycete mycelium they are both, nevertheless, part of the sporothallus and each may be presumed to have a role in the regulation of the respective dikaryons. The extent to which the descent of the latter from the former is reflected in the structure, biochemistry and physiology of the respective septal complexes remains to be investigated.

Despite the ostiole and various types of ascal dehiscence, spore dispersal from perithecia is still inefficient. The apothecium, however, essentially everts the perithecium and changes the role of the gametothallus from producing containers for a limited sporothallus to one of producing underlying supports for it. This facilitates, or at least is accompanied by, two major advances in the cup fungi: one is the shift to the extensive production of air-borne meiospores (made possible by the exposed hymenium) and the associated very marked reduction in the dependence on conidia for dissemination; the other is the formation of much larger ascocarps that can support extensive, albeit still dependent, ascogenous mycelia.

The transition from the ascomycetes to the basidiomycetes is particularly difficult to model. Part of this transition is a specific difference in the apparent wall structure of the two groups. In the ascomycetes the septa generally appear homogeneous, yeast bud scars are continuous with the daughter cells, and yeast cell walls have a negative colour reaction with Diazonium Blue B (DBB), while in basidiomycetes the septa are trilaminate with a light layer between two dark ones, yeast bud scars are frayed, and yeast cell walls give a positive DBB reaction (Kreger-van Rij & Veenhuis, 1971; van der Walt & Hopsu-Havu, 1976; see also Moore, 1980*b*; Heath, Ashton & Rethoret, 1982). This fundamental difference in the septal ultrastructure appearance may be a maturation effect. *Neurospora crassa* septa from 5-day-old cultures studied by Hunsley & Gooday (1974) appeared as two dark lamellae separated by an electron-lucent area and the micrograph in their Fig. 15 is directly comparable, for example, to Butler & Bracker's (1970) micrographs of developing septa in *Thanatephorus cucumeris.* What is particularly significant about these observations is their implication that the early ontogenetic stages of septagenesis are common to both the ascomycetes and basidiomycetes, i.e. that the ingrowing septagenetic membrane initially produces a three-layered structure and that the generally observed differences in septal morphology are a consequence of later events. In the ascomycetes it would appear that the bounding electropaque layers undergo further development and become converted to additional electrolucent material (mostly chitin?) of the type found in the central layer (see Brenner & Carroll, 1968). I would suggest that the basidiomycete septum, on the other hand, may be neotonous. That is, that the liberation of the sporothallus and the associated evolutionary changes in the basic haplo-dikaryotic life cycle of these fungi may have been effected by retaining an essentially juvenile septum.

In the basidiomycetes, the d/p septum is fully expressed and has two basic states. In the monokaryon, mycelial with a few notable exceptions, it is labile and subject to degradation following plasmogamy and the cytoplasmic events of nuclear migration. The dikaryotic hyphae that ensue, however, have fulfilled septa that are stable. This latter, secondary, mycelium is the culmination of the septomycete sporothallus. It has a unique compatability system that permits virtually complete outbreeding accompanied by the capacity to form large, geotropic fruit bodies composed of variously modified tertiary hyphae which can produce and disperse spores in unprecedented numbers. The evolutionary appearance and radiation of the higher fungi (septomycetes) must be co-eval with that of the seed plants because, as Corner (1964) remarks, 'The fungi of today are those of coniferous and flowering forest that adapt themselves to other kinds of vegetation. Ferns and their like do not decay rapidly. The thickness of the Coal Measures, composed of the remains of fern-like plants, indicates that even at this stage of the world's history there were not many organisms able to remove their detritus. Sphagnum bogs are another example. Conifers, too, with resinous tissue are not so easily decayed as flowering plants. It seems indeed that it could not have been until their advent that the fungus world expanded. Flowering plants employed the animal and propagated the fungus. The flowering forest is both eminently edible and putrescible, wherein lies the other half of its success. Its remains do not accumulate. They are rotted down by fungus, animal, and bacterium and reincorporated in the soil.' The basidiomycetes, particularly the lignin-degrading, white-rot species, are among the major contributors to this breakdown process.

The nomenclature of the major taxa mentioned in the text follows that of Moore (1974, 1978, 1980*b*).

The challenge of its structure

The septum

Three basic structures constitute the d/p septal complex (Fig. 7.2): the barrel-shaped dolipore, the occlusions within the pore, and the parenthesomes on either side that delimit the pore domain. The morphology of the septum itself is essentially the same in monokaryons, dikaryons, and diploids (Casselton, Lewis & Marchant, 1971) throughout the Basidiomycota. (There is, however, some debate on which method of fixation gives the truest picture; see Hoch & Howard, 1981). Girbardt (1979), studying *Coriolus versicolor*, has shown that special cytoplasmic elements accumulate within the incipient septal area of the hyphal wall, that very quickly after nuclear separation there appears a peripheral band of 4–7-nm microfilaments, and that almost immediately after this band forms, the septum is initiated. Its ingrowth could be similar to that of the hyphal tip. In both sequences there is active membrane increase, associated vesicles (Raudaskoski, 1970; Grove, 1978; Hoch & Howard, 1980) and the deposition of primary wall material to the outside. At the hyphal tip this material is deposited on a constantly expanding dome (Wessels, Sietsma & Sonnenberg, 1983) while in septum formation (Fig. 7.3) it is deposited as a layer within the limited lumen formed by an invaginating membrane. Craig *et al.* (1979) labelled developing septa with *N*-acetyl-*N*[1-^{3}H]glucosamine and showed that the ingrowing pore lip was the most active site of chitin synthesis. Wessels & Sietsma (1981) pointed out the similarity between regenerating protoplasts, in which pure chitin is often one of the first components to appear, and the septagenetic invaginating membrane, in which homogeneous chitin is the first material to be deposited. This initial phase of development ends near the cell centre when the contracting pore lip starts to inflate into the mould of the dolipore. The next phase of development is the deposition of secondary wall material. This material, unlike that of the primary septal component, is deposited on all of the cell membrane and forms a continuous electron-dense layer on the lateral walls, along both sides of the electrolucent central disc, and into the developing dolipore (Fig. 7.4*a*). A probable third phase of development is indicated from an examination of a number of published electron micrographs

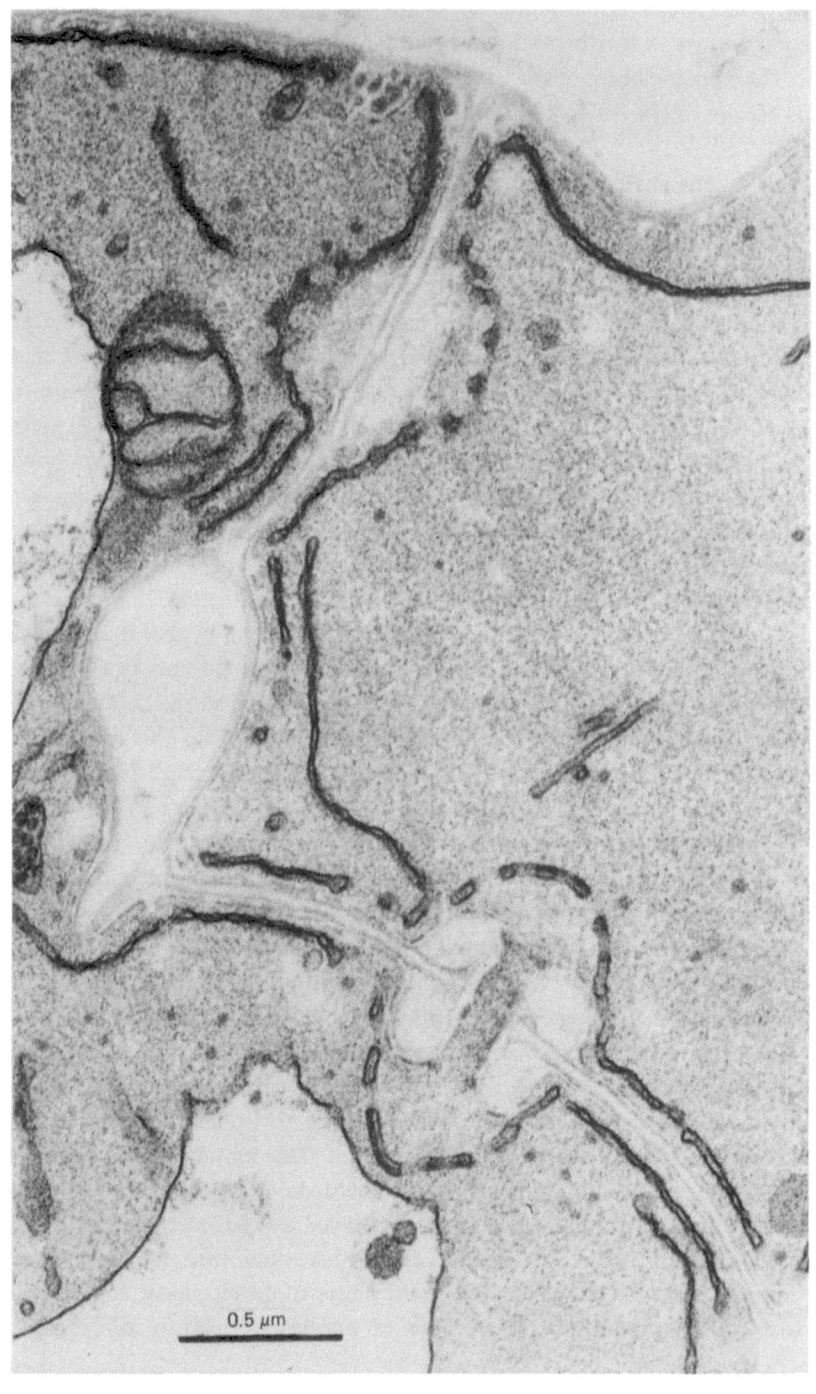

0.5 μm

(Setliff, MacDonald & Patton, 1972; Moore, 1975; Flegler, Hooper & Fields, 1976; Gull, 1976; Tu, Kimbrough & Aldrich, 1977; van der Valk, Marchant & Wessels, 1977; Wessels & Sietsma, 1979; Dykstra, 1982). These micrographs and diagrams (Moore, 1975; van der Valk *et al.*, 1977; Wessels & Sietsma, 1979) (Figs 7.3 and 7.4) show the central light layer extending most of the way into the dolipore, the dark bounding material

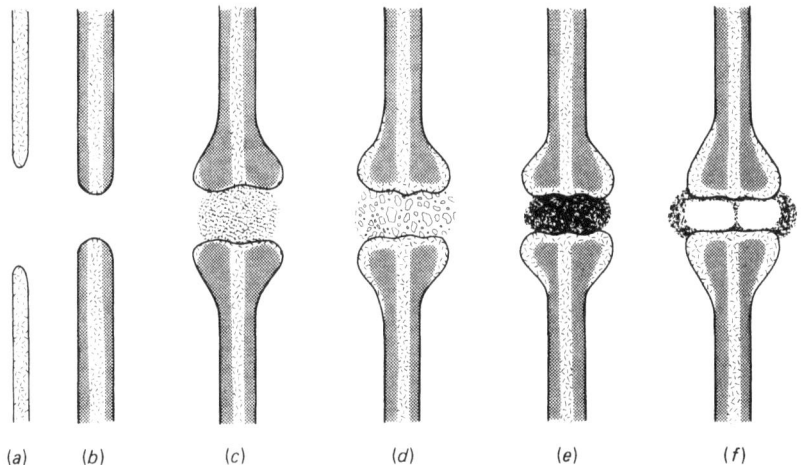

(*a*)   (*b*)   (*c*)   (*d*)   (*e*)   (*f*)

Fig. 7.3. Speculative diagram of dolipore formation. (*a*) Invaginating cytokinetic membrane as it approaches the centre. (*b*) The septum has become three-layered and the lip has closed upon a region of dense cytoplasm within the pore. (*c*) The impeded invaginating lip has flared to form two fronts, the material within the pore has become more compacted, and additional light material is starting to be deposited in the cusps of the developing dolipore and its middle. (*d*) There is now a jacket of light material around the dolipore and the material in the centre is becoming discernible with the electron microscope. (*e*) The centre material has become electropaque and compressed into a solid 8-shape. (*f*) The centre mass has split and been largely extruded to the orifices as a pair of occlusions; each moiety has a shape like a champagne cork and the pair of opposed bases produces either a narrow dark band in the centre of the dolipore or, if they separate, a light band. (Redrawn from Moore, 1975.)

Fig. 7.2. Section through a young clamp of *Polyporus biennis* showing typical dolipore/parenthesome (d/p) septum of the clamp (above) and the hypha (below). The parenthesomes are regularly perforate (note upper septum) and have several continuities with the extensive wall endoplasmic reticulum (ER). The septal shoulders and centre of the dolipore are three-layered and appear similar to juvenile ascomycete septa (Hunsley & Gooday, 1974), a condition that may be a phylogenetic result of neoteny.

continuing into and occupying something like two-thirds to three-quarters of the dolipore, and the outer portion of the dolipore, composed of electrolucent material. Fitting an ontogeny to these static pictures (Fig. 7.3) one could envisage, first, the 'firming up' of the light primary layer as invagination ceases (*a, b*), followed by the deposition of the dark material on both sides of the central layer and into the expanded lip (*c, d*), and, finally, (*e, f*), the renewed deposition of electrolucent material by that portion of the plasmalemma bounding the dolipore (this last is clearly evident in Moore, 1975 – Fig. 8). This final layer gives the dolipore its mature shape. It should be noted that the dolipore micrographs do not show a sharp demarcation between the core of secondary dark material and the jacket of light material; rather, the core–jacket interface appears, in section, to be variously dendroid or reticular.

There are differences in wall composition between monokaryons and dikaryons. One would intuitively suspect this from the fact that septa of

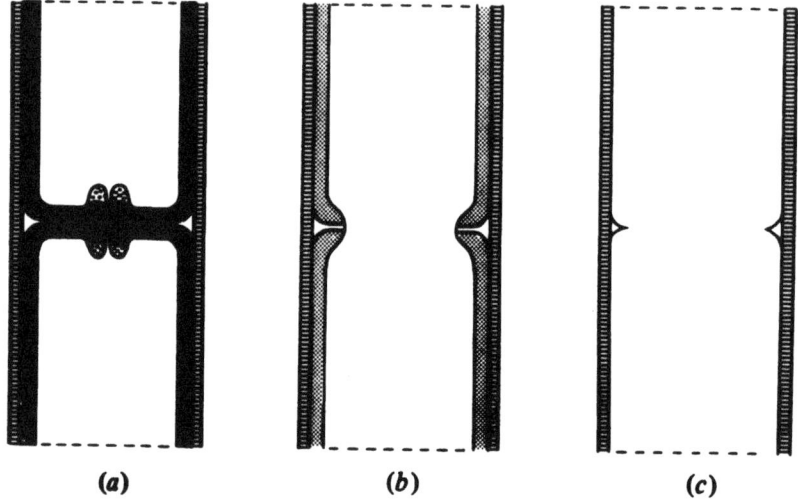

(a)   *(b)*   *(c)*

Fig. 7.4. Wall architecture in *Schizophyllum commune* illustrating dissolution of septa in monokaryotic hyphal preparations by *R*-glucanase and
chitinase. (*a*) before enzyme treatment: the central chitinous plate of the
septum is indicated as white; the dotted area indicates the septal swelling
around the pore; black corresponds to *R*-glucan–chitin; the hatched area
represents *S*-glucan. (*b*) After moderate treatment with the enzymes the
septum is dissolved and some *R*-glucan–chitin has disappeared from the
lateral wall. This situation probably corresponds to that occurring *in vivo*
during nuclear migration. (*c*) Prolonged enzyme treatment leaves only the
S-glucan layer and a chitinous ring from the perimeter of the septum.
(From Wessels & Sietsma, 1979.)

monokaryons break down in advance of nuclear migration (Giesy & Day, 1965; Mayfield, 1974) while there is no evidence of comparable dissociation of septa of dikaryons. Marchant (1978) chemically analysed the walls of a dikaryon of *Coprinus cinereus*, the walls of its constituent monokaryons and those of the stipe, and found marked differences in the ratios of the major components. (See Maret (1972) for a comprehensive analysis of the cell walls of an ascomycete.) The most detailed information on septal composition comes from studies of *Schizophyllum commune* (Wessels & Sietsma, 1979). These studies show the central layer to be a plate of compact chitin, the bounding layers and secondary wall to be composed of an *R*-glucan–chitin mixture, and the outer, primary, hyphal wall to be composed of *S*-glucan (Fig. 7.4*a*). The septa of monokaryons are readily attacked by a mixture of chitinase and *R*-glucanase (Fig. 4*b/c*) but the septa of the dikaryon are not affected by this treatment. In another study, Casselton *et al.* (1971) found that while common *A* heterokaryons had self-destructing septa, the common *A* diploid did not; further, these latter septa in mating with compatible monokaryons were indurate and prohibited nuclear migration. (Note: the dikaryotic d/p septal complex is, however, disassembled in the course of conidia formation in *Pleurotus cystidiosus*; Moore, 1977.) These sophisticated studies confirm that there is a critical difference between the dikaryotic sporothallus and its constituent haploids (see also Casselton, 1978). Comparable studies are now required of other basidiomycetes to provide some idea of the range of similarities and differences around these prime examples.

The occlusions

The occlusions occupy the channels of the dolipore and all but those of the Tremellineae have a similar morphology. In near-median sections there is a dark band in the middle of the channel, light areas on either side, and dark material in the orifices; this banding was noted by Moore & McAlear (1962) in the paper naming the dolipore and parenthesome and various subsequent authors have also noted this general pattern, e.g. Khan & Kimbrough (1979), Desole (1982). An integrated interpretation of these bands was given by Moore & Marchant (1972) based on a perfect median section. Their model (Fig. 7.5) visualised the occlusions as being shaped like champagne corks oriented base to base within the pore channel. It was later proposed (Moore, 1975) that this dual structure arose as a consequence of an impediment to cytokinesis in the centre of the cell. But is there any ultrastructural evidence for such cell division-associated material? First, 'the cytoplasm of most eukaryotic cells contains a cytoskeletal fabric formed of *microtubules* and of various types of *microfilaments*' (DeRobertis & DeRobertis, 1980; see also Marx, 1983). In plant and animal cells these become particularly evident during cell division. In plants (Hepler & Jackson, 1968; Fineran, Wild & Ingerfeld, 1982) the cell plate is formed by the coalescence of a plane of centrifugally fusing vesicles that arise in the post-telophase microtubular array; the microtubules occur in small clusters and within each cluster there is a local accumulation of an amorphous electropaque material. In animal cells there is a cleavage furrow formed by the indrawing of the plasma membrane and

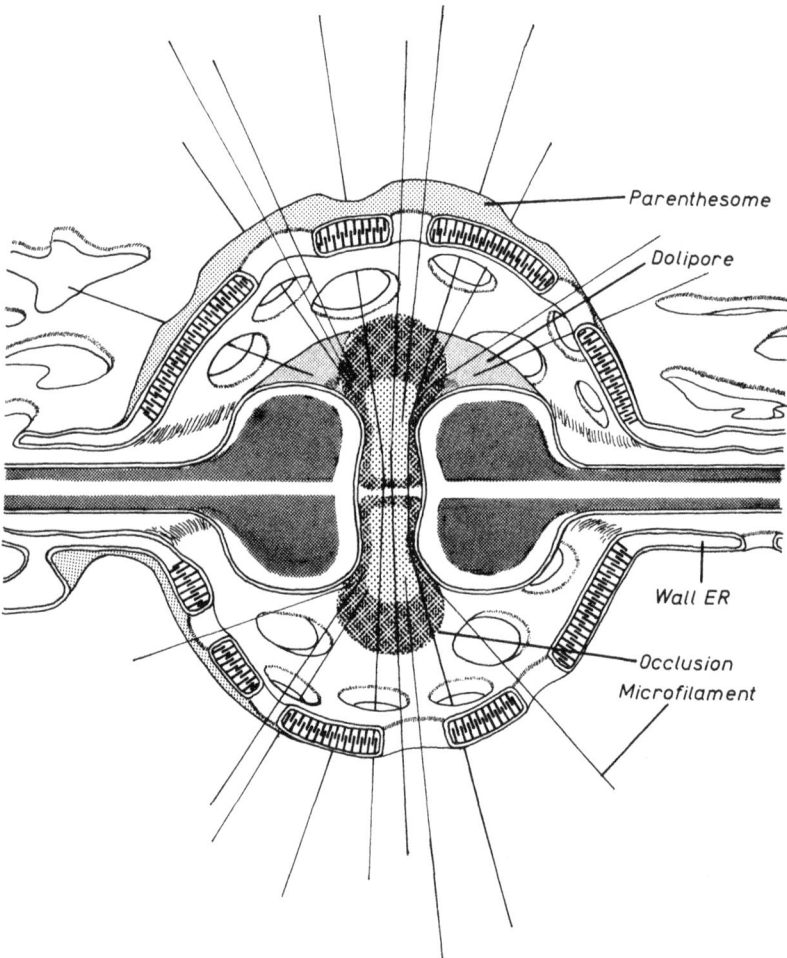

Fig. 7.5. Interpretation of homobasidiomycete d/p septum. (From Moore & Marchant, 1972.)

by centripetally fusing vesicles (Buck & Tisdale, 1962). The dense central midbody (Robbins & Gonatas, 1964) is composed of a compacted bundle of post-telophase microtubules and vesicle-derived osmiophilic material; the elimination of the midbody may be affected by symmetrical invagination of the plasma membrane in a sort of pincer movement. Microfilaments have been reported to participate in dolipore septagenesis (Moore & Marchant, 1972; Patton & Marchant, 1978a; Traquair & McKeen, 1978; Khan & Kimbrough, 1979; Hoch & Howard, 1980; Desole, 1982). It seems reasonable to postulate that the pursing of the pore could cause microfilament fasciculation. The greatest compaction would be within the closing pore circle and this could become of such a density that, though growth briefly continued, further membrane ingress was denied: the consequence would be the outline of the dolipore. As visualised (Fig. 7.3), the closing cytokinetic membrane (*a*) encounters an obstacle to further ingrowth (*b*) and flattens against it (*c*). This has two consequences. One is that continued membrane increase produces two growth fronts, one on either side of the obstacle, thereby establishing the mould of the dolipore (*c–e*). The second is that pressure on the impeding obstacle compacts it (*b–e*) and causes electropaque material to be extruded to either side (*f*). This substance could be similar to the microtubular associated material of plant and animal cell division; there is also the comparison with ascomycete septa (Curry & Kimbrough, 1983).

This model can also help to make more sense of the anomalous occlusions in the Tremellineae. These fungi are characterised by vesiculate parenthesomes tethered by microfilaments/-tubules to occlusions whose orificial component appears as a striated band (see Fig. 7.10). In the centre of the pore channel, however, there is still the median dark band with light areas on either side. The initial ontogeny could be the same with, perhaps, the addition/retention of the parenthesome microfilaments/-tubules. The extrusion of the compacted material would be combed through these resistant elements and be deposited at the orifices as a layered honeycomb.

The parenthesomes

The parenthesomes, the third component of the d/p septal complex, show the greatest variation in structure. They occupy and define the cytoplasmic domain on either side of the dolipore and they have partial continuity with the ER skirt that lies along the disc of the septum (Fig. 7.2). At the edge of the dolipore the two ER membranes diverge slightly and within the resultant widened lumen there generally appears an interposed layer that usually lies closer to the outer membrane (Bracker

& Butler, 1963; Ellis, Rogers & Mims, 1972; Moore & Marchant, 1972; Setliff *et al.*, 1972; Moore, 1975, 1978, 1984) (see Fig. 7.11). This central layer has been variously interpreted. Moore & Marchant (1972) thought it might be the juncture of some kind of nap arising from the inner surfaces of the bounding membranes (Fig. 7.5) while Moore & Patton's (1975) micrographs and interpretation suggested it was an interposed membrane. Sometimes the centre line is absent or poorly defined. Most papers, in some way or another, comment on this interposed layer and variously identify it; Tu & Kimbrough (1978) attach taxonomic significance to its several patterns in the *Rhizoctonia* complex. None of the available interpretations, however, is particularly satisfactory from the point of view of cell biology.

The parenthesome is not an easy organelle to study. Although there are four per clamp connection, hyphal cells are relatively long and good median sections are difficult to accumulate (e.g. Patton & Marchant, 1978*b*). Further, parenthesomes are small and cup- to saucer-shaped and although they have a fairly consistent appearance for any given sample there is the unanswered question of their responses to various fixation protocols, as well as the question of whether the internal structure for a given species is fixed or changes with time. No one, as far as I am aware, has isolated and studied parenthesome cell fractions electron microscopically or biochemically. All the information we have at present comes from transmission electron micrographs of thin sections. As already indicated, the internal structure results from a dilation of the ER lumen. It is hard to see how this increase in lumen width could induce either the formation of an internal membrane or a reinforcement image of interdigitated tips of processes arising on the inner surfaces of the bounding unit membranes. There is, however, an intermembrane phenomenon that mimics what we see in the parenthesome. This is the gap junction (Larsen, 1977; Peracchia, 1980) which forms between juxtaposed plasma membranes of animal cells (Figs 7.6 and 7.7*a*). In gap junctions the opposing membranes'... are glued to each other in discrete spots where intramembrane particles, protruding from opposite membrane surfaces, come in contact with each other to bridge the gap' (Peracchia, 1980, p. 102). These particles are seen in electron micrographs of freeze-fractured gaps as matching bumps and pits (Figs 7.7*b,c* and 7.8). Even though it is difficult to visualise any similarity in function between gap junctions (cell-to-cell communication) and parenthesomes (?) the molecular basis of the model is attractive. Present knowledge of biological membranes sees them as a kind of fluid mosaic of lipid and integral proteins (see DeRobertis & DeRobertis, 1980). It is not difficult to imagine that there could be proteins that were inserted into the respective ER membranes and/or attached to their inner surfaces which could give rise to the observed parenthesome structure. In perforate parenthesomes (Fig. 7.5) the interposed layer generally extends to the pore boundary (Moore & Patton, 1975; Gull, 1976; Khan & Kimbrough, 1979) while in imperforate parenthesomes it is frequently absent in the centre, resulting in a tented profile (see Fig. 7.11). Patton & Marchant (1978*b*) carried out a mathematical analysis of d/p structure using nine parameters and presented the results in a summary dendrogram. A very important contribution of this study is that by printing respective sets of micrographs to the same magnification they graphically illustrated the generally unappreciated variations in d/p dimensions.

The possible ontogeny of perforate parenthesomes was elucidated by Moore (1975): ribbons of ER arise from the nuclear envelope and come to lie free in the cytoplasm in paired parallel orientation; differentiation starts in the centre of the array and results in fenestration across both ER elements; further development results in conjoined parenthesomes that then become separated. In some undetermined way the parenthesomes, still with extensive ER continuities, assume their positions next to the dolipore (Fig. 7.2). Although this study proposed ontogenies for dolipore and

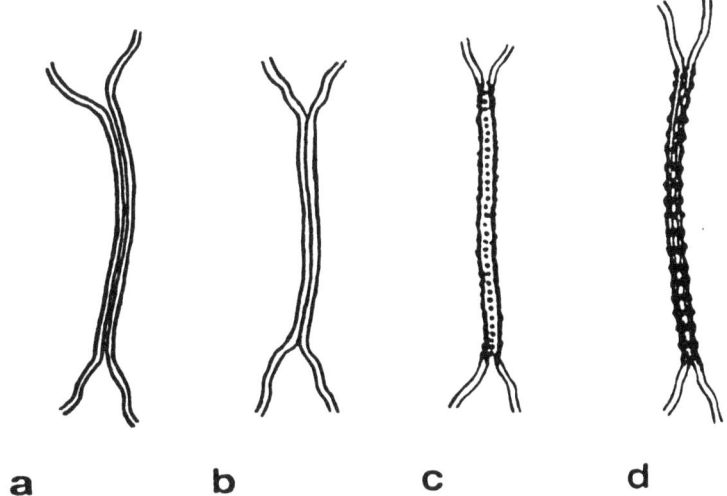

a   **b**   **c**   **d**

Fig. 7.6. Diagrams of thin-section profiles of apparent gap junctions bearing a remarkable and suggestive similarity to parenthesome profiles. (*a*) Classical profile with a 2–4-nm 'gap'. (*b*) Profile of junction fixed with permanganate. (*c*) Profile with intercellular densities with a periodicity of 8 nm. (*d*) Beaded profile produced by adding hydrogen peroxide to the fixative. (From Peracchia, 1980.)

188

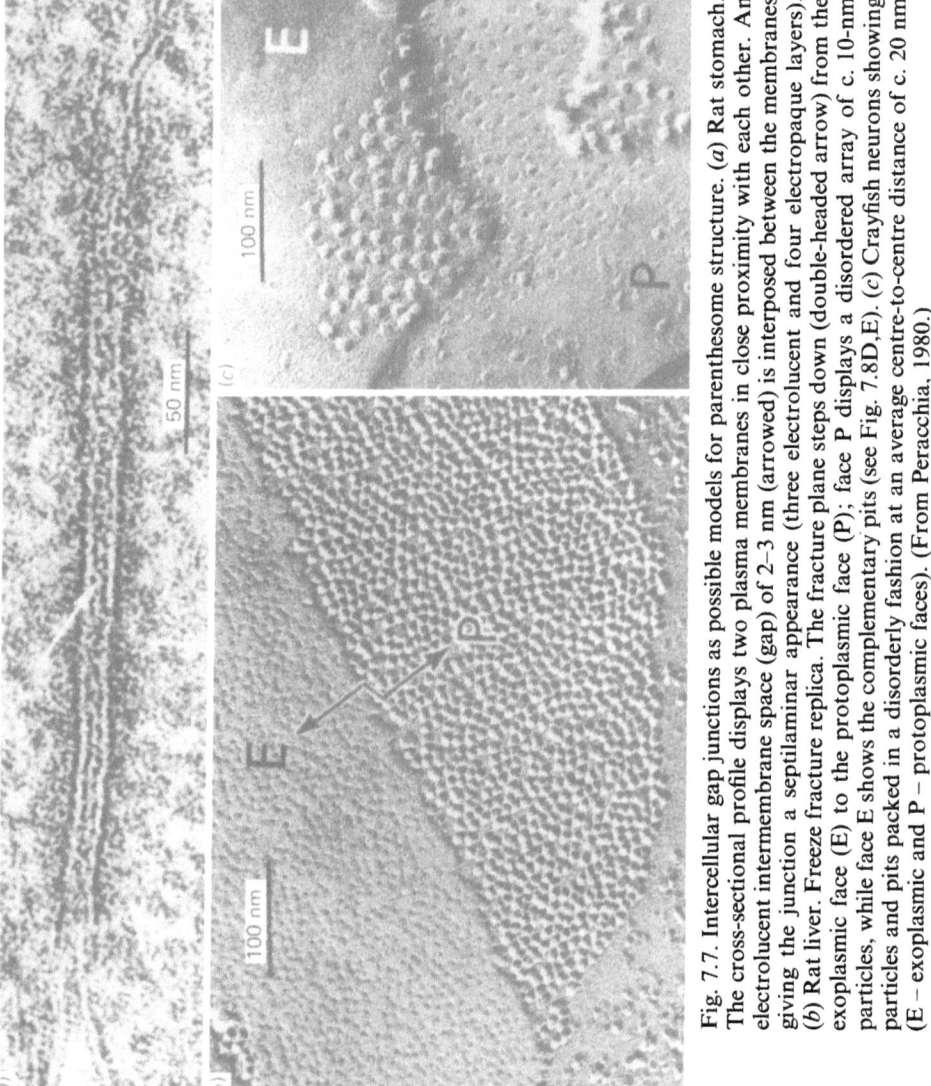

Fig. 7.7. Intercellular gap junctions as possible models for parenthesome structure. (a) Rat stomach. The cross-sectional profile displays two plasma membranes in close proximity with each other. An electrolucent intermembrane space (gap) of 2–3 nm (arrowed) is interposed between the membranes giving the junction a septilaminar appearance (three electrolucent and four electropaque layers). (b) Rat liver. Freeze fracture replica. The fracture plane steps down (double-headed arrow) from the exoplasmic face (E) to the protoplasmic face (P); face P displays a disordered array of c. 10-nm particles, while face E shows the complementary pits (see Fig. 7.8D,E). (c) Crayfish neurons showing particles and pits packed in a disorderly fashion at an average centre-to-centre distance of c. 20 nm (E – exoplasmic and P – protoplasmic faces). (From Peracchia, 1980.)

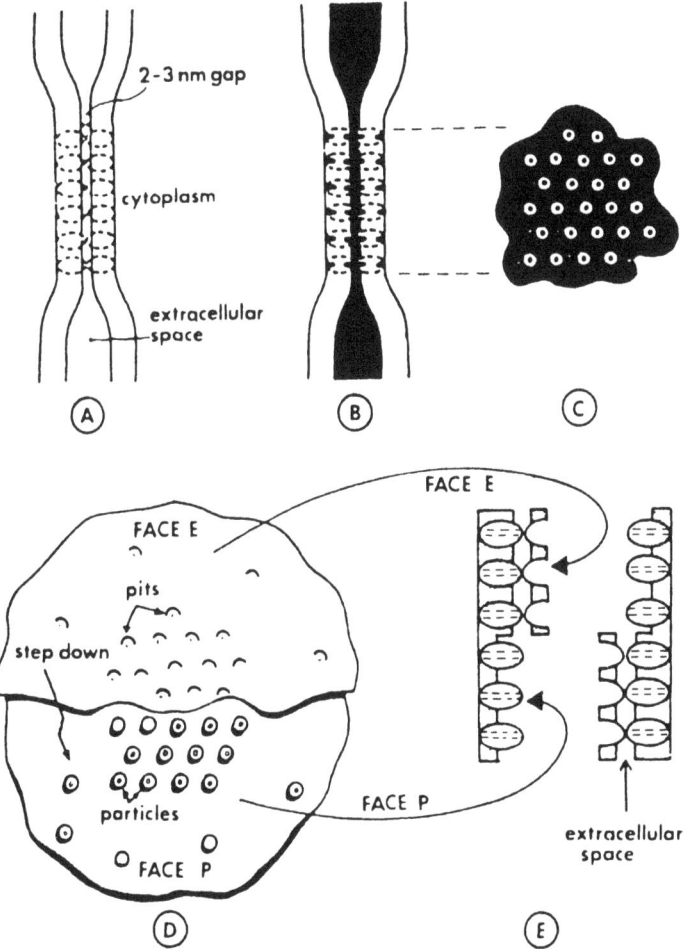

Fig. 7.8. Interpretation of gap junction structure in positively (A) and negatively (B,C) stained sections and in freeze fracture (D,E). The junctions are shown in profile (A,B,E) and in face view (C,D). In (D) the fracture plane steps down from face E (exoplasmic) to face P (protoplasmic) following the path shown in (E). Each gap junction membrane contains a polygonal array of intramembrane particles matching precisely and binding extracellularly with a similar array in the adjoined membrane. The intramembrane particles are not usually seen in positively stained sections, a reason why the membranes appear separated by a continuous gap 2–3 nm wide (A). The gap appears electropaque when colloidal lanthanum is mixed with the fixative, as the tracer penetrates between the membranes filling the extracellular space (B). By surrounding the particle protrusions, lanthanum produces face-view images of electrolucent rings (the particle protrusions) surrounded by an electropaque network (C). In freeze fracture (D,E), the membranes split down the middle exposing a good portion of the intramembrane particles on the protoplasmic leaflet (face P) and their complementary pits on the exoplasmic leaflet (face E). (Compare with Fig. 7.7.) (From Peracchia, 1980.)

parenthesome formation it gave no indication of the possible coordination of the two sequences to achieve the final d/p complex. An additional comment: in most published electron micrographs ER/parenthesome continuities are relatively infrequent or absent – the parenthesomes lying seemingly free in the cytoplasm. Inasmuch as they apparently arise from the ER, it may be that as cells age the ER component decreases until it is lost. Dykstra (1982) shows a pair of parenthesomes lying free in the degenerated peripheral material of an ejected glebal mass of *Sphaerobolus stellatus*, which suggests that parenthesome structure, whatever it is, is quite durable (see also Todd & Aylmore: Chapter 9).

Outer caps

Finally, in addition to these fundamental components of the d/p septum, several papers have reported the presence of outer caps. Gull (1976) critically analysed the septa of the outer hymenium, the subhymenium, and the central trama of gills of *Agrocybe praecox*. In the first and third of these tissues he found regular dolipores and parenthesomes. In the subhymenium, however, there occurred, additionally, an ER-bound,

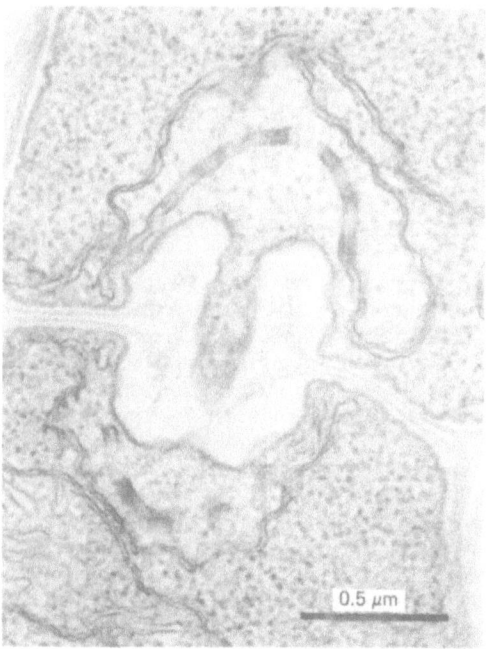

0.5 μm

Fig. 7.9. D/p septum with outer caps from fruit body primordium of *Coprinus cinereus*. (From van der Valk & Marchant, 1978.)

ribosome-free, thick lens of cytoplasm next to the parenthesome; the ER had continuities with the edge of the parenthesome and it also had scattered gaps in its perimeter. Where septa straddled the hymenium/sub-hymenium transition region, the outer caps appeared associated only with the latter cells. Similar outer caps have been shown by Thielke (1972) and Craig, Newsam & Gull (1977) for *Agaricus bisporus*. van der Valk & Marchant (1978) found outer caps associated with the septa of the short cells comprising the core of fruit body primordia of *Coprinus cinereus* (Fig. 7.9). The impression these micrographs convey is that the skirt of the parenthesome-generating ER may have become reflexed over the centrally differentiated area. Arita (1979) found subhymenial outer caps in *Pholiota adiposa* and *P. nameko* to be non-membranous, electron-dense zones consisting of microgranules about 20 nm in diameter. Wells (1978) has observed similar granular outer caps in the hymenium of *P. terrestris*. Considerably more reports of this additional parenthesome component are necessary, however, before its place in the scheme of the d/p septal complex can be properly evaluated.

The challenge of its morphology in taxonomy

Ascomycete versus basidiomycete septa

Septal ultrastructure is coming to be recognised as a major indicator of taxonomic affinities. The simple, large-pored septum with associated membrane-bound, paracrystalline Woronin bodies is acknowl-edged to be the most consistent indicator that a species is an ascomycete (Brenner & Carroll, 1968; McKeen, 1971; Wergin, 1973; Bronchart & Demoulin, 1975; von Arx, van der Walt & Liebenberg, 1981; Thomas & Jackson, 1982; Wimble & Young, 1983), while the d/p septum equally denotes that a species is a basidiomycete. The latter often have perforate parenthesomes, indicating that their teleomorphs are homobasidiomycetes e.g.: *Moniliophthora roreri* (Evans *et al.*, 1978); teleomorph probably *Crinipellis perniciosa* (Evans, 1981); *Sclerotium hydrophilum* (Mordue, 1983); *Dendrosporomyces prolifer* (Nawawi, Webster & Davy, 1977); *Nia vibrissa* and *Digitatispora marina* (Brooks, 1975); hyphae of fungal gardens cultivated by *Acromyrmex octospinosus* ants of Guadeloupe (Angeli-Papa & Eymè, 1979); and unique nodular bodies on raspberry root tips (Harris & Mackenzie, 1982). Mycorrhizal studies have shown mycobionts to have ascomycete- (Bonfante-Fasolo & Gianinazzi-Pearson, 1979; Thomas & Jackson, 1982), homobasidiomycete- (Nieudorp, 1972), or heterobasidio-mycete-type (Strullu & Gourret, 1974; Bonfante-Fasolo, 1980) septa (see Fig. 7.10).

Perforate versus *imperforate parenthesomes*

The literature of the past 20 years has shown that within the basidiomycetes the d/p septum is a conservative structure of taxonomic importance. Its morphologies (Fig. 7.10) are generally restricted to, and thus key characters of, specific taxa. The regularly perforate parenthesome (type O_1/P_1) occurs only in the superclass Homobasidiomycia (see Wilsenach & Kessel, 1965; Ellis *et al.*, 1972; Moore, 1975, 1984; Hanlin, 1978; Patton & Marchant, 1978a; Wells, 1978; Arita, 1979). It has been found without exception in the agarics and the Gasteromycetes. In the aphyllophores, however, there are two anomalies. One is the genus *Hirschioporus* (Polyporaceae). Traquair & McKeen (1978) showed *H. (Polyporus) pargemenus* to have imperforate parenthesomes (type O_1/P_2). Subse-

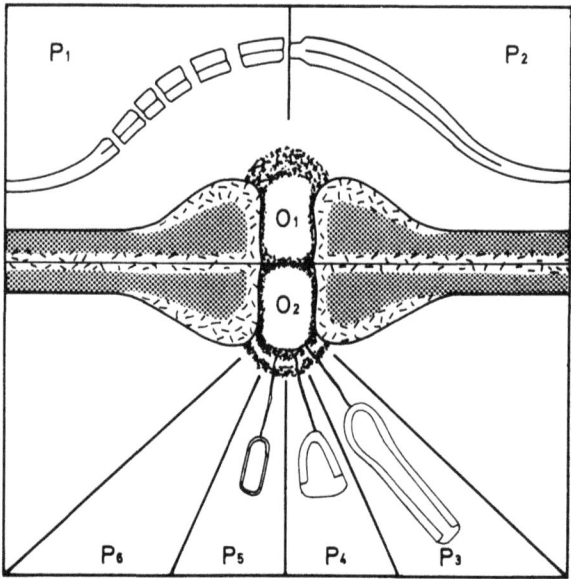

Fig. 7.10. Summary diagram of d/p septal types found in Basidiomycota. The diagram is divided top and bottom between the two types of septal occlusion found in the dolipore orifice: O_1 septa have an occluding granule while O_2 septa are characterised by a striated band. Associated with these types are several forms of parenthesome: P_1, regularly perforate; P_2, imperforate; P_3, P_4 and P_5, vesiculate; P_6, absent. Fungi generally characterised by these several variations of the d/p septum are as follows: O_1/P_1 – superclass Homobasidiomycia; O_1/P_2 – most jelly fungi; O_2/P_3 – *Sirobasidium magnum*; O_2/P_4 – *Tremella* spp., *Filobasidium capsuligenum* and *Hyalodendron*; O_2/P_5 – *Wallemia sebi*; O_2/P_6 – *F. floriforme*, *Filobasidiella*, *Stilbotulasnella* and *Trichosporon beigelii*. Not drawn to scale. (From Moore, 1980b).

quently, Moore (1984) examined other species of the genus: *H.* (*Polyporus*) *abietinus* (type), *H.* (*Irpex*) *fusco-violaceus*, *H.* (*Lenzites*) *laricinus*, and *H.* (*Coriolus*) *subchartaceus*. All of these species also had imperforate parenthesomes (Fig. 7.11). These species at one time or another were classified in five different genera but they came to be considered congeneric on the basis of gross and light-microscopic characters. The correctness of placing these species together is very strongly supported by the fact that they share the O_1/P_2 type of parenthesome. There may be other less obvious species whose possible admission to the genus could be confirmed or denied by septal analysis. The parenthesome morphology of *Hirschioporus* sets it markedly apart from other polypores and may have phylogenetic implications. The second anomaly is in the Hymenochaetaceae. In the subfamily Hymenochaetoideae there is an aggregate of genera whose type-species have imperforate parenthesomes (Moore, 1980*a*). These, and the approximate number of species they represent, are: *Inonotus hispidus* (40), *Onnia circinata* (5), and *Phellinus torulosus* (80). Fiasson (1983) uses

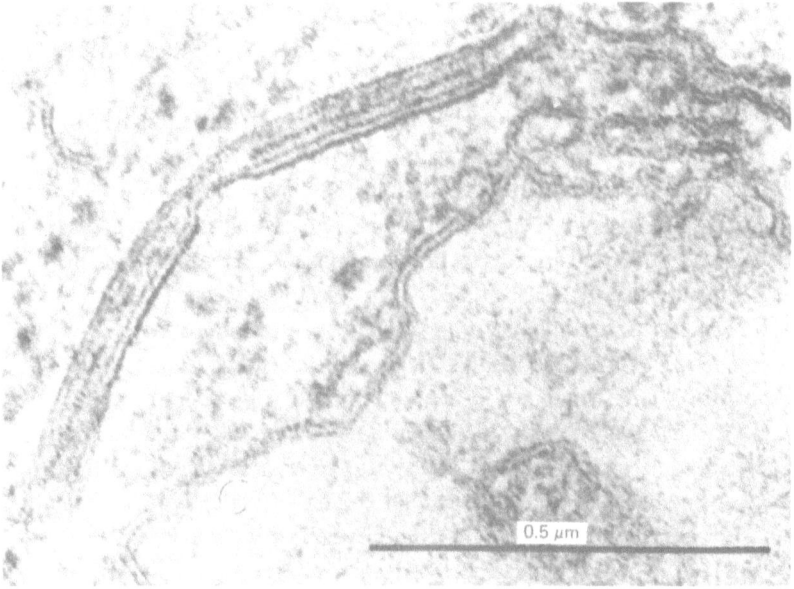

0.5 μm

Fig. 7.11. Imperforate parenthesome of *Hirschioporus abietinus* (Polyporales) showing the gap in the middle layer that gives this type of parenthesome a characteristic tented profile; note the parenthesome–ER continuity at the upper right and compare the bounding membranes of the parenthesome with the plasma membrane lining the dolipore (lower right) and with those of the ER. (From Moore, 1984.)

this character of imperforate parenthesomes as one of the criteria for his tribe Phellineae which includes *Phylloporia* (five species) in addition to the genera already mentioned.

The imperforate, O_1/P_2, type of parenthesome is generally characteristic of the superclass Heterobasidiomycia, specifically the Auriculariales (*sensu stricto*), including the basidia (McLaughlin, 1980), the Dacrymycetales, the sub-order Exidiinae, and the Tulasnellales (see Moore, 1978, 1980*b*; McLaughlin, 1978; Khan & Kimbrough, 1980*a*; Bandoni & Oberwinkler, 1982). These fungi are also characterised by basidiospores that form hyphae when they germinate.

The interface between the two superclasses of the Basidiomycota has long been a difficult problem. Traditionally, the one has been characterised by homobasidia that change little after meiosis and by basidiospores that germinate directly into hyphae, while the other has been characterised by various types of heterobasidia that undergo post-meiotic differentiation, usually with septation, and by basidiospores capable of reproducing secondary spores on sterigma-like outgrowths. The genus *Ceratobasidium* (Ceratobasidiaceae), however, straddles this dichotomy – it has homobasidia (i.e. there are no septa at the bases of the so-called epibasidia) and basidiospores that can germinate by repetition. As a consequence it has been variously interpreted. Lentz (in Petersen, 1971, p. 108) considers it to be part of the Aphyllophorales but Christiansen (in Petersen, 1971, pp. 263–5) considers it to be in the Tulasnellales, while Rogers (in Petersen, 1971, p. 249) points out that 'there is an unbroken series of forms connecting species that are heterobasidiomycetous in all respects with loose, tufted, merely sterigma bearing species that are unquestionably pellicularias and unquestionably homobasidiomycetous'. In addition, however, *Ceratobasidium* has perforate parenthesomes (Patton & Marchant, 1978*b*; Tu & Kimbrough, 1978) and this is also true for *Uthatobasidium* and *Waitea* which are in the same family (Tu & Kimbrough, 1978). Perforate parenthesomes and homobasidia are also characteristic of *Corticium* (Patton & Marchant, 1978*b*) and *Botryobasidium* and *Athelia* (Tu & Kimbrough, 1978) in the closely related family Corticiaceae. *Dictyonema*, a basidiolichen thought to be close to *Athelia* (which also has lichenised species), also has perforate parenthesomes (Slocum, 1980). Rogers (in Petersen, 1971, p. 249) further comments that 'The breaking up of this series into a number of genera has brought no advantage in the practical matter of finding names for specimens and has served only to obscure the series. It exists, nevertheless, and it is here that the Homo-and Heterobasidiomycetes are united.' Perhaps as, in another context, there are the secotioid Gasteromycetes (Smith, 1973; Watling, 1978; Thiers, 1984) and the gasteroid Russulales (Pegler & Young, 1979), we should be less procrustean and think of these boundary species as tulasnellate aphyllophores.

Vesiculate parenthesomes

Vesiculate parenthesomes, type O_2/P_{3-5} (Fig. 7.10), are restricted in their distribution to the suborder Tremellineae (Moore, 1978, 1979, 1980b); these genera are also characterised by yeast haplophases (*Cryptococcus* and *Bullera* when identified). *Tremella* and similar genera are, as well, characterised by cruciately septate hypobasidia (see Fig. 7.15a) and probably, therefore, constitute a natural group. Besides the free-living species with large fruit bodies there are also a number of reduced species that parasitise other higher fungi (Bezerra & Kimbrough (1978) include a complete list of parasitic *Tremella* spp.; see also Oberwinkler & Bandoni, 1981, 1983).

There is a second, heterogeneous, group brought together by septal morphology and blastic anamorphs but which have holobasidia. The family Filobasidiaceae is characterised by *Cryptococcus* anamorphs (Moore & Kreger-van Rij, 1972; Kwon-Chung, Bennett & Rhodes, 1982) and by clavate basidia that bear either a corona of spores (*Filobasidium*; Fig. 7.15e) or chains of spores from four loci (*Filobasidiella*; Fig. 7.15d). The dolipore septa have bipartite occlusions (Fig. 7.10) and the parenthesomes are either present (O_2/P_4) – *Filobasidium capsuligenum* (Moore & Kreger-van Rij, 1972) – or absent (O_2/P_8) – *F. floriforme* (Moore & Kreger-van Rij, 1972), *Filobasidiella neoformans* (Kwon-Chung & Popkin, 1976) and *F. depauperatus* (Khan & Kimbrough, 1980a; Samson, Stalpers & Weijman, 1983). The last-named species is homothallic, monokaryotic, and without clamps; it was originally described as an *Aspergillus* but its morphological similarity to heterothallic species of *Filobasidiella* and its septal morphology (Khan, Kimbrough & Kwon-Chung, 1981) show it to belong to this genus. Three other genera have been ascribed to the family: *Chionosphaera* (Fig. 7.15b) has been transferred by Oberwinkler & Bandoni (1982b) to their ambiguous Atractiellales; *Cystofilobasidium* (Oberwinkler *et al.*, 1983a) based on *Rhodosporidium capitatum* (Nakagiri & Tubaki, 1981) while perhaps meriting generic rank is probably nonetheless still a sporidiomycete (see Fig. 7.14); *Rogersiomyces* (Fig. 7.15c) has not been examined electron microscopically (it lacks clamps and the spores germinate directly). The family Syzygosporaceae (Jülich, 1982) (= Carcinomycetaceae) has been intensively studied by Oberwinkler & Bandoni (1982a). It comprises three genera, *Carcinomyces*, *Christiansenia* (see also Oberwinkler *et al.*, 1984) and *Syzygospora*, that are parasites of homobasidiomycetes and which have dolipore septa that apparently lack parenthesomes.

Finally, there are three other genera that probably belong here. One is *Hyalodendron*, a dark, yeast-like hyphomycete which produces both arthric and catenate conidia. Martinez (1979) has examined its one species, *H. lignicola*, and found that the septa are occluded by an electron-dense layer similar to that of *Tremella* and *Filobasidium* and have arches of small vesicles on either side but which lack microfilaments. de Hoog (1979) found the species to be physiologically similar to *Bullera alba*. The second genus is *Stilbotulasnella*. Bandoni & Oberwinkler (1982) characterise this monotypic, synnematous genus as having basidiospores and conidia that germinate by budding, dolipore septa that lack parenthesomes (in marked contrast to the prominent imperforate parenthesomes of a *Tulasnella* sp. they also depict), and occasional collapsed basidia that appear to be partially cruciate-septate. What weighs more heavily for them in the naming and family classification of the genus, however, are the tulasnellaceous basidia which they compare with those of *Pseudotulasnella*, a genus which also has partially cruciate-septate basidia. However, *P. guatemalensis* has septa at the base of the epibasidia (the defining character of the Tulasnellaceae) while they are not indicated for *S. conidiophora*. The third genus is *Wallemia*. Terracina (1974) showed that the septa of *W. sebi*, the only species, flared around the pore, that in the opening there appeared bands of light and dark material, and that associated with these were ER elements radiating into the cytoplasm. I have also examined these septa: the small associated vesicles are composed of single membranes (i.e. there is no inner parenthesome cupule) and they are tethered by microfilaments to the bipartite occlusions (Fig. 7.10, O_2/P_5).

The challenge of its reduction and absence
In the Tremella *complex*

In the Tremellineae parenthesome reduction and loss correlates with probable phylogenetic descent. The fortunate occurrence of regular vesiculate parenthesomes in *Filobasidium capsuligenum*, in combination with the basidial and anamorphic characters, allows the latter traits, in combination with dolipore septa without parenthesomes, to serve as identifying markers.

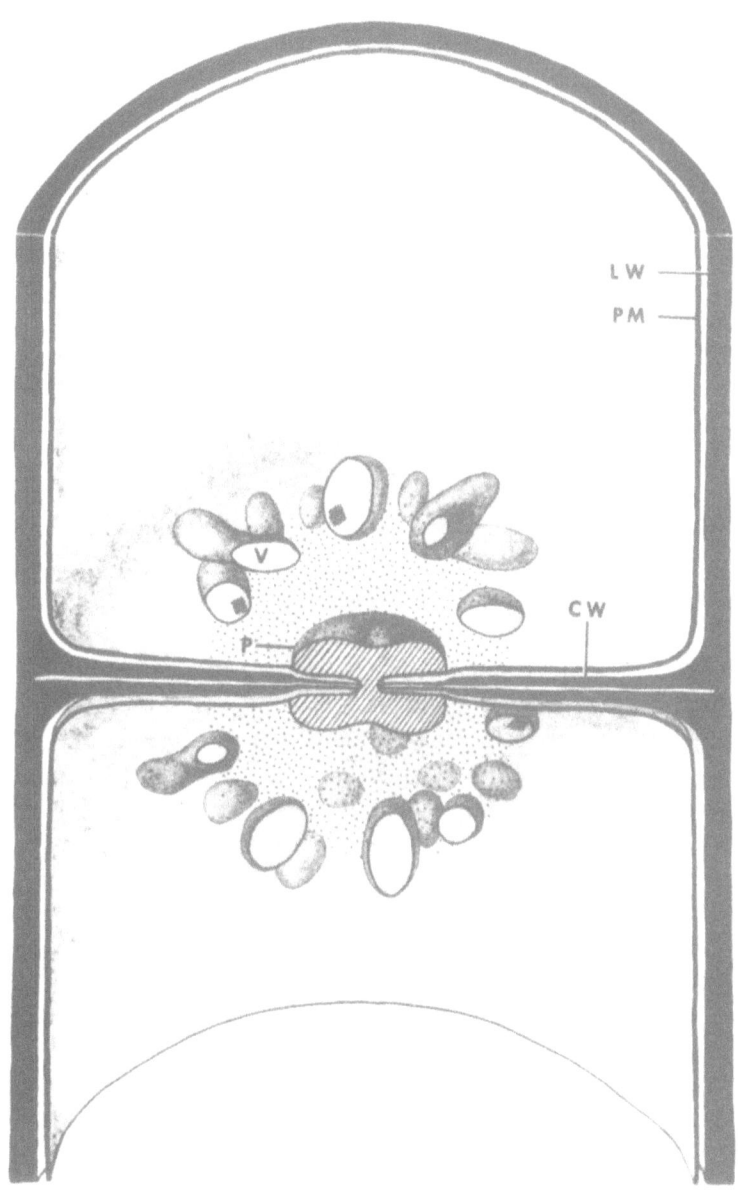

Fig. 7.12. Diagrammatic interpretation of a representative rust septum. The pore is blocked by a pulley-shaped structure (P) that is surrounded by a matrix of cytoplasm (stippled area) more dense than elsewhere in the cell. Organelles are usually excluded from this matrix and it is often surrounded by vesicles (V), some of which contain crystalline bodies. This reduced septal morphology may be the phylogenetic remains of a d/p septal system. (CW, cross wall; LW, lateral wall; PM, plasma membrane.) (From Littlefield & Bracker, 1971.)

In the auricularialian complex

The sub-class Stichophragmobasidiomycetidae represents another complex in which parasitic descent and/or reduction of the fruit body has been accompanied by septal degradation. In the Uredinales the septa of several genera have been examined (Ehrlich, Ehrlich & Schafer, 1968; Littlefield & Bracker, 1971; Coffey, Palevitz & Allen, 1972; Jones, 1973; Mims, 1977; Littlefield & Heath, 1979; Khan, Kimbrough & Webb, 1982). The following ontogeny seems to emerge from these studies: at the end of cytokinesis the membrane lip, instead of becoming inflated into a dolipore, closes on itself and the excess membrane folds back along either side of the septum; fusion in the centre results in a pulley-shaped membrane matrix. On either side of the pore there appears a dome of fine-textured, homogeneous cytoplasm whose ill-defined boundary is not membrane-delimited but which is impinged on by a number of vesicles that may contain paracrystalline inclusions (Fig. 7.12). As the septum ages the membrane matrix in the pore becomes an electron-dense, pulley-shaped plug. I would interpret this morphology as reflecting a partially degraded d/p complex in which the membrane matrix in the pore is a relic of the dolipore and the domes of cytoplasm and their partially embedded vesicles on either side are the vestiges of a parenthesome system.

Dykstra (1974) has examined five species of *Septobasidium* and shown that while there is some dense material (non-pulley-shaped) plugging the pore, there is no evidence of a specialised cytoplasm/vesicle complex on either side of it. Particularly noteworthy, however, is the massive thickness of the electropaque layers on either side of the thin, electrolucent middle layer. In profile it is possible to recognise three zones (see Khan & Kimbrough, 1982): (1) the central zone in which the three layers are about the same thickness and at whose very centre is the small pore and its associated pair of dense granules, (2) a shoulder zone that thickens abruptly and then rounds into (3) a peripheral zone. In wide hyphae the shoulder zone appears like a torus but in narrower cells it is continuous with the peripheral zone. Similar trizonate septa also occur in *Atractogloea* (Oberwinkler & Bandoni, 1982c) and *Eocronartium* (Khan & Kimbrough, 1980b, 1982; Jülich, 1982).

This assemblage of fungi has traditionally been comprised of three well-defined orders. The Septobasidiales and Uredinales still retain their circumscriptions but the Auriculariales are starting to be dismembered. Characteristic d/p septa with imperforate parenthesomes have been reported only in *Auricularia* (McLaughlin, 1978, 1980), *Hirneola* (Moore, 1978; Tu & Kimbrough, 1978), and *Mylittopsis* (Oberwinkler & Bandoni,

1982*b*). Genera in the order for which dolipore septa are absent are *Eocronartium* (see above) and (Oberwinkler & Bandoni, 1982*b*) *Helicobasidium* (Fig. 7.13), *Herpobasidium, Jola, Kriegera (Xenogloea), Mycogloea, Neotyphula, Paraphelaria, Phleogena, Platygloea* and *Stilbum.*

The new synnematous order Atractiellales (Oberwinkler & Bandoni, 1982*b,c*) has simple septa, holo- and parallei-septate basidia, and passively discharged spores that germinate directly or by budding. The Hoehnelomycetaceae is particularly ambiguous because *Agaricostilbum* seems to be a ustomycete (see below), *Atractogloea* has trizonate septa, and *Atractiella* has thin septa.

The Auriculariales, *sensu stricto*, must be regarded as the stem group from which this possible bouquet of species arose. The partially reduced septal morphology of the rusts mark them out as a coherent assemblage separate from the other species with simpler septa. Oberwinkler (1982), discussing the Uredinales, comments that '...(iv) it seems likely that obligate parasitism and a high rate of asexual reproduction are primitive characters that do not commonly occur among the higher basidiomycetes.

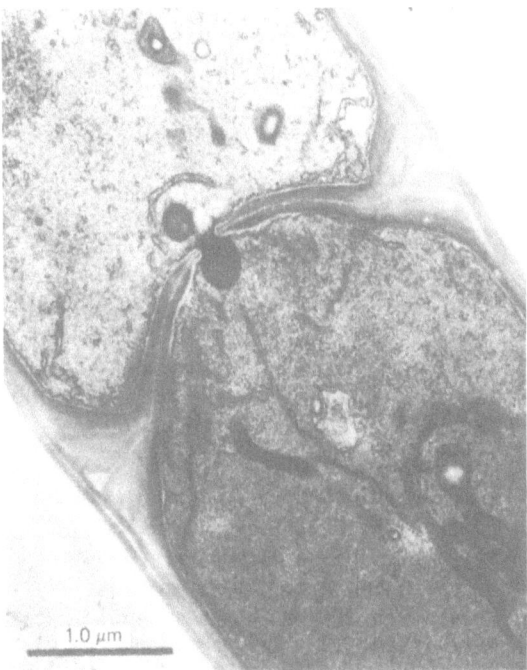

1.0 μm

Fig. 7.13. Septal structure of *Helicobasidium mompa*. (Courtesy of Bourett & McLaughlin.)

For these reasons it is appropriate to place the rusts before other basidiomycetes in a linear system of classification.' I would emphasise again, however, that parasitism is a *derived* condition, that, to a greater or lesser degree, it is invariably one of degenerative specialisation, and that, in a line of descent, obligate parasitism is more advanced than facultative (see Moore, 1971). The trizonate septa of the Septobasidiales is unusual enough to be diagnostic and, if so, then both *Atractogloea* and *Eocronartium* would belong in this order. The other, thin, reduced septal morphology may, or may not, be taxonomically significant.

In the Ustomycota

The smuts and their allies are alien to the mainstream of the basidiomycetes: the modes of formation and germination of their meiocytes are unique (Fischer & Holton, 1957; Donk, 1972a,b, 1973a,b/c; Zambettakis, 1973, 1978; Ramberg & McLaughlin, 1981; Ingold, 1983a; (compare with Ingold, 1983b; Gold & Mendgen, 1984); the septa are trilaminate, small pored, often with sharp rims, and lack parenthesomes (Moore, 1972; Johnson-Reid & Moore, 1972; Deml, 1977; Deml & Oberwinkler, 1981; Deml, Nebel & Oberwinkler, 1981; Ramberg &

	100	105	110	115
	:	:	:	:
Ustilago violacea	5'CACUGUGCUGCCGCAGGU OH			
Rhodosporidium toruloides	5'CACUGUGCUGCCGCAGGU OH			
Sporobolomyces salmonicolor	5'CACUGUGCUGCCGCAGGU OH			
Filobasidium floriforme	5'UCCUAGG-UGCUGUGGUU OH			
Filobasidium capsuligenum	5'UCCUAGG-UGCUGUGGUU OH			
Tremella mesenterica	5'UCCUGGG-UGCUGUGGUU OH			
Schizophyllum commune	5'UCCUGGG-UGCUGUGGUU OH			
Bjerkandera adusta	5'UCCUAGG-UGCUGUGGUU OH			

Fig. 7.14. A portion of the nucleotide sequence of the 5S rRNAs of eight species of basidiomycetes. Note that the ustomycetes, the first three species, have an insertion in the 107 position that is absent from the homo- and heterobasidiomycete sequences of the basidiomycetes proper. (From Walker & Doolittle, 1982.)

McLaughlin, 1981; Nakagiri & Tubaki, 1981); and their 5S rRNA has an insertion in the 107 position that is absent in the basidiomycetes proper (Fig. 7.14). Also, while there are no more than 13 nucleotide differences within each of these clusters, there is an average difference of 42 nucleotides between them (Walker & Doolittle, 1982; see also Templeton, 1983). Appreciating the singularity of these fungi, Moore (1972) put them into their own division, the Ustomycota (from *ustos* – L., burn) and recognised two classes (1980*b*): the free-living Sporidiomycetes (see Summerbell, 1983) and the Ustomycetes (smuts and Graphiolales) (Cole, 1983; Oberwinkler *et al.*, 1983*b*), and termed their meiocyte a ustospore; I propose here that the ustospore germ hypha, which may be one or more and bears lateral or terminal sporidia, be termed a **ustidium** (pl. -ia). This treatment removes the ustomycetes from an unsatisfactory, shadowy, para-auricularian position and admits a serious consideration of their relationship with the rest of the septomycetes: could they be intermediate between the ascomycetes and the basidiomycetes proper; a second line of basidiomycete descent from the ascomycetes; or, perhaps, another line of reduction from the homobasidiomycetes, paralleling that of the heterobasidiomycetes? These questions take on immediacy from Walker's (personal communication) observations that the 5S rRNA sequences of *Exobasidium*, *Trichosporon* and *Tilletia* are quite similar to each other, that the sequence of *Chionosphaera apobasidialis* is clearly allied to the sequences from doliporous groups, that the *Tilletiaria* (Septobasidiales; Moore, 1980*b*) sequence diverges about equally from the sequence clusters of, respectively, the doliporous group and the *Exobasidium–Trichosporon–Tilletia* group, and that the sequence from *Rhizoctonia crocorum* and *Agaricostilbum palmicolum* share the insertion position 107 with those from the ustomycetes. The first of these observations is interesting because the septa of *Exobasidium* spp. (and *Microstroma juglandis*; Fig. 7.15*g*) are simple with thick dark layers on either side of the thin light one (Blanz, 1978) (similar non-d/p septa also occur in *Dicellomyces* (Exobasidiales) and *Coniodyctium chevalieri*, a cryptobasidiaceous species (Oberwinkler & Bandoni, 1982*b*)) and because *Trichosporonoides* is said to differ only slightly from *Trichosporon* (de Hoog, 1979) and has been shown to have septa that are rather doliporous (Haskins, 1975; Martinez, 1979); *Moniliella* is also more or less in this complex (de Hoog, 1979) and, too, has rather doliporous septa (Martinez, 1979). Walker's last observation has particular significance because *R. crocorum* is the generic type and because, according to Oberwinkler & Bandoni (1982*b*), *Agaricostilbum* is the only genus outside the Ustomycota to have ustilago-type meiospore production (ustidia

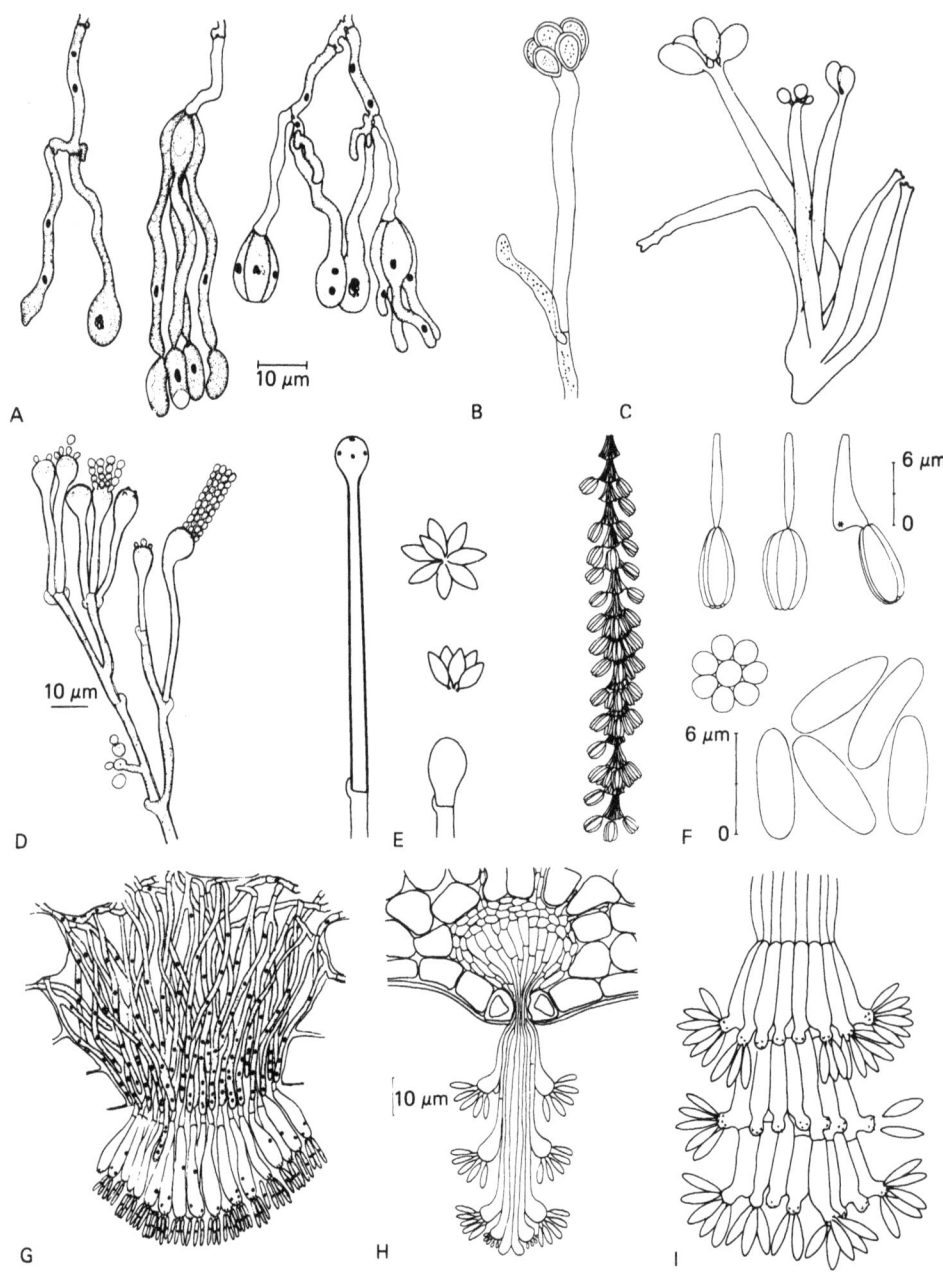

Fig. 7.15. (A–E) Heterobasidiomycetes and possibly related species. (A) Developmental stages of phragmobasidia in *Exidia nucleata*, representative of those Tremellales with gelatinous fruit bodies (from Wells, 1964). (B–E) Reduced species with holobasidia similar to the probasidia in (A). (B) *Chionosphaera apobasidialis* (Atractiellales) has simple, non-d/p septa (from Cox, 1976). (C) *Rogersiomyces okefenokeenis*

producing indeterminate numbers of sporidia), an indication that it probably belongs in the Sporidiomycetes.

Closing statement

In the basidiomycetes proper (Basidiomycota) a number of phylogenetic lines have been postulated to lead from the agarics to the several other major assemblages (aphyllophores, Gasteromycetes and jelly fungi). The corollary of these proposals is that descendent groups are the more advanced; e.g. that the hetero- rather than the homobasidiomycetes are the Higher Fungi (Petersen, 1971). Another consequence of these schemes is that perforate parenthesomes have to be the general condition. Imperforate parenthesomes occur in all of the major orders of jelly fungi and appear to have arisen within the homobasidiomycetes in the aphyllophores, probably at least twice, unless a phylogenetic relationship can be found between *Hirschioporus* and the sub-set of genera in the Hymenochaetoideae. Vesiculate parenthesomes are the most advanced as they occur only in the Tremellineae. The dolipore is present in all of these taxa, even in the last named when parenthesomes are no longer present.

More extreme reductions occur in the Stichophragmobasidiomycetidae: d/p septa with imperforate parenthesomes are found only in the Auricu-

---

(Filobasidiaceae?), septal morphology unknown (from Crane & Schoknecht, 1978). (D) *Filobasidiella neoformans* (Filobasidiaceae), dolipore septa without parenthesomes (from Kwon-Chung, 1975). (E) *Filobasidium capsuligenum* (Filobasidiaceae) showing a basidial initial, a mature basidium with spore scars, and two views of floriform clusters of basidiospores, the d/p septa have vesiculate parenthesomes like *Tremella* (drawn from micrographs in de Miranda, 1972). (F–I) Species that may be basidiomycetes. (F) *Articularia quercina*, a parasite on the lower surface of oak leaves, produces multi-tiered pendants with each tier apparently composed of a bundle of scapula-like cells packed together radially. Each flattened cell appears to have two meristematic cusps, the one facing outwards developing spores while the inner one (*) develops another similar cell. The sporogenous cusp produces eight, seven around one, elongate, closely appressed (basidio-?)spores that evidently cohere and do not become floriform (adapted from Morris, 1963). (G) *Microstroma juglandis* has been reported to have basidiomycete-type budding and trilamellar septa (Blanz, 1978). It is a leaf parasite (*Carya* and *Juglans*) and its fine hyphae emerge at the stomata as a cushion of apobasidia(?) bearing six to eight hyaline sterigmate spores (from Wolf, 1929). (H, I) *Helostroma alba* is like *Microstroma* in having somatic hyphae that are intercellular and become concentrated in stomatal chambers but it differs explicitly in that the sporogenous cells from several tiers (see Moore, 1982); from von Arx (1970) and Carmichael *et al.* (1980), respectively.

lariales, *sens. str.* (three genera at present); all of the other species so far examined in this complex lack assured dolipores and parenthesomes. Three patterns can be discerned: the 'ghost' of a d/p system in the Uredinales; the trizonate configuration of the Septobasidiales; and others (variously simple, ambiguous, and not very meaningful taxonomically).

The Ustomycota also have simple septa but their taxonomic distinctness and unity is marked out by the nature of their ustospores and ustidia and by their unique 5S rRNA sequences.

Electron microscopy has, by demonstrating in known forms that there is a fundamental difference between septa of ascomycetes and basidiomycetes and, within the latter, between the d/p septa of homobasidiomycetes, jelly fungi, and the Tremellineae, facilitated the assignment of otherwise anomalous forms to one of these taxa. There are other species whose spore or hyphal forms are suggestively basidiomycetous (e.g. Hornby (1984); Fig. 7.15*f*/*h*) that await future ultrastructural and biochemical analysis.

Finally, there are the very basic questions of how is the dolipore formed and what is the cell biology and developmental biology of the parenthesome. I have advanced some novel suggestions about both of these problems that I hope will provoke future discussions and investigations.

References

Angeli-Papa, J. & Eymè, J. (1979). Le champignon cultivé par la "fourmi-manioc", *Acromyrmex octospinosus* Reich en Guadeloupe; résultats préliminaires sur le mycélium en culture pure et sur l'infrastructure des hyphes. *Comptes Rendus Hebdomadaires des Séances de l'Académie de Sciences, Paris*, Série D, **281**, 21–4.

Arita, I. (1979). Cytological studies on *Pholiota*. *Reports of the Tottori Mycological Institute*, **17**, 1–118.

Arx, J. A. von (1970). A revision of the fungi classified as *Gloeosporium*. *Bibliotheca Mycologica*, **34**, 1–203.

Arx, J. A. von, Walt, J. P. van der & Liebenberg, N. V. D. W. (1981). On *Mauginiella scaettae*. *Sydowia*, **34**, 42–5.

Bandoni, R. J. & Oberwinkler, F. (1982). *Stilbotulasnella*: a new genus in the Tulasnellaceae. *Canadian Journal of Botany*, **60**, 1875–9.

Bezerra, J. L. & Kimbrough, J. W. (1978). A new species of *Tremella* parasitic on *Rhytidhysterium rufulum*. *Canadian Journal of Botany*, **56**, 3021–33.

Beckett, A. (1981). The ultrastructure of septal pores and associated structures in the ascogenous hyphae and asci of *Sordaria humana*. *Protoplasma*, **107**, 127–47.

Blanz, P. (1978). Uber die systematische Stellung der Exobasidiales. *Zeitschrift für Mykologie*, **44**, 91–107.

Bonfante-Fasolo, P. (1980). Occurrence of a basidiomycete in living cells of mycorrhizal hair roots of *Calluna vulgaris*. *Transactions of the British Mycological Society*, **75**, 320–5.

Bonfante-Fasolo, P. & Gianinazzi-Pearson, V. (1979). Ultrastructural aspects of endomycorrhiza in the Ericaceae. I. Naturally infected hair roots of *Calluna vulgaris* L. Hull. *New Phytologist*, **83**, 739–44.

Bracker, C. E. & Butler, E. E. (1963). The ultrastructure and development of septa in hyphae of *Rhizoctonia solani*. *Mycologia*, **55**, 35–8.

Brenner, D. M. & Carroll, G. C. (1968). Fine-structural correlates of growth in hyphae of *Ascodesmis sphaerospora*. *Journal of Bacteriology*, **95**, 658–71.

Bronchart, R. & Demoulin, V. (1975). Septum ultrastructure of *Ostracoderma torrendii*. *Canadian Journal of Botany*, **53**, 1549–53.

Brooks, R. D. (1975). The presence of dolipore septa in *Nia vibrissa* and *Digitatispora marina*. *Mycologia*, **67**, 172–4.

Buck, R. C. & Tisdale, J. M. (1962). An electron microscopic study of the development of the cleavage furrow in mammalian cells. *Journal of Cell Biology*, **13**, 117–25.

Burnett, J. H. (1968). *Fundamentals of Mycology*. London: Edward Arnold.

Butler, E. E. & Bracker, C. E. (1970). Morphology and cytology of *Rhizoctonia solani*. In *Rhizoctonia solani, Biology and Cytology*, ed. J. R. Parmeter, Jr, pp. 32–51. Berkeley: University of California Press.

Carmichael, J. W., Kendrick, W. B., Conners, I. L. & Sigler, L. (1980). *Genera of Hyphomycetes*. Edmonton: The University of Alberta Press.

Casselton, L. A. (1978). Dikaryon formation in higher basidiomycetes. In *The Filamentous Fungi*, vol. 3, *Developmental Mycology*, ed. J. E. Smith & D. R. Barry, pp. 275–97. London: Edward Arnold.

Casselton, L. A., Lewis, D. & Marchant, R. (1971). Septal structure and mating behaviour of common *A* diploid strains of *Coprinus lagopus*. *Journal of General Microbiology*, **66**, 273–8.

Coffey, M. D., Palevitz, B. A. & Allen, P. J. (1972). The fine structure of two rust fungi, *Puccinia helianthi* and *Melampsora lini*. *Canadian Journal of Botany*, **50**, 231–40.

Cole, G. T. (1983). *Graphiola phoenicis*: a taxonomic enigma. *Mycologia*, **75**, 93–116.

Corner, E. J. H. (1964). *The Life of Plants*. Cleveland: The World Publishing Company.

Cox, D. E. (1976). A new homobasidiomycete with anomalous basidia. *Mycologia*, **68**, 481–510.

Craig, G. D., Newsam, R. J. & Gull, K. (1977). Subhymenial branching and dolipore septation in *Agaricus bisporus*. *Transactions of the British Mycological Society*, **69**, 337–44.

Craig, G. D., Newsam, R. J., Gull, K. & Wood, D. A. (1979). An ultrastructural and autoradiographic study of stipe elongation in *Agaricus bisporus*. *Protoplasma*, **98**, 15–29.

Crane, J. L. & Schoknecht, J. D. (1978). *Rogersiomyces*, a new genus in the Filobasidiaceae (Homobasidiomycetes) from an aquatic habitat. *American Journal of Botany*, **65**, 902–6.

Curry, K. J. & Kimbrough, J. W. (1983). Septal structures in apothecial tissues of the Pezizaceae (Pezizales, ascomycetes). *Mycologia*, **75**, 781–94.

Deml, G. (1977). Feinstrukturelle Merkmalsanalysen an Ustilaginales-Arten. *Zeitschrift für Pilzkunde*, **43**, 291–303.

Deml, G., Nebel, M. & Oberwinkler, F. (1981). Light and scanning electron microscopic studies of spore formation in *Ustilago pustulata* and *U. scabiosae*. *Canadian Journal of Botany*, **59**, 122–8.

Deml, G. & Oberwinkler, F. (1981). Studies in Heterobasidiomycetes. Part 4. Investigations on *Entorrhiza casparyana* by light and electron microscopy. *Mycologia*, **73**, 392–8.

DeRobertis, E. D. P. & DeRobertis, E. M. F. (1980). *Cell and Molecular Biology.* Philadelphia: Saunders College.

Desole, S. (1982). Die Entwicklung der Dolipore von *Coprinus radiatus* (Bolt.) Fr. *Bibliotheca Mycologica*, **88**, 1–86 (+60 plates).

Donk, M. A. (1972*a*). The heterobasidiomycetes: A reconnaissance – I. A restricted emendation. *Proceedings. Nederlandse akademie van wetenschappen. Amsterdam.*, series C, **75**, 365–75.

Donk, M. A. (1972*b*). The heterobasidiomycetes: A reconnaissance – II. Some problems connected with the restricted emendation. *Proceedings. Nederlandse akademie van wetenschappen. Amsterdam.*, series C, **75**, 376–90.

Donk, M. A. (1973*a*). The heterobasidiomycetes: A reconnaissance – III. How to recognize a basidiomycete? *Proceedings. Nederlandse akademie van wetenschappen. Amsterdam.*, series C, **76**, 1–22.

Donk, M. A. (1973*b/c*). The heterobasidiomycetes: A reconnaissance – IV/V. *Proceedings. Nederlandse akademie van wetenschappen. Amsterdam.*, series C, **76**, 109–125/126–140.

Dykstra, M. J. (1974). Some ultrastructural features in the genus *Septobasidium*. *Canadian Journal of Botany*, **52**, 971–2.

Dykstra, M. J. (1982). A cytological examination of *Sphaerobolus stellatus* fruiting bodies. *Mycologia*, **74**, 44–53.

Ehrlich, M. A., Ehrlich, H. G. & Schafer, J. F. (1968). Septal pores in the Heterobasidiomycetidae, *Puccinia graminis* and *P. recondita*. *American Journal of Botany*, **55**, 1020–7.

Ellis, T. T., Rogers, M. A. & Mims, C. W. (1972). The fine structure of the septal pore cap in *Coprinus stercorarius*. *Mycologia*, **64**, 681–8.

Evans, H. C. (1981). Pod rot of cacao caused by *Moniliophthora (Monilia) roreri*. *Phytopathological Papers*, **24**, 1–44. Kew: Commonwealth Mycological Institute.

Evans, H. C., Stalpers, J. A., Samson, R. A. & Benny, G. L. (1978). On the taxonomy of *Monilia roreri* an important pathogen of *Theobroma cacao* in South America. *Canadian Journal of Botany*, **56**, 2528–32.

Fiasson, J.-L. (1983). Taxonomie phylétique des hyménochétacées porées d'Europe, spécialement du genre *Phellinus*. Lyon: Docteur Sciences Naturelles thése, Université Claude Bernard.

Fineran, B. A., Wild, D. J. C. & Ingerfeld, M. (1982). Initial wall formation in the endosperm of wheat, *Triticum aestivum*: a reevaluation. *Canadian Journal of Botany*, **60**, 1776–95.

Fischer, G. W. & Holton, C. S. (1957). *Biology and Control of the Smut Fungi*. New York: The Ronald Press Company.

Flegler, S. L., Hooper, G. R. & Fields, W. G. (1976). Ultrastructural and cytochemical changes in the basidiomycete septum associated with fruiting. *Canadian Journal of Botany*, **54**, 2243–53.

Giesy, R. M. & Day, P. R. (1965). The septal pores of *Coprinus lagopus* in relation to nuclear migration. *American Journal of Botany*, **52**, 287–93.

Girbardt, M. (1979). A microfilamentous septal belt (FSB) during induction of cytokinesis in *Trametes versicolor* (L. ex Fr.). *Experimental Mycology*, **3**, 215–28.

Gold, R. E. & Mendgen, K. (1984). Cytology of teliospore germination and basidiospore formation in *Uromyces appendiculatus* var. *appendiculatus*. *Protoplasma*, **119**, 150–5.

Grove, S. N. (1978). The cytology of hyphal tip growth. In *The Filamentous Fungi*, vol. 3, *Developmental Mycology*, ed. J. E. Smith & D. R. Berry, pp. 28–50. London: Edward Arnold.

Gull, K. (1976). Differentiation of septal ultrastructure according to cell type in the basidiomycete, *Agrocybe praecox*. *Journal of Ultrastructural Research*, **54**, 89–94.

Hanlin, R. T. (1978). Septum structure in *Spiniger meineckellus*. *American Journal of Botany*, **65**, 471–6.

Harris, D. C. & Mackenzie, K. A. D. (1982). Nodular bodies of an unidentified basidiomycete on raspberry roots. *Transactions of the British Mycological Society*, **79**, 530–4.

Haskins, R. H. (1975). Septa ultrastructure and hyphal branching in the pleomorphic imperfect fungus *Trichosporonoides oedocephalis*. *Canadian Journal of Botany*, **53**, 1139–48.

Heath, I. B., Ashton, M.-L. & Rethoret, K. (1982). Mitosis and the phylogeny of *Taphrina*. *Canadian Journal of Botany*, **60**, 1696–725.

Hepler, P. K. & Jackson, W. T. (1968). Microtubules & early stages of cell-plate formation in the endosperm of *Haemanthus katherinae* Baker. *Journal of Cell Biology*, **38**, 437–46.

Hoch, H. C. & Howard, R. J. (1980). Ultrastructure of freeze-substituted hyphae of the basidiomycete *Laetisaria arvalis*. *Protoplasma*, **103**, 281–97.

Hoch, H. C. & Howard, R. J. (1981). Conventional chemical fixations induce artifactual swelling of dolipore septa. *Experimental Mycology*, **5**, 167–72.

Hoog, G. S. de (1979). The taxonomic position of *Moniliella, Trichosporonoides* and *Hyalodendron* – an essay. In *Studies in Mycology*, No. 19, *The Black Yeasts, II: Moniliella and Allied Genera*, ed. G. S. de Hoog, pp. 81–90. Baarn: Centraalbureau voor Schimmelcultures.

Hornby, D. (1984). *Akenomyces costatus* sp. nov. and validation of *Akenomyces* Arnaud. *Transactions of the British Mycological Society*, **82**, 653–64.

Hunsley, D. & Gooday, G. W. (1974). The structure and development of septa in *Neurospora crassa*. *Protoplasma*, **82**, 125–46.

Ingold, C. T. (1983a). The basidium in *Ustilago*. *Transactions of the British Mycological Society*, **81**, 573–84.

Ingold, C. T. (1983b). A view of the basidium. *Bulletin of the British Mycological Society*, **17**, 82–94.

Johnson-Reid, J. A. & Moore, R. T. (1972). Some ultrastructural features of *Rhodosporidium toruloides* Banno. *Antonie van Leeuwenhoek*, **38**, 417–35.

Jones, D. R. (1973). Ultrastructure of septal pore in *Uromyces dianthi*. *Transactions of the British Mycological Society*, **61**, 227–35.

Jülich, W. (1982). Notes on some basidiomycetes (Aphyllophorales and Heterobasidiomycetes). *Persoonia*, **11**, 421–8. (Issued 16 August)

Khan, S. K. & Kimbrough, J. W. (1979). Ultrastructure of septal pore apparatus in the lamellae of *Nematoloma puiggarii*. *Canadian Journal of Botany*, **57**, 2064–70.

Khan, S. K. & Kimbrough, J. W. (1980a). Septal ultrastructure in some genera of the Tremellaceae. *Canadian Journal of Botany*, **58**, 55–60.

Khan, S. K. & Kimbrough, J. W. (1980b). Ultrastructure and taxonomy of *Eocronatrium*. *Canadian Journal of Botany*, **58**, 642–7.

Khan, S. K. & Kimbrough, J. W. (1982). A reevaluation of the basidiomycetes based upon septal and basidial structures. *Mycotaxon*, **15**, 103–20.

Khan, S. K. & Kimbrough, J. W. & Kwon-Chung, K. J. (1981). Ultrastructure of *Filobasidiella arachnophila*. *Canadian Journal of Botany*, **59**, 893–7.

Khan, S. K., Kimbrough, J. W. & Webb, P. G. (1982). The fine structure of septa and haustoria of *Cronartium quercuum* formae speciales *fusiforme* on *Quercus rubra*. *Mycologia*, **74**, 809–19.

Kreger-van Rij, N. J. W. & Veenhuis, M. (1971). A comparative study of the cell wall structure of basidiomycetous and related yeasts. *Journal of General Microbiology*, **68**, 87–95.

Kwon-Chung, K. J. (1975). A new genus, *Filobasidiella*, the perfect state of *Cryptococcus neoformans. Mycologia*, **67**, 1197–200.

Kwon-Chung, K. J., Bennett, J. E. & Rhodes, J. C. (1982). Taxonomic studies on *Filobasidiella* species and their anamorphs. *Antonie van Leeuwenhoek*, **48**, 25–38.

Kwon-Chung, K. J. & Popkin, T. J. (1976). Ultrastructure of septal complex in *Filobasidiella neoformans* (*Cryptococcus neoformans*). *Journal of Bacteriology*, **126**, 524–8.

Larsen, W. J. (1977). Structural diversity of gap junctions. A review. *Tissue and Cell*, **9**, 373–94.

Littlefield, L. J. & Bracker, C. E. (1971). Ultrastructure of septa in *Melampsora lini. Transactions of the British Mycological Society*, **56**, 181–8.

Littlefield, L. J. & Heath, M. C. (1979). *Ultrastructure of Rust Fungi*. New York: Academic Press.

McKeen, W. E. (1971). Woronin bodies in *Erysiphe graminis* DC. *Canadian Journal of Microbiology*, **12**, 1557–60.

McLaughlin, D. J. (1978). Ultrastructure of the hymenium of *Auricularia polytricha. Mushroom Science*, **10**, 219–29.

McLaughlin, D. J. (1980). Ultrastructure of the metabasidium of *Auricularia fuscosuccinea. American Journal of Botany*, **67**, 1225–35.

Marchant, R. (1978). Wall composition of monokaryons and dikaryons of *Coprinus cinereus. Journal of General Microbiology*, **106**, 195–9.

Maret, R. (1972). Chimie et morphologie submicroscopique des parois cellulaires de l'ascomycète *Chaetomium globosum. Archiv für Mikrobiologie*, **81**, 68–90.

Martinez, A. T. (1979). Ultrastructure of *Moniliella, Trichosporonoides* and *Hyalodendron*. In *Studies in Mycology*, No. 19, *The Black Yeasts, II*: Moniliella *and Allied Genera*, ed. G. S. de Hoog, pp. 50–7. Baarn: Centraalbureau voor Schimmelcultures.

Marx, J. L. (1983). Organizing the cytoplasm. *Science*, **222**, 1109–11.

Mayfield, J. E. (1974). Septal involvement in nuclear migration in *Schizophyllum commune. Archiv für Mikrobiologie*, **95**, 115–24.

Mims, C. W. (1977). Ultrastructure of teliospore formation in the cedar-apple rust fungus *Gymnosporangium juniperi-virginianae. Canadian Journal of Botany*, **55**, 2319–29.

Miranda, L. R. de (1972). *Filobasidium capsuligenum* nov. comb. *Antonie van Leeuwenhoek*, **38**, 91–9.

Moore, R. T. (1965). The ultrastructure of fungal cells. In *The Fungi, an Advanced Treatise*, vol. 1, *The Fungal Cell*, ed. G. C. Ainsworth & A. S. Sussman, pp. 95–118. New York: Academic Press.

Moore, R. T. (1971). An alternative concept of the Fungi based on their ultrastructure. In *Recent Advances in Microbiology*, ed. A. Perez-Miravete & D. Pelaez, pp. 49–64. Mexico City: Libreria International.

Moore, R. T. (1972). Ustomycota, a new division of higher fungi. *Antonie van Leeuwenhoek*, **38**, 567–84.

Moore, R. T. (1974). Proposal for the recognition of super ranks. *Taxon*, **23**, 650–2.

Moore, R. T. (1975). Early ontogenetic stages of dolipore/parenthesome formation in *Polyporus biennis. Journal of General Microbiology*, **87**, 251–9.

Moore, R. T. (1977). Dolipore disjunction in *Antromycopsis broussonetiae* Pat. *Experimental Mycology*, **1**, 92–101.

Moore, R. T. (1978). Taxonomic significance of septal ultrastructure with particular reference to the jelly fungi. *Mycologia*, **70**, 1007–24.

Moore, R. T. (1979). Septal ultrastructure in *Sirobasidium magnum* and its taxonomic implications. *Antonie van Leeuwenhoek*, **45**, 113–18.

Moore, R. T. (1980*a*). Taxonomic significance of septal ultrastructure in the genus *Onnia* Karsten (Polyporineae/Hymenochaetaceae). *Botaniska Notiser*, **133**, 169–75.

Moore, R. T. (1980*b*). Taxonomic proposals for the classification of marine yeasts and other yeast-like fungi including the smuts. *Botanica Marina*, **23**, 361–73.

Moore, R. T. (1982). The correct typification of *Microstroma* Niessl and *Helostroma* Pat. *Mycotaxon*, **14**, 13–16.

Moore, R. T. (1984). Taxonomic implications of septal ultrastructure in the Aphyllophorales. *Wissenschaftliche Arbeiten aus dem Burgenland*. (in press)

Moore, R. T. & Kreger-van Rij, N. J. W. (1972). Ultrastructure of *Filobasidium* Olive. *Canadian Journal of Microbiology*, **18**, 1949–51.

Moore, R. T. & McAlear, J. H. (1962). Fine structure of mycota 7. Observations on septa of ascomycetes and basidiomycetes. *American Journal of Botany*, **49**, 86–94.

Moore, R. T. & Marchant, R. (1972). Ultrastructural characterization of the basidiomycete septum of *Polyporus biennis*. *Canadian Journal of Botany*, **50**, 2463–9.

Moore, R. T. & Patton, A. M. (1975). Parenthesome fine structure in *Pleurotus cystidiosus* and *Schizophyllum commune*. *Mycologia*, **67**, 1200–5.

Mordue, J. E. M. (1983). Dolipore septa in *Sclerotium hydrophilum*. *Transactions of the British Mycological Society*, **81**, 654–5.

Morris, E. F. (1963). The synnematous genera of the Fungi Imperfecti. *Western Illinois University Series in the Biological Sciences*, **3**, 1–143.

Nakagiri, A. & Tubaki, K. (1981). A taxonomic study of *Rhodosporidium capitatum*. *Canadian Journal of Botany*, **61**, 1898–1905.

Nawawi, A., Webster, J. & Davey, R. A. (1977). *Dendrosporomyces prolifer* gen. et sp. nov., a basidiomycete with branched conidia. *Transactions of the British Mycological Society*, **68**, 59–63.

Nieudorp, P. J. (1972). Some observations with light and electron microscope on the endotrophic mycorrhiza of orchids. *Acta Botanica Néerlandica*, **21**, 128–44.

Oberwinkler, F. (1982). The significance of the morphology of the basidium in the phylogeny of basidiomycetes. In *Basidium and Basidiocarp. Evolution, Cytology, Function, and Development*, ed. K. Wells and E. K. Wells, pp. 9–35. New York: Springer-Verlag.

Oberwinkler, F. & Bandoni, R. J. (1981). *Tetragoniomyces* gen. nov. and Tetragoniomycetaceae fam. nov. (Tremellales). *Canadian Journal of Botany*, **59**, 1034–40.

Oberwinkler, F. & Bandoni, R. J. (1982*a*). Carcinomycetaceae: a new family in the Heterobasidiomycetes. *Nordic Journal of Botany*, **2**, 501–16.

Oberwinkler, F. & Bandoni, R. J. (1982*b*). A taxonomic survey of the gasteroid, auricularioid Heterobasidiomycetes. *Canadian Journal of Botany*, **60**, 1726–50.

Oberwinkler, F. & Bandoni, R. J. (1982*c*). *Atractogloea*: a new genus in the Hoehnelomycetaceae (Heterobasidiomycetes). *Mycologia*, **74**, 634–9.

Oberwinkler, F. & Bandoni, R. J. (1983). *Trimorphomyces*: a new genus in the Tremellaceae. *Systemic and Applied Microbiology*, **4**, 105–13.

Oberwinkler, F., Bandoni, R. J., Bauer, R., Deml, G. & Kisimova-Horovitz, L. (1984). The life-history of *Christiansenia pallida*, a dimorphic, mycoparasitic heterobasidiomycete. *Mycologia*, **76**, 9–22.

Oberwinkler, F., Bandoni, R. J., Blanz, P. & Kisimova-Horovitz, L. (1983*a*).

Cystofilobasidium: a new genus in the Filobasidiaceae. *Systematic and Applied Microbiology*, **4**, 114–22.

Oberwinkler, F., Bandoni, R. J., Blanz, P., Deml, G. & Kisimova-Horovitz, L. (1983*b*). Graphiolales: Basidiomycetes parasitic on palms. *Plant Systematics and Evolution*, **140**, 251–77.

Olive, L. S. (1944). Development of the perithecium in *Aspergillus fischeri* Wehmer, with a description of crozier formation. *Mycologia*, **36**, 266–75.

Patton, A. M. & Marchant, R. (1978*a*). An ultrastructural study of septal development in hyphae of *Polyporus biennis*. *Archives of Microbiology*, **118**, 271–8.

Patton, A. M. & Marchant, R. (1978*b*). A mathematical analysis of dolipore/parenthesomes structure in basidiomycetes. *Journal of General Microbiology*, **109**, 335–49.

Pegler, D. N. & Young, T. W. K. (1979). The gasteroid Russulales. *Transactions of the British Mycological Society*, **72**, 353–81.

Peracchia, C. (1980). Structural correlates of gap junction permeation. *International Review of Cytology*, **66**, 81–146.

Petersen, R. H. (ed.) (1971). *Evolution in the Higher Basidiomycetes*. Knoxville: University of Tennessee Press.

Ramberg, J. E. & McLaughlin, D. J. (1981). Ultrastructural study of promycelial development and basidiospore initiation in *Ustilago maydis*. *Canadian Journal of Botany*, **58**, 1548–61.

Raudaskoski, M. (1970). Occurrence of microtubules and microfilaments, and origin of septa in dikaryotic hyphae of *Schizophyllum commune*. *Protoplasma*, **70**, 415–22.

Robbins, E. & Gonatas, N. K. (1964). The ultrastructure of a mammalian cell during the mitotic cycle. *Journal of Cell Biology*, **21**, 429–63.

Rosing, W. C. (1981). Ultrastructure of septa in *Chaetomium brasiliense* (Ascomycotina). *Mycologia*, **73**, 1204–7.

Samson, R. A., Stalpers, J. A. & Weijman, A. C. M. (1983). On the taxonomy of the entomogenous fungus *Filobasidiella arachnophila*. *Antonie van Leeuwenhoek*, **49**, 447–56.

Setliff, E. C., MacDonald, W. L. & Patton, R. F. (1972). Fine structure of the septal pore apparatus in *Polyporus tomentosus*, *Poria latemarginata*, and *Rhizoctonia solani*. *Canadian Journal of Botany*, **50**, 2559–63.

Slocum, R. D. (1980). Light and electron microscopic investigations in the Dictyonemataceae (basidiolichens). II. *Dictyonema irpicinum*. *Canadian Journal of Botany*, **58**, 1005–15.

Smith, A. H. (1973). Agaricales and related secotioid Gasteromycetes. In *The Fungi, an Advanced Treatise*, vol. IVb, *A Taxonomic Review with Keys: Basidiomycetes and Lower Fungi*, ed. G. C. Ainsworth, F. K. Sparrow & A. S. Sussman, pp. 421–50. New York: Academic Press.

Steffens, W. L. & Jones, J. P. (1983). Ascus and ascospore development in *Eleutherascus peruvianus*. 2. Nuclear and early ascus ontogeny. *Canadian Journal of Botany*, **61**, 1599–617.

Strullu, D.-G. & Gourret, J.-P. (1974). Ultrastructure et évolution de champignon symbiotique des racines de *Dactylorhiza maculata* (L.) Verm. *Journal de Microscopie*, **20**, 285–94.

Summerbell, R. C. (1983). The heterobasidiomycetous yeast genus *Leucosporidium* in an area of temperate climate. *Canadian Journal of Botany*, **61**, 1402–10.

Swart, H. J. (1969). Proliferating croziers in *Asterina*. *Transactions of the British Mycological Society*, **53**, 322–3.

Templeton, A. R. (1983). Systematics of basidiomycetes based on 5S rRNA sequences and other data. *Nature*, **303**, 731–2.

Terracina, F. C. (1974). Fine structure of the septum in *Wallemia sebi*. *Canadian Journal of Botany*, **52**, 2587–90.

Thielke, C. (1972). Die Dolipore der Basidiomyceten. *Archiv für Mikrobiologie*, **82**, 31–7.

Thiers, H. D. (1984). The secotioid syndrome. *Mycologia*, **76**, 1–8.

Thomas, G. W. & Jackson, R. M. (1982). *Complexipes moniliformis* – ascomycete or zygomycete? *Transactions of the British Mycological Society*, **79**, 149–86.

Traquair, J. A. & McKeen, W. E. (1978). Ultrastructure of the dolipore septum in *Hirschioporus paragamenus* (Polyporaceae). *Canadian Journal of Microbiology*, **24**, 767–71.

Trinci, A. P. J. (1979). The duplication cycle and branching in fungi. In *Fungal Walls and Hyphal Growth*, ed. J. H. Burnett & A. P. J. Trinci, pp. 319–58. Cambridge University Press.

Tu, C. C. & Kimbrough, J. W. (1978). Systematics and phylogeny of fungi in the *Rhizoctonia* complex. *Botanical Gazette*, **139**, 454–66.

Tu, C. C., Kimbrough, J. W. & Aldrich, H. C. (1977). Cytology and ultrastructure of *Thanatephorus cucumeris* and related taxa of the *Rhizoctonia* complex. *Canadian Journal of Botany*, **5**, 2419–36.

Valk, P. van der & Marchant, R. (1978). Hyphal ultrastructure in fruit-body primordia of the basidiomycetes *Schizophyllum commune* and *Coprinus cinereus*. *Protoplasma*, **95**, 57–72.

Valk, P. van der, Marchant, R. & Wessels, J. G. H. (1977). Ultrastructural localization of polysaccharides in the wall and septum of the basidiomycete *Schizophyllum commune*. *Experimental Mycology*, **1**, 69–82.

Walker, W. F. & Doolittle, W. F. (1982). Redividing the basidiomycetes on the basis of 5S rRNA sequences. *Nature*, **299**, 723–4.

Walt, J. P. van der & Hopsu-Havu, V. K. (1976). A colour reaction for the differentiation of ascomycetous and hemibasidiomycetous yeasts. *Antonie van Leeuwenhoek*, **42**, 157–63.

Watling, R. (1978). From infancy to adolescence: advances in the study of the higher fungi. *Transactions of the Botanical Society of Edinburgh*, **42** (suppl., *Cryptogamic Centenary Symposium 1975*), 61–73.

Wells, K. (1964). The basidia of *Exidia nucleata*. II. Development. *American Journal of Botany*, **51**, 360–70.

Wells, K. (1978). The fine structure of septal pore apparatus in the lamellae of *Pholiota terrestris*. *Canadian Journal of Botany*, **56**, 2915–24.

Wergin, W. P. (1973). Development of Woronin bodies from microbodies in *Fusarium oxysporum* f. sp. *lycopersici*. *Protoplasma*, **76**, 249–60.

Wessels, J. G. H. & Sietsma, J. H. (1979). Wall structure and growth in *Schizophyllum commune*. In *Fungal Walls and Hyphal Growth*, ed. J. H. Burnett & A. P. J. Trinci, pp. 27–48. Cambridge University Press.

Wessels, J. G. H. & Sietsma, J. H. (1981). Fungal cell walls: a survey. In *Encyclopedia of Plant Physiology*, new series vol. 13B, *Plant carbohydrates*, II. *Extracellular Carbohydrates*, ed. W. Tanner & F. A. Loewus, pp. 352–94. Berlin: Springer-Verlag.

Wessels, J. G. H., Sietsma, J. H. & Sonnenberg, S. M. (1983). Wall synthesis and assembly during hyphal morphogenesis in *Schizophyllum commune*. *Journal of General Microbiology*, **129**, 1607–16.

Wilsenach, R. & Kessel, M. (1965). On the function and structure of the septal pores of *Polyporus rugulosus*. *Journal of General Microbiology*, **40**, 397–400.

Wimble, D. B. & Young, T. W. K. (1983). Septum structure in *Dactylella lysipaga*. *Mycologia*, **75**, 174–5.

Wolf, F. A. (1929). The relationship of *Microstroma* (Bereng.) Sacc. *Journal of the Elisha Mitchell Scientific Society*, **43**, 97–9.

Zambettakis, Ch. (1973). Recherches sur la germination des téliospores des Ustilaginales. I. Différents modes de germination selon l'espèce et le milieu utilisé. *Bulletin de la Société Mycologique de France*, **89**, 253–75.

Zambettakis, Ch. (1978). La sexualité chez les Ustilaginales. *Revue de Mycologie*, **41**, 469–91; **42**, 13–39 & 113–42.

8
Dikaryon formation

LORNA A.CASSELTON AND
ANDROULLA ECONOMOU

*School of Biological Sciences, Queen Mary College, University of London,
Mile End Road, London E1 4NS*, UK

Introduction

In the typical life cycle of most higher basidiomycetes, two different vegetative states can be distinguished, the *monokaryon* (or *homokaryon*) and the *dikaryon*. The monokaryon is the primary mycelium which develops on germination of a single sexual spore. The dikaryon is the secondary mycelium which predominates in nature and is the mycelium on which the highly differentiated fruit bodies develop. These two mycelia have different functional roles and in the two species which are the main subject of this review they are morphologically quite distinct because in the dikaryon there are clamp connections present at most septa.

The two species *Coprinus cinereus* and *Schizophyllum commune* have been used extensively for studies concerned with understanding the genetic control of dikaryon formation and there are many structural and biochemical changes which can be correlated (see Raper, 1966; Casselton, 1978; Raper, 1983). A dikaryon is generally derived from the interaction between two genetically different monokaryons. This interaction is a very important one in terms of a population because it represents the stage at which sexual compatibility is recognised. In any natural population the essence of sexual reproduction is to maintain genetic heterogeneity and we find barriers which act to restrict self-fertilisation and thereby promote cross-fertilisation. These are the *breeding systems* or *incompatibility mechanisms*. A monokaryon interaction which successfully gives rise to a dikaryon is a compatible one which will lead ultimately to fusion of genetically different nuclei, meiosis and genetic recombination.

Table 8.1. *Multi-allelic incompatibility control in higher basidiomycetes*

System	Genes	Single mating	Meiotic segregation	Inbreeding restriction (%)	Outbreeding potential[a]		
					alleles A	B	%
Bipolar	$A_1A_2A_3\cdots A_n$	$A_1 \times A_2$	$0.5\ A_1$ $\qquad$ $0.5\ A_2$	50	2		50
					5		80
					10		90
					20		95
Tetrapolar	$A_1A_2A_3\cdots A_n$ $B_1B_2B_3\cdots B_n$	$A_1B_1 \times A_2B_2$	$0.25\ A_1B_1$ $\quad$ $0.25\ A_2B_2$ $0.25\ A_1B_2$ $\quad$ $0.25\ A_2B_1$	25	2	2	25
					5	5	64
					10	10	81
					20	20	90

[a] Outbreeding potential calculated from: 1 gene $(n_A - 1)/n_A$; 2 genes $(n_A n_B - n_A - n_B + 1)/n_A n_B$ (Koltin, Stamberg and Lemke, 1972).

Breeding systems and mating type genes

The genes which determine sexual compatibility are termed the mating type genes. They are of considerable interest genetically because in these basidiomycetes they are seen to be regulatory genes which control the different events involved in dikaryon morphogenesis.

The efficiency of a breeding system is measured in terms of how well it restricts inbreeding and, as a consequence, how much this then imposes a restriction on outbreeding. In ascomycete fungi where there are only two alleles at a single mating type gene, the population falls into two cross-compatible groups. Inbreeding restriction can be maintained without a corresponding reduction in outbreeding potential if there are many different alleles of the mating type genes all of which are cross-compatible. This is the situation that is found in the higher basidiomycetes. The only interaction between monokaryons which is incompatible is when alleles are the same. There are two basic breeding systems in these fungi referred to as bipolar and tetrapolar. The important features of these two systems are summarised in Table 8.1.

In the bipolar system there is a single mating type gene, A, with multiple alleles. Any two different alleles give compatibility. From a single fruit body the progeny fall into two cross-compatible groups, hence there is a 50% restriction on selfing. In terms of outbreeding potential it requires only ten alleles for this to reach 90%. In the tetrapolar system there are two genes, A and B, each with multiple alleles. For compatibility it is now required that alleles of both genes are different. A and B are in different chromosomes and assort independently at meiosis, thus four mating types are derived from a single fruit body in equal numbers. This allows inbreeding restriction to be reduced to 25%. There is a little more restriction on outbreeding potential with two genes to satisfy but it still only requires 20 alleles of each gene to reach 90%.

Studies with the tetrapolar species *Schizophyllum commune* have shown that the mechanism exists to create much more allelic variability at A and B than is actually required (Papazian, 1954; Raper, Baxter & Ellingboe, 1960; Raper, Baxter & Middleton, 1958). It is more appropriate to refer to A and B as factors rather than genes because, in *S. commune* at least, the specificity of each factor is not determined by a single gene, but by two closely linked genes designated α and β. α and β appear to be alternatives in determining specificity because each A and B specificity derives from its particular combination of α and β. Assuming that A_1 is $\alpha_1\beta_1$ and A_2 is $\alpha_2\beta_2$, recombination can yield two new $\alpha\beta$ combinations, $\alpha_1\beta_2$ and $\alpha_2\beta_1$, which each have a different A specificity and are compatible with A_1 and

A_2. This two-gene structure provides a very economical way of generating a large number of A and B specificities with only a few α and β alleles. Nine α and 32 β alleles of A and nine α and nine β alleles of B have been identified in *S. commune* (see Koltin, Stamberg & Lemke, 1972) giving a potential 288 A and 81 B specificities. These numbers are far in excess of the actual specificities identified in worldwide samples.

The two-gene structure of A and B has been demonstrated for *Collybia velutipes* (Takemaru, 1961) but studies with other fungi are few. A in *Coprinus cinereus* has been shown to have a similar two-gene structure (Day, 1963a) but attempts to demonstrate recombination in B were unsuccessful (Haylock, Economou & Casselton, 1980). Since so few alleles are required for high outbreeding potential there would seem to be little pressure to realise the potential that exists to create such numbers of allelic specificities. The evidence from *Schizophyllum commune* implies, however, that both α and β DNA sequences contribute equally to the structure of the mating type gene product. In different species these sequences may not be so easily separable by recombination or even lie within the confines of a single gene. It should be remembered that close linkage of α and β is essential since recombination generates a compatible allele and therefore reduces inbreeding restriction. If α and β were unlinked, all the advantage of the tetrapolar system would be lost (Koltin, Stamberg & Lemke, 1972) and inbreeding potential would rise to 56% !

To have multiple alleles of the mating type genes is clearly a successful strategy for maintaining near unrestricted recombination of other genes in the population but it imposes increased complexity on the mating type genes themselves. The mating type genes have to recognise a very large number of non-self alleles in a compatible interaction. It is non-self alleles that are compatible rather than self alleles being incompatible. This is evident from studies with diploids and aneuploids. It is possible to make a single mating type factor heterozygous either in a somatic diploid monokaryon (Casselton, 1965; Casselton & Lewis, 1966) or as a consequence of irregular meiotic division (Raper & Oettinger, 1962). A_1/A_2 is compatible with an A_1 or A_2 monokaryon and similarly B_1/B_2 is compatible with a B_1 or B_2 monokaryon. It simply requires a different factor to be present in one of the monokaryons for successful dikaryon formation; the common factor does not prevent it.

Morphogenesis

Fig. 8.1 gives an outline of the essential events occurring in dikaryon formation (see also Todd & Aylmore: Chapter 9). The first step is the fusion of two compatible monokaryotic hyphae. This is followed by a phenomenon first described by Buller (1931). The fusion cells initially contain both nuclei. A nucleus from one monokaryon then starts to migrate through the cells of the other monokaryon. This can be a reciprocal event so migration can occur through cells of both participants in the mating. Eventually the migrating nuclei reach the apical cells which then contain two genetically different nuclei which will now remain associated to constitute the dikaryotic pair. These two nuclei will eventually fuse in the basidia of the fruit body, but in the vegetative dikaryon, they remain in close association as two haploid nuclei in what is essentially a diploid cell.

All further cell divisions in the dikaryon involve a complex series of

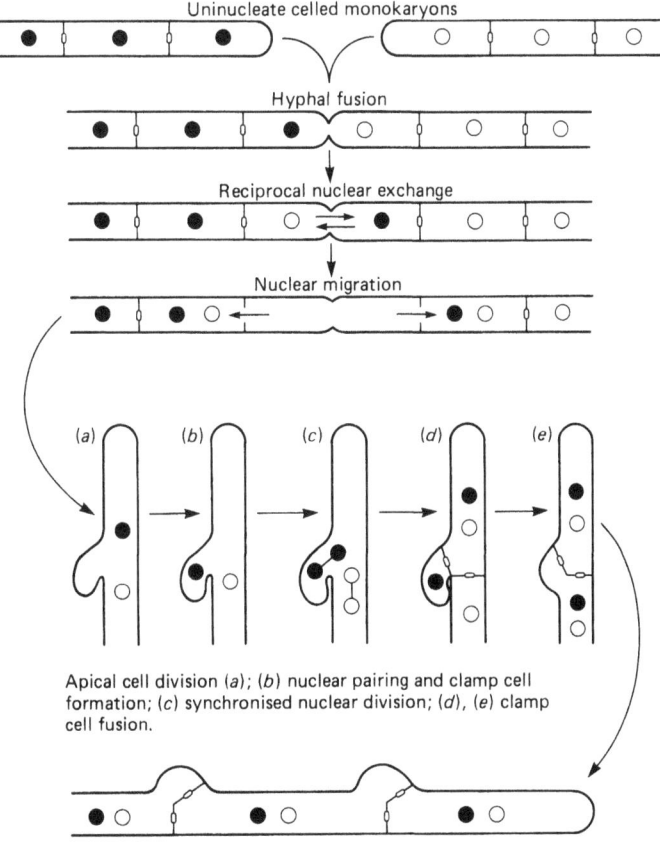

Apical cell division (a); (b) nuclear pairing and clamp cell formation; (c) synchronised nuclear division; (d), (e) clamp cell fusion.

Binucleate celled dikaryon

Fig. 8.1. Schematic representation of the morphogenetic sequence leading to dikaryon formation from two compatible monokaryons.

Table 8.2. *Characteristics of four types of homokaryon interaction in tetrapolar species*

Genetic constitution	Morphogenetic events		Result
	Operating	Blocked	
$A_1B_1 \times A_2B_2$ compatible	Nuclear migration Nuclear pairing Clamp cell formation Synchronised nuclear division Clamp cell fusion	—	Dikaryon
$A_1B_1 \times A_1B_2$ common A	Nuclear migration	Nuclear pairing Clamp cell formation Synchronised nuclear division	Common A Heterokaryon
$A_1B_1 \times A_2B_1$ common B	Nuclear pairing Clamp cell formation Synchronised nuclear division	Nuclear migration Clamp cell fusion	Common B Heterokaryon
$A_1B_1 \times A_1B_1$ common AB	—	Nuclear migration Nuclear pairing Synchronised nuclear division Clamp cell formation	Common AB Heterokaryon

events in which the clamp connection is formed. A clamp cell develops on the side of the apical cell. One nucleus passes into the clamp cell and both nuclei then divide synchronously. New septa are laid down, cutting off one nucleus in the clamp cell and one in the sub-apical cell. The clamp cell fuses with the sub-apical cell and its nucleus then passes into this cell to restore a binucleate condition.

In bipolar forms, all these events, if they occur, are controlled by a single gene. In tetrapolar forms it has been shown that part of the sequence is controlled by *A* and part by *B*. Although compatibility and hence dikaryon formation is only possible when both *A* and *B* factors are different, there is no barrier to cell fusion between incompatible monokaryons. Such fusions can give rise to heterokaryons, mycelia in which two different nuclei are present but not in the regular dikaryotic association. It may be necessary to force the association by introducing gene mutations which impose complementing growth requirements.

Three incompatible associations are possible depending on which factor is common: common *AB*, common *B* and common *A*. Table 8.2 summarises the characteristics of these heterokaryons compared with a normal dikaryon and is based on studies with both *Coprinus cinereus* (Swiezynski & Day, 1960*a*,*b*) and *Schizophyllum commune* (see Raper, 1966). Where only one factor is common, part of the morphogenetic sequence leading to dikaryon formation is expressed. The partial sequence operating in heterokaryon formation is controlled by the factor which is different and the sequence which is blocked by the common factor.

In a common *B* mating (where *A* is different), there is no nuclear migration to precede the formation of a heterokaryon. However, once formed, many tip cells of the heterokaryon have both nuclei. Clamp cells develop and a synchronised division of the two nuclei occurs, but following septation (see Fig. 8.1*d*), the clamp cell cannot fuse so its nucleus becomes trapped in what is called a false clamp. It is evident that different *A*s allow clamp cell formation and synchronised nuclear division. Different *B*s must be required for clamp cell fusion and also the early nuclear migration following cell fusion. In a common *A* mating where *B* is different, nuclear migration follows cell fusion but because *A* functions are blocked, no clamp cells develop and there is no nuclear pairing or synchronised nuclear division.

B functions are particularly interesting because they can be divided into mating functions and post-mating functions. *B* mating function is the extensive nuclear migration which precedes dikaryon formation. That this occurs is interesting in view of the complex dolipore septum separating each cell (Moore & McAlear, 1962; see also R. T. Moore: Chapter 7). The size of the pore and the parenthesome membrane afford a considerable barrier to nuclear movement. As first demonstrated by Giesy & Day (1965) using the electron microscope, nuclear migration is associated with the disruption of the septum to allow nuclei to pass freely. Illustrated in Fig. 8.2 are a typical dolipore septum in the monokaryon of *Coprinus cinereus* and a disrupted septum in a mycelium in which nuclear migration is occurring. It is interesting to note that the disrupted septum has a multivesicular body associated with it. This is commonly observed in both *C. cinereus* (Casselton, 1978) and *Schizophyllum commune* (Marchant & Wessels, 1974) and suggests that these vesicles contain the enzymes involved in septum disruption.

Once the dikaryon is established, septal disruption no longer occurs, indeed it is important for the maintenance of the binucleate condition of the dikaryotic cells that the septa remain intact. There is genetic evidence to suggest that one of the post-mating functions of *B* is to switch off septal disruption. Somatic diploids heterozygous for the *B* factor are fully compatible with monokaryons having either of its two different *B* factors, or any other, but compatibility is only expressed in terms of donating nuclei that can migrate through these homokaryotic cells to establish a dikaryon (Casselton, Lewis & Marchant, 1971). There is no migration through the diploid cells and the septa are no longer susceptible to disruption. The presence of two different *B* factors in the same cell, a condition normally found in every dikaryotic cell, results in a switch from mating function to post-mating function. There is still a need for nuclear migration in the dikaryon to release the clamp nucleus by clamp cell fusion; this is a post-mating function. A possible mechanism for switching off *B* mating

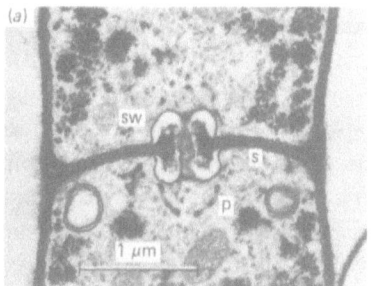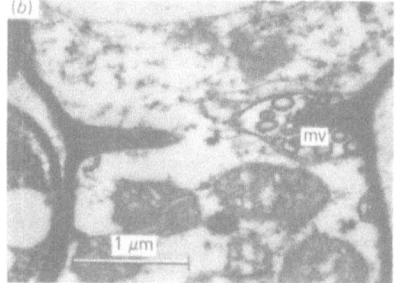

Fig. 8.2. (*a*) Dolipore septum in a monokaryon of *Coprinus cinereus*. (p, parenthesome membrane; s, septum; sw, septal swelling.) (*b*) Disrupted septum in a mycelium in which nuclear migration is occurring. (mv, multivesicular body.)

function is a change in septal structure as first suggested by Casselton *et al.* (1971); and Wessels & Marchant (1974) showed that monokaryotic septa were more susceptible to *in vitro* enzymic dissolution than dikaryotic septa.

Mating type gene mutations

Mutations in the mating type factors are of interest because they clearly show the regulatory function of these genes. The technique for selecting mutations in *A* and *B* was first introduced by Parag (1962) and has yielded mutations in both factors of *Schizophyllum commune* (Koltin, 1968; Raper, Boyd & Raper, 1965; Raudaskoski *et al.*, 1976; Koltin *et al.*, 1979) and *Coprinus cinereus* (Day, 1963*b*; Haylock, Economou & Casselton, 1980). It makes use of the fact that heterokaryons with a common *A* or *B* factor are generally sterile. After treatment of the heterokaryon with a mutagen, subsequent development of a fruit body is visual evidence that a true dikaryon has formed. Formation of a true dikaryon can only result from mutation in the common mating type factor to give compatibility. A modification of this technique devised by Haylock, Economou & Casselton (1980) was to look for evidence of nuclear migration in a common *B* interaction. Because they select for the expression of compatibility, both techniques can only yield mutations which either give new alleles, or break down compatibility control.

Mutation has only ever yielded mutants which have lost compatibility control and become self-compatible. It has never given rise to new alleles. This suggests that the DNA sequences which determine allelic variability may be significantly different. Concomitant with self-compatibility is the constitutive expression of the morphogenetic events regulated by the mutated gene. For example, mutation in *A* leads not only to self-compatibility (it requires only a different *B* for dikaryon formation), the mutant monokaryon resembles in all respects a common *B* heterokaryon with different *A* factors. *A* controls synchronised nuclear division and associated clamp cell formation and, as seen in Fig. 8.3*a*, the mutant *A* monokaryon has clamp cells present at every septum and inside each clamp the migrating nucleus is trapped. It requires the operation of *B* function to get clamp cell fusion to release the trapped nucleus and this can be achieved by introducing a mutation in *B*. There is now constitutive expression of *B* function which fuses the clamps. The double mutant seen in Fig. 8.3*b* has a single haploid nuclear genotype but has binucleate cells and true clamp connections and resembles in all respects a true dikaryon (Raper, Boyd & Raper, 1965; Haylock, Economou & Casselton, 1980).

This similarly also extends to the numbers of proteins which it synthesises (De Vries, Hoge & Wessels, 1980).

In *Coprinus cinereus*, mutation in B gives rise to a monokaryon which is morphologically indistinguishable from a normal non-mutant monokaryon. It can, moreover, donate and accept migrating nuclei in matings with any monokaryon having a different A factor, including those with the same B mutation and the original B factor in which the mutation occurred. In trying to determine what type of regulatory control the mating type genes effect, somatic diploid monokaryons have been constructed which are heterozygous for B_{1mut}/B_1 and compared for mating behaviour with the corresponding B_1/B_1 non-mutant and B_{1mut}/B_{1mut} mutant diploid monokaryons (Economou & Casselton, unpublished). In the heterozygous diploid, the non-mutant B could not restore compatibility control, this diploid was still self-compatible and could form a dikaryon with B_1 and B_{1mut}. However, unlike the homozygous non-mutant or mutant diploid which had normal monokaryon mating behaviour, it could only donate nuclei; B mating function had been switched off. The mutant allele had recognised B_1 as being a different factor. This suggests that the product of the mating type gene may be acting as a positive regulator of post-mating function (clamp cell fusion) but normally interacts with the product of another factor to act as a negative regulator of mating functions. Such dual regulatory roles of mating and post-mating functions are seen in the

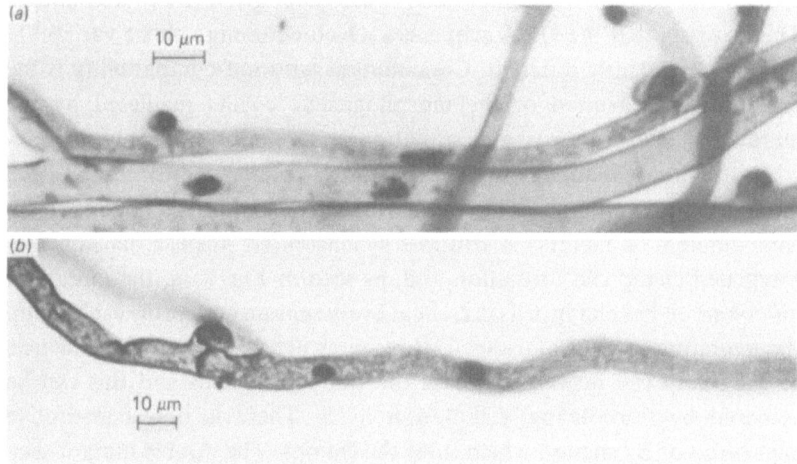

Fig. 8.3. Effect of mutation in the A and B incompatibility factors. (*a*) A_{mut} monokaryon with uninucleate cells and false clamps. (*b*) $A_{mut}B_{mut}$ double mutant with binucleate cells and true clamp connections.

best-analysed mating type genes, those of the ascomycete *Saccharomyces cerevisiae* (see Nasmyth, 1982).

In *Schizophyllum commune* mutation in *B* leads to development of a grossly abnormal monokaryon in which mating functions are constitutively expressed (Parag, 1962). Disruption of septa (Koltin & Flexer, 1969) is associated with high levels of lytic enzymes (Wessels, 1969) which also remove storage glucan from the cell wall. This induction of high levels of lytic enzymes may be in response to an *in vivo* deficiency in mitochrondrial function reported by Hoffman & Raper (1971, 1974) which leads to inefficient utilisation of substrate carbon source. Because of abnormal growth of mutant *B* monokaryons of *S. commune*, it is relatively easy to recognise and recover secondary mutations within the B factor which restore normal growth by loss of constitutive *B* function (Raper, Boyd & Raper, 1965; Raper & Raudaskoski, 1968; Raper & Raper, 1973). Such mutations always lead to varying degrees of loss of function. This is indeed what would be expected if mutations were accumulating to prevent the normal function of a positive regulator.

It seems likely that the tetrapolar incompatibility system has been derived from the bipolar as a more efficient system for restricting inbreeding. The rather different effects of mutation in *B* in *Coprinus cinereus* and *Schizophyllum commune* may well reflect a slightly different distribution of functions controlled by the two mating type factors.

Nuclear migration and inheritance of mitochondria

The nuclear migration which precedes dikaryon formation is not essential, dikaryon formation can occur without it; but it does appear to be advantageous in nature, as seen in the non-clamp-forming *Coprinus patouillardii* described by Kemp (1980). This species no longer requires *B* function in clamp cell fusion, and compatability with different alleles of a single *A* factor is sufficient for dikaryon formation without nuclear migration. The *B* factor is still present in the genome, however, and when different alleles are present nuclear migration occurs. Whatever its biological significance, there is certainly one very important consequence of nuclear migration which is that it imposes uniparental inheritance of cytoplasmic components of the cell, in particular the mitochondria.

In organisms that have differentiated gametes it is common to find that the mitochondrial genome of the zygote is exclusively derived from the maternal parent. In fungi this is evident in *Neurospora crassa* where the first mitochondrial gene mutations were described (Mitchell & Mitchell, 1952). Such mutations always show complete maternal inheritance. Despite the fact that the essential sexual interaction in basidiomycetes, dikaryon formation, is between purely vegetative hyphae, nuclear migration can just as effectively exclude any mitochondrial contribution from the donor parent. Although migration of donor nuclei involves widespread disruption of septa in recipient cells, there is no exchange and accompanying migration of mitochondria.

Illustrated in Fig. 8.4 are the reciprocal dikaryons formed on mating monokaryons of *C. cinereus* which differ in alleles of a mitochondrial gene. The mutation, [*acu*-10], causes a cytochrome oxidase deficiency resulting in a severe respiratory defect and poor growth (Casselton & Condit, 1972). Following hyphal fusion and reciprocal exchange of nuclei, nuclear migration has led to the establishment of two distinct reciprocal dikaryons having identical nuclei but only the mitochondria of the recipient mycelium.

If there is no exchange of mitochondria during dikaryosis, it would seem that there is no opportunity for recombination of mitochondrial genes. Recombination does occur, but recombinant mitochondria can only be found in the region of hyphal anastomosis (Baptista-Ferreira, Economou & Casselton, 1983). To show this, monokaryons with different mitochon-

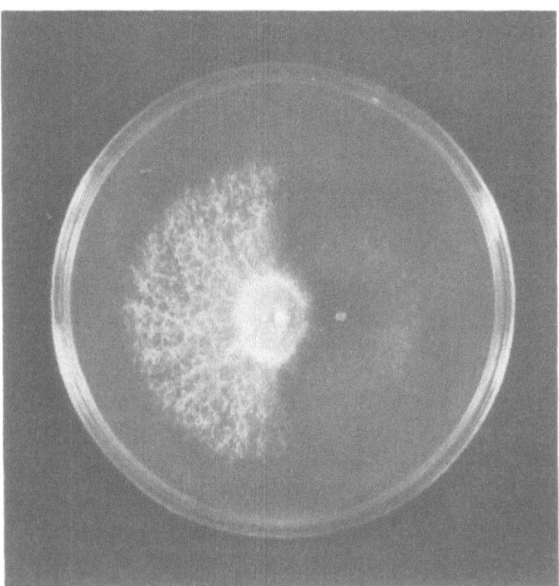

Fig. 8.4. Distinctive reciprocal dikaryons of *Coprinus cinereus* derived by mating a monokaryon with normal mitochondria (left) and a mutant with defective mitochondria (right).

Table 8.3. *Comparison of J, H and H/J recombinant mitochondrial DNA by means of restriction with enzymes EcoR1 and PstI. Fragment sizes are given in kilobases. Fragments of interest are shown in italics.*

	EcoR1 fragments			PstI fragments		
	J	H	H/J	J	H	H/J
			23.36		*15.49*	*15.49*
	22.13	22.13		*14.26*		
	11.87	11.87	11.87	10.47	10.47	10.47
	3.89	3.89	3.89	5.50	5.50	5.50
	2.09	2.09	2.09	3.37	3.37	3.37
	1.05	1.05	1.05	*3.16*		*3.16*
	0.92	0.92	0.92	2.82	2.82	2.82
				2.25	2.25	2.25
					1.93	
Total size	41.95	41.95	43.18	41.83	41.83	43.06

drial (mt) gene mutations were used so that it was possible to screen for recombinant genotypes. The two mutations used were [*acu*-10] (illustrated in Fig. 8.4) which can be distinguished by the mutant's inability to utilise acetate as sole carbon source for growth, and [*cap*-1] which confers resistance to the mitochondrial ribosome inhibitor chloramphenicol.

Neither [*acu*-10] nor [*cap*-1] causes a detectable change in the mitochondrial (mt) DNA but in the experiment reported here, we were able to demonstrate recombination at the molecular level using a naturally

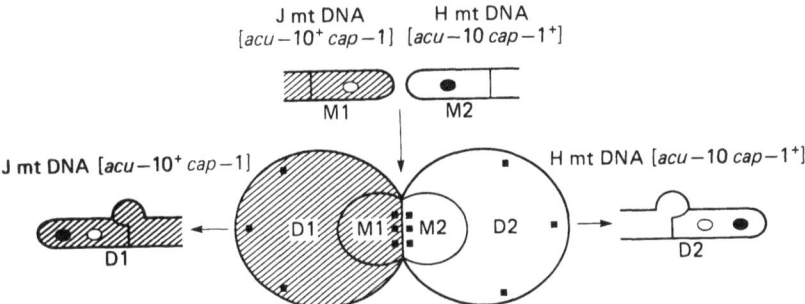

Fig. 8.5. Cross designed to select for recombination between two mitochrondrial genes. Cross: J [*acu*-10+*cap*-1] × H [*acu*-10 *cap*-1+] where J and H represent the DNA types and [*acu*-10/*acu*-10+] and [*cap*-1/*cap*-1+] alleles of the mitochondrial genes. (M1, M2, parent monokaryons; D1, D2, reciprocal dikaryons; ■, sampling sites.)

Table 8.4. *Segregation of mitochondrial genes during dikaryosis in the cross:* H[acu-*10 cap-1*$^+$] × J(acu-*10*$^+$ *cap-1*] *illustrated in Fig. 8.5. Data from two independent crosses (1) and (2).*

		Somatic segregation			
		Parental genotypes		Recombinant genotypes	
Cytoplasm of parent		acu-10 cap-1$^+$	acu-10$^+$ cap-1	acu-10$^+$ cap-1$^+$	acu-10 cap-1
[acu-10$^+$ cap-1]	(1)	0	60	0	0
(D1)	(2)	0	60	0	0
[acu-10 cap-1$^+$]	(1)	60	0	0	0
(D2)	(2)	60	0	0	0
Junction	(1)	0	117	3	0
	(2)	0	117	3	0

Entries show the numbers of mycelium samples exhibiting the indicated genotype. Refer to text and Fig. 8.5 for details.

occurring restriction enzyme site polymorphism. The two DNA types are designated H and J. As summarised in Table 8.3, there is no difference in the sizes of fragments when H and J DNAs are restricted with *EcoRI* but restriction with *PstI* shows that there is a difference in the relative position of the two restriction sites. H DNA yields two fragments of 15.49 and 1.93 Kb which are replaced by two fragments of 14.26 and 3.16 Kb in digests of J DNA.

The strategy for setting up the cross and DNA type and mitochondrial genotype of the participants is illustrated in Fig. 8.5. Following reciprocal dikaryon formation, three samples were taken from each dikaryon to confirm that these had only the recipient mt genome, and six samples from the junction region where the initial hyphal fusions occurred between mated monokaryons. This junction region is not dikaryotic but samples develop into dikaryon on further growth. All samples were grown into colonies to promote any somatic segregation, then 20 samples from each were tested for mt genotype. The results of these tests are shown in Table 8.4.

As expected, only the recipient mt genotype was recovered in samples taken from the two reciprocal dikaryons showing that there had been no exchange of mitochondria during nuclear migration. From the junction region the majority of samples had the [*acu*$^+$*cap-1*] parental genotype but from two separate crosses three recombinant genotypes were detected. Mt

DNA was extracted from one of these recombinants and a *Pst I* digest of this and mt DNA from the two reciprocal dikaryons is compared in Fig. 8.6. It will be seen that the mt DNAs recovered from the two reciprocal dikaryons were identical to their respective parental mt DNAs whereas the recombinant had a new restriction pattern. It had the H 15.49 Kb fragment together with the J 3.16 Kb fragment. Recombination between [*acu*-10] and [*cap*-1] generated a recombinant mt genome some 1.23 Kb larger than either of the parental genomes. This can be seen in both the *Eco RI* and *Pst I* restrictions (Table 8.3). The most likely difference between H and J mt DNA is a small translocation which becomes duplicated in the recombinant.

It is evident that recombination of mitochondrial genomes can occur in *Coprinus cinereus* but it has required careful genetic selection to show it in a region of hyphal interaction which is not normally dikaryotic. A possible significance of nuclear migration in nature is that it ensures that nuclear exchange for sexual reproduction is successful even in the presence of cytoplasmically determined incompatibilities.

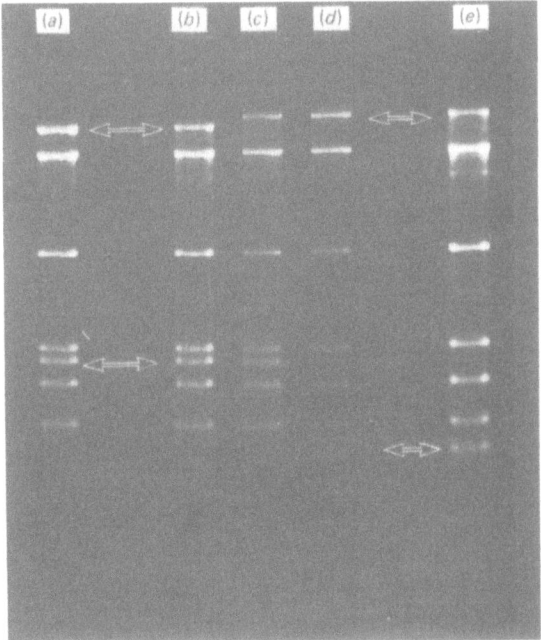

Fig. 8.6. *Pst I* restriction of mitochondrial DNA. (*a*) J standard DNA; (*b*) [*acu*-10⁺*cap*-1] dikaryon DNA; (*c*) [*acu*-10⁺*cap*-1⁺] recombinant DNA; (*d*) [*acu*-10 *cap*-1⁺] dikaryon DNA; (*e*) H standard DNA.

References

Baptista-Ferreira, J. L. C., Economou, A. & Casselton, L. A. (1983). Mitochondrial genetics of *Coprinus*: recombination of mitochondrial genomes. *Current Genetics*, **7**, 405–7.

Buller, A. H. R. (1931). *Researches on Fungi*, vol. 4. London: Longmans, Green & Co.

Casselton, L. A. (1965). The production and behaviour of diploids of *Coprinus lagopus*. *Genetical Research*, **6**, 190–208.

Casselton, L. A. (1978). Dikaryon formation in higher basidiomycetes. In *The filamentous Fungi*, vol. 3, ed. J. E. Smith & D. R. Berry, pp. 275–97. London: Edward Arnold.

Casselton, L. A. & Condit, A. (1972). A mitochondrial mutant of *Coprinus lagopus*. *Journal of General Microbiology*, **72**, 521–7.

Casselton, L. A. & Lewis, D. (1966). Compatibility and stability of diploids in *Coprinus lagopus*. *Genetical Research*, **8**, 61–72.

Casselton, L. A., Lewis, D. & Marchant, R. (1971). Septal structure and mating behaviour of common *A* diploids of *Coprinus lagopus*. *Journal of General Microbiology*, **66**, 273–8.

Day, P. R. (1963a). The structure of the *A* mating-type factor in *Coprinus lagopus*: wild alleles. *Genetical Research*, **4**, 323–5.

Day, P. R. (1963b). Mutations affecting the *A* mating-type locus in *Coprinus lagopus*. *Genetical Research*, **4**, 55–65.

De Vries, O. M. H., Hoge, J. H. C. & Wessels, J. G. H. (1980). Translation of RNA from *Schizophyllum commune* in a wheat germ and rabbit reticulocyte cell-free system. *Biochimica et Biophysica Acta*, **607**, 373–8.

Giesy, R. M. & Day, P. R. (1965). The septal pores of *Coprinus lagopus* (Fr.) *sensu* Buller in relation to nuclear migration. *American Journal of Botany*, **52**, 287–94.

Haylock, R. W., Economou, A. & Casselton, L. A. (1980). Dikaryon formation in *Coprinus cinereus*: selection and identification of *B* factor mutants. *Journal of General Microbiology*, **121**, 17–26.

Hoffman, R. M. & Raper, J. R. (1971). Genetic restriction of energy conservation in *Schizophyllum*. *Science*, **171**, 418–19.

Hoffman, R. M. & Raper, J. R. (1974). Genetic impairment of energy conservation in development of *Schizophyllum*. Efficient mitochondria in energy-starved cells. *Journal of General Microbiology*, **82**, 67–75.

Kemp, R. F. O. (1980). Bifactorial incompatibility without clamp connections in *Coprinus patouillardii* group. *Transactions of the British Mycological Society*, **74**, 355–60.

Koltin, Y. (1968). The genetic structure of the incompatibility factors of *Schizophyllum commune*. Comparative studies of primary mutations in the *B* factor. *Molecular and General Genetics*, **102**, 196–203.

Koltin, Y. & Flexer, A. S. (1969). Alteration of nuclear migration in *B*-mutant strains of *Schizophyllum commune*. *Journal of Cell Science*, **4**, 739–49.

Koltin, Y., Stamberg, J., Bawnik, N., Tamarkin, R. & Werczberger, R. (1979). Mutational analysis of natural alleles in and affecting the *B* incompatibility factor of *Schizophyllum*. *Genetics*, **93**, 383–91.

Koltin, Y., Stamberg, J. & Lemke, P. A. (1972). Genetic structure and evolution of the incompatibility factors in higher fungi. *Bacterial Reviews*, **36**, 156–71.

Marchant, R. & Wessels, J. G. H. (1974). An ultrastructural study of septal dissolution in *Schizophyllum commune*. *Archives of Microbiology*, **96**, 115–24.

Mitchell, M. B. & Mitchell, H. K. (1952). A case of "maternal" inheritance in *Neurospora crassa*. *Proceedings of the National Academy of Sciences, USA*, **38**, 442–9.

Moore, R. T. & McAlear, J. H. (1962). Fine structure of Mycota. 7. Observations on septa of Ascomycetes and Basidiomycetes. *American Journal of Botany*, **49**, 86–94.

Nasmyth, K. A. (1982). Molecular genetics of yeast mating type. *Annual Review of Genetics*, **16**, 439–500.

Papazian, H. (1954). Exchange of incompatibility factors between the nuclei of a dikaryon. *Science*, **119**, 691–3.

Parag, Y. (1962). Mutations in the *B* incompatibility factor of *Schizophyllum commune*. *Proceedings of the National Academy of Sciences, USA*, **48**, 743–50.

Raper, C. A. (1983). Controls for development and differentiation of the dikaryon in basidiomycetes. In *Secondary Metabolism and Differentiation in Fungi*, ed. J. W. Bennett & A. Ciegler, pp. 195–238. New York and Basel: Marcel Dekker, Inc.

Raper, C. A. & Raper, J. R. (1973). Mutational analysis of a regulatory gene for morphogenesis in *Schizophyllum*. *Proceedings of the National Academy of Sciences, USA*, **70**, 1427–31.

Raper, J. R. (1966). *Genetics of sexuality in higher fungi*. New York: The Ronald Press.

Raper, J. R., Baxter, M. G. & Ellingboe, A. H. (1960). The genetic structure of the incompatibility factors of *Schizophyllum commune*: the *A* factor. *Proceedings of the National Academy of Sciences, USA*, **46**, 833–42.

Raper, J. R., Baxter, M. G. & Middleton, R. B. (1958). The genetic structure of the incompatibility loci in *Schizophyllum*. *Proceedings of the National Academy of Sciences, USA*, **53**, 889–900.

Raper, J. R., Boyd, D. H. & Raper, C. A. (1965). Primary and secondary mutations at the incompatibility loci in *Schizophyllum*. *Proceedings of the National Academy of Sciences, USA*, **53**, 1324–32.

Raper, J. R. & Oettinger, M. T. (1962). Anomolous segregation of incompatibility factors in *Schizophyllum commune*. *Revista de Biologia*, **3**, 205–21.

Raper, J. R. & Raudaskoski, M. (1968). Secondary mutations at the *B* incompatibility locus of *Schizophyllum*. *Heredity*, **23**, 109–17.

Raudaskoski, M., Stamberg, J., Bawnik, N. & Koltin, Y. (1976). Mutational analysis of natural alleles at the *B* incompatibility factor of *Schizophyllum commune*: 2 and 6. *Genetics*, **83**, 507–16.

Swiezynski, K. M. & Day, P. R. (1960*a*). Heterokaryon formation in *Coprinus lagopus*. *Genetical Research*, **1**, 114–28.

Swiezynski, K. M. & Day, P. R. (1960*b*). Migration of nuclei in *Coprinus lagopus*. *Genetical Research*, **1**, 129–39.

Takemaru, T. (1961). Genetic studies on fungi. X. The mating system in Hymenomycetes and its general mechanism. *Biological Journal Okayama University*, **7**, 133–211.

Wessels, J. G. H. (1969). Biochemistry of sexual morphogenesis in *Schizophyllum commune*: effect of mutations affecting the incompatibility system on cell-wall metabolism. *Journal of Bacteriology*, **98**, 697–704.

Wessels, J. G. H. & Marchant, R. (1974). Enzymic degradation of septa in wall preparations from a monokaryon and a dikaryon of *Schizophyllum commune*. *Journal of General Microbiology*, **83**, 359–68.

9

Cytology of hyphal interactions and reactions in *Schizophyllum commune*

N. K. TODD AND R. C. AYLMORE

Department of Biological Sciences, Washington Singer Laboratories, University of Exeter, Exeter EX4 4QG, UK

Introduction

Gregory (1984) in his elegant address to the British Mycological Society meeting on 'The ecology and physiology of fungal mycelium' outlined the history of ideas about the fungal mycelium (see also Buller, 1931, 1933) and stressed the need for more basic cytological work on living material. At this same meeting, we presented findings on the fusion between living hyphae of *Coriolus versicolor* (Aylmore & Todd, 1984a). This, and other work on mycelial interactions, was made possible by a technique which combines both light and electron microscopy and allows the continuous observation of living hyphae up to moments before fixation for ultrastructural studies (Aylmore, 1983; Aylmore & Todd, 1984b). We have now extended our work to examine the cytology of various phenomena in *Schizophyllum commune*. Here we present observations on the reaction of septa in response to physical damage of hyphae, hyphal and clamp cell fusion, and the behaviour of nuclei following fusion.

Septal sealing

Somatic incompatibility in many species of higher fungi is a post-fusion event and often involves the lysis and degeneration of the protoplasm within the fused compartments (e.g. *Neurospora* (Garnjobst & Wilson, 1956); *Rhizoctonia* (Flentje & Stretton, 1964; Parmeter *et al.*, 1969); *Phanerochaete* (Rayner *et al.*, 1984)). Such interactions are generally localised to the fusion segment and immediate cells. It seems likely therefore, that septa are responsible for containing this type of reaction by a plugging mechanism. Indeed, in ascomycetous fungi, plugging mechanisms are well documented (Reichle & Alexander, 1965; Brenner & Carroll, 1968; Furtado, 1971; McKeen, 1971; Wergin, 1973; Trinci & Collinge,

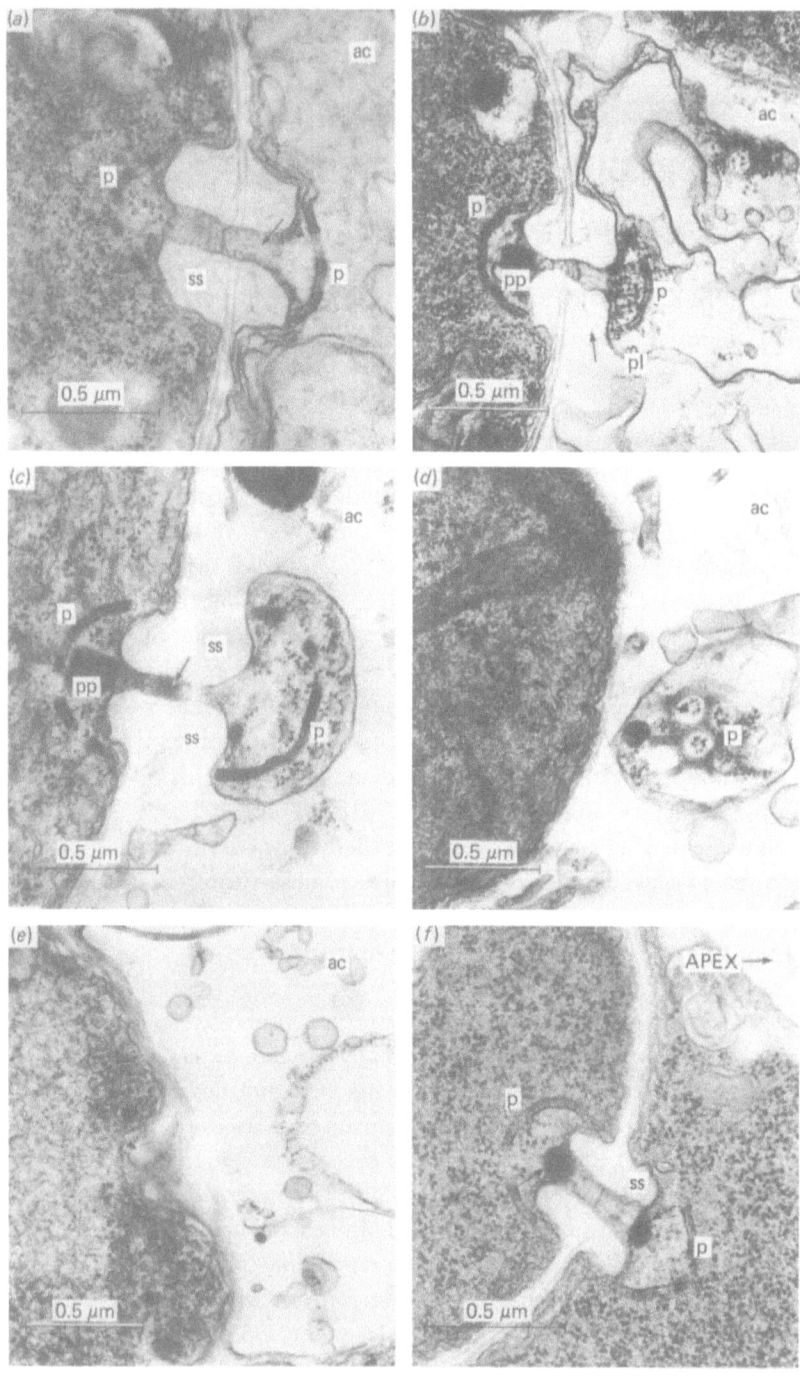

1973; Brouchart & Demoulin, 1975; Collinge *et al.*, 1978) and, in basidiomycetes, the highly differentiated septal apparatus, the dolipore (R. T. Moore: Chapter 7), is often found blocked by electron-dense material (Bracker & Butler, 1963; Koltin & Flexer, 1969; Casselton *et al.*, 1971; Moore & Marchant, 1972; Setliff *et al.*, 1972; Craig *et al.*, 1977). Here, in *S. commune*, we have employed physical means (using a tungsten needle to injure individual apical compartments) to mimic such cellular damage and have studied the role of the dolipore in containing the reaction. The technique involved puncturing the leading hyphae of mycelia grown on cellophane membranes. These were then processed and sectioned longitudinally for electron microscopy using methods described previously (Aylmore & Todd, 1984*b*). Septa present in damaged hyphae were examined and compared to those found in undamaged regions of the same mycelium, these latter serving as controls.

The septa in the apical regions of undamaged leading hyphae showed typical dolipore structure, with their pore channels unobstructed (Fig. 9.1*a*). However, when septa in punctured hyphae were examined, they showed a response which followed a consistent train of events. In specimens fixed within 30 s, instantaneous plugging, which prevented any

Fig. 9.1. (*a–e*) Transmission electron micrographs showing the septa of apical compartments (ac) of leading dikaryotic hyphae of *Schizophyllum commune*. (*a*) An undamaged actively growing hypha. Note the typical dolipore structure with septal swellings (ss) and parenthesomes (p). The cross wall has a layered appearance with two bands of dense-staining material sandwiching a non-staining region. The pore channel (arrowed) is unoccluded. (*b*) Specimen fixed within 30 s of apical compartment puncture. An electron-dense plug (pp) occludes one end of the pore channel. Note the intact parenthesomes (p). The plasmalemma (pl) has detached to expose regions of the septal swelling (arrowed). Little cytoplasm has escaped into the disrupted cell. (*c*) Within 5 min the septal swellings (ss) are completely exposed and begin to degenerate. Note the pore plug (pp) with electron-dense material (arrowed) extending into the pore channel. The parenthesomes (p) on both sides remain intact. (*d*) Near-median longitudinal section of hypha fixed 15 min after apical compartment puncture. The parenthesome (p) present in the empty cell has detached from the cross wall. (*e*) 1 h after puncture the detached apical apparatus in the empty compartment has degenerated. On the other side of the cross wall the septal swellings, now devoid of the parenthesome, have deformed to give a single pad-like structure (arrowed) with staining properties similar to the dense material present in the rest of the cross wall. (*f*) The second septum behind a punctured apical compartment (direction of apex indicated), fixed 30 s after damage. Note the electron-dense plugs (arrowed) occluding either end of the pore channel leaving the central region clear. The septal swellings (ss) and parenthesomes (p) remain intact.

significant loss of cytoplasm, was accomplished by the formation of electron-dense plugs at the mouth of the pore channel (Fig. 9.1*b*). In the following 15–20 min, the plug was consolidated in the pore channel (Fig. 9.1*c*). At the same time there was a progressive cleavage, away from the cross wall, of the septal swelling present in the empty damaged cell (Fig. 9.1*c,d*). Throughout this the parenthesomes remained intact. After 1 h the detached septal apparatus degenerated leaving a plain surface on this side of the cross wall. In the penultimate cell, however, the septal apparatus behaved in a different way with the swelling deforming and fusing over and around the pore plug (Fig. 9.1*e*). This pad-like structure assumed staining properties similar to the outer layers found in normal sections through cross walls (see Fig. 9.1*a*). Interestingly, examination of the second septum back from a damaged apex, fixed within 30 s, also showed an occlusion reaction, with electron-dense plugs forming either end of the pore channel leaving the central region clear (Fig. 9.1*f*).

From these observations, septal sealing seems to be a two-stage process, the first stage being the instantaneous plugging of the pore channel, with the septal swellings and parenthesomes apparently playing no part in these initial events. The pore plugs appear to form *in situ*, as if by a process of coagulation, and closely resemble structures reported by other workers (Bracker & Butler, 1963; Koltin & Flexer, 1969; Casselton *et al.*, 1971; Moore & Marchant, 1972; Setliff *et al.*, 1972; Craig *et al.*, 1977). The second stage starts within several minutes of damage being inflicted and entails the consolidation of the plug within the channel and the detachment and eventual degeneration of the septal apparatus present in the ruptured compartment. At the same time a more permanent seal is formed on the other side of the cross wall. The resulting pad-like structures have been seen before and in situations involving senescence of hyphal compartments (Bracker & Butler, 1963; Wells, 1964).

The detachment of the septal swellings as described here lends support to the view that they may be labile in nature, having different chemical and physical properties to the rest of the cross wall (for discussion see Bracker & Butler, 1963, 1964; Wells, 1964; Bracker, 1967; Moore & Marchant, 1972; Wessels & Sietsma, 1979). How, and to what extent, the plugging stimulus is transmitted through hyphae is not known. During regeneration and growth of the damaged hyphae, the plugging of septa not in direct contact with the ruptured cell may be reversible. We are currently investigating the role of septal sealing during expression of somatic incompatibility and the way in which this may contain the reaction in fused segments.

Hyphal fusion

The fusion process in *Schizophyllum commune* appears to be similar to that observed in *Coriolus versicolor* (Aylmore & Todd, 1984a). Briefly, this involves the opening of a single pore which undergoes regular and symmetrical enlargement until it reaches the full hyphal diameter (Fig. 9.2d,e). The majority of successful fusions in *S. commune* were of the tip-to-side type, although signs of telemorphotic and zygotropic behaviour were more obvious in this fungus than in *C. versicolor* grown under the same conditions (Fig. 9.2b). Most hyphal contacts did not result in fusion; rather, bifurcation or deflection of the growing apex occurred (Fig. 9.2a). Interestingly, whereas successful fusions were comparatively rare events, when found they were often clustered in the same area of the mycelium. This is clearly seen in Fig. 9.2f, where it is possible to distinguish particular hyphae participating in fusion at very high frequency, giving a ladder-like appearance. It should be remembered that these observations were made in the growing periphery of the colony.

In the accepted model of hyphal extension (Bartnicki-Garcia, 1973), some of the vesicles present in the apex are presumed to contain wall lytic enzymes. However, attempts to examine in detail the role of the apical apparatus during anastomosis were hampered by the difficulty in distinguishing the earlier stages of tip-to-side fusions from non-fusing cell contacts. Thus ultrastructural examination to date in *S. commune* and *C. versicolor* (Aylmore & Todd, 1984a) has given little clue as to the behaviour of the apical vesicles. Therefore, in an attempt to observe the very early stages of the fusion process, we have examined clamp connection formation, an event in which anastomosis is virtually guaranteed.

The observations are shown in Fig. 9.3 and they illustrate several interesting features. The apical region of the clamp cell outgrowth resembles that of extending vegetative hyphae (Fig. 9.3a), containing numerous discrete membrane-bound vesicles. These become displaced in the direction of curvature of the clamp cell soon after its emergence (Fig. 9.3a,b). It appears that these structures are only present until a stage, before mitosis, when the walls have just touched, where they accumulate at the point of contact (Fig. 9.3c,d). Ultrastructural evidence for the onset of wall dissolution was only found after disappearance of the apical vesicles, completion of mitosis and during later stages of septum formation (see Fig. 9.3e–i). As in hyphal anastomosis, clamp cell fusion involves localised wall lysis and enlargement of a single fusion pore (Fig. 9.3j).

Drawing these observations together, we can speculate on a possible role for apical vesicles in the fusion process. Their accumulation and sudden

Fig. 9.2. Micrographs showing dikaryotic hyphae of *Schizophyllum commune*. (*a*) Photomicrograph showing bifurcation of a non-fusing tip-to-side contact. (*b*) Photomicrograph showing attracted growth of a branch apex preceding tip-to-side contact (arrowed). This eventually fused. (*c*) Electron micrograph showing fully mature fusion bridge (F) formed from the tip-to-tip contact of two branch initials. Two nuclei (n) are present close to the site of anastomosis. (*d*) Photomicrograph showing hyphae fusing at F. (*e*) Electron micrograph of a median section through the fusion shown in (*d*). The cytoplasm is continuous through a single fusion pore (fp). The remains of the fusion cross wall (cw) are evident. (*f*) A ladder-like appearance created by a high frequency of fusions (F) between two parallel hyphae.

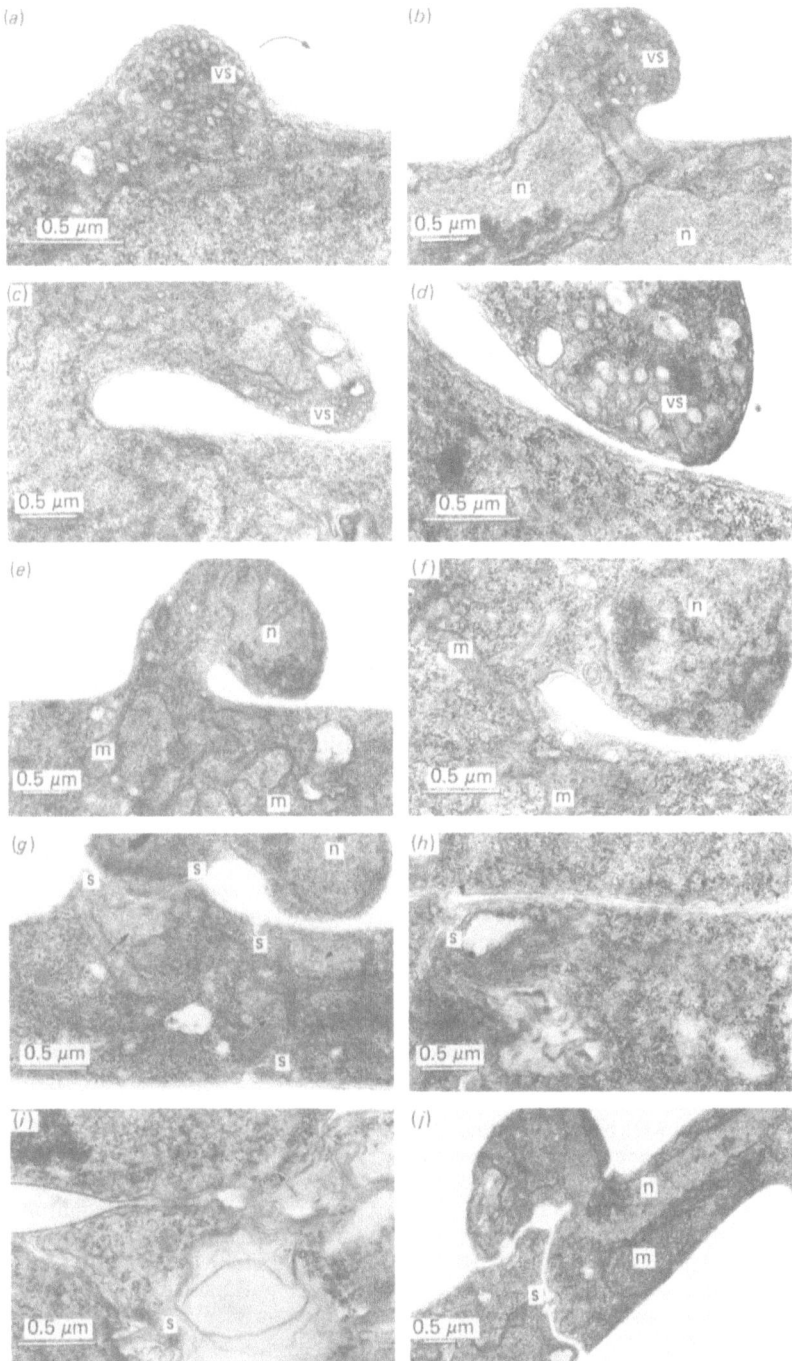

Fig. 9.3. For caption see p. 238

disappearance at points of contact may be evidence that they discharge their contents in the wall prior to dissolution. Indeed, this would be consistent with previous micrographs showing that vesicles can be found at points of wall lysis during various fusion events (Hawker & Beckett, 1971; Harvey, 1975; Rijkenberg & Truter, 1975; van der Valk & Marchant, 1978) and provide circumstantial evidence for the notion that the apical apparatus contains wall-lytic enzymes. It is interesting that although they are released soon after contact, visible signs of wall lysis are delayed. This may explain the absence of convincing evidence on this point in our studies of fusing hyphae in *S. commune* and earlier in *C. versicolor* (Aylmore & Todd, 1984*a*).

Clamp connection formation may provide a good system for studying the regulation of a developmental pathway in that the biochemically opposite processes of wall *lysis* during fusion and *synthesis* during septum formation are occurring in close spatial and temporal proximity. Indeed, Girbardt's elegant studies on how sites of septum formation are controlled have already thrown some light on this aspect (Girbardt, 1979).

Post-fusion events

Our work on the cytology of post-fusion events in *Coriolus versicolor* produced some interesting findings concerning the behaviour of nuclei which show novel patterns of intracellular migration, disintegration

---

Fig. 9.3. Clamp connection development and fusion in dikaryons of *Schizophyllum commune*. (*a*) An aggregate of vesicles (vs) at the site of clamp cell initiation. Note the apparent displacement of the vesicles in the direction of future curvature (arrowed). (*b*) A more mature clamp initial. Vesicles (vs) occur in the extending apex. Two nuclei (n) are present in the hypha. (*c*), (*d*) Clamps fixed soon after contact with the parent hypha, just before the onset of mitosis. Vesicles (vs) are present, accumulated in the apex at the point of wall appression. (*e*), (*f*) Specimens fixed just after mitosis and entry of the nucleus (n) into the clamp. Vesicles are now absent from the apex in the region of contact with main hypha. Note the numerous mitochondria (m). (*g*) Septa (s) forming by annular ingrowth. Note the regions of amorphous material bounded by a double membrane (arrowed) associated with these structures. The nucleus (n) is present in the clamp. No vesicles are evident in the clamp apex and there is no sign of wall lysis. (*h*) As in (*g*), septa (s) are forming. Note that the region of contact between the clamp and wall of the main hypha (between arrows) show no sign of lysis. Vesicles are absent from this region. (*i*) With the dolipore septum almost fully formed (s), wall lysis and clamp fusion has occurred at point arrowed. (*j*) Migration of nucleus (n) through enlarging fusion pore. Note the forming dolipore septum (s) and enlarged mitochondrion (m) in the penultimate cell.

and division. In *C. versicolor* a nuclear replacement reaction usually follows fusion in both 'self' (intramycelial) and 'non-self' (intermycelial) situations. The basics of this are shown in Fig. 9.4. For the work on *S. commune*, however, we have examined only 'self' fusions occurring within both dikaryotic and monokaryotic mycelia. In the case of the latter, detailed examination of ten fusions failed to give any consistent pattern. The events were variable, ranging from persistence of the binucleate condition in the fusion segment to partial disintegration and fragmentation of one of the nuclei *but* on no occasion was the complete nuclear

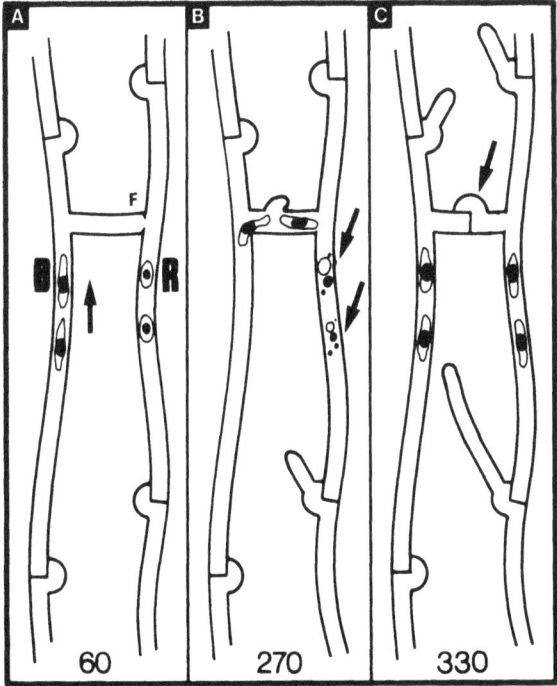

Fig. 9.4. The nuclear replacement reaction in *Schizophyllum commune* following fusion of genetically identical dikaryotic hyphae. The numbers indicate minutes after fusion. (A) Tip-to-side fusion (F) results in a transient tetranucleate compartment. The nuclei of the donor cell (D) begin migration towards the site of anastomosis while those in the recipient cell (R) round up and remain stationary. (B) Following fragmentation and degeneration of the recipient cell nuclei (arrowed), the donor pair stabilise position close to the fusion and begin conjugate mitosis. (C) The fused segment is converted into two binucleate compartments, separated by an intercalary clamp connection (arrowed). Both compartments possess nuclei derived from the original donor cell. (Diagram not to scale.)

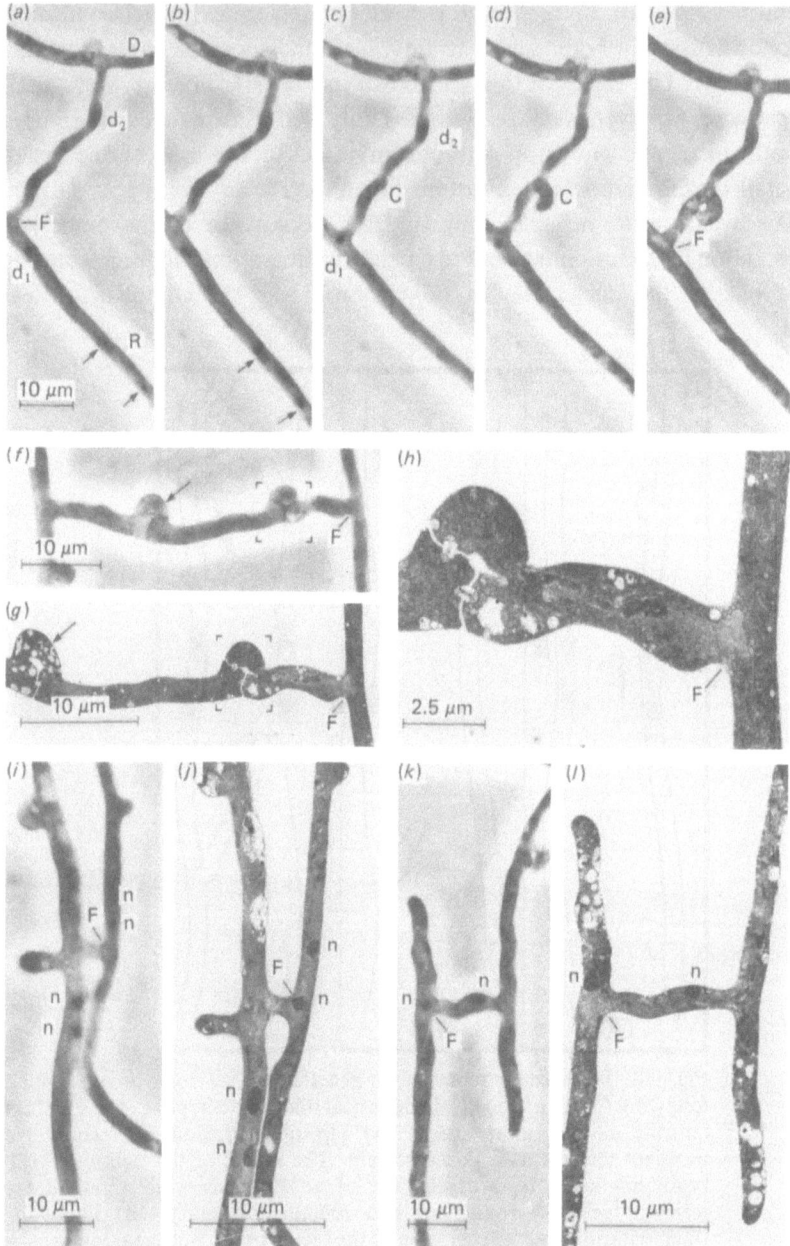

Fig. 9.5. Fusions between genetically identical dikaryotic hyphae of *Schizophyllum commune*. D = donor; R = recipient. (*a–e*) Photomicrographs showing tip-to-side fusion at (F). The donor cell nuclei (d_1 and d_2) migrate towards the anastomosis site, one entering the recipient cell. A

replacement found. On the other hand, the reaction of nuclei in most dikaryotic fusions (we examined 18 in all) was identical to that found previously in *C. versicolor*; only rarely was exceptional behaviour noted. Here we focus attention on the cytology of dikaryotic fusions and the details of the nuclear degeneration process.

The essential features of the replacement reaction in dikaryons are shown in Fig. 9.5. The participating cells display a donor–recipient relationship in which both nuclei of the recipient cell degenerate and are replaced by a normal conjugate division of the donor pair, the original tetranucleate fusion cell being converted into two binucleate compartments which remain separated by a clamp connection (see Fig. 9.5*a–e*). The intercalary clamps arising in this way can grow out in either direction (Fig. 9.5*a–g*), and possess typical structures with normal septal apparatus (Fig. 9.5*h*). The two principal stages, before and after degeneration of the nuclei, are clearly evident in the light micrographs and their confirmatory electron micrographs shown in Fig. 9.5*i–l*.

The studies on the degeneration of nuclei provided further details of the process. Under phase optics they appeared to undergo cycles of enlargement and contraction before fragmenting into several dense spherical bodies which eventually faded from view. Throughout this process, the nucleoli became diffuse within the nucleus accompanied by changes in their contrast properties. In later stages, it was sometimes possible to distinguish fragments of dense nucleolar-like material outside the nuclear envelope, free in the cytoplasm, often persisting for several hours (Fig. 9.6*a–f*). At the ultrastructural level the most obvious feature was the unusually dense-staining material in the nucleus or fragments of nuclei, possibly reflecting abnormal condensation of chromatin (see Figs 9.6*k*, 9.7*c,g*). This was accompanied by detachment of the chromatin mass from the nuclear envelope (Figs 9.6*j*, 9.7*g*) which was often associated with several layers

---

clamp initial (C) then forms which is followed by normal mitosis and clamp-connection formation. The clamp on the fusion bridge is in reverse orientation in relation to growth of the original branch. Fragments of recipient cell nuclei are arrowed in (*a*) and (*b*). (*f–h*) Photomicrograph (*f*) and electron micrographs (*g*, *h*) showing intercalary clamp formation following tip-to-side fusion at F. The clamp forming after the nuclear replacement (in brackets) has opposite orientation to that which formed at the base of the branch before fusion occurred (arrowed). Septa of intercalary clamps possess typical dolipore structure (*h*). (*i–l*) Photomicrographs (*i*, *k*) and their corresponding electron micrographs (*j*, *l*). These show tetranucleate (*i, j*) and binucleate (*k, l*) fusion compartments before and after nuclear degeneration, respectively. Points of fusion (F) and positions of nuclei (n) are indicated.

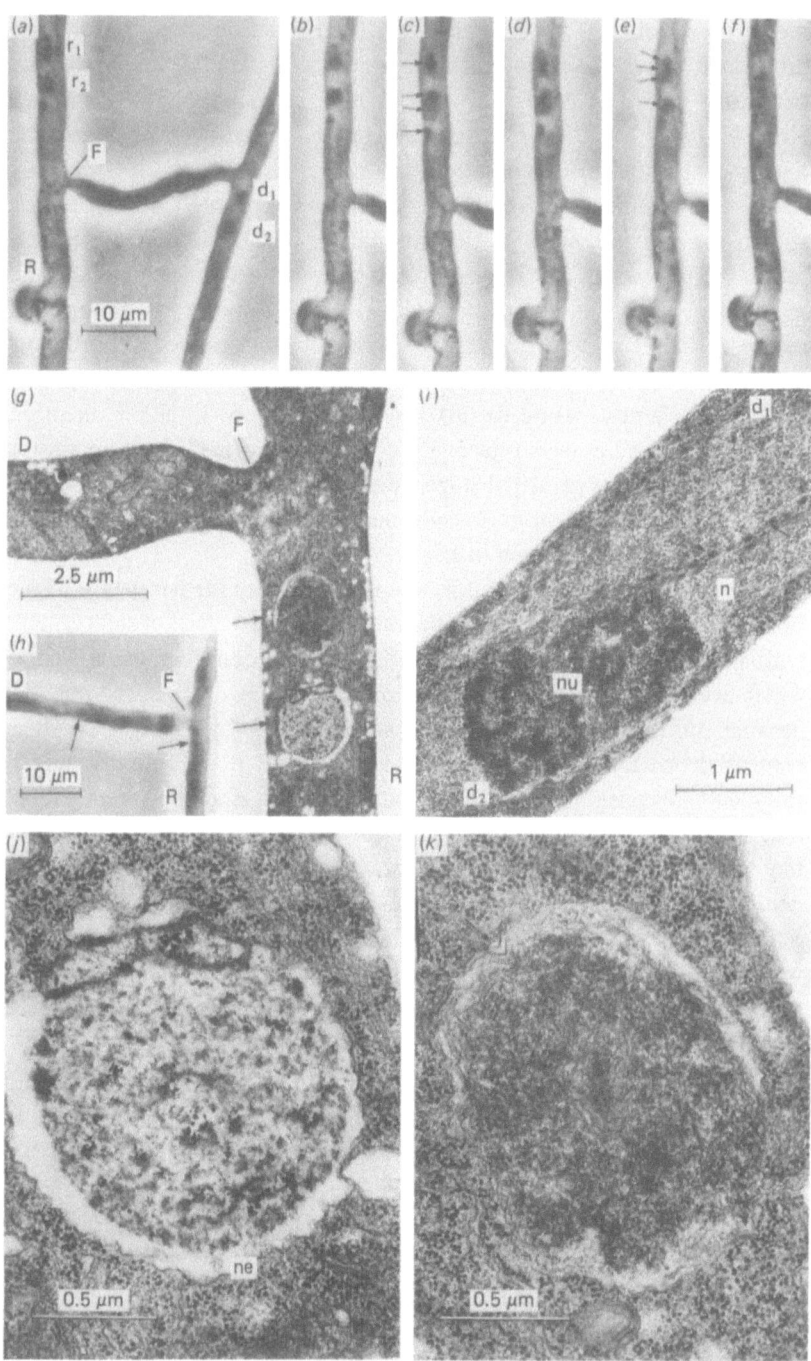

of membrane (Figs. 9.6*k*, 9.7*g*). Electron microscopy confirmed the presence of naked chromatin in the cytoplasm at the site of degeneration (Fig. 9.7*g,i*). The integrity of the cytoplasm surrounding these nuclei was maintained and the reaction was nuclear-specific.

From these observations, it is clear that the degeneration process in *S. commune* closely resembles that in *C. versicolor*. This work and the early reports of Bensaude (1918, cited in Papazian, 1958) and Noble (1937) suggests that fusion-induced nuclear degeneration is a widespread phenomenon in vegetative hyphae of many basidiomycetes. In addition to *S. commune*, these include *Coprinus cinereus* (Bensaude, 1918), *Typhula trifolii* (Noble, 1937), *Coriolus versicolor* (Aylmore & Todd, 1984*a*), *Hypholoma fasciculare* (Aylmore, unpublished) and possibly *Phanerochaete velutina* (Ainsworth, unpublished). Despite all these observations, we are no nearer to elucidating the mechanism underlying this process and obvious questions remain: (*a*) why does it occur – is it purely to maintain a constant number of nuclei per compartment, (*b*) what triggers nuclear breakdown – the nucleus-specific agent responsible appears to act transiently and in a unilateral fashion, (*c*) how are genetically identical nuclei sharing a common cytoplasm able to behave so differently? These and other issues concerning similar behaviour of fungal nuclei were discussed fully in our original article on *C. versicolor* (Aylmore & Todd, 1984*a*). The biochemistry of this process will be difficult to work out since fusion is a comparatively rare event that cannot, as yet, be induced synchronously *en*

Fig. 9.6. Micrographs showing the behaviour of nuclei following fusion in dikaryons of *Schizophyllum commune*. (*a–f*) A series of photomicrographs showing donor cell nuclei (d$_1$ and d$_2$) and degeneration of nuclei in the recipient cell (R) following fusion at F. Both recipient cell nuclei (r$_1$ and r$_2$) fragment into several spherical bodies of different density (arrowed in (*c*) and (*e*). (*g*, *h*) Photomicrograph (*h*) and electron micrograph (*g*) showing hyphae fused at F. The position of the nuclei present in both donor (D) and recipient (R) cells at the time of fixation is arrowed in (*h*). (*i*) A section through the donor nuclei d$_1$ and d$_2$. Note the typical appearance with a single granular nucleolus (nu) and uniform nucleoplasm (n). Patches of electron-dense material (arrowed) are associated with the nuclear envelopes in the region of contact. (*j*, *k*) Enlargements of the recipient cell nuclei shown in (*g*). The nucleus in (*j*) appears abnormal, the diffuse-staining chromatin apparently shrinking away from the nuclear envelope (ne). In (*k*) several layers of membrane (arrowed) are associated with the surface of the nucleus. The chromatin is amorphous and densely stained. In both micrographs the nuclei show circular profiles and lack an obvious nucleolar region. The cytoplasm surrounding the structures has normal appearance.

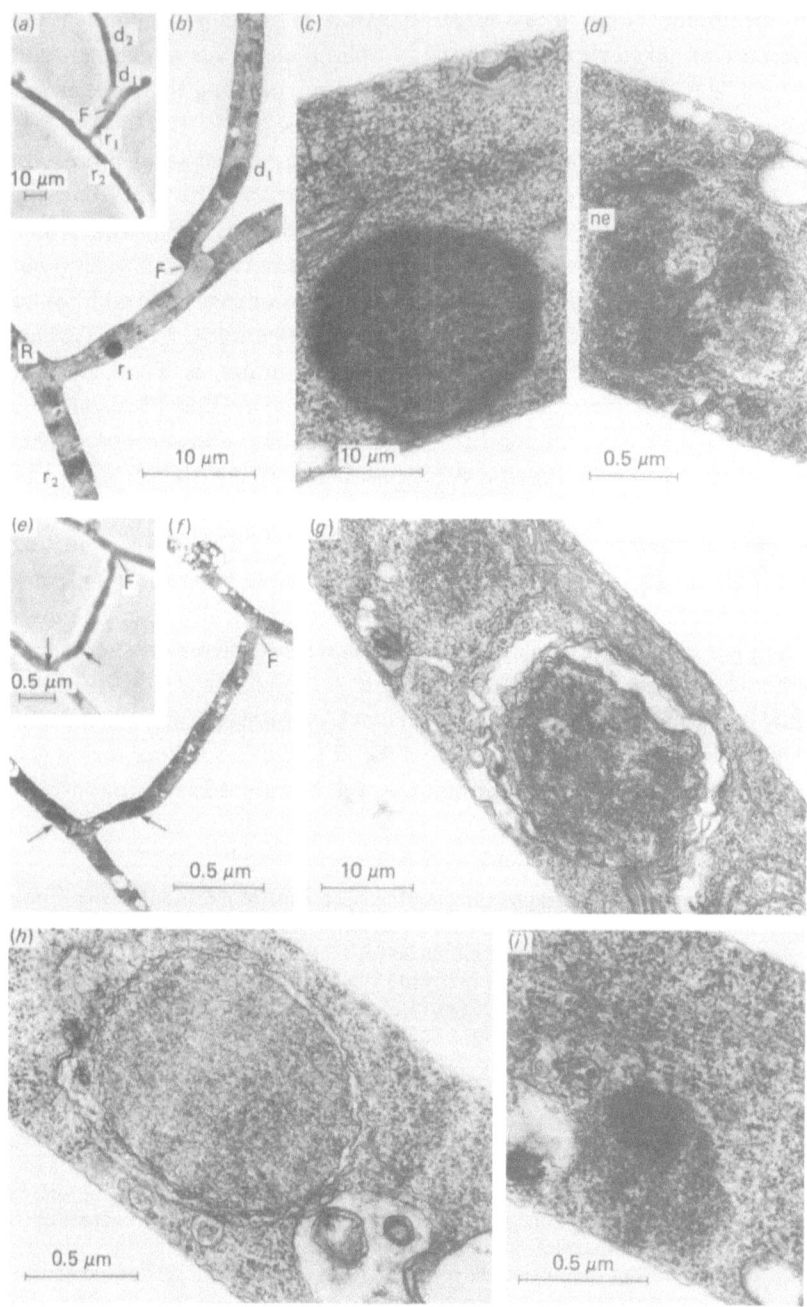

masse. However, the use of metabolic inhibitors to affect the process and enzyme cytochemistry may assist our understanding.

Conclusions

In this chapter we have provided the first full description of septal sealing, further information regarding the fusion process, particularly the role of the apical apparatus, and comprehensive details of nuclear behaviour following anastomosis in *Schizophyllum commune*. It is generally assumed that hyphae of filamentous fungi are functionally coenocytic, with each cell in communication with its neighbour through the septal pore. In ascomycetous species, which generally possess simple pores, the compartments comprising the actively growing margin of the colony (the so-called peripheral growth zone see Trinci, 1978*a,b*) are thought to act effectively as a single unit with wholesale translocation of protoplasm along the hyphae towards the extending apices. Whilst a fundamentally similar process is likely to occur in basidiomycetes, the extent of metabolic cooperation and exchange between individual compartments is not so clearly understood. The highly elaborate septal apparatus present in these fungi may provide potential for greater control of intercompartment communication and we could speculate that the basidiomycete hyphae show a greater tendency towards a more cellular organisation than is often supposed. Like ascomycetous species, basidiomycetes have the capacity to seal septa and isolate regions of hypha (Fig. 9.1). Although we looked at the response of septa to drastic damage, it is likely that similar

---

Fig. 9.7. Micrographs showing behaviour of nuclei following fusion in dikaryons of *Schizophyllum commune*. (*a*, *b*) Photomicrograph (*a*) and electron micrograph (*b*) showing fusion at F. The positions of the donor cell nuclei (d_1 and d_2) and recipient cell nuclei (r_1 and r_2) are indicated. Nucleus d_1 is just entering the recipient cell (R). (*c*) The remains of nucleus r_1 from (*b*) are seen as a very densely staining spherical fragment, devoid of delimiting membrane. (*d*) Nucleus r_2 from (*b*). Note the dense chromatin within the nuclear envelope (ne). (*e*, *f*) Photomicrograph (*e*) and electron micrograph (*f*) showing a fusion at F. The donor cell nuclei are arrowed. (*g*, *h*) The recipient cell nuclei from the fusion shown in (*e*). Micrograph (*g*) shows the condensed chromatin apparently shrinking away from the surrounding envelope which is associated with several layers of membrane. Note the distinct region of chromatin-like material free in the cytoplasm (arrowed). (*h*) This nucleus has a circular profile with homogeneous nucleoplasm lacking a typical nucleolar region. (*i*) Naked chromatin (arrowed) found in the region between the degenerating nuclei shown in (*g*) and (*h*). The cytoplasm surrounding these nuclei appears normal.

plugging reactions could modify compartment exchange during normal mycelial development and differentiation (for discussion see Gull, 1978).

It is of interest that the observations of 'self' fusions in dikaryotic mycelia again focus attention on the status of each compartment within the hypha. While fusion leads to an increase in the numbers of nuclei occurring within a single compartment, it has little influence on the actual proportions of nuclear–cytoplasmic material. Yet an elaborate pattern of behaviour has evolved to restore nuclear numbers and strictly maintain cellular integrity following fusion. Such behaviour would hardly seem necessary if all nuclei within a hypha were simply contributing to the same cytoplasmic system. Perhaps then, the replacement reaction implies some degree of compartmental autonomy.

In conclusion, the form and function of septa and the way they influence nucleus–cytoplasmic interactions is of primary importance to our understanding of colony development. A cytological reappraisal, along the lines described in this chapter, of the various patterns of nuclear distribution and hyphal organisation, branching and mitotis, in a range of filamentous fungi should prove extremely useful in this respect. We feel that the time is ripe for a re-examination of the cell concept in filamentous fungi and a new synthesis of information regarding the development and function of the vegetative mycelium.

Acknowledgement. We thank the Science and Engineering Research Council for financial support. We wish to acknowledge Dr P. H. Gregory for his encouragement and Dr D. Niederpruem for helpful and amusing discussions.

References

Aylmore, R. C. (1983). *Hyphal fusion in* Coriolus versicolor. Ph.D. thesis, University of Exeter.

Aylmore, R. C. & Todd, N. K. (1984*a*). Hyphal fusion in *Coriolus versicolor*. In *The Ecology and Physiology of the Fungal Mycelium*, Symposium of the British Mycological Society, ed. D. H. Jennings & A. D. M. Rayner, pp. 103–25. Cambridge University Press.

Aylmore, R. C. & Todd, N. K. (1984*b*). A microculture chamber and improved method for combined light and electron microscopy of filamentous fungi. *Journal of Microbiological Methods*, **2**, 317–22.

Bartnicki-Garcia, S. (1973). Fundamental aspects of hyphal morphogenesis. In *Microbial Differentiation*, 23rd Symposium of the Society for General Microbiology, ed. J. O. Ashworth & J. E. Smith, pp. 245–67. Cambridge University Press.

Bensaude, M. (1918). *Récherches sur le cycle évolutif et la sexualité chez les Basidiomycètes.* Thesis, Nemours.

Bracker, C. E. (1967). Ultrastructure of Fungi. *Annual Review of Phytopathology*, 5, 343–74.

Bracker, C. E. & Butler, E. E. (1963). The ultrastructure and development of septa in hyphae of *Rhizoctonia solani. Mycologia*, 55, 35–58.

Bracker, C. E. & Butler, E. E. (1964). Function of the septal pore apparatus in *Rhizoctonia solani* during protoplasmic streaming. *Journal of Cell Biology*, 21, 152–7.

Brenner, D. M. & Carroll, G. C. (1968). Fine-structural correlates of growth in hyphae of *Ascodesmis sphaerospora. Journal of Bacteriology*, 95, 658–71.

Brouchart, R. & Demoulin, V. (1975). Septum ultrastructure of *Ostracoderma torrendii. Canadian Journal of Botany*, 53, 1549–53.

Buller, A. H. R. (1931). *Researches on Fungi*, vol. 4. London: Longmans Green.

Buller, A. H. R. (1933). *Researches on Fungi*, vol. 5. London: Longmans Green.

Casselton, L. A., Lewis, D. & Marchant, R. (1971). Septal structure and mating behaviour of common-A diploid strains of *Coprinus lagopus. Journal of General Microbiology*, 66, 273–8.

Collinge, A. J., Miles, E. A. & Trinici, A. P. J. (1978). Ultrastructure of *Penicillium chrysogenum* hyphae from colonies and chemostat cultures. *Transactions of the British Mycological Society*, 70, 401–8.

Craig, G. D., Newsam, R. J., Gull, K. & Wood, D. A. (1977). Subhymenial branching and dolipore septation in *Agaricus bisporus. Transactions of the British Mycological Society*, 69, 337–44.

Flentje, N. T. & Stretton, H. M. (1964). Mechanisms of variation in *Thanatephorus cucumeris* and *T. praticolus. Australian Journal of Biological Sciences*, 17, 686–704.

Furtado, J. S. (1971). The septal pore and other ultrastructural features of the Pyrenomycete *Sordaria fimicola. Mycologia*, 63, 104–13.

Garnjobst, L. & Wilson, J. F. (1956). Heterocaryosis and protoplasmic incompatibility in *Neurospora crassa. Proceedings of the National Academy of Sciences, USA*, 42, 613–18.

Girbardt, M. (1979). A microfilamentous septal belt (FSB) during induction of cytokinesis in *Trametes versicolor* (L. ex Fr.). *Experimental Mycology*, 3, 215–28.

Gregory, P. H. (1984). The First Benefactor's Lecture. The Fungal Mycelium: an historical perspective. *Transactions of the British Mycological Society*, 82, 1–11.

Gull, K. (1978). Form and function of septa in filamentous fungi. In *The Filamentous Fungi*, vol. 3, ed. J. E. Smith & D. R. Berry, pp. 78–93. London: Edward Arnold.

Harvey, I. C. (1975). Development and germination of chlamydospores in *Pleiochaeta setosa. Transactions of the British Mycological Society*, 64, 489–95.

Hawker, L. E. & Beckett, A. (1971). Fine structure and development of the zygospore of *Rhizopus sexualis* (Smith) Callen. *Philosophical Transactions of the Royal Society of London (B)*, 263, 71–100.

Koltin, Y. & Flexer, A. S. (1969). Alteration of nuclear distribution in B-mutant strains of *Schizophyllum commune. Journal of Cell Science*, 4, 739–49.

McKeen, W. E. (1971). Woronin bodies in *Erysiphe graminis* DC. *Canadian Journal of Microbiology*, 17, 1557–60.

Moore, R. T. & Marchant, R. (1972). Ultrastructural characterisation of the basidiomycete septum of *Polyporus biennis. Canadian Journal of Botany*, 50, 2463–9.

Noble, M. (1937). The morphology and cytology of *Typhula trifolii* (Rostr.). *Annals of Botany* (N.S.), **1**, 67–98.

Papazian, H. P. (1958). The genetics of Basidiomycetes. *Advances in Genetics*, **9**, 41–69.

Parmeter, J. R., Sherwood, R. T. & Platt, W. D. (1969). Anastomosis grouping among isolates of *Thanatephorus cucumeris*. *Phytopathology*, **59**, 1270–8.

Rayner, A. D. M., Coates, D., Ainsworth, A. M., Adams, T. J. H., Williams, E. N. D. & Todd, N. K. (1984). The biological consequences of the individualistic mycelium. In *The Ecology and Physiology of the Fungal Mycelium*, Symposium of the British Mycological Society, ed. D. H. Jennings & A. D. M. Rayner, pp. 509–40. Cambridge University Press.

Reichle, R. E. & Alexander, J. V. (1965). Multiperforate septations, Woronin bodies and septal plugs in *Fusarium*. *Journal of Cell Biology*, **24**, 489–96.

Rijkenberg, F. H. J. & Truter, S. J. (1975). Cell fusion in the aecium of *Puccinia sorghi*. *Protoplasma*, **83**, 233–46.

Setliff, E. C., MacDonald, W. L. & Patton, R. F. (1972). Fine structure of the septal pore apparatus in *Polyporus tomentosus*, *Poria latemarginata*, and *Rhizoctonia solani*. *Canadian Journal of Botany*, **50**, 2559–63.

Trinci, A. P. J. (1978*a*). Wall and hyphal growth. *Science Progress (Oxford)*, **65**, 75–99.

Trinci, A. P. J. (1978*b*). The duplication cycle and vegetative development in moulds. In *The Filamentous Fungi*, vol. 3, ed. J. E. Smith & D. R. Berry, pp. 132–63. London: Edward Arnold.

Trinci, A. P. J. & Collinge, A. J. (1973). Structure and plugging of septa of wild type and spreading colonial mutants of *Neurospora crassa*. *Archiv für Mikrobiologie*, **91**, 355–64.

Valk, P. van der & Marchant, R. (1978). Hyphal ultrastructure in fruit-body primordia of the basidiomycetes *Schizophyllum commune* and *Coprinus cinereus*. *Protoplasma*, **95**, 57–72.

Wells, K. (1964). The basidia of *Exidia nucleata*. 1. Ultrastructure. *Mycologia*, **56**, 327–41.

Wergin, W. P. (1973). Development of Woronin bodies from microbodies in *Fusarium oxysporum* f.sp. *lycopersici*. *Protoplasma*, **76**, 249–60.

Wessels, J. G. H. & Sietsma, J. H. (1979). Wall structure and growth in *Schizophyllum commune*. In *Fungal Walls and Hyphal Growth*, Symposium of the British Mycological Society, ed. J. H. Burnett & A. P. J. Trinci, pp. 27–48. Cambridge University Press.

10
Morphogenesis of vegetative organs

A. D. M. RAYNER, K. A. POWELL,* W. THOMPSON†
and D. H. JENNINGS†

*School of Biological Sciences, *Department of Materials Science, University of
Bath, Claverton Down, Bath BA2 7AY, UK*

†Department of Botany, The University, PO Box 147, Liverpool L69 3BX, UK

Introduction

Besides reproductive fruit bodies, agarics, as well as many other
fungi, elaborate a wide variety of plectenchymatous structures which serve
a purely vegetative function. Production of these vegetative organs is a
response to the changing demands encountered by a dynamic mycelium
growing under heterogeneous conditions, allowing different parts of the
colony to fulfil separate and sometimes contradictory roles (Rayner,
Watling & Frankland: Chapter 1). Knowledge of the factors controlling
morphogenesis of vegetative organs is therefore crucial to our understanding of the biological and ecological consequences of the capacity of
fungal colonies to differentiate and to switch between alternative modes of
functioning (Gregory, 1984). Nonetheless, by comparison with reproductive
fruit bodies, it is perhaps fair to suggest both that the morphogenesis of
vegetative organs has been neglected and that the significance of these
structures has not been as widely appreciated as it should have been. Partly
as a result of lack of information, it has been necessary to use non-agaricoid
fungi to illustrate particular points in the following discussion. We feel that
many of the central issues concerning morphogenesis of vegetative organs
are essentially similar to those in fruit body development, so that each field
of study ought to benefit mutually. Rather than providing an exhaustive
review, our approach, therefore, has been to focus on these issues, using
some of our own recent work for illustration. For relevant reviews of
previous work the reader is referred to Butler (1966), Willetts (1973),
Watkinson (1979) and Cooke (1983).

Types of vegetative organ

To begin with, it is necessary to outline the range of vegetative organs produced by agarics. Traditionally, several different types of plectenchymatous vegetative organs have been recognised in fungi, including cords (=strands), rhizomorphs, sclerotia, pseudosclerotia, microsclerotia and bulbils, stromata, pseudorhiza and ectomycorrhizal sheaths. Superficially, there seem to be obvious functional and morphological distinctions between these types. However, as will be revealed below, and as with so much biological classification, such distinctions may often be a matter only of degree, the types intergrading in a continuous spectrum of forms and behaviour. Indeed the distinction, implicit in the title of this chapter and our introductory comments, between purely reproductive and vegetative functioning is by no means absolute. If, as therefore seems likely, we are really observing variations on a theme of mycelial aggregation, there is every reason to expect common underlying mechanisms of control of that aggregation.

Linear organs

A primary morphological distinction which can be made is between linear and non-linear organs. Linear organs include all those structures variously described as cords, threads, strands, rhizomorphs and pseudorhiza. They consist of aggregations of predominantly parallel, longitudinally aligned hyphae which may be additionally cross-linked by narrow thin- or thick-walled hyphae. Within the structure there can be a wide variety of hyphal types (Hornung & Jennings, 1981) while the structure itself may also differentiate into a distinct outer crust with a high proportion of wide-vessel hyphae within the core. As discussed in Chapter 1, the principal role of these structures is probably in allowing exit from and/or recolonisation of substrata, accompanied by mobilisation of resources which the latter contain.

We believe an important distinction must be made between truly *migratory organs*, capable of autonomous extension from a food base, and those cases where there is initially diffuse mycelial extension. While this distinction is academically important, in reality there is considerable intergradation between the two types of behaviour, even within a single fungus. This variation probably reflects the relative ease with which transitions between diffuse and aggregated structures can occur. In turn these transitions may depend solely on the degree of apical control, that is the extent of coordination between the faster growing marginal hyphae during extension from a food base (Figs 10.1–10.4). At one extreme of the spectrum are true rhizomorphs, in the sense defined by Garrett (1963) and best known in *Armillaria*. These are truly migratory, extending from a highly organised apical growing point (Fig. 10.2) at a rate faster than that of any associated diffuse mycelium and producing endogenous lateral branches subject to considerable apical dominance.

Progressive loss of apical control over extension, with concomitant loss of apical dominance and coherence of apical hyphae (Figs. 10.3, 10.4), leads to the forms illustrated in Fig. 10.1 – that is, first to increased lateral

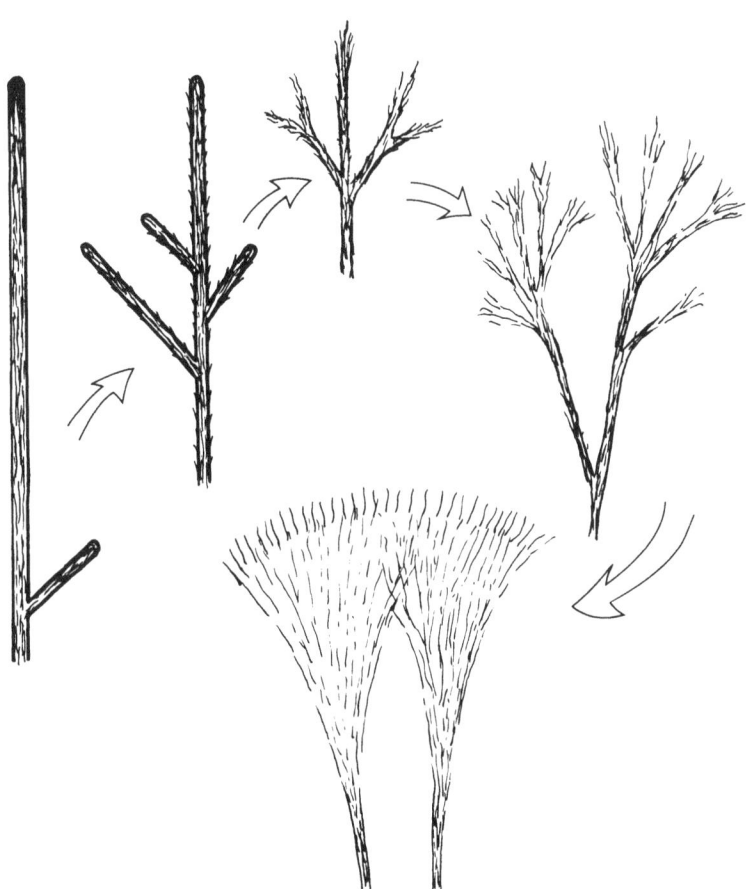

Fig. 10.1. Diagram illustrating the spectrum of mycelial outgrowth patterns resulting in production of linear organs. Progression from strongly rhizomorphic outgrowth (far left) to diffuse outgrowth followed by consolidation (bottom centre) is associated with loss of apical control over extension of marginal hyphae resulting in increased branching and loss of apical coherence.

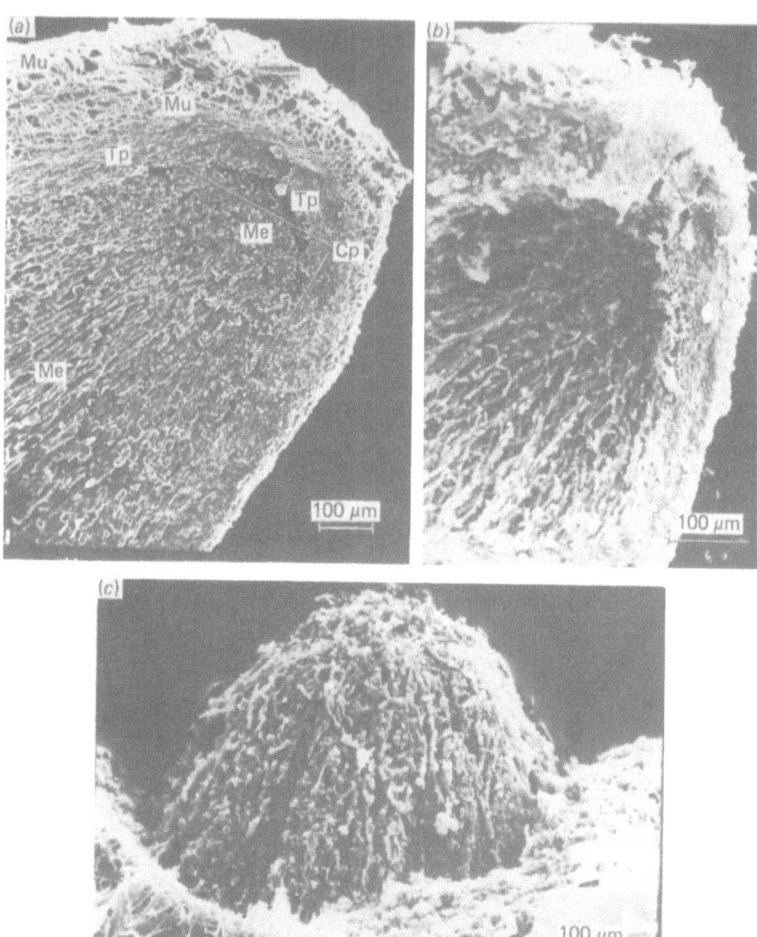

Fig. 10.2. (*a*) Scanning electron micrograph of one half of a longitudinally bisected apex of a rhizomorph of *Armillaria bulbosa* obtained by incubation of naturally colonised wood. The hyphae are organised into distinct layers, including an outer mucilaginous region (Mu), a tightly packed layer (Tp) overarching the apical dome and a central medullary region (Me) containing loosely packed, predominantly axially aligned hyphae with swollen compartments. The tightly packed layer appears to form a cross-latticed tissue over the apical dome, and at the junction with the medullary region there is a distinctive cleavage plane (Cp), presumably reflecting the different textural arrangements between the two tissues. (*b, c*) Scanning electron micrographs of tissue layers subtending the cleavage plane, seen in face view after the apical tightly packed layer has been fractured away from the medullary dome. The basically filamentous organisation of each layer is clearly shown: (*b*) tightly packed layer; (*c*) medullary layer.

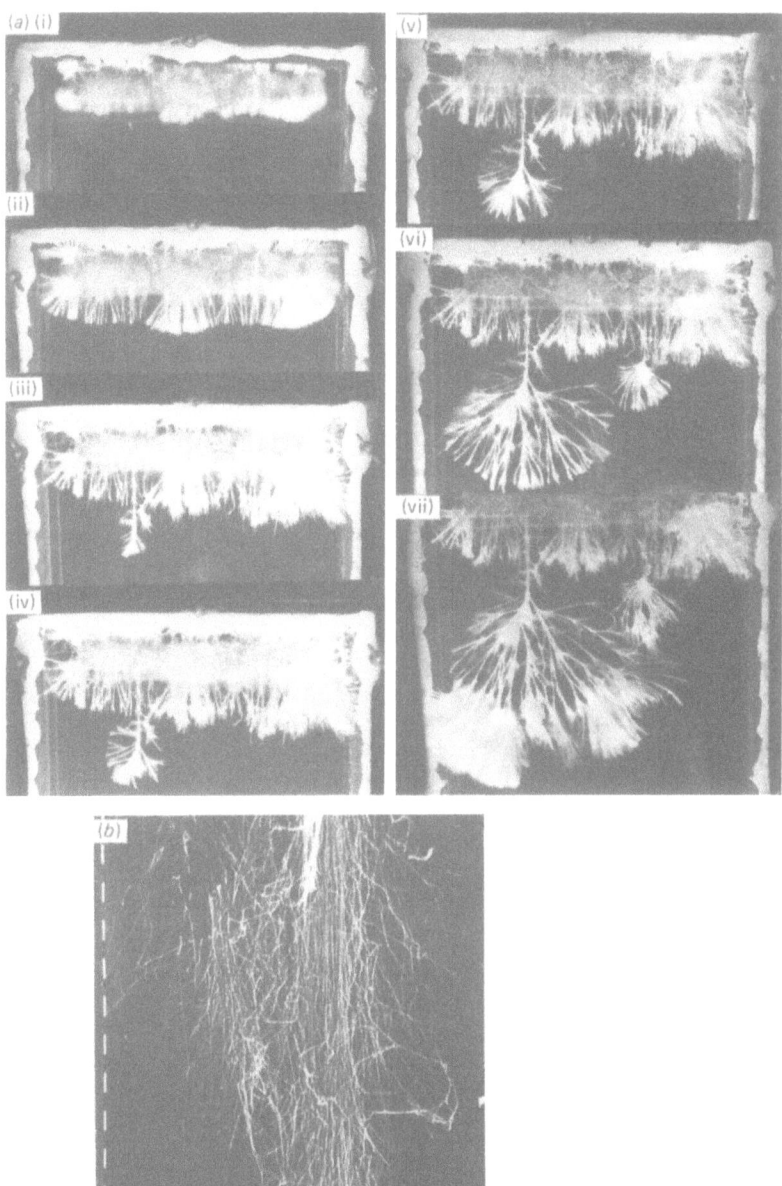

Fig. 10.3. (*a*) Mycelial cords of *Phallus impudicus* extending from colonised beech blocks onto a perspex platform in a sterile chamber. (i) 8 days' growth; (ii) 15 days' growth; (iii) 41 days' growth; (iv) 50 days' growth; (v) 56 days' growth; (vi) 64 days' growth; (vii) 73 days' growth. (*b*) Scanning electron micrograph of mycelial front of *Phallus impudicus* after 30 days' growth in a perspex chamber as in (*a*), scale markers represent 100 μm.

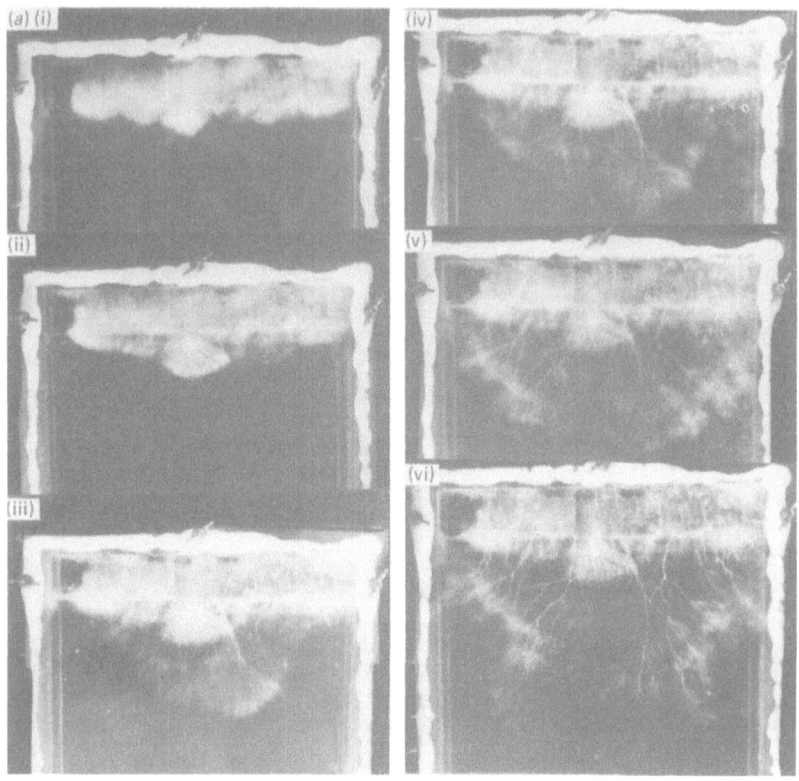

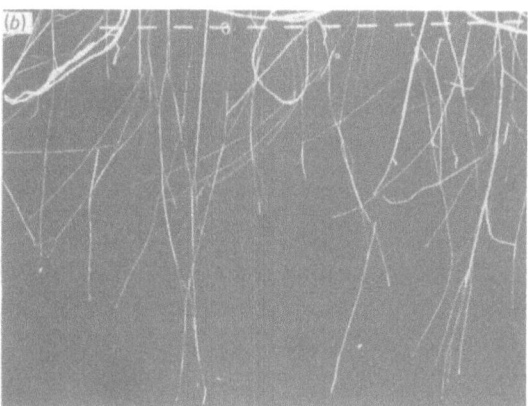

Fig. 10.4. (*a*) Mycelium and cords of *Phanerochaete velutina* extending from colonised beech blocks onto a perspex platform in a sterile chamber. (i) 8 days' growth; (ii) 16 days' growth; (iii) 28 days' growth; (iv) 34 days' growth; (v) 42 days' growth; (vi) 56 days' growth. (*b*) Scanning electron micrograph of mycelial front of *Phanerochaete velutina* after 30 days' growth in a perspex chamber as in (*a*), scale markers represent 100 μm.

branching, then to apically extending 'fan' mycelia, and finally to extension of diffuse mycelium followed by consolidation of linear organs. At this end of the spectrum the best-known example is *Serpula lacrimans* (Brownlee & Jennings, 1982). Decreased supply of exogenous nutrients or increased concentrations of exogenous morphologically active and/or inhibitory substances (see below) will all tend to push development towards the rhizomorphic end of the spectrum.

Possible mechanisms relating to the coordination of marginal hyphae into rapidly extending collateral systems will be discussed later in connection with slow-dense/fast-effuse transitions. However, what is shown in Fig. 10.1 readily describes the variations between and within species. The more 'rhizomorphic' a fungus, the more readily it will produce linear aggregations, even at high nutrient levels on agar media. By the same token, fungi at the opposite end of the spectrum would not be expected to produce linear organs readily at high nutrient levels. We would predict that by suitably adjusting external conditions it should be possible to alter morphogenesis progressively from one stage to another, particularly in the direction from rhizomorph to diffuse mycelium. It is therefore satisfying that exactly this pattern is exhibited by *Armillaria* species when producing subcortical mycelium in the cambial region of infected woody hosts. At the other end of the spectrum, the normally diffuse mycelium of *S. lacrimans* produces at its margin, under low-nutrient conditions, fans of fast-growing mycelium, the so-called 'point growth' (Coggins *et al.*, 1980), the fast growth rate of which has been compared, on a previous occasion, to that of a rhizomorph (Jennings, 1982).

Continuous variation in the origin and structure of vegetative linear organs raises the question of the terminology used to describe them. Currently, the usual practice is to distinguish true *rhizomorphs* from other structures which are variously delineated as cords, strands, fans, threads or syrrotia, doubt then always arising as to which is most apt to describe a particular case. Perhaps the best solution is to choose one or two terms, such as cords and rhizomorphs, and to qualify these where necessary by adjectives such as 'apically dominant', 'apically branched', 'apically spreading' and 'apically diffuse' to cover the range of morphogenetic patterns illustrated in Fig. 10.1. This will be our approach, using *rhizomorphs* in the sense of Garrett (1963; see above) and *cords* for all other types.

Another problem of nomenclatural demarcation concerns the fact that linear organs may not always be strictly *vegetative* in function. Often they serve as connections to fruit bodies, an extreme case being the pseudorhiza of *Oudemansiella radicata* and *Termitomyces* spp. which, respectively, connect aerial fruit bodies to buried roots and fungus gardens. At what stage can pseudorhiza be distinguished from cords or rhizomorphs connected to the base of fruit bodies? Here the issues are blurred even further in rhizomorphic species of *Marasmius* (Hedger: Chapter 2), where the melanised stipe of the fruit body is clearly homologous with the rhizomorph and which, if ceasing to extend apically, may develop a pileus instead. A rigorous study of *Marasmius* rhizomorphs appears to have been neglected, resulting in a serious gap in our present knowledge.

Non-linear organs

A variety of vegetative organs are produced and remain *in situ*, the types most commonly recognised being sclerotia, microsclerotia, pseudosclerotia, stromata and ectomycorrhizal sheaths. However, the former four types intergrade, having the common feature of being composed of greater or smaller masses of mycelium bounded by a crust which is often pseudoparenchymatous, and differing only in size and whether or not they contain extraneous material or reproductive structures. Sclerotia, which contain neither reproductive structures nor extraneous material, are, together with stromata, not well known in agarics (but see Moore, 1981; Rayner *et al.*: Chapter 1) and will not be discussed further in any detail. In the case of pseudosclerotia, which do incorporate extraneous material (Campbell, 1934), the outer crusts (pseudosclerotial plates) may be immersed in the substratum, whence they appear as lines in cross-section which have been termed 'zone lines', although lines of similar gross appearance can develop for a variety of other reasons (Rayner & Todd, 1982). Terminological problems are again apparent in the suggestion that pseudosclerotial plates of *Armillaria* and *Xylaria* are homologous with the melanised outer crusts of rhizomorphs and perithecial stromata, respectively (Campbell, 1933, 1934).

Exogenous control of morphogenesis
Abiotic factors

A wide variety of abiotic factors affect morphogenesis of vegetative organs. In some cases such factors have been attributed a role as principal determinants of morphogenesis but since they cannot themselves explain the intrinsic capacity for differentiation, it seems more reasonable to view their effects in terms of interactions with endogenous control mechanisms (see below). For example, perhaps related to their role as connectives between nutrient depots, differentiation of mycelial cords has often been attributed to low availability of external nutrients, which causes aggregation of hyphal branches in response to their own exudates (Day, 1969; Garrett, 1970; Watkinson, 1971). Whereas this model is attractive for specific examples, such as *Serpula lacrimans*, it does not seem to provide a basis for explaining the variation both within and between fungal species in their capacity to produce linear organs, unless variation in the extent of nutrient exudation and translocation is invoked. In fact, there are arguments in favour of taking the view that *high* nutrient levels cause *de*-differentiation (cf. earlier discussion of interchange between types of linear organs).

Apart from major nutrients, a wide range of chemical stimuli have been implicated in the initiation of vegetative organs. Ethanol, certain amino acids, indole acetic acid, aminobenzoic acid and various oils and fatty acids have all been reported to elicit rhizomorph formation in *Armillaria* (Weinhold, Hendrix & Raabe, 1962; Weinhold, 1963; Pentland, 1965; Weinhold & Garraway, 1966; Moody, Garraway & Weinhold, 1968; Garraway, 1970; Sortkjaer & Allerman, 1972). However, the underlying mechanism and ecological significance of such stimuli is not clear. By contrast, apparent volatile or diffusible stimuli emanating from natural substrata may, as suggested by Rayner *et al.* (Chapter 1), fulfil an important ecological role in inducing formation and directed growth of linear organs. Where a directive stimulus is operating, this would be expected to enhance apical control of extension, that is, to push development towards the rhizomorphic pattern illustrated in Fig. 10.1, and preliminary observations suggest that this may indeed be the case (Payne, Rayner and Thompson, unpublished).

Studies on the effects of moisture conditions and aeration on morphogenesis of vegetative organs have so far mostly concerned the pseudosclerotial plates (PSPs) of certain wood-decaying fungi, and the rhizomorphs of *Armillaria*. In the former case, a probable role for these factors was indicated by the production of PSPs immediately below a freshly cut surface of colonised wood, access of air and drying out of the surface layers being obvious concomitants of cutting (Lopez-Real, 1975; Lopez-Real & Swift, 1975, 1977). However, experimental studies failed to confirm a role for desiccation or oxygenation as *stimuli* for PSP formation, and Lopez-Real & Swift (1977) concluded that injury to hyphae elicited the response. Aeration has, by contrast, been implicated as an important factor governing rhizomorph extension. Thus, Smith & Griffin (1971) working with *Armillaria elegans* demonstrated that optimal growth of the rhizomorph required a high partial pressure of oxygen within the apex, but a low one outside – high external oxygen levels promote melanisation and inhibit extension. We shall return to this point.

With respect to low water potential, there is evidence that this often promotes 'point growth' phenomena in wood-decaying basidiomycetes (Clarke, Jennings & Coggins, 1980; Jennings, 1982; Boddy, personal communication).

Light has not so far been shown to affect *morphogenesis* of vegetative organs *per se*, although it does inhibit extension of mycelial cords in soil tubes (Thompson & Rayner, 1982, 1983) and, presumably via effects on the mycelium, it affects the timing of production of sclerotia by many fungi (Humpherson-Jones & Cooke, 1977). This is perhaps consistent with the fact that vegetative organs are usually produced in darkness. However, where vegetative organs become reproductive, as in pseudorhizas and *Marasmius* rhizomorphs, it seems likely that light plays an important role in completing the morphogenetic sequence from aggregation to reproduction.

Purely physical factors also affect morphogenesis. Injury caused by cutting has already been mentioned in relation to PSP production and is very commonly observed to accelerate aggregation and consequent formation of a wide variety of vegetative organs. The presence of a permeable physical barrier, such as cellophane, between the mycelium and a medium which would not otherwise support their formation, has been shown to elicit cord formation in *Calvatia sculpta* (Bellotti & Couse, 1980).

Biotic factors

Interactions with other organisms, and particularly other fungi, frequently result in differentiation of vegetative organs. In some cases, diffusible or volatile chemical factors may induce telemorphotic and perhaps even chemotropic responses. For example, ethanol produced by *Aureobasidium pullulans* has been implicated in rhizomorph initiation in *Armillaria* (Pentland, 1965), and *Penicillium* species cause remarkable telemorphotic induction and apparent directed growth of cords of *Phallus impudicus* (see Rayner & Webber, 1984).

As discussed by Rayner *et al.* (Chapter 1), many agarics probably possess a combative ecological strategy involving either defence of captured domain or secondary capture of domain from other fungi (replacement/ secondary resource capture). The production of PSPs may have a particular role in defence, providing an effective physical obstacle to hyphal or mycelial penetration – and the influence of injury inferred by Lopez-Real & Swift (1977; see above) would seem relevant in this context. In fact, PSPs are regularly formed, both in culture and in nature, in response to interactions both between species and between somatically incompatible genotypes of the same species (Rayner, 1978; Rayner & Todd, 1979, 1982).

The ability of many basidiomycetes, including agarics, to replace other fungi in culture is often related to the production at the interaction front of a region of bulked-up mycelium, which may or may not differentiate into linear organs prior to invasion, as shown for *Collybia peronata* in Fig. 10.5(*a,b*). An unusual case (Fig. 10.5*c*) is exhibited by *C. dryophila* which constitutively produces cords in culture, and reverts to extension of appressed mycelium during replacement of fungi such as *Clitocybe flaccida*.

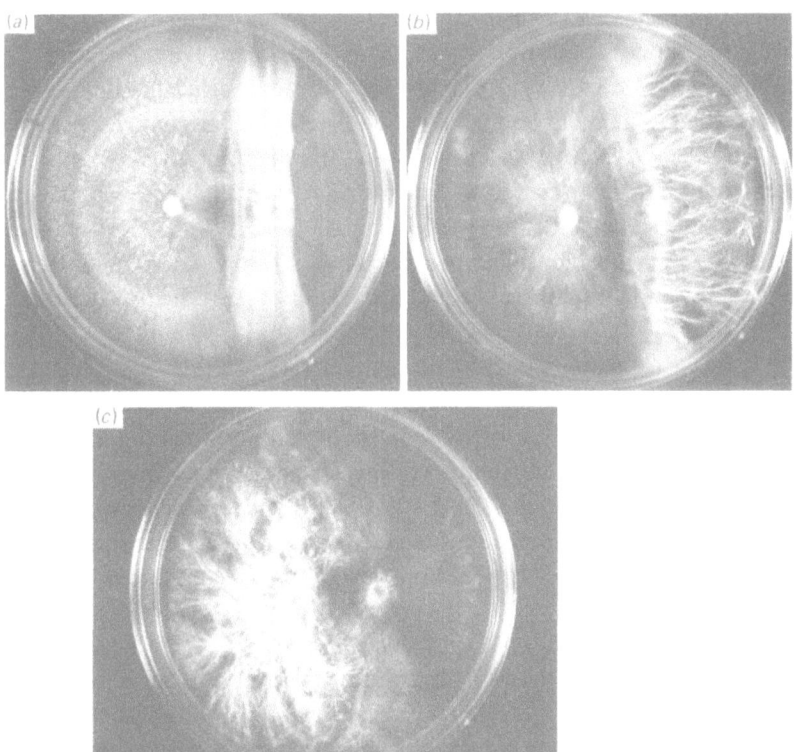

Fig. 10.5. Replacement interactions between certain litter-inhabiting agarics on malt agar. (*a*) Replacement of *Collybia butyracea* (right) by a dense mycelial front of *C. peronata*. (*b*) Replacement of *Clitocybe nebularis* (right) by *Collybia peronata*, with the normally appressed mycelium of the latter producing cords. (*c*) Replacement of *Clitocybe flaccida* (right) by *Collybia dryophila*, with the normal cord-forming growth of the latter reverting to diffuse, appressed mycelium. (Photographs by A. G. Mitchell, from Cooke & Rayner, 1984.)

What triggers production of these invasive fronts of mycelium is not certain, but it seems plausible that they provide an effective means of combatting staling products produced by opposing fungi in the medium (cf. discussion of slow-dense/fast-effuse transitions below). It is interesting here that Park (1963) described production of cords of the ascomycete *Xylaria polymorpha* in response to diffusates emanating from cultures of other fungi.

Endogenous control of morphogenesis

As we have already intimated, the effects of exogenous factors cannot in themselves explain the intrinsic capacity of agarics and other fungi to produce vegetative organs. Here, three fundamental issues are as follows. First, how can the switch from diffuse mycelial development to collateral development and aggregation be effected? Secondly, once such switches have been operated, by what mechanism does the ordered structure of some vegetative organs generate? Finally, what signals maintain development along particular pathways? We shall explore these questions in relation to some specific examples below. These fundamental issues may apply equally to the morphogenesis of reproductive structures.

Switch mechanisms

In the terminology of Gregory (1984), differentiation of vegetative organs involves a change of *mode* in the mycelium, fulfilling a different function and obeying different laws. This change of mode involves one or both of two basic elements, *collateral growth* of hyphae, which is particularly important in the development of migratory or linear organs, and *proliferation and compaction* of hyphal branches to form a continuous tissue.

Ordering of the mycelial front: slow-dense/fast-effuse transitions

During development of a radiating mycelium on a semi-solid medium such as agar, it is often noticeable that initial establishment from a spore involves growth in all directions with branching of hyphae of similar diameter often occurring at right-angles (e.g. Prosser, 1983). If growth were to continue in this way, the expected result would be a comparatively slowly extending, densely branched colony, subject to autoinhibition by accumulated staling products in the medium. However, in many fungi maturation of the colony is associated with production of a spreading marginal mycelial front (or fronts) which overrides any such autoinhibition, consistent with a change in functional mode from establishment to exploration and capture of a domain (Rayner *et al.*: Chapter 1).

The production of spreading marginal zones may be achieved in a variety of ways (e.g. Steele & Trinci, 1975; Bull & Trinci, 1977; Prosser, 1983; Butler, 1984) and may be virtually continuous with prior development or occur abruptly (when we describe them as slow-dense/fast-effuse transitions) either from localised foci or from a broad mycelial front. In the latter case, when there is alternation between 'staled' submerged mycelium and spreading surface mycelium, which may be due to endogenous or exogenous factors, the result is rhythmic colony development which is, in turn, often related to rhythmic patterns of sporulation (Lysek, 1984). Often, extension of spreading mycelial fronts is sustained by wide leader hyphae which extend faster than their more attenuate subordinate branches. Alternatively, there may be formation of submerged or aerial hyphae which under- or overarch staled marginal hyphae, so that the colony advances rhythmically or by a series of 'leapfrogs' (Butler, 1984; Lysek, 1984).

From the point of view of the morphogenesis of linear organs, a vital feature in the development of spreading mycelial fronts is that there is usually a change to *acute-angled branching* of the marginal hyphae, such that they extend in virtually parallel array, that is, by *collateral growth*. The hyphae are thus brought into exactly that orientation which would seem to be a precondition for development of linear organs; understanding of the mechanisms controlling the transition to collateral growth must therefore be a vital consideration in morphogenetic studies. That this is indeed so is manifest in certain cord-forming fungi exhibiting what has been termed 'point-growth' (Coggins *et al.*, 1980). Here there is an abrupt transition from localised foci in the margins of slowly extending, densely branched colony forms which are auto- and allo-inhibitory, resulting in rapid collateral extension of effuse mycelial sectors, which in some cases may aggregate directly into cords or rhizomorphic structures (Fig. 10.6). Indeed, as we have already suggested, it is easy to envisage the whole spectrum of forms illustrated in Fig. 10.1 arising via such a process. Evidence for endogenous control of the process is provided below and is further apparent from the stability of fast-effuse colony types following sub-culture (Coggins *et al.*, 1980; Rayner, unpublished).

Genetic control. Although to our knowledge there is no published account of genetic control of morphogenesis, several observations indicate that the intrinsic capacity to produce vegetative organs is controlled by genetic

262

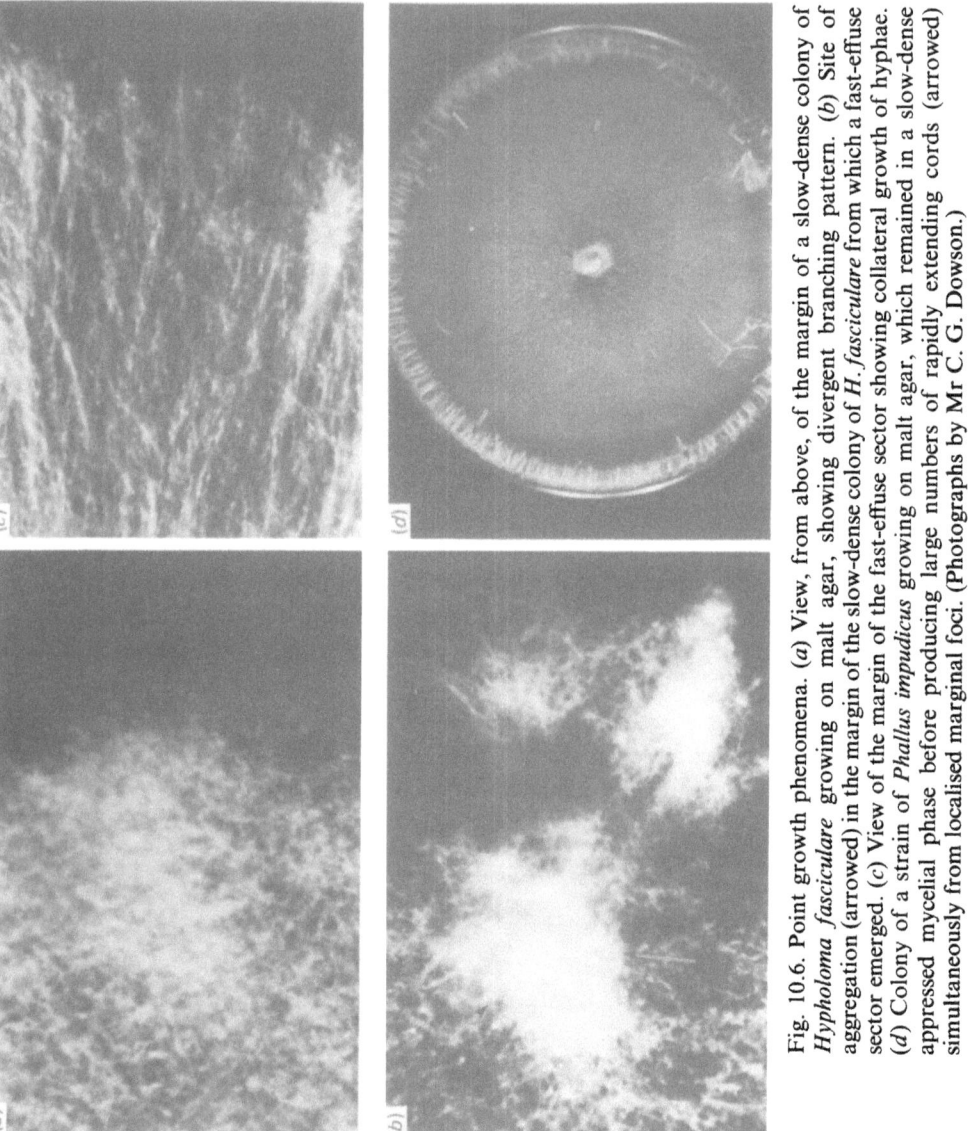

Fig. 10.6. Point growth phenomena. (*a*) View, from above, of the margin of a slow-dense colony of *Hypholoma fasciculare* growing on malt agar, showing divergent branching pattern. (*b*) Site of aggregation (arrowed) in the margin of the slow-dense colony of *H. fasciculare* from which a fast-effuse sector emerged. (*c*) View of the margin of the fast-effuse sector showing collateral growth of hyphae. (*d*) Colony of a strain of *Phallus impudicus* growing on malt agar, which remained in a slow-dense appressed mycelial phase before producing large numbers of rapidly extending cords (arrowed) simultaneously from localised marginal foci. (Photographs by Mr C. G. Dowson.)

mechanisms. Perhaps the most obvious concerns the fact that this capacity is a particular feature of secondary mycelia or dikaryons, as opposed to homokaryons. For example, homokaryons of *Armillaria* species rarely, if ever, produce pseudosclerotial plates in culture, and rhizomorph formation is less regularly observed than in diploids (Rishbeth, personal communication). Such differences may well be related to distinctions between the growth form of homokaryons and secondary mycelia. Within the context of our earlier comments, these may often amount to a difference between slow-dense and fast-effuse colony forms, as occurs in *Coprinus cinereus* (Casselton, 1978).

If controlling genetic elements exist, a logical mode of action would be on collateral growth and/or proliferation or aggregation of branches, and any mechanism which modulates expression of the slow-dense/fast-effuse colony dimorphism might be an effective switch. Some evidence here points to expression of certain homogenic incompatibility factors (B-factors and their equivalents) which, when 'unstabilised', for example, in the absence of complementary A-factors, result in unilateral or bilateral inhibition of colony extension and proliferation of abnormal branches (Rayner *et al.*, 1984). In some instances, as appears to be the case in *Coniophora puteana* (Ainsworth, personal communication), such inhibition appears to precede point-growth of cord-forming mycelium from restricted colony margins in both mating-compatible and incompatible pairings between homo-karyons.

Given the probability of endogenous control of mode transitions resulting in development of vegetative organs, it seems likely, by analogy with similar systems and its suggested role in fruit body initiation, that cyclic AMP is involved. So far we know of no published evidence for this, but in a preliminary study of *Serpula lacrimans* a ten-fold increase in cyclic AMP levels was detected in differentiating regions of the colony as compared with older regions (Thompson, Macdonald & Jennings, unpublished).

Induction of rhizomorphic organs in Stereum hirsutum

Recent studies with *Stereum hirsutum*, a non-rhizomorphic basidiomycete (Aphyllophorales) which usually colonises wood via air-borne basidiospores, have provided further evidence for endogenous control mechanisms and links between slow-dense/fast-effuse transitions and involvement of mating compatibility factors. A spontaneous change occurring during storage of a homokaryotic culture resulted in a dramatic transformation in its properties, such that it rapidly produced farinaceous hymenial surfaces and showed abnormal mating characteristics (Coates & Rayner, 1985). This culture was able to transform other homokaryons, apparently via a cytoplasmically transmissible factor having strong affinities with the mating-compatibility loci, such that they acquired the fruiting and mating properties. Basidiospore progeny from one transformed strain gave rise to two types of colony morphology, slow-dense and fast-effuse, the former always spontaneously reverting by producing fast-effuse sectors. Certain reverted slow-dense colonies produced tumour-like mycelial aggregations or 'mounds' – similar to those which have been described in *Schizophyllum commune* (Gaber & Leonard, 1981) – and these in turn

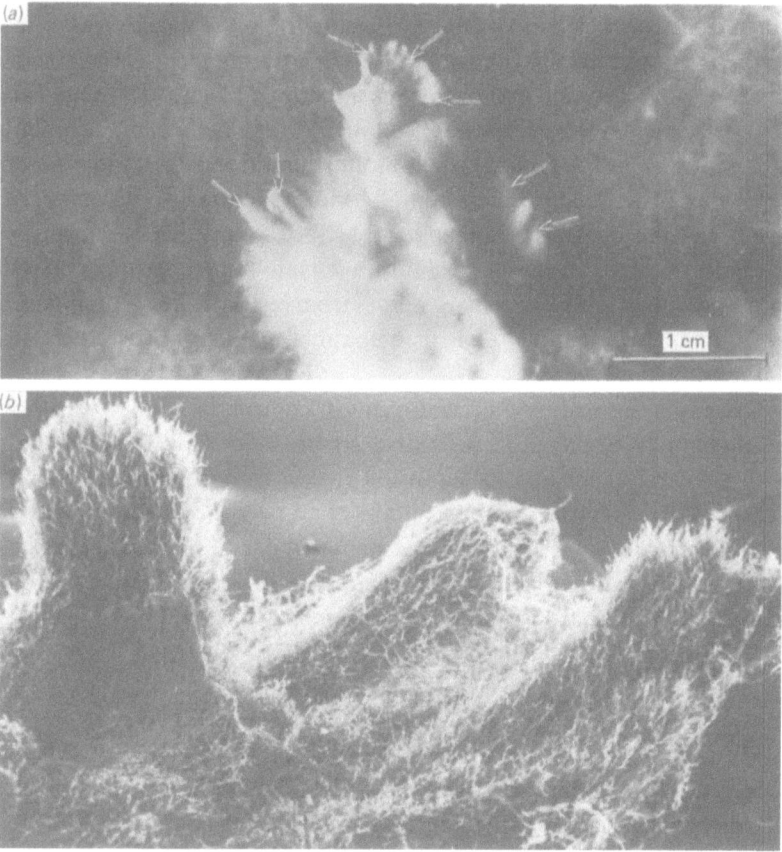

Fig. 10.7. (*a*) Outgrowth of rhizomorphic organs (arrowed) in a mound-producing strain of *Stereum hirsutum* from a site of mound formation above a colony of *Penicillium* (Photograph by D. Coates.) (*b*) Scanning electron micrograph of the organs shown in (*a*).

produced basidiospore progeny which developed mounds. Certain of the mound-producing strains have been found – in one case in the presence of a *Penicillium* contaminant, and in others where aggregations developed in submerged mycelium – to produce remarkably rhizomorph-like organs (Fig. 10.7; Coates & Rayner, 1985). These organs resembled those of *Phallus impudicus* and *Tricholomopsis platyphylla*, for example, but nothing similar has previously been reported in *Stereum hirsutum*, so far as we are aware. Their production from localised sites of aggregation and association with a probable slow-dense/fast-effuse transition (tissue subcultures from mounds produce slow-dense colonies) seems highly significant.

Ordering of structure in Armillaria *rhizomorphs*

In studying the structural ordering of vegetative organs, the rhizomorphs of *Armillaria* which, as we have pointed out, are the most clearly definable entities, provide an obvious starting point. Despite this, cytological studies of these rhizomorphs until recently have been surprisingly limited (Motta, 1967, 1969*a*, *b*, 1971; Schmid & Liese, 1970; Wolkinger, Plank & Brunegger, 1975). More recently, Powell & Rayner (1983, 1984 and unpublished) have made a series of light microscope and scanning and transmission electron microscope studies of vigorously growing rhizomorphs of *Armillaria bulbosa* obtained by incubation of naturally colonised wood samples, while Granlund, Jennings & Veltkamp (1984) have made a detailed scanning electron microscope study of cultured rhizomorphs of *A. mellea*. Results from these and earlier studies provide the basis for the following discussion.

Median longitudinal sections through the apices of vigorously growing rhizomorphs were quite variable in appearance (see Motta, 1969*a,b*, 1971; Motta & Peabody, 1982; Powell & Rayner, 1983; Fig. 10.2). However, there were common features. Thus, there was an outer mucilage-producing layer, underneath which were one or two layers, variable in thickness, of tightly packed cells which gave way, sometimes abruptly to an inner medullary region, containing swollen vacuolated and often multinucleate cells (see Gooday: Chapter 12) surrounded by extensive air or mucilage-filled spaces. This medullary region formed a central channel through the rhizomorph and, as maturation occurred, was increasingly traversed by narrow, thick-walled densely staining fibre hyphae filled with protoplasm. In young rhizomorphs it was often possible to dissect out from this central channel bundles (Fig. 10.8) composed of elongated vacuolated vessel-like hyphae interspersed with fibre hyphae. In some cases curious balloon-like

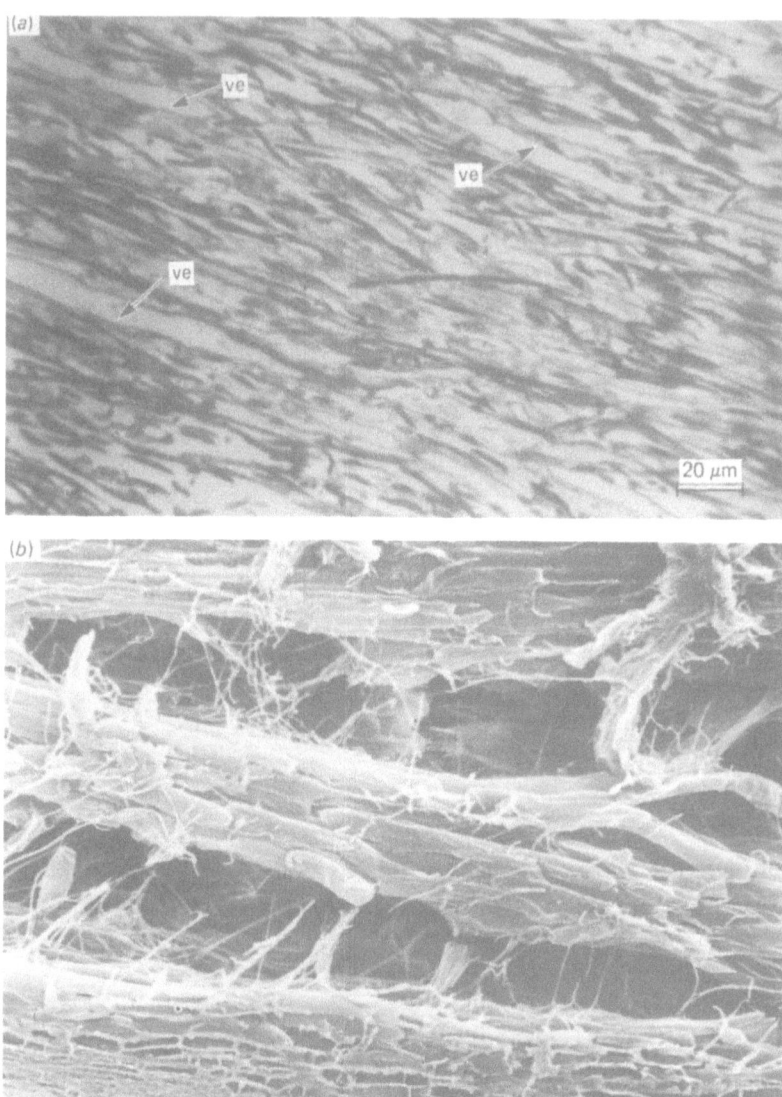

Fig. 10.8. (*a*) Light micrograph of a toluidine blue-stained longitudinal thin section through medullary tissue of a rhizomorph of *Armillaria bulbosa* showing longitudinally aligned vessel-like hyphae (ve) of wide diameter interspersed by narrow, strongly stained fibre hyphae. (*b*) Scanning electron micrograph showing an axial medullary bundle and connective fibre hyphae within the central channel of a rhizomorph of *A. mellea*. (From Granlund *et al.*, 1974.)

structures were present (Granlund *et al.*, 1984). These were about 20 μm in diameter, considerably wider than the vessel hyphae found extensively in mature rhizomorphs (Eamus *et al.*, 1985) and in cords of *Serpula lacrimans* (Hornung & Jennings, 1981) where they are usually about 10 μm in diameter, but less wide than in cords of other fungi such as *Tricholomopsis platyphylla* (Thompson & Rayner, 1983).

Production of the outer mucilage layer (Fig. 10.2 *a*) was characteristic of actively extending rhizomorphs and was lost if outgrowth into an oxygenated environment occurred, whereupon extension usually soon ceased. In the presence of high exogenous nutrient levels, outgrowth of fringing mycelium was sometimes observed, often quite close to the apex (Fig. 10.9 *b*), and disorganised masses of hyphae sometimes developed from apices which had ceased extension (Fig. 10.9 *c*). However, these features did not occur in other circumstances. Within the context of our previous

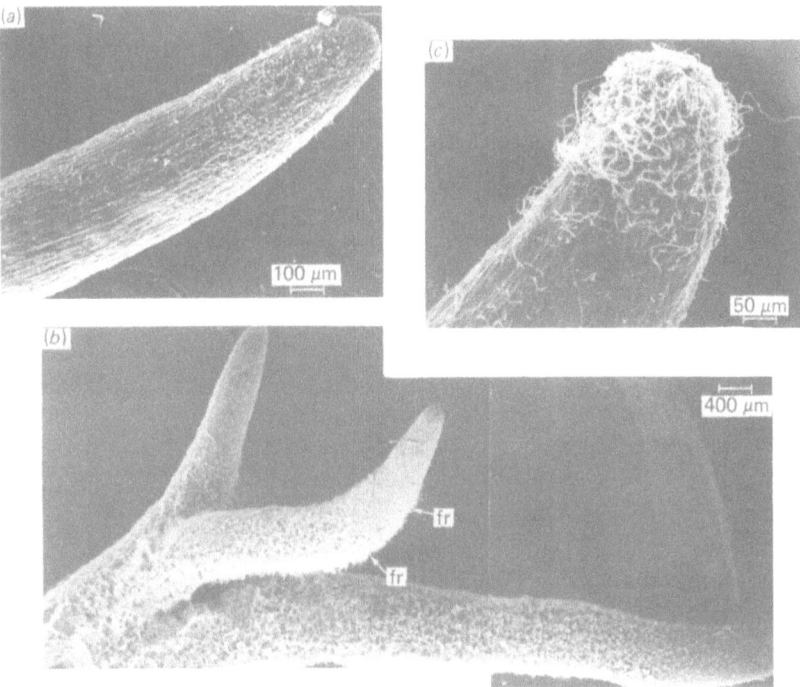

Fig. 10.9. Scanning electron micrographs of rhizomorphs of *Armillaria mellea*. (*a*) An aerial rhizomorph showing lack of mucilage layer and easily discernible outer hyphal layer. (*b*) Rhizomorphs grown in submerged liquid culture showing abundant outgrowth of fringing mycelium (fr). (*c*) Outgrowth of a disorganised mass of hyphae from an apex which has ceased extension. (From Granlund *et al.*, 1984.)

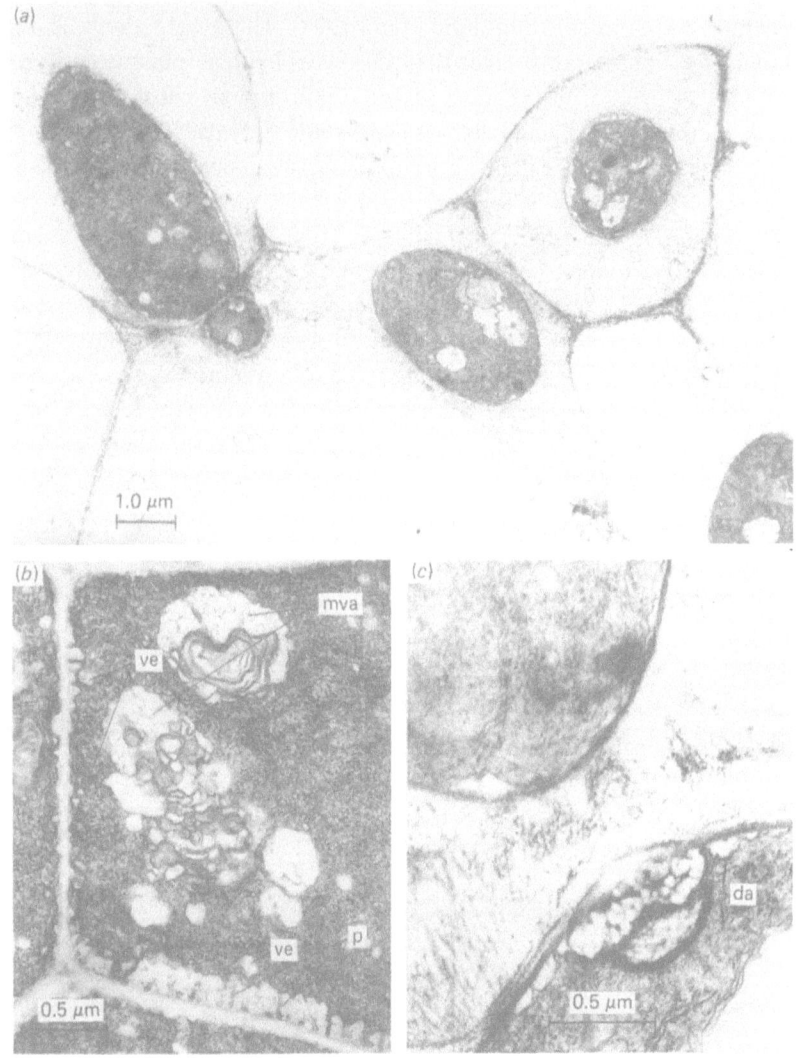

Fig. 10.10. Transmission electron micrographs illustrating patterns of
mucilage production in rhizomorphs of *Armillaria bulbosa*. (*a*) Hyphae
of the outer mucilage layer surrounded by capsules containing radially
aligned mucilaginous strands. (*b*) Part of a hypha in closely packed tissue
underlying the mucilage layer, showing apparently mucilage-filled vesicles
(ve) coalescing with the plasma membrane and discharging their contents
into a paramural space traversed by membrane-bound protuberances (p).
Large whorls of intracytoplasmic membrane are associated with mucilage-
filled vacuoles (mva). (*c*) Hyphal compartments cleaved apart, apparently
due to extrusion of mucilage into a septal region. Part of the dolipore
apparatus (da) is still apparent in one of the compartments. (From Powell
& Rayner, 1983.)

discussion of the origin of rhizomorphs, the fringing mycelium and disorganised outgrowth can be regarded as equivalent to undifferentiated mycelial margin. That is not to say that the fringing mycelium and disorganised apices are not without importance to the functioning of the rhizomorphs. The fringing mycelium may aid nutrient capture, particularly phosphorus, in a manner akin to hairs and the hyphae of endophytic mycorrhiza extending from a root (Jennings, 1982, 1984), while the disorganised apices produce air pores (Smith & Griffin, 1971; Granlund *et al.*, 1984).

The process of mucilage production has been studied electron microscopically by Powell & Rayner (1983). Within the mucilage layer itself, widely dispersed hyphae with dense protoplasmic contents were embedded, the mucilage appearing to radiate from them (Fig. 10.10 *a*). Mucilage production began in tightly packed tissue immediately below the mucilage layer (Fig. 10.10 *b*). Here, mucilage-containing vesicles – possibly derived from large vacuoles – coalesced with the plasmalemma creating a mucilage-filled space between the plasmalemma and all parts of the hyphal wall, including the septal plate, traversed by numerous membrane-bound protoplasmic protuberances. This space, together with the protuberances, disappeared following apparent partial or complete digestion of the wall concomitant with release, in a radiate pattern, of mucilage to the exterior. Importantly, mucilage released into septal regions apparently could result in cleavage of compartments into separate cellular entities (Fig. 10.10 *c*), perhaps implying a role in plasticisation and part in lubrication of the apical region. An alternative, or accompanying, function of mucilage production, which has been proposed, is as an adaptation to unfavourable carbon: nitrogen ratios, allowing excess carbon, brought in by the translocation stream, to be removed from the growing region (Jennings, 1982). Mucilage is also produced in cords of *Serpula lacrimans* (Watkinson, 1984), as well as in fruit bodies (e.g. Williams, Beckett & Read: Chapter 18) and in sclerotia (Willetts, 1978). A wide variety of functions therefore seems likely under different circumstances.

Another layer of cells which could well play an important role in rhizomorph morphogenesis, occurred in the apical dome immediately adjacent to the medullary region (Powell & Rayner, 1984). This circum-medullary layer, up to several cells wide, could extend 0.75 mm or more back from the apex in *A. bulbosa* rhizomorphs, whence it was lost progressively to the outside. Cells in this layer appeared to be very active and were characterised by large numbers of mitochondria with distinctive outer membranes and by axial bundles, up to 1.4 μm wide, of microfilaments

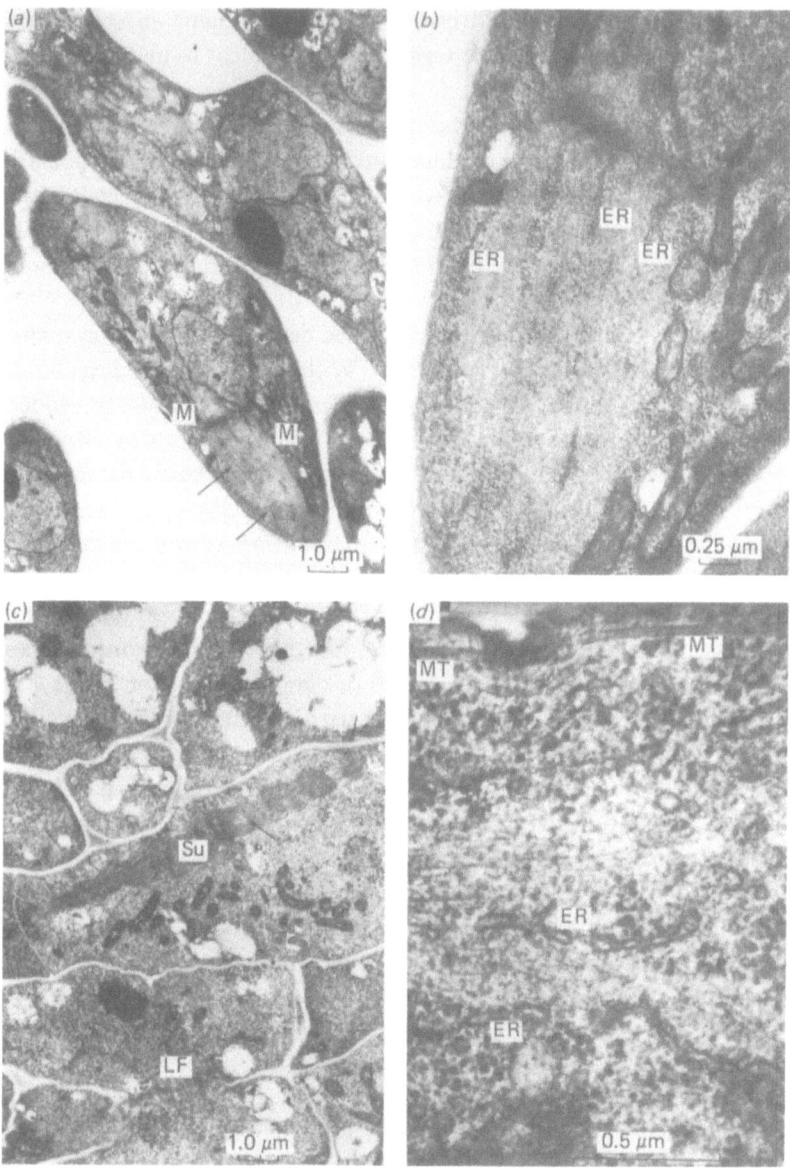

Fig. 10.11. Electron micrographs of the circum-medullary region in the apices of *Armillaria bulbosa* rhizomorphs. (*a*) Structures seen at low magnification, with large bundles of microfilaments present (arrowed) together with clusters of associated axially aligned mitochondria (M). (*b*) Enlargement of bundle shown in (*a*) illustrating the microfilaments, 4–9 nm in diameter, and associated endoplasmic reticulum (ER). (*c*) A heavily stained preparation showing longitudinally and obliquely sectioned bundles (arrowed) which in one case appears to have traversed a site of lateral fusion (LF) between adjacent compartments and in another forms a large (0.96-μm wide) sinuous unit (Su) extending for at least 9.5 μm longitudinally. (*d*) Part of a bundle with associated endoplasmic reticulum (ER) and microtubule (MT). (From Powell & Rayner, 1984.)

4–9 nm in diameter (Fig. 10.11). The mitochondria appeared to become vacuolate in cells on either side of this layer, and this may explain the origin of the vacuolate medullary cells.

Microfilament bundles with essentially similar characteristics to those of *A. bulbosa* have been seen in a wide range of eukaryotes, including plants, animals, Myxomycota and Eumycota (Wohlfarth-Bottermann, 1964; Partharasarthy & Muhlethaler, 1972; Allen, Lowry & Sussman, 1974; Anderson & Zachariah, 1974; Beckett, 1976; Stebbings & Hyams, 1979). Perhaps significantly, reports of their occurrence in fungi often concern aggregated structures: apothecia in *Ascobolus stercorarius* (Anderson & Zachariah, 1974); stromata, ascogenous hyphae and perithecial wall cells in various Xylariaceous ascomycetes (Beckett, 1976); rhizomorphs in *Sphaerostilbe repens* (Botton & Dexheimer, 1977); and cords and fruit body primordia in *Agaricus sylvicola* (Angeli-Papa & Eyme, 1978). In several cases they have been likened to the P protein in phloem sieve-tubes of higher plants, particularly clear evidence being provided by Beckett (1976) who, using high magnification, was able to distinguish truly tubular elements of 10–13 nm diameter composed of sub-units arranged in a double helix. These contrasted with fibrous elements which were present only in the ascospores of *Xylaria polymorpha* and *X. longipes*. The occurrence of microfilament bundles may therefore be further evidence of anatomical parallels between a range of types of vegetative and reproductive organs in higher fungi. However, bundles have been described in the *snowflake* morphological mutant of *Neurospora crassa* (which has a slow-dense colony morphology), where evidence has been provided that they are composed of actin-like protein, as well as occasionally in wild-type strains (Allen, Lowry & Sussman, 1974; Allen & Sussman, 1978). Small bundles have also been observed in dikaryotic hyphae of *Schizophyllum commune* (Raudaskoski, 1970); microfilaments are associated with dolipore septa (see Moore: Chapter 7) and have been observed forming a network in hyphal tips (Gooday, 1983).

Various functions have been attributed to microfilament bundles. However, two which seem significant in relation to their occurrence in the circum-medullary zone of *Armillaria* rhizomorph apices are cytoplasmic streaming, direct evidence for which has been obtained in the alga *Nitella* (Shimmen & Tazawa, 1982), and provision of cytoskeletal support in elongating cells (Anderson & Zachariah, 1974). The apparent considerable activity of the circum-medullary zone, indicated by the large number of mitochondria, is also consistent with the suggestion mentioned previously, that maintenance of an oxygen gradient from the medulla to the exterior of the rhizomorph apex is important for continued extension.

These observations clearly impinge on the mechanism underlying the polarised extension of rhizomorphs. Motta (1969a) has suggested, from the apparent orientation of 'cells' in rhizomorph apices and perhaps by analogous behaviour to the histogens of the apical meristems of higher plants (Watkinson, 1979), that extension is due to an apical centre containing actively dividing initials which give rise to the various other layers. However, no such centre is apparent in the highly developed rhizomorphs of the ascomycete *Sphaerostilbe repens*, where apically extending hyphae are discernible in the tips (Botton & Dexheimer, 1977). Furthermore, this concept is difficult to reconcile with our suggestion of morphological and behavioural continuity between types of linear organs, and with known patterns of branching and partitioning between hyphal compartments which are fundamentally different from cell division and separation in plant tissues. Our own observations appear to be consistent with the presence of an interdigitating mass of apically extending, septating, branching and anastomosing hyphae forming an intricate cross-latticed tissue overarching the apical dome, but being in predominantly parallel array further back. Such a tissue, *in section*, might well provide an impression of central apical initials giving rise to axially aligned tissues, but the three-dimensional information provided by scanning electron microscopy (see Figs 10.2, 10.9) is seemingly in accord with the basically filamentous organisation of rhizomorph apices.

As an alternative to meristematic development, we suggest that rhizomorph extension in *Armillaria* is by a more recognisably fungal strategy. This is perhaps analogous to the balanced lysis mechanism which has been proposed for hyphal extension (Bartnicki-Garcia, 1973), and involves the driving forward of a plasticised apical dome by forward pressure generated within a tube with rigidified side walls and compensated for by branching and growth of the intercalated apical hyphae. Plasticisation could be facilitated by mucilage production, which disrupts the continuity of the hyphal mesh which covers the dome; rigidification is brought about by melanisation and compaction of an outer crust; forward pressure could be provided via osmotically driven flow through the medullary region. This last possibility introduces our final consideration.

Role of translocation in relation to structure and growth of linear organs

The arguments for long-distance translocation of nutrients through mycelia being brought about by a hydrostatic pressure gradient generated osmotically by the absorption of solutes from the substratum have been rehearsed previously (Jennings *et al.*, 1974; Jennings, 1982, 1984). The arguments have been based on the demonstration of water-flow through mycelia, the similar velocity of translocation irrespective of the radiotracer used in the determination, and the characteristics of inhibition of translocation by low solute potential and metabolic inhibitors presented to the source of the translocate. Until recently, direct demonstration of the driving force – a turgor gradient – has been lacking. This has now been rectified (Eamus & Jennings, 1984) with studies on (1) relatively undifferentiated mycelia and cords of *Serpula lacrimans*, (2) laboratory and field-grown cords of *Phallus impudicus* and *Phanerochaete velutina*, and (3) field-grown rhizomorphs of *Armillara mellea*. In all cases, a gradient of water potential and turgor potential existed between the older and the younger extending regions. By investigating the flow of water through rhizomorphs of *A. mellea* and cords of *Phallus impudicus* it has been possible to determine their hydraulic conductivity. From these and other data it has been possible to show that the flow of water and solutes through these linear structures and through mycelium of *S. lacrimans* is of the correct order of magnitude to support known rates of growth.

In the natural environment, one must not assume that all nutrients for extension growth must be translocated from a source increasingly distant from the extending organ or mycelium. While there is some information available on the ability of rhizomorphs and cords to absorb nutrients from the environment through which they grow (Morrison, 1975; Granlund, 1983), more information on this matter is required (Jennings, 1982). We have pointed out earlier that the peripheral hyphae of a rhizomorph may be important in scavenging nutrients, particularly phosphorus, from the surrounding soil.

It seems certain, however, that the bulk of the combined carbon reaching the extending part of the mycelium must be translocated from the food source through the linear structure connecting the two. In this respect there is now unambiguous evidence that rhizomorphs of *A. mellea* can translocate 0.27–1.84 nmol glucose equivalents of carbohydrate $cm^{-2} s^{-1}$ at velocities of 3.5 cm h^{-1} towards, and 1.2 cm h^{-1} from, the apex (Granlund, Jennings & Thompson, 1985). Given that translocation of combined carbon must be occurring along cords and rhizomorphs in nature, there must be a turgor gradient along them. The longer the distance from source to sink, i.e. the extending hyphae, the more susceptible will translocation be to the dissipation of this gradient through water loss from the translocation pathway. Jennings *et al.* (1974) pointed out that the development of a rind or outer cortex of thick-walled hyphae will prevent the gradient from being dissipated since the rind will reduce evaporation and restrict lateral expression of the hydrostatic pressure.

This pressure can not only drive translocation but also contributes to the positive turgor which is necessary for extension growth of the linear structure. Nevertheless we now know (Thompson, Eamus & Jennings, 1985) that, certainly over the short term and when mycelium is growing over a non-absorptive surface, growth of *S. lacrimans* is not dependent upon the turgor gradient driving translocation. Indeed, water can be made to leave the mycelium at the source of translocation by an unfavourable solute potential in the medium without any change in the rate of mycelial extension. Translocation ceases and the turgor gradient along the axis of growth of the mycelium is reversed, yet growth continues. In contrast, when the solute potential in the medium at the food source of a rhizomorph of *A. mellea* is considerably decreased there is little effect on translocation (Granlund *et al.*, 1985). Both sets of observations support a view that the mycelia of these two basidiomycetes are able to buffer themselves against the effects of unfavourable external water potentials. In view of the fact that translocation in the rhizomorphs of *A. mellea* is much less susceptible to changes in the solute potential of the external medium than is that in young cords of *S. lacrimans*, and in keeping with the fact that the former are more bulky than the latter, we would argue that the greater the aggregation of hyphae into a linear structure the more resistant it will be to the external environment, particularly with respect to water loss. Thus, the more bulky the linear structure the better able it will be to translocate and to maintain the forward direction of the hydrostatic pressure such that it can be manifested in growth. Thus we are now beginning to obtain a measure of the physiological parameters behind the views expressed earlier on the organised growth of individual hyphae within a rhizomorph.

Conclusion

We hope to have demonstrated that the vegetative organs of agarics and other higher fungi provide exciting and, as yet, surprisingly unfulfilled prospects for studies of morphogenesis. The issues underlying their development have much in common with those relating to fruit bodies, with which they have several significant structural and functional parallels. However, their relative simplicity may be expected to facilitate both the conduct and interpretation of experimental work.

Acknowledgements. We thank the SERC for financial support of the Electron Optics Unit at Bath University.

References

Allen, E. D., Lowry, R. J. & Sussman, A. S. (1974). Accumulation of microfilaments in a colonial mutant of *Neurospora crassa*. *Journal of Ultrastructure Research*, **48**, 455–64.

Allen, E. D. & Sussman, A. S. (1978). Presence of an actin-like protein in mycelium of *Neurospora crassa*. *Journal of Bacteriology*, **135**, 713–16.

Anderson, R. H. & Zachariah, K. (1974). On the structure, function and distribution of filament bundles in apothecial cells of the fungus *Ascobolus stercorarius*. *Journal of Ultrastructure Research*, **46**, 375–92.

Angeli-Papa, J. & Eyme, J. (1978). Ultrastructural changes during development of *Agaricus bisporus* and *Agaricus sylvicola*. In *The Biology and Cultivation of Edible Mushrooms*, ed. S. T. Chang & W. A. Hayes, pp. 52–81. New York, San Francisco & London: Academic Press.

Bartnicki-Garcia, S. (1973). Fundamental aspects of hyphal morphogenesis. In *Microbial Differentiation* (23rd Symposium of the Society for General Microbiology, April 1973), ed. J. M. Ashworth & J. E. Smith, pp. 245–67. Cambridge University Press.

Beckett, A. (1976). Fibrous and tubular inclusions in four Xylariaceous fungi. *Protoplasma*, **89**, 279–90.

Bellotti, R. A. & Couse, N. L. (1980). Induction of mycelial strands in *Calvatia sculpta*. *Transactions of the British Mycological Society*, **74**, 19–25.

Botton, B. & Dexheimer, J. (1977). The ultrastructure of the rhizomorphs of *Sphaerostilbe repens*, B. & Br. *Zeitschrift für Pflanzenphysiologie*, **85**, 429–43.

Brownlee, C. & Jennings, D. H. (1982). The pathway of translocation in *Serpula lacrimans*. *Transactions of the British Mycological Society*, **79**, 401–7.

Bull, A. T. & Trinci, A. P. J. (1977). The physiology and metabolic control of fungal growth. *Advances in Microbial Physiology*, **15**, 1–84.

Butler, G. M. (1966). Vegetative structures. In *The Fungi: An Advanced Treatise*, vol. 2, ed. G. C. Ainsworth & A. S. Sussman, pp. 83–112. London: Academic Press.

Butler, G. M. (1984). Colony ontogeny in basidiomycetes. In *The Ecology and Physiology of the Fungal Mycelium*, ed. D. H. Jennings & A. D. M. Rayner, pp. 53–71. Cambridge University Press.

Campbell, A. H. (1933). Zone lines in plant tissues. I. The black lines formed by *Xylaria polymorpha* (Pers.) Grev. in hardwoods. *Annals of Applied Biology*, **20**, 123–45.

Campbell, A. H. (1934). Zone lines in plant tissues. II. The black lines formed by *Armillaria mellea* (Vahl.) Quel. *Annals of Applied Biology*, **21**, 1–22.

Casselton, L. A. (1978). Dikaryon formation in higher basidiomycetes. In *The Filamentous Fungi*, vol. 3, *Developmental Mycology*, ed. J. E. Smith & D. R. Berry, pp. 275–97. London: Edward Arnold.

Clarke, R. W., Jennings, D. H. & Coggins, C. R. (1980). Growth of *Serpula lacrimans* in relation to water potential of substrate. *Transactions of the British Mycological Society*, **75**, 271–80.

Coates, D. & Rayner, A. D. M. (1985). Evidence for a cytoplasmically transmissible factor affecting recognition and somato-sexual differentiation in the basidiomycete *Stereum hirsutum*. *Journal of General Microbiology*, **131**, 207–19.

Coates, D. & Rayner, A. D. M. (1985). Induction of rhizomorphic organs in the non-rhizomorphic fungus, *Stereum hirsutum*, via a mobile regulatory factor. *Transactions of the British Mycological Society*. (in press)

Coggins, C. R., Hornung, U., Jennings, D. H. & Veltkamp, C. J. (1980). The phenomenon of 'point growth' and its relation to flushing and strand formation in mycelium of *Serpula lacrimans*. *Transactions of the British Mycological Society*, **75**, 69–76.

Cooke, R. C. (1983). Morphogenesis of sclerotia. In *Fungal Differentiation*, ed. J. E. Smith, pp. 397–418. New York: Marcel Dekker.

Cooke, R. C. & Rayner, A. D. M. (1984). *The Ecology of Saprotrophic Fungi*. London & New York: Longman.

Day, S. C. (1969). *The morphogenesis of mycelial strands in the timber dry rot fungus, Merulius lacrymans* (*Wulf.*), *Fr*. Ph.D. thesis, University of Cambridge.

Eamus, D. & Jennings, D. H. (1984). Determination of water, solute and turgor potentials of mycelium of various basidiomycete fungi causing wood decay. *Journal of Experimental Botany*, **35**, 1782–6.

Eamus, D., Thompson, W., Cairney, J. W. G. & Jennings, D. H. (1985). Internal structure and hydraulic conductivity of basidiomycete translocating organs. *Journal of Experimental Botany*. (in press)

Gaber, R. F. & Leonard, T. J. (1981). Unilateral internuclear gene transfer and cell differentiation in *Schizophyllum*. *Nature*, **291**, 342–4.

Garraway, M. O. (1970). Rhizomorph initiation and growth in *Armillaria mellea* promoted by *o*-aminobenzoic and *p*-aminobenzoic acids. *Phytopathology*, **60**, 61–5.

Garrett, S. D. (1963). *Soil Fungi and Soil Fertility*. Oxford: Pergamon Press.

Garrett, S. D. (1970). *Pathogenic Root-Infecting Fungi*. Cambridge University Press.

Gooday, G. W. (1983). The hyphal tip. In *Fungal Differentiation*, ed. J. E. Smith, pp. 315–56. New York: Marcel Dekker.

Granlund, H. I. (1983). *Nutrient uptake and translocation by rhizomorphs of* Armillaria mellea. Ph.D. thesis, University of Liverpool.

Granlund, H. I., Jennings, D. H. & Thompson, W. (1985). Translocation of solute along rhizomorphs of *Armillaria mellea*. *Transactions of the British Mycological Society*, **84**, 111–19.

Granlund, H. I., Jennings, D. H. & Veltkamp, K. (1984). Scanning electron microscope studies of *Armillaria melea*. *Nova Hedwigia*, **39**, 85–100.

Gregory, P. H. (1984). The fungal mycelium: an historical perspective. *Transactions of the British Mycological Society*, **82**, 1–11.

Hornung, U. & Jennings, D. H. (1981). Light and electron microscopical observations of surface mycelium of *Serpula lacrimans*: stages of growth and hyphal nomenclature. *Nova Hedwigia*, **34**, 101–26.

Humpherson-Jones, F. M. & Cooke, R. C. (1977). Morphogenesis in sclerotium-forming fungi. I. Effects of light on *Sclerotinia sclerotiorum*, *S. delphinii* and *S. rolfsii*. *New Phytologist*, **78**, 171–80.

Jennings, D. H. (1982). The movement of *Serpula lacrimans* from substrate to substrate over nutritionally inert surfaces. In *Decomposer Basidiomycetes; their Biology and Ecology*, ed. J. C. Frankland, J. N. Hedger & M. J. Swift, pp. 91–108. Cambridge University Press.

Jennings, D. H. (1985). Salt relations of cells, tissues and roots. In *Plant Physiology: A Treatise*, vol. 7, ed. F. C. Steward. New York: Academic Press. (in press)

Jennings, D. H., Thornton, J. D., Galpin, M. F. J. & Coggins, C. R. (1974). Translocation in fungi. In *Transport at the Cellular Level* (28th Symposium of the Society for Experimental Biology), ed. M. A. Sleigh & D. H. Jennings, pp. 139–56. Cambridge University Press.

Lopez-Real, J. M. (1975). Formation of pseudosclerotia ('zone lines') in wood decayed by *Armillaria mellea* and *Stereum hirsutum*. I. Morphological aspects. *Transactions of the British Mycological Society*, **64**, 465–71.

Lopez-Real, J. M. & Swift, M. J. (1975). Formation of pseudosclerotia ('zone lines') in wood decayed by *Armillaria mellea* and *Stereum hirsutum*. II. Formation in relation to the moisture content of the wood. *Transactions of the British Mycological Society*, **64**, 473–81.

Lopez-Real, J. M. & Swift, M. J. (1977). Formation of pseudosclerotia ('zone lines') in wood decayed by *Armillaria mellea* and *Stereum hirsutum*. III. Formation in relation to composition of gaseous atmosphere in wood. *Transactions of the British Mycological Society*, **66**, 321–5.

Lysek, G. (1984). Physiology and ecology of rhythmic growth and sporulation in fungi. In *The Ecology and Physiology of the Fungal Mycelium*, ed. D. H. Jennings & A. D. M. Rayner, pp. 323–42. Cambridge University Press.

Moody, A. R., Garraway, M. O. & Weinhold, A. R. (1968). Stimulation of rhizomorph production in *Armillaria mellea* with oils and fatty acids. *Phytopathology*, **58**, 1060–1.

Moore, D. (1981). Developmental genetics of *Coprinus cinereus*: genetic evidence that sclerotia and carpophores share a common pathway of initiation. *Current Genetics*, **3**, 145–50.

Morrison, D. J. (1975). Ion uptake by rhizomorphs of *Armillaria mellea*. *Canadian Journal of Botany*, **53**, 48–51.

Motta, J. J. (1967). A note on the mitotic apparatus in the rhizomorph meristem of *Armillaria mellea*. *Mycologia*, **59**, 370–5.

Motta, J. J. (1969*a*). Cytology and morphogenesis in the rhizomorph of *Armillaria mellea*. *American Journal of Botany*, **56**, 610–19.

Motta, J. J. (1969*b*). Somatic nuclear division in *Armillaria mellea*. *Mycologia*, **61**, 873–86.

Motta, J. J. (1971). Histochemistry of the rhizomorph meristem of *Armillaria mellea*. *American Journal of Botany*, **58**, 80–7.

Motta, J. J. & Peabody, D. C. (1982). Rhizomorph cytology and morphogenesis in *Armillaria tabescens*. *Mycologia*, **74**, 671–4.

Park, D. (1963). Evidence for a common fungal growth regulator. *Transactions of the British Mycological Society*, **46**, 541–8.

Partharasathy, M. Y. & Muhlethaler, K. (1972). Cytoplasmic microfilaments in plant cells. *Journal of Ultrastructure Research*, **38**, 46–62.

Pentland, G. D. (1965). Stimulation of rhizomorph development of *Armillaria mellea* by *Aureobasidium pullulans* in artificial culture. *Canadian Journal of Microbiology*, **11**, 345–50.

Powell, K. A. & Rayner, A. D. M. (1983). Ultrastructure of the rhizomorph apex in *Armillaria bulbosa* in relation to mucilage production. *Transactions of the British Mycological Society*, **81**, 529–34.

Powell, K. A. & Rayner, A. D. M. (1984). Occurrence of bundles of microfilaments in circum-medullary cells underlying the apical dome of *Armillaria bulbosa* rhizomorphs. *Transactions of the British Mycological Society*, **83**, 217–21.

Prosser, J. I. (1983). Hyphal growth patterns. In *Fungal Differentiation*, ed. J. E. Smith, pp. 357–96. New York: Marcel Dekker.

Raudaskoski, M. (1970). Occurrence of microtubules and microfilaments, and origin of septa in dikaryotic hyphae of *Schizophyllum commune*. *Protoplasma*, **70**, 415–22.

Rayner, A. D. M. (1978). Interactions between fungi colonising hardwood stumps and their possible role in determining patterns of colonization and succession. *Annals of Applied Biology*, **89**, 131–4.

Rayner, A. D. M., Coates, D., Ainsworth, A. M., Adams, T. J. H., Williams, E. N. D. & Todd, N. K. (1984). The biological consequences of the individualistic mycelium. In *The Ecology and Physiology of the Fungal Mycelium*, ed. D. H. Jennings & A. D. M. Rayner, pp. 509–40. Cambridge University Press.

Rayner, A. D. M. & Todd, N. K. (1979). Population and community structure and dynamics of fungi in decaying wood. *Advances in Botanical Research*, **7**, 333–420.

Rayner, A. D. M. & Todd, N. K. (1982). Population structure in wood-decomposing basidiomycetes. In *Decomposer Basidiomycetes: Their Biology and Ecology*, ed. J. C. Frankland, J. N. Hedger & M. J. Swift, pp. 109–28. Cambridge University Press.

Rayner, A. D. M. & Webber, J. F. (1984). Interspecific mycelial interactions – an overview. In *The Ecology and Physiology of the Fungal Mycelium*, ed. D. H. Jennings & A. D. M. Rayner, pp. 383–417. Cambridge University Press.

Schmid, R. & Liese, W. (1970). Feinstruktur der Rhizomorphs von *Armillaria mellea*. *Phytopathologisches Zeitschrift*, **84**, 352–9.

Shimmen, T. & Tazawa, M. (1982). Cytoplasmic streaming in the cell model of *Nitella*. *Protoplasma*, **112**, 101–6.

Smith, A. M. & Griffin, D. M. (1971). Oxygen and the ecology of *Armillaria elegans*. Heim. *Australian Journal of Biological Sciences*, **24**, 231–62.

Sortkjaer, O. & Allermann, K. (1972). Rhizomorph formation in fungi. I. Stimulation by ethanol and acetate and inhibition by disulfiram of growth and rhizomorph formation in *Armillaria mellea*. *Physiologia Plantarum*, **26**, 376–80.

Stebbings, H. & Hyams, J. S. (1979). *Cell Motility*. London & New York: Longman.

Steele, G. C. & Trinci, A. P. J. (1975). Morphology and growth kinetics of hyphae of differentiated and undifferentiated mycelia of *Neurospora crassa*. *Journal of General Microbiology*, **91**, 362–8.

Thompson, W., Eamus, D. & Jennings, D. H. (1985). Water flux through mycelium of *Serpula lacrimans*. *Transactions of the British Mycological Society*. (in press)

Thompson, W. & Rayner, A. D. M. (1982). Structure and development of mycelial cord systems of *Phanerochaete laevis* in soil. *Transactions of the British Mycological Society*, **78**, 193–200.

Thompson, W. & Rayner, A. D. M. (1983). Extent, development and functioning of mycelial cord systems in soil. *Transactions of the British Mycological Society*, **81**, 333–45.

Watkinson, S. C. (1971). The mechanism of mycelial strand induction in *Serpula lacrimans*: a possible effect of nutrient distribution. *New Phytologist*, **70**, 1079–88.

Watkinson, S. C. (1979). Growth of rhizomorphs, mycelial strands, coremia and sclerotia. In *Fungal Walls and Hyphal Growth*, ed. J. H. Burnett & A. P. J. Trinci, pp. 93–113. Cambridge University Press.

Watkinson, S. C. (1984). Morphogenesis of the *Serpula lacrimans* colony in relation to its function in nature. In *The Ecology and Physiology of the Fungal Mycelium*, ed. D. H. Jennings & A. D. M. Rayner, pp. 165–84. Cambridge University Press.

Weinhold, A. R. (1963). Rhizomorph production by *Armillaria mellea* induced by ethanol and related compounds. *Science*, **142**, 1065–6.

Weinhold, A. R. & Garraway, M. O. (1966). Nitrogen and carbon nutrition of *Armillaria mellea* in relation to growth promoting effects of ethanol. *Phytopathology*, **56**, 108–12.

Weinhold, A. R., Hendrix, F. F. & Raabe, R. D. (1962). Stimulation of rhizomorph growth of *Armillaria mellea* by indole-3-acetic acid and figwood extract. *Phytopathology*, **52**, 757.

Willetts, A. J. (1978). Sclerotium formation. In *The Filamentous fungi*, vol. 3, *Developmental Mycology*, ed. J. E. Smith & D. R. Berry, pp. 197–213. London: Edward Arnold.

Wohlfarth-Bottermann, K. E. (1964). Differentiation of the ground cytoplasm and their significance for the generation of the motive force of amoeboid movement. In *Primitive Motile Systems in Cell Biology*, ed. R. D. Allen & N. Kamiya, pp. 79–109. New York: Academic Press.

Wolkinger, F., Plank, S. & Bruneggar, A. (1975). Rasterelektronenmikroskopische Untersuchungen von *Armillaria mellea*. *Phytopathologisches Zeitschrift*, **84**, 352–9.

11
Developmental characters of agarics

ROY WATLING

Royal Botanic Garden, Edinburgh EH3 5LR, UK

Introduction

Agaricology is one of the few remaining disciplines in which the development of the organisms concerned plays such a minor role in the overview of the group. The father of present-day agaricology is undoubtedly Elias Fries who worked in the early nineteenth century, when the classification of the flowering plants had already been consolidated by Linnaeus (1753) and the development of flowering structures was well understood. The classification of the agarics is therefore far behind that of vascular plants; a surprising fact since toadstools are as colourful, as bizarre in shape and smell, and exploit a wide range of biological lifestyles. The field botanist still only looks at fungi casually and many early floras ignored the fungi because of the difficulty of their identification.

Whatever the reason for this neglect of development, agaricologists have a long way to go; no zoologist of any standing would attempt a discussion on classification without call on developmental studies in existent and fossil taxa. Yet many agaricologists happily describe new taxa with only a casual knowledge or interest in agaric development.

With the introduction of the ideas of Nannfeldt (1932) and Luttrell (1951), it was appreciated that not all ascomycetes produced their asci, or even the fruit body, in the same way. Perhaps because the ascomycetologist has to use the microscope continuously the transition was made painlessly. But not for agarics!

In this volume the colloquial term 'fruit body' has been adopted to describe the basidiomycete sexual sporing structure. The preferred term is basidioma; other alternatives (and definitions of other terms used here) are given by Hawksworth, Sutton & Ainsworth (1983).

It is undoubtedly correct for one to be cautious in using fruit body

281

Table 11.1. *Development in the agarics arranged in the classification adapted by Dennis, Orton & Hora (1960). Arrows indicate where some modification to the classification is necessary; asterisks indicate genera in which additional species have been analysed by Watling & Sweeney; † indicates those added to Reijnders' 1963 list; and ● genera with species examined by both Reijnders and Watling & Sweeney. Abbreviations: gymno = gymnocarpic; -angio = -angiocarpic; iso = isocarpic; -stipito = -stipitocarpic; hymeno = hymenocarpic; pileo = pileocarpic; R. = Reijnders; C. = Corner.*

	Number of species analysed	Development
Hydnaceae		
Hydnum rufescens	1*	gymno
Cantharellaceae		
Cantharellus & *Craterellus*	11 (9†)	gymno (C.1966)
Boletaceae		
Xerocomus	5 (2*)	gymno, paravelang
Pulveroboletus	3	metavelangio (C.1972)
●*Suillus*	8	{ gymno / pilangio / mixangio / gymno
Leccinum	2 (1*; 1†)	pilangio (C.1972)
Austroboletus	1†	metavelangio? (C.1972)
Boletellus	2†	mixangio (C.1972)
Boletochaete	1†	gymno (C.1972)
Tylopilus	2†	gymno (C.1972)
Gyroporus	2	{ gymno / metavelangio
Gyrodon (≡ Uloporus)	1	gymno
Boletinus	4	pilangio
Strobilomyces	2 (1*)	metavelangio
Heimiella	1†	gymno (C.1972)
Paxillaceae		
Paxillus	2 (1*)	pilangio
Gomphidiaceae		
Gomphidius (incl. *Chroogomphus*)	3	metavelangio
Russulaceae		
Russula	8 (1*; 3†)	{ gymno / mixangio (R.1976)
Lactarius	7 (4*; 1†)	{ gymno / mixangio (R.1976)
Hygrophoraceae		
Hygrophorus (sg.) *Hygrophorus*	7 (1†)	mixangio gymno (R.1983*a*)

	Number of species analysed	Development
(sg.) *Camarophyllus*	2 (1*)	gymno
(sg.) *Hygrocybe*	7 (3*)	gymno
(sg.) *Hygrotrama*	1†	monovelangio (R.1983*b*)
Pleurotaceae		
Pleurotus	3 (1*)	{ gymno { metavelangio
Hohenbuehelia ←	1*	gymno
Resupinatus ←	2 (1*)	gymno
● *Pleurotellus* ←	2*	gymno
Phyllotopsis ←	1	gymno
Geopetalum ←	1	gymno
● *Lentinus/Panus*	6 (1*; 3†)	{ mixangio { gymno (C.1981)
(*Lentinula*: see Leatham: Chapter 17)	1*	metavelangio
Lentinellus (= Lentinellaceae)	2 (1*)	gymno
● *Panellus* ←	2 (1†)	gymno (R.1983*a*)
● *Tectella* ←	1†	bivelangio (R.1983*c*)
● *Schizophyllum* (= Schizophyllaceae) ←	1	gymno
Polyporaceae		
● *Polyporus* (incl. *Microporus*)	3 (1*)	gymno
Entolomataceae (Rhodophyllaceae)	5 (2*)	see text. gymno
Pluteaceae (Volvariaceae)		
Volvariella	5 (1*)	bulbangio
● *Pluteus*	4	{ gymno { paravelangio
Amanitaceae		
● *Amanita*	6	bivelangio
Limacella	2 (1†)	bivelangio (R.1979*c*)
Termitomyces (incl. *Podabrella*) ←	2	bivelangio
Agaricaceae		
Tribe Agariceae		
Agaricus (see Wood: Chapter 15)	5 (1*)	bivelangio
Melanophyllum	1	bivelangio
Tribe Lepioteae		
Lepiota		
Lepiota	7 (1*)	⎫
● *Leucocoprinus*	3 (1*; 1†)	⎬ bivelangio (R.1975)
Macrolepiota	2 (1*)	⎪
Leucoagaricus	5 (4*; 1†)	⎭
Cystoderma	2	monovelangio
Drosella (≡ *Chamaemyces*)	1†	monovelangio (R.1975)

Table 11.1. (*cont.*)

	Number of species analysed	Development
Bolbitiaceae		
● *Bolbitius*	3 (1*)	paravelangio
● *Conocybe*	18 (15*)	{ paravelangio gymnangio
● *Agrocybe*	4 (2*)	paravelangio
(*A. cylindrica*)	1†	bivelangio (R.1971*a*)
Coprinaceae		
● *Coprinus*	37 (10*; 9†) see Table 11.3	Predominantly bivelangio (R.1979*b*; also see R.1974*b*)
Psathyrella		
sg. *Psathyrella*	3 (2*)	bivelangio
● sg. *Psathyra*	9 (1*)	monovelangio
● *Lacrymaria*	2	metavelangio
● *Panaeolus* (incl. *Panaeolina*)	7 (3*)	paravelangio
Strophariaceae		
● *Stropharia*	4 (1*)	bivelangio
● *Hypholoma*	9 (3*)	{ bivelangio monovelangio
● *Psilocybe*	10 (2*)	{ bivelangio monovelangio
Melanotus	1	bivelangio
Cortinariaceae		
Cortinarius	22 (7*; 8†)	⎧ bivelangio monovelangio (R.1974, R.1976, R.1979*a*) ⎭
Leucocortinarius	1†	bivelangio (R.1976)
Rozites	2	bivelangio
● *Phaeolepiota*	1	monovelangio
Flocculina (= *Flammulaster*)	3 (1*)	bivelangio
● *Tubaria*	3 (2*)	monovelangio
Gymnopilus	5 (3*)	{ monovelangio paravelangio
Galerina	4 (2*)	monovelangio
● (*Kuehneromyces mutabilis*)	1†	bivelangio (R.1971*b*))
Pholiota	13 (7*)	bivelangio
● *Hebeloma*	5 (3*)	monovelangio
Naucoria (incl. *Simocybe*)	2	monovelangio paravelangio
● *Inocybe*	10 (2*; 1†)	monovelangio (R.1974*a*)
Ripartites	1	gymnangio

	Number of species analysed	Development
Triocholomataceae		
Tribe Tricholomeae		
Tricholoma		{ paravelangio
Tricholoma s. st.	11 (3*; 3†)	{ monovelangio (R.1983*a*)
Calocybe	1*	gymno
Tricholomopsis	3 (1*)	{ paravelangio
		{ gymno
Lyophyllum	1*	gymno
Melanoleuca	1*	gymno
Squamanita	1	monovelangio
Tribe Clitocybeae		
Clitocybe	4	gymno
Tephrocybe	1*	gymno
Armillaria	3 (1*)	{ monovelangio
		{ metavelangio
Leucopaxillus	2 (1*)	gymno
Cantharellula	1	gymno
● *Hygrophoropsis*	1†	gymno (R.1983*a*)
● *Laccaria*	2 (1*)	monovelangio
Biannularia	1	bivelangio
Tribe Collybieae		
Collybia (excl. *Tephrocybe*)	7 (4*)	gymno
Nyctalis	2	gymno
● *Oudemansiella*	4	{ monovelangio
		{ bivelangio
		{ paravelangio
● *Flammulina* (see Williams *et al.*: Chapter 18)	1	paravelangio
● *Mycena*	26 (10*)	{ monovelangio
		{ gymno
		{ stipitangio
		{ pilangio
Myxomphalia	1*	gymno
Omphalina	5 (4*)	gymno
Marasmius (incl. *Marasmiellus*)	8 (2*; 2†)	paravelangio (R.1983*a*)
Physalacria	1	gymno
● *Micromphale*	1	paravelangio
Trogia	3	gymno (C.1966)
Calyptella (reduced agaric)	1	gymno
Clitopilaceae ←		
Clitopilus	1*	gymno?
● *Lepista*	1	gymno
Rhodocybe	1	pilangio
Rhodotus	1	paravelangio

ontogeny in taxonomy, but if proven to be useful then a very important tool is available to us. Brefeld (1877, 1881) was probably the first mycologist to describe fully the development of an agaric in his exceptional series of treatises; he was followed by de Bary (1884), but it was left to Fayod (1889) to draw attention to the different ways in which different taxonomic groups of agarics develop. Seventy five years later, the study was placed on a much firmer footing, almost entirely by the work of a single man, A. F. M. Reijnders. Indeed this chapter is dedicated to this man with whom the study of agaric ontogeny has become so clearly associated.

It is true that others since Fayod, e.g. Beer (1911), Atkinson (1914, 1915, 1916) and Kühner (1925, 1926, 1929), had examined the development of individual species but it was left to Reijnders to carry out a wide survey. By 1963 he had studied 156 fungi bringing together his results spanning 25 years of effort (Reijnders, 1948, 1952); to this impressive list he added many more species from the literature bringing the total to 232, 57% of which he had studied himself. Subsequent work by Reijnders (1971 *et seq.*) and Corner (1966, 1972), and our own studies in Edinburgh have added observations on a further 190 species. Thus from an (under)estimated 4850 species of Agaricales, Boletales and Russulales (Hawksworth, Sutton & Ainsworth, 1983), we only have information on 9%, a small proportion indeed (Table 11.1).

During the last few years we have been able to add an additional parameter to this study by marrying information from primordia collected in the field with that from pure cultures in the laboratory.

The 232 (including two polypores and eight gasteromycetes) dealt with by Reijnders (1963) cover 95 genera and include 63 taxa which are considered type-species for their genus and six additional segregate genera. The information is at the moment rather biased in favour of European species, only 30 of the species studied being sub-tropical and still fewer (about 4.5%) exclusively North American. We have examined in Edinburgh 177 species including 131 not previously studied by Reijnders. This later work extends our knowledge to a further 25 genera (Table 11.1).

Terminology

Three groups of terms were introduced by Reijnders (1948) to describe the various types of development shown by agarics and similar fungi. The main group circumscribes the position of the hymenium in relation to the presence or absence of any protective tissues. The other two groups of terms refer to (1) the relationship between the gills and the surrounding sterile tissues and (2) those tissues concerned with the initial development of the hymenium. Readers are referred to illustrations of these developmental patterns to be found in Reijnders (1963) and/or Watling (1978).

The major developmental terms are an elaboration and expansion of those used by de Bary (1884), Fayod (1889) and Maire (1902):

> **gymnocarpic**, where the hymenium appears and develops to maturity on the surface of the fruit body
>
> **angiocarpic** where the hymenium is enclosed
>
> **hemiangiocarpic** where the hymenium first appears within a cavity but becomes exposed before its maturity by the dehiscence of an enclosing membrane.

Gilbert (1931) doubted the distinctions between these terms, but Kühner (1938) accepted them and restricted the use of hemiangiocarpic to where the hymenium is surrounded by a veil until the spores mature and it then becomes exposed. Reijnders (1948) introduced the term velangiocarpic for this phenomenon, which Kühner (1938) had previously described as hemiangiocarpic, and further sub-divided it into four, based on the number and the location of the protective membranes (veils), i.e. primarily angiocarpic in the sense of Reijnders:

(*a*) **monovelangiocarpic**
(*b*) **paravelangiocarpic**
(*c*) **bivelangiocarpic**
(*d*) **metavelangiocarpic**

In the monovelangiocarpic type the veil envelops the primordium; in the paravelangiocarpic type the veil is exposed only at adolescence and protects the hymenium, and in some cases may be considerably reduced (when it is described as gymnangiocarpic). In the bivelangiocarpic type, in addition to the enveloping membrane there is tissue which protects the developing gills (i.e. the classic diagrammatic representation of the 'agaric'). In the metavelangiocarpic type there is a union of secondary structures which arise from the pileus and/or stipe to form a universal veil.

In contrast there are other types of development where the hymenium is either naked at the moment of initiation (*e* below), or the hymenium is protected by tissue other than a specialised primary veil. These are:

(*e*) **gymnocarpic**
(*f*) **pilangiocarpic**
(*g*) **mixangiocarpic**
(*h*) **stipitangiocarpic**

Mixangiocarpy is similar in some ways to the metavelangiocarpic state;

after the initial gymnocarpic phase, tissue proliferates from the pileus margin and/or stipe base to embrace the hymenium. This therefore is an extension of gymnocarpy where the hymenium is initiated in predetermined areas on the surface of the fruit body and secondarily covered. The pilangiocarpic state is where the protection is offered by downward growth of the margin of the pileus; stipitangiocarpic is where protection is offered by upward growth from the stipe base. Thus, (*f*), (*g*) and (*h*) above are considered secondary angiocarpic. Reijnders (1963) introduced bulbangiocarpic to describe the rather specialised state found in *Volvariella*, which superficially resembles the stipitangiocarpic condition but differs in that the primordial pileus is completely enveloped.

Two further terms, although they might be considered redundant, need to be mentioned for completeness. These are **hypovelangiocarpic**, where the veil development whether in mono- or paravelangiocarpy is only slightly produced, and **pseudoangiocarpic** (Kühner, 1925) where the gymnocarpic hymenium is later protected by the downward and inward curling of the pileus-margin, such as in *Lactarius rufus* and *Russula fellea*.

The three terms concerned with the relation of the hymenium to the surrounding tissue are:

(*i*) **levhymenial**, where the gills push down into a preformed gill-cavity

(*j*) **rupthymenial**, as (*i*) but gills are formed by linking of two hymenial palisades (see Rosin et al.: Chapter 13)

(*k*) **schizohymenial**, where the gills are differentiated from the background tissue.

Although a schizohymenial development is clear cut (Atkinson, 1914*c*), separation of the other two is very difficult; indeed Atkinson (1916) did not see any evidence to support the sub-division into three categories. Although Levine (1914), Kühner (1926) and Reijnders (1948) recognised the first two categories, it is appreciated that intermediates exist and that critical interpretation is necessary to decide which set of tissues in one group of agarics is strictly parallel to those in another group. Indeed although this is a very fundamental point, it is at the moment difficult to interpret the available data.

The terms which cover the order in which the various tissues develop are as follows:

(*l*) **stipitocarpy**, young primordium stipe-like, a group or bundle (fascicle) of hyphae lacking an apical area of differentiated cells

(*m*) **pileocarpy**, young primordium pileus-like, an apical interlaced mass of hyphae (*n*) **pileostipitocarpy**, both the features above are more or less of an equal level of differentiation

(*o*) **hymenocarpy**, where a circular zone of imminent hymenial tissue is found, without accompanying differentiation of pileus or stipe

(*p*) **hymenopileocarpy**, where hyphae of the hymenium are observed as the first differentiation of the pileus takes place

(*q*) **isocarpy** where apparently simultaneous differentiation takes place in all the major organs.

With these terms one is able to categorize the developmental types exhibited in the agarics.

Developmental types within the agaricoid fungi

The main categories of developmental pattern exhibited in the agaricoid fungi have been overshadowed in the past in the non-taxonomic literature by emphasis on laboratory 'organisms' which are rather specialised; e.g. *Coprinus cinereus*, with its deliquescent equilateral gills, dimorphic/trimorphic basidia and ephemeral nature, and the rather isolated *Schizophyllum commune*, better placed in the Aphyllophorales with *Polyporus* and *Lentinus*. With the introduction, particularly in the Far East, of species other than *Agaricus bisporus* into cultivation either for food or recreational drugs, the range of developmental types demonstrable in laboratory subjects can be expanded, e.g. *Flammulina velutipes* with reduced inner veil, *Stropharia ferrii* and *Psilocybe cubensis* with strongly developed inner veil, *Pholiota nameko* with double velar system, and *Volvariella esculenta* with a basal volva. To this group can be added *Pleurotus ostreatus* and its relatives, and the unrelated *Lentinula edodes*, 'shi-itake'.

Because of the widespread use of the classification adopted by Dennis, Orton & Hora (1960), together with its familiarity in Britain and the ease by which it can be consulted, the distribution of developmental types is discussed within this taxonomic framework. Examples from each of the major families of agarics, based on information both of Reijnders (1963 *et seq.*) and particularly from work completed in Edinburgh, are given to demonstrate how the developmental knowledge is of use in confirming or modifying classifications.

The Cantharellaceae because of their low, radiating, flabellate (fan-shaped), branched gill-folds, are generally linked with the Agaricales although their affinities are more with the Aphyllophorales; indeed Corner (1966) found it difficult to draw a distinction between the clavarioid

Table 11.2. *Agarics additional to Reijnders' studies and compilation*
(1963). Data from Watling & Sweeney (unpublished). Material in E.
Abbreviations as in Table 11.1.

Species analysed	Development of hymenium	Development of tissues
Hydnaceae		
Hydnum rufescens	gymno	stipito
Boletaceae		
Boletus armeniacus	gymno	stipito
B. porosporus	gymno	stipito
Leccinum versipelle	gymno → pilangio	stipito
Strobilomyces polypyramis	metavelangio	stipito
Paxillaceae		
Paxillus atrotomentosus	?metavelangio	?
Russulaceae		
Lactarius deterrimus	gymno → pilangio	stipito
L. glyciosmus	gymno	stipito
L. pubescens	gymno → pilangio	stipito
L. torminosus	gymno → pilangio	stipito
Russula fellea	gymno	stipito
Hygrophoraceae		
Hygrocybe lilacina	gymno	stipito
H. nivea	gymno	stipito
H. psittacina	gymno	stipito
H. aff. *ceracea*	gymno	stipito
Pleurotaceae		
Lentinellus cochleatus	gymno	stipito
(= Lentinellaceae)		
Pleurotus flabellatus	gymno	stipito
Polyporaceae		
Lentodium squamulosum	metavelangio	stipito
Polyporus squamosus	gymno	stipito
Entolomataceae		
Entoloma abortivum	gymno?	pileostipito
Nolanea verna	gymno	pileostipito
Pluteaceae		
Volvariella esculenta	bulbangio	pileo
Agaricaceae		
Agaricus vaporarius	bivelangio	iso
Lepiotaceae		
Lepiota xanthophylla	bivelangio	pileostipito
Leucoagaricus bresadolae	bivelangio	pileostipito
L. carneifolius	bivelangio	pileostipito
Leucoagaricus sp. (2 undet.)	bivelangio	pileostipito
Leucocoprinus birnbaumii	bivelangio	pileostipito
Macrolepiota bohemica	bivelangio	iso

Table 11.2. (*cont.*)

Species analysed	Development of hymenium	Development of tissues
Bolbitiaceae		
Agrocybe firma	paravelangio	hymeno
A. acericola	paravelangio	hymeno
Bolbitius titubans	paravelangio	hymeno
Conocybe coprophila	paravelangio	hymeno
C. pubescens	paravelangio	hymeno
Coprinaceae		
Panaeolus antillarum	paravelangio	hymeno
P. phaleanarum	paravelangio	hymeno
P. sub-balteatus	paravelangio	hymeno
Psathyrella coprophila	paravelangio	stipito
P. polycystis	paravelangio	hymeno
P. spadiceogrisea	bivelangio	iso?
Strophariaceae		
Hypholoma marginatum	bivelangio	pileostipito
H. subericaeum	bivelangio	pileostipito
H. elongatum	bivelangio	pileostipito
Melanotus vorax	bivelangio	pileo
Psilocybe cyanescens	monovelangio	stipito
P. fimetaria	bivelangio?	pileostipito
P. merdaria (incl. gasteroid form)	bivelangio	pileostipito
Stropharia cyanea	bivelangio	hymeno
Cortinariaceae		
Cortinarius alboviolaceus	bivelangio	pileostipito
C. anomalus	bivelangio	pileostipito
C. callisteus	bivelangio	pileostipito
C. lepidopus	bivelangio	pileostipito
C. paleaceus	bivelangio	pileostipito
Cortinarius spp. (2 undet.)	bivelangio	pileostipito
Flocculina aff. *carpophila*	bivelangio	stipito
Galerina ampullaceocystis	monovelangio	stipito
G. mycenopsis	monovelangio	stipito
Gymnopilus penetrans	monovelangio	stipito
G. hybridus	monovelangio?	stipito
G. junonius	monovelangio	stipito
Hebeloma cavipes	monovelangio	pileo
H. populinum	monovelangio	stipito
Hebeloma sp. (*H. fragile?*)	monovelangio	stipito
Inocybe fastigiata	monovelangio	stipito
I. leptocystis	monovelangio	stipito
Pholiota alnicola	bivelangio	pileostipito
P. highlandensis	bivelangio	pileostipito
P. lubrica	bivelangio	pileostipito
P. pusilla	bivelangio	pileostipito
P. scamba	bivelangio	pileostipito
P. terrestris	bivelangio	pileostipito
P. tuberculosa	bivelangio	pileostipito
Tubaria conspersa	bivelangio	stipito
Tubaria sp. (cf. *hiemalis*)	bivelangio	stipito

Table 11.2. (*cont.*)

Species analysed	Development of hymenium	Development of tissues
Tricholomataceae		
Armillaria bulbosa	metavelangio	stipito
Clitopilus prunulus	gymno?	stipito
Collybia cirrhata	gymno	stipito
C. cookei	gymno	stipito
C. maculata	gymno	stipito
C. peronata	gymno	stipito
Hohenbuehelia geogina	gymno	stipito
Laccaria bicolor	monovelangio	stipito
Lentinula edodes	metavelangio	pileostipito
Lyophyllum connatum	gymno	stipito
Marasmius hudsonii	paravelangio	stipito
M. epiphyllus	paravelangio	stipito
Melanoleuca sp.	gymno	stipito
Mycena fibula	gymno	stipito
M. galericulata	mixangio	stipito
M. galopus	mixangio	stipito
M. pullata?	mixangio	stipito
M. purpureofusca	mixangio	stipito
M. veneta	stipitangio	pileostipito
Mycena spp. (4 undet.)	mixangio	stipito
Myxomphalia maura	gymno	stipito
Omphalina cupulatoides	gymno	stipito
O. ericetorum	gymno	stipito
O. hudsoniana	gymno	stipito
O. luteovitellina	gymno	stipito
Resupinatus cyphelliformis	gymno	pileostipito
Strobilurus tenacellus	gymno	stipito
Tephrocybe palustris	gymno?	stipito
Tricholoma imbricatum	monovelangio	stipito
T. vaccinium	monovelangio	stipito
T. virgatum	monovelangio	stipito
Tricholomopsis decora	metavelangio	stipito
Reduced form: *Calyptella campanella*	gymno	stipito

Clavariadelphus and *Cantharellus*, the type-genus of the family. Relationships have also been suggested with the Hydnaceae (see Table 11.2) and the Clavulinaceae (Petersen, 1967), both of which are also gymnocarpic. Patouillard (1900) considered the Aphyllophorales gymnocarpic, and the Agaricales hemiangiocarpic, yet included *Cantharellus* in the latter. Those members so far studied are all gymnocarpic (Reijnders, 1963; Corner, 1966) and our own observations confirm this opinion. Gomphaceae is undoubtedly the same (Corner, 1966) and is related to the gymnocarpic, clavarioid genus *Ramaria*. All are stipitocarpic, a condition which is considered primitive, and this, coupled with the presence of stichic (cylindrical) basidia, refers the chantarelles to the Aphyllophorales. The cantharelloid fungi are undoubtedly primitive and may well be similar in structure and development to those primitive basidiomycetes which gave rise to both the agarics and polypores recognised today.

The boletes, despite their tubular hymenophore, are generally accepted as related to the agarics; indeed the Gomphidiaceae and Paxillaceae, families with true gills, have been placed with them in the Boletales. Several patterns of development are displayed by members of this group, including velate and non-velate behaviour but, although veils are formed in the majority of cases, they are secondary in origin and therefore only superficially resemble the veils exhibited by *Agaricus* spp. and *Amanita* spp.

In the genus *Boletus* as circumscribed by Dennis *et al.* (1960) several natural genera, e.g. *Suillus* and *Leccinum*, can be separated out (Rayner, Watling & Frankland: Chapter 1); but even in *Suillus* apparent anomalies exist where annulate, velate and exannulate species occur side by side. Such a phenomenon can be explained by all species exhibiting a basic gymnocarpic development which leads in some to protection by a fluffy floccose veil left in the mature fruit body as a marginal roll of tissue on the pileus, and in others to protection by a strong membrane which forms a ring, and usually a marginal, appendiculate veil, in the mature specimens. In others no veil develops at all and the downward curling of the pileus margin affords the only protection to the developing hymenium. Thus the members of the genus are unified by a common primary pattern of development but diverge in respect to their secondary patterns.

In *Gomphidius* and its segregate *Chroogomphus* (Gomphidiaceae) all examples studied are metavelangiocarpic in a similar way to the boletes *Strobilomyces* spp. and *Pulveroboletus ravenelii* (Table 11.2), whilst *Paxillus involutus* (Paxillaceae) is pilangiocarpic, parallel to some species of *Suillus*. Pilangiocarpic development is rare elsewhere in the agarics.

Results from additional boletes to those already studied (Table 11.2) extend the observations to the genus *Leccinum* and confirm the general theme, i.e. variations on a basic gymnocarpic pattern. However, whilst the development recorded for the two European members of the unusual genus *Gyroporus* is metavelangiocarpic (Reijnders, 1963), Corner (1972) considers the genus gymnocarpic on the basis of Malaysian collections.

Much discussion has surrounded the delimitation of *Pulveroboletus* as to whether it should be restricted to the type-species (*P. ravenelii*) which is metavelangiocarpic, or expanded to include several taxa many of which have contrasting macromorphological features. By doing the latter the range of developmental types found in the genus is also extended. Similar differences, often rather major, are experienced with some authors' treatments of other bolete groupings but many of the anomalies found have been solved by a more radical and restricted approach to the boletes adopted by Pegler & Young (1981). Boletes and their allies generally develop relatively rapidly for their size and, although bulky, produce basidiospores from a very early stage, often before the pileus has fully differentiated. They are stipitocarpic, a trait considered primitive.

It is relevant at this point to mention *Gastroboletus*, since in this single genus there are connections to all the known major bolete genera; in this secotioid genus a range of developmental types is found, from naked margins to where the pileus margin extends to enclose the gleba. One unusual feature of *Tylopilus humilis* is its apparent trend towards a gasteroid habit with the stipe often being poorly developed and eccentric, the tubes sometimes permanently covered by the cuticle of the pileus, and the fruit bodies often remaining below the ground until fully mature.

The Russulaceae, with several gasteromycetoid members, exhibits variations in developmental type but on a unifying theme. The northern temperate species appear to be universally stipitocarpic and gymnocarpic, some species becoming almost pilangiocarpic by secondary downward curling of the pileus margin (e.g. *Lactarius deterrimus*, *L. pubescens* and *L. torminosus*; Table 11.2). The non-temperate species studied by Heim (1937) appear, in contrast, to be mixangiocarpic and are placed in Sect. Lactariopsidei. A similar division is found in *Russula* between southern-hemisphere taxa and the more familiar European constituents; *R. emetica* and *R. olivacea* (Reijnders, 1963) and *R. fellea* (Watling & Sweeney, unpublished) are pileostipitocarpic and gymnocarpic, whilst the African *R. annulata* and *R. radicans* (Heim, 1937) are mixangiocarpic/pilangiocarpic. Both species studied by Heim are in Sect. Pelliculariae, a section known from Central and East Africa, Madagascar, the Lesser Antilles and South America. He records an exannulate form of *R. annulata* which indicates how unwise it is to have rigid categories; or does it show that more studies are required?

The heteromerous flesh of the Russulaceae (which has groups of inflated cells among filamentous hyphae) has been studied by Reijnders (1976) and Watling & Nicol (1980) and is considered to be a way in which the fruit body rapidly expands, thus taking maximum advantage of transiently favourable environmental conditions. *Macowanites*, a gasteroid member of the family, is gymno- or slightly pilangiocarpic (Bucholtz, 1903) in keeping with the other members of the family so far studied (see also *Hydnangium* below).

In the Hygrophoraceae a basic gymnocarpic pattern is also demonstrable. The family possesses three sub-genera, *Camarophyllus*, *Hygrocybe* and *Hygrophorus*; the first two are strictly gymnocarpic, a character confirmed by the recent analysis of *H.* aff. *ceracea*, *H. lilacina*, *H. nivea* and *H. psittacina* (Table 11.2). Indeed these two sub-genera are now united in a single genus, *Hygrocybe*. In those species of sub-genus *Hygrophorus* studied so far, a tendency towards a mixangiocarpic state is found, and this state is correlated with the presence of bilateral, and not regular, hymenophoral trama and the ability to form sheathing mycorrhizas. This trend is substantiated by Reijnders' (1983*a*) studies on *H. pudorinus* (see also Reijnders, 1963). The segregation of *Hygrotrama* as a separate genus is supported by the developmental pattern of *H. atropuncta*; it is neither gymno- nor mixangiocarpic (Reijnders, 1983*a*); it is monovelangiocarpic and therefore quite different to those hygrophoraceous species already discussed.

The pleurotoid fungi

The Pleurotaceae in Dennis, Orton & Hora (1960) is a miscellany of taxa brought together by virtue of their common lignicolous habit, lateral stipe and pale or white spore-print. Pegler's (1972) reassessment of the group rightly emphasises the importance of the presence or absence of inflating hyphae, and he recognises several natural groupings. It is true that there is a basic gymnocarpic pattern exhibited in these lignicolous agarics but this, coupled with stipitocarpy, represents the primitive nature of their overall development and probably also their ecological preferences. *Pleurotus dryinus*, related to *Polyporus*, is metavelangiocarpic and in this way resembles *Lentinula edodes* but this latter fungus has inflating hyphae and belongs to the Tricholomataceae. The genera *Resupinatus*, *Hohenbuehelia* and *Pleurotellus* are all gymnocarpic but on anatomical grounds could not be related to *Polyporus*. *Schizophyllum commune*, on the other hand, takes an isolated position (Watling & Sweeney, 1971) although the development superficially resembles that illustrated for *Lachnella subfalcispora* (Aeger, 1983), a cyphelloid member of the Tricholomataceae.

Lentinus tigrinus is mixvelangiocarpic (Kühner, 1925) but in one growth-form an enclosing veil is retained intact even into maturity (Rosinski & Robinson, 1968) giving a gasteromycetoid fruit body. Such structures have been called *Lentodium squamulosum* and grow, often intermixed or at the site of the normal form; no environmental advantage can be hypothesized for this 'form' although the results from intensive field work might offer some clues.

Lentinellus is also placed in the Pleurotaceae by Dennis *et al.* (1960) but differs in many fundamental characters; it is related to the Auriscalpiaceae and like it is gymnocarpic. Studies have been completed on *L. cochleatus* (Table 11.2) and *L. flabelliformis* (= *omphalodes*) confirming a relationship with the clavarioid genus *Clavicorona*; indeed *L. cochleatus* can frequently be found in the field as *Clavaria*-like tongues which are covered in asexual spores before a hymenium is formed.

The 'Rhodosporae' of classical texts

The Entolomataceae await analysis. Little or nothing can be added to the observations of Douglas (1918). No rings or veils have thus far been found in the family; members are apparently gymnocarpic. Even the smallest fruit bodies of *Nolanea staurospora* found in the field show no sign of a veil or the former presence of such a structure; but were the primordia studied young enough?

One family in the old classical texts, which was associated with the rhodophylloid fungi (Entolomataceae) by virtue of the pinkish spore-print, was the pluteoid agarics. After a period of separation the two groups have been reunited, based on common characters of the spore-wall (Kühner, 1980). The Pluteaceae (*Pluteus* and *Volvariella*: Volvariaceae of Dennis, Orton & Hora (1960)) are a paradox, as within the family several distinctive characters are exhibited which at first sight might be thought to be self-exclusive or unrelated.

Thus in *Pluteus*, groups of species may be recognised based on whether they possess cellular or filamentous pileipelli (outer layers of fruit body cap); two species of each of the two groups have been studied, in the first group one is pilangiocarpic and a second probably gymnocarpic, whilst in the second group one is gymnocarpic and one paravelangiocarpic (Reijnders, 1963). Obviously more work is required!

Although rather common and widespread, *Pluteus* has not been grown in culture and I cannot add anything to the observations already published, although it might be supposed that *Pluteus* is a derived taxon specialised to rather ephemeral fruiting on wood and woody debris.

In contrast, bulbangiocarpic development is unique to *Volvariella*; it has been recorded in four species (Reijnders, 1963) and is now demonstrable in a fifth, *V. esculenta* (Table 11.2). Such a development gives protection to the growing primordium into a relatively late period of its differentiation. The volva characterising the mature fruit body is superficially similar to that found in *Amanita*, although the two are not homologous, the latter being part of a bivelangiocarpic pattern (see below).

The inverse hymenophoral trama unique to the Pluteaceae, although important in defining the family, is, we believe, probably over-exaggerated in that in young primordia the trama is regular and becomes inverse by inflation and expansion of the gills. This is an example where a developmental end product is useful in classification but only careful analysis reveals its full usefulness in judging relationships. Thus, the family is more closely related to those agarics with regular tramas than to those exhibiting bilateral tramas where the curvature of the elements is more fundamental, such as *Amanita*.

Amanitaceae, the true and parasol mushrooms

The Amanitaceae (two constituent genera) have been related to *Volvariella* (Singer, 1962), although members possess bilateral hymenophoral tramas, a secondary configuration (Bas, 1969), and bivelangiocarpic development in all the species so far studied – seven out of a possible 200! In all species studied, however, the development is schizohymenial, a feature unknown outside the family. As developing fruit bodies of the genus are commonly found in the field and may be quite large in the primordial state, field observations are available for a much larger number of species. The structure of the outer protective membrane (universal veil) may be membranous, mealy or viscid, features made use of in taxonomy, but whatever its form it protects the developing fruit body for a much longer period than in many other agarics, thus reducing the period during which the hymenium is exposed to the environment. The hymenium appears to be exposed to coincide almost with peak spore production. The pileocarpic, schizohymenial, concentrated, bivelangiocarpic characters of *Amanita* are considered to be advanced. *Termitomyces*, a genus frequently associated with *Amanita*, is discussed by Rayner *et al.* in Chapter 1; the two species studied are bivelangiocarpic (Heim, 1940, 1943).

The Agaricaceae and Lepiotaceae can be considered together as the developmental patterns exhibited by their main constituent genera are parallel. Indeed in many morphological characters they are close, leading several authors to unite them into a single family; the Lepiotaceae differs primarily in the lack of pigment in the spore-wall. The developmental states of members of the Lepiotaceae appear to vary considerably, or do we have genera incorrectly placed? The stipitocarpic, poorly annulate genera (*Cystoderma*, *Phaeolepiota* and *Chamaemyces*) contrast markedly with the isocarpic, pileocarpic or hymenocarpic, prominent bivelangiocarpy of the core-genera, e.g. *Agaricus* and *Macrolepiota* where the well-developed inner tissues (lipsanenchyma) leave a prominent ring. Whereas Reijnders (1975) sought a link between *Agaricus* and these three genera, no link is necessary as they doubtfully belong to the Agaricaceae and on anatomical characters alone would be placed elsewhere by many authors (see Table 11.2); for example, *Cystoderma* is monovelangiocarpic with a pileipellis composed of velar spherocytes. The structure of the pileipellis of other Lepiotaceae can be traced in a series from a strict trichoderm, through a hymenoderm (as in *Leucocoprinus*), to the true palisododerm of *Macrolepiota* (Reijnders, 1975). *Squamanita* is related to *Cystoderma* but the flesh exhibits a more compacted form, a feature Reijnders considered advanced (1975). Thus the exceptions prove the rule giving uniformity to the Agaricaceae.

The strong bivelangiocarpic and non-stipitocarpic condition of the true mushrooms lends itself easily to modification, and by strengthening and restricted expansion the secotioid condition could result. *Endoptychum* is one such gasteromycetoid member of the Agaricaceae; the development of this genus has been studied by Conrad (1915) and Lohwag (1924).

Coprinaceae (inky caps) and relatives

Two usually closely linked families are the Bolbitiaceae and Coprinaceae, brought together, perhaps erroneously, by their general similarity in shape and ephemeral nature, reflections of their habitat preferences and not genetic relationships. Eighteen species in the Bolbitiaceae, in addition to those listed by Reijnders (1963), have been studied. The findings lead one to believe there is a basic paravelangiocarpic/gymnangiocarpic and hymenocarpic pattern (Watling, 1975). The development of the lipsanenchyma (tissue remaining after differentiation of hymenophore, stipe and pileus margin) is critical in determining whether a ring is left in the mature fruit body or not, and its structure in determining whether it is left as a membrane or as filamentous elements. This is in marked contrast to the Coprinaceae where there appears to be a basic bivelangiocarpy. It is true, however, that Reijnders (1971a) found bivelangiocarpic development in *Agrocybe cylindrica*, but here the pattern of tissues is unusual, the lipsanenchyma being replaced by hyphae emanating from the pileus, gill and stipe. This may be a result of modifications to environmental conditions, i.e. growing on standing trees, and comparison should be made with the xerophytic structures exhibited by other lignicolous fungi, e.g. hairy surfaces (Rayner *et al.*: Chapter 1). It is interesting to record that the silvicolous members of the Bolbitiaceae are velate and comparatively thick fleshed for such small agarics, whereas praticolous species have no velar elaboration and grow rapidly using as little building material as possible for the spores they produce. One can easily postulate an evolutionary link between *Conocybe* and *Gastrocybe* where the pileus does not expand and the fruiting period is reduced to a few pre-dawn hours.

The Coprinaceae is split into three sub-families; the first, Panaeoloideae, would appear to express the same range of developmental characters as *Conocybe* (Bolbitiaceae), i.e. paravelangiocarpic → gymnangiocarpic, based on species studied by Reijnders (1963) and three by Watling & Sweeney (unpublished) (Table 11.1). In the second, Psathyrelloideae, *Lacrymaria* is metavelangiocarpic, whereas although the nine taxa examined by Reijnders (1963) and *Psathyrella spadiceogrisea*, *P. candolleana* and *P. pennata* (Watling & Sweeney, unpublished) are bivelangiocarpic, the *P. gracilis* and the *P. coprophila* groups have been found to be monovelangiocarpic (Watling & Sweeney, unpublished). It is interesting to speculate that *P. conopilea*, studied by Reijnders (1948), and *P. polycystis* and *P. fimetaria* studied by Watling & Sweeney (unpublished) would in older literature have been placed in *Psathyrella* whereas *P. spadiceogrisea*, *P. candolleana* and *P. pennata* would have been placed in the genus *Psathyra*, now synonymised with *Psathyrella*. Thus, as has been found in other groups, anatomical characters go 'hand in hand' with morphological characters. Not only is the presence or absence of a particular tissue in the field important, but also how that tissue is constructed and how it may be altered by environmental conditions prevailing at the time of expansion of the fruit bodies. Thus the collection of *P. pennata* studied by Watling & Sweeney had a well-formed annulus in the adult specimens; comparison should be made with the studies by Kits van Waveren (1971) and Watling (1971 a) in the Bolbitiaceae.

The range of developmental type found in the Psathyrelloideae is parallel to that found in the third sub-family (Coprinoideae). With the ease by which many *Coprinus* species can be maintained in culture, a special study was undertaken to add to Reijnder's observations (1979 b). Table 11.3 summarises the developmental patterns known in *Coprinus*, expressed within the presently accepted taxonomic structure (Orton & Watling, 1979). It shows that members are typically bivelangiocarpic, but for a few exceptions in Section *Pseudocoprinus*; i.e. *C. plicatilis* and its allies are gymnangiocarpic, and *C. stellatus* and *C. bisporus* are recorded as paravelangiocarpic (Kühner, 1926; Reijnders, 1963). Closer examination of the last two show that evidence of an outer veil, although rudimentary, can be observed on the stipe in young primordia. It would appear that a

Table 11.3. *Development of species of* Coprinus *with particular reference to Sect.* Pseudocoprinus. *All species show bivelangiocarpic development except the* C. plicatilis *group and* C. bisporus. *The latter is not fully described in the literature (Kühner, 1926b) and judging from the closely related* C. congregatus *(Kemp, 1970) both should be parallel to the* C. ephemerus *group with scanty lipsanenchyma and universal veil not clothing the entire primordium. Asterisks show material examined by Watling & Downie (unpublished). Classification follows Orton & Watling (1979). Authorities where possible also follow this publication; if not authorities are added. Abbreviations as in Table 11.1.*

	Species analysed within stirps and sections	Previously analysed species and developmental type of analysed species
Section 1 Coprinus		
Stirps 5	Picaceus	*C. cubensis* Berk. & Curt.
	C. kimurae	iso; bivelangio
Stirps 6	Lagopus	*C. radiatus*; *C. lagopus*; *C. macrorhizus*; *C. 'macrocephalus'*
	C. cinereus	iso; bivelangio rupthymenial; see Rosin, et al.: Chapter 13; Gooday: Chapter 12
Stirps 7 & 8	Friesii & Tigrinellus = Herbicolae Pilát & Svrček	*C. phaeosporus*; *C. brassicae*
	C. gonophyllus	iso; bivelangio
Stirps 9	Filamentifer	
	C. filamentifer s. Bell	iso; almost monovelangio
	C. xenobius	prob. iso; bivelangio
Section 2 Micaceus		
Stirps 12	Domesticus	*C. radians*; *C. tomentosus s.* Chow
	C. domesticus	pileostipito; bivelangio; rupthymenial
	C. ellisii	pileostipito, bivelangio; rupthymenial
Stirps 15	Niveus	*C. niveus*; *C. ephemeroides*; *C. patouillardii*
	C. cordisporus	pileostipito; bivelangio; levhymenial
	C. corthurnatus agg.	pileostipito → iso; bivelangio
Stirps 16	Narcoticus	*C. stercorarius*; *C. narcoticus*
	C. sclerotiger	pileostipito → iso; bivelangiocarpic; levhymenial
	C. tuberosus	?, bivelangio; levhymenial

Table 11.3. (*cont.*)

Species analysed within stirps and sections	Previously analysed species and developmental type of analysed species
Section 3 Pseudocoprinus	
Stirps 17 Disseminatus	
C. disseminatus	soon iso
C. hexagonosporus	bivelangio almost
C. curtus	monovelangio;
**C. heptemerus*	rupthymenial; see Reijnders (1963, 1975)
Stirps 18 Ephemerus	
C. ephemerus	iso; bivelangio
C. stellatus	(lipsanenchyma scanty, universal veil clothing part of stipe only); rupthymenial ($=$?*C. miser* (Reijnders, 1963))
**C. congregatus*	as in *C. stellatus*; see Ross: Chapter 14
Stirps 21 Hemerobius	*C. plicatilis*
**C. miser*	iso; gymnangio; poss. rupthymenial

general pattern for the group is adhered to. Perhaps in Section *Pseudocoprinus*, because of the very delicate and ephemeral nature of its members, we see the enveloping secondary veil becoming redundant.

The sections of *Coprinus* are based on the structure of the universal veil; e.g. *C. cinereus* group ($=C. lagopus$ of many geneticists: Pinto-Lopes & Almeida, 1970) – sausage-shaped velar cells (Lanatuli); *C. friesii* group – thick-walled, angular, often contorted, pigmented hyphae (Herbicolae). In the larger members of the genus one can see how the bivelangiocarpic condition may have led to the evolution of *Montagnites* found in arid regions of the world; here the universal veil forming a sheathing skin at first restricts water loss but later contracts to pull the modified gills into an arched position and so expose the hymenium.

The predominant pattern of tissue differentiation in Bolbitiaceae and Coprinaceae varies from hymenocarpic (all bolbitioid agarics and *Panaeolus* spp. studied) to hymenopileocarpic (or even isocarpic). *Lacrymaria velutina* is pileostipitocarpic in parallel to *C. lagopus* and its allies. Further studies would be of value in this field.

Geophiloid agarics

The Strophariaceae, when restricted to those agarics displaying a purple-brown spore-print and filamentous pileipellis, comprises four major genera, the status of each having been questioned on more than one occasion by more than one authority. The species of *Hypholoma* and *Stropharia* studied (four additional species to those tabulated by Reijnders (1963), i.e. *H. marginatum*, *H. subericaeum*, *H. udum* and *Stropharia cyanea* (Table 11.2) are all bivelangiocarpic. *S. merdaria* and its allies are now placed in *Psilocybe* close to the annulate *P. fimetaria* and *P. cubensis*; they are bivelangiocarpic also. It would appear, however, that the non-annulate species are monovelangiocarpic, e.g. *P. semilanceata*, but it could be hypothesised that, like the Bolbitiaceae, fruit bodies which are protected by the surrounding vegetation and which grow quickly can dispense with elaborate velar material. The pleurotoid *Melanotus vorax* is bivelangio-carpic as are the gasteromycetoid forms of *P. merdaria* studied by Watling (1971*b*). Examination of primordia accompanying herbarium material of members of the gasteroid genus *Galeropsis* and *Weraroa copro-phila* show that these conform to the bivelangiocarpic pattern. The majority of *Stropharia* species exhibit a tendency to hymenocarpy, whilst *Hypholoma* species exhibit pileostipitocarpy.

Cortinariaceae and Tricholomataceae

The two remaining families to be discussed, the Cortinariaceae and Tricholomataceae, between them possess something of the order of 2000 taxa. Both are rather heterogeneous, the latter possibly containing several primitive and/or reduced species. In fact the latter family is more difficult to interpret as many patterns are expressed therein, whether chemical, physiological, anatomical or morphological.

The Cortinariaceae is divided into two subfamilies, Cortinarieae and Inocybeae, which, although based primarily on spore-print colour, can be distinguished by the fact that the species so far studied (exclusively temperate species) are all monovelangiocarpic in the Inocybeae, whereas the *Cortinarius* group generally are bivelangiocarpic with the veils developed to different degrees, e.g. arachnoid in *Cortinarius* (five species), membranous in *Rozites* (one species) and mucilaginous in *Pholiota* (six species). Thirteen additional species from the three genera have been analysed (Table 11.2); *Cortinarius infractus* is said to be monovelangiocarpic (Douglas, 1916) contrasting with Reijnders' (1974, 1979*a*) information. In the Inocybeae, *Naucoria escharoides* (= *Alnicola* Singer), *Hebeloma* (two species) and *Inocybe* (nine species, seven listed by Reijnders, 1963) are all monovelangiocarpic. *Naucoria centunculus* is paravelangiocarpic (Reijnders, 1963) therefore exhibiting quite a different pattern to *Naucoria sensu stricto* supporting its placement in the genus *Simocybe*. In fact based on anatomical and morphological features Romagnesi (1962) links *Simocybe* to *Agrocybe*, i.e. Bolbitiaceae and paravelangiocarpic. *Galerina*, a genus of generally small agarics which have long been confused with members of the genus *Conocybe* (Bolbitiaceae), is, in contrast to *Cortinarius*, monovelangiocarpic and it might be postulated that it represents a specialisation for growing in moss-cushions, *Sphagnum* beds, etc. where relatively rapid growth and protection from the close 'turf' removes the necessity for extensive velar development. *Gymnopilus* is monovelangiocarpic except for the small *G. fulgens*, a not infrequent species in montane communities which is rather isolated in the genus; it is gymnocarpic (Reijnders, 1963).

Tubaria (three species) and *Flammulaster* (two species; = *Flocculina*) are two small genera, undoubtedly related by virtue of their similar spore morphology, tissue arrangement, etc., and it would appear that bivelangiocarpic development unites the two although the inner veil is much reduced in *Tubaria*. Although Reijnders (1963) interprets Walker's (1919) observations on *T. furfuracea* as monovelangiocarpic, my observations agree with Walker's illustrations and I interpret this accordingly as bivelangiocarpic.

Several hypogeous members of the Cortinariaceae are known and it can be speculated how easily the bivelangiocarpic condition can lead to an angiocarpic state as in *Thaxterogaster*, parallel to *Cortinarius*, and *Nivatogastrium*, parallel to *Pholiota*. In each case, not only the anatomical characters are parallel but also the ecological preferences. Intermediates which still give spore deposits if the velar tissue is physically removed have been described from montane, xerophytic habitats by Thiers & Smith (1969) and Watling (1980).

Undoubtedly the basic theme of development exhibited in the Tricholomataceae is one of gymnocarpy associated with stipitocarpy, e.g. *Calocybe*, *Cantharellula*, *Clitocybe*, *Collybia*, *Hygrophoropsis*, *Leucopaxillus* and *Omphalina* (Table 11.1), indicating a rather primitive condition. Modifications of this theme or rather different patterns are expressed in some of the major genera within the family. Thus, in the large genus *Mycena* (about 200 taxa), those in the small section Basipedes possess the distinctive character of stipitangiocarpy; the three species studied are undoubtedly closely related to the other taxa in the genus but further developmental studies may give support to the recognition of the genus *Pseudomycena* first proposed in 1930 but later considered to be *Mycena*.

The section is represented in many different parts of the world in different vegetational types, perhaps themselves united by their very seasonal environments. The species can be recognised immediately in the field by the disc at the stipe base. The rest of the 22 species of *Mycena* studied, nine more being added to Reijnders' (1963) tabulation by Watling & Sweeney (Table 11.2), are gymnocarpic or develop a very slight hint of an outer veil, i.e. monovelangiocarpic; the primordia are pileostipitocarpic or in a few species stipitocarpic.

Mycena fibula is gymnocarpic and stipitocarpic; it is now placed in *Rickenella* and grows in moss-cushions, i.e. protected by vegetation. *Mycena*, along with *Collybia* and *Marasmius*, comprise the main litter-decomposers of the world's forests. In contrast to *Collybia* (see above) *Marasmius* (including *Marasmiellus* and *Micromphale*) is paravelangio- and pileostipitocarpic; indeed developmental differences may be the way to separate these otherwise superficially similar genera.

The smaller genera of the Tricholomataceae, e.g. *Oudemansiella* and *Tricholomopsis*, show contrasting patterns which may have to be recognised in any future studies on relationships. *Oudemansiella* includes species with bivelangiocarpic (three species) and paravelangiocarpic (one species) development. At least two of these have in the past been placed in separate genera, a position which might warrant review, although they all are united by their common isocarpy. Similarly in *Tricholomopsis*, *T. rutilans* is paravelangiocarpic and *T. platyphylla* is gymnocarpic; both species are stipitocarpic. *T. platyphylla* has in fact been separated on other criteria by Kotlaba & Pouzar (1972) as *Megacollybia* (Rayner *et al.*: Chapter 1). In *Armillaria* the species are basically bivelangiocarpic with the exannulate *A. tabescens* lacking extensive development of the outer veil.

The pleurotoid members of the Tricholomataceae, e.g. *Geopetalum*, *Panellus*, *Phyllotopsis*, *Resupinatus* and *Hohenbuehelia*, are spread over several sub-families and sections of the family (see above). The primordium in all species is at first symmetrical but as the fruit body develops further it becomes bilateral, presumably to accommodate the growth on vertical or inclined woody surfaces. Many species have gelatinised tissue which prevents or reduces water loss (Rayner *et al.*: Chapter 1).

The biotrophs in the family belong particularly to the genera *Tricholoma* and *Laccaria* (Rayner *et al.*: Chapter 1). Both are velate, although in mature species this may not be very evident; the first is monovelangiocarpic (Reijnders, 1963) and the latter metavelangiocarpic or hypoangiocarpic (Beer, 1911, *fide* Reijnders, 1948). The fungicolous necrotroph *Nyctalis*, a genus in which the fruit body is reduced to a massive asexual structure, is gymno- and stipitocarpic; it is related to *Calocybe* (see above). *Laccaria* has recently been placed in the family Hydnangiaceae (Kühner, 1980), a family named after the gasteroid genus *Hydnangium*. Although undoubtedly related, there does not appear to be extant any morphological intermediates linking the agaric life-form of *Laccaria* with the hypogeous *Hydnangium*. Unfortunately Fischer's (1925) work on *H. stephensii* is of little help except in confirming that the Russulales have a basic gymnocarpic pattern, for *H. stephensii* is a *Zelleromyces* (Russulales) not a *Hydnangium*!

Now that this anomaly has been resolved it is of great interest to ask why the Tricholomataceae does not appear to have any secotioid relatives; could it be that because the bivelangiocarpic condition is rare, opportunities have not arisen for those species possessing the character to develop the gasteroid facies?

Finally, mention should be made of some pale-spored agarics whose affinities have been in dispute for generations. Unfortunately developmental studies do not allow decisions to be made except in the case of *Lepista* which is similar to its close relative *Clitocybe*. *Hygrophoropsis*, although white-spored, has been associated with *Paxillus*, and therefore the Boletales. Its development is gymnocarpic as in Boletaceae but the gill structure is cantharelloid (see Corner, 1966, and Reijnders, 1983a). The genus is traditionally connected to *Clitocybe* (Dennis, Orton & Hora, 1960); although bilateral the hymenophoral trama in detail is in keeping neither with the boletes nor with *Clitocybe*. *Rhodocybe* (one species) is pilangiocarpic, therefore superficially resembling some of the Boletales; *Ripartites* (one species), often associated with Paxillaceae, is gymnangiocarpic as found in the Bolbitiaceae. The lignicolous *Rhodotus palmatus* also resembles the Bolbitiaceae in development; it is a species which is isolated in many characters, especially in its unusual spore-wall morphology (Kühner, 1980). Neither *Ripartites* nor *Rhodotus* is related to *Bolbitius*, and both are examples of convergence.

Conclusion

Only a small number of agarics have been studied from a developmental point of view but, it is expected that if the classification is anything approaching natural, there should be very good correlation, and when species are apparently anomalous further study should reveal their true relationship. The classification of the agarics and related fungi is in a fluid state, so much so that new approaches are still being proposed, e.g.

Kühner (1980), Pegler & Young (1981). It is therefore not surprising that species which appear to be anomalous in one system, based upon non-ontogenetic criteria, are homologous when using a second system.

Many of the differences between modern and classical texts involve species superficially close in macromorphology but with radically different developmental patterns, e.g. members of the genera *Tricholoma* and *Tricholomopsis*, a fact which exposes the earlier conservative approach (Rayner *et al.*: Chapter 1). The modern classification, which has received so much adverse criticism since its introduction (Singer, 1951 *et seq.*), is supported by the ontogenetic studies above, whilst application of developmental data to a Friesian system, so popular still with many, gives far less correlation with all the data available.

The New check list of British Agarics and Boleti (Dennis, Orton & Hora, 1960) was based primarily but not entirely on Singer (1951), e.g. *Naucoria* includes *Simocybe*, a taxonomic judgement which can now be more clearly reassessed with the developmental studies available.

Surely the ultimate aim must be to include developmental data in family and generic circumscription although we are a long way from doing so at the moment. Yet the bulbangiocarpic and pileocarpic state defines *Volvariella*, the schizohymenial state the Amanitaceae, and iso-, pileo- or hymenocarpy with prominent bivelangiocarpic development characterises the core-genera of the Agaricaceae. Developmental studies have been of assistance in confirming the closeness of fungi traditionally classified in quite different taxa because of the over-emphasis of a single character, e.g. the close relationship now accepted between the agaricoid *Mycena* and discoid (cyphelloid) *Calyptella*.

Although one must guard against making generalisations (Reijnders, 1983*a*) it is very probable that in the future when we have learnt to interpret the jigsaw we will be able to read the direction of the morphological series, to enquire into evolutionary pathways and to attempt predictions as to primitive *versus* reduced or advanced characters, and to unravel the reticulation of characters found in the agarics on which their classification is based (see Reijnders, 1979*b*). Periodically it is important to summarise and analyse the information available within a small group (Boletales: Pegler & Young, 1981; Singer, 1981), and equally it is useful to stop and assess the facts on a much broader front.

Further study will undoubtedly also show positive correlations with ecological preferences, some of which are hinted at in the text from field observations. Ecology and taxonomy of organisms are intimately linked and the developmental patterns outlined above represent a picture of the strategy of development adopted in this group of fungi for colonisation of various substrates or exploitation of different environments. When the developmental patterns expressed by groups of agarics are analysed in the future and expanded to include many other taxa I have every confidence that positive correlation with ecology will be found. It is hoped that the data offered herein will act as a base-line from which to carry out further studies.

Acknowledgements. I am particularly grateful to Dr A. F. M. Reijnders, Amersfoot, for his encouragement over the years, to Mrs J. Sweeney for unselfish assistance in the laboratory, and to Miss J. Downe for her help in sectioning *Coprinus* primordia. I am also deeply grateful to M. J. Richardson for his helpful criticism of the text.

References

Aeger, R. (1983). Typusstudien on Cyphelloiden Pilzen IV. *Mitteilungen Botanische Staatssammlungen München*, **19**, 163–334.

Atkinson, G. F. (1914*a*). The development of *Agaricus campestris*. *The Botanical Gazette*, **42**, 241–64.

Atkinson, G. F. (1914*b*). The development of *Armillaria mellea*. *Mykologische Zentralblatt*, **4**, 113–21.

Atkinson, G. F. (1914*c*). The development of *Amanitopsis vaginata*. *Annales Mycologici*, **12**, 369–92.

Atkinson, G. F. (1914*d*). The development of *Agaricus arvensis* and *A. comtulus*. *American Journal of Botany*, **1**, 3–22.

Atkinson, G. F. (1914*e*). The development of *Lepiota clypeolaria*. *Annales Mycologici*, **12**, 346–56.

Atkinson, G. F. (1915). Morphology and development of *Agaricus rodmani*. *Proceedings of the American Philosophical Society*, **1954**, 309–43.

Atkinson, G. F. (1916). Development of *Lepiota cristata* and *L. seminuda*. *Memoirs of the New York Botanical Garden*, **6**, 209–88.

Bas, C. (1969). Morphology and subdivision of *Amanita* and a monograph of its section *Lepidella*. *Persoonia*, **5**, 285–579.

Beer, R. (1911). Notes on the development of the carpophore of some Agaricaceae. *Annals of Botany*, **25**, 683–9.

Brefeld, O. (1877). *Botanische Untersuchungen über Schimmelpilze*, No. 3, 266 pp.

Brefeld, O. (1881). *Botanische Untersuchungen über Schimmelpilze*, No. 4, 191 pp.

Bucholtz, F. (1903). Zur Morphologie und Systematik der Fungi hypogaei. *Annales Mycologici*, **1**, 152–74.

Conrad, H. S. (1915). The structure and development of *Secotium agaricoides*. *Mycologia*, **7**, 94–104.

Corner, E. J. H. (1966). *A monograph of Cantharelloid fungi*. *Annals of Botany*, Memoir 2, 225 pp.

Corner, E. J. H. (1972). Boletus *in Malaysia*. Singapore: Govt. Printer.

Corner, E. J. H. (1981). The agaric genera *Lentinus, Panus* and *Pleurotus. Beihefte Nova Hedwigia*, **69**, 1–169.

de Bary, A. (1884). *Vergleichende Morphologie und Biologie der Pilze*. Leipzig: USW.

Dennis, R. W. G., Orton, P. D. & Hora, F. B. (1960). New check list of British Agarics and Boleti. *Transactions of the British Mycological Society*, **43**, suppl., 1–225.

Douglas, G. E. (1916). A study of development in the genus *Cortinarius. American Journal of Botany*, **3**, 319–35.

Douglas, G. E. (1918). The development of some exogenous species of Agarics. *American Journal of Botany*, **5**, 36–54.

Fayod, V. (1889). Prodrome d'une histoire naturelle des Agaricinés. *Annales des Sciences Naturelles*, Botanique Série, **7–9**, 179–411.

Fischer, E. (1925). Zur Entwicklungsgeschichte der Fruchtkorper der Secotiaceae. *Festchrift Carl Schröter, Geobotanische Institut Ruebel in Zurich*, **3**, 571–82.

Gilbert, E. J. (1931). *Les Bolets*. Paris: Lechavellier.

Hawksworth, D. L., Sutton, B. C. & Ainsworth, G. C. (1983). *Ainsworth and Bisby's Dictionary of Fungi*, 7th edn. Kew: Commonwealth Mycological Institute.

Heim, R. (1937). Les Lactario – Russulés à anneau: Ontogénie et Phylogénie. *Revue de Mycologie, NS*, **2**, 4–17, 61–75, 109–17.

Heim, R. (1940). Etudes descriptives et experimentales, sur les Agarics termitophiles d'Afrique tropicale. *Memoires de l'Académie des Sciences de l'Institut de France*, **64**, 1–74.

Heim, R. (1943). Nouvelles études descriptives sur les Agarics termitophiles d'Afrique tropicale. *Archives Museum National d'Histoire Naturelle*, Series 6, **18**, 107–66.

Kemp, R. F. O. (1970). Interspecific sterility in *Coprinus bisporus, C. congregatus* and other basidiomycetes. *Transactions of the British Mycological Society*, **54**, 488–9.

Kits van Waveren, E. (1971). The genus *Conocybe* subgen. *Pholiotina*. 1. The European annulate species. *Persoonia*, **6**, 119–65.

Kotlaba, F. & Pouzar, Z. (1972). Taxonomic and nomenclatural notes on some Macromycetes. *Ceská Mykologie*, **26**, 217–21.

Kühner, R. (1925). Le développement de *Lentinus tigrinus* Bull. *Comptes rendus hebdomadaires des séances de l'Académie des Sciences*, **180**, 137.

Kühner, R. (1929). Le développement et la position taxonomique de l'*Agaricus* 177–81.

Kühner, R. (1926b). Contribution à l'étude des Hyménomycètes et specialement des Agaricacés. *Le Botaniste*, **17**, 1–215.

Kühner, R. (1929). Le développement et la position taxonomique de l'*Agaricus disseminatus* Pers. *Le Botaniste*, **20**, 147–56.

Kühner, R. (1938). *Le Genre Mycena*. Paris: Lechavellier.

Kühner, R. (1980). Les Hyménomycètes agaricoides. Etude générale et classification. *Bulletin de la Société Linnéene de Lyon*, **49** (Suppl.), 1–1027.

Levine, M. (1914). The origin and development of the lamellae in *Agaricus campestris* and in certain species of *Coprinus. American Journal of Botany*, **9**, 509–33.

Linnaeus, C. (1753). *Species plantarum*. Holmiae.

Lohwag, H. (1924). Entwicklungsgeschichte und Systematische. Stellung van *Secotium agaricoides. Osterreichische Botanische Zeitschrift*, **73**, 177–334.

Luttrell, E. S. (1951). Taxonomy of the Pyrenomycetes. *University of Missouri Studies*, **24**(3), 1–120.

Maire, R. (1902). Recherches cytologiques et taxonomiques sur les Basidiomycètes. *Bulletin de la Société Mycologique de France*, **18**, Suppl., 1–212.

Nannfeldt, J. A. (1932). *Studien über die morphologie und systematik der nicht-lichenisierten Inoperculaten Discomyceten*. Uppsala: Norblads Bokhandel.

Orton, P. D. & Watling, R. (1979). *British Fungus Flora, Agarics and Boleti*, vol. 2, *Coprinaceae*, Coprinus. Edinburgh: HMSO.

Patouillard, N. (1900). *Essai taxonomique sur les familles et les genres des Hyménomycètes.* Lons-le-Saunier: Luciend Declume.

Pegler, D. N. (1972). Lentineae (Polyporaceae), Schizophyllaceae et espèces lentinoides et pleurotoides des Tricholomataceae. *Flore Illustré Champignons d'Afrique Centrale*, **1**, 1–26.

Pegler, D. N. & Young, T. (1981). A natural arrangement of the Boletales, with reference to spore morphology. *Transactions of the British Mycological Society*, **76**, 103–46.

Petersen, R. H. (1967). Notes on Clavarioid fungi VI. Two new species and notes on the origin of *Clavulina*. *Mycologia*, **59**, 39–46.

Pinto-Lopes, J. & Almeida, M. G. (1970). 'Coprinus lagopus' a confusing name as applied to several species. *Portugaliae Acta Biologica*, (*B*), **11**, 167–204.

Reijnders, A. F. M. (1948). Etudes sur le développement et l'organisation histologiques des carpophores dans les Agaricales. *Recueil Travaux Botaniques Néerlandais*, **41**, 213–396.

Reijnders, A. F. M. (1952). Recherches sur le développement des carpophores dans les Agaricales. *Medelingen, Nederlandsche Mycologische Vereeniging*, **30**, 1–116.

Reijnders, A. F. M. (1963). *Les problèmes du développement des carpophores dans Agaricales et de quelques groupes voisins.* Hague: Dr W. Junk.

Reijnders, A. F. M. (1971*a*). The veil of *Agrocybe aegerita*. *Acta Botanica Neerlandica*, **20**, 299–304.

Reijnders, A. F. M. (1971*b*). The development of *Kuehneromyces mutabilis*. *Acta Botanica Neerlandica*, **20**, 305–8.

Reijnders, A. F. M. (1974*a*). Le développement de deux espèces de Cortinariaceae et la nature du bulb primordial. *Bulletin de la Société Linnéene de Lyon*, Suppl., 355–64.

Reijnders, A. F. M. (1974*b*). *Coprinus auricomus* (Pat.) et ses voiles. *Bulletin de la Société Mycologique de France*, **90**, 223–30.

Reijnders, A. F. M. (1975). The development of three species of the Agaricaceae and the ontogenetic pattern of this family as a whole. *Persoonia*, **8**, 307–19.

Reijnders, A. F. M. (1976). Recherches sur le développement et l'histogénèse dans les Asterosporales. *Persoonia*, **9**, 65–83.

Reijnders, A. F. M. (1979*a*). On carpophore development in the genera *Cortinarius, Dermocybe* and *Leucocortinarius*. *Sydowia*, **8**, 335–48.

Reijnders, A. F. M. (1979*b*). Developmental anatomy of *Coprinus*. *Persoonia*, **10**, 383–424.

Reijnders, A. F. M. (1979*c*). Le développement de *Limacella glioderma*. *Persoonia*, **10**, 301–8.

Reijnders, A. F. M. (1983*a*). Supplementary notes on basidiocarp ontogeny in agarics. *Persoonia*, **12**, 1–20.

Reijnders, A. F. M. (1983*b*). Le développement d'*Hygrotrama atropuncta* (Pers.: Fr.) Sing. *Cryptogamie Mycologie*, **4**, 71–8.

Reijnders, A. F. M. (1983*c*). Le développement de *Tectella patellaris* (Fr.) Murr. et la nature des basidiocarps cupuliformes. *Bulletin de la Société Mycologique de France*, **99**, 109–26.

Romagnesi, H. (1962). Les *Naucoria* du groupe *Centunculus*. *Bulletin de la Société Mycologique de France*, **78**, 337–58.

Rosinski, M. A. & Robinson, A. D. (1968). Hybridization of *Panus tigrinus* and *Lentodium squamulosum*. *American Journal of Botany*, **55**, 242–6.

Singer, R. (1951). The Agaricales in modern taxonomy. *Lilloa*, **22** (1949), 1–832.

Singer, R. (1962). *Agaricales in Modern Taxonomy*, 2nd edn. Weinheim: Cramer.

Singer, R. (1981). Notes on bolete taxonomy. *Persoonia*, **11**, 269–302.

Thiers, H. D. & Smith, A. H. (1969). Hypogeous Cortinarii. *Mycologia*, **61**, 526–36.

Walker, L. B. (1919). The development of *Pluteus admirabilis* and *Tubaria furfuracea*. *The Botanical Gazette*, **68**, 1–21.

Watling, R. (1971 *a*). Developmental studies on *Conocybe* with particular reference to the annulate species. *Persoonia*, **6**, 281–9.

Watling, R. (1971 *b*). Polymorphism in *Psilocybe merdaria*. *New Phytologist*, **70**, 307–26.

Watling, R. (1975). Studies on fruit-body development in the Bolbitiaceae and the implications of such work. *Nova Hedwigia*, Beihefte **51**, 319–46.

Watling, R. (1978). From infancy to adolescence: advances in the study of higher fungi. *Transactions of the Botanical Society of Edinburgh*, **42**, Suppl., 61–73.

Watling, R. (1980). A hypogeous *Cortinarius* from Kashmir. *Notes from the Royal Botanic Garden, Edinburgh*, **38**, 357–62.

Watling, R. & Nicol, H. (1980). Sphaerocysts in *Lactarius rufus*. *Transactions of the British Mycological Society*, **75**, 331–3.

Watling, R. & Sweeney, J. (1971). Observations on *Schizophyllum commune*. *Sabouraudia*, **12**, 214–26.

12
Elongation of the stipe of
Coprinus cinereus

G.W.GOODAY

*Department of Microbiology, Marischal College, University of Aberdeen,
Aberdeen AB9 1AS, UK*

Introduction

The stipe of the agaric fruit body serves but one function, to place the cap in the best position for release of the basidiospores. In order to do this it responds to environmental stimuli as it develops and has the mechanical properties and transport properties to support and nurture the development of the pileus. *Coprinus* species have been widely used in the study of stipe development, as this process is often particularly dramatic amongst them, and several species can be grown readily in the laboratory.

As pointed out by Buller (1909), Percy Bysshe Shelley was clearly referring to the stipe of *Coprinus* when he wrote of agarics:

> Their mess rotted off them, flake by flake,
> Till the thick stalk stuck like a murderer's stake,
> Where rags of loose flesh yet tremble on high,
> Infecting the winds that wander by.

There has been confusion over the taxonomy of *Coprinus* species, which is now being clarified (Pinto-Lopes & Almeida, 1970; Moore, Elhiti & Butler, 1979). Thus studies using *Coprinus cinereus* have been published under the names *Coprinus lagopus*, *Coprinus macrorhizus* f. *microsporus* and *Coprinus fimetarius*. It is clear that much published experimental work has used incorrectly identified strains of *Coprinus* species, and without the cultures we cannot be certain what was used (Pinto-Lopes & Almeida, 1971). The major species in current use in laboratory experiments are *C. cinereus*, *C. radiatus* and *C. congregatus*. It is also clear that (1) there are some physiological attributes that are shown to different extents by different species (such as cap/stipe interdependency, and photomorphogenetic responses) and (2) published differences between 'different' species

311

now known to belong to the same taxon need to be looked at anew as possible differences between genetic stocks or between experimental protocols.

Most work has used wild-type isolates, but Takemaru & Kamada (1972) treated a dikaryon of *C. cinereus* with ultraviolet light or *N*-methyl-*N'*-nitro-*N*-nitrosoguanidine and obtained a range of developmental variants. Particularly relevant to this account are an '*elongationless*' strain, in which stipe cells expand in girth instead of in length (Kamada & Takemaru, 1977*a*) and a '*sporeless*' strain, in which at 28 °C meiosis is disrupted and basidia disintegrate (Tani, Kuroiwa & Takemaru, 1977).

The *Coprinus* stipe

The stages of development of fruit bodies have been described in detail for *Coprinus cinereus* by Matthews & Niederpruem (1972, 1973), Morimoto & Oda (1973), Stewart & Moore (1974), Lu (1974), Blayney & Marchant (1977), van der Valk & Marchant (1978) and Moore, Elhiti & Butler (1979); for *C. radiatus* by Borriss (1934*a*) and Eilers (1974); and for *C. congregatus* by Manachère (1970), Manachère *et al.* (1983), Robert (1977*a*) and Bret (1977) (see also Chapter 14). Fruit body formation requires massive relocation of material from the vegetative mycelium (Madelin, 1956, 1960; Robert, 1977*b*). Chanter & Thornley (1978) modelled the process of fruit body formation by *Agaricus bisporus* and calculated that for a single flush, 45.5% of the fungal dry mass was in the fruit bodies. This is close to the experimental value of 44% obtained for *C. radiatus* by Madelin (1956) (see also Chapter 17, p. 410).

The morphology and cytology of stipe elongation have been investigated for *C. cinereus* by Cox & Niederpruem (1975) and Kamada & Takemaru (1977*a*); for *C. radiatus* by Borriss (1934*a*), Hafner & Thielke (1970) and Eilers (1974); and for *C. congregatus* by Bret (1977). It is clear from all of these studies that elongation of the stipe is not uniform over its length. The greatest elongation is shown by the mid-upper portion, i.e. that enclosed initially by the cap, and the apex and base show little elongation. The rapid increase in length is chiefly the result of cell elongation (Kamada & Takemaru, 1977*a*) but Eilers (1974) reported a contribution by cell division in *C. radiatus*, with cell numbers doubling and cells increasing six-to eight-fold in length. In analysing stipe elongation of *Agaricus bisporus*,

Fig. 12.1. A clump of fruit bodies of *Coprinus atramentarius* pushing up through an asphalt path in Oxfordshire, photographed on three successive days.

Craig, Gull & Wood (1977) concluded that it was chiefly the result of cell elongation in the lower stipe, but of cell division and cell elongation in the upper stipe.

In his detailed discussion of the biology of *Coprinus* species, Buller (1909, 1924, 1931) records the characteristic features of the stipe. The mature stipe is a thin-walled hollow cylinder, in keeping with the ephemeral nature of the fruit bodies, having to be formed quickly and as strongly and economically as possible. Thus the stipe, although somewhat brittle, is very strong in its longitudinal axis, as shown by elongating fruit bodies thrusting their way through their substrates, usually dung, and through obstacles such as asphalt paths (Fig. 12.1), and by fruit bodies remaining quite motionless in windy conditions: 'Those with the longest stipes e.g. *Coprinus comatus* scarcely sway on very windy days, whilst most fruit bodies remain practically unstirred even during gales' (Buller, 1909). Buller (1931) measured the lifting power of fruit bodies of *C. sterquilinus* by loading weights into a tube supported above them. One fruit body lifted 204 g, and the area of the solid part of the stipe was 0.045 square inches (29 mm^2), giving an upwards pressure of the elongating stipe of approximately two-thirds of an atmosphere (7×10^4 N m^{-2}).

The upwards pressure comes from the turgor pressure of the individual cells making up the wall of the hollow stipe. These cells are allowed to elongate by the synthesis of new wall material throughout their length (see below). The osmotic value of the cells has been estimated by immersing stipes in solutions of increasing concentration and measuring the point of onset of incipient plasmolysis. Thus Kamada & Takemaru (1977a), using sucrose, determined that the osmotic value of stipe cells stayed almost constant throughout elongation at 0.45–0.50 M. As these cells were estimated to increase eight times in length and volume, the osmotic pressure was clearly actively maintained as water entered the cell. The cells of the primordium have dense cytoplasm and small vacuoles, but become highly vacuolated during elongation with the cytoplasm restricted to the cylindrical periphery and to either end of the cell (Moore, Elhiti & Butler, 1979). The solutes contributing to the maintenance of osmotic pressure have not been identified. As stipes will elongate dipping in water or in a moist atmosphere (Gooday, 1974), these must be chiefly organic metabolites. Trehalose accumulated in elongating stipes, while glucose content declined (Rao & Niederpruem, 1969; Gooday, 1982) but as the water content was increasing, Gooday (1982) estimated that the concentration of trehalose plus glucose declined from 90 to 25 mM. Rao & Niederpruem (1969) could detect no polyols in stipes. Calculations from results of Robert

(1977*a*) show that the methanol-soluble carbohydrate content of the stipe of *C. radiatus* declined from 32% to 21% during its final rapid elongation. Ewaze, Moore & Stewart (1978) and Moore, Elhiti & Butler (1979) reported low and declining contents of amino nitrogen compounds and urea in stipes, although they implicated these as osmotically active solutes during development of the cap.

Kamada & Takemaru (1977*a*) measured the values of three parameters representing mechanical properties of the stipe cell wall of *C. cinereus*: shrinkage in hypertonic mannitol solution, which they took as a measure of the elastic extensibility of the cell wall; extensibility of methanol-killed segments with a 10-g load in a tensile tester; and T_0, minimum stress-relaxation time following this stretching. All three parameters showed a good positive correlation with the elongation rate the stipe would have undergone in the hour after harvesting. Kamada & Takemaru point out that the results with T_0 are the reverse of those obtained with higher plants, where a decrease in T_0 signifies cell wall loosening which causes cell extension. Incubating stipes in increasing concentrations of aerated mannitol or sucrose solutions in buffer for 90 min resulted in progressive inhibition of elongation, with progressive shrinkage occurring above 0.25 M solute. Capacity for subsequent elongation on transfer to buffer, extensibility and T_0 value all decreased with increasing solute concentration, suggesting that, as for higher plant systems, extensibility of the stipe is an expression of the history of wall extension as well as of the capacity of the wall to extend.

Protein, glycogen and stipe elongation

The total protein content of the stipe of *Coprinus cinereus* increases during development (Kamada, Miyazaki & Takemaru, 1976; Moore, Elhiti & Butler, 1979; Gooday, 1982; Fig. 12.2, Table 12.1). The percentage protein content however shows little change. Gooday (1982), from measurements of 34 stipes ranging from 1 to 25 mg dry weight and 5–125 mm long, reported a linear relation between protein content and length, corresponding to a content of 13.4% of dry weight. Calculations from the results of Moore, Elhiti & Butler (1979), who used a different analytical method and different strains, show protein content rising slightly between stages II and V, from 16.2 to 20.5%, with the proportion of this as water-soluble protein falling from 14.4 to 9.4%. Calculations from results of Kamada, Miyazaki & Takemaru (1976) show a 1.75-fold increment in protein between 3 and 18 h for stipes of their wild-type strain (Table 12.1), and 1.6, 3.6 and 2.4-fold increases, respectively, for their

Table 12.1. *Fold increases during stipe elongation of* Coprinus cinereus
from 3 to 18h of development

	Wild-type	Wild-type decapitated[a]	Elongationless mutant	Sporeless mutant
Length	6.9	—	2.5	10.0
Fresh weight	4.1	—	3.8	8.1
Dry weight	3.7	—	3.8	5.5
DNA	1.65	1.8	2.1	2.5
RNA	3.4	4.8	2.3	4.0
Protein	1.75	2.4	1.6	3.6

[a] Decapitated at 9 h.
Data calculated from results of Kamada, Miyazaki & Takemaru (1976).

elongationless and *sporeless* mutants, and the wild-type decapitated at 9 h. In the latter two cases, this build-up of protein might represent material that otherwise would be translocated into the developing pileus. One qualitative change in protein during elongation has been reported by Blayney & Marchant (1977) who described membrane-bound proteinaceous inclusions in the cytoplasm, which increased in size between stage II and stage III and then decreased in size to stage V. This correlated with a decline in particulate protein to stage V of cell-free extracts. These inclusions, probably crystals of one protein, may represent a storage protein with a particular function, or an enzyme that is mobilised during

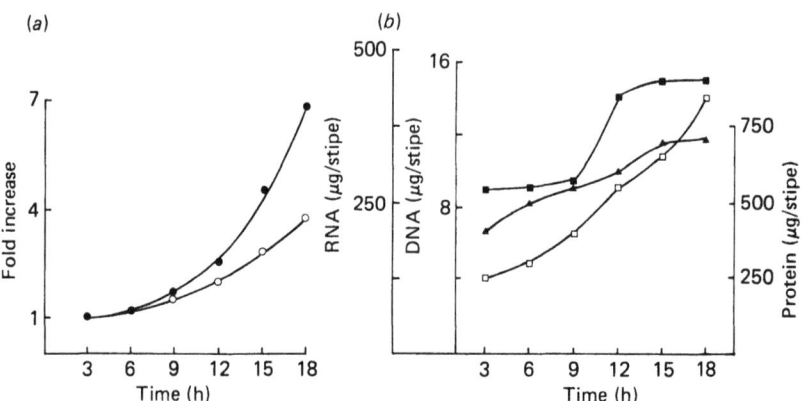

Fig. 12.2. Stipe elongation in *Coprinus cinereus*. (*a*) Incremental increases in length (●) and dry weight (○). (*b*) Contents of DNA (■), RNA (□) and protein (▲). (Redrawn from Kamada, Miyazaki & Takemaru, 1976.)

elongation but, as emphasised by Moore, Elhiti & Butler (1979) they cannot represent a major source of precursors for development.

Robert (1977*a*) presented results for protein contents of stipes of *C. congregatus* during development. Stipes of the smallest primordia had a protein content of 16.8% which declined to 11.3% just before rapid elongation, and again to 8.9% during elongation.

Glycogen acts as a mobilisable source of carbon reserve for development in *Coprinus*: in fruit body formation, when glycogen disappears from swollen vegetative cells of the mycelium (Madelin, 1960); and from cells of the subhymenium during basidium differentiation to reaccumulate in the developing spores (Bonner *et al.*, 1957; McLaughlin, 1974). Jirjis & Moore (1976) characterised the glycogen of *Coprinus cinereus* and showed it to be similar to that from other sources. It usually appears as characteristic rosettes in electron micrographs of stipe cells (Matthews & Niederpruem, 1973; McLaughlin, 1974; Blayney & Marchant, 1977; Moore, Elhiti & Butler, 1979; Gooday, 1982). Glycogen content in young stipes of *C. cinereus* was found to be very variable from 0.5 to 12% dry weight (Gooday, 1982; *cf.* table 2 of Moore, Elhiti & Butler, 1979), falling to a more uniform content at less than 1% at Stage V. The variability in glycogen content of young stipes suggests that it is a transient source of carbohydrate, utilised if it has been accumulated, but not a prerequisite for development. A similar non-essential role for glycogen was suggested by Haines & Ashworth (1974) during morphogenesis of *Dictyostelium discoideum*, when fruit bodies developed whether or not 'slug' cells had accumulated glycogen. Nevertheless, glycogen was rapidly depleted from the *Coprinus* stipe during development, but concomitant with an accumulation in the cap (Moore, Elhiti & Butler, 1979).

Moore, Elhiti & Butler (1979) conclude 'that neither glycogen nor protein, the two most readily identified reserve materials, can be allocated a centrally important function in stipe elongation'.

Cell wall metabolism and stipe elongation

Analyses of cell walls of *Coprinus cinereus* have shown them to be composed of chitin, glucans with α-$(1 \rightarrow 4)$, β-$(1 \rightarrow 3)$ and β-$(1 \rightarrow 6)$ linkages, and glycans/glycoproteins containing xylomannans (Marchant, 1978; Schaeffer, 1977; Kamada & Takemaru, 1977*b*, 1983; Bottom & Siehr, 1979, 1980). Apart from the preponderance of α-$(1 \rightarrow 4)$ glucan in place of α-$(1 \rightarrow 3)$ glucan, this is typical of the wall compositions of basidiomycetes (Bartnicki-Garcia, 1968; Wessels & Sietsma, 1981). Marchant (1978) reported considerable differences in the compositions of walls of monokaryons, dikaryons and stipes, the chitin content being highest in stipe walls (43.3% w/w).

Chitin is thus a major component of the stipe walls, and more attention has been paid to it than other components of the wall. During the period of rapid stipe elongation, chitin content increased linearly with stipe length, remaining at about 11% of the total stipe dry weight (Gooday, 1979, 1982). Excised stipes, which elongated endotrophically in a moist atmosphere to an extent dependent on their size at excision (Gooday, 1974), showed an increase in chitin percentage during elongation (Gooday, 1979). Thus, chitin must be made in this situation at the expense of other cell components. Polyoxin D, a specific inhibitor of chitin synthase, efficiently inhibited elongation (Gooday, de Rousset-Hall & Hunsley, 1976). Kamada & Takemaru (1977*b*) reported an increase in chitin content of isolated cell walls from 12% to 20% during elongation, but as the dry weight per mm length decreased during this period the amount of chitin proved to be directly proportional to stipe length. Chitin contents in their *elongationless* mutant showed a similar trend but of much smaller magnitude. These observations, coupled with the very high specific activity of chitin synthase in stipes (Montgomery, Adams & Gooday, 1984), point to the importance of chitin as a structural component during elongation. Kamada & Takemaru (1977*a, b*), in a study of mechanical properties of the stipe, have found good negative correlations between percentage chitin content of the wall and both shrinkage and extensibility of the cells, i.e. the higher the chitin content the less the cells shrank in hypertonic mannitol solution and the less they could be stretched by a tensile tester.

The molecular architecture of the stipe wall of *Coprinus cinereus* has not been investigated in detail, but by analogy with other basidiomycete cell walls that have – *Schizophyllum commune* (Hunsley & Burnett, 1970; van der Valk, Marchant & Wessels, 1977) and *Agaricus bisporus* (Michalenko, Hohl & Rast, 1976) – the chitin is reckoned to be in the inner layers while the glucans span the thickness of the wall. Very importantly, Sietsma & Wessels (1979, 1981) and Wessels, Sietsma & Sonnenberg (1983) provided evidence for covalent linkages between chitin and glucans in *S. commune* and *C. cinereus*.

The chitin in *C. cinereus* is in the form of microfibrils, 7–25 nm in diameter, which can often be seen to be at least 1 μm long (Gow & Gooday, 1983). These dimensions are very similar to those determined for chitin microfibrils in mature walls of *Neurospora crassa* (Burnett, 1979). In the stipe cells of *C. cinereus* the microfibrils were strongly oriented as shallow helices (Gooday, 1979). This was demonstrated by polarised light microscopy and by analysis of electron micrographs of shadowcast preparations. Initial analysis indicated that the helices were right-handed, but these observations proved to be fortuitous, as more extensive analyses suggest that the helices are right-handed or left-handed (Gooday, Kamada & Takemaru, unpublished). Either way, the pitch of the helix at between 80 and 90 degrees to the vertical axis is maintained throughout elongation of the cell. As autoradiographs clearly indicate uniform intercalary deposition of chitin during elongation (Gooday, 1982), the conclusion is that the new microfibrils must be inserted between pre-existing ones.

Changes in wall components other than chitin, i.e. chiefly glucans, during elongation have been reported by Kamada & Takemaru (1977b, 1983). Thus, their fraction I (soluble in hot dilute acid; $(1 \rightarrow 3)$, $(1 \rightarrow 4)$ and $(1 \rightarrow 6)$ links) declined in amount, in $(1 \rightarrow 4)$ links and in average molecular weight; fraction II (alkali-soluble, precipitated by dilute acid; $(1 \rightarrow 3)$-linked glucose, $(1 \rightarrow 3)$-linked mannose, xylose) increased in amount and average molecular weight but its xylomannan content fell dramatically; the major glucan fraction, III (alkali-soluble; $(1 \rightarrow 3)$ and $(1 \rightarrow 6)$ links) increased from 12% to 17% and then decreased to 8%. The alkali-soluble residue $((1 \rightarrow 3)$-and $(1 \rightarrow 6)$-linked glucose) was treated with strong alkali and then nitrous acid to remove chitin, and fractionated into a solubilised fraction, IVa, and fraction IVb, solubilised with further alkali treatment. During elongation, fraction IVa showed a fall in molecular weight and an increase in the ratio of β-$(1 \rightarrow 3)$ to β-$(1 \rightarrow 6)$ linkages.

Fraction II was negatively correlated with shrinkage and extensibility of stipes, as was chitin, whereas fractions III and IV were positively correlated; i.e. the higher their percentage contents in cell walls, the more the cells shrank in hypertonic mannitol solution and extended in the tensile tester (Kamada & Takemaru, 1977a, b). Fraction I showed no correlation.

Wall autolytic enzymes can be implicated in some of these changes in composition during elongation. Cell wall preparations from elongating stipes are rich in chitinase, β-$(1 \rightarrow 3)$-glucanase and β-$(1 \rightarrow 6)$-glucanase activities (Kamada, Fujii & Takemaru, 1980). Kamada, Hamada & Takemaru (1982) allowed these walls to autolyse and found that they digested 40% of their dry weight to neutral sugars, chiefly glucose, in 24 h at 37 °C. A further 8% of protein equivalents was lost, with a total autolysis of 55% of the dry weight. Of the glucan-rich fractions characterised by Kamada & Takemaru (1977b, 1983), fractions I, III and IV were highly susceptible to autolysis, whereas fraction II was not. There was an obvious relation between activities of soluble glucanases, wall-bound chitinase, autolysis rate and stipe elongation rate. The clear implication is that wall autolysis is associated with the rapid elongation of the stipe. However, despite considerable chitinase activities in the wall fractions and a high chitin content in the wall, *N*-acetylglucosamine only constituted 7% of the liberated neutral sugars. This suggests that only certain sites on the chitin microfibrils may be attacked readily. This appears to contrast with the situation in autolysing gill tissue, where chitinase has a major role in wall digestion (Iten, 1970; Iten & Matile, 1970; Bush, 1974; Miyake, Takemaru & Ishikawa, 1980).

Cox & Niederpruem (1975) describe an amorphous brown gel in the stipe lumen that disappears during elongation, implicating it as a nutrient reserve, but its nature remains unknown. It disappeared first from the elongating mid-region, then from the apical region, and finally from the basal region.

Phototropism and geotropism

Phototropism is a common phenomenon in sporulating structures of fungi (Banbury, 1959). Young elongating fruit bodies of some, perhaps all, *Coprinus* species are positively phototropic (Fig. 12.3). Thus Buller (1909, 1931) described growth towards light by the stipes of the small

Fig. 12.3. Phototropism and geotropism of a *Coprinus* fruit body, formed underneath a flat roofing sheet of compressed straw in a garage. Initially the stipe grew ageotropically towards the light, then it became negatively geotropic so that the cap became horizontal.

coprophilous species *C. niveus* and *C. curtus*, by intercalary elongation of the region just beneath the pileus. During this period there was no response to gravity, but negative geotropism was acquired and phototropism lost as the pileus began to expand. Buller interpreted this behaviour in terms of these coprophilous fungi growing out from dung. In the case of the larger *C. sterquilinus*, Buller (1931) described the formation of a solid stipe base, and the inhibition of elongation of this base by light, so that the hollow stipe of the adult fruit body could then be formed at the surface of the dung. The youngest fruit bodies, with pileus only just differentiating, were not responsive to gravity or light, simply growing away from the substrate, but they soon acquired both tropic responses to elongate up from dung angled towards the light at up to 10 degrees. Buller interpreted this behaviour as the response of a large fruit body needing to be close to vertical to maintain its stability. Phototropism of stipes has been observed in *C. stiriacus*, *C. radiatus*, *C. sterquilinus* and *C. domesticus* by Knoll (1909), Borriss (1934*b*), Jeffreys & Greulach (1956) and Chapman & Fergus (1973), respectively; in *Amanita* species by Streeter (1909); in *Schizophyllum commune* by Schwalb & Shanler (1974); in *Polyporus brumalis* by Plunkett (1961); and in *Pleurotus ostreatus* by Eger-Hummel (1980). Plunkett (1961) concluded that the stipe of *Polyporus brumalis* was positively phototropic throughout development, masking geotropism if both stimuli were acting, but that as the reception and response to light were in the region of elongation, shading by the pileus could modulate the phototropism. Experiments with shading suggested that the site of reception of the light was the site of response, i.e. the elongating zone of the stipe. Schwalb & Shanler (1974), showing that phototropism was a response to unilateral blue light, suggested that the same receptor pigments were responsible for photomorphogenesis and phototropism. Eger-Hummel (1980) described *P. ostreatus* as being unique in having a second phototropic response during its development: stages II and III stipes reacted positive-phototropically; then stage IV stipes were negatively geotropic; then the pileus margins that received more light grew faster and towards the light, giving asymmetrical caps in unilateral light. Eger-Hummel (1980) described the isolation of mutants with aberrant tropic responses.

Carlile (1970) and Dennison (1979) described two mechanisms of positive phototropism: (1) the lens effect, when the far side of a cell is stimulated to grow, as in *Phycomyces*; and (2) directed growth towards the light resulting from an asymmetrical distribution of growth substrates in multicellular systems. In the relatively translucent hollow stipe of *Coprinus* there is no evidence for growth substances (see below), and so a possible mechanism is that each cell in the elongating zone is reacting by the lens mechanism and bending. This requires much greater elongation by, and thus diversion of nutrients to, the far side of the stipe, perhaps mediated by 'stretch receptors' such as have been invoked in the response of *Phycomyces* to gravity (Dennison & Roth, 1967).

Geotropism is widespread among fungi (Banbury, 1962). Buller (1909, 1924, 1931) presented detailed accounts of negative geotropism in several species of *Coprinus*. He emphasised that it is a characteristic of this genus for the pileus and gills not to be geotropic, but for the stipe to be strongly so through all or at least the latter part of its elongation. The site of the geotropism is the upper portion, i.e. the site of elongation, so that the axis of the stipe just below the pileus is exactly vertical (Fig. 12.3). If a fruit body is turned from vertical to horizontal, asymmetrical growth of the stipe rapidly returns the apex of the stipe to the vertical. Buller (1924), by pinning down the cap of a horizontal excised fruit body of *Coprinus atramentarius*, observed that the lower side of the horizontal stipe elongated more than the upper, with the result that the base of the stipe moved into a vertically upright 'upside down' position. Buller (1909, 1924, 1931) described the phenomenon of 'geotropic swinging', whereby stipes of fruit

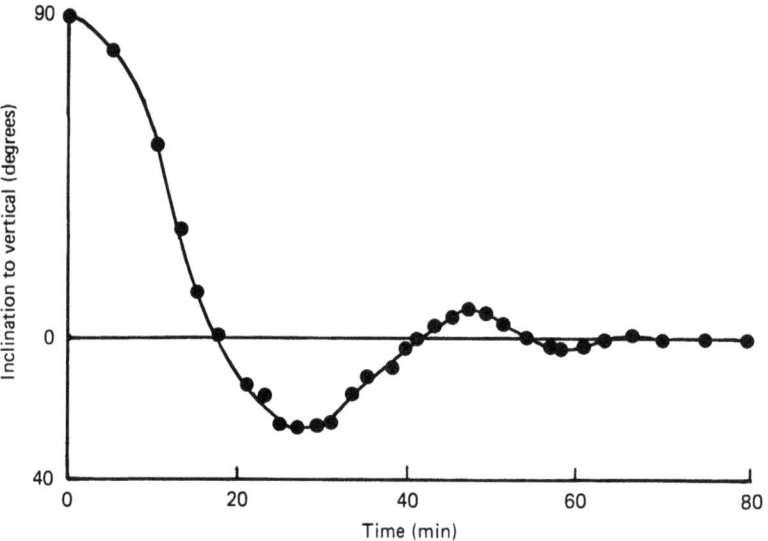

Fig. 12.4. 'Geotropic swinging' of a fruit body of *Coprinus plicatiloides* (= *Coprinus curtus*). At time zero the fruit body was turned from a vertical to a horizontal position. The inclination of the axis of the apex of the stipe to the vertical was measured at intervals. (Redrawn from Buller, 1909.)

bodies of *Coprinus* species that had been placed horizontally overshot in their negative geotropism so that their response was a series of damped oscillations finally resulting in exact vertical growth (Fig. 12.4). This is also visible in the time-lapse photographs and drawings of *C. stiriacus* by Knoll (1909), who also gives results of experiments in which he longitudinally split most of the stipe leaving it joined at its base. Each piece independently showed geotropism.

Cox & Niederpruem (1975) incubated excised stipes vertically upside down, and observed them elongating around in a U-shape so that the stipe apex again pointed upwards. Cutting off the stipe apex made no difference to the response. Eger-Hummel (1980) recorded geotropic curvatures of up to 180 degrees for stipes of *Pleurotus ostreatus*, but Plunkett (1961) showed that exactly inverted fruit bodies of *Polyporus brumalis* did not turn up, concluding that initiation of geotropism requires a departure from the vertical.

The cytology of geotropism of stipes of *C. cinereus* was investigated by the late G. H. Banbury (personal communication). Excised stipes of 10–50 mm that had been laid on their side in a moist atmosphere raised their apices as they elongated. Examination of these cells by light and electron microscopy of transverse thin sections showed that when horizontal, the distribution of cell contents was displaced so that the vacuole occupied most of the upper part, and the cytoplasm was concentrated in the lower part. It is tempting to see this cytological asymmetry as a biochemical asymmetry, with a greater supply of precursors to the lower wall. It could thus be a direct cause for an increase in rate of wall synthesis on the lower wall and a decrease on the upper, directly giving rise to the observed response of each cell bending upwards as it grows. Once vertical, the central position of the vacuole would be restored, but if this process were slow in comparison to the growth of the cell, then overshoot would occur, resulting in the 'geotropic swinging' described by Buller (1909, 1931). The cause of the asymmetry of the vacuole in horizontal cells is unclear. It is unlikely to be a consequence of a difference in density between vacuole and cytoplasm, otherwise cytoplasm and hence growth would be expected to be confined to the lower portions of vertical cells, and this is not observed. Rather, it may represent a rearrangement of the cytoskeleton.

As discussed above for phototropism, as well as each elongating cell showing negative geotropic growth, the stipe tissue itself in that region must also show asymmetrical growth, with the lower cells elongating and perhaps dividing at a faster rate than the upper cells. This implies communication between the cells, with directional flow of precursors to the regions of most growth. Again, we may invoke the 'stretch receptors' discussed above.

Growth hormones

Stipe elongation in agarics is dependent, to varying degrees, on the presence of the pileus, but despite much effort the regulatory factor(s) has yet to be characterised (Gruen, 1982). This phenomenon is less marked among *Coprinus* species than other genera such as *Agaricus* and *Flammulina*. Gruen (1982) provides an extensive review of a range of experiments by different authors with *Coprinus* species. Some experiments used excised fruit bodies, others were left attached. Thus, stipes of *C. sterquilinus* and *C. cinereus* grew by the same amount with or without decapitation (Gräntz, 1898; Jeffreys & Greulach, 1956; Gooday, 1974); whereas those of *C. radiatus* and *C. congregatus* grew less after decapitation (Borriss, 1934*a*; Eilers, 1974; Bret, 1977). Evidence for stimulation of stipe elongation by a growth factor or hormone in these latter cases comes from Borriss (1934*a*) and Hagimoto & Konishi (1959) who reported that curvature of the elongating stipes of *C. radiatus* and *Agaricus bisporus* can result from partial decapitation or asymmetrical replacement of pileus material or extracts, and Eilers (1974) who reported increased elongation of excised stipes of *C. radiatus* with apical application of cap tissue. Banbury (personal communication) suggested that at least some of the growth stimulation of the stipe in such experiments is a result of nutrients being translocated back from the cap to the stipe, i.e. for its full elongation the stipe needs a nutrient supply into its apex as well as into its base. Additionally in some cases, in the intact attached fruit body the presence of the cap may aid the supply of nutrients to the stipe, perhaps by transpiration. Gooday (1974) and Bret (1977) have shown that the final length of elongated excised stipes of *C. cinereus* and *C. congregatus* was dependent on the size at excision, and Gooday (1974) that this could be increased to nearly equal that of intact fruit bodies by a supply of nutrients to the base of the stipe.

Nuclear numbers in stipe cells

Some of the cells of fruit bodies of many basidiomycetes are multinucleate. Harper (1902) concluded that this departure from the binucleate state of the vegetative dikaryon arose in cells of the stipe and pileus that were concerned with support and transport, and was not shown by cells directly concerned with the production of basidiospores. Kniep

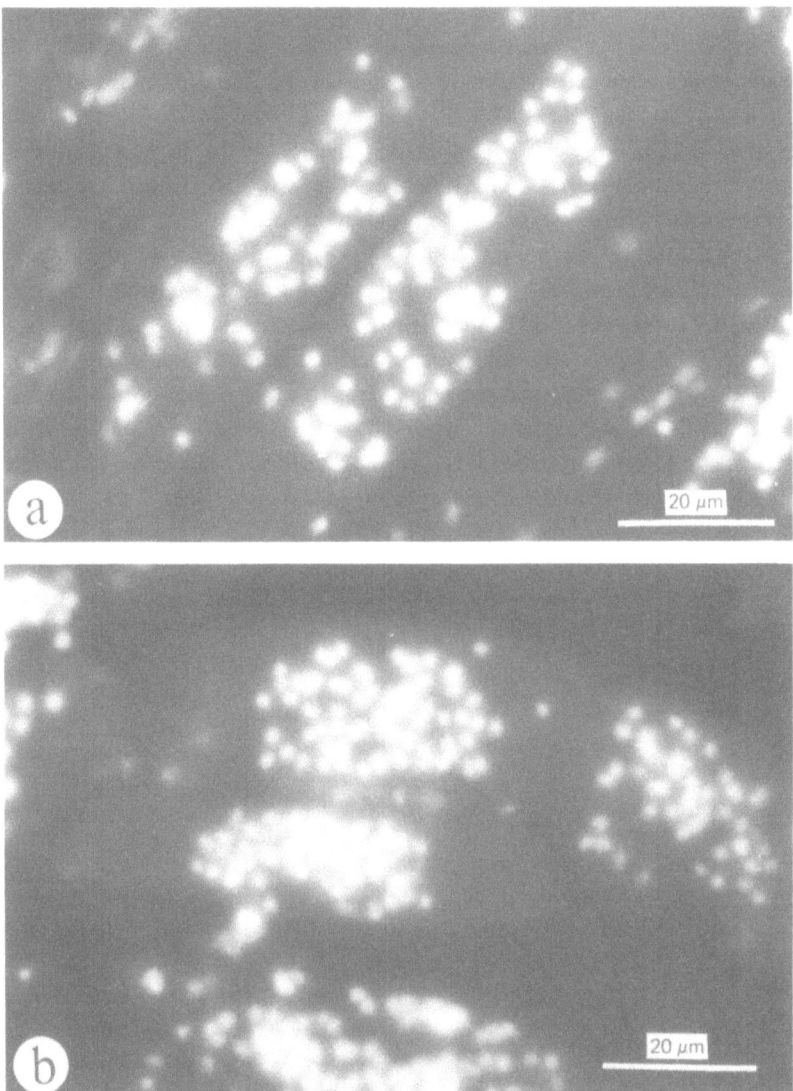

Fig. 12.5. Stipe cortex cells from *Coprinus cinereus*, with their DNA stained with DAPI. Both preparations are from stipes 18 mm tall, showing between 65 and 80 nuclei per cell with differential focusing. (From Stephenson & Gooday, 1984.)

(1913) and Lu (1974) described multinucleate cells in the stipes of *Coprinus nycthemerus* and *C. cinereus*, respectively. Estimates of maximum numbers of nuclei per stipe cell include 36 in *Agaricus bisporus* (Evans, 1959), 56 in *Clavaria sphagnicola* (Geitler, 1965), and 230–80 in *Coprinus radiatus* (Hafner & Thielke, 1970). Hafner & Thielke (1970) suggest that this number in stipe cells of *C. radiatus*, at about 256, represents seven synchronous divisions from the original dikaryon.

Using the fluorescent stain DAPI, we have investigated nuclear numbers in stipe cells of primordia and young fruit bodies of *C. cinereus* (stages I, II, III of Moore, Elhiti & Butler, 1979). Characteristically the vegetative cells of *C. cinereus* are dikaryotic, maintaining this state by clamp connection formation during mitosis. Prior to rapid elongation, however, the nuclei in the stipe cortex cells undergo repeated divisions (Fig. 12.5; Stephenson & Gooday, 1984). Thus stipe cells of fruit bodies 16.6–20.5 mm tall had between 32 and 156 nuclei each. In stipes between 2 and 16 mm

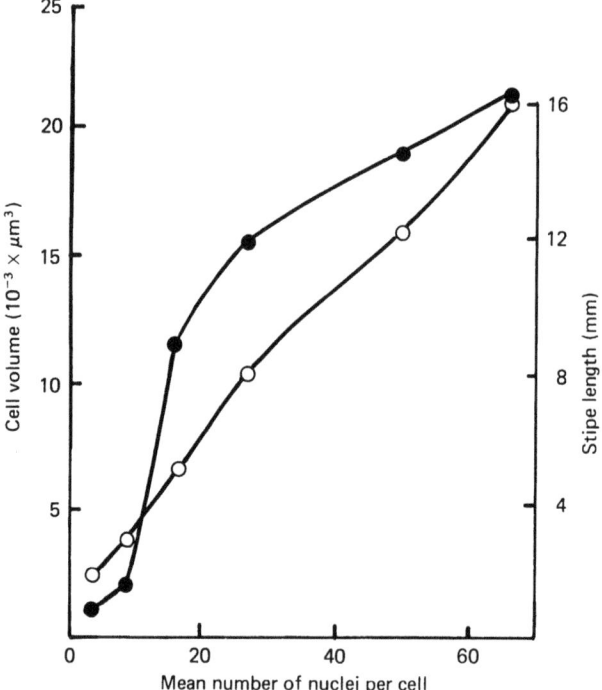

Fig. 12.6. Nuclear numbers in stipe cells of *Coprinus cinereus* in relation to cell volume (●) and stipe length (○). (Drawn from results of Stephenson & Gooday, 1984.)

long, stipe length was directly proportional to mean nuclear number (Fig. 12.6).

Although cell size also increased with increasing nuclear number, there was a jump in the cell volume per nucleus between stipes of length 3 and 5 mm, from 230 to 720 μm^3/nucleus (Fig. 12.6). Although there may be synchronous divisions in cells at early stages (Lu, 1974), detailed counts of nuclear numbers in older cells gave no evidence for synchrony (Stephenson & Gooday, 1984). Cells from slightly older stipes showed even higher numbers of nuclei, but with so many it was not possible to count them accurately by epifluorescence.

Thus each stipe cell accumulated hundreds of nuclei before its rapid several-fold elongation. These nuclei appeared smaller than their progenitors. Thus, at stipe lengths between 2 and 18 mm, the mean nuclear diameter declined from 916 to 723 nm (Stephenson & Gooday, 1984), corresponding to a decrease in nuclear volume from 0.40 to 0.20 μm^3.

Kamada et al. (1976) presented analyses of dry weight and DNA, RNA and protein contents of the stipe of Coprinus cinereus (Fig. 12.2). These record a rise in DNA content between 9 and 12 h of development, corresponding to the 3 h preceding the onset of rapid elongation. In contrast, RNA and protein contents increased more steadily. Fold increases in wild-type strains were 1.65, 3.4 and 1.75, respectively, for DNA, RNA and protein (Table 12.1). Because of differences in expressing results and in the growth conditions it is difficult to compare these results with those given earlier on nuclear number. Nevertheless the relatively small increase in DNA recorded in Table 12.1 suggests that the bulk of nuclear division had occurred before these samples were taken.

The much greater fold increases in RNA than in DNA, for all three rapidly elongating systems (wild-type stipes, decapitated stipes and *sporeless* mutants), suggest a role for the preceding proliferation, i.e. that the very active metabolic process of rapid stipe elongation may require a more rapid rate of RNA synthesis than that in vegetative cells.

References

Banbury, G. H. (1959). Phototropism in lower plants. In *Encyclopedia of Plant Physiology*, vol. 17 (i), ed. W. Ruhland, pp. 530–78. Berlin: Springer-Verlag.

Banbury, G. H. (1962). Geotropism in lower plants. In *Encyclopedia of Plant Physiology*, vol. 17 (ii), ed. W. Ruhland, pp. 344–77. Berlin: Springer-Verlag.

Bartnicki-Garcia, S. (1968). Cell wall chemistry, morphogenesis, and taxonomy of fungi. *Annual Review of Microbiology*, **22**, 87–108.

Blayney, G. P. & Marchant, R. (1977). Glycogen and protein inclusions in elongating stipes of *Coprinus cinereus*. *Journal of General Microbiology*, **98**, 467–76.

Bonner, J. T., Hoffman, A. A., Marioka, W. T. & Chiquoine, A. D. (1957). The distribution of polysaccharides and basophilic substances during the development of the mushroom *Coprinus*. *Biological Bulletin, Marine Biological Laboratory, Woods Hole, Mass.*, **122**, 1–6.

Borriss, H. (1934*a*). Beiträge zur Wachstums und Entwicklungsphysiologie der Fruchtkörper von *Coprinus lagopus*. *Planta, Berlin*, **22**, 28–69.

Borriss, H. (1934*b*). Uber den Einfluss äusserer Factoren auf Wachstum und Entwicklung der Fruchtkörper von *Coprinus lagopus*, *Planta, Berlin*, **22**, 644–84.

Bottom, C. B. & Siehr, D. J. (1979). Structure of an alkali-soluble polysaccharide from the hyphal wall of the basidiomycete *Coprinus macrorhizus* var. *microsporus*. *Carbohydrate Research*, **77**, 169–81.

Bottom, C. B. & Siehr, D. J. (1980). Structure and composition of the alkali-insoluble cell wall fraction of *Coprinus macrorhizus* var. *microsporus*. *Canadian Journal of Biochemistry*, **58**, 147–53.

Bret, J. P. (1977). Respective role of cap and mycelium on stipe elongation of *Coprinus congregatus*. *Transactions of the British Mycological Society*, **68**, 262–9.

Buller, A. H. R. (1909). *Researches on Fungi*, vol. 1. London: Longmans, Green & Co.

Buller, A. H. R. (1924). *Researches on Fungi*, vol. 3. London: Longmans, Green & Co.

Buller, A. H. R. (1931). *Researches on Fungi*, vol. 4. London: Longmans, Green & Co.

Burnett, J. H. (1979). Aspects of the structure and growth of hyphal walls. In *Fungal Walls and Hyphal Growth*, ed. J. H. Burnett & A. P. J. Trinci, pp. 1–25. Cambridge University Press.

Bush, D. A. (1974). Autolysis of *Coprinus comatus* sporophores. *Experientia*, **30**, 984.

Carlile, M. J. (1970). The photobiology of fungi. In *Photobiology of Microorganisms*, ed. P. Halldal, pp. 309–43. London: Academic Press.

Chanter, D. O. & Thornley, J. H. M. (1978). Mycelial growth and the initiation and growth of sporophores in the mushroom crop; a mathematical model. *Journal of General Microbiology*, **106**, 55–65.

Chapman, E. S. & Fergus, C. L. (1973). An investigation of the effects of light on basidiocarp formation of *Coprinus domesticus*. *Mycopathologia*, **51**, 315–26.

Cox, R. J. & Neiderpruem, D. J. (1975). Differentiation in *Coprinus lagopus*. III, Expansion of excised fruit-bodies. *Archives of Microbiology*, **105**, 257–60.

Craig, G. D., Gull, K. & Wood, D. A. (1977). Stipe elongation in *Agaricus bisporus*. *Journal of General Microbiology*, **102**, 337–47.

Dennison, D. S. (1979). Phototropism. In *Physiology of Movements, Encyclopedia of Plant Physiology*, vol. 7, ed. W. Haupt & M. E. Feinleib, pp. 506–66. Berlin: Springer-Verlag.

Dennison, D. S. & Roth, C. C. (1967). *Phycomyces* sporangiophores: fungal stretch receptors. *Science*, **156**, 1386–8.

Eger-Hummel, G. (1980). Blue-light photomorphogenesis in mushrooms (basidiomycetes). In *The Blue Light Syndrome*, ed. H. Seager, pp. 555–62. Berlin: Springer-Verlag.

Eilers, F. I. (1974). Growth regulation in *Coprinus radiatus*. *Archives of Microbiology*, **96**, 353–64.

Evans, H. J. (1959). Nuclear behaviour in the cultivated mushroom. *Chromosoma*, **10**, 115–35.

Ewaze, J. O., Moore, D. & Stewart, G. R. (1978). Co-ordinate regulation of enzymes involved in ornithine metabolism and its relation to sporophore morphogenesis in *Coprinus cinereus*. *Journal of General Microbiology*, **107**, 343–57.

Geitler, L. (1965). Zur Cytologie von *Clavaria* s. str. *Osterreichische Botanische Zeitschrift*, **112**, 543–50.

Gooday, G. W. (1974). Control of development of excised fruit bodies and stipes of *Coprinus cinereus*. *Transactions of the British Mycological Society*, **62**, 391–9.

Gooday, G. W. (1979). Chitin synthesis and differentiation in *Coprinus cinereus*. In *Fungal Walls and Hyphal Growth*, ed. J. H. Burnett & A. P. J. Trinci, pp. 203–23. Cambridge University Press.

Gooday, G. W. (1982). Metabolic control of fruit body morphogenesis in *Coprinus cinereus*. In *Basidium and Basidiocarp: Evolution, Cytology, Function and Development*, ed. K. Wells & E. K. Wells, pp. 157–73. New York: Springer-Verlag.

Gooday, G. W., de Rousset-Hall, A. & Hunsley, D. (1976). Effect of polyoxin D on chitin synthesis in *Coprinus cinereus*. *Transactions of the British Mycological Society*, **67**, 193–200.

Gow, N. A. R. & Gooday, G. W. (1983). Ultrastructure of chitin in hyphae of *Candida albicans* and other dimorphic and mycelial fungi. *Protoplasma*, **115**, 52–8.

Gräntz, F. (1898). *Uber den Einfluss des Lichtes auf die Entwicklung einiger Pilze*. Thesis, Universität Leipzig.

Gruen, H. E. (1982). Control of stipe elongation by the pileus and mycelium in fruitbodies of *Flammulina velutipes* and other Agaricales. In *Basidium and Basidiocarp: Evolution, Cytology, Function and Development*, ed. K. Wells & E. K. Wells, pp. 125–55. New York: Springer-Verlag.

Hafner, L. & Thielke, C. (1970). Kernzahl und Zellgrösse im Fruchtkörperstiel von *Coprinus radiatus* (Solt.) Fr. *Berichte Deutsche Botanische Gesellschaft*, **83**, 27–31.

Hagimoto, H. & Konishi, M. (1959). Studies on the growth of fruit body of fungi. I. Existence of a hormone active to the growth of fruit body in *Agaricus bisporus* (Lange) Sing. *Botanical Magazine, Tokyo*, **72**, 359–66.

Haines, B. D. & Ashworth, J. M. (1974). The control of saccharide synthesis during development of myxamoebae of *Dictyostelium discoideum* containing different amounts of glycogen. *Biochemical Journal*, **142**, 317–25.

Harper, R. A. (1902). Binucleate cells in certain hymenomycetes. *Botanical Gazette*, **33**, 1–25.

Hunsley, D. & Burnett, J. H. (1970). The ultrastructural architecture of the walls of some fungi. *Journal of General Microbiology*, **62**, 203–18.

Iten, W. (1970). Zur Funktion hydrolytischer Enzyme bei der Autolyse von *Coprinus*. *Berichte Schweizerische Botanische Gesellschaft*, **79**, 175–98.

Iten, W. & Matile, P. (1970). Role of chitinase and other lysosomal enzymes of *Coprinus lagopus* in the autolysis of fruiting bodies. *Journal of General Microbiology*, **61**, 301–9.

Jeffreys, D. B. & Greulach, V. A. (1956). The nature of tropisms of *Coprinus sterquilinus*. *Journal of the Elisha Mitchell Scientific Society*, **72**, 153–8.

Jirjis, R. I. & Moore, D. (1976). Involvement of glycogen in morphogenesis of *Coprinus cinereus*. *Journal of General Microbiology*, **95**, 348–52.

Kamada, T., Fujii, T. & Takemaru, T. (1980). Stipe elongation during basidiocarp maturation in *Coprinus macrorhizus*; changes in activity of cell wall lytic enzymes. *Transactions of the Mycological Society of Japan*, **21**, 359–67.

Kamada, T., Hamada, Y. & Takemaru, T. (1982). Autolysis *in vitro* of the stipe cell wall in *Coprinus macrorhizus*. *Journal of General Microbiology*, **128**, 1041–6.

Kamada, T., Kurita, R. & Takemaru, T. (1978). Effects of light on basidiocarp maturation in *Coprinus macrorhizus*. *Plant & Cell Physiology*, **19**, 263–75.

Kamada, T., Miyazaki, S. & Takemaru, T. (1976). Quantitative changes of DNA, RNA, and protein during basidiocarp maturation in *Coprinus macrorhizus*. *Transactions of the Mycological Society of Japan*, **17**, 451–60.

Kamada, T. & Takemaru, T. (1977*a*). Stipe elongation during basidiocarp maturation in *Coprinus macrorhizus*; mechanical properties of stipe cell wall. *Plant & Cell Physiology*, **18**, 831–40.

Kamada, T. & Takemaru, T. (1977*b*). Stipe elongation during maturation in *Coprinus macrorhizus*; changes in polysaccharide composition of stipe cell walls during elongation. *Plant & Cell Physiology*, **18**, 1291–300.

Kamada, T. & Takemaru, T. (1983). Modification of cell-wall polysaccharides during stipe elongation in basidiomycete *Coprinus cinereus*. *Journal of General Microbiology*, **129**, 703–9.

Kniep, H. (1913). Über die Herkunft der Kernpaare im Fruchtkörper von *Coprinus nycthemerus* Fr. *Botanische Zeitung*, **5**, 610–37.

Knoll, F. (1909). Untersuchungen über Längenwachstum und Geotropismus der Fruchtkörperstiele von *Coprinus stiriacus*. *Sitzungsberichte Akademie Wissenschaften in Wien*, **118**, 573–633.

Lu, B. C. (1974). Meiosis in *Coprinus*. V. The role of light on basidiocarp initiation, mitosis and hymenium differentiation in *Coprinus lagopus*. *Canadian Journal of Botany*, **52**, 299–306.

McLaughlin, D. J. (1974). Ultrastructural localization of carbohydrate in the hymenium and subhymenium of *Coprinus*. Evidence for the function of the Golgi apparatus. *Protoplasma*, **82**, 341–64.

Madelin, M. F. (1956). Studies on the nutrition of *Coprinus lagopus* Fr. especially affecting fruiting. *Annals of Botany*, **20**, 307–30.

Madelin, M. F. (1960). Visible changes in the vegetative mycelium of *Coprinus lagopus* Fr. at the time of fruiting. *Transactions of the British Mycological Society*, **43**, 105–10.

Manachère, G. (1970). Recherches physiologiques sur la fructification de *Coprinus congregatus* Bull ex Fr.; action de la lumière; rythme de production de carpophores. *Annales des Sciences, Botanique et Biologie Végetale, Paris*, Série 12, **11**, 1–96.

Manachère, G., Robert, J. C., Durand, R., Bret, J. P. & Fevre, M. (1983). Differentiation in the Basidiomycetes. In *Fungal Differentiation: a Contemporary Synthesis*, ed. J. E. Smith, pp. 481–514. New York: Marcel Dekker.

Marchant, R. (1978). Wall composition of monokaryons and dikaryons of *Coprinus cinereus*. *Journal of General Microbiology*, **106**, 195–99.

Matthews, T. R. & Niederpruem, D. J. (1972). Differentiation in *Coprinus lagopus*. I. Control of fruiting and cytology of initial events. *Archiv für Mikrobiologie*, **87**, 257–68.

Matthews, T. R. & Niederpruem, D. J. (1973). Differentiation in *Coprinus lagopus*. II. Histology and ultrastructural aspects of developing primordia. *Archiv für Mikrobiologie*, **88**, 169–90.

Michalenko, G. O., Hohl, H. R. & Rast, D. (1976). Chemistry and architecture of the mycelial wall of *Agaricus bisporus*. *Journal of General Microbiology*, **92**, 251–62.

Miyake, H., Takemaru, T. & Ishikawa, T. (1980). Sequential production of enzymes and basidiospore formation in fruiting bodies of *Coprinus macrorhizus*. *Archives of Microbiology*, **126**, 201–5.

Montgomery, G. W. G., Adams, D. J. & Gooday, G. W. (1984). Studies on the purification of chitin synthase from *Coprinus cinereus*. *Journal of General Microbiology*, **130**, 291–7.

Moore, D., Elhiti, M. M. Y. & Butler, R. D. (1979). Morphogenesis of the carpophore of *Coprinus cinereus*. *New Phytologist*, **83**, 695–722.

Morimoto, N. & Oda, Y. (1973). Effects of light on fruit body formation in a basidiomycete, *Coprinus macrorhizus*. *Plant and Cell Physiology*, **14**, 217–25.

Pinto-Lopes, J. & Almeida, M. G. (1970). '*Coprinus lagopus*' a confusing name as applied to several species. *Portugaliae Acta Biologica* (*B*), **11**, 167–204.

Plunkett, B. E. (1961). The change of tropism in *Polyporus brumalis* stipes and the effect of directional stimuli or pileus differentiation. *Annals of Botany*, N.S., **25**, 206–23.

Rao, P. S. & Niederpruem, D. J. (1969). Carbohydrate metabolism during morphogenesis of *Coprinus lagopus* (*sensu* Buller). *Journal of Bacteriology*, **100**, 1222–8.

Robert, J. C. (1977*a*). Fruiting of *Coprinus congregatus*: biochemical changes in fruit-bodies during morphogenesis. *Transactions of the British Mycological Society*, **68**, 379–87.

Robert, J. C. (1977*b*). Fruiting of *Coprinus congregatus*: relationship to biochemical changes in the whole culture. *Transactions of the British Mycological Society*, **68**, 389–95.

Schaeffer, H. P. (1977). An alkali-soluble polysaccharide from the cell walls of *Coprinus lagopus*. *Archives of Microbiology*, **113**, 79–82.

Schwalb, M. N. & Shamler, S. (1974). Phototropic and geotropic responses during the development of normal and mutant fruit bodies of the basidiomycete *Schizophyllum commune*. *Journal of General Microbiology*, **82**, 209–12.

Sietsma, J. H. & Wessels, J. G. H. (1979). Evidence for covalent linkage between chitin and β-glucan in a fungal wall. *Journal of General Microbiology*, **114**, 99–108.

Sietsma, J. H. & Wessels, J. G. H. (1981). Solubility of $(1-3)$-β-D/$(1-6)$-β-D-glucan in fungal walls: importance of presumed linkage between chitin and glucan. *Journal of General Microbiology*, **125**, 209–12.

Stephenson, N. A. & Gooday, G. W. (1984). Nuclear numbers in the stipe cells of *Coprinus cinereus*. *Transactions of the British Mycological Society*, **82**, 531–4.

Stewart, G. R. & Moore, D. (1974). The activities of glutamate dehydrogenase during mycelial growth and sporophore development in *Coprinus lagopus*, (*sensu* Lewis). *Journal of General Microbiology*, **83**, 73–81.

Streeter, S. G. (1909). The influence of gravity on the direction of growth of *Amanita*. *Botanical Gazette*, **48**, 414–26.

Takemaru, T. & Kamada, T. (1972). Basidiocarp development in *Coprinus macrorhizus*. I. Induction of developmental variations. *Botanical Magazine, Tokyo*, **85**, 51–7.

Tani, K., Kuroiwa, T. & Takemaru, T. (1977). Cytological studies on sporeless mutant in the basidiomycete *Coprinus macrorhizus*. *Botanical Magazine, Tokyo*, **90**, 235–45.

Valk, P. van der & Marchant, R. (1978). Hyphal ultrastructure in fruit-body primordia of the basidiomycetes *Schizophyllum commune* and *Coprinus cinereus*. *Protoplasma, Berlin*, **95**, 57–72.

Valk, P. van der, Marchant, R. & Wessels, J. G. H. (1977). Ultrastructural localization of polysaccharides in the wall and septum of the basidiomycete *Schizophyllum commune*. *Experimental Mycology*, **1**, 69–82.

Wessels, J. G. H. & Sietsma, J. H. (1981). Fungal cell walls: a survey. In *Plant Carbohydrates II, Encyclopedia of Plant Physiology*, vol. 13B, ed. W. Tanner & F. A. Loewus, pp. 352–94. Berlin: Springer-Verlag.

Wessels, J. G. H., Sietsma, J. H. & Sonnenberg, A. S. M. (1983). Wall synthesis and assembly during hyphal morphogenesis in *Schizophyllum commune*. *Journal of General Microbiology*, **129**, 1607–16.

13
Differentiation and pattern formation in the fruit body cap of *Coprinus cinereus*

ISABELLE V. ROSIN, JACQUELINE HORNER and
DAVID MOORE
Department of Botany, The University, Manchester M13 9PL, UK

Introduction

Over the past few years we have gained considerable insight into the metabolism involved in development of the fruit body cap of *Coprinus cinereus*. Most of this work has made use of homogenates of whole caps for biochemical analyses and there is now an urgent need to relate the findings to differentiation of individual cells and the creation of patterns that characterise the different tissues of the cap. The biochemical data identify specific enzyme regulatory mechanisms associated with developmental changes in the fruit body. However, proper understanding of the ways in which this regulation is integrated endogenously during morphogenesis requires detailed knowledge of the relationships between individual cells and of the differentiation processes leading to establishment of tissue domains. The existing literature is silent on the former and confusing and contradictory on the latter. In this chapter we review the biochemical data and indicate how we are attempting to extend the analysis to the cellular level. We concentrate, though, on an account of cap development derived from microscopical observations that leads to a unifying interpretation of tissue differentiation in the fruit body of this species and provides the basis for hypotheses about the integration of control systems during morphogenesis.

Metabolic control of morphogenesis

It appears that in *C. cinereus* the nitrogen metabolism of the developing cap has the most direct bearing on its morphology, though there is considerable interplay with carbohydrate and carboxylic acid metabolisms. The developmental sequence of the fruit body has been divided into a series of stages to facilitate analysis. This has been done in a number of ways by different authors over the years (see Moore, Elhiti & Butler, 1979). The scheme used here is illustrated in Fig. 13.1.

Attempts to characterise the metabolism of the developing fruit body have concentrated on enzyme surveys and metabolite measurements. These investigations have revealed four enzymes which show large increases in activity in the developing cap while remaining at low levels (or declining in activity) in the stipe. These four enzymes are NADP-linked glutamate dehydrogenase (NADP–GDH), glutamine synthetase (GS), ornithine acetyltransferase (OAT) and ornithine carbamyltransferase (OCT). A fifth enzyme, urease, showed the reverse behaviour, being absent from the cap though present in the stipe and, indeed, being constitutive in mycelium (Table 13.1). Amplification of tricarboxylic acid (TCA) cycle activity is signalled by the observations that succinate dehydrogenase also shows much greater activity in the cap than in the stipe, and isocitrate dehydrogenase activity increases in both parts of the fruit body as development

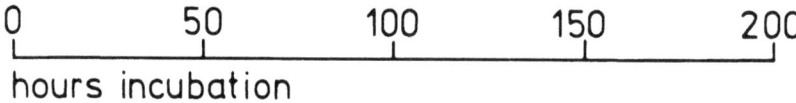

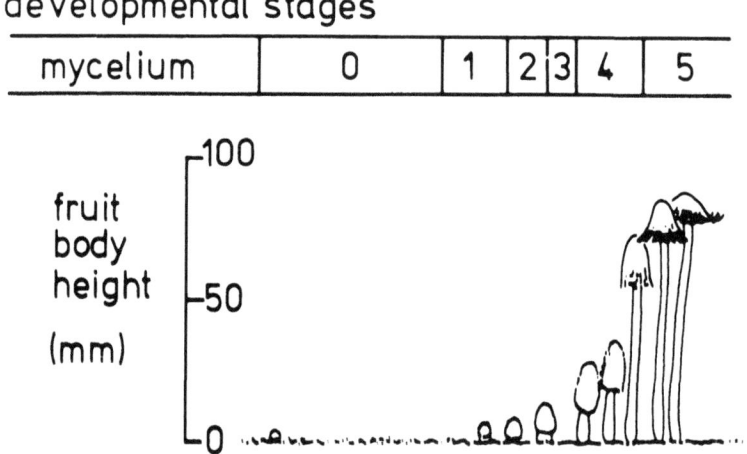

Fig. 13.1. Descriptions of developmental stages in *Coprinus cinereus* in relation to incubation period. Standardised culture conditions were described by Moore & Ewaze (1976) and the different stages are described by Moore, Elhiti & Butler (1979).

Table 13.1. *Some enzyme activities in developing fruit bodies of* Coprinus cinereus

Enzyme	Tissue	Fruit body developmental stage		
		3	4	5
NADP-linked glutamate dehydrogenase	cap	150	267	830
	stipe	12	15	103
Glutamine synthetase	cap	300	950	2067
	stipe	85	290	550
Ornithine acetyltransferase	cap	16	37	34
	stipe	8	7	5
Ornithine carbamyltransferase	cap	353	693	1240
	stipe	153	436	310
Urease	cap	55	70	38
	stipe	143	570	4300
NAD-linked glutamate dehydrogenase	cap	270	553	1183
	stipe	363	483	1287

Enzyme activities are shown as pmol substrate used min^{-1} $(mg\ protein)^{-1}$ for urease, and as nmol substrate used min^{-1} $(mg\ protein)^{-1}$ for all other enzymes.

proceeds (Stewart & Moore, 1974; Moore & Ewaze, 1976; Ewaze, Moore & Stewart, 1978) (Table 13.2).

Consideration of the metabolic pathways in which these enzymes are concerned leads to the inference that the increases in activity provide for amplification of the urea cycle, leading to accumulation of urea. Feeding of $[U^{-14}C]$-citrulline to live tissue slices confirms that urea synthesis and accumulation occurs *in vivo* in the cap but not in the stipe (the latter, of course, having a high urease activity). Urea is not the only compound which is accumulated. Arginine content increases by a factor of four (concentration by a factor of two) as the primordial cap develops to maturity, though both content and concentration decline in the stipe during this time. This situation can be interpreted as a means by which arginase activity is regulated. In *Coprinus* this enzyme has a K_m of 100 mM and it can be calculated that the flux through the arginase reaction is likely to increase by a factor of two to three in the cap while declining in the stipe as development proceeds from stage 3 (immature fruit body, meiosis ending but spore formation only just starting) to stage 5 (mature fruit body discharging spores). Indeed the arginine accumulation will lead to at least a six-fold greater flux through the arginase reaction in the cap than in the stipe even though there is little difference between the *in vitro* measurements

Table 13.2. *Some enzymes and metabolites in developing fruit bodies of* Coprinus cinereus

Enzyme or metabolite	Tissue	Fruit body developmental stage		
		3	4	5
NADP-linked glutamate dehydrogenase	cap	47	420	780
	stipe	29	11	20
Succinate dehydrogenase	cap	12	27	62
	stipe	28	16	10
Isocitrate dehydrogenase	cap	50	150	200
	stipe	90	200	220
Glycogen:				
μg per fruit body		1857	1361	97
% in cap		97	96	86
Ammonium	cap	32	11	9
	stipe	90	40	22

Enzyme activities are shown as nmol substrate used min^{-1} (mg protein)$^{-1}$; the amount of ammonium is shown as μmol (g fresh weight)$^{-1}$.

of arginase activity of the two tissues. This further strengthens the view that the urea cycle is specifically amplified in developing cap tissues.

The level of arginine increases whether quantified in terms of tissue fresh weight or dry weight; but while urea content on a dry weight basis increases by a factor of 2.5, the urea concentration (on a fresh weight basis) is essentially unchanged during cap development (Ewaze, Moore & Stewart, 1978). Among the compounds assayed urea was the only one to behave like this. The conclusion was drawn that during cap development urea accumulation drives water into the cells osmotically.

There is certainly a need for considerable water uptake during the later stages of cap development, for the hymenial cells particularly become greatly inflated (Moore *et al.*, 1979). This cell expansion is absolutely central to the whole morphogenesis of the developing cap. Mature gill hymenia are largely made up of paraphyses and these cells increase most dramatically in size. Expansion of these cells therefore increases the area of the gill plate but since this is bounded on its outer edge (i.e. the edge furthest from the stipe) by an inextensible, but flexible, layer of outer cap tissue, the increase in gill area is accommodated by a curling of the gill away from the stipe (Moore *et al.*, 1979). Thus the changes in cap morphology which characterise the maturation process can be accounted for by inflation of hymenial cells, and this depends on osmotic influx of water driven by the substrate accumulations which are a consequence of the metabolic shift discussed above.

Experimentation with vegetative cultures has shown that the basic features of NADP–GDH are that induction occurs when acetyl-CoA accumulates in the mycelium in the virtual absence of ammonium (Moore, 1981). This correlates extremely well with observations made on fruit bodies (Table 13.2). Assays of normal fruit bodies show that glycogen is accumulated to high levels in the fruit body cap, but is metabolised as the cap matures (Moore *et al.*, 1979). During this process both isocitrate dehydrogenase and succinate dehydrogenase are elevated in activity in the cap, implying enhanced metabolism through the TCA cycle (Moore & Ewaze, 1976). There is a good evidence that the TCA cycle in *Coprinus cinereus* proceeds through the glutamate decarboxylation loop (Ewaze *et al.*, 1978). Thus 2-oxoglutarate amination is a necessary component of TCA cycle reactions and if the TCA cycle activity is to be amplified, this amination step must be amplified too. However, the cap always contains less free ammonium than does the stipe and the concentration of this metabolite declines drastically as the primordium develops into the mature fruit body (Ewaze *et al.*, 1978). Metabolism of up to 2 mg of glycogen per fruit body with little available ammonium could obviously lead to accumulation of acetyl-CoA to levels sufficient to induce NADP–GDH and associated enzymes because of the overall rate-limiting effect at the 2-oxoglutarate amination step normally carried out by NAD-linked glutamate dehydrogenase.

We believe that the NADP–GDH and glutamine synthetase together form an ammonium-scavenging system which is induced to safeguard the requirements of both the TCA and urea cycles under these metabolic conditions in the fruit body; NADP–GDH having a ten-fold higher affinity for ammonium than the NAD-linked enzyme (Al-Gharawi & Moore, 1977). Such an enzyme combination is a rather novel way of going about this task, but there is evidence that at least one other basidiomycete, *Sporotrichum*, employs a similar system (Buswell, Ander & Eriksson, 1982). In this discussion, however, we do not wish to take the metabolic description any further. Clearly, we have identified a system of enzymes which serve a distinct purpose during cap development. These enzymes exhibit a variety of levels of control; some are quite evidently derepressed by controls which must operate at the gene level, one is repressed in an apparently coordinated way, and there is substrate-level regulation too, including an allosteric control of NADP–GDH activity which is apparently related to accumulation of the substrate and which therefore again implies that ammonium availability is rate-limiting for the TCA cycle (Al-Gharawi & Moore, 1977). These various phenomena are evidently integrated during development so as to provide for the maturation of the fruit body cap. Our major interest now is in the means by which the control of this metabolism is integrated in the differentiated cells of the fruit body cap.

Cytochemistry of developing gill tissues

Although the metabolism described seems to be specifically concerned with the expansion phase of fruit body maturation, the derepression of NADP–GDH (which is used as a model representative of this whole metabolic shift) occurs in quite young primordia. To understand the regulatory events themselves, we must look at the primordium and its mode of differentiation. The first efforts to examine the behaviour of individual cells were made some years ago using frozen sections of fruit body tissues (Elhiti, Butler & Moore, 1979). The sections were stained with tetrazolium and revealed a very distinct difference between the NAD-linked GDH (which is always present in cap and stipe tissues) and the NADP-linked enzyme, activity of which is found in caps but not stipes. The cytochemical observations paralleled those derived from assays of tissue homogenates. They also seem to show a new phenomenon, for the developing activity of NADP–GDH did not increase uniformly in all cells. Isolated islands of hymenial tissue showed positive enzyme staining. As successively older tissues were examined those islands enlarged and eventually coalesced, implying that the steadily increasing activity of NADP–GDH recorded in homogenates prepared from successively older primordia reflected a steady increase in the population of cells able to express this enzyme activity. These experiments used material frozen and sectioned at $-20\,^{\circ}\text{C}$. It is potentially possible to apply the tetrazolium reaction to living material. The technique involves stripping individual gills (or even individual hymenia) from primordia and incubating in a solution containing tetrazolium salt. After about 30 min, various patterns of staining are observed, and the most favourable give the impression of successful staining of individual cells to reveal NADP–GDH activity. Unfortunately, this staining reaction is non-specific (i.e. it is not dependent on the substrates of the glutamate dehydrogenase reaction included in the reaction mixture). However, it *is* interesting that for some reason something reacts differentially with tetrazolium in these immature hymenia. Nevertheless it is quite clear that frozen sections stained with tetrazolium salts reveal the localisation of NADP–GDH activity in basidia (Elhiti *et al.*, 1979); that in very young tissues only some of the basidia stain, and in older tissues all of the basidia (indeed, eventually all of the cells) stain; and that these reactions *are* substrate-specific. So the question is, how is the regulatory signal communicated between these cells? Any attempt to answer this question will require detailed histological information.

Development of fruit body structure

To derive a developmental description at the biochemical and molecular levels we need an account of the histological structure and development of the fruit body. At the start of spore discharge, the mature cap encloses the top of the stipe and comprises a set of vertically oriented gills radiating inwards, towards the stipe, from the thin outer layer of cap tissue. As spore release proceeds, the cap margin curls outwards, the cap opens like a parasol as splits penetrate the tramal layer between adjacent hymenia, and the gills (or, at least, the gill remnants since autodigestion is occurring too) are eventually brought to an almost horizontal orientation. This terminating phase in development is well documented and the description is non-controversial. The enzyme control mechanisms are exercised in very young primordial tissues and the published literature lacks the clarity required for an unambiguous description of the way tissue domains are initially laid down. The difficulty seems to lie in the variety of developmental patterns observable in agarics (Watling: Chapter 11), in a controversy over interpretations which took place in the early decades of this century, and, in the fact that *C. cinereus* does not appear in the most authoritative account of *Coprinus* spp. (Reijnders, 1979) and has also been frequently misidentified over the years (Pinto-Lopes & Almeida, 1970; Moore *et al.*, 1979).

Accounts in the current literature

The existing literature claims that 'the lamellae originate as downward projecting salients of the palisade hymenophore fundament, in a series radiating outward towards the margin of the pileus, *the younger portions of the salients being towards the margin of the pileus and continuing to arise in a centrifugal direction*, following up the progressive development in the same direction of the palisade hymenophore, cavity, and pileus margin' (Atkinson, 1916, describing various species of *Coprinus*; the stress is ours). Thoughts on the direction of widening of the gill are variable. The prevalent belief is that the gills expand *towards* the stipe owing to growth of the trama hyphae at the inner edge (i.e. the edge closest to the stipe) and to intercalary growth of elements into the hymenium (Brefeld, 1877; De Bary, 1887; Atkinson, 1916; Reijnders, 1948, 1963). Levine (1914) contradicts the popular belief that widening of the gill occurs from the outer to the inner edge of the gill (i.e. towards the stipe): 'The length of the ridges increases by the formation of new palisade cells from above...'; in this context the outer surface of the cap is 'above', so Levine seems to be indicating a widening from the inner, stipe-adjacent edge towards the outer edge.

Two different modes of development have been described for the origin of lamellae. Schmitz (1842) was one of the first to observe the presence of a general annular cavity around the stipe, the roof of which was lined with a continuous palisade layer, this latter being the young hymenophore. These observations were confirmed in Brefeld's (1877) study on *Coprinus lagopus*, Hoffman's (1860) work on *C. fimetarius* and Atkinson's (1906, 1914) descriptions of *Agaricus* spp. The second method of development was originally thought to be of limited applicability (Atkinson, 1914) until Levine (1914) reported it in *C. micaceus* and then in other species and concluded that this course of development prevailed in most Agaricaceae. According to Levine no annular prelamellar cavity is found. Instead, the palisade layer develops a series of groups or ridges which then elongate and split; halves of adjacent ridges unite to form the lamellae. Levine maintained that the protenchyme tissue between the ridged groups of palisade cells is continuous with the underlying stipe tissue from the earliest stage.

Atkinson (1916) refuted Levine's work in his treatise on *C. comatus*, *C. atramentarius* and *C. micaceus*. He remains adamant that, in Agaricaceae, there is first a general annular cavity with a continuous palisade layer which grows 'outward in a centrifugal direction over the undersurface of the pileus, following the centrifugal growth of the latter'. According to Atkinson, the unequal growth of areas of the palisade layer gives rise to folds which are the fundaments of the lamellae: as these widen in the cavity, they reach the underlying stipe or the fundamental plectenchyma surrounding the stipe. The stipe tissues and the gill trama therefore come to appear continuous. Atkinson argues that it is this secondary attachment of the gill trama to the stipe which led Levine to his 'erroneous' conclusions. Similarly, Chow (1934) reports palisade pockets with the interlying tissues continuous with the stipe, but, again, attributes this feature to the growth of the gills into the stipe.

Observations made since those already discussed have shown that both modes of development occur, but in different species (Reijnders, 1963, 1979). Owing to Levine's demonstration that cavities widened due to fixation artifacts and points-of-stress tearing, Reijnders fixed his material in very mild Flemming solution. His results indicate that certain species do not have a continuous palisade layer or a general annular cavity and that pileus and stipe hyphae in those species are intimately intermingled in the regions between the palisade ridges. Reijnders (1979) examined a number of *Coprinus* species and found some to be rupthymenial (gill differentiation proceeding away from the stipe) while others were levhymenial (gills differentiate towards the stipe).

A considerable portion of the existing literature thus strongly suggests

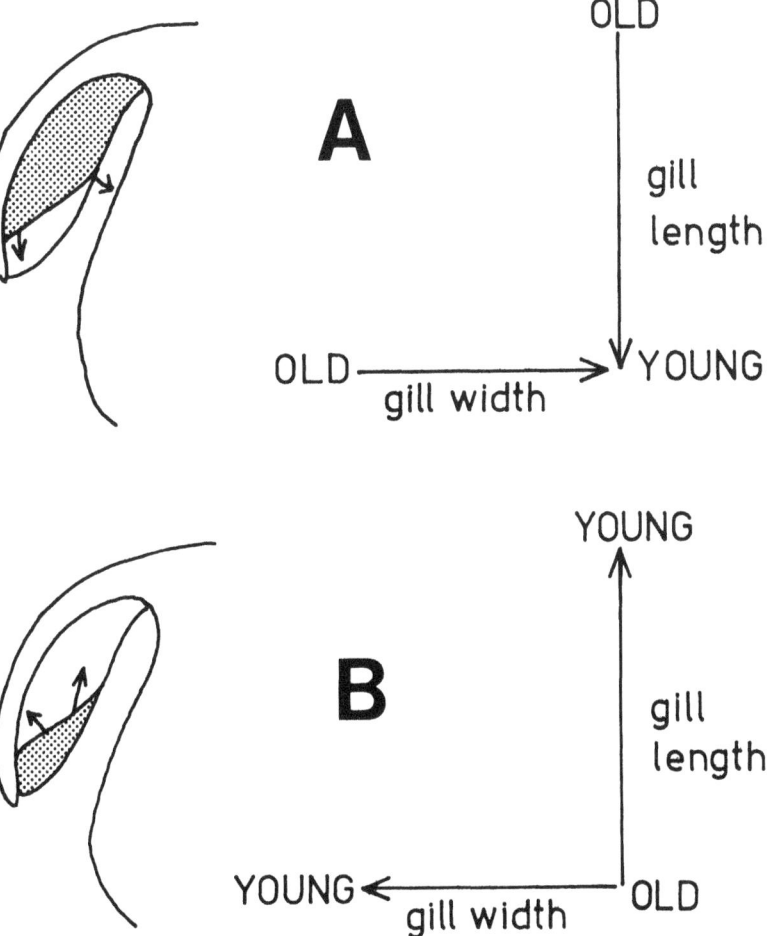

Fig. 13.2. Developmental polarity of the *Coprinus* fruit body cap. A depicts polarity of differentiation where development is levhymenial; B shows polarity in the rupthymenial mode of development.

that early development of the gills takes place in a direction towards the cap margin and towards the stipe. This implies that the youngest tissue will be that which is adjacent to the stipe and that which is closest to the cap margin (Fig. 13.2A). But what of the later stages of maturation? In *Coprinus cinereus*, nuclear fusion is immediately followed by chromosome pairing. This process is asynchronous, with the basal part of the gill (i.e. that at the cap margin) initiating the events. The difference, though, between the margin and the apex is less than an hour (Raju & Lu, 1970). Nevertheless, the implication is that it is cap margin tissue which first reaches the stage of development at which meiosis is initiated. Chow (1934) notes that the maturation of the basidia follows the same general order in *Coprinus* species and begins at the interior–inferior margin of each lamella. Thus, it is commonly agreed and a matter of simple observation that the production and pigmentation of spores is initiated at the edge of the gill closest to the stipe and the 'wave' of pigmentation travels from that edge towards the outer edge of the cap and from the cap margin towards the apex. These observations indicate that events associated with maturation progress in an upward direction, from the cap margin to the apex, and across the gill, from the inner edge (adjacent to the stipe) to the outer. Consequently, the oldest part of the gill (i.e. the part most advanced in development) appears to be at the cap margin *and* at the edge closest to the stipe (Fig. 13.2B).

There is quite obviously a clear contradiction in these accounts, and they could imply that morphogenetic polarity changes direction by 180 degrees at some stage during development. To determine whether this is so, one needs to know whether *Coprinus cinereus* shows a rupthymenial or levhymenial mode of origin of the hymenophore. Unfortunately, this species has often been misidentified and work has been published in the past under such names as *C. lagopus*, *C. fimetarius* and *C. macrorhizus* (Pinto-Lopes & Almeida, 1970). Furthermore, Reijnders (1979) showed that *C. macrorhizus* and *C. lagopus* differed in their mode of hymenophore origin, but he did not examine *C. cinereus*. We have therefore examined the situation in this species for ourselves.

Observations on the differentiation of Coprinus cinereus

Our observations were made on serial transverse sections cut from young primordia at three different stages of development – stages 1, 2 and 3 (see Fig. 13.1). The material was fixed in 5.75% glutaraldehyde, post-fixed in osmium tetroxide, then dehydrated through an alcohol series and embedded in a low-viscosity resin (Spurr, 1969). Serial sections 2 μm thick

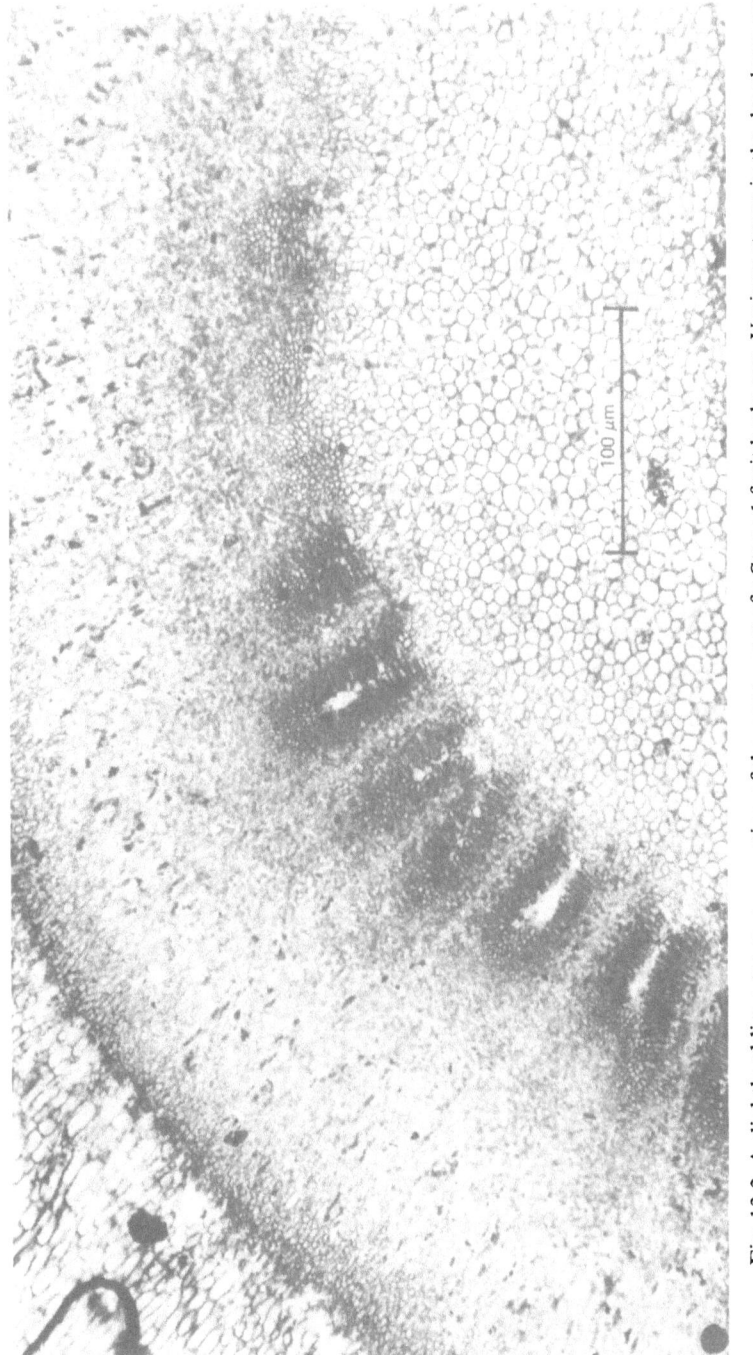

Fig. 13.3. A slightly oblique transverse section of the upper part of a Stage 1 fruit body cap. Various stages in the development of the gill cavities can be seen in this section. Note the absence of a general annular cavity and the continued connection between the stipe tissues and the tissue located between developing hymenia.

were cut on an ultramicrotome and mounted on glass microscope slides for observation with the light microscope. Sections were stained with a solution of 1% toluidine blue in 1% boric acid.

Examination of these serial sections shows that stipe differentiation can be recognised by the increased diameter and vertically parallel arrangement of the stipe cells. From the very earliest point at which such cells can be seen, the outer layers of the stipe are composed of hyphae which are thoroughly intermingled with the undifferentiated hyphae of the cap regions. The first evidence of cap differentiation is the formation of a wave-like contour around the stipe comprised of arched groups of cells (Figs 13.3 and 13.4). These are the palisade ridges, separated by protenchyme hyphae extending into the peripheral tissue of the stipe. Such 'Levine ridges' occur at the apical end of the gill and at the time they are first observed there is no evidence of a general annular cavity. The gill cavity

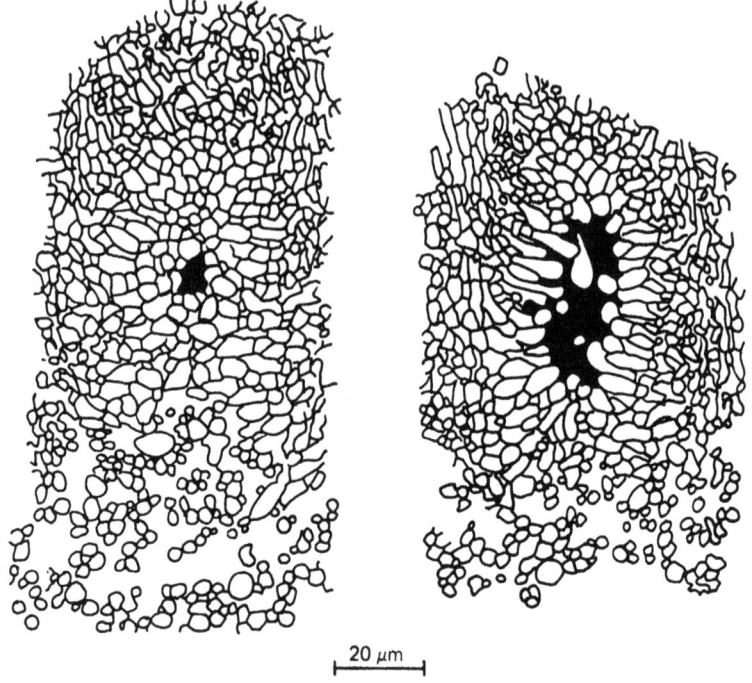

20 μm

Fig. 13.4. Two successive stages in development of the gill cavity (black). The more loosely organised tissue in the bottom of the drawings corresponds in position to the 'annular cavity' region. Tracings made from photographs of transverse sections.

develops within the palisade ridge as its differentiation proceeds towards the outer edge of the cap. At about this time the region between the differentiating palisades and the stipe proper becomes more open and may represent the annular cavity region, but it is still traversed by a large number of intact hyphae. Moreover, the palisade layers (which clearly differentiate into cell plates that are to become the hymenia) have proto-tramal regions which are in full communication with the outer context of the stipe (Fig. 13.3). This means that the hymenium is discontinuous at the inner edge of each primary gill and the gill is physically connected to the stipe. In a stage 1 primordium this hymenial discontinuity at the inner edge of each gill is seen in all cross-sections down the length of the cap. Such connections between the primary gills and the stipe do not represent a secondary attachment; they arise as a result of the particular way in which the gill differentiates. In older stages, the hymenium discontinuity becomes less apparent with the formation of hymenial cells over the inner (stipe-adjacent) edge of the gill; and in older stages still, the appearance of an annular cavity is accompanied by tearing of the gills away from the stipe.

The discontinuity of the hymenium, forming an open inner edge to the gill, indicates that *C. cinereus* shares the rupthymenial mode of hymenophore development described for many other *Coprinus* species (Reijnders, 1979). With this mode of development the gill is envisaged as widening *towards* the periphery (outer edge) of the cap as a differentiating front moves into, and differentiates from, the protenchyme. Since the widest parts of the gills are those at the cap margin, it follows that the differentiating front is also moving *upwards*, towards the apex of the cap. Consequently, the morphogenetic polarities which initially establish the different tissue domains are the same as those that characterise the later, maturational, changes.

Detailed histological structure of the gill and the relationships of its component cells

While the protenchyme hyphae are continuously supplementing the palisade front along a defined acropetal polarity, the older palisades, near the inner edge, are differentiating into basidia. As a result of each ridge splitting in half behind the lateral advance of the palisade front, the gill cavity is formed, flanked by a hymenium of basidia (Fig. 13.4).

The young hymenium consists of poorly differentiated basidia and of conspicuous cystidia. The latter grow from the subhymenium and trama branches (Fig. 13.5). Cystidia insert into the opposite hymenium and their

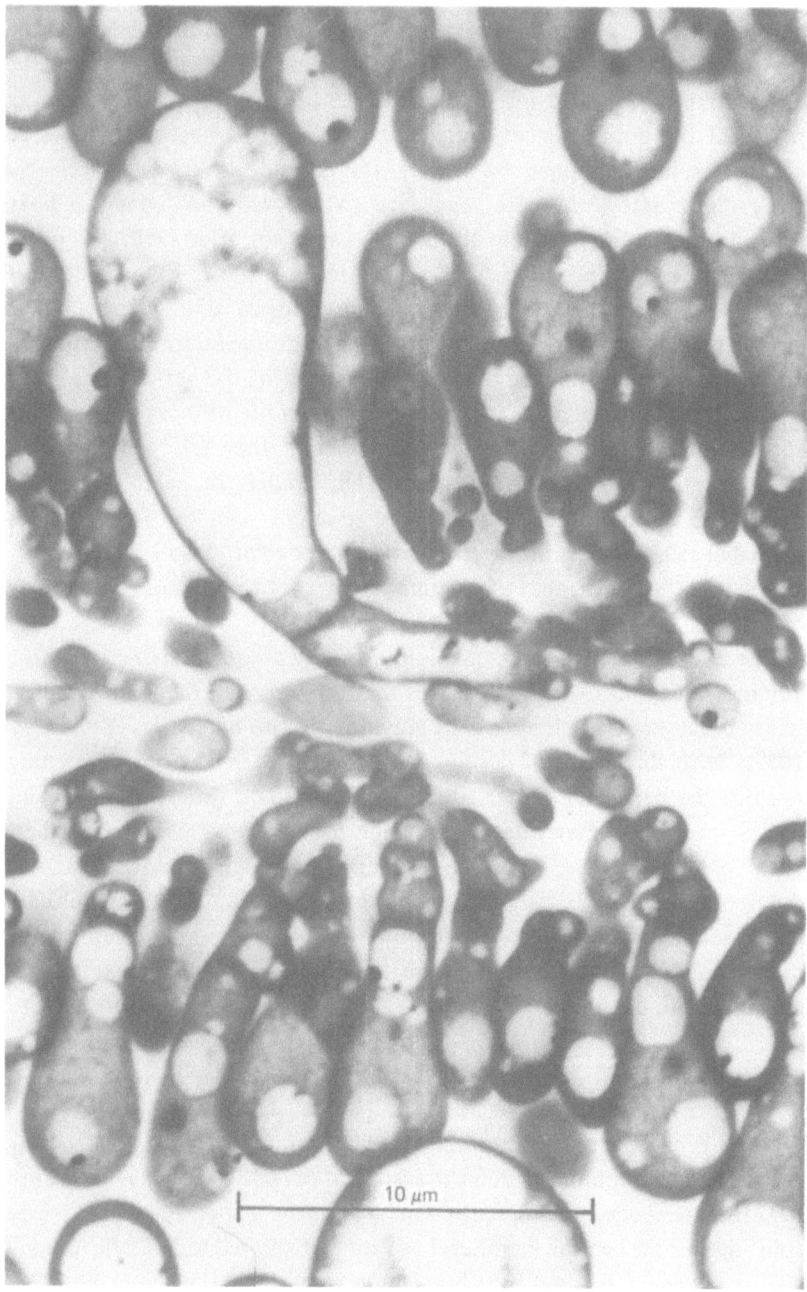

Fig. 13.5. Transverse section of the hymenium of a Stage 2 primordium. A cystidium emerges directly from the tramal hyphae. Paraphyses emerge as branches from sub-basidial cells and insert into the basidial layer.

turgidity prevents hymenium layers from touching and hindering sterigma and spore growth and spore discharge (Buller, 1924; Chow, 1934).

Most of the literature dealing with basidial ontogeny appears to ignore the initial stages of basidial formation and to picture basidia developing as outgrowths from subhymenial segments. In *C. cinereus*, basidia originate as slightly swollen chromophilic cells at the apex of protenchyme or pileus–trama hyphae. This apical differentiation follows the pattern of most basidiomycetes (Sundberg, 1978). Towards stage 2, cystidial cells and the apices of the basidia enlarge. The expansion of basidial apices is often associated with fruit bodies in which the hyphae of the trama, subhymenium and hymenium show similar expansion (Oberwinkler, 1982). The basidia are dimorphic with a slight preponderance of short basidia – a situation analogous to that in *C. sterquilinus* and *C. lagopus* (Buller, 1924).

During the early stages of hymenium development it is noticeable that adjacent basidia arise at the apex of sister branches from a parent hypha. This branching occurs at a distance away from the young hymenium and the intermediate cells give rise to subhymenial branches and to paraphyses. Paraphyses develop as outgrowths of sub-basidial cells during stage 2. Assuming that the chromophily of the tissue represents the distribution of carbohydrates, then at stage 1 the carbohydrate accumulation is observed in basidia and at stage 2 in the subhymenium and particularly the emerging paraphyseal branches (Fig. 13.5). This distribution appears to reflect the zone of the lamella which is in active 'extension' growth. In *C. micaceus* and *C. lagopus*, Chow (1934) observed a chromophilic pattern analogous to that found here in *C. cinereus* but concluded that basidia grew from the subhymenium and into a layer of paraphyses. We do not agree with this interpretation, believing that our observations show quite clearly that paraphyses emerge *from* sub-basidial cells and insert *into* a layer of young basidia. Our identification of the cells which insert into the existing layer as paraphyses is the basis of our identification of the cells that differentiate from the original palisade as 'poorly differentiated basidia'. Pukkila (see Chapter 22) puts the commencement of paraphyseal insertion at about the time that meiosis starts; with metaphase I of meiosis being about coincident with the start of the main phase of paraphyseal expansion. Statistical analysis of the geometrical relations between hymenial cells suggests that basidial numbers do not increase after the gill lamella is formed. About 60% of the paraphyseal population insert in this initial wave at about the time of meiosis, the rest inserting as gill maturation proceeds.

Elongation and enlargement of the gill is clearly dependent on this increase in the paraphyseal cell population, but more especially on an enormous increase in the constituent cell volumes. Paraphyses expand as much as 2.5 times their original volume, this expansion accounting for half the width of the gill, and most probably for the same proportion of its length. Basidial expansion accounts for the remaining increase in size. The ingress of water into the hymenium elements produces much vacuolation (Moore *et al.*, 1979). The exact identity of an osmoticum has not been established but, as discussed earlier, amplification of urea cycle activity, occurring specifically in the cap tissue, could contribute to such a force (Ewaze *et al.*, 1978).

The determination of developmental pattern

So far, although the direction of gill formation has been suggested, no attempts have been made to explain how this spatial and temporal sequence might be achieved. However, some discussion of a more general nature is in order. The tissue has an initial hyphal mass which is relatively homogeneous, but it then alters to form a well-defined spatial pattern. Some kind of specification has occurred. The specification of development occurs along two axes of polarity. These are from base to apex, and from the inner to the outer surface. The polarity is a unit vector which determines the direction in which the cells will become organised. The specification of information along these two axes of polarity results in the formation of a palisade layer and thus precipitates the establishment of basic tissue domains and the pattern of the cells within those domains.

The most popular ideas concerning the mechanism of polarity come from work on animal systems in which polarity is regarded as being determined by the concentration gradient of a substance, the so-called 'morphogen'. According to Wolpert (1969), this information is positional information and 'largely determines with respect to the cell's genome and development history the nature of its molecular differentiation'. Models of the regulation of cell differentiation have been analysed by Meinhardt & Gierer (1974) who illustrate how two-dimensional patterns very similar to those observed in the agaric hymenium may be generated in response to activators and inhibitors capable of diffusing through the tissues.

The concepts of positional information and polarity axes are capable of providing a basic understanding as to how the gill pattern develops, the specification of positional information being determined by the two axes of polarity. We presume that hyphal cells interpret this positional information and differentiate into palisades. These are the fundamental building blocks of the hymenium and differentiate initially into basidia and cystidia. The differentiating agaric gill seems, on the face of it, to be an ideal candidate for interpretation along the lines of a diffusion gradient theory. The gill *does* have two developmental axes to which differentiation might be referred, and the hymenial cells *do* become positionally differentiated in a manner apparently analogous to epidermal cell layers (the usual classic examples) in higher plants and animals. There are problems, though. We are dealing with a structure whose construction is based on the hyphal organisation. As Read (1983) has put it, in the fungi '...cellular polarity resulting from polarised differentiation of the cytoplasm and/or the cell wall has only been found in hyphal like elements'. Dependence on hyphal organisation is well illustrated by the above description of the origin of paraphyses as branches formed beneath the hyphal tip cell which is differentiating into a basidium (other examples are mentioned by Reijnders & Moore: Chapter 27). Differentiation of the hymenium, therefore, which has the appearance and function of a plate-like layer of cells, owes a great deal to the *linearised* differentiation of the components which eventually come to comprise that cell layer. Much of the final pattern observable in the mature hymenial cell layer thus arises as a consequence of the sequential differentiation of compartments in the hyphae which terminate in the young hymenium. Nevertheless, some definition of positional significance in the hymenial layer must be involved since the basidium–paraphysis relation is an organised one, and cystidia certainly differentiate at regular intervals. In a broader sense, gill lamellae are organised at a constant distance from one another (Burnett, 1968) and the formation of other domains in the cap are also organised in ways that imply coordination of a spatially dispersed cell population may be achieved by the dissemination of regulatory signals.

This highlights another problem which exists in the application of currently favoured models relating to the definition of positional information. All of these models have been developed primarily from studies of differentiating animal tissues, though they can be applied to at least some higher plant systems. All of the models which have been developed depend on the existence of lateral cytoplasmic communications between cells in the differentiating tissue which extend over many cell diameters. Although the ultrastructure of many fungal structures has been carefully studied, there appears to be no evidence for any form of lateral communication between neighbouring hyphal compartments other than via lateral anastomoses. Nothing like gap junctions or plasmodesmata has ever been reported in structures which have arisen by hyphal aggregation. This does not have disabling consequences for theoretical treatment of tissue pattern formation, but it does mean that a cytoplasmic route for coordinating regulatory signals might be excluded. Any regulatory molecule must be excreted across the membrane and wall of the 'sending' cell and absorbed across the wall and membrane of the 'responding' cell. This implies that membrane- and wall-associated processes, and perhaps transport through an extracellular matrix (Williams *et al.* : Chapter 18), may be the rate-limiting steps in the determination of fungal tissue organisation.

Acknowledgements. We thank the SERC for a Research Grant during the tenure of which some of the work reported here was carried out, and Dr N. D. Read for constructive criticism of this chapter.

References

Al-Gharawi, A. & Moore, D. (1977). Factors affecting the amount and the activity of the glutamate dehydrogenases of *Coprinus cinereus*. *Biochimica et Biophysica Acta*, **496**, 95–102.

Atkinson, G. F. (1906). The development of *Agaricus campestris*. *Botanical Gazette*, **42**, 241–69.

Atkinson, G. F. (1914). The development of *Agaricus arvensis* and *A. comtulus*. *American Journal of Botany*, **1**, 3–22.

Atkinson, G. F. (1916). Origin and development of the lamellae in *Coprinus*. *Botanical Gazette*, **61**, 89–103.

Brefeld, O. (1877). *Botanische Untersuchungen über Schimmelpilze*, vol. 3. Leipzig: Felix.

Buller, A. H. R. (1924). *Researches on Fungi*, vol. 3. London: Longman, Green & Co.

Burnett, J. H. (1968). *Fundamentals of Mycology*. London: Edward Arnold.

Buswell, J. A., Ander, P. & Eriksson, K.-E. (1982). Ligninolytic activity and levels of ammonia assimilating enzymes in *Sporotrichum pulverulentum*. *Archives of Microbiology*, **133**, 165–71.

Chow, C. H. (1934). Contribution à l'étude du développement de coprins. *Le Botaniste*, **26**, 89–233.

De Bary, A. (1887). *Comparative morphology and biology of the fungi, mycetozoa and bacteria*. Oxford: Oxford University Press.

Elhiti, M. M. Y., Butler, R. D. & Moore, D. (1979). Cytochemical localization of glutamate dehydrogenases during carpophore development in *Coprinus cinereus*. *New Phytologist*, **82**, 153–7.

Ewaze, J. O., Moore, D. & Stewart, G. R. (1978). Co-ordinate regulation of enzymes involved in ornithine metabolism and its relation to sporophore morphogenesis in *Coprinus cinereus*. *Journal of General Microbiology*, **107**, 343–57.

Hoffman, H. (1860). Beiträge zur Entwicklungsgeschichte und Anatomie der Agaricinen. *Botanische Zeitung*, **18**, 389–404.

Levine, M. (1914). The origin and development of lamellae in *Coprinus micaceus*. *American Journal of Botany*, **1**, 343–56.

Meinhardt, H. & Gierer, A. (1974). Applications of a theory of biological pattern formation based on lateral inhibition. *Journal of Cell Science*, **15**, 321–46.

Moore, D. (1981). Evidence that the NADP-linked glutamate dehydrogenase of *Coprinus cinereus* is regulated by acetyl-CoA and ammonium levels. *Biochimica et Biophysica Acta*, **661**, 247–54.

Moore, D. & Ewaze, J. O. (1976). Activities of some enzymes involved in metabolism of carbohydrate during sporophore development in *Coprinus cinereus*. *Journal of General Microbiology*, **97**, 313–22.

Moore, D., Elhiti, M. M. Y. & Butler, R. D. (1979). Morphogenesis of the carpophore of *Coprinus cinereus*. *New Phytologist*, **83**, 695–722.

Oberwinkler, F. (1982). The significance of the morphology of the basidium in the phylogeny of basidiomycetes. In *Basidium and Basidiocarp*, ed. K. Wells & E. K. Wells, pp. 9–35. New York: Springer-Verlag.

Pinto-Lopes, J. & Almeida, M. G. (1970). '*Coprinus lagopus*' a confusing name as applied to several species. *Portugaliae Acta Biologica (B)*, **11**, 167–204.

Raju, N. B. & Lu, B. C. (1970). Meiosis in *Coprinus*. III. Timing of meiotic events in *Coprinus lagopus* (*sensu* Buller). *Canadian Journal of Botany*, **48**, 2183–6.

Read, N. D. (1983). Structural features of multicellular differentiation in the higher fungi. *Abstracts of the Third International Mycological Congress, Tokyo*, p. 603.

Reijnders, A. F. M. (1948). Etudes sur le développement et l'organisation histologique des carpophores dans les Agaricales. *Recueil Travaux Botanique de la Néerlande*, **41**, 213–396.

Reijnders, A. F. M. (1963). *Les Problèmes du Développement des Carpophores dans les Agaricales et de quelques Groupes Voisins*. The Hague: Junk.

Reijnders, A. F. M. (1979). Developmental anatomy of *Coprinus*. *Persoonia*, **10**, 383–424.

Schmitz, J. (1842). Mykologische Beobachtungen, als Beiträge zur Lebens- und Entwicklungsgeschichte einiger Schwämme aus der Klasse der Gasteromyceten und Hymenomyceten. *Linnaea*, **16**, 141–215.

Spurr, A. R. (1969). A low-viscosity epoxy resin embedding medium for electron microscopy. *Journal of Ultrastructural Research*, **26**, 31–43.

Stewart, G. R. & Moore, D. (1974). The activities of glutamate dehydrogenases during mycelial growth and sporophore development in *Coprinus lagopus* (*sensu* Lewis). *Journal of General Microbiology*, **83**, 73–81.

Sundberg, W. J. (1978). Hymenial cytodifferentiation in basidiomycetes. In *The Filamentous Fungi*, vol. 3, ed. J. E. Smith & D. R. Berry, pp. 298–314. London: Edward Arnold.

Wolpert, L. (1969). Positional information and the spatial pattern of cellular differentiation. *Journal of Theoretical Biology*, **25**, 1–47.

14

Determination of the initial steps in differentiation in *Coprinus congregatus*

IAN K. ROSS

Department of Biological Sciences, University of California at Santa Barbara, Santa Barbara, California 93106, USA

Introduction

The structure of the multicellular fruit bodies of the Agaricales has been studied in great detail. They are of commercial interest (Flegg: Chapter 24; Wu: Chapter 25) and may also represent a means of investigating differentiation and integration in multicellular organisms. Much of the work on fruit body development starts with the primordium (see Manachère, 1980) because this is the first visible indication of fruiting, but to find the events regulating fruit body formation one must look much earlier than this. In general, agarics produce fruit bodies on secondary mycelia which may be dikaryon, diploid homokaryon, or heterokaryon (Anderson & Ullrich, 1979; Ullrich, 1973; Viigalyse & Miller, 1983). Although homokaryotic fruiting has been reported, this is the exception and merely emphasises the fact that developmental events are controlled by genetic events; if the genes are expressed, by whatever trigger, developmental events will follow.

In the basidiomycetes, there is no morphological differentiation of gametes as there is in animals, most plants, and many lower fungi. Higher fungi have only vegetative, somatic, hyphal fusions. The sequence of events that is assumed to cause developmental genes to be expressed is the formation of the secondary mycelium by the somatic fusion of two haploid homokaryotic mycelia or cells. The act of somatogamy sets the stage for subsequent differentiation. However, according to Wessels' group (Zantinge, Dons & Wessels, 1979; Zantinge, Hoge & Wessels, 1981), there are no major changes in mRNA populations in the transition from homokaryon to dikaryon, nor are there many protein changes that could not be accounted for by post-translational modifications.

Where, then, are the gene products required for differentiation? Their absence might be surprising if one considers the transition from homokaryon to dikaryon to be an act of differentiation. If, however, somatogamy is cell fusion, and the dikaryon is regarded as a growing fusion cell (or zygote, if a diploid is formed), new gene expression is not needed to maintain the filamentous, apically extending, branching habit of the thallus, whether homokaryon or dikaryon. Some small changes may be required to cause modification of wall lysing and synthesising enzymes to obtain the clamp connection, but such would be few and difficult to find. To use an animal analogy, after sperm and egg fusion to form a zygote, the requirement for the expression of new genes comes later, at gastrulation – most animal zygotes can divide to form morulae or blastulae without any new mRNAs being formed. In agarics, the formation of primordia may be equated with gastrulation (Ross, 1979); it is here that new gene expression is required and found (Hoge, Springer & Wessels, 1982; Hoge *et al.*, 1982; Dons *et al.*, 1984).

It is unusual to find a dikaryon capable of producing primordia immediately after formation or transfer to fresh medium. A period of growth is required before the mycelium is competent to initiate differentiation (Ross, 1982*b*). This attainment of competence, or maturation, is essentially the time required for the accumulation of the molecular apparatus necessary for the cell to receive and to respond to stimuli. As shown by Axelrod, Gealt & Pastushok (1973), the attainment of competence is a genetically controlled event and mutants may be obtained with longer or shorter times of attainment. Once a mycelium has attained developmental competence, it can receive and respond to external stimuli; in fact, it is the first ability to respond that delimits the attainment of maturity in this context.

The first questions in the search for initial events in differentiation are: what stimulus is required, at what age is a mycelium able to respond to that stimulus, and what parts of the mycelium are able to respond? A second series of questions would ask what is the molecular apparatus that receives the stimulus and how is this reception translated into genomic activity?

There are at least three approaches to the molecular biology of differentiation in higher fungi. One, elegantly conducted by Wessels' group, is to go into the genome and isolate genes that are expressed differentially during developmental stages, as in their recent cloning of a gene that is expressed only during primordium formation in *Schizophyllum commune* (Dons *et al.*, 1984; see also Chapter 21). This approach uncovers genes, and their protein gene products, but does not necessarily indicate their function and how they are regulated. Given adequate transforming systems, it should eventually be possible to sequence such genes, insert man-made mutations, replace them in the organism, and see what malfunction occurs. A second approach, used with effect by Uno & Ishikawa, for example, is to study changes in the biochemistry of the mycelium after a developmental trigger has been received. In this way, Uno & Ishikawa (1982), showed that cyclic AMP is involved in events following light reception in *Coprinus cinereus*, possibly as an activator of protein kinases. Another approach is to delimit as accurately as possible the actual number, kind and timing of steps in the differentiation process, to determine which are genetically controlled and which are epigenetic, and to uncouple these steps so that one will not automatically trigger the next. In this way, gene activation initiated by a step can be isolated in space and time and any genes purified or cloned as a result of that step-trigger may automatically be assumed to be part of that step.

Another reason for attempting to find the initial events in differentiation is that they are probably pathways common to several taxa and the genes regulating these pathways may represent conserved sets of genes reflecting phylogenetic relationships.

Earlier reports of the localisation of fruit body initiation in *Coprinus congregatus* (Ross, 1982*b*; Durand, 1983) had shown that for this fungus, as with any others, light and nutrient depletion were the two primary stimuli for the induction of primordia (see Moore, Elhiti & Butler, 1979). Neither of these stimuli was effective unless the mycelium was at least 3 days old, and they were only effective if the light was received by a young, growing, peripheral growth zone. Subsequent work (Ross, 1982*c*) suggested that phenoloxidase may be involved in part of the attainment of competence and therefore may be implicated in differentiation regulation (see Leslie & Leonard, 1979; Wood, 1980; Leatham & Stahman, 1981). The description of a layer of specialised thick-walled cells that forms under the agar surface after light induction but prior to primordium formation provided another marker for investigating the initial events in differentiation (Henderson & Ross, 1983; Henderson, Elliott & Ross, 1983). The work to be described below indicates that there are at least two light-requiring steps prior to the first appearance of primordia, that the second of these steps not only establishes the developmental pathway to be followed but also causes inhibition of subsequent initiation, and that both induction and inhibition kinetics are genetically controlled and vary among different strains.

Two terms are used in this study that may require definition here: *zones* and *fixation*. A zone represents the area capable of receiving or competent to receive and to respond to an initial stimulus, in this case the first light exposure. In our hands, a zone corresponds to one day's growth of a Petri dish culture, so that a competent zone is equivalent to the peripheral growth zone, the zone of youngest hyphal tips. Fixation is the activation of the initial events in developmental programming which commit the cell to a particular developmental pathway. This definition does not specify what the initial events are, it merely postulates that they exist and that their activation is required for a cell to become committed. A fixed cell is committed to a predictable developmental pathway but may not necessarily enter that pathway then, or in the future. Only if other, subsequent, conditions are favourable, can the fixed cell actually enter that developmental pathway. Such cells are morphologically indistinguishable from other cells, they show no signs of having become committed, and may remain in that state with no overt change in morphology for several days.

Light induction: number of steps

Details of the two strains of *Coprinus congregatus* used in this study, one from California and one from Scotland, have been described by Ross (1982*a*, *b*, *c*) and Evers & Ross (1983). Cultures of the Californian strain, dikaryon (13×16), when exposed to single 12 h periods of white light formed primordia in response to contact with the edge of the Petri dish (Ross, 1982*c*). Primordia only formed after edge contact, and then in a single zone at the site of the peripheral growth zone at the time of the light exposure (Fig. 14.1). If, instead of a single 12-h exposure, shorter and multiple exposures were given, multiple zones of primordia could form (Fig. 14.1). The results depended on the number of light exposures, their duration and timing. A single exposure to either low ($c.$ 2.5 μE m^{-2} s^{-1}) or high ($c.$ 50 μE m^{-2} s^{-1}) irradiance given for less than 1 h did not result in primordium formation (Fig. 14.1). If a second exposure of similar length was given between 3 and 12 h after the first, primordia formed in the zone given the first exposure and subsequent zones were inhibited (Fig. 14.2). If the period between short light exposures was extended to 24 h or more, multiple zones of primordia could form (Figs 14.1, 14.2). Cells fixed by the first exposure were still able to respond with a period between short light exposures as long as 10 days.

The first exposure to a short light period appears to commit the cells to a developmental pathway but cannot exert any influence on cells in other, subsequent, zones. It requires a second light exposure after at least

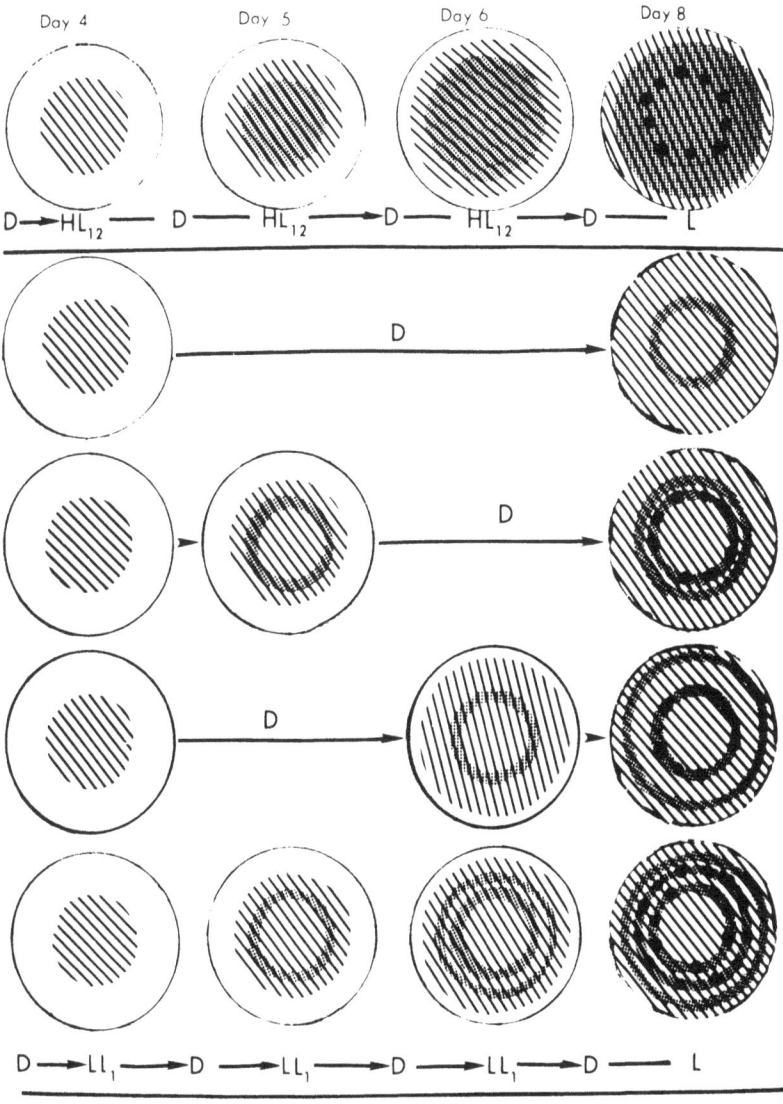

Fig. 14.1. Effect of single and multiple short light exposures on primordium formation. Top: control, exposed to alternate 12 h light/12 h dark.
One zone of primordia forms, the rest are inhibited. Bottom: results of
giving a 1-h exposure on day 4, followed by either none, one or two more
exposures as indicated. Dots represent primordia, dark rings represent
pigment bands, shading represents mycelial growth.

DURATION and TIMING of LIGHT EXPOSURES

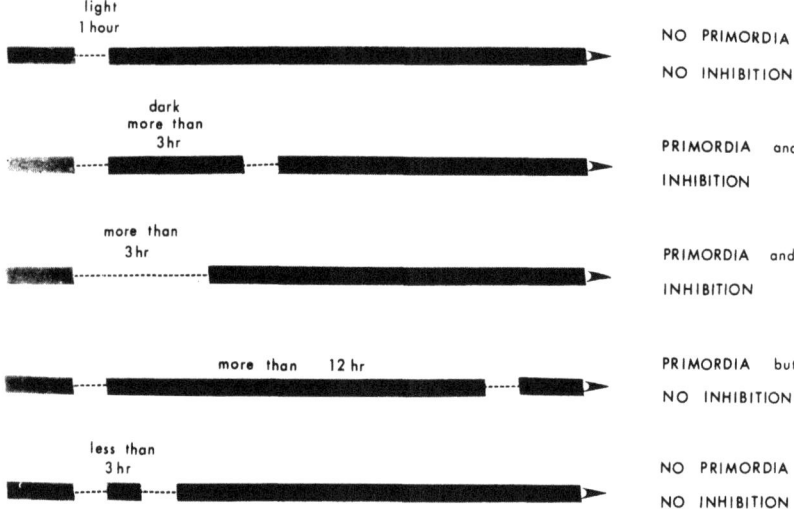

Fig. 14.2. Effect of duration and timing of light exposures on primordium formation. All exposures equivalent to $50\ \mu E\ m^{-2}\ s^{-1}$ for the time indicated.

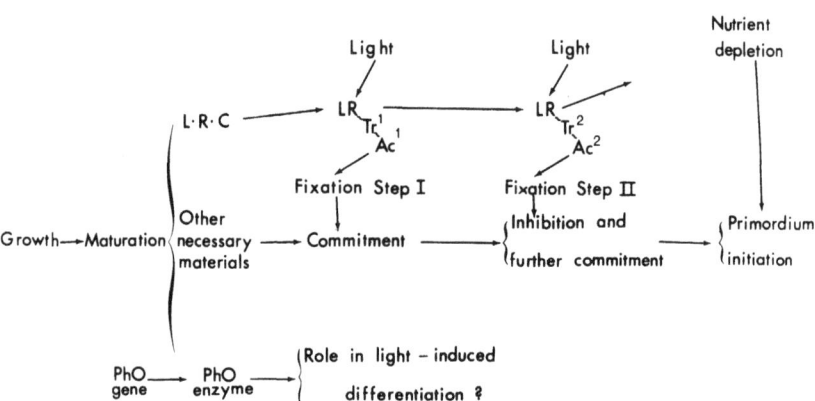

Fig. 14.3. Model of possible interactions involved in primordium initiation. LRC = light receptor complex; LR = light receptor; Tr = transducer (see Fig. 14.7); Ac = activator (see Fig. 14.7); PhO = phenol oxidase. See text for details.

a 3-h delay to cause inhibition of subsequent zones. Although the first exposure can commit the zone of cells to a future morphogenetic event, it requires a second exposure to permit this morphogenesis to occur. This is summarised in Fig. 14.3. The first light exposure activates Fixation Step I, and the second, Fixation Step II. This model is discussed further below.

Light induction: inhibition studies

To determine how extensive the inhibition of subsequent zones was after a full (two-step) induction, the dikaryon (13×16) was grown in an alternate 12 h light/12 h dark regime (L/D) in long 'race tubes' (Fig. 14.4) and in large (150-mm) Petri dishes. In both cases, primordia began to form in zone 3 (the youngest capable of becoming competent, Ross, 1982b) at the time that the peripheral growth zone was 40 mm away from zone 3 (approximately 4 days growth). In the race tubes, zones of primordia formed every 40 mm (Fig. 14.4) throughout the length of the tubes. In 100-mm Petri dishes, the mycelium would only have about 40 mm of radial growth after zone 3 before the edge of the dish is reached. The apparent requirement for edge contact before primordia can form may, therefore, merely be due to the fact that an induced zone can inhibit up to three subsequent growth zones. If the mycelium can grow beyond that distance, subsequent zones are released from the inhibition.

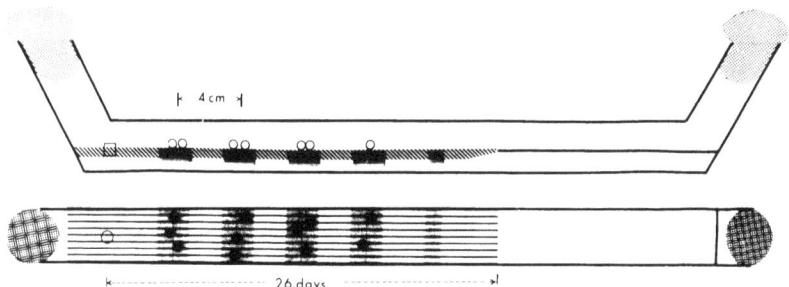

SEQUENTIAL PRIMORDIUM FORMATION IN L_{12} / D_{12} REGIME

IN

RACE TUBES

Fig. 14.4. Effect of a light/dark regime on sequential primordium formation during growth in race tubes. ○ and ● = primordia formed in upper and lower diagrams respectively; □ = point of inoculation in upper, ○ = point of inoculation in lower diagram.

Light induction and inhibition: strain variation

The Scottish strain reacted to light with different timing and results. The Scottish dikaryon (22 × 28), grown under L/D conditions, began forming primordia in zone 2 on the fifth day, long before edge contact. Instead of a single zone forming under L/D regimes (22 × 28), dikaryons formed a broad zone covering three to four growth zones. There was no apparent inhibition of subsequent zones.

Interstrain crosses between cultures 16, 22 and 28 gave intermediate responses. The dikaryons (16 × 22) and (16 × 28) grew somewhat faster than the Californian strain, and much faster than the Scottish strain. Cultures matured and could be induced after only 36 h of growth, a very clear-cut distinction in L/D grown cultures. L/D-grown (16 × 22) and (16 × 28) dikaryons formed multiple zones of primordia, approximately one growth zone apart, suggesting that the inhibitory effect of an induced zone is weak, or can extend only into the next growth zone. Typically, there was a sharp line between the beginning of the first primordium zone and the immature area in the centre. The first zone of primordia extended partly over zone 2 into zone 3. The rings of primordia in zones 4, 5 and 6 were clearly defined. Under similar culture conditions, the Californian dikaryon (13 × 16) would have but one zone of primordia, in zone 3.

Another major difference between the Scottish and the Californian strains is that no matter when the Scottish dikaryons received the first light exposure, previously grown zones did not lose their ability to sporulate, so that a culture exposed on day 5 produced a defined ring of primordia on zone 5, and scattered to dense primordia extending into zone 3–2. Breeding studies are under way to see if these differences are due to single or multiple gene effects.

Phenoloxidase and light reception

The phenoloxidase (PhO) of *Coprinus congregatus* is a membrane-localised enzyme that is not secreted into the medium as are most fungal laccases. The enzyme has a substrate specificity similar to laccases (Ross, 1982c), and cannot oxidise tyrosine or DOPA. Preliminary work has shown that intact protoplasts of *C. congregatus* contain active PhO that can be inhibited by the addition of concanavalin-A (con-A) (Choi & Ross, unpublished). Removal of the con-A with α-methyl mannoside restores activity of the PhO. Fractionation of protoplasts or intact cells in Percoll gradients reveals that the phenoloxidase activity co-bands with the membrane fractions (Swords & Ross, unpublished), while treatment of the gradients with digitonin causes disruption of the membrane fractions and a distribution of the PhO activity throughout the gradient. This association of the PhO with the membrane may be significant in view of the suspected membrane location of fungal blue-light receptors.

It is possible to predict which cells of a dark-grown culture will respond to light merely by assaying the levels of PhO in the different growth zones (Ross, 1982c). Under normal nutritional conditions, PhO levels were highest in the youngest growth zone, the zone of developmental competence. As the mycelium grew, this zone moved outwards with the growing hyphal tips (Ross, 1982b). Once a zone had received light, it became pigmented (melanised) prior to the formation of primordia.

When cultures were grown on low-nutrient media (50% normal), only one zone usually formed and it did not move out with the growth of the mycelium. This can be correlated with the PhO activity in that the zone that ultimately became the only one to form primordia was also the only one that attained a high (threshold) level of PhO activity (Ross, 1982c). No other zone of the mycelium, whether produced before or after this zone, attained similar levels of PhO activity.

The PhO level in zones fully induced with 12 h light dropped to basal

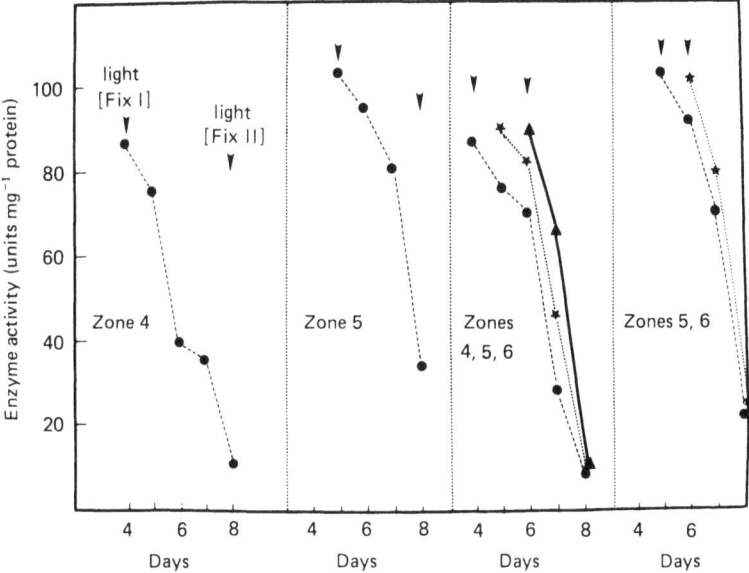

Fig. 14.5. Phenoloxidase levels after Fixation Step I light exposure. Far left panel shows the decrease in activity is independent of receipt of the second Fixation Step exposure.

levels long before primordia began to form, suggesting that PhO could not be involved directly in the actual morphogenesis of the primordia (Ross, 1982c). To determine the relation of PhO activity to the two Fixation Steps, cultures of dikaryon (13 × 16) were exposed to a 5-min light exposure after varying times of growth (Fig. 14.5). In all cases, the levels of extractable PhO began to drop after the first light exposure. Since cells can remain in Fixation Step I for several days, it was possible to wait until the PhO levels had dropped almost to zero before giving the second Fixation Step light exposure. Even though PhO levels at the time of the second exposure were extremely low, full fixation, inhibition and subsequent primordium formation occurred. If PhO is involved in the differentiation process, it must be at the first light receptor step, Fixation Step I.

The first visible result of light induction is the formation of a pigmented area extending under the induced zone, accompanied by the formation of the special thick-walled cells (Ross, 1982b; Henderson & Ross, 1983). Primordia form only above this pigmented area. The pigmented area is the result of PhO activity and develops soon after the levels of PhO begin to drop following a light exposure. It must be emphasised that prior to exposure to light there is no visible pigmentation of the mycelium, the youngest parts of the mycelium (i.e. the only parts competent to receive light) have the highest level of extractable PhO activity, and melanisation does not begin until after the induced cells have become developmentally programmed.

Nutritional studies

Studies designed to uncouple steps in differentiation by nutritional means have just commenced. As shown in Fig. 14.6, glucose has an interesting effect. Contrary to the results obtained by Manachère (1970), Robert (1978) and Durand & Jacques (1983) with their strain, the Californian strain of *C. congregatus* does not fruit well in the presence of glucose and does not fruit on the malt extract medium of Manachère (1970). It fruits best on a soluble starch medium (Difco Emerson's YpSs). If glucose is used as the sole carbon source, primordia do not form. In 1% or 2% glucose alone, PhO levels are lower than normal and fall rapidly after light exposure, and pigmentation is precocious. In 0.25% glucose alone, PhO levels are very high, but the level does not fall after light exposure, pigmentation does not occur, and primordia do not form. When soluble starch is the sole carbon source, all behaviour patterns are normal. If 1% or 2% glucose is added to a 1% soluble starch medium, primordia do not form, pigmentation is precocious, PhO does not reach normal levels, but the fall in activity is normal. If 0.25% glucose is added to 1% starch, PhO levels are normally high, pigmentation occurs with a drop of enzyme activity level, and primordia form, though only to 10% of normal numbers.

The effects of different nitrogen sources on the differentiation programme are being investigated. In Emerson's YpSs medium, nitrogen is supplied as 4 g yeast extract per litre. When the yeast extract was reduced to 1 g l⁻¹ and supplemented by 1 g l⁻¹ of asparagine, growth and zone inhibition were normal and primordium formation was enhanced. When sodium nitrate was substituted for asparagine, all growth rates, numbers of primordia and inhibition effects were normal, but the inducible zone did not move out with the peripheral hyphae. Primordia and pigment only formed in zones 2–3, no matter when the culture was exposed to light. This response is completely different from that of the asparagine-grown cultures. In fact, this is the same apparent response that the mycelia make to low nutrients (Ross, 1982*b*). The asparagine and the sodium nitrate responses occur in both 10 g l⁻¹ and 15 g l⁻¹ of soluble starch, so it is doubtful if the carbon in the 1 g l⁻¹ of asparagine is contributing to the different response by affecting the carbon:nitrogen ratio.

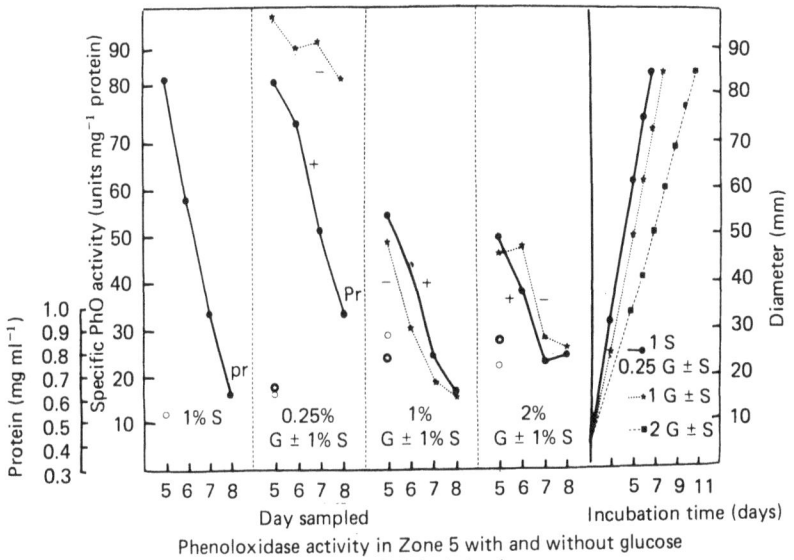

Phenoloxidase activity in Zone 5 with and without glucose
after Zone 5 light induction

Fig. 14.6. Phenoloxidase activity in zone 5 with and without glucose after zone 5 has been exposed to 12 h light. See text for details. G = glucose; S = starch; + = glucose+starch; − = glucose alone; Pr,pr = primordia formed.

Different nitrogen sources, supplied at molar equivalents of nitrogen at the same pH, cause quite variable results. Although the reasons for this variability are not known, they do offer a means of controlling the extent and location of inducible zones, a point that will be of considerable value in later analyses.

Light reactions of a second *Coprinus* species

A second species of *Coprinus*, as yet unidentified and termed Cd, has been studied. This species produces copious quantities of an extracellular phenoloxidase. Unlike *C. congregatus*, Cd is completely growth-inhibited by light, yet still requires light for sporulation. The response of Cd hyphae to light exposure of less than 3 h was complete cessation of growth. Such cultures formed primordia (in the zone of peripheral hyphae at the time of light exposure) several days later, after edge contact. In this respect, Cd is similar to *C. congregatus*. However, if light inhibition of growth was maintained for 6–9 h, primordia began to form only a few hours after dark growth recommenced, even though the mycelium had several days of growth before reaching the edge of the dish. Superficially, it would seem that Cd is different from *C. congregatus* in its reaction to light, but an analysis of the sequence of events suggests that both species have very similar initiation regulation.

Conclusions

The early events in fruit body initiation have been examined. It is very important to delimit what happens and how many discrete steps are involved before morphogenesis actually occurs. As Turian (1983) has pointed out, differential gene expression, although important in cell differentiation, does not of itself explain how the initial commitment is made. Only by knowing the exact sequence of biochemical and molecular steps will it be possible to assign specific genes to specific events, and to interpret mutant studies.

These studies have shown that many complex molecular interactions occur long before a morphological indication of development is visible. Anything that affects these interactions will modify or prevent the normal course of differentiation. There are several points to consider here. One is the accumulation of the molecular apparatus that will receive the light stimulus and transduce it into genomic and cellular activity. This accumulation corresponds to the attainment of competence and in *Coprinus congregatus* requires 2–3 days depending on the strain, suggesting a genetic regulation. The composition of this molecular accumulation is still obscure. Durand & Jacques (1982) suggested that because all the light responses, enhancing and inhibitory, in *C. congregatus* had exactly the same action spectrum, they were probably all regulated by the same light receptor. This further suggests that the differing effect of light falling on the same light receptor must be due to different components of the light receptor complex (LRC). The idealised model of an LRC shown in fig. 14.7 implies that the actual light receptor molecule remains in the membrane, and that either the transducer or the activator changes between successive developmental events. This concept fits the evidence presented on the two Fixation Steps, indicated in the model shown in Fig. 14.3. Maturation would be in part the insertion into the membrane of the first LRC, with nothing overt happening until light is received. Once light has been received a series of events would begin that would lead to the activation of Fixation Step I and the disruption of LRC-1. The reformation of an active LRC (LRC-2) is postulated to require gene activity and the formation of new LRC members, either the transducer or the activator, or both. The new gene

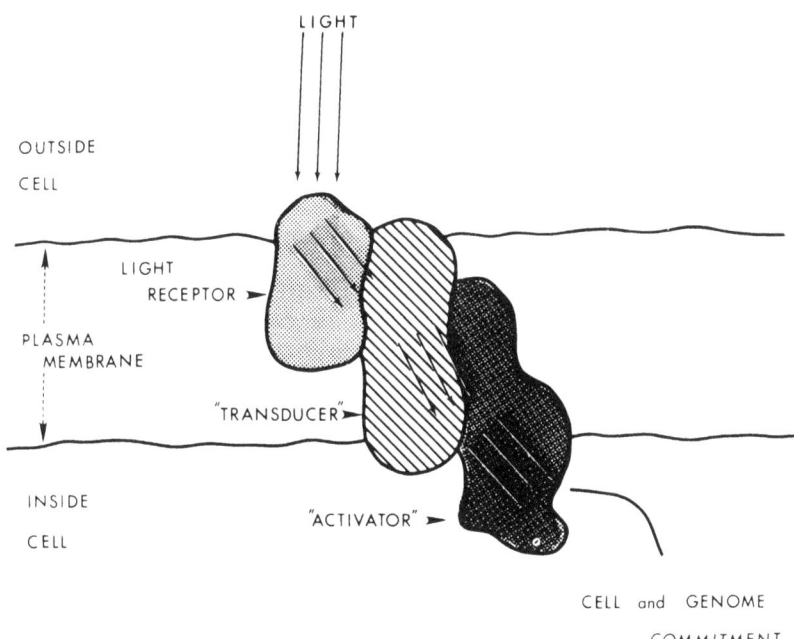

Fig. 14.7. Idealised model of a light receptor complex involved in primordium initiation.

activity is suggested by the 3-h time period between effective short light exposures and by Durand's (1983) finding that RNA synthesis inhibitors and cycloheximide added to cultures before or after photoinduction inhibited the formation of primordia.

The role of phenoloxidase (PhO) in this system is not clear, but must either be at the first Fixation Step, or not at all. Pigmentation, the overt sign of PhO activity, does not occur in young mycelia unless light is received. Then pigmentation occurs prior to the formation of visible primordia, and concurrently with the formation of thick walled cells in the substrate. Primordia are found only over pigmented areas, yet by the time primordia form, the actual levels of PhO are down to almost zero. It may be argued that PhO plays a secondary role, that of the formation of pigment and cells necessary for primordial development by a parallel pathway also triggered, or activated, by the first Fixation Step, but is not involved in differentiation initiation. On the other hand, the consistent correlation of PhO levels and the predictable behaviour of the cells prior to light exposure could also be taken as evidence that PhO is involved in the molecular apparatus responsible for the first Fixation Step, and that the pigment and thick-walled cells are merely fortuitous and secondarily useful end products of the mechanism of removal of a highly active enzyme from the cell. This suggests the possibility that PhO may be the activator

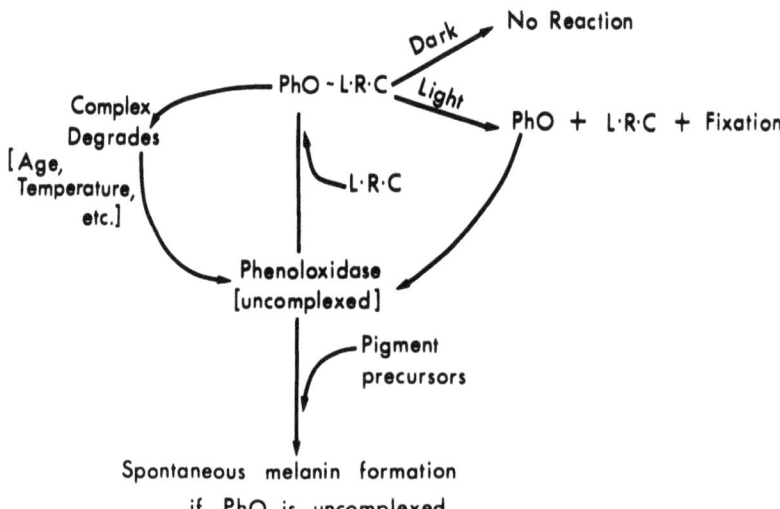

Fig. 14.8. Postulated roles and behaviour of the phenoloxidase of *Coprinus congregatus* in light activation and pigmentation.

in the LRC, and that the complexing of PhO with the LRC prevents any oxidising activity (see Fig. 14.8). The membrane location of the PhO supports this contention. The disruption of the LRC by the reception of light could be sufficient to activate PhO, resulting in the presence of an enzyme capable of forming free radicals, quinones and other highly reactive compounds in the cell. It may be to the cell's advantage to get rid of such an enzyme, once it has performed its initial triggering activities, by incorporating it into non-reactive pigment. The cytotoxicity of melanin intermediates is well known. Miranda, Botti & DiCola (1984) have shown that early intermediates in tyrosinase-mediated melanin synthesis can bind directly to DNA *in vitro*. This suggests that PhO reaction products have a mechanism for regulation of gene expression.

This speculation will not be resolved until both detailed mutant studies are done, and the enzyme is purified and immunological localisation studies performed. The analysis so far has shown where not to expect Pho involvement, and has led to the discovery of the various steps that do exist.

There is one more important aspect of Fixation Step I. The reception of the first light exposure can commit a cell to a future pathway and can remain dormant in that cell for several days. After the first light exposure, it is postulated that LRC-1 disrupts, genes are activated that make the new components of LRC-2, and PhO is activated. The interesting point is that the new LRC remains in the original cells and does not move out with the new hyphal growth as LRC-1 must. It is known that if only Fixation Step I is activated, subsequent growth zones may also become fixed, but if both fixation steps are activated in a single zone, subsequent zones are inhibited. Consequently, even though Fixation Step I can result in the formation of LRC-2, this act does not prevent the formation/activation of LRC-1 in the next growth zone. But if Fixation Step II is activated, this can prevent the formation or activation of LRC-1 in the next zone.

The inhibition of subsequent zones by a Step II-fixed zone appears to be under genetic control as indicated by the differential inhibition among Scottish and Californian isolates. This then represents another area for mutation and gene expression studies. Since the critical parameter appears to be the distance of the fixed zone from the next inducible zone, it is possible that fixation may regulate the translocation of glucose to glycogen storage areas that form under the induced zones (Fig. 14.9). This could be of considerable importance because in the Californian isolates, with three zones of inhibition between inducible zones, fruit bodies that develop are uniformly large, whereas, in the Scottish clones, with only one zone of inhibition or less, fruit bodies are uniformly small. By regulating the

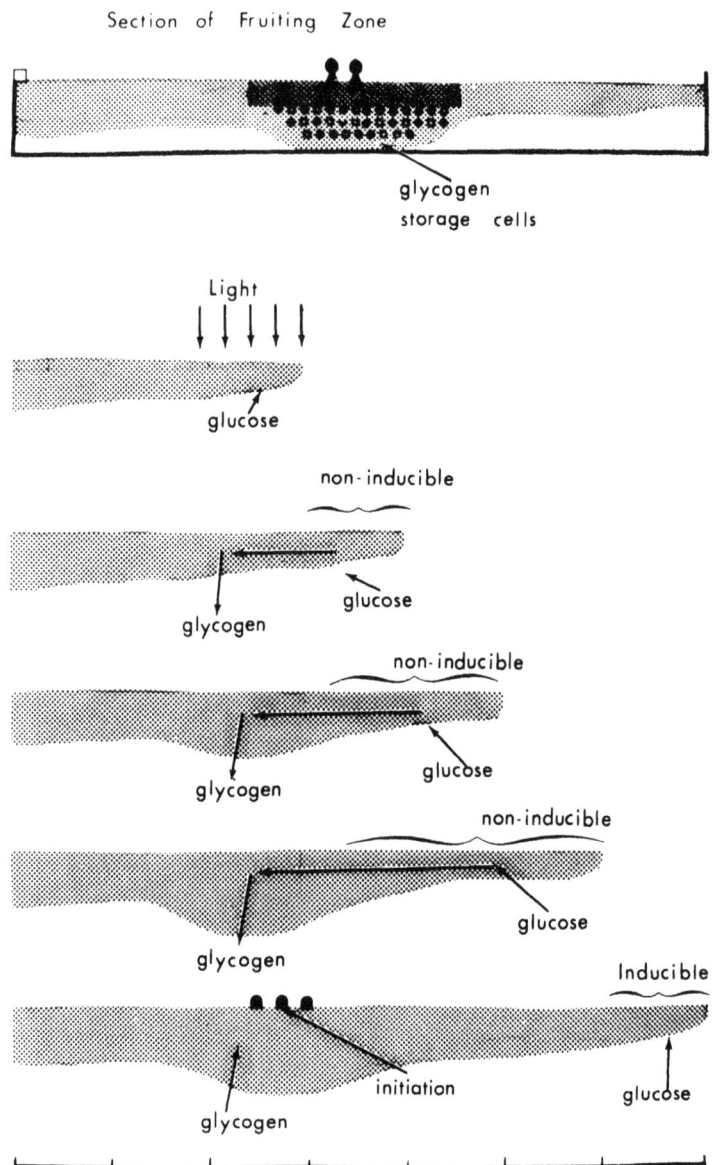

POSSIBLE   TRANSLOCATION   EVENTS

Fig. 14.9. Possible translocation events regulated by an induced zone leading to primordium development in *Coprinus congregatus*. See text for details.

number of zones that are inhibited, the size of the fruit body may be controlled.

The hypothesis is that as long as translocation is occurring, those cells involved in translocation are inhibited from becoming fixed. Once translocation ceases, and the glucose has been stored as insoluble glycogen, the cells of the fixed zone would then effectively become glucose-deficient. Such reduction of glucose could well be the trigger for the cyclic AMP activity reported by Uno & Ishikawa that might lead to the actual morphogenesis of the primordia. This is reflected in the behaviour of the Cd species. Since growth is completely inhibited by light, it is probable that uptake and translocation are also inhibited. If the light exposure is for only for a short time, 3 h or less, fixation may occur, but glucose levels would still be high. If the cessation of vegetative growth goes on for 6–9 h, then glucose levels may drop below the level required to activate cyclic AMP and induce actual primordium formation. Those cells are actually in a situation of glucose deprivation because of the vegetative growth inhibition. It is obvious then, that commitment to the developmental pathway leading to primordium formation can occur long before the total carbohydrates in the medium are exhausted. The initial events, therefore, are independent of exogenous carbohydrates, but are dependent on endogenous amounts.

It must be remembered that the strain of *Coprinus congregatus* used does not differentiate in the presence of free glucose. Yet, starch, the best carbohydrate found for this strain, supplies only glucose. The results with 1% or 2% glucose, precocious and heavy pigmentation, suggest that those levels of glucose were not inhibiting PhO formation, but were inhibiting the postulated complexing of PhO with the LRC, a complexing that prevents pigment formation and keeps extractable levels of PhO high. The results with 0.25% glucose that showed very high PhO levels but no light response or primordium formation, suggest that this level of glucose does not interfere with PhO formation but does affect the LRC in such a way that the PhO is not activated by light, by preventing the synthesis either of the light receptor itself, or of the transducer. The addition of starch to 0.25% glucose results in some fixation and primordium formation. It can be concluded that the degradation of starch, when supplied alone, occurs slowly enough so that glucose levels are not so high as to inhibit fixation. Since starch may also act as an absorbant, it is possible that low levels of glucose in the medium may be absorbed on to the starch and not be freely available.

Two further points deserve comment. First, the primordia that formed on light-inhibited hyphae of species Cd did so in a narrow ring that represented the extreme tips of the peripheral hyphae at the time the light exposure began. This means that the receptive part of the growing thallus is the most actively growing part. As Gooday (1983) has pointed out, the last 100 μm of the hyphal tip is an extremely dynamic area, supposedly functioning only as the site of apical growth extension. It now seems probable that this area of great lytic and synthetic enzyme activity is also the site of the light receptor complex and its attendant functions. Secondly, the different modes of inhibition of subsequent zones by induced zones in the Californian and Scottish clones has important consequences on the eventual fruit body maturation. Californian dikaryons form fruit bodies from each induced zone in sequence in large plates or race tubes. The Scottish and Cal/Scot dikaryons form fruit bodies simultaneously from any or all induced zones. Subsequent flushes may also come from any zone.

The regulation of the Californian clones seems to be that of distance. Because an induced zone inhibits the next three growth zones, the induced zone will be induced to differentiate primordia and complete maturation before the next induced zone is ready. In the other dikaryons, the inhibitory effect on initiation extends only for one zone, and this does not seem to prevent maturation in different zones. The Scottish and Cal/Scot dikaryons seem to represent the situation found in *Agaricus bisporus* where primordia form after casing and then mature in sequential flushes. In the Scottish and Cal/Scot dikaryons all the primordia that form are present on a Petri dish before any begin to mature into a fruit body. Subsequent flushes arise from pre-existing primordia. Maturation of fruit bodies in these dikaryons thus represents a selection, from a large number of primordia, of those that will develop at a particular time, but does not prevent some of those remaining from developing later.

This suggests that the regulation of the initiation of primordia is separate from the regulation of which primordia will subsequently develop. Although primordium formation is essential for later fruit body development, the two processes are not necessarily regulated by the same mechanisms. Manachère and Robert (see Manachère *et al.*, 1983, for review) have shown that excised maturing fruit bodies placed on plates with primordia will inhibit development of those primordia. This work shows that the initiation of primordia (the completion of the initial events in commitment to differentiation) can inhibit the ability of subsequent cells to initiate primordia.

The work described here has set the stage for a detailed molecular analysis of the initial events in differentiation in *Coprinus congregatus*.

Knowing the sequence of the various steps in the developmental pathway will make it possible to seek mutants that affect specific steps. Screening for mutants of developmental processes, whether in homokaryotic fruiters or normal dikaryons, is a time-consuming process. By knowing the timing and the location of particular events, appropriate cultural and environmental triggers may be applied that will separate in space and time the activation of specific genes. It should therefore be somewhat easier to identify a particular mutant or a particular mRNA with a specific developmental process.

Acknowledgements. I would like to thank Lorraine E. Henderson for SEM work, Choi Hyoung Tae, Katherine M. M. Swords and Jeffrey Holmes for their work on phenoloxidases and protoplasts, and Gail Turner, Scott Benson, Keith Rayburn, Pam House, Laura Sugino, Rebecca Wilks, Gabrielle Adams and Robert Marks for their technical assistance in all the tedious parts of the research, and for their continued enthusiasm and support. This research was funded in part by the Faculty Research Committee, UCSB, by the College of Creative Studies, and by President's Undergraduate Research Fellowships to G. Turner, L. Sugino, R. Wilks, G. Adams and R. Marks.

References

Anderson, J. B. & Ullrich, R. C. (1979). Biological species of *Armillaria mellea* in North America. *Mycologia*, **72**, 402–14.

Axelrod, D. E., Gealt, M. & Pastushok, M. (1973). Gene control of developmental competence in *Aspergillus nidulans*. *Developmental Biology*, **34**, 9–15.

De Vries, O. M. H. & Wessels, J. G. H. (1972). Release of protoplasts from *Schizophyllum commune* by lytic enzyme preparations from *Trichoderma viride*. *Journal of General Microbiology*, **73**, 13–22.

Dons, J. J. M., Springer, J., De Vries, S. C. & Wessels, J. G. H. (1984). Molecular cloning of a gene abundantly expressed during fruiting body initiation in *Schizophyllum commune*. *Journal of Bacteriology*, **157**, 802–8.

Durand, R. (1983). Effects of inhibitors of nucleic acid and protein synthesis on light-induced primordial initiation in *Coprinus congregatus*. *Transactions of the British Mycological Society*, **81**, 553–8.

Durand, R. & Jacques, R. (1982). Action spectra for fruiting of the mushroom *Coprinus congregatus*. *Archives of Microbiology*, **132**, 131–4.

Evers, D. C. & Ross, I. K. (1983). Isozyme patterns and morphogenesis in higher basidiomycetes. *Experimental Mycology*, **7**, 9–16.

Gooday, G. W. (1983). The Hyphal Tip. In *Fungal Differentiation*, ed. J. E. Smith, pp. 315–56. New York: Marcel Dekker.

Henderson, L. E., Elliott, M. & Ross, I. K. (1983). Sclerotium formation in *Coprinus congregatus*. *Mycologia*, **75**, 738–41.

Henderson, L. E. & Ross, I. K. (1983). Ultrastructural studies of vegetative and fruiting mycelia of *Coprinus congregatus*. *Mycologia*, **75**, 634–43.

Hoge, J. H. C., Springer, J. & Wessels, J. G. H. (1982). Changes in complex RNA during fruit-body initiation in the fungus *Schizophyllum commune*. *Experimental Mycology*, **6**, 233–43.

Hoge, J. H. C., Springer, J., Zantinge, B. & Wessels, J. G. H. (1982). Absence of differences in polysomal RNA's from vegetative monokaryotic and dikaryotic cells of the fungus *Schizophyllum commune*. *Experimental Mycology*, **6**, 225–32.

Leatham, G. F. & Stahman, M. A. (1981). Studies on the laccase of *Lentinus edodes*: specificity, localization, and association with the development of fruiting bodies. *Journal of General Microbiology*, **125**, 147–57.

Leslie, J. F. & Leonard, T. J. (1979). Monokaryotic fruiting in *Schizophyllum commune*: phenoloxidases. *Mycologia*, **71**, 1082–5.

Manachère, G. (1970). Recherches physiologique sur la fructification de *Coprinus congregatus* Bull. ex Fr.: action de la lumière; rhythme de production de carpophores. *Annals des Sciences Naturelles Botanique et Biologique Végetale*, **11**, 1–96.

Manachère, G. (1980). Conditions essential for controlled fruiting of macromycetes – a review. *Transactions of the British Mycological Society*, **75**, 255–70.

Manachère, G., Robert, J.-C., Durand, R., Bret, J.-P. & Fevre, M. (1983). Differentiation in the Basidiomycetes. In *Fungal Differentiation*, ed. J. E. Smith, pp. 481–514. New York: Marcel Dekker.

Miranda, M., Botti, D. & Di Cola, M. (1984). Possible genotoxicity of melanin synthesis intermediates: tyrosinase reaction products interact with DNA *in vitro*. *Molecular and General Genetics*, **193**, 395–9.

Moore, D., Elhiti, M. M. Y. & Butler, R. D. (1979). Morphogenesis of the carpophore of *Coprinus cinereus*. *New Phytologist*, **83**, 695–722.

Robert, J. C. (1977). Fruiting of *Coprinus congregatus*: relationship to biochemical changes in the whole culture. *Transactions of the British Mycological Society*, **68**, 389–95.

Ross, I. K. (1979). *Biology of the Fungi*. New York: McGraw-Hill.

Ross, I. K. (1982*a*). The genetic basis for the meiotic disorder in *Coprinus congregatus*. *Current Genetics*, **5**, 53–6.

Ross, I. K. (1982*b*). Location of carpophore initiation in *Coprinus congregatus*. *Journal of General Microbiology*, **128**, 2755–62.

Ross, I. K. (1982*c*). The role of laccase in carpophore initiation in *Coprinus congregatus*. *Journal of General Microbiology*, **128**, 2763–70.

Turian, G. (1982). Concepts of fungal differentiation. In *Fungal Differentiation*, ed. J. E. Smith, pp. 1–18. New York: Marcel Dekker.

Ullrich, R. C. (1973). Sexuality, incompatibility, and intersterility in the biology of the *Sistotrema brinkmannii* aggregate. *Mycologia*, **65**, 1234–49.

Uno, I. & Ishikawa, T. (1982). Biochemical and genetic studies on the initial events of fruitbody formation. In *Basidium and Basidiocarp*, ed. K. Wells & E. K. Wells, pp. 113–24. New York & Berlin: Springer-Verlag.

Vilgalyse, R. & Miller, O. K. (1983). Biological species in the *Collybia dryophila* group in North America. *Mycologia*, **75**, 707–22.

Wood, D. A. (1980). Production, purification and properties of extracellular laccase of *Agaricus bisporus*. *Journal of General Microbiology*, **117**, 327–38.

Zantinge, B., Dons, H. & Wessels, J. G. H. (1979). Comparison of Poly(A)-containing

RNA's in different cell types of the lower eukaryote *Schizophyllum commune.*
European Journal of Biochemistry, **101**, 251–60.

Zantinge, B., Hoge, J. H. & Wessels, J. G. H. (1981). Frequency and diversity of RNA
sequences in different cell types of the fungus *Schizophyllum commune.*
European Journal of Biochemistry, **102**, 1–9.

15

Production and roles of extracellular enzymes during morphogenesis of basidiomycete fungi

D. A. WOOD

Glasshouse Crops Research Institute, Worthing Road, Littlehampton, West Sussex BN17 6LP, UK

Introduction

The molecular and cellular processes in the production of extra-cellular enzymes by basidiomycete fungi are probably similar to such processes occurring in the production of similar molecules in more well-defined biochemical systems, such as enzyme excretion by mammalian secretory tissues or by bacteria (Byrde, 1982; Priest, 1983). The evidence from these latter systems is that proteins destined for export are synthesised from messenger RNA (mRNA) molecules whose sequences are longer than the final gene product. This mRNA is processed at various stages to produce the final translated gene product (Priest, 1983). At the structural level, mRNA synthesised in the cell nucleus is exported to the endoplasmic reticulum for translation. The translated product then passes via Golgi bodies into storage vesicles and via the cell membrane for externalisation or directly into vesicular bodies which fuse with the cell membrane and then externalise (Lampen, 1974). The evidence from other systems is that all exported mRNAs contain a sequence which codes for a stretch of mostly hydrophobic amino acids, called the signal peptide. This signal peptide allows excreted proteins to pass through membranes where proteolytic processing occurs and the final form of the protein is externalised (Byrde, 1982; Priest, 1983).

The evidence that basidiomycete fungi, or fungi in general, employ a similar system, is based on a consideration of the homologies of cell structure in all eukaryotes and the evolutionary conservation of biochemical pathways and cellular organisation. The knowledge of basidiomycete cell structure, as studied by electron microscope techniques, is now extensive

375

(Beckett, Heath & McLaughlin, 1974; Grove, 1978). These ultrastructural studies have not been matched by corresponding biochemical studies on the mechanisms of excretion of macromolecules by basidiomycetes or other fungi (Cohen, 1980). The importance of such biochemical studies of enzyme excretion by basidiomycete fungi can be justified by the significance of basidiomycetes in biodegradation and nutrient cycling in terrestrial environments (Frankland, 1982; Rayner, Watling & Frankland: Chapter 1). A better understanding of the mechanisms of resource utilisation, coupled with further ecological studies of this important microbial group, is of considerable importance for fundamental studies and for applied studies whose aim is the manipulation of these organisms in natural or man-made environments.

Many basidiomycete fungi in natural environments or in systems where they are commercially exploited grow in lignocellulosic substrates (Montgomery, 1982; Wood, 1984). The outstanding biochemical feature of these types of resource is that the sources of carbon, and in many cases also of nitrogen, are in the form of macromolecules, often in an insoluble form. These substrates will contain lignin, cellulose, hemicellulose and protein, and, for those lignocellulosic substrates previously colonised by microorganisms, the residual biomass of such microorganisms (Fermor & Wood, 1981). To capture and utilise these nutrients requires the export from the basidiomycete hyphae of a range of extracellular enzymes which act to depolymerise such substrates into lower-molecular-weight compounds which the mycelium can then assimilate and utilise for growth and reproduction (Montgomery, 1982; Kirk, 1983). All of the major groups of basidiomycetes, including the agarics, must utilise extracellular enzymes for at least some stages of their life cycle to gain entry to living or dead plant material and to colonise such material successfully (Rayner *et al.*: Chapter 1).

It is known that many basidiomycetes are capable of the simultaneous production of several hydrolytic and oxidative enzymes (King, 1966; Eriksson, 1981; Kirk, 1983; Wood, 1984). The combined action of these enzymes is probably responsible for the bulk of the observed degradation of complex lignocellulosic substrates such as wood, leaf litter or compost. The cultivated mushroom *Agaricus bisporus* has, for example, been shown to be able to excrete the following enzymes: endocellulase, exocellulase, β-glucosidase, laccase, protease(s), xylanase(s), laminarinase, lipase, DNase, RNase, glucosaminidase and *N*-acetyl-muramidase (Wood, 1984). This list is unlikely to be the total capacity of this organism for extracellular enzyme production.

Control and localisation of enzyme production

A fundamental problem in the excretion of extracellular depolymerising enzymes in relation to both basidiomycete nutrition and microbial nutrition generally is that of control of product formation. Regulatory and structural mechanisms must be imposed on the production, excretion and activity of such enzymes to ensure maximal cell economy (Priest, 1983). Since the products of hydrolytic enzyme activity are likely to be growth substrates for competitor microorganisms, the fungus excreting such enzymes must have available regulatory controls to ensure that a significant part of the nutrient from degraded substrate is channelled into its own growth. The mechanisms of enzyme induction, repression, catabolite repression and end-product inhibition clearly have a role here and have been shown to be operative in regulating enzyme activity in cultures of basidiomycetes growing on lignocellulosic substrates (Manning & Wood, 1983). It is also likely that there must be some degree of control over enzyme *localisation*. Fragmentary evidence is available which suggests that certain extracellular enzymes excreted by basidiomycete fungi remain associated with the hyphal surfaces (Dickerson & Baker, 1979; Leatham & Stahmann, 1981; Highley, Palmer & Murmanis, 1983; Palmer, Murmanis & Highley, 1983). Localisation of extracellular enzymes at the hyphal surface in sheaths or similar structures would confer significant advantages on the producer fungus when colonising lignocellulosic substrates. In addition to limiting microbial competition for low-molecular-weight substrates surface localisation may protect against loss of enzyme activity by desiccation and may serve to determine some degree of structural organisation of enzyme activity comparable to that found within the cell. Further work is required to investigate extracellular enzyme localisation in basidiomycete cultures growing in lignocellulosic substrates. There is also some evidence available that extracellular enzyme production is not uniform within fungal colonies (Yanagita & Nomarchi, 1967) but these types of studies have not been extended to cultures growing on solid substrates with the use of modern cytochemical techniques, including immunocytochemistry.

Modulation of extracellular enzyme activity and basidiomycete morphogenesis

Morphogenesis in fungi, and in organisms generally, is associated with differential genetic activity (Griffin, 1981; Wright, 1978). Associated with morphogenesis, and often used to monitor it, are changes in the levels of activity of enzymes. The bulk of enzymological work in morphogenesis of fungal systems has been concerned with changes in *intracellular* activity (Smith & Berry, 1974; Loomis, 1982). In basidiomycete fungi what is also of interest is that there are now some well-studied examples of *extracellular* enzymes whose activities are regulated in association with morphogenesis and often with the development of fruit bodies.

Extracellular laccase and basidiomycete morphogenesis

Several published examples indicate that changes in the activity of extracellular polyphenoloxidases, particularly laccases, are associated with fruit body construction in basidiomycetes. Leonard (1971) showed that total culture laccase activity of *Schizophyllum commune* is associated with the ability of homokaryotic cultures to fruit. Fertile mycelial cultures possess high levels of laccase activity, but in sterile mycelia laccase activity is absent. Further studies showed that laccase activity of whole cultures exhibits a temporal relation to the formation of fruit bodies in *S. commune*. Laccase activity increases until mature spore-bearing fruit bodies are produced and then declines rapidly. In cultures grown in continuous dark conditions, only stage I and stage II fruit bodies are produced and laccase activity continues to increase (Leonard & Philips, 1973). Assays of laccase in the culture medium (extracellular enzyme) and of intracellular laccase showed that the change in total activity is mostly due to change in the activity of the extracellular enzyme, and that the profile of activity change resembles that of the total culture, with a large decline in activity occurring when mature fruit bodies are produced; no decline in activity occurs in cultures grown in continuous light (Philips & Leonard, 1976). Experiments with mixed extracts of extracellular and intracellular enzyme preparations provided evidence for the presence of an inhibitor of laccase activity. Assays of protein content of the extracellular medium were used to show that the fluctuations of laccase activity are not merely due to changes in extracellular protein content (Philips & Leonard, 1976).

A similar type of modulation of extracellular laccase activity has also been observed in cultures of the cultivated white mushroom growing in commercial straw substrates (Turner, 1974). Laccase activity increases up to the stage of production of the first group ('flush') of fruit bodies and then measurable laccase activity declines rapidly. Wood & Goodenough (1977) confirmed these observations and produced further evidence that the source and modulation of laccase activity is not due to other microorganisms present in compost cultures of *Agaricus bisporus* and that the enzyme is under some form of developmental regulation. Cultures where no 'casing' layer is applied to the culture surface fail to fruit but laccase activity remains high indicating that modulation of activity is not due merely to culture age or exhaustion of particular nutrients. Axenically fruited cultures of *A. bisporus* produce laccase in comparable quantities to non-axenic cultures, indicating that *A. bisporus* mycelium is the source of laccase activity. Loss of laccase activity is found only in those axenic cultures where fruit bodies form *and* develop beyond 1–2 cm in height. Laccase activity in cultures of a strain of *A. bisporus* which fails to fruit remains at a high level. Activity in a strain which forms immature fruits which fail to develop beyond 1–2 cm tall declines when these are formed. Wood & Goodenough (1977) concluded that, taken together, these pieces of evidence indicated laccase activity to be under developmental regulation but that there was insufficient proof that 'the changes observed in enzyme activity or properties are directly correlated with the differentiation process or are merely secondary events of the developmental sequence' (Mandelstam, 1976; Wood & Goodenough, 1977; see also Chapter 14).

The molecular properties of the forms of *A. bisporus* laccase prior to fruiting, when enzyme activity is at a high level (HA), and after fruiting, when enzyme activity has declined by 10- to 20-fold (LA), have been compared (Wood, 1980a). Both HA and LA enzymes are stable on extraction from compost cultures and stable on storage at room temperature, or after repeated freezing and thawing. Passage of HA or LA through gel filtration columns, extended dialysis or ultrafiltration produces no significant activation or inhibition, suggesting that loss of activity cannot be accounted for by production of a loosely bound inhibitor (Philips & Leonard, 1976). Many of the physicochemical properties of HA and LA are the same, including molecular weight, the pH activity profile, substrate specificity and Michaelis constant (Wood, 1980a). Large differences are found in the electrophoretic profiles of HA and LA on both non-denaturing and denaturing (sodium dodecyl sulphate) polyacrylamide gel electrophoresis indicating that significant structural changes have taken place but with apparently little change in molecular weight. Since the purified enzyme from malt extract-grown cultures is a glycoprotein, these electrophoretic changes could be due to alterations in either the protein or glycan moieties of the enzyme. Both forms of the enzyme cross-react in identical fashion with antibodies prepared against extensively purified enzyme obtained from malt extract-grown cultures (Wood, 1980b). Recent work shows, however, that the two forms of enzyme show only *partial* identity when cross-reacted against antibodies raised against either purified HA or LA enzyme forms from compost cultures (Wood & Matcham, unpublished).

Laccase represents a major protein produced by *A. bisporus* cultures, comprising some 2% of cell protein (Wood, 1980*b*). Since it is resolvable from other extracellular enzymes on electrophoresis, this property can be used to quantitate the amount of *enzyme protein* in extracts. The availability of antisera to purified *A. bisporus* laccase also allows an estimate of the quantity of enzyme protein. The use of these methods shows that enzyme protein loss in fruiting cultures is about 60–70% whereas *enzyme activity* loss is 90–95%. Wood (1980*a*) concluded that this indicated that inactivation of laccase was occurring, followed by enzyme degradation.

Enzyme inactivation systems

In recent years enzyme inactivation has been shown to be a third major means of regulating enzyme activity (Thurston, 1972; Switzer, 1977). Inactivation has been defined as 'the irreversible loss *in vivo* of catalytic activity in the physiologically significant reaction of an enzyme' (Switzer, 1977). This is clearly distinguishable from the other two major methods of activity control which are control by affecting rate of enzyme synthesis and control by binding of ligands. Switzer (1977) has outlined two basic types of inactivation control: (1) modification inactivation, where enzyme protein remains intact and activity loss is caused by change in the physical state of the enzyme or by attachment of modifying groups; and (2) degradative inactivation, where there is cleavage of a limited number of peptide bonds (distinct from complete protein breakdown and turnover). Switzer's (1977) review illustrates many examples of both classes of inactivation in a variety of microbial systems, including fungi, and gives criteria for establishing what type of inactivation is occurring. It is noteworthy that the cultural conditions leading to enzyme inactivation are shifts in carbon or nitrogen metabolism associated with, for example, the stationary phase of growth, glucose addition or starvation, nitrogen addition or starvation, or inducer removal.

The other major conditions leading to enzyme inactivation are morphogenetic processes such as spore formation or germination in bacteria, yeasts or filamentous fungi. Since microbial differentiation may be a consequence of nutrient limitation or changes in nutrient status, the second type of cultural condition may well be equivalent to the first.

The rationale for enzyme inactivation is similar to that of other cellular control systems, namely that of control of cell economy. Inactivation of enzymes can be shown to prevent 'futile' cycling of metabolites, to divert metabolic pathways, and spare metabolites and energy when shifts to new media occur (Switzer, 1977).

There is still insufficient evidence to determine what type of inactivation occurs to the *Agaricus bisporus* laccase, but since the enzyme is extracellular the implication is that the inactivating system is also located externally. Further work on the protein chemistry and metabolism of the enzyme should reveal what type of system is utilised to inactivate laccase and the nature of the inactivating system itself. Examples of *extracellular* enzyme inactivation are rare but, interestingly, one has been observed in cultures of a basidiomycete. Friebe & Holldorf (1975), working with an extracellular β-glucanase of basidiomycete QM 806, found this enzyme is synthesised *de novo* on glucose depletion from the medium. Addition of glucose to derepressed cultures produces rapid inactivation of glucanase activity, but this inactivation is blocked by the concomitant addition of cycloheximide, indicating that the inactivating system for the glucanase requires protein synthesis. The inactivation system produces loss of enzyme protein although activity loss was not measured in the extracts.

Intracellular inactivation systems have also been found in fruiting cultures of basidiomycetes. Schwalb (1974, 1977), working with *Schizophyllum commune*, has observed inactivation of various enzymes of glucose metabolism during fruit body formation. Evidence was obtained that the loss in activity is due to the *de novo* production of a specific protease at a particular stage of fruiting development.

Roles of laccase in basidiomycete morphogenesis

Various functions have been suggested for laccase activity or modulation of activity during development but definitive proof as to its role(s) is still lacking.

Leonard (1971) suggested that the laccase activity of *S. commune* is associated with sites of pigmentation within colonies where fruiting bodies are due to be formed but he made no comment on the correlation of *modulation* of activity with development of mature fruit bodies. Wood (1980a) has suggested that since laccase represents a major export of protein by the mycelium and enzyme protein loss occurs at fruiting, degraded laccase protein might be reassimilated as amino acids by the mycelium at fruiting, when nitrogen demand on the mycelium is at a high level. An alternative, or complementary, suggestion is that the metabolites produced by laccase activity, which could include quinones, might be inhibitory to fruit body development. One approach to this would be to generate mutants lacking laccase activity and examine their fruiting competence. This type of approach has been used to show that laccase may play a role in lignin biodegradation (Ander & Eriksson, 1976), but the evidence for the direct role of laccase in lignin biodegradation is still the subject of some controversy (Haars & Hutterman, 1980). It should be possible to determine the metabolic fate of laccase in fruit body-forming cultures of *Agaricus* by the use of radiolabelling techniques.

In addition to evidence that extracellular laccase is involved in the primordium initiation of basidiomycete fruit bodies, it has been proposed that laccase may continue to act on the hyphal surfaces throughout fruit body development. Leatham & Stahmann (1981) show evidence that the bulk of the fruit body laccase of *Lentinus edodes* is extracellular and located at the cell surface. They propose that since the pH of the cell surface may be close to that for optimal laccase activity, the enzyme could be functional in this environment. They further suggest that since laccase can produce reactive quinone compounds by its action on phenols, these compounds could be involved in oxidative polymerisations with cell surface components such as carbohydrates or proteins (Leatham, King & Stahmann, 1980). These reactions might serve chemically to cross-link adjacent hyphae and thus perform a constructional function in assembling previously spatially divergent hyphae into aggregated tissue (Bu'lock, 1967). Recent molecular approaches analysing the patterns of polypeptides synthesised during fruit body formation have shown that specific polypeptides are synthesised at the aggregation stage of fruit body formation in *S. commune* (De Vries & Wessels, 1984). Certain of these polypeptides were both associated with cell walls and excreted by the hyphae and it has been speculated that these might represent polypeptides that were part of the polyphenoloxidase protein previously examined by Leonard (1971).

Modulation of endocellulase activity during fruiting of Agaricus bisporus

It has been shown that, in addition to modulation of laccase activity, extracellular endocellulase activity also shows dramatic changes during fruiting of *A. bisporus*. In contrast to laccase activity, endocellulase activity of *A. bisporus* remains at a low level prior to fruiting but increases remarkably at fruit body formation (Turner *et al.*, 1975; Wood & Goodenough, 1977). By using the criteria outlined previously, Wood & Goodenough (1977) were able to show that this activity also originated from *A. bisporus* mycelium, was developmentally regulated and was associated with fruit body enlargement.

Re-examination of this work has shown that the situation is even more interesting than a single switch in enzyme activity levels. By increasing culture sampling frequency, Claydon (unpublished) has shown that extracellular endocellulase activity remains in direct proportion to the quantity of fruit body biomass produced during each 'flush' of mushrooms and, further, that each flush is associated with a fresh peak of cellulase activity. By controlling the harvesting strategy, either by picking mushrooms continuously at very early developmental stages or by allowing the mushrooms in particular flushes to complete development to the stage of mature senescent fruit bodies, it is possible to show that endocellulase remains in direct proportion to harvested mushroom biomass. Thus the fruit body biomass is in some way responsible for the activity of an extracellular hydrolytic enzyme produced in the underlying substrate. This finding has significant implications in respect of carbon flow from the substrate into the mycelium (Hammond: Chapter 16). Models of mushroom cropping have assumed constant inputs of substrate into the mycelium (Chanter, 1979) but the findings shown above suggest this may be unlikely at least for substrate derived from cellulose, which should consist of cellobiose or glucose. Control by fruit body biomass of enzymes regulating carbon input to the mycelium would be an elegant mechanism for ensuring that carbon reserves depleted from the mycelium during fruit body construction are balanced by carbon input via extracellular hydrolase activity. However, the evidence available indicates that other excreted hydrolases, such as xylanases or proteases of *Agaricus*, are not modulated in activity when fruiting occurs (Wood & Goodenough, 1977).

Hammond (Chapter 16) suggests that cellulase activity may be controlled by the intracellular mycelial levels of soluble carbohydrates such as glucose or trehalose. *A. bisporus* endocellulase is a classical cellulose-induced, catabolite-repressed type of fungal cellulase (Manning & Wood, 1983). Thus concentration changes of sugars within the mycelium due to their export to the developing fruit bodies would be a control mechanism for enzyme production. What is of importance is that it is now possible to connect, at the biochemical level, the mechanisms producing flux of carbon from substrate, through mycelium and into the developing fruit bodies, and to connect this to periodicity of fruiting cycles. We know little as yet of the molecular changes associated with the enzyme activity level changes of cellulase, but comparable studies to those described for *Agaricus* laccase should reveal whether enzyme activity increases and decreases are due to synthesis, degradation, inhibitor binding or inactivation, or combinations of these.

Conclusions

Three significant points emerge from this review of extracellular enzymes and basidiomycete morphogenesis. First, we still know very little about how basidiomycete cultures in laboratory media or natural environments produce, regulate and localise their extracellular enzyme production to ensure maximal cell economy of substrate utilisation. This is important to establish, since in the microbial world basidiomycetes are amongst the slowest growing members in nutrient-rich environments. Nevertheless, it is clear that part of their dominance in lignocellulosic substrates is due to their efficient degradation of substrate. The mechanisms by which these fungi limit competitor organism exploitation of substrates they themselves are colonising is worthy of more study (Rayner *et al.*: Chapter 1).

Secondly, it seems that at least some basidiomycetes can regulate the activity of certain extracellular enzymes during morphogenesis. These regulations are produced using a variety of control systems. Such regulatory systems would repay further study to throw light on problems of resource allocation and utilisation during basidiomycete morphogenesis (Rayner *et al.*: Chapter 1).

Thirdly, the example of *Agaricus* endocellulase modulation indicates that fruit body development and mycelial metabolism are tightly coupled, even to the extent that developing fruit body tissue can influence the activity of an extracellular enzyme that is spatially and metabolically separate. In the light of these findings, an investigation would be justified to examine whether cultures of basidiomycetes, such as *Coprinus* or *Schizophyllum*, grown on insoluble substrates also regulate extracellular enzyme activity during morphogenesis. Control of flushing response is a highly desirable aim in commercial mushroom growing (Manachère, 1980). Since *Coprinus* and *Schizophyllum* both have well-defined genetic analysis systems and a considerable body of information on their intracellular metabolisms is available, they might provide ideal model systems for manipulating the flushing response by genetic or chemical means. In any event the studies with *Agaricus* enzyme modulation have revealed fascinating examples of the intimate and necessary connection between the basidiomycete mycelium and the fruit bodies that develop on it (Reijnders & Moore: Chapter 27).

Acknowledgements. The author thanks Dr N. Claydon and Mr S. Matcham for useful discussion and permission to use unpublished data.

References

Ander, P. & Eriksson, K. E. (1976). The importance of phenol oxidase activity in lignin degradation by the white rot fungus *Sporotrichum pulverulentum*. *Archives for Microbiology*, **109**, 1–8.

Beckett, A., Heath, I. B. & McLaughlin, D. J. (1974). *An Atlas of Fungal Ultrastructure*. London: Longman.

Bu'lock, J. D. (1967). *Essays on Biosynthesis and Microbial Development. Rutgers University Institute of Microbiology E. R. Squibb Lectures on Chemistry of Microbial Products*. New York: John Wiley. (See especially pp. 1–18.)

Byrde, R. J. W. (1982). Presidential address. Fungal 'pectinases' from ribosome to plant cell wall. *Transactions of the British Mycological Society*, **79**, 1–14.

Chanter, D. O. (1979). Harvesting the mushroom crop: a mathematical model. *Journal of General Microbiology*, **115**, 79–87.

Cohen, B. L. (1980). Transport and utilization of proteins by fungi. In *Micro-organisms and Nitrogen Sources*, ed. J. W. Payne, pp. 411–30. Chichester: John Wiley.

De Vries, O. H. M. & Wessels, J. G. H. (1984). Patterns of polypeptide synthesis in non-fruiting monokaryons and a fruiting dikaryon of *Schizophyllum commune*. *Journal of General Microbiology*, **130**, 145–54.

Dickerson, A. G. & Baker, R. C. F. (1979). The binding of enzymes to fungal β-glucans. *Journal of General Microbiology*, **112**, 67–75.

Eriksson, K.-E. (1981). Fungal degradation of wood components. *Pure and Applied Chemistry*, **53**, 33–43.

Fermor, T. R. & Wood, D. A. (1981). Degradation of bacteria by *Agaricus bisporus* and other fungi. *Journal of General Microbiology*, **126**, 377–87.

Frankland, J. C. (1982). Biomass and nutrient cycling by decomposer basidiomycetes. In *Decomposer Basidiomycetes, Their Biology and Ecology*, ed. J. C. Frankland, J. N. Hedger & M. J. Swift, pp. 241–61. Cambridge University Press.

Friebe, B. & Holldorf, A. W. (1975). Control of extracellular β-1,3-glucanase activity in a basidiomycete species. *Journal of Bacteriology*, **122**, 818–25.

Griffin, D. H. (1981). *Fungal Physiology*. New York: John Wiley.

Grove, S. N. (1978). The cytology of hyphal tip growth. In *The Filamentous Fungi*, vol. 3, ed. J. E. Smith & D. R. Berry, pp. 28–50. London: Edward Arnold.

Haars, A. & Hutterman, A. (1980). Function of laccase in the white rot fungus *Fomes annosus*. *Archives for Microbiology*, **125**, 233–7.

Highley, T. L., Palmer, J. G. & Murmanis, L. (1983). Decomposition of cellulose by *Poria placenta*. *Holzforschung*, **37**, 179–84.

King, N. J. (1966). The extracellular enzymes of *Coniophora cerebella*. *Biochemical Journal*, **100**, 784–92.

Kirk, T. K. (1983). Degradation and conversion of lignocelluloses. In *The Filamentous Fungi*, vol. 4, *Fungal Technology*, ed. J. E. Smith, D. R. Berry & B. Kristiansen, pp. 266–95. London: Edward Arnold.

Lampen, J. O. (1974). Movement of extracellular enzymes across cell membranes. *Symposium of the Society for Experimental Biology*, **28**, 351–74.

Leatham, G. F., King, V. & Stahmann, M. A. (1980). *In vitro* protein polymerisation by quinones or free radicals generated by plant or fungal oxidative enzymes. *Phytopathology*, **70**, 1134–40.

Leatham, G. F. & Stahmann, M. A. (1981). Studies on the laccase of *Lentinus edodes*: specificity, localization and association with the development of fruiting bodies. *Journal of General Microbiology*, **125**, 147–57.

Leonard, T. J. (1971). Phenoloxidase activity and fruiting body formation in *Schizophyllum commune*. *Journal of Bacteriology*, **106**, 162–7.

Leonard, T. J. & Philips, L. E. (1973). Study of phenoloxidase activity during the reproductive cycle in *Schizophyllum commune*. *Journal of Bacteriology*, **114**, 7–10.

Loomis, W. F. (1982). The Development of *Dictyostelium discoideum*. New York: Academic Press.

Manachère, G. (1980). Conditions essential for controlled fruiting of macromycetes. A review. *Transactions of the British Mycological Society*, **75**, 255–70.

Mandelstam, J. (1976). Bacterial sporulation: a problem in the biochemistry and genetics of a primitive developmental system. *Proceedings of the Royal Society (B)*, **193**, 89–106.

Manning, K. & Wood, D. A. (1983). Production and regulation of extracellular endocellulase by *Agaricus bisporus*. *Journal of General Microbiology*, **129**, 1839–47.

Montgomery, R. A. P. (1982). The role of polysaccharidase enzymes in the decay of wood by basidiomycetes. In *Decomposer Basidiomycetes, their Biology and Ecology*, ed. J. C. Frankland, J. N. Hedger & M. J. Swift, pp. 51–65. Cambridge University Press.

Palmer, J. G., Murmanis, L. & Highley, T. L. (1983). Visualization of hyphal sheath in wood-decay hymenomycetes. I. Brown-rotters. *Mycologia*, **75**, 995–1004.

Philips, L. E. & Leonard, T. J. (1976). Extracellular and intracellular phenoloxidase activity during growth and development in *Schizophyllum*. *Mycologia*, **68**, 268–78.

Priest, F. G. (1983). Enzyme synthesis: regulation and process of secretion by micro-organisms. In *Microbial Enzymes and Biotechnology*, ed. W. M. Fogarty, pp. 319–66. London & New York: Applied Science Publishers.

Schwalb, M. N. (1974). Changes in activity of enzymes metabolizing glucose 6-phosphate during development of the basidiomycete *Schizophyllum commune*. *Developmental Biology*, **40**, 84–9.

Schwalb, M. N. (1977). Developmentally regulated proteases from the basidiomycete *Schizophyllum commune*. *Journal of Biological Chemistry*, **252**, 8435–40.

Smith, J. E. & Berry, D. R. (1974). An introduction to the biochemistry of fungal development. London: Academic Press.

Switzer, R. L. (1977). The inactivation of microbial enzymes *in vivo*. *Annual Review of Microbiology*, **31**, 135–57.

Thurston, C. F. (1972). Disappearing enzymes. *Process Biochemistry*, **7**, 18–20.

Turner, E. M. (1974). Phenoloxidase activity in relation to substrate and development stage in the mushroom *Agaricus bisporus*. *Transactions of the British Mycological Society*, **63**, 541–7.

Turner, E. M., Wright, M., Ward, T., Osborne, D. J. & Self, R. (1975). Production of ethylene and other volatiles and changes in cellulase and laccase during the life cycle of the cultivated mushroom *Agaricus bisporus*. *Journal of General Microbiology*, **91**, 167–76.

Wood, D. A. (1980*a*). Inactivation of extracellular laccase during fruiting of *Agaricus bisporus*. *Journal of General Microbiology*, **117**, 339–45.

Wood, D. A. (1980*b*). Production, purification and properties of extracellular laccase of *Agaricus bisporus*. *Journal of General Microbiology*, **117**, 327–38.

Wood, D. A. (1984). Microbial processes in mushroom cultivation; a large scale solid substrate fermentation. *Journal of Chemical Technology and Biotechnology*, **34B**, 232–40.

Wood, D. A. & Goodenough, P. W. (1977). Fruiting of *Agaricus bisporus*. Changes in extracellular enzyme activities during growth and fruiting. *Archives of Microbiology*, **114**, 161–5.

Wright, B. E. (1978). Concepts of differentiation. In *The Filamentous Fungi*, vol. 3, *Developmental Mycology*, ed. J. E. Smith & D. R. Berry, pp. 1–7. London: Edward Arnold.

Yanagita, T. & Nomarchi, Y. (1967). Kinetic analysis of the region of protease formation in the hypha of *Aspergillus niger*. *Journal of General and Applied Microbiology*, **13**, 227–35.

16

The biochemistry of *Agaricus* fructification

J.B.W.HAMMOND

Glasshouse Crops Research Institute, Worthing Road, Littlehampton, West Sussex, BN17 6LP, UK

Introduction

Agaricus fructification is an interesting study in fungal differentiation, involving the aggregation of vegetative hyphae into compact fruit body initials which further differentiate and grow to form the mature fruit body. In *Agaricus*, fructification is interesting not only to biologists but also to commercial growers, since the exact timing and yield of a mushroom crop depend on the control of these processes. At present some control is achieved by empirical manipulation of the environment. Understanding of the metabolic mechanisms underlying fruiting could lead to the development of more reliable methods of control and the production of better quality mushrooms which can be stored for longer periods after harvest.

Certain aspects of the biochemistry of fruit body initiation and growth have been studied in some detail in other basidiomycetes (see Manachère *et al.*, 1982, for review). Work on *Agaricus* has been restricted since techniques are not yet available for the initiation and growth of fruit bodies under axenic, defined conditions (Wood, 1979). This means that study of the mycelium which supports the fruit body during its development is very difficult. However, despite these difficulties evidence has accumulated for the existence of major differences, especially in the area of carbohydrate metabolism, between the metabolism of *A. bisporus* fruit bodies and mycelium. It is these differences that will be considered here since it seems likely that a full understanding of the biochemistry of fructification will only be possible when we consider the role of the mycelium in the differentiation and growth of the fruit body.

Fruit body formation

Some of the differences between mycelial and fruit body metabolism are detectable at the early aggregation stage of fruit body growth. In our studies we have tried to use these metabolic 'markers' for fruit body metabolism in developing an understanding of the processes which control fruit body initiation and growth. Additionally, we have examined the importance of these processes in determining the commercial quality of the fruit body.

Because I shall mention mostly carbohydrate metabolism and fructification, I shall be presenting a personal, but I hope not too biased, view of the biochemistry of *Agaricus* fructification. The carbohydrate metabolism of *Agaricus* during fructification has been of interest for some years, and we are now beginning to arrive at some sort of picture of how the metabolism changes during fruiting. Work on other fungi has demonstrated the importance of carbohydrate metabolism in differentiation (e.g. *Schizophyllum commune* (Wessels, 1965); yeast (Panek, 1963; Panek & Bernardes, 1983); *Dictyostelium discoideum* (Rosness & Wright, 1974)), and the same is probably true of fruit body initiation and growth in *Agaricus.*

The transition between *Agaricus* mycelium and fruit body appears to take place in at least two easily recognisable steps (Couvy, 1972). First, primordia which are competent to form fruit bodies are initiated and develop. Then the primordia are released for differentiation and further growth into fruit bodies. There is some evidence that the two steps are distinct and may be separated by a considerable period of time (see also Ross: Chapter 14, p. 370). A developmental variant of *A. bisporus*, B430, which produced structures which were indistinguishable physically, and so far biochemically, from normal primordia, was isolated (Elliott & Wood, 1978). These structures, however, developed no further, even on compost, and so may have been blocked in the metabolic pathway which enables the primordium to develop into a growing fruit body.

During a commercial crop, fruit bodies are produced heavily with a relatively synchronous emergence time at approximately weekly intervals, provided the fruit bodies are harvested at the normal commercial stage (buttons; stages 2–4 in Hammond & Nichols, 1976). When harvesting is delayed, the emergence of the next flush is also delayed (Cooke & Flegg, 1965). Flegg (1979) reported that in a small-scale cropping experiment, the number of visible primordia was at its maximum before or during the first flush of fruit body growth. Little or no further production of primordia was seen; on the contrary there was an overall decline in the number of primordia over the remaining 30 days and two flushes of the crop. Thus, later flushes of fruit body growth appear to be due to the release of already formed primordia for further growth and development.

Various experiments have led to the conclusion that *A. bisporus* mycelium produces an initiation-inhibiting substance (Long & Jacobs, 1974; Wood, 1979), this hypothetical substance being removed by the microbial flora of the casing material (normally a peat/chalk mixture in the UK) under commercial conditions, or by activated charcoal under axenic conditions (Long & Jacobs, 1974). The initiation-inhibiting substance has not been identified but Long & Jacobs concluded that it was a compound of low volatility. Cyclic AMP, which is associated with fruit body initiation in *Coprinus* (Uno & Ishikawa, 1973 *a,b*) had no effect on the formation of initials in *A. bisporus* (Wood, 1979). However, the cyclic AMP-synthesising enzyme adenylate cyclase is present in *A. bisporus* fruit bodies (Hintermann & Parish, 1979), so it is still possible that cyclic AMP plays some part in the fruiting process.

The effect of the removal of the hypothetical fruiting inhibiting substance on hyphal metabolism is still largely unexplored. However, we can speculate that cell wall metabolism may be modified to allow branching and the aggregation of hyphae. Elaboration of intracellular membranes and vesicles also appears to be associated with the early stages of fruit body initiation (Eyme & Couvy, 1977; Reisinger, Desbiens & Olah, 1979). These changes may be associated with cell wall metabolism.

Differences in carbohydrate metabolism in mycelium and fruit body

In terms of chemical composition, the main difference that has been observed between *A. bisporus* mycelium and fruit body is in the mannitol content. Vegetative mycelium typically contained 1.5–4.5% dry weight as mannitol, and mycelium supporting fruit bodies contained levels at the lower end of this scale (Hammond & Nichols, 1976). Fruit bodies accumulated mannitol; at the aggregate stage they normally had twice the mannitol content of their supporting mycelium, and accumulation continued throughout growth until the mature fruit body typically contained 25–35% mannitol as dry weight, and sometimes as much as 50% (Rast, 1965; Hammond & Nichols, 1976). Thus, an understanding of mannitol function and metabolism appears to be a good starting point for developing an understanding of fruit body growth.

Labelling experiments suggested that mannitol was turned over at a low rate in the fruit body (Hammond & Nichols, 1977), although senescing fruit bodies either metabolised it or exported it back to the mycelium

(Hammond & Nichols, 1976). This suggests that mannitol is an unlikely storage material, and also that it is not used as an intermediate in an NAD/NADP transhydrogenase system, since this would involve turnover.

Mannitol is synthesised from fructose in *A. bisporus*, which is reduced by mannitol dehydrogenase using NADPH as cofactor (Edmundowicz & Wriston, 1963; Ruffner *et al.*, 1978). The enzyme shows a high degree of specificity, and no other mannitol-metabolising enzymes have been reported in *A. bisporus*. The enzyme was not appreciably inhibited in the mannitol synthesis direction by either NADP or mannitol, however there was considerable inhibition of mannitol oxidation at *in vivo* levels of NADPH (Morton, Dickerson & Hammond, unpublished). This, together with the fact that the K_{eq} is far towards mannitol (Edmundowicz & Wriston, 1963), suggests that the system is heavily loaded against mannitol oxidation under normal cellular conditions. However, mannitol appears to be used as a respiratory substrate in harvested fruit bodies, its conversion to fructose presumably being regulated by the rate of NADPH utilisation in other pathways (Hammond & Nichols, 1975).

The enzyme was not subject to allosteric regulation (Morton, Dickerson & Hammond, unpublished), but it was not clear whether the reaction was close to equilibrium, and some form of regulation of the reaction may take place (Hammond, unpublished). There is evidence for some form of osmoregulatory control of mannitol synthesis, in that although the mannitol content of fruit bodies increased dramatically on a dry weight basis, the concentration of the polyol on a total cell water basis remained relatively constant between 150 and 200 mM during growth (Hammond & Nichols, 1976).

Mannitol accumulation is likely to occur in the vacuole, and mannitol content appears to be highest in those fruit body tissues which are most highly vacuolated. Possible mechanisms for the osmoregulation of mannitol synthesis have not been studied, but could involve transfer of mannitol or its precursors across the tonoplast into the vacuole.

The accumulation of mannitol in fruit bodies suggests that its function is one which is linked to the difference in the growth mode of fruit bodies and mycelium. *A. bisporus* fruit bodies, like those of other agarics, grow mainly by the insertion of new cell wall material along the length of the hyphae (Craig, Gull & Wood, 1977). In vegetative mycelium, hyphal growth only occurs at the tip (Burnett, 1976). *A. bisporus* cell wall composition changes between mycelium and fruit body, and it has been suggested that the changes are to allow extension growth to occur in the fruit body (Vincent-Davies, 1972). The function of mannitol is possibly to attract water into the fruit body hyphae and create the hydrostatic pressure necessary for extension of the hyphae. The resultant turgor pressure would also function in supporting the fruit body.

A similar function has been suggested for urea in the expanding cap of *Coprinus* (Ewaze, Moore & Stewart, 1978; see also Chapter 13), although the concentrations reported were a fraction of those seen for mannitol in *A. bisporus*. Urea concentrations in *A. bisporus* were found to be considerably lower than those of mannitol, and it seems unlikely that urea has an osmotic role in this organism (Hammond, 1979). Unfortunately we have been unable to correlate rates of growth (as dry weight increments over a 2-day period) of individual whole fruit bodies with the concentrations of mannitol detected in them, but increment in dry weight may not be a true reflection of extension growth. However, Parrish, Beelman & Kneebone (1976) reported that mannitol content of fruit bodies was positively correlated with the flush of the commercial crop in which they were growing, suggesting some relation between mannitol content and fruit body growth. In addition, cell wall properties must also affect growth rate.

In vegetative mycelium, although some hydrostatic force appears to be necessary to produce hyphal tip growth (Burnett, 1976), it is presumably only necessary at the growing point. Thus even if mannitol fulfilled the same function as that proposed for it in the fruit body, it would only be required at the tip, and the overall mannitol content of the mycelium would be expected to be lower.

Another change which may be linked with mannitol accumulation also occurred at a fairly early stage of fruit body formation. Glucose 6-phosphate dehydrogenase (G6PD), the first enzyme of the pentose phosphate pathway and one of the major sites of NADPH formation in the cell, exhibited higher activity in fruit body extracts than in those of vegetative mycelium (Hammond, 1977). Dütsch & Rast (1977) demonstrated that NADPH synthesised in the pentose phosphate pathway was used in mannitol synthesis. Radiorespirometry and specific labelling showed that the higher G6PD level in fruit bodies was accompanied by a greater participation of the pentose phosphate pathway in hexose oxidation (Hammond, 1977). This must make more NADPH available for mannitol and perhaps lipid and other biosyntheses, and pentoses for nucleic acid synthesis, in the fruit body.

Thus, some obvious changes in metabolism occurred during the phase of initiation and formation of fruit body primordia; doubtless other changes also occurred. The further development of the primordium into a growing and differentiating fruit body appears to depend on the reception of a stimulus which originates from the mycelium. This is suggested by the fact that flushes of fruit body growth are relatively synchronous in the commercial system, i.e. a large number of primordia start to develop at the same time. However, there is undoubtedly some feedback mechanism from fruit bodies to the mycelium, since changes in harvesting practice for fruit bodies already growing influenced the time of appearance of the newly developing primordia. Our results suggest that the stimulus for release of primordia for further development is linked with carbohydrate metabolism.

Interest in the effect of flushing on carbohydrate levels arose as a development of work on carbohydrates in growing fruit bodies. Fruit bodies at two defined stages of development were sampled daily throughout several flushes in a commercial crop and analysed. The results showed that trehalose and glycogen levels were maximal at the time of flush emergence, i.e. the time at which primordia were being released for growth into mature fruit bodies. The levels of both carbohydrates declined to minima of 10–25% of the levels seen at the peak, as the fruit bodies in the flush developed to stages 2–4, 3–4 days later. As the flush reached this stage of growth and was harvested, the levels of trehalose and glycogen reached a minimum and then increased again towards the maximum at the emergence of the next flush. This pattern was seen in fruit bodies sampled at both stages of development (Hammond & Nichols, 1979; Hammond, 1981).

Mannitol behaved differently, showing smaller fluctuations and, although levels were maximal at about the time of flush emergence in stage 1 fruit bodies, in stage 2 fruit bodies they were higher during later flush growth (Hammond & Nichols, 1979; Hammond, 1981). Thus while trehalose and glycogen changes seemed to occur over the whole 'culture', changes in mannitol were linked to the growth of individual fruit bodies, mannitol levels being highest in those fruits growing in phase with the flush.

At about the same time that we were doing this work, D. Chanter of the Biomaths Department at GCRI devised a mathematical model of flushing (Chanter, 1979). In the model, mycelium absorbs a hypothetical substrate from the compost at a constant rate. This then accumulates in the mycelium until a threshold level is reached at which initiation and growth of fruit bodies occurs. This consumes the substrate, and the substrate level in the mycelium falls until the initiation and growth of new fruit bodies stops. After the fruit bodies in the flush have been picked, substrate again accumulates in the mycelium until another flush is initiated. This model behaves in a similar way to a real crop when parameters such as harvesting time are altered.

The behaviour of the mycelial substrate in the model appeared very similar to that of trehalose and glycogen in our fruit body samples. Some evidence existed that these two carbohydrates supplied a significant proportion of the substrate for fruit body growth in that their peak levels were related to the ultimate productivity of the emerging flush.

The means by which trehalose and glycogen are accumulated is fairly easy to understand in terms of the Chanter model of fruit body initiation. The mycelium absorbs compounds from the compost released by the action of cellulase and other lytic enzymes (Wood & Goodenough, 1977) and converts carbohydrates to glucose in order to synthesise trehalose and glycogen.

Recent work has shown that the input of carbohydrate into the mycelium may not be constant throughout the cropping period. The amount of extracellular cellulase activity found in the compost appears to be very tightly linked to fruit body growth (Claydon, unpublished). Thus high levels of cellulase activity were present when fruit bodies were growing, and low levels when few fruit bodies were present. This observation has a number of corollaries. First, it means that in terms of the Chanter model, a constant input of hexose into the mycelium cannot be assumed; it seems that growth of fruit bodies stimulated cellulase activity, leading to greater glucose absorption during the flush. However, since cellulase activity is so tightly linked to fruit body growth, it is likely that the only effect on the flush pattern would be a lengthening of the period over which initiation could take place.

Secondly, it is possible that cellulase production was controlled by the intracellular level of trehalose or glucose. Extracellular cellulase production in *Trichoderma viride* is known to be controlled by glucose and disaccharides (Nisizawa *et al.*, 1971) and *A. bisporus* cellulase activity was reduced by glucose in the medium (Manning & Wood, 1983). The latter effect may be due to absorption of the hexose and repression of enzyme synthesis. Thus the high levels of trehalose observed between flushes could repress cellulase production, while the consumption of hexose during fruit body growth and respiration may lower mycelial levels to the point where repression was lifted and cellulase secreted in order to maintain some minimal hexose level.

Considered in isolation, the increase in cellulase could mean a large increase in the carbon:nitrogen ratio of the metabolite input to mycelium during flush growth. Whether this is the case, or whether other extracellular enzymes responsible for breakdown and absorption of N-compounds also increase during fruit body growth, remains to be seen. Additionally, the probability that glucose absorption is linked to fruit body growth means that fruit body construction relies to a significant extent on an external substrate, rather than endogenous storage materials.

Fruit body carbohydrate metabolism

In order to assess the possibility that the changes in trehalose were leading to changes in intracellular metabolism, a study of some enzymes of carbohydrate metabolism in fruit bodies sampled through the flushing cycle has been carried out. The results show that the extracted activity of G6PD varied by up to 20-fold during the flushing cycle; peak activity occurred at the same time as the maximum trehalose and glycogen levels. Mannitol dehydrogenase and phosphoglucose isomerase activities also varied over the sampling period, but did not show such a consistent pattern or as large variations as G6PD (Hammond, 1981).

The results from the carbohydrate assays during flushing (Hammond & Nichols, 1979; Hammond, 1981) and from earlier work (Hammond & Nichols, 1976) suggested that mannitol synthesis was highest at the early stages of growth of those fruit bodies growing in phase with the flush. The high level of trehalose and glycogen, coupled with a high G6PD activity, allowing a high pentose phosphate pathway flux and NADPH production, would provide the substates for mannitol production at the time of flush initiation.

Evidence that the higher G6PD levels may be linked to a greater pentose phosphate pathway activity at flush emergence was obtained by determination of intermediate and coenzyme levels in fruit bodies. The mass action ratio (MAR) of the G6PD + lactonase reaction calculated from these results gave a significant positive correlation with the level of trehalose in the tissue (Hammond, unpublished). This suggests that there is a greater flux through the G6PD + lactonase reaction at times when trehalose levels were high and, by implication, at flush emergence. However, the maximal *in vitro* activity of G6PD was approximately ten times greater than that necessary to give the observed pentose phosphate pathway flux. This may mean that there was some inhibition of the G6PD reaction *in vivo*, or that the reaction was not rate limiting for the pathway and that variations in its extracted activity did not affect the overall rate of the pathway. The former alternative seems most likely at the present time, especially in view of the fact that liver G6PD is thought to be almost completely inhibited by NADPH *in vivo* (Eggleston & Krebs, 1974).

The next question is whether the level of trehalose directly controlled G6PD activity, and how this could occur. Although it is possible that synthesis or activation of the enzyme could be regulated by trehalose levels, application of trehalose to excised fruit bodies did not affect G6PD activity. Our data suggest that the signal to increase G6PD activity came from the mycelium. As described above there was a synchronous increase in G6PD in fruit bodies at two stages of development. When fruit bodies were sampled through the stages of their growth, those which started to grow at flush emergence showed a decline in G6PD activity in parallel with that seen when fruit bodies at a fixed stage of development were sampled as described above. Thus a decline in G6PD activity with growth appears to be the normal pattern. However, when fruit bodies starting growth 3–4 days later were sampled as they developed, a different pattern was seen. Here, G6PD activity increased until stage 4 of development (when the gills were exposed) and then declined. Although this pattern differed from that seen in fruit bodies growing in phase with the flush, the variations were still in parallel with those seen in fruit bodies sampled at a fixed stage of development over the same period (Minamide & Hammond, unpublished).

Thus, variations in G6PD activity in fruit bodies were controlled by the stage of the flushing cycle rather than the stage of growth of the fruit body. The variation in trehalose level in growing fruit bodies appears to be influenced by the stage of the flushing cycle in a similar way to G6PD (Minamide & Hammond, unpublished). Thus G6PD activity may be linked to trehalose levels under all conditions of fruit body growth.

Recent work has indicated that increases seen in G6PD activity at flush emergence were due to the synthesis of new enzyme protein, as determined by immunoassay. There was an apparent increase in molecular specific activity of G6PD at this time. One form of G6PD has been purified from fruit bodies and is present at all stages of the flushing cycle. During the period of maximal G6PD activity, another form was also present which showed little or no cross-reaction with antiserum prepared against the 'normal' G6PD (Minamide & Hammond, unpublished). This lack of antigenicity may be due to the isozyme being the product of a different gene, which would allow this form to be regulated separately from the 'normal' G6PD.

Further evidence for the control of fruit body biochemistry by the flushing cycle was found when the DNA content of gill tissue in growing fruit bodies was examined. Fruit bodies growing at the normal time, in phase with the flush, showed a doubling in DNA content of the gill tissue just before sporulation. This was presumably associated with meiosis. However, fruit bodies developing out of phase with the flush showed no doubling of gill DNA, although gill development, basidia and spores appeared normal (Minamide & Hammond, unpublished). These data suggest that meiosis will proceed only when the fruit body receives a stimulus related to the progress of the flushing cycle, from the mycelium.

Summary

From the work described here it is possible to construct a hypothesis describing the biochemical changes which accompany fruit body initiation, and the development of primordia into mature fruit bodies. First, both fruit body initiation and further growth appear to be linked to an increased capacity for mannitol synthesis and accumulation. Mannitol accumulation may be a mechanism to move water into the fruit body in order to supply the hydrostatic power necessary for cell extension. Increased G6PD synthesis is associated with fruit body initiation and early growth, and thus with mannitol accumulation. However, whether an increased flux of the G6PD reaction is primarily responsible for greater mannitol synthesis *via* an increased NADPH:NADP ratio is not clear.

Once primordia have formed, flushing is associated with accumulating trehalose and glycogen in the mycelium and primordia, accompanied by a rise in G6PD. When a critical level is reached, mannitol accumulation may be sufficient to power elongation of the cells and allow fruit body growth to proceed. The high pentose phosphate pathway activity could supply the NADPH for mannitol synthesis as well as the NADPH and pentoses necessary for synthesis of the nucleic acids, lipids and other substances needed during growth.

This sort of mechanism could help to explain why only a proportion of the primordia is released for growth at each flush. Trehalose, glycogen and G6PD levels peak over a period of 1 day or less, thus maximal mannitol synthesis can only occur over a short period. If a primordium has wall properties which make it more resistant to extension, this would prevent it from developing at that time. As the flushes proceed there may be some long-term wall lytic activity in the primordia which weakens the cell walls and makes them more susceptible to extension. It is clear that G6PD synthesis is closely associated with the early stages of growth of the fruit body from the primordium, and by determining the mechanism by which this synthesis is stimulated it should be possible to characterise the factors responsible for flush formation. It also seems that the flushing cycle, in addition to being the endogenous rhythm in which fruit body growth from initials is started, may influence other events in fruit body development. Thus, normal fruit body development is a function not only of the age of the individual fruit body, but also of the status of the mycelium and of other fruit bodies. This brings me back to my first point: that the importance of the mycelium in fruit body development should not be overlooked.

References

Burnett, J. H. (1976). *Fundamentals of Mycology*, 2nd edn. London: Edward Arnold.

Chanter, D. O. (1979). Harvesting the mushroom crop: a mathematical model. *Journal of General Microbiology*, **115**, 79–87.

Cooke, D. & Flegg, P. B. (1965). The effect of stage of maturity at picking on the flushing of crops of the cultivated mushroom. *Journal of Horticultural Science*, **40**, 207–12.

Couvy, J. (1972). Etude de l'induction de la fructification chez *Agaricus bisporus* (Lange) Sing. (= *Psalliota hortensis*): action du glucose. *Comptes Rendus Hebdomadaire des Séances l'Académie des Sciences, Série D*, **274**, 2475–7.

Craig, G. D., Gull, K. & Wood, D. A. (1977). Stipe elongation in *Agaricus bisporus*. *Journal of General Microbiology*, **102**, 337–47.

Dütsch, G. A. & Rast, D. (1972). Biochemische Beziehung zwischen Mannitbildung und Hexosemonophosphatzyklus in *Agaricus bisporus*. *Phytochemistry*, **11**, 2677–81.

Edmundowicz, J. M. & Wriston, J. C. (1963). Mannitol dehydrogenase from *Agaricus campestris*. *Journal of Biological Chemistry*, **238**, 3539–41.

Eggleston, L. V. & Krebs, H. A. (1974). Regulation of the pentose phosphate cycle. *Biochemical Journal*, **138**, 425–35.

Elliott, T. J. & Wood, D. A. (1978). A developmental variant of *Agaricus bisporus*. *Transactions of the British Mycological Society*, **70**, 373–81.

Ewaze, J. O., Moore, D. & Stewart, G. R. (1978). Coordinate regulation of enzymes involved in ornithine metabolism and its relation to sporophore morphogenesis in *Coprinus cinereus*. *Journal of General Microbiology*, **107**, 343–57.

Eyme, J. & Couvy, J. (1977). Ultrastructure des hyphes de l'*Agaricus sylvicola* et de l'*Agaricus bisporus* au cours des premiers stades de l'initiation fructifère. *Mushroom Science*, **9**, 1–25.

Flegg, P. B. (1979). Effects of competition on the development of the mycelium, mycelial aggregates and sporophores of *Agaricus bisporus*. *Scientia Horticulturae*, **11**, 141–9.

Hammond, J. B. W. (1977). Carbohydrate metabolism in *Agaricus bisporus*: oxidative pathways in mycelium and sporophore. *Journal of General Microbiology*, **102**, 245–8.

Hammond, J. B. W. (1979). Changes in composition of harvested mushrooms (*Agaricus bisporus*). *Phytochemistry*, **18**, 415–18.

Hammond, J. B. W. (1981). Variations in enzyme activity during periodic fruiting of *Agaricus bisporus*. *New Phytologist*, **89**, 419–28.

Hammond, J. B. W. & Nichols, R. (1975). Changes in respiration and soluble carbohydrates during the post-harvest storage of mushrooms (*Agaricus bisporus*). *Journal of the Science of Food and Agriculture*, **26**, 835–42.

Hammond, J. B. W. & Nichols, R. (1976). Carbohydrate metabolism in *Agaricus bisporus* (Lange) Sing.: Changes in soluble carbohydrates during growth of mycelium and sporophore. *Journal of General Microbiology*, **93**, 309–20.

Hammond, J. B. W. & Nichols, R. (1977). Carbohydrate metabolism in *Agaricus bisporus* (Lange) Imbach: Metabolism of [¹⁴C] labelled sugars by sporophores and mycelium. *New Phytologist*, **79**, 315–25.

Hammond, J. B. W. & Nichols, R. (1979). Carbohydrate metabolism in *Agaricus bisporus*: Changes in non-structural carbohydrates during periodic fruiting (flushing). *New Phytologist*, **83**, 723–30.

Hintermann, R. & Parish, R. W. (1979). Determination of adenylate cyclase activity in a variety of organisms: evidence against the occurrence of the enzyme in higher plants. *Planta*, **146**, 459–61.

Long, P. E. & Jacobs, L. (1974). Aseptic fruiting of the cultivated mushroom *Agaricus bisporus*. *Transactions of the British Mycological Society*, **63**, 99–107.

Manachère, G., Robert, J. C., Durand, R., Bret, J. P. & Fevre, M. (1982). Differentiation in the Basidiomycetes. In *Fungal Differentiation*, ed. J. E. Smith, pp. 481–514. New York & Basel: Marcel Dekker.

Manning, K. & Wood, D. A. (1983). Production and regulation of extracellular endocellulase by *Agaricus bisporus*. *Journal of General Microbiology*, **126**, 1839–47.

Nisizawa, T., Suzuki, H., Nakayama, M. & Nisizawa, K. (1971). Inductive formation of cellulase by sophorose in *Trichoderma viride*. *Journal of Biochemistry* (*Tokyo*), **70**, 375–85.

Panek, A. (1963). Function of trehalose in baker's yeast (*Saccharomyces cerevisiae*). *Archives of Biochemistry and Biophysics*, **100**, 422–5.

Panek, A. & Bernardes, E. J. (1983). Trehalose: its role in germination of *Saccharomyces cerevisiae*. *Current Genetics*, **7**, 393–7.

Parrish, G. K., Beelman, R. B. & Kneebone, L. R. (1976). Relationship between yield and mannitol content during the crop cycle of cultivated mushrooms. *HortScience*, **11**, 32–3.

Rast, D. (1965). Zur stoffwechselphysiologischen Bedeutung von Mannit und Trehalose in *Agaricus bisporus* (eine gaschromatographische Studie). *Planta*, **64**, 81–93.

Reisinger, O., Desbiens, O. & Olah, G. M. (1979). Transformation du substrat organique utilisé pour la culture industrielle d'*Agaricus bisporus*. Etude ultrastructurale preliminaire et hypotheses de travail. *Mushroom Science*, **10**, 287–302.

Rosness, P. A. & Wright, B. E. (1974). *In vivo* changes of cellulose, trehalose and glycogen during differentiation of *Dictyostelium discoideum*. *Archives of Biochemistry and Biophysics*, **164**, 60–72.

Ruffner, H. P., Rast, D., Tobler, H. & Karesch, H. (1978). Purification and properties of mannitol dehydrogenase from *Agaricus bisporus* sporocarps. *Phytochemistry*, **17**, 865–8.

Uno, I. & Ishikawa, T. (1973*a*). Purification and identification of the fruiting inducing substances in *Coprinus macrorhizus*. *Journal of Bacteriology* **113**, 1240–8.

Uno, I. & Ishikawa, T. (1973*b*). Metabolism of adenosine 3′,5′-cyclic monophosphate and induction of fruiting bodies in *Coprinus macrorhizus*. *Journal of Bacteriology*, **113**, 1249–55.

Vincent-Davies, S. (1972). *Relationship between* Mycogone perniciosa *Magnus and its host* Agaricus bisporus (*Lange*) Sing *the cultivated mushroom*. Ph.D. thesis, University of Bath.

Wessels, J. G. H. (1965). Morphogenesis and biochemical processes in *Schizophyllum commune* Fr. *Wentia*, **13**, 1–113.

Wood, D. A. (1979). Studies on primordium initiation in *Agaricus bisporus* and *Agaricus bitorquis* (syn. *edulis*). *Mushroom Science*, **10**, 565–86.

Wood, D. A. & Goodenough, P. W. (1977). Fruiting of *Agaricus bisporus*. Changes in extracellular enzyme activities during growth and fruiting. *Archives of Microbiology*, **114**, 161–5.

17
Growth and development of *Lentinus edodes* on a chemically defined medium

GARY F.LEATHAM

USDA Forest Products Laboratory, PO Box 5130, Madison, Wisconsin 53705, USA

Introduction

The single largest bioconversion process utilising wood is currently the cultivation of *Lentinus* (= *Lentinula*) *edodes*. The Orient cultivates more than US $1 billion of this edible mushroom each year (Royse & Schisler, 1980; San Antonio, 1981; Leatham, 1982). This lignin-degrading basidiomycete (Leatham & Kirk, 1983) is cultivated primarily out of doors on hardwood logs. Growers are now beginning to cultivate this valuable food in other parts of the world (Farr, 1983) and are trying to grow it in environmentally controlled chambers or on new media such as lignocellulosic particles (Hans *et al.*, 1981).

For many growers, however, fruiting is not reliable. The same problem occurs when attempting to cultivate other edible mushrooms under new environmental conditions or on new media. This points to the need for research on the requirements for fungal development and the mechanism by which development is regulated. Unfortunately, the complexity of fungal development, the difficulty in fruiting many of the edible mushrooms in the laboratory, and the frequent use of non-defined media have slowed research progress and limited our understanding of their development.

Recently, a chemically defined medium was developed that rapidly fruits *L. edodes* (Leatham, 1983). Using this medium and a fractionation technique, the author studied the growth and development of *L. edodes*. The areas studied were carbon and nitrogen source uptake and utilisation, soluble protein content, and several extractable enzymes that may be important in development. These results are described here.

Presented first is a brief review of what is currently known about the roles of these enzymes and of carbon and nitrogen source uptake in the growth and development of basidiomycetes including *L. edodes*. This is

403

followed by a discussion of the new studies with the chemically defined medium.

Carbon and nitrogen

Uptake, mobilisation and utilisation of carbon and nitrogen sources are of special importance in the development of fungi. Carbon/nitrogen source ratios can influence whether or not basidiomycetes (Plunkett, 1953; Wessels, 1965) including *L. edodes* (Ando, 1974; Leatham, 1983) produce fruit bodies. It is likely that the carbon and nitrogen sources have separate effects. In a wide range of fungi, nitrogen source depletion precedes and is perhaps essential for the development of fruit bodies. When *L. edodes* was grown on a semi-defined medium, exogenous carbon source was necessary to support the growth of fruit bodies (Tokimoto & Kawai, 1975). Once fruit body growth is underway, rapidly expanding fruit bodies can act as a major sink for carbon and nitrogen compounds, and probably other nutrients (Plunkett, 1953; Madelin, 1956; Wessels, 1965; Gruen & Wong, 1982).

Enzymes

Organisms require enzymes to utilise nutrients and to regulate growth and development. To carry out its role, each enzyme must be in the appropriate location, i.e. extracellular, cell wall-bound (De Vries & Wessels, 1984), periplasmic, and/or intracellular. To function, each enzyme requires the correct substrate(s) and an acceptable pH. Many classes of enzymes are important to fungal growth and development; those reviewed here include proteases, phosphatases, phenoloxidases, and cell wall-associated polysaccharidases and saccharidases.

Proteases are commonly produced by nitrogen-limited cultures in an attempt to scavenge protein nitrogen (North, 1982). Proteases are developmentally regulated in *Schizophyllum commune* (Schwalb, 1977). During the onset of development in *S. commune* the internal accumulation of proteolytically stable enzymes apparently shifts metabolism from vegetative growth to development (Schwalb, 1978). Proteases are also likely to be involved in the development of other fungi. Protease inhibitors profoundly influence the fruiting of *L. edodes* (Terashita, Kono & Murao, 1978; Terashita *et al.*, 1981).

Phosphatases are developmentally regulated in fungi and myxomycetes (Loomis, 1969; O'Day & Horgen, 1974). Intracellular alkaline phosphatases are present during hyphal tip growth in fungi (Meyer, Parish & Hohl, 1976). Extracellular acid phosphatases are often present in the periplasmic space or the cell wall of fungi, yeast and filamentous bacteria (Lampen, 1968; Field & Schekman, 1980; Poirier & Holt, 1983 *a,b*). They are thought to function in cell wall formation (Bojović-Cvetić & Vujicić, 1982; Poirier & Holt, 1983 *c*).

Phenoloxidases, such as laccase, are developmentally regulated in basidiomycetes (Leonard & Phillips, 1973; Wood & Goodenough, 1977; Ross, 1982) including *L. edodes* (Leatham & Stahmann, 1981). Besides polymerising phenols which then form pigments, strengthen cell walls, and protect against pathogens, extracellular phenoloxidases probably have other roles in vegetative growth and development (Leatham & Stahmann, 1981; Leatham, unpublished).

Polysaccharidases and saccharidases have many functions during the growth and development of basidiomycetes. Those relevant to this study include extracellular functions related to basidiomycete cell wall polysaccharides. Water-insoluble (skeletal) cell wall polysaccharides in basidiomycetes include chitin microfibrils, $(1 \rightarrow 3)$-and/or $(1 \rightarrow 6)$-β-D-glucans, and $(1 \rightarrow 3)$-α-D-glucans (Rosenberger, 1976; Sietsma & Wessels, 1977; Novaes-Ledieu & Garcia Mendosa, 1981; Wessels & Sietsma, 1981). Water-soluble (non-skeletal) cell wall polysaccharides include homo- or hetero-D-galactans and -D-mannans (Barreto-Bergter & Gorin, 1983). The fruit body cell wall of *L. edodes* contains water-insoluble polysaccharides similar to those described above (Shida, Uchida & Matsuda, 1978; Shida *et al.*, 1981) and a water-soluble L-fuco-D-manno-$(1 \rightarrow 6)$-α-D-galactan (the L-fucose and D-mannose linkages are not yet known; Shida, Haryu & Matsuda, 1975). During hyphal growth or development, synthesis of new cell wall polysaccharides, cross-linking (Sietsma & Wessels, 1979, 1981; Sonnenberg, Sietsma & Wessels, 1982), and lytic modifications in existing cell wall polysaccharides occur (Bartnicki-Garcia, 1973; Gooday & Trinci, 1980; Wessels, Sietsma & Sonnenberg, 1983; Kamada & Takemaru, 1983). Lysis can be extensive. Older cell wall polysaccharides can serve as a source of new cell wall material (Bartnicki-Garcia, 1973; Fèvre, 1977).

Those polysaccharidases and saccharidases produced by basidiomycetes that are capable of degrading insoluble cell wall polysaccharides and/or their intermediate degradation products, include laminarinases, β-D-glucosidases, chitinases and N-acetyl-β-D-glucosaminidases (Wilson & Niederpreum, 1967; Kamada, Fuji & Takemaru, 1980; Miyake, Takemaru & Ishikawa, 1980; Ishikawa, Oki & Senba, 1983). By analogy, the other appropriate enzymes capable of degrading water-soluble cell wall polysaccharides are probably also produced (e.g. in *L. edodes*: α-D-glucanase, L-fucosidase, D-mannosidase and α-D-galactosidase). An earlier study with

L. edodes grown on a lignocellulosic medium suggested that laminarinase, β-D-glucosidase, N-acetyl-β-D-glucosaminidase, β-D-mannosidase (but not α-D-mannosidase) and α-D-galactosidase were all important during development (Leatham, unpublished).

Results and discussion

Lentinus edodes ATCC strain 48085 (hetero-dikaryon) was grown on the chemically defined medium developed by Leatham (1983); the composition is given in Table 17.1.

A new fractionation technique was developed to help study the uptake of nutrients and to determine the general location of soluble proteins (Fig. 17.1). The technique was sequential and involved collection of the growth medium, limited homogenisation of the mycelia, and finally extensive homogenisation and breakage of the cells. From these steps were collected the extracellular, shockable, intracellular, and particulate fractions. Water used during the homogenisations extracted soluble components and, by dilution of residual components, resulted in minimal contamination of subsequent fractions. What follows is an explanation of how the technique fractionates and the likely sources of the components in each fraction.

Vacuum filtration of whole cultures on glass fibre filters did not extract the aerial mycelia nor damage the cultures. Thus, the extracellular fraction contained components in the growth medium and also extracellular metabolites and enzymes secreted into it. Limited homogenisation of the mycelium in a Waring blender effectively wetted the extracellular surfaces of the aerial mycelium and only partially fragmented the mycelium. Thus, the shockable fraction contained components extracted from the aerial mycelium and components released during partial fragmentation, i.e. from the following locations: cell wall, periplasmic space, cell membrane (extracellular side), fragile cells (e.g. hyphal tips), and vesicles accumulated near the cell membrane (e.g. undergoing extracellular transport). Cell breakage of the mycelial fragments in the Braun homogeniser was essentially complete. Thus, the intracellular fraction contained components extracted only after extensive cell breakage. The final vacuum filtration of the final homogenate collected insoluble cell debris. Thus, the particulate fraction contained components not previously extracted (solubilised). No attempt was made to fractionate further (or characterise) the particulate fraction (Poirier & Holt, 1983c).

In this technique there was obvious concern as to the extent of cell breakage during preparation of the shockable fraction – this would result

Table 17.1. *Composition of the chemically defined medium used to grow and fruit* Lentinus edodes (*Leatham, 1983*)

Component	Final concentration (per litre)[a]
D-Glucose	25.0 g
D-Glucuronic acid	4.0 g
L-Glutamic acid	2.5 g
KH_2PO_4	2.0 g
$MgSO_4 . 7H_2O$	2.0 g
$MnSO_4 . 5H_2O$	43.9 mg
$CaCl_2 . 2H_2O$	36.7 mg
$ZnSO_4 . 7H_2O$	22.0 mg
$Fe(NH_4)_2(SO_4)_2 . 6H_2O$	14.1 mg
$CuSO_4 . 6H_2O$	784.0 μg
$CoCl_2 . 2H_2O$	81.0 μg
$NiCl_2 . 6H_2O$	81.0 μg[b]
$Na_2MoO_4 . 2H_2O$	51.0 μg
$SnCl_2 . 2H_2O$	38.0 μg[b]
Salicylic acid	1.0 mg
i-Inositol	1.0 mg
Thiamine HCl	1.0 mg
p-Aminobenzoic acid	100.0 μg
Nicotinic acid	100.0 μg
Pyridoxine HCl	100.0 μg
Riboflavine	100.0 μg
Sodium pantothenate	100.0 μg
Biotin	30.0 μg
Cyanocobalamin	10.0 μg
Folic acid	10.0 μg

[a] Distilled/deionised water was used as solvent. The medium was adjusted to pH 4.0 with KOH, filter sterilised through 0.22 μm-size filters, and stored at 4 °C until use.
[b] These unusual trace elements were added because they stimulate fruiting (Leatham, 1985).

in intracellular components contaminating the shockable fraction. By using an intracellular component as a marker, it was possible to approximate an upper limit for the extent of cell breakage. In doing so it was necessary to assume that (1) all of the protein in the shockable fraction was derived from cell breakage, and (2) young and middle-aged mycelia were equally fragile. Based on protein content in the intracellular and shockable fractions from 6- and 34-day-old cultures, apparently no more than 23%

cell breakage occurred during preparation of the shockable fractions. Thus, unless a residual component was incompletely extracted, the shockable and other fractions were at least 77% free of components from other culture fractions.

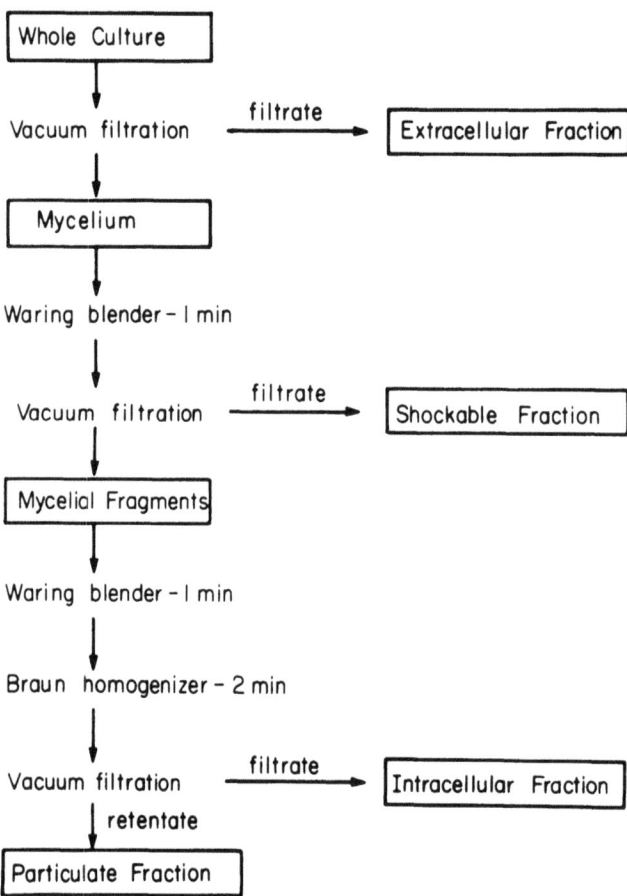

Fig. 17.1. Technique for fractionating whole cultures grown in the chemically defined medium. Homogenisations and cell breakage were in distilled/deionised water at 1–5 °C. Homogenisations were at low speed in an Eberbach model 8580 semi-micro stainless steel blender assembly and Waring blender using 25 ml total sample volume per 25-ml culture. Cell breakage was at 4000 rpm in a Braun 70-ml glass homogenising cell and Braun model MSK homogeniser using 30 ml total sample volume and 20 ml of 0.1–0.2-mm-diameter glass beads per culture. Glass fibre filters were used during all vacuum filtrations to collect mycelial fragments or to remove cell fragments and glass beads from filtrates.

Growth, growth stages and development

Analysis of growth and development of *L. edodes* on the chemically defined medium has revealed the growth stages and extent of development

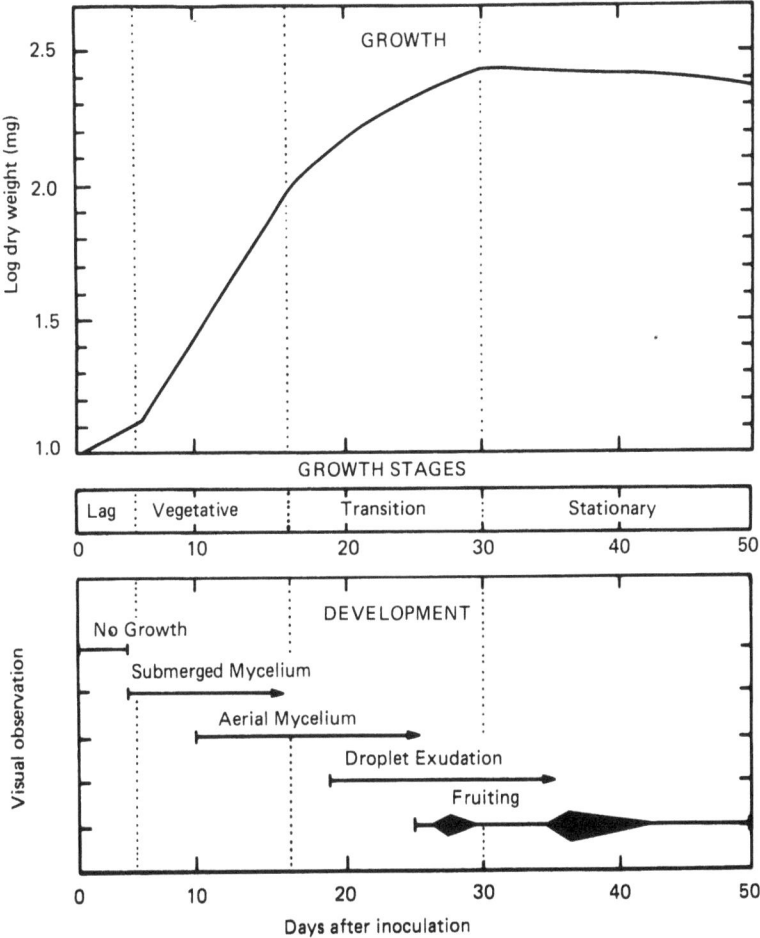

Fig. 17.2. Growth, growth stages and the development of cultures of *L. edodes* in a chemically defined medium (Leatham, 1983). Cultures were incubated at 22 ± 1 °C and 80% relative humidity with illumination at 200–400 lux from 40-watt Sylvania Gro-Lux fluorescent bulbs on a 9 h light/15 h dark cycle (Leatham & Stahmann, 1981). *Upper graph*: values for growth are expressed as the log of the average dry weight (mg) for mycelia from triplicate cultures dried at 60 °C. *Middle graph*: changes in growth were used to assign the growth stages. *Lower graph*: growth stages are delineated (vertical dotted lines). Development was determined by visual inspection of the cultures.

of the fungus at different ages (Fig. 17.2). Early in the vegetative growth stage, only submerged mycelia grew, aerial mycelia forming later. During the transition stage, the rate of growth slowed to one-half and then to one-fourth of the initial rate. In spite of this, before growth finally stopped the mycelial weight had doubled during the transition stage.

Droplet exudation from tiny primordia buried within the mycelium was the first outward sign of development. All cultures exuded droplets. Exudation began on day 18 (during the transition stage) and was most active through day 30. Following the active exudation period, approximately 40% of the cultures fruited. Fruiting began on day 25 (near the end of the transition stage) and continued through day 50. The bulk of fruiting usually happened at two times – near days 27 and 36 – but was most frequent near day 36. Each culture usually fruited only once and produced one or two fruit bodies. Although other possibilities exist, this pattern suggests that *L. edodes* may have fruited as a result of the depletion of an essential nutrient.

Once fruiting was initiated, fruit body expansion was rapid and a considerable portion of the total fungal mass was committed to fruit body formation. Fruit bodies produced by individual cultures commonly grew to full weight within 2–5 days and contained up to 48% of the total colony dry weight.

Uptake and utilisation of carbon source

D-Glucose is the major carbon source in the chemically defined medium. Throughout the transition stage, at least 20% of the initial D-glucose remained in the extracellular fraction (Fig. 17.3). Only near day 45 was the D-glucose depleted. These observations have important implications. Carbon limitation does not trigger either the transition stage or primordia formation. However, as Tokimoto & Kawai suggested (1975), depletion of carbon source perhaps best explains the failure of cultures to fruit. Since primordia produce fruit bodies by expansion, it is probable that carbon limitation stops development by blocking primordium expansion.

L. edodes is efficient in utilising D-glucose as a carbon source. The maximal theoretical yield of cells for an aerobic organism grown on glucose is near 0.5 g cells g^{-1} glucose consumed (Righelato, 1975). Here, the cell yield of *L. edodes* on D-glucose was near the theoretical limit. At the end of the vegetative growth stage (day 16), 0.6 g of cells had been produced per gram of D-glucose consumed (Fig. 17.3). At the time of

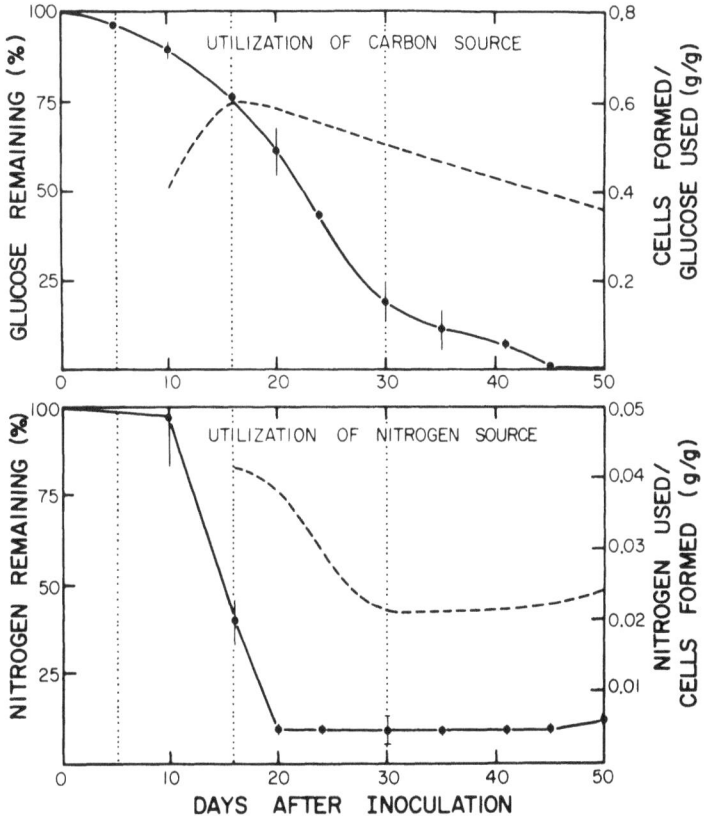

Fig. 17.3. Uptake and utilisation of carbon and nitrogen sources by cultures grown in the chemically defined medium. *Upper graph*: carbon source uptake was determined by measuring the amount (%) of D-glucose remaining in the extracellular fraction (solid line) of duplicate cultures with a Beckman Model 2 Glucose Analyzer equipped with a D-glucose oxidase/O_2 probe using D-glucose as standard. Cell yields (g cells formed g^{-1} glucose utilised) in cultures (dashed line) were calculated using the growth data in Fig. 17.2. *Lower graph*: nitrogen source uptake was determined by measuring the amount (%) of total nitrogen (initially L-glutamate) remaining in the extracellular fraction (solid line) of duplicate cultures by the micro-Kjeldahl procedure using acetanilide as standard. The nitrogen contents of cells (g nitrogen utilised g^{-1} cells formed) in the cultures (dashed line) were calculated using the growth data in Fig. 17.2. Growth stages (see Fig. 17.2) are delineated (vertical dotted lines).

D-glucose depletion (day 45), 0.39 g of cells had been produced per gram of D-glucose consumed.

Uptake and utilisation of nitrogen source

L-Glutamic acid is the only significant nitrogen source in the chemically defined medium. In contrast to D-glucose, L-glutamic acid was rapidly depleted from the extracellular fraction (Fig. 17.3). By the beginning of the transition stage (day 20), more than 90% of the total extracellular nitrogen had been taken up by the mycelium. Because the nitrogen level did not change after day 20, the L-glutamic acid was probably depleted. The nitrogen remaining in the medium after day 20 was probably fungal protein. Based on these observations, it is likely that nitrogen source depletion triggered the transition stage. Because of the depletion of utilisable extracellular nitrogen, growth during the transition stage might occur at the expense of stored nitrogen reserves, which perhaps includes nitrogen liberated from protein and/or nucleic acids (Legerton & Weiss, 1979).

The medium composition is known to influence the efficiency by which certain strains of *L. edodes* utilise L-glutamic acid as a nitrogen source (Kawamura & Goto, 1980). With the medium used here, this strain of *L. edodes* utilised L-glutamic acid efficiently. Fungal cells are characteristically 4–7% nitrogen by dry weight. In the current study, the cell yield of *L. edodes* was remarkably high. At the end of the vegetative growth stage through the onset of the transition stage (days 16–20) the cells contained 4% nitrogen (Fig. 17.3). Due to the approximate doubling in weight (and the absence of available extracellular nitrogen source) during the transition stage (day 16–30), this value decreased until it reached 2.2% nitrogen. Failure to grow further suggests depletion of stored nitrogen source reserves.

These results indicate that development is likely to be regulated by nitrogen limitation. Droplet exudation occurred as the utilisable extracellular nitrogen was depleted. Fruiting occurred with the probable depletion of reserve nitrogen sources. These results also explain how carbon:nitrogen source ratios may regulate development in basidiomycetes that produce abundant or large fruit bodies. Adequate nitrogen source is undoubtedly required to produce a mycelium of sufficient mass to support development. However, nitrogen source depletion is not possible without sufficient carbon source to utilise the nitrogen source. Carbon:nitrogen source ratios may also be important to fungi such as *L. edodes* growing on wood. The low nitrogen content of wood undoubtedly ensures rapid nitrogen limitation.

Protein production and localisation and resistance to proteases

A mechanism is needed to explain how nitrogen limitation may regulate the development of *L. edodes*. Because the patterns for the production, stability and turnover of protein may be involved, they were investigated. With the chemically defined medium, maximum soluble protein was produced during the vegetative growth stage (Fig. 17.4). Consistent with the low nitrogen source content of the medium, little protein accumulated in the extracellular fraction. Instead, the majority accumulated in the intracellular and shockable fractions. Nearly equivalent amounts of protein were in these two fractions. The exact cellular location and function(s) of the protein are not known and deserve study. Recently,

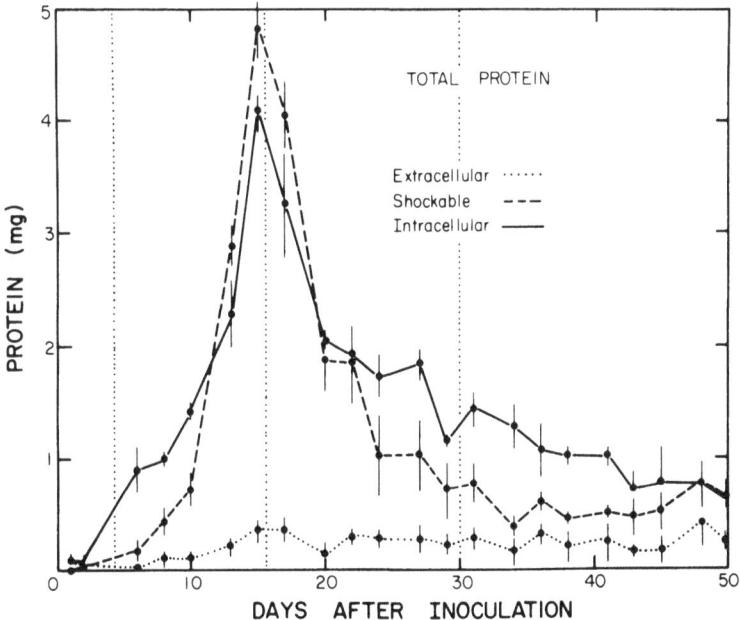

Fig. 17.4. Production and localisation of protein in cultures grown in the chemically defined medium. The protein contents of cultures were determined on fractions from triplicate cultures by the Coomassie blue dye method (Spector, 1978). Values are given for the total protein contents per culture (mg) in the extracellular (dotted line), shockable (dashed line) and intracellular (solid line) fractions. Growth stages (see Fig. 17.2) are delineated (vertical dotted lines).

De Vries and Wessels (1984) suggested that cell wall protein in *S. commune* is involved in fruiting.

With the onset of the transition stage in *L. edodes* (day 16), a marked loss in total soluble protein occurred (Fig. 17.4). As the utilisable extracellular nitrogen was depleted (Fig. 17.3), protein levels in the intracellular and shockable fractions decreased to 20% and 11% of their

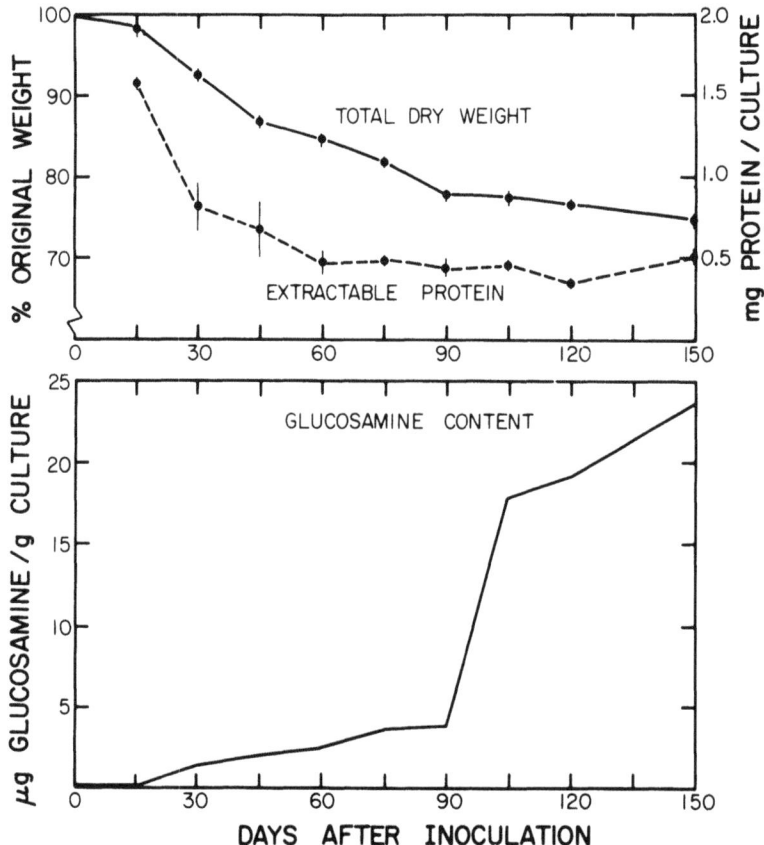

Fig. 17.5. Accumulation of glucosamine by nitrogen-limited cultures during degradation of a shredded oak wood/oatmeal medium. *Upper graph*: total dry weights (%) remaining after different lengths of incubation were determined on triplicate whole cultures (medium and fungus) dried at 60 °C (solid line). Extractable protein (%) remaining was determined in distilled-water extracts from triplicate cultures by the Coomassie blue dye method (Spector, 1978) (dotted line). *Lower graph*: D-glucosamine contents were determined on acid-hydrolysed samples from pooled triplicate cultures by the method of Gurusiddaiah, Blanchette & Shaw (1978).

original levels, respectively. In contrast, the low protein level in the extracellular fraction remained nearly constant. These data suggest that a massive reorganisation of soluble protein occurs during the transition stage. Late in the life cycle, enzymes essential for vegetative growth are certainly not as important as are enzymes essential for fruiting. Proteases and other developmentally regulated enzymes are likely to be induced in these nitrogen-depleted cultures. Thus, only those proteins that are either stable to proteolysis or rapidly resynthesised will accumulate.

Because these data suggest that significant net protein turnover occurs during the transition stage, it is important to determine the fate of the nitrogen from scavenged protein. Major non-protein sinks for nitrogen in developing fungi may include DNA for spore (nucleus) formation, D-glucosamine for chitin (cell wall) synthesis, or urea for osmotic pressure (needed for fruit body pileus expansion in *Coprinus cinereus* (Moore, Elhiti & Butler, 1979)). When cultures of *L. edodes* were grown on a lignocellulosic medium (Leatham, unpublished), analysis of acid-hydrolysed samples showed that after depletion of utilisable nitrogen source (protein), the cultures rapidly accumulated D-glucosamine, resulting in a net five-fold increase (Fig. 17.5). The form of the D-glucosamine is not yet known but it may be in reserve for the later production of fruit bodies (Tokimoto & Fukuda, 1981). If the same process occurs in cultures grown on the chemically defined medium, this may account for a significant amount of the soluble protein lost during the transition stage. However, for protein turnover and development to occur, the proteases and the other necessary enzymes must be present.

Production, localisation and possible roles of proteases,
phosphatases, laccase and saccharidases

Proteases. Proteases are produced by *L. edodes* and are in distinct locations. Cultures began to exhibit significant protease activities late in the vegetative growth stage (Fig. 17.6). These peaked during the transition stage near day 20 (after nitrogen source depletion had occurred) and were in high titres throughout the rest of the incubation. Proteases with pH optima acceptable for *in vivo* function were located in each of the culture fractions. The extracellular fraction contained acid protease but little detectable neutral or alkaline protease. The intracellular fraction contained neutral and alkaline proteases but little detectable acid protease. The shockable fraction contained all three classes of protease. Essentially non-synchronous fluctuations in protease titres began during the transition stage and the extracellular acid protease peaked near the end of the incubation. These data suggest that the three classes of proteases are distinct and that their titres are developmentally regulated. The appearance, relative abundance and widespread localisation of proteases suggest that significant rates of protein turnover are indeed possible in nitrogen-depleted cultures.

As reported for *S. commune* (Schwalb, 1977, 1978), protein turnover and

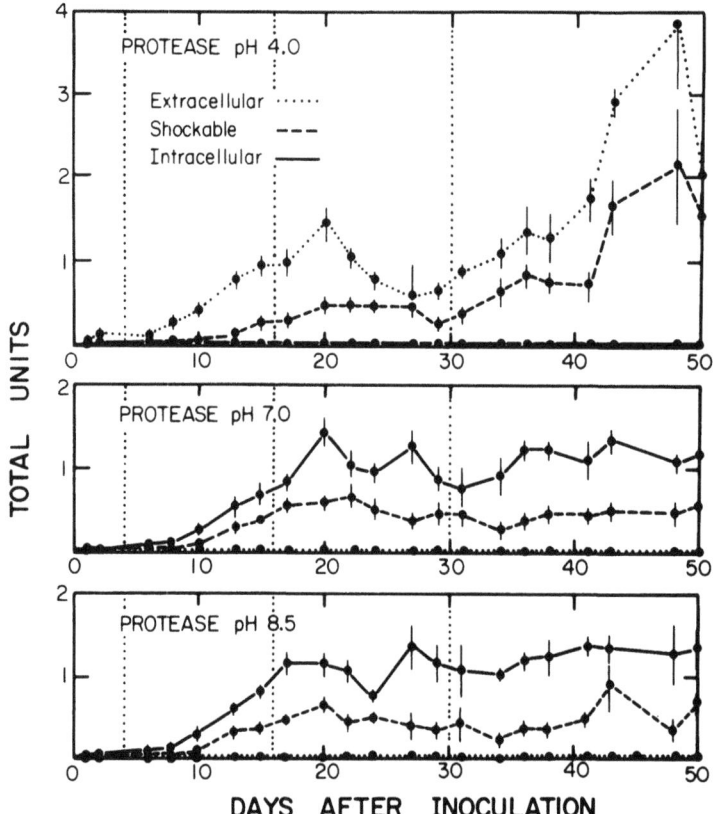

Fig. 17.6. Production and localisation of acid (upper graph), neutral (middle graph), and alkaline proteases (lower graph) in cultures grown in the chemically defined medium. Titres were determined at 22 °C on fractions from triplicate cultures in 50mM pH 4.0, 7.0, or 8.5 buffer (sodium acetate (HCl), piperazine-N,N'-bis[2-ethane-sulphonic acid] (NaOH), tri(hydroxymethyl)amino-methane (HCl), respectively) with 2.5 mg ml^{-1} hide powder azure as substrate (increase in A_{595}). Values are given for the total enzyme units (one unit equals 0.1 optical density change min^{-1}.) per culture in the extracellular (dotted lines), shockable (dashed lines) and intracellular (solid lines) fractions. Growth stages (see Fig. 17.2) are delineated (vertical dotted lines).

Table 17.2. *Stability of total protein and enzymes in fractions from 24-day-old cultures of* Lentinus edodes *during storage for 4 days at 4 °C*

Component	pH of assay[c]	Percent of original remaining in the culture fraction after storage		
		Extracellular[a] fraction	Shockable[a] fraction	Intracellular[a] fraction
Total protein[b]		95	31[d,i]	45[d,i]
Enzymes	*pH of assay[c]*			
Acid protease	4.0	109[d]	96[d]	NA
Neutral protease	7.0	NA	108[d]	89[d]
Alkaline protease	8.5	NA	108[d]	136[d]
Acid phosphatase	4.0	102	93[d]	13[i]
Neutral phosphatase	7.0	NA	72[d]	29[i]
Alkaline phosphatase	8.5	NA	57	63[d]
Laccase	4.0	98[d]	65[d]	59
β-D-Glucosidase	4.0	74[d]	81[d]	48[i]
N-Acetyl-β-D-Glucosaminidase	4.0	81[d]	101[d]	78
α-D-Galactosidase	4.0	92	101[d]	NA
β-D-Mannosidase	4.0	82	97[d]	NA

[a] The final average pH of the culture fractions were 3.7, 5.4 and 9.3, respectively. Endogenous proteases are present in all fractions. Pooled triplicate culture fractions were assayed.
[b] Determined by the Coomassie blue dye method (Spector, 1978).
[c] Methods for enzyme assays are given in the legends for Figs 17.6 through 17.10. NA: Activity normally not present in this culture fraction.
[d] Dominant site for this particular protein/enzyme.
[i] Marked instability of the protein/enzyme in this culture fraction. Protease inhibitors decrease the extent of net protein disappearance during storage.

the relative stability of enzymes to proteolysis may be important to the development of *L. edodes*. Studies on protein and enzyme stability in fractions from nitrogen-depleted cultures stored for 4 days at 4 °C showed that both the total soluble protein and specific enzymes in the extracellular fraction were resistant to proteolysis by endogenous proteases (Table 17.2). In contrast, the bulk of the total protein in the shockable and intracellular fractions was not stable during storage, 55–69% of it was apparently degraded. Even so, several enzymes in these latter two fractions were markedly stable, especially when in the culture fraction of their dominance. These included proteases, phosphatases, laccase, and certain saccharidases (Table 17.2). This stability may help account for their relatively high titres in these nitrogen-depleted cultures.

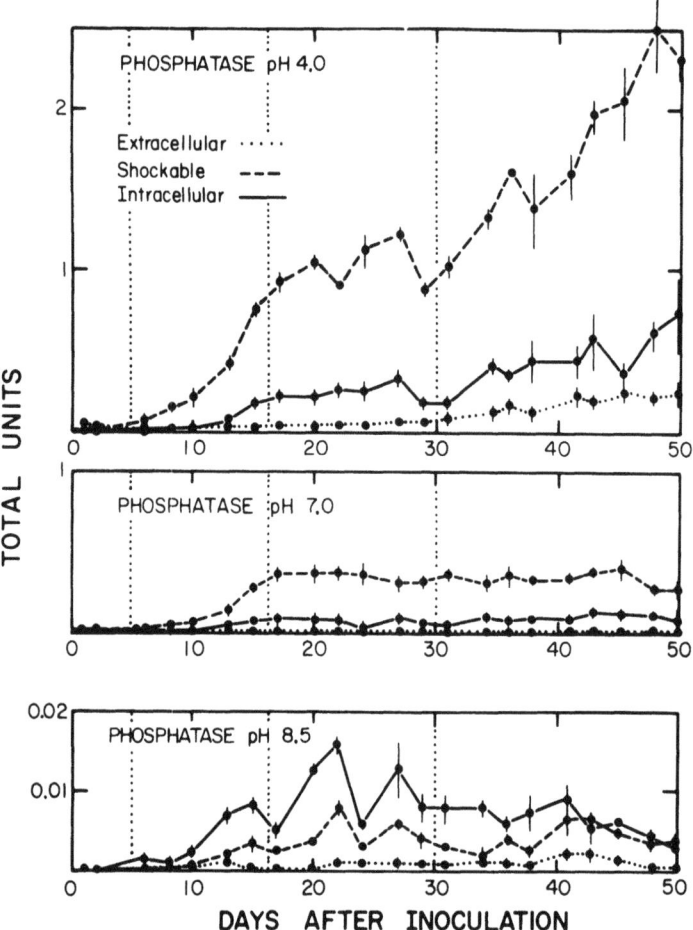

Fig. 17.7. Production and localisation of acid (upper graph), neutral (middle graph), and alkaline phosphatases (lower graph) in cultures grown in the chemically defined medium. Titres were determined at 22 °C on fractions from triplicate cultures in 50mM pH 4.0, 7.0, or 8.5 buffer (sodium acetate (HCl), piperazine-N,N'-bis[2-ethane-sulphonic acid] (NaOH), or tri(hydroxymethyl)amino-methane (HCl), respectively) with 3.3mM p-nitrophenol-phosphate ester as substrate (increase in A_{400}). Values are given for the total enzyme units (international units) per culture in the extracellular (dotted lines), shockable (dashed lines), or intracellular (solid lines) fractions. Growth stages (see Fig. 17.2) are delineated (vertical dotted lines).

Proteases are known to have functions in addition to protein turnover. Regulation by proteolytic activation of enzymes is important in many eukaryotes (Marzluf, 1981). Chitin synthetases of yeast (Cabib & Ulane, 1973) and basidiomycetes (e.g. *Agaricus bisporus* (Hänseler, Nyhlén & Rast, 1983)) are known to be proteolytically activated. Many proteins produced by developing cultures of *S. commune* are known to undergo post-translational modification (De Vries *et al.*, 1980). Certainly some may undergo proteolytic modifications including proteolytic activation. However, other modifications are also possible including glycosylation and phosphorylation. Yeast fructose-1,6-bis-phosphatase, a key metabolic enzyme, is regulated by proteolytic cleavage and by phosphorylation/dephosphorylation (Tortora *et al.*, 1981).

Phosphatases. Phosphatases are produced by *L. edodes* and are in distinct locations (Fig. 17.7). Like the protease activities, phosphatase activities became significant late in the vegetative growth stage and remained in high titre throughout the rest of the incubation period. The phosphatase in highest titre was acid phosphatase, followed by moderate and low titres of neutral and alkaline phosphatases, respectively. Unlike the proteases, relatively little phosphatase was found in the extracellular fraction. Acid and neutral phosphatases were located predominantly in the shockable fraction. Alkaline phosphatase was located predominantly in the intracellular fraction and was at least 20 times lower in titre than the acid or neutral phosphatases. The titres of acid and alkaline (but not neutral) phosphatases underwent marked fluctuations. Acid phosphatase titres fluctuated during the transition stage and afterwards, whereas alkaline phosphatase titres fluctuated significantly only during the transition stage. These data suggest that, like the proteases, the three classes of phosphatases are both distinct and developmentally regulated.

As discussed earlier, little is known about the roles for phosphatases in fungi. These studies suggest that the phosphatases of *L. edodes* have more than one role. Since the extracellular pH of the primordia and fruit bodies of *L. edodes* is maintained near 4.0 (Leatham & Stahmann, 1981), it is likely that the relatively abundant acid phosphatase has a role in the cell wall region. The natural substrate(s) for this enzyme is not known and thus deserves study. Perhaps phosphoprotein, phospho-β-D-glucan (e.g. myco-laminarin in *Phytophthora* (Wang & Bartnicki-Garcia, 1980)), or another phosphopolysaccharide (e.g. phospho-α-D-mannan in *Hansenula* (San-Blas & Cunningham, 1974)) is the natural substrate.

Laccase. Laccase active at pH 4.0 is produced by cultures of *L. edodes* grown on the chemically defined medium (Fig. 17.8). Confirming earlier work with cultures grown on non-defined media (Leatham & Stahmann, 1981), the laccase is in distinct locations, and the titre in each location changes with the growth stage. Laccase was initially excreted into the medium. Later, as the (non-pigmented) aerial mycelium formed, the laccase titre increased in the shockable fraction. The greatest increase occurred during formation of the pigmented primordia and fruit bodies. Because the laccase was extracellular in these aerial structures, it was found predominantly in the shockable fraction. These data suggest that the laccase, too, is developmentally regulated and, as discussed earlier (Leatham & Stahmann, 1981), plays an extracellular role in the cell wall region. The

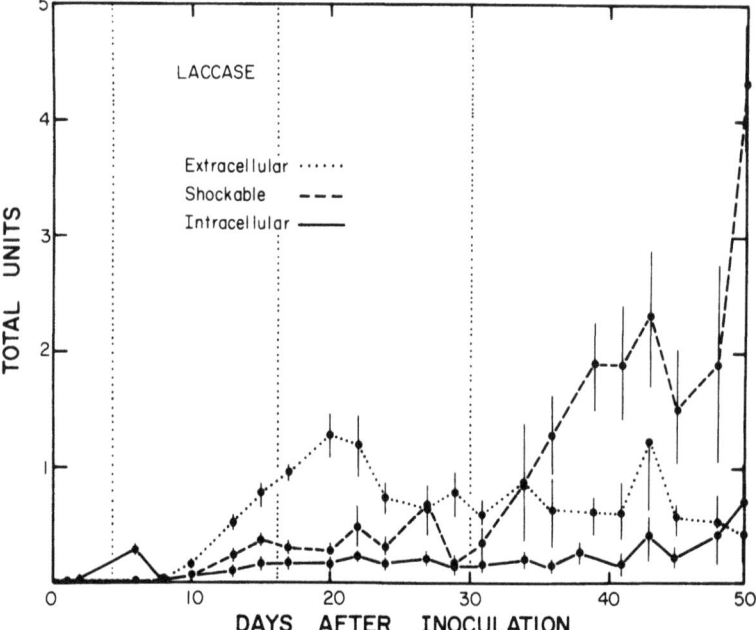

Fig. 17.8. Production and localisation of laccase in cultures grown in the chemically defined medium. Titres were determined at 22 °C on fractions from triplicate cultures in 50mM pH 4.0 sodium acetate (HCl) buffer with 1mM *o*-toluidine as substrate (increase in A_{600}, Leatham & Stahmann, 1981). Values are given for the total enzyme units (international units) per culture in the extracellular (dotted line), shockable (dashed line), and intracellular (solid line) fractions. Growth stages (see Fig. 17.2) are delineated (vertical dotted lines).

current study shows that intracellular roles are unlikely; little laccase was detected in the intracellular fraction at any growth stage (Fig. 17.8).

Saccharidases. Saccharidases active at pH 4.0 are produced by *L. edodes* and are in distinct locations (Figs 17.9 and 17.10). They included β-D-glucosidase, *N*-acetyl-β-D-glucosaminidase, α-D-galactosidase and β-D-mannosidase. As with the proteases, phosphatases and laccase, saccharidase titres became significant late in the vegetative growth stage and remained high throughout the incubation period. The enzyme in

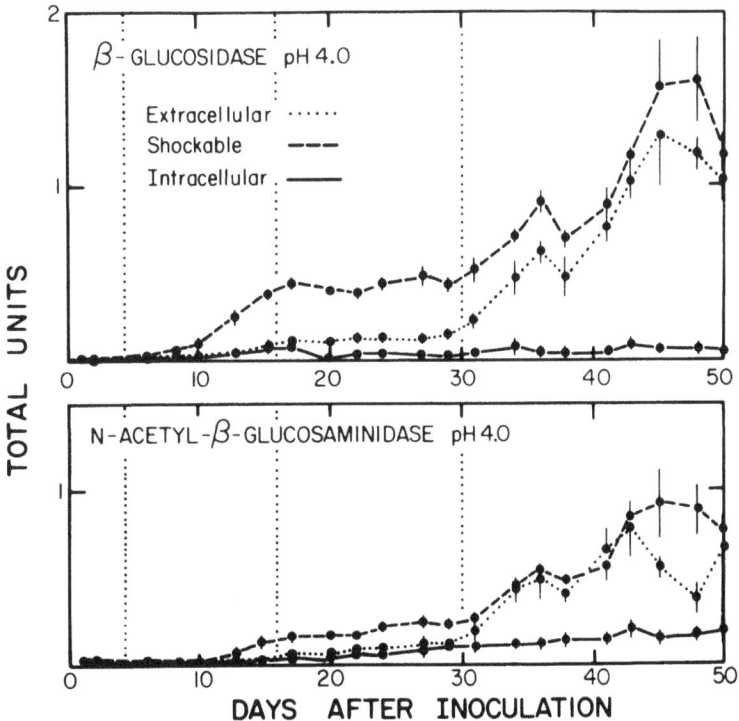

Fig. 17.9. Production and localisation of β-D-glucosidase (upper graph) and *N*-acetyl-β-D-glucosaminidase (lower graph) in cultures grown in the chemically defined medium. Titres were determined at 22 °C on fractions from triplicate cultures in 50mM pH 4.0 sodium acetate (HCl) buffer with 3.3mM *p*-nitrophenol-β-D-glucopyranoside or *p*-nitrophenol-2-acetami-do-2-deoxy-β-D-glucopyranoside as substrate, respectively (increase in A_{400}). Values are given for the total enzyme units (international units) per culture in the extracellular (dotted lines), shockable (dashed lines) and intracellular (solid lines) fractions. Growth stages (see Fig. 17.2) are delineated (vertical dotted lines).

highest titre was β-D-glucosidase, followed by N-acetyl-β-D-glucosamini-
dase. During the transition period, the saccharidases were in highest titre
in the shockable fraction. After the transition period, however, their titres
also increased in the extracellular fraction. With the exception of
β-D-mannosidase, the saccharidase titres all markedly increased as the
D-glucose (carbon source) in the growth medium became depleted (compare
Fig. 17.3 with Figs 17.9 and 17.10). During all growth stages only low and
moderately low activities of β-D-glucosidase and N-acetyl-β-D-gluco-
saminidase, respectively, were found in the intracellular fraction. In contrast,

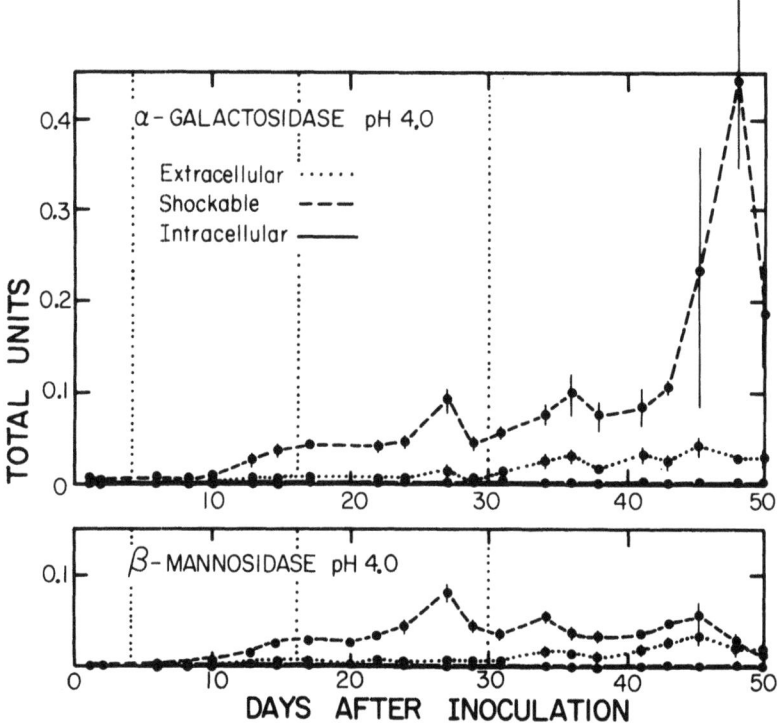

Fig. 17.10. Production and localisation of α-D-galactosidase (upper
graph) and β-D-mannosidase (lower graph) in cultures grown in the
chemically defined medium. Titres were determined at 22 °C on fractions
from triplicate cultures in 50mM pH 4.0 sodium acetate (HCl) buffer
with 3.3mM p-nitrophenol-α-D-galactopyranoside or p-nitrophenol-β-
D-mannopyranoside as substrate, respectively (increase in A_{400}). Values
are given for the total enzyme units (international units) per culture in
the extracellular (dotted lines), shockable (dashed lines) and intracellular
(solid lines) fractions. Growth stages (see Fig. 17.2) are delineated
(vertical dotted lines).

essentially no α-D-galactosidase or β-D-mannosidase activity was present in the intracellular fraction (Fig. 17.10). Based on these data it is possible that these saccharidases have roles in the cell wall of *L. edodes*. Their natural substrates may be the insoluble cell wall β-D-glucans, chitin and the water-soluble L-fuco-D-manno-α-D-galactan of *L. edodes*, or their partial degradation products. The low *N*-acetyl-β-D-glucosaminidase and β-D-glucanase activities in the intracellular fraction may be important. Chitin microfibrils are the innermost structural polymer in the cell wall of *L. edodes* and they are embedded in extensively branched β-D-glucans (Shida *et al.*, 1981). During collection of the shockable fraction these enzymes may be incompletely extracted from these insoluble substrates.

Summary and conclusions

The studies summarised above have shown that during growth on the chemically defined medium, *L. edodes* (1) efficiently uses D-glucose as carbon source; (2) efficiently uses L-glutamic acid as nitrogen source; (3) depletes the nitrogen source before the carbon source; (4) enters a transition stage probably triggered by nitrogen source depletion; (5) produces proteases; (6) slows in growth rate; (7) loses more than 80% of its total soluble protein; (8) produces and accumulates protease-resistant enzymes in specific locations, with appropriate pH optima, including proteases, phosphatase, laccase, β-D-glucosidase, *N*-acetyl-β-D-gluco-saminidase, α-D-galactosidase and β-D-mannosidase; and (9) produces fruit bodies as net growth (increased dry mass) finally stops. Based on these observations, it is likely that nitrogen limitation regulates development in *L. edodes* and that, in part, the regulatory mechanism involves the new synthesis and the proteolytic turnover and/or modification of key enzymes.

The sequential fractionation technique developed for this study was effective in localising soluble components in the extracellular, shockable or intracellular fractions. Those in the shockable fraction clearly merit further study. During the vegetative growth stage, half of the total soluble protein was in the shockable fraction. Afterwards, acid phosphatase, laccase, β-D-glucosidase, *N*-acetyl-β-D-glucosaminidase, α-D-galactosidase and β-D-mannosidase were in highest titre in the shockable fraction. These enzymes presumably were extracted from the cell wall region and all may have roles related to development.

Further use of fractionation/localisation techniques and chemically defined media will facilitate research into the roles of these and other enzymes in fungal development. The practical application of this knowledge should result in more reliable industrial methods for producing edible mushrooms.

Acknowledgements. The author thanks K. R. Bogart (Microbiologist, Forest Products Laboratory) for excellent technical assistance.

References

Ando, M. (1974). Fruit-body formation of *Lentinus edodes* (Berk.) Sing. on the artificial media. *Mushroom Science*, **9**, 415–22.

Barreto-Bergter, E. & Gorin, P. A. J. (1983). Structural chemistry of polysaccharides from fungi and lichens. *Advances in Carbohydrate Chemistry and Biochemistry*, **41**, 67–103.

Bartnicki-Garcia, S. (1973). Fundamental aspects of hyphal morphogenesis. *Symposia of the Society for General Microbiology*, **23**, 245–67.

Bojović-Cvetić, D. & Vujicić, R. (1982). Acid phosphatase localization and distribution in *Aspergillus flavus. Transactions of the British Mycological Society*, **79**, 137–41.

Carib, E. & Ulane, R. (1973). Chitin synthetase activating factor from yeast, a protease. *Biochemical and Biophysical Research Communications*, **50**, 186–91.

De Vries, O. M. H., Hoge, J. H. C. & Wessels, H. G. H. (1980). Translation of RNA from *Schizophyllum commune* in wheat germ and rabbit reticulocyte cell-free system. Comparison of *in vitro* and *in vivo* products after two-dimensional gel electrophoresis. *Biochimica et Biophysica Acta*, **563**, 100–12.

De Vries, O. M. H. & Wessels, J. G. H. (1984). Patterns of polypeptide synthesis in non-fruiting monokaryons and a fruiting dikaryon of *Schizophyllum commune. Journal of General Microbiology*, **130**, 145–54.

Farr, D. F. (1983). Mushroom industry: diversification with additional species in the United States. *Mycologia*, **74**, 351–60.

Fèvre, M. (1977). Subcellular localization of glucanase and cellulase in *Saprolegnia monoica* Pringsheim. *Journal of General Microbiology*, **103**, 287–95.

Field, C. & Schekman, R. (1980). Localized secretion of acid phosphatase reflects the pattern of cell surface growth in *Saccharomyces cerevisiae. Journal of Cell Biology*, **86**, 123–8.

Gooday, G. W. & Trinci, A. J. P. (1980). Wall structure and biosynthesis in fungi. *Symposia of the Society for General Microbiology*, **30**, 207–51.

Gruen, H. E. & Wong, W. M. (1982). Distribution of cellular amino acids, protein, and total organic nitrogen during fruitbody development in *Flammulina velutipes*. II. Growth on potato-glucose solution. *Canadian Journal of Botany*, **60**, 1342–51.

Gurusiddaiah, S., Blanchette, R. A. & Shaw, C. G. (1978). A modified technique for the determination of fungal mass in decayed wood. *Canadian Journal of Forestry Research*, **8**, 486–90.

Han, Y. H., Ueng, W. T., Chen, L. C. & Cheng, S. (1981). Physiology and ecology of *Lentinus edodes* (Berk.) Sing. *Mushroom Science*, **11**, 623–58.

Hänseler, E., Nyhlén, L. E. & Rast, D. M. (1983). Isolation and properties of chitin synthetase from *Agaricus bisporus* mycelium. *Experimental Mycology*, **7**, 17–30.

Ishikawa, H., Oki, T. & Senba, Y. (1983). Changes in the activities of extracellular enzymes during fruiting of the mushroom *Lentinus edodes* (Berk.) Sing. *Mokuzai Gakkaishi*, **29**, 280–7.

Kamada, T., Fuji, T. & Takemaru, T. (1980). Stipe elongation during basidiocarp maturation in *Coprinus macrorhizus*: changes in activity of cell wall lytic enzymes. *Transactions of the Mycological Society of Japan*, **21**, 359–67.

Kamada, T. & Takemaru, T. (1983). Modifications of cell-wall polysaccharides during stipe elongation in the basidiomycete *Coprinus cinereus*. *Journal of General Microbiology*, **129**, 703–9.

Kawamura, N. & Goto, M. (1980). Changes in the activities of some TCA cycle enzymes in *Lentinus edodes* mycelia after replacement of the culture medium with a glutamate medium. *Transactions of the Mycological Society of Japan*, **21**, 523–6.

Lampen, J. O. (1968). External enzymes of yeast: their nature and formation. *Antonie van Leeuwenhoek Journal of Microbiology*, **34**, 1–18.

Leatham, G. F. (1982). Cultivation of shiitake, the Japanese forest mushroom, on logs: a potential industry for the United States. *Forest Products Journal*, **32**, 29–35.

Leatham, G. F. (1983). A chemically defined medium for the fruiting of *Lentinus edodes*. *Mycologia*, **75**, 905–8.

Leatham, G. F. (1985). Stimulatory effect of nickel or tin on the fruiting of *Lentinus edodes*. *Transactions of the British Mycological Society*. (in press)

Leatham, G. F. & Kirk, T. K. (1983). Regulation of ligninolytic activity by nutrient nitrogen in white-rot basidiomycetes. *FEMS Microbiology Letters*, **16**, 65–7.

Leatham, G. F. & Stahmann, M. A. (1981). Studies on the laccase of *Lentinus edodes*: specificity, localization and association with the development of fruiting bodies. *Journal of General Microbiology*, **125**, 147–57.

Legerton, T. L. & Weiss, R. L. (1979). Mobilization of sequestered metabolites into degradative reactions by nutritional stress in *Neurospora*. *Journal of Bacteriology*, **138**, 909–14.

Leonard, T. J. & Phillips, L. E. (1973). Study of phenoloxidase activity during the reproductive cycle in *Schizophyllum commune*. *Journal of Bacteriology*, **114**, 7–10.

Loomis, W. F., Jr (1969). Developmental regulation of alkaline phosphatase. *Journal of Bacteriology*, **100**, 417–22.

Madelin, M. F. (1956). Studies on the nutrition of *Coprinus lagopus* Fr., especially as affecting fruiting. *Annals of Botany, New Series*, **20**, 307–30.

Marzluf, G. A. (1981). Regulation of nitrogen metabolism and gene expression in fungi. *Microbiological Reviews*, **45**, 437–61.

Meyer, R., Parish, R. W. & Hohl, H. R. (1976). Hyphal tip growth in *Phytophthora*: gradient distribution and ultrahistochemistry of enzymes. *Archives of Microbiology*, **110**, 215–24.

Miyake, H., Takemaru, T. & Ishikawa, T. (1980). Sequential production of enzymes and basidiospore formation in fruiting bodies of *Coprinus macrorhizus*. *Archives of Microbiology*, **126**, 201–5.

Moore, D., Elhiti, M. M. Y. & Butler, R. D. (1979). Morphogenesis of the carpophore of *Coprinus cinereus*. *New Phytologist*, **83**, 695–722.

North, M. J. (1982). Comparative biochemistry of the proteinases of eukaryotic microorganisms. *Microbiological Reviews*, **46**, 308–40.

Novaes-Ledieu, M. & Garcia Mendoza, C. (1981). The cell walls of *Agaricus bisporus* and *Agaricus campestris* fruiting body hyphae. *Canadian Journal of Microbiology*, **27**, 779–87.

O'Day, D. H. & Horgen, P. A. (1974). The developmental patterns of lysosomal enzyme activities during Ca^{++} induced sporangium formation in *Achlya bisexualis*. I. Acid phosphatase. *Developmental Biology*, **39**, 116–24.

Plunkett, B. E. (1953). Nutritional and other aspects of fruit body production in pure cultures of *Collybia velutipes* (Curt.) Fr. *Annals of Botany, New Series*, **17**, 193–216.

Poirier, T. P. & Holt, S. C. (1983*a*). Acid and alkaline phosphatases of *Capnocytophaga* species. I. Production and cytological localization of the enzymes. *Canadian Journal of Microbiology*, **29**, 1350–60.

Poirier, T. P. & Holt, S. C. (1983*b*). Acid and alkaline phosphatases of *Capnocytophaga* species. II. Isolation, purifications, and characterization of the enzymes from *Capnocytophaga ochracea*. *Canadian Journal of Microbiology*, **29**, 1361–8.

Poirier, T. P. & Holt, S. C. (1983*c*). Acid and alkaline phosphatases of *Capnocytophaga* species. III. The relationship of the enzymes to the cell wall. *Canadian Journal of Microbiology*, **29**, 1369–81.

Righelato, R. C. (1975). Growth kinetics of mycelial fungi. In *The Filamentous Fungi*, vol. 1, *Industrial Mycology*, ed. J. E. Smith & D. R. Berry, pp. 79–103. London: Edward Arnold.

Rosenberger, R. F. (1976). The cell wall. In *The Filamentous Fungi*, vol. 2, *Biosynthesis and Metabolism*, ed. J. E. Smith & D. R. Berry, pp. 328–44. New York: Wiley.

Ross, I. K. (1982). The role of laccase in carpophore initiation in *Coprinus congregatus*. *Journal of General Microbiology*, **128**, 2763–70.

Royse, D. J. & Schisler, L. C. (1980). Mushrooms – their consumption, production and culture development. *Interdisciplinary Science Reviews*, **5**, 324–32.

San Antonio, J. P. (1981). Cultivation of the shiitake mushroom. *Horticulture*, **16**, 151–6.

San-Blas, G. & Cunningham, W. L. (1974). Structure of the cell wall and exocellular mannans from the yeast *Hansenula holstii*. I. Mannans produced in phosphate-containing medium. *Biochimica et Biophysica Acta*, **354**, 233–46.

Schwalb, M. N. (1977). Developmental regulated proteases from the basidiomycete *Schizophyllum commune*. *Journal of Biological Chemistry*, **252**, 8435–9.

Schwalb, M. N. (1978). Regulation of fruiting. In *Genetics and Morphogenesis in the Basidiomycetes*, ed. M. N. Schwalb & P. G. Miles, pp. 135–65. New York: Academic Press.

Shida, M., Haryu, K. & Matsuda, K. (1975). On the water-soluble heterogalactan from the fruit-bodies of *Lentinus edodes*. *Carbohydrate Research*, **41**, 211–18.

Shida, M., Uchida, T. & Matsuda, K. (1978). A $(1 \rightarrow 3)$-α-D-glucan isolated from the fruit bodies of *Lentinus edodes*. *Carbohydrate Research*, **60**, 117–27.

Shida, M., Ushioda, Y., Nakajima, T. & Matsuda, K. (1981). Structure of the alkali-insoluble skeletal glucan of *Lentinus edodes*. *Journal of Biochemistry*, **90**, 1093–100.

Sietsma, J. H. & Wessels, J. G. H. (1977). Chemical analysis of the hyphal wall of *Schizophyllum commune*. *Biochimica et Biophysica Acta*, **496**, 225–39.

Sietsma, J. H. & Wessels, J. G. H. (1979). Evidence for covalent linkages between chitin and β-glucan in a fungal wall. *Journal of General Microbiology*, **114**, 99–108.

Sietsma, J. H. & Wessels, J. G. H. (1981). Solubility of $(1 \rightarrow 3)$-β-D-/$(1 \rightarrow 6)$-β-D-glucan in fungal walls: importance of presumed linkage between glucan and chitin. *Journal of General Microbiology*, **125**, 209–12.

Spector, T. (1978). Refinement of the Coomassie blue method of protein quantitation. *Analytical Biochemistry*, **86**, 142–6.

Sonnenberg, A. S. M., Sietsma, J. H. & Wessels, J. G. H. (1982). Biosynthesis of alkali-insoluble cell wall glucan in *Schizophyllum commune* protoplasts. *Journal of General Microbiology*, **128**, 2667–74.

Terashita, T., Kono, M. & Murao, S. (1978). Effect of streptomyces pepsin inhibitor on the fruit body formation of a few basidiomycetes. *Hakkokogaku Kaishi*, **56**, 175–81.

Terashita, T., Oda, K., Kono, M. & Murao, S. (1981). Purification and some properties of carboxyl proteinase in mycelium of *Lentinus edodes*. *Agricultural and Biological Chemistry*, **45**, 1929–35.

Tokimoto, K. & Fukuda, M. (1981). Relation between mycelium quantity and fruit-body yield in *Lentinus edodes* bed-logs. *Taiwan Mushrooms*, **5**, 1–5.

Tokimoto, K. & Kawai, A. (1975). Nutritional aspects on fruit-body development in replacement culture of *Lentinus edodes* (Berk.) Sing. *Report of the Tottori Mycological Institute (Japan)*, **12**, 25–30.

Tortora, P., Birtel, M., Lenz, A.-G. & Holzer, H. (1981). Glucose-dependent metabolic interconversion of fructose-1,6-bis-phosphatase in yeast. *Biochemical and Biophysical Research Communications*, **100**, 688–95.

Wang, M. C. & Bartnicki-Garcia, S. (1980). Distribution of mycolaminarans and cell wall β-glucans in the life cycle of *Phytophthora*. *Experimental Mycology*, **4**, 269–80.

Wessels, J. G. H. (1965). Biochemical processes in *Schizophyllum commune*. *Wentia*, **13**, 1–113.

Wessels, J. G. H. & Sietsma, J. H. (1981). Fungal cell walls: a survey. In *Encyclopedia of Plant Physiology*, vol. 13B, ed. W. Tanner & F. A. Loewus, pp. 352–94. Heidelberg: Springer-Verlag.

Wessels, J. G. H., Sietsma, J. H. & Sonnenberg, A. S. M. (1983). Wall synthesis and assembly during hyphal morphogenesis in *Schizophyllum commune*. *Journal of General Microbiology*, **129**, 1607–16.

Wilson, R. W. & Niederpruem, D. J. (1967). Control of β-glucosidases in *Schizophyllum commune*. *Canadian Journal of Microbiology*, **13**, 1009–20.

Wood, D. A. & Goodenough, P. W. (1977). Fruiting of *Agaricus bisporus*. Changes in extracellular enzyme activities during growth and fruiting. *Archives of Microbiology*, **114**, 161–5.

18

Ultrastructural aspects of fruit body differentiation in *Flammulina velutipes*

M.A.J.WILLIAMS, A.BECKETT and N.D.READ
Department of Botany, University of Bristol, Woodland Road, Bristol BS8 1UG,
UK

Introduction
Flammulina velutipes is a common agaric of the family Tricholo-mataceae. The centrally stipitate fleshy fruit body is light brown in colour and is commonly found in the winter months of the year growing on dead stumps and logs of a wide range of deciduous trees, only occasionally growing on living trees. The fruit body is frost-tolerant and can survive repeated freezing and thawing (Stewart, 1918; Ingold, 1978, 1981). The fungus is edible, and is grown commercially in the Far East (Tonomura, 1978).

F. velutipes fruits readily in culture and has been used as an experimental system to investigate environmental effects on fruit body production. It has also been used extensively by Gruen and co-workers for studying the physiology and biochemistry of fruiting, especially the relationship of the pileus to stipe elongation. Table 18.1 summarises published work on fruit body development in *F. velutipes*.

The study of the developmental morphology of *F. velutipes* is largely restricted to the early work of Biffen (1898) and Moss (1923). These studies, though meticulous, were restricted by the available methods of investigation. This chapter describes the preliminary results of a study of the developmental morphology of the fruit body of *F. velutipes* using correlated techniques for microscopy.

The cultured material used throughout this study was derived from four wild strains of *F. velutipes* isolated from different sites in the Bristol area. The fungus was grown in glass Petri dishes on a 0.5% malt extract agar medium supplemented with finely ground elm sawdust. Incubation was at 25 °C in darkness for 7 days followed by a temperature of 18 °C in continuous light. Primordia were formed about 40 days after inoculation.

Table 18.1. *Summary of previous work on aspects of fruit body structure and development in* Flammulina velutipes

Aspect studied	References
Developmental anatomy	Biffen (1898); Moss (1923); Ingold (1980*a*)
Genetics and cytology	Aschan (1954*a*); Pihakaski (1972); Wong & Gruen (1977)
Physiology	Gruen (1969, 1976, 1979, 1982); Gruen & Wu (1972*a*, *b*); Kitamoto & Gruen (1976)
Environmental effects	
Nutrition	Hawker (1942); Plunkett (1953); Aschan (1954*b*); Aschan-Åberg (1958); Gruen & Wong (1982*a*, *b*)
Light	Aschan(1954*b*);Plunkett(1956);Aschan-Åberg(1960)
Aeration	Plunkett (1956); Long (1966)
Temperature	Aschan-Åberg (1958, 1960); Ingold (1978, 1981)
pH	Plunkett (1953); Aschan (1954*b*)
Competition	Gruen (1983)
Taxonomy	Bas (1983)
Reviews	Tonomura (1978); Ingold (1980*b*)

The preparatory procedures used for microscopy are summarised in Table 18.2. Further details of the techniques in relation to interpretation of the results will be presented where relevant.

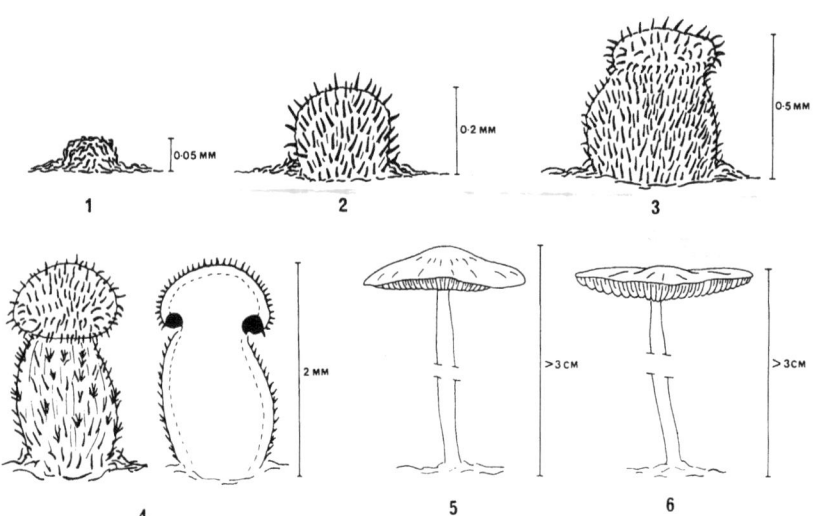

Fig. 18.1. Diagrammatic representation of the developmental stages of the *Flammulina velutipes* fruit body.

Table 18.2. *Preparative methods for microscopy*

Type of microscopy	Methods of preparation
Light microscopy	Glutaraldehyde/osmium tetroxide fixed. Resin-embedded sections stained with: 1. Methylene blue/Azure II (O'Brien & McCully, 1981) or 2. Periodic acid-Schiff's reagent (PAS) (Jensen, 1962)
Transmission electron microscopy	Glutaraldehyde/OsO_4-fixed. Resin-embedded sections stained with lead citrate (Reynolds, 1963) and uranyl acetate.
Scanning electron microscopy	1. Glutaraldehyde/OsO_4-fixed. Critical point-dried (Read, Porter & Beckett, 1983) 2. Cryo-fixed, freeze-fractured and freeze-dried (Read & Beckett, 1985) 3. Cryo-fixed and examined frozen-hydrated (Beckett, Porter & Read, 1982; Read *et al.*, 1983) 4. Cryo-fixed, freeze-fractured and examined frozen-hydrated (Beckett & Porter, 1982; Beckett *et al.*, 1982; Read & Beckett, 1985) 5. Cryo-fixed and partially freeze-dried (etched) on cold stage in microscope. Examined partially freeze-dried.

The sequence of development

A sequence of stages observed macroscopically in the development of a fruit body of *F. velutipes* is illustrated in Fig. 18.1. The first gross morphological indication of fruit body initiation was the formation of a hyphal aggregation in a much-branched region of the dikaryotic mycelium. This aggregation consisted of widely spaced, interwoven, branched hyphae (Fig. 18.1, stage 1; Fig. 18.2*a*). 'Aspergilloid hyphae' (Ingold, 1980*a*) were sometimes found in its vicinity (not illustrated). As the aggregation enlarged in diameter the outermost hyphae gave rise to numerous tapering cystidial elements, uniformly distributed over the surface, which gave the young primordium a spiky appearance (Fig. 18.1, stage 2; Fig. 18.2*b*).

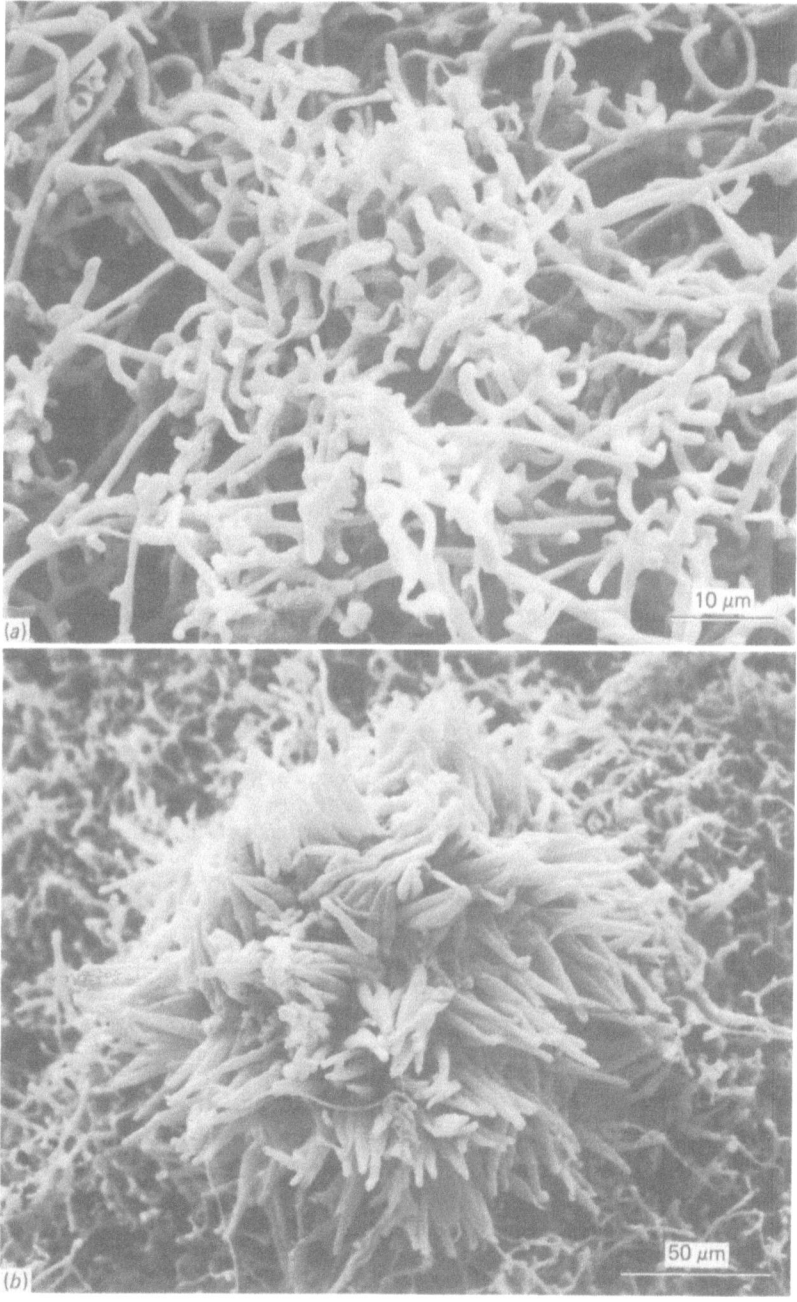

Fig. 18.2. (*a*) Scanning electron micrograph of a hyphal aggregation on the culture surface (Fig. 18.1, stage 1). Note the highly branched nature of the surrounding mycelium and of the aggregation itself. Critical point-dried. (*b*) Scanning electron micrograph of a young primordium completely covered in tapering cystidia (Fig. 18.1, stage 2). Critical point-dried.

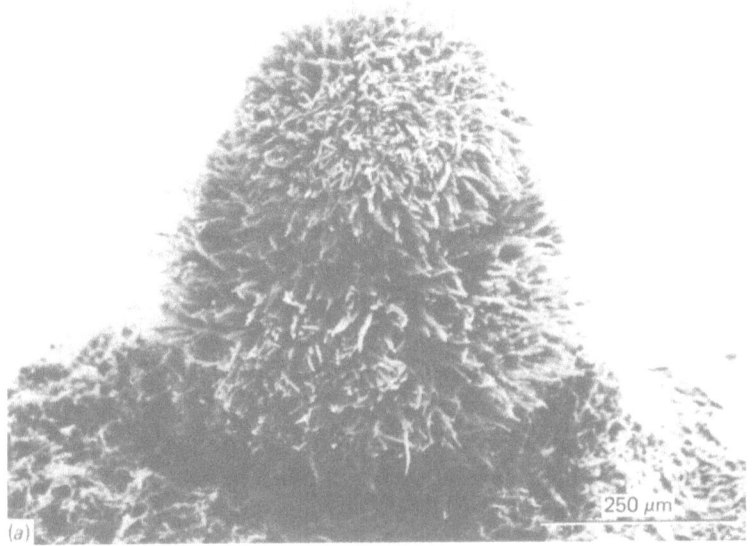

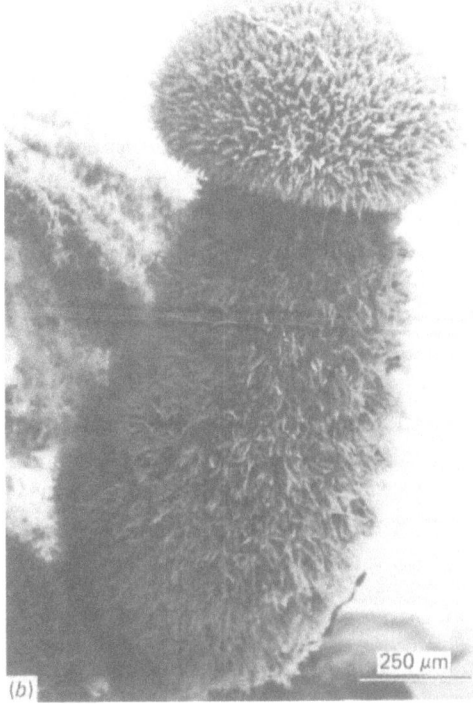

Fig. 18.3. (*a*) Scanning electron micrograph of a primordium showing the first indication of pileal differentiation towards the tip (Fig. 18.1, stage 3). Critical point-dried. (*b*) Scanning electron micrograph of a young fruit body showing distinct differentiation into stipe and pileus (Fig. 18.1, stage 4). Note uniform distribution of cystidia on the pileus. Critical point-dried.

With further increase in size the primordium became slightly elongated about its vertical axis and at the apex the rudimentary pileus became differentiated (Fig. 18.1, stage 3; Fig. 18.3a). Cystidia continued to be produced over the entire primordium surface.

Further expansion and elongation resulted in a primordium with the typical form of a mushroom in miniature (Fig. 18.1, stage 4; Fig. 18.3b). Sectioned material of this stage showed the basic organisation of tissues into a stipe, pileus and hymenium (albeit rudimentary) as found in the mature fruit body. Increase in size up to this stage was a slow process but once the basic form had differentiated, rapid elongation and expansion resulted in the mature fruit body form (Fig. 18.1, stage 5). The inrolled edge of the pileus curved back to expose the gills. Spore discharge commenced and as it continued the convex upper surface of the pileus flattened out (Fig. 18.1, stage 6).

Primordia developing in any one culture did not show synchrony of development and their number varied in different cultures. The number of fruit bodies produced in culture was apparently not dependent on the strain of fungus used. Under the same incubation conditions different cultures of the same strain produced different numbers of primordia that matured at different rates. The number of fruit bodies that fully matured in a culture was also variable. Some primordia and young fruit bodies at any stage of development aborted, that is, they ceased to develop further (see also Ingold, 1980a). The size of the primordium was not an absolute indication of its stage of development because primordia of the same dimensions could have been at different stages of development. Whether *F. velutipes* forms sclerotia in culture, as do various species of *Coprinus* (see Henderson, Elliott & Ross, 1983, and references cited therein), was not clear because sclerotia may have been indistinguishable from the early stages of primordium development.

The structural organisation of the developing fruit body

For convenience, the sectioned fruit body primordia at stage 4 of Fig. 18.1 have been divided into labelled sites (Fig. 18.4). The organisation and form of the cells within these areas have been studied and the details and illustrations below relate specifically to them.

The stipe

In sectioned and freeze-fractured material the tissue of the stipe was composed of mainly elongated, longitudinally aligned elements arranged parallel to each other (Figs 18.5, 18.6a, b and 18.7a). Hyphal branching was not common. At the base of the stipe this regular arrangement was not apparent (not illustrated).

Structurally, two distinct zones could be recognised in the stipe tissue (Fig. 18.5*a*) as described by Biffen (1898) and Moss (1923). The outer, cortical region (Fig. 18.4, site 1b) was composed of narrow, closely packed elements which stained densely. Seen in the transmission electron microscope (TEM) (Fig. 18.5*b*), these outer cortical cells had electron-dense contents. The outermost layer of the cortex (Fig. 18.4, site 1a) gave rise to the cystidia on the surface of the stipe (see Fig. 18.8*c*). The elements of the inner zone of the stipe (Fig. 18.4, site 1c) were broader, less tightly

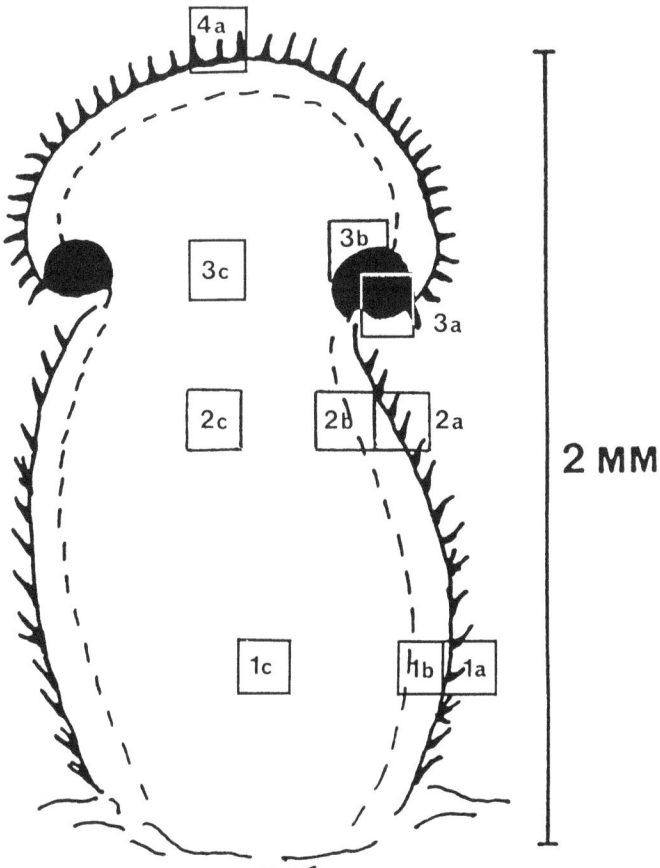

Fig. 18.4. Diagram of a sectioned stage 4 fruit body showing the locations of tissues studied by electron microscopy and indicated by site numbers in the text and legends. Site 3b is not illustrated in this chapter.

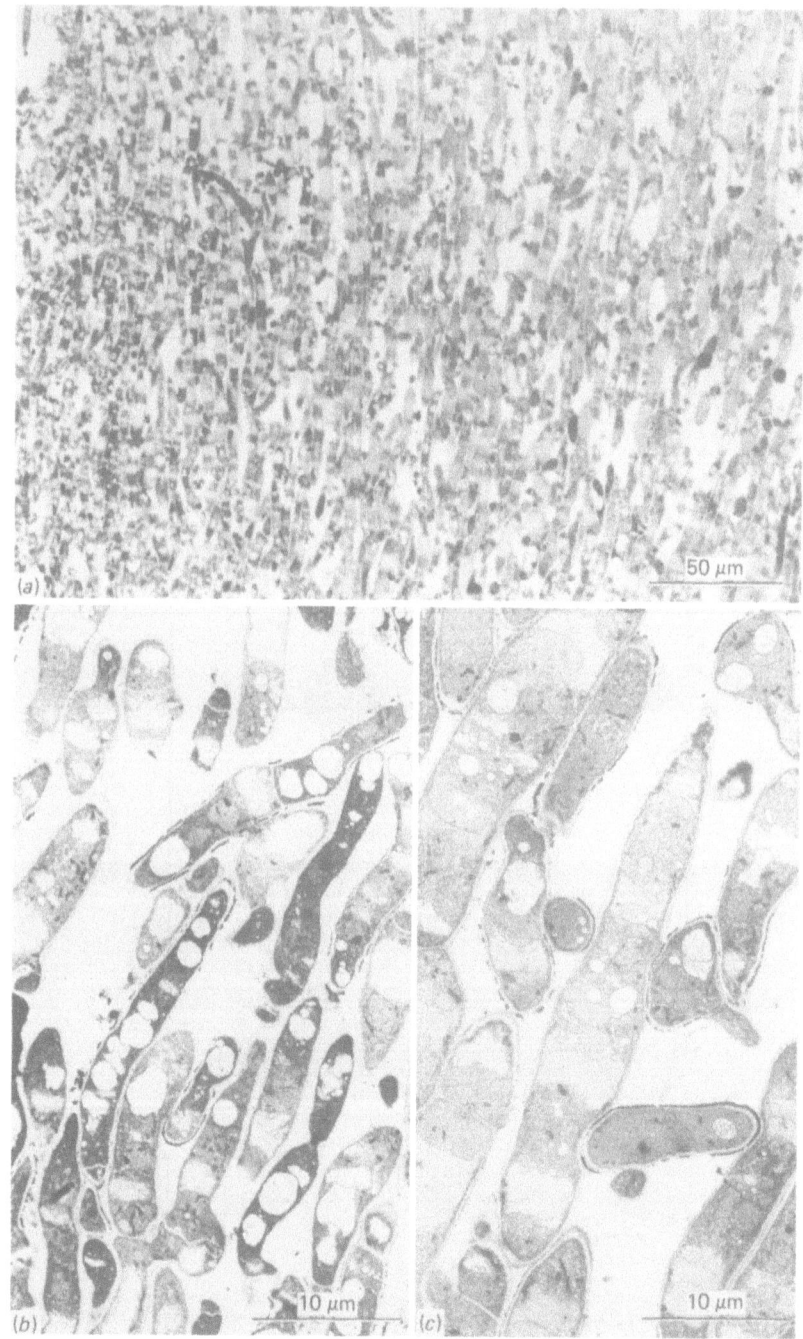

packed, and the cell contents stained less densely (Fig. 18.5 *a, c*). There were occasional cells oriented transversely to the longitudinal axis of the stipe. In longitudinal and transverse fractures of the stipe the different arrangements of the elements in the two zones were also apparent in the scanning electron microscope (Fig. 18.6). The close packing of narrow elements in the cortex coupled with the looser packing of broader elements in the central stipe region, which is often hollow in mature fruit bodies, is the optimal distribution of elements to resist bending stresses (Buller, 1909).

The pileus

At the junction of the stipe and pileus, the outermost elements of the stipe curved slightly towards the centre of the fruit body. In this region (Fig. 18.4, site 3c) the regular arrangement of the elongated stipe cells was lost (Fig. 18.7 *a, c*). The elements of the pilear context were not regularly arranged. Hyphae ran in all directions, were probably interwoven (see below) and, like the stipe elements, were embedded in extracellular material. Some of the elements in this zone were very wide and some were branched (Fig. 18.7 *c*).

The elements of the developing hymenium were narrow in diameter and had densely staining contents (Fig. 18.7 *a, b*).

There was a diffuse layer of tissue in the angle between the developing gill and the stipe in which some cellular elements had little or no contents. These apparently necrotic cells were embedded in extracellular material (Fig. 18.7 *a*). This evanescent layer was described by Moss (1923) and on the basis of his description Reijnders (1963) classed the development of *F. velutipes* as paravelangiocarpic (Watling: Chapter 11).

The pileus, like the stipe, had a densely packed cortical layer of elements, the pileipellis. The outer elements of this zone (Fig. 18.4, site 4a) gave rise to the pileocystidia (see Fig. 18.9 *a*).

Fig. 18.5. (*a*) Brightfield light micrograph of a longitudinal section through a young stipe at stage 4 (Fig. 18.1). Dense-staining cortical elements are on the left side, central elements on the right. Stained with methylene blue/Azure II. (*b*) Transmission electron micrograph of a longitudinal section of elements in the cortical region of the stipe (Fig. 18.4, site 1b). (*c*) Transmission electron micrograph of a longitudinal section through elements in site 1c (Fig. 18.4). Note that the elements are wider in diameter and have less-densely staining contents than those in (*b*). Note transverse elements, some in cross-sectional view.

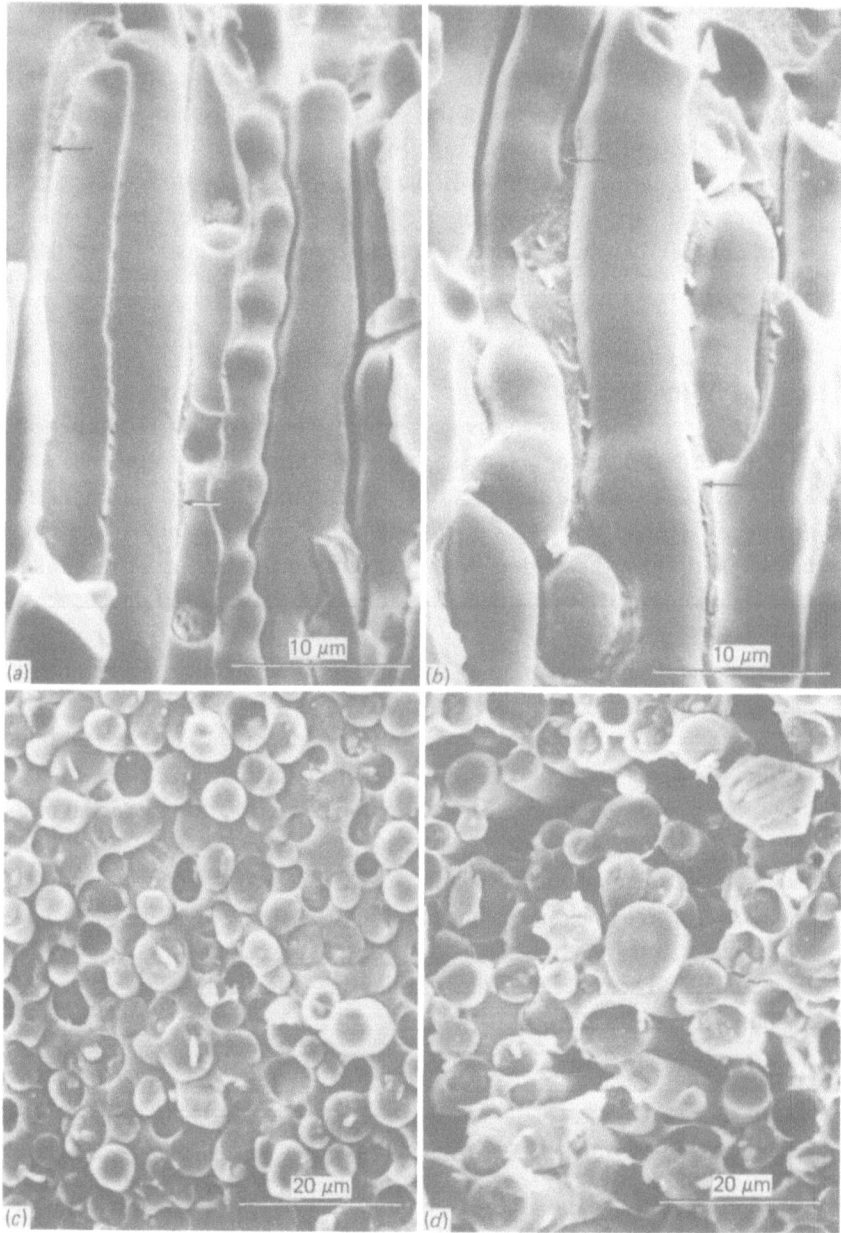

Fig. 18.6. (*a*) Scanning electron micrograph of a longitudinal fracture through the cortex of the stipe (Fig. 18.4, site 2b). (Compare with Fig. 18.5*b*.) Arrows point to extracellular material. Freeze-fractured, frozen-hydrated. (*b*) Scanning electron micrograph of a longitudinal fracture through the central region of the stipe (Fig. 18.4, site 2c). (Compare also

Cystidia

Cystidial elements were present on all external surfaces of the primordium from an early stage. The form of these cystidia was constant, comprising a bulbous base and tapered apex (Fig. 18.8c, d). Occasionally they were branched (not illustrated). At stage 4 (Fig. 18.1), the spatial distribution of the cystidia on the surface of the primordium was not uniform. On the stipe, cystidia were grouped in distinct clumps (Fig. 18.8e), while on the pileus, they formed a continuous covering (Fig. 18.3b). Whether the cystidia occurred on the stipe, the top surface of the pileus or amongst the basidia of the hymenium their origin was always from the underlying undifferentiated hypha-like elements (Fig. 18.8e). On the stipe, cystidia arose from longitudinally orientated elements (Fig. 18.8e) while on the pileus they arose from the irregular elements of the pileipellis (Fig. 18.9a, b).

The cystidia may have a role in mucilage secretion (see next section).

Extracellular material

Extracellular material with a mucilaginous appearance was present both externally in association with cystidia (Figs 18.7a, 18.8a, c, 18.9) and internally between hypha-like elements of the primordium (Figs 18.6a, b, 18.7a, b, 18.10c, d).

At all stages primordia developing on the culture surface sometimes had glistening droplets associated with them. These were described as 'exudation droplets' by Plunkett (1951). The amount of this apparently mucilaginous material was variable.

Cystidia on the stipe and pileus were embedded in a layer of extracellular material which stained weakly with the periodic acid-Schiff's reagent (Fig. 18.9a). This material sometimes formed a shroud covering the whole surface of the primordium with occasional cystidia poking through (Fig. 18.8a, b). The presence of mucilaginous droplets at the apices of cystidia suggests that cystidia may actually secrete extracellular material which then covers the outer surfaces. Biffen (1898) considered that cystidia might

with Fig. 18.5c.) Note the wide diameter of these elements compared with those in (a), the less-dense packing and the large amounts of extracellular material present (arrowed). Freeze-fractured, frozen-hydrated. (c) Scanning electron micrograph of a transverse fracture through the elements of the stipe cortex (Fig. 18.4, site 2b). Note close packing of elements. Freeze-fractured, frozen-hydrated. (d) Scanning electron micrograph of a transverse fracture through the elements of the central region of the stipe (Fig. 18.4, site 2c). Note wider diameters of elements and their looser packing compared with (c). Freeze-fractured, frozen-hydrated.

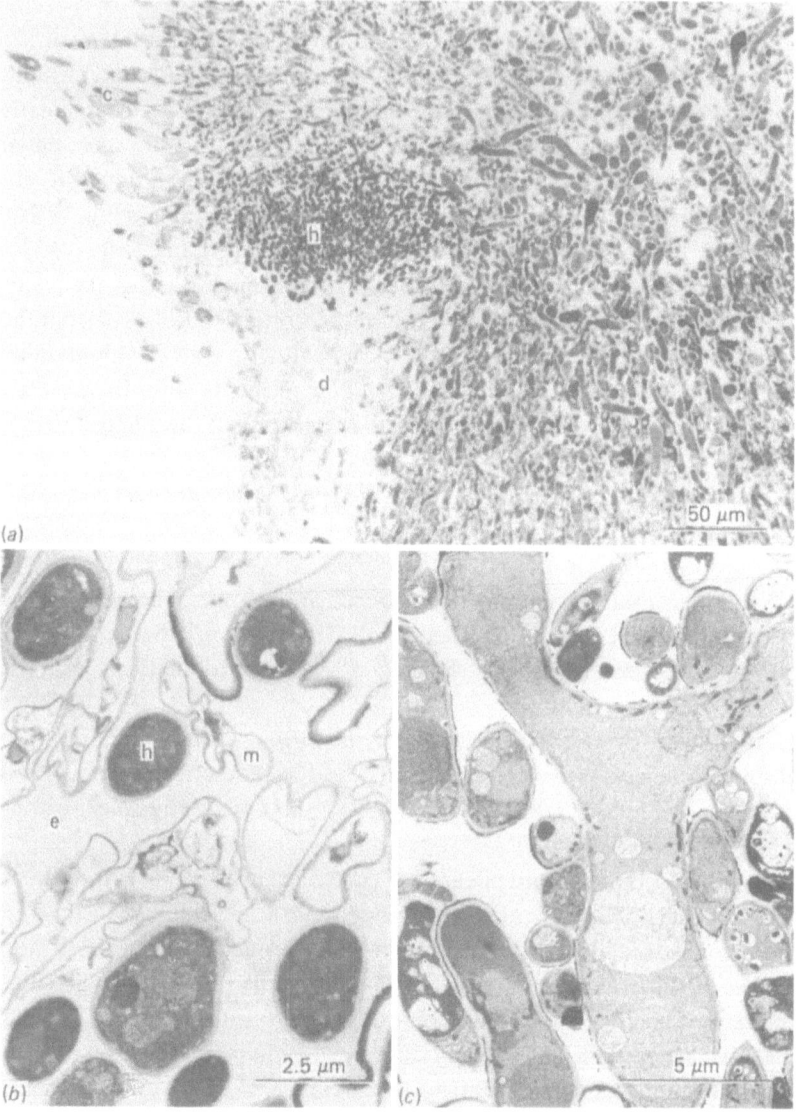

Fig. 18.7. (*a*) Brightfield light micrograph of a longitudinal section through part of the transition zone between stipe (below) and pileus (above). Note: irregular arrangement of pilear context compared with stipe; rudimentary hymenium (h); pileocystidia (c); and the diffuse layer (d) in the angle of the pileus and stipe, consisting largely of extracellular material with a few rudimentary hymenial elements and moribund cells embedded in it. Stained with methylene blue/Azure II. (*b*) Transmission electron micrograph of a longitudinal section through the diffuse region in (*a*) (Fig. 18.4, site 3a). Note: densely staining elements of the hymenium (h); moribund, walled elements with little or no contents (m); and the extracellular material (e). (*c*) Transmission electron micrograph of a longitudinal section through the pilear context (Fig. 18.4, site 3c). Note that elements seen in different orientations suggests interweaving of hyphae.

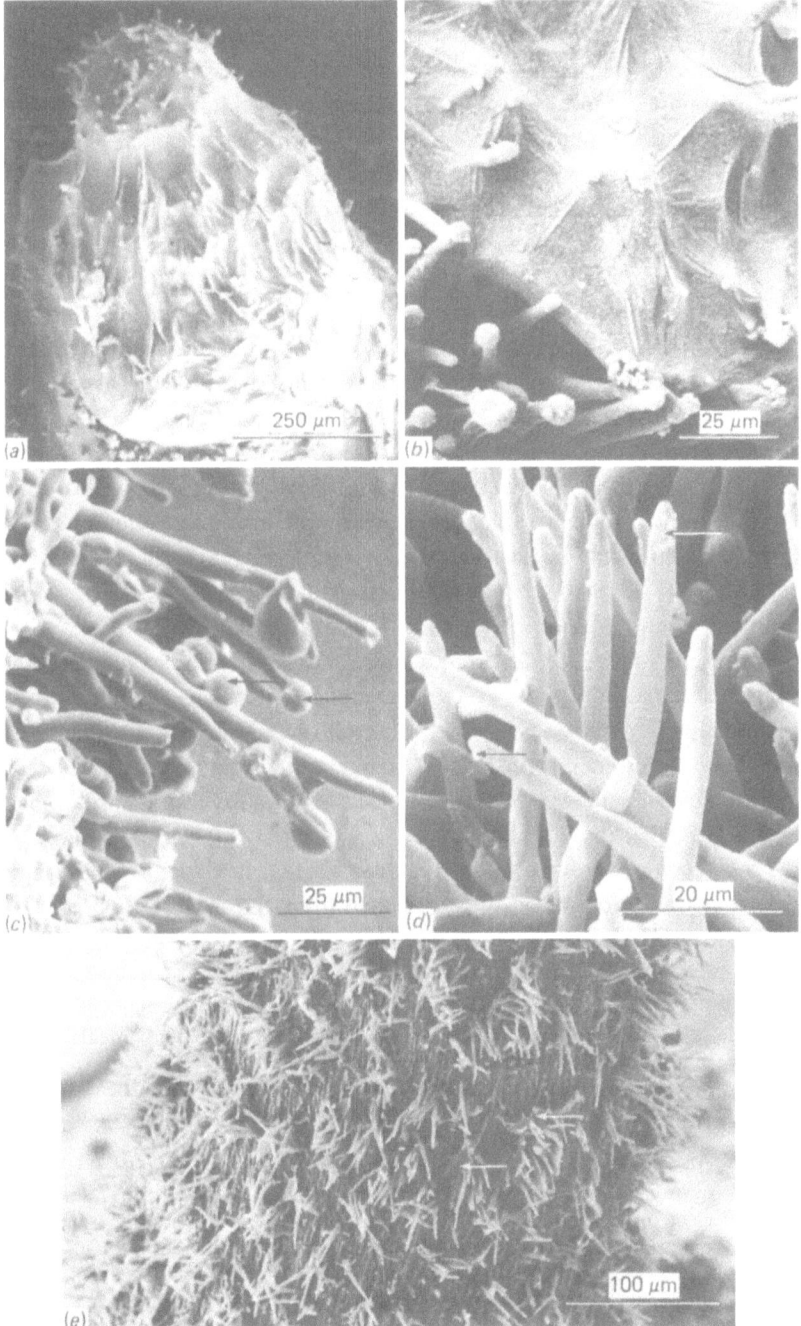

Fig. 18.8. For caption see p. 442.

produce mucilage but was unable to substantiate this; he also suggested that they might play a role in 'transpiring' water. Further evidence for the secretory role of cystidia is the large amount of vesiculate smooth endoplasmic reticulum (ER) seen in sections of cystidial tips (Fig. 18.9c). Large quantities of smooth ER are associated with higher plant glandular cells having a secretory function (Schnepf, 1974).

Ingold (1980a) speculated that a possible role for the cystidia in *F. velutipes* might be to suck into the fruit body aqueous fluid rich in organic substances, and to concentrate it by the excretion of excess water. However, our results do not support the interpretation of the cystidial droplets as being composed only of water. For example, in frozen-hydrated specimens many cystidia had droplets of extracellular material associated with them (Fig. 18.8c) and these droplets remained when superficial ice was removed by etching (not illustrated). In addition, in specimens which had been subjected to the action of solvents during preparation for scanning and transmission electron microscopy, insoluble components of both the external and internal extracellular material remained (Figs 18.7a, b, 18.8d, c, 18.9, 18.10c, d). In thin sections viewed with the TEM, these remnants often had a fibrous nature (Figs 18.7b, 18.9c, 18.10c, d).

In material prepared for transmission electron microscopy, Beckett (unpublished) found the external extracellular material to have a continuous, regular boundary layer which covered the whole primordium. This layer had a trilaminar appearance which resembles biological membranes prepared by the same method. It could be homologous with the trilaminar pellicle described by McLaughlin (1982) as overlying the hymenium of *Coprinus cinereus*. However, it is not clear whether the layer in *F. velutipes* was an artifact produced during specimen preparation.

Considerable amounts of extracellular material were present in the

---

Fig. 18.8. (a) Scanning electron micrograph of an entire primordium (Fig. 18.1, stage 4) shrouded in a dense layer of extracellular mucilaginous material. Frozen-hydrated. (b) Detail of extracellular layer over the pileocystidia. Creases in this coat suggest that it has dried slightly. Frozen-hydrated. (c) Scanning electron micrograph of a longitudinal fracture of the edge of the stipe (Fig. 18.4, site 2a) showing origin of the stipe cystidia. Note the drops of extracellular material associated with the cystidia (arrowed). Freeze-fractured, frozen-hydrated. (d) Scanning electron micrograph of some pileocystidia. Note remnants of extracellular material on the cystidial tips after chemical immersion fixation and dehydration (arrowed). Critical point-dried. (e) Scanning electron micrograph of the outer region of the stipe (Fig. 18.4, site 1a) showing cystidia in clumps and the longitudinal arrangement of the underlying elements. Note remnants of extracellular material (arrowed). Critical point-dried.

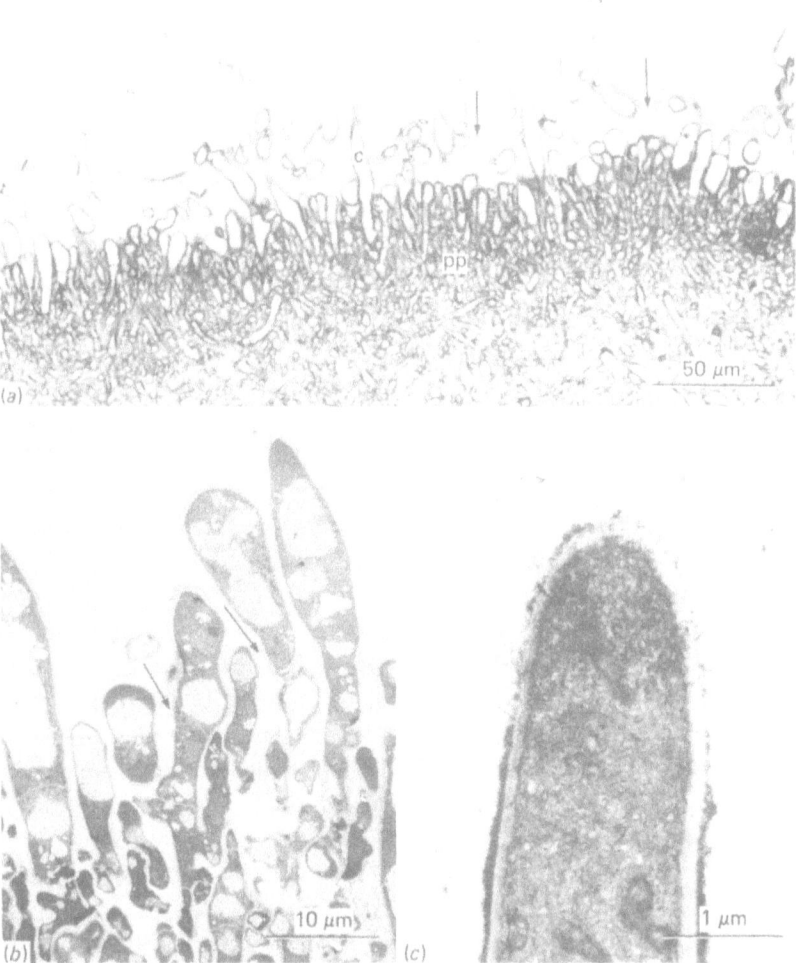

Fig. 18.9. (*a*) Brightfield light micrograph of a longitudinal section through the upper surface of the pileus (Fig. 18.4, site 4a). Note: densely packed tissue of the pileipellis (pp); pileocystidia (c); and extracellular material around the cystidia (arrowed). Also note that the walls of the cellular elements are strongly stained while the extracellular material is not. Stained with PAS. (*b*) Transmission electron micrograph of a longitudinal section through the pileocystidia. Note dense cytoplasm towards the cystidial tips and the extracellular material around the cystidia (arrowed). (*c*) Transmission electron micrograph of a longitudinal section through the tip of a pileocystidium. Note the very dense cytoplasm containing much vesicular smooth endoplasmic reticulum, and the fibrous remnants of extracellular material surrounding the apex of the cystidium.

interhyphal spaces in the tissues of the stipe (Figs 18.6*a*, *b*, 18.10*c*, *d*) and pileus (Fig. 18.7*a*, *b*). Extracellular mucilaginous material has also been reported as occurring between the cells of developing primordia in *C. cinereus* (van der Valk & Marchant, 1978), *Agaricus bisporus* and *A. sylvicola* (Angeli-Papa & Eyme, 1978), and *Schizophyllum commune* (Niederpruem & Wessels, 1969; van der Valk & Marchant, 1978).

The origin of the extracellular material within the fruit body is not known.

Although the function(s) of the extracellular material is(are) not clear, the following is a list of possible roles:

1. *Protection*. It may provide a protective coat helping to maintain a relatively stable environment around the primordium in adverse or changing conditions. Furthermore, it may act as a natural cryoprotective agent under freezing conditions.

2. *Chemical transport*. The extracellular material between elements of the stipe and pileus provides an uninterrupted medium around cells of the whole fruit body. As such it may provide a medium for the transmission of chemicals involved in cellular communication, particularly in relation to the formation of an organised fruit body (as suggested for *C. cinereus* by McLaughlin, 1982; see also Chapter 13). It may also be involved in the transport of nutrients and serve a similar role to that of the apoplast of plants. Schütte (1956) and Littlefield (1966) studied translocation in fruit bodies of *F. velutipes*, but they did not determine whether it occurred extracellularly.

3. *Lubrication*. The extracellular material may allow adjacent cells to slide against each other and, if necessary, elongate at different rates with respect to each other. A similar role, as a lubricant, has been suggested for the perithecial mucilage produced by *Sordaria humana* during fruit body morphogenesis (Read & Beckett, 1985).

4. *Hyphal binding* (see next section).

Structural integrity of the young fruit body

The developing primordium was essentially prosenchymatous being composed of readily identifiable, discrete hyphal elements of which most were not tightly appressed together. These elements must, nevertheless, be held together to maintain the primordium as an organised structure. Anatomical features which may help to maintain its structural integrity include:

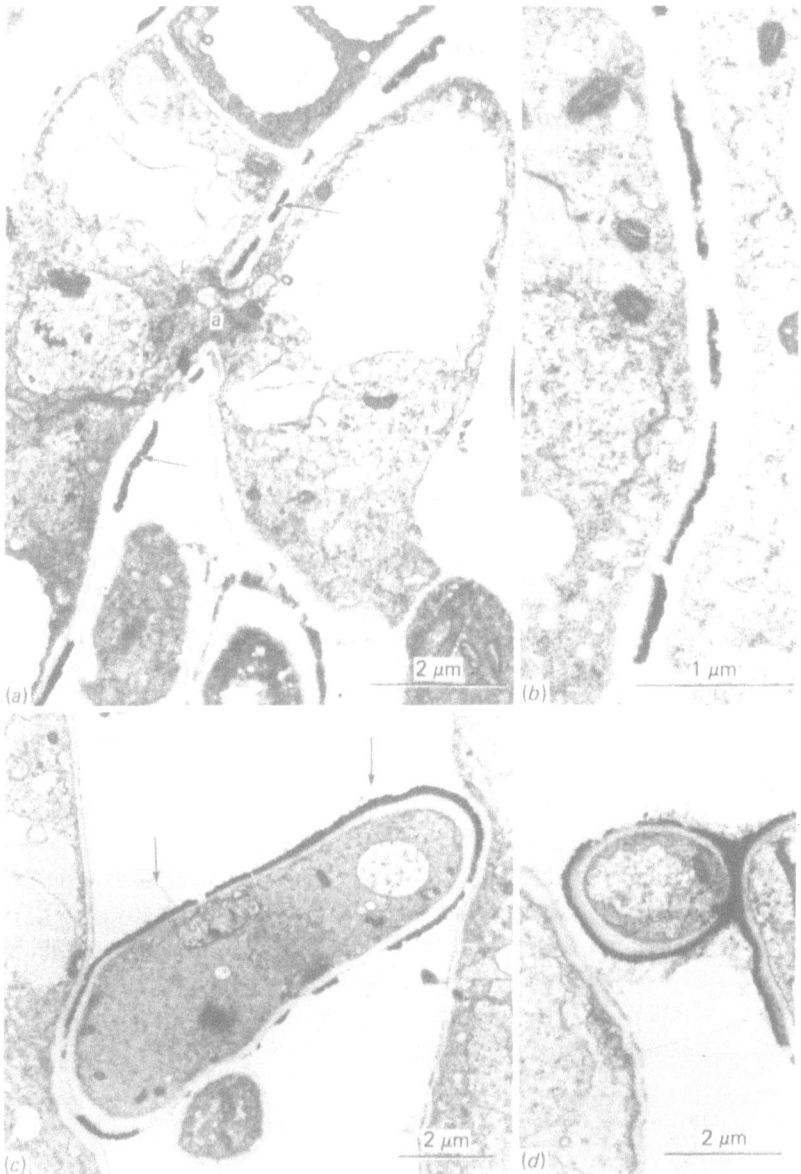

Fig. 18.10. (*a*) Transmission electron micrograph (detail of Fig. 18.5*b*) showing longitudinal section through the cortical zone of the stipe (Fig. 18.4, site 1b). There is a hyphal anastomosis (*a*) between two adjacent elements. Note the very-electron-dense material lying outside the cell wall of one of the elements and between the walls of the two anastomosed elements (arrowed). (*b*) Transmission electron micrograph showing a section through part of two adjacent elements in the stipe (Fig. 18.4, site

1. *Hyphal interweaving.* The extent of hyphal interweaving within a primordium is not known. The fact that its component elements, particularly in the pilear context, occurred in different orientations in any one plane of section (Figs 18.5, 18.7*a*, *c*, 18.9*a*) indicates possible interweaving. The elements which were orientated transversely across the longitudinal axis of the stipe are perhaps more suggestive of interweaving. However, only as a result of laborious serial sectioning can hypha! interweaving be proven.

2. *Hyphal anastomoses.* Occasional anastomoses occurred between adjacent hyphal elements in the stipe (Fig. 10*a*) but were not observed in the pileus. Anastomoses will inevitably provide the primordium with a more rigid, tightly bound structure but, as Ingold (1980*a*) also reported, such anastomoses are only rarely found in primordia of *F. velutipes*. In contrast, van der Valk & Marchant (1978) have described anastomoses as common in parts of the primordia of *C. cinereus* and *S. commune*.

3. *Hyphal cohesion.* In two recent studies (Read, 1983; Read & Beckett, 1985), a simple classification was proposed which divided all cellular elements within developing perithecia of *Sordaria humana* into two basic types: (1) discrete hyphae and hypha-like elements which exhibit a pronounced longitudinal type of growth pattern; and (2) coherent elements which arise by the cohesion of the cell walls of adjacent hyphae or hypha-like elements which become firmly fixed to each other. Sliding movements are not possible between adjacent coherent elements. In sectioned perithecia the coherent elements were usually easily recognisable because they had a pseudoparenchymatous appearance (Read & Beckett, 1985). Most of the developing primordium of *F. velutipes* was clearly composed of hyphae and hypha-like elements as was indicated by its prosenchymatous appearance in sectioned material (Figs 18.5, 18.7*a*, 18.9*a*, *b*). The presence of

---

1*c*). Between the two walls there is a large amount of electron-dense material. (*c*) Transmission electron micrograph (detail of Fig. 18.5*c*) showing a section through a transverse element in the central zone of the stipe (Fig. 18.4, site 1*c*). Note depressions in the walls of the neighbouring elements suggesting tight association. Also note electron-dense material lying outside the wall of the element and fibrous remnants of material in the space between elements (arrowed). (*d*) Transmission electron micrograph showing section through a transverse element similar to that in (*c*). Note large electron-dense deposits around the outside of the cell walls and fibrous extracellular material between the elements.

coherent elements was less obvious. Certainly some hyphae were in very close contact with adjacent hyphae. This close association commonly involved an electron-dense, discontinuous, outer wall layer as viewed with the TEM (Figs 18.5*b*, *c*, 18.7*c*, 18.10). However, this layer was also present where close cell–cell associations were not evident (Figs 18.5*b*,*c* 18.7*c*, 18.10*c*,*d*). It is not possible from these results alone to conclude whether these hyphae were firmly fixed together. Attempts to identify coherent elements by teasing apart living primordial hyphae have proved inconclusive because of the problems of preventing mechanical damage during dissection (Williams, unpublished). Examination of micrographs presented in other ultrastructural studies of basidiomycete fruit body primordia has been similarly inconclusive with regard to the interpretation of coherent elements.

4. *Extracellular material.* Depending on its stickiness, this will provide some degree of cohesion between adjacent hyphae.

Structure of the young fruit body in relation to its development

Fruit body development involves the coordinated and integrated growth and differentiation of many elements, all of which originate from hyphae of the undifferentiated mycelium. These vegetative hyphae become aggregated and differentiate to fulfil different roles in the fruit body.

The pronounced longitudinal growth pattern exhibited by the hyphae and hypha-like elements allows the apices of these elements to 'move' from one location to another within developing fruit bodies (Read & Beckett, 1985). These growth 'movements' in basidiomycete fruit bodies may be considerable, perhaps in some cases up to several centimetres in distance. As a result, there is the potential for the differentiation of part of a cellular element at a site which is sometimes considerably removed from that at which the element originated. In this respect, comparisons may be made with the morphogenetic movements exhibited by animal cells and by cellular slime mould cells. In these cases, however, whole cells migrate from one location to another.

The virtual (if not complete) lack of coherent tissues in the primordium of *F. velutipes*, compared with other multicellular fungal structures such as perithecia, may be attributable to different modes of development. The rapid expansion of the mushroom may be facilitated by the essentially loosely bound nature of its tissues which allows unimpeded elongation and expansion of the component elements. Coherent fungal tissues (e.g. the perithecium peridium) often provide a rigid, protective layer around hyphae and hypha-like elements. Such tissues would probably inhibit the rate of fruit body expansion.

Acknowledgements. We thank Dr David J. McLaughlin for critically reading the manuscript. Thanks are also due to the Shell Trust for Higher Education for a scholarship award to M.A.J.W. A.B. thanks the Science and Engineering Research Council for a research grant (GR/B/04914) and the Royal Society for a Scientific Investigations grant.

References

Angeli-Papa, J. & Eyme, J. (1978). Ultrastructural changes during development of *Agaricus bisporus* and *Agaricus sylvicola*. In *The Biology and Cultivation of Edible Mushrooms*, ed. S. T. Chang & W. A. Hayes, pp. 53–81. New York: Academic Press.

Aschan, K. (1954*a*). Some facts concerning the incompatibility groups, the dicaryotization and the fruitbody production in *Collybia velutipes*. *Svensk Botanisk Tidskrift*, **48**, 603–25.

Aschan, K. (1954*b*). The production of fruit bodies in *Collybia velutipes*. I. Influence of different culture conditions. *Physiologia Plantarum*, **7**, 571–91.

Aschan-Åberg, K. (1958). The production of fruit bodies in *Collybia velutipes*. II. Further studies on the influence of different culture conditions. *Physiologia Plantarum*, **11**, 312–28.

Aschan-Åberg, K. (1960). The production of fruit bodies in *Collybia velutipes*. III. Influence of the quality of light. *Physiologia Plantarum*, **13**, 276–9.

Bas, C. (1983). *Flammulina* in western Europe. *Persoonia*, **12**, 51–66.

Beckett, A. & Porter, R. (1982). *Uromyces viciae-fabae* on *Vicia faba*: scanning electron microscopy of frozen-hydrated material. *Protoplasma*, **111**, 28–37.

Beckett, A., Porter, R. & Read, N. D. (1982). Low temperature scanning electron microscopy of fungal material. *Journal of Microscopy*, **125**, 193–9.

Biffen, R. H. (1898). On the biology of *Agaricus velutipes*, Curt. (*Collybia velutipes*, P. Karst.). *Journal of the Linnean Society: Botany*, **34**, 147–62.

Buller, A. H. R. (1909). *Researches on Fungi*. London: Longmans, Green and Co.

Gruen, H. E. (1969). Growth and rotation of *Flammulina velutipes* fruitbodies and the dependence of stipe elongation on the cap. *Mycologia*, **61**, 149–66.

Gruen, H. E. (1976). Promotion of stipe elongation in *Flammulina velutipes* by a diffusate from excised lamellae supplied with nutrients. *Canadian Journal of Botany*, **54**, 1306–15.

Gruen, H. E. (1979). Control of rapid stipe elongation by the lamellae in fruit bodies of *Flammulina velutipes*. *Canadian Journal of Botany*, **57**, 1121–35.

Gruen, H. E. (1982). Control of stipe elongation by the pileus and mycelium in fruitbodies of *Flammulina velutipes* and other Agaricales. In *Basidium and Basidiocarp. Evolution, Cytology, Function and Development*, ed. K. Wells & E. K. Wells, pp. 125–55. New York: Springer-Verlag.

Gruen, H. E. (1983). Effects of competition among *Flammulina velutipes* fruitbodies on their growth. *Mycologia*, **75**, 604–13.

Gruen, H. E. & Wong, W. M. (1982*a*). Distribution of cellular amino acids, protein, and total organic nitrogen during fruitbody development in *Flammulina velutipes*. I. Growth on sawdust medium. *Canadian Journal of Botany*, **60**, 1330–41.

Gruen, H. E. & Wong, W. M. (1982*b*). Distribution of cellular amino acids, protein, and total organic nitrogen during fruitbody development in *Flammulina velutipes*. II. Growth on potato-glucose solution. *Canadian Journal of Botany*, **60**, 1342–51.

Gruen, H. E. & Wu, S. (1972*a*). Dependence of fruitbody elongation on the mycelium in *Flammulina velutipes*. *Mycologia*, **64**, 995–1007.

Gruen, H. E. & Wu, S. (1972*b*). Promotion of stipe elongation in isolated *Flammulina velutipes* fruitbodies by carbohydrates, natural extracts and amino acids. *Canadian Journal of Botany*, **50**, 803–18.

Hawker, L. E. (1942). The effect of vitamin B_1 on the concentration of glucose optimal for the fruiting of certain fungi. *Annals of Botany*, N.S., **6**, 631–6.

Henderson, L. E., Elliott, M. & Ross, I. K. (1983). Sclerotium formation in *Coprinus congregatus*. *Mycologia*, **75**, 738–41.

Ingold, C. T. (1978). Survival of *Flammulina velutipes* in severe frost and the liberation of spores under freezing conditions. *Bulletin of the British Mycological Society*, **12**, 86–7.

Ingold, C. T. (1980*a*). Mycelium, oidia and sporophore initials in *Flammulina velutipes*. *Transactions of the British Mycological Society*, **75**, 107–16.

Ingold, C. T. (1980*b*). *Flammulina velutipes*. *Bulletin of the British Mycological Society*, **14**, 112–18.

Ingold, C. T. (1981). *Flammulina velutipes* in relation to drying and freezing. *Transactions of the British Mycological Society*, **76**, 150–2.

Jensen, W. A. (1962). *Botanical Histochemistry*. San Francisco: W. H. Freeman.

Kitamoto, Y. & Gruen, H. E. (1976). Distribution of cellular carbohydrates during development of the mycelium and fruitbodies of *Flammulina velutipes*. *Plant Physiology*, **58**, 485–91.

Littlefield, L. (1966). Translocation of phosphorus-32 in sporophores of *Collybia velutipes*. *Physiologia Plantarum*, **19**, 264–70.

Long, T. J. (1966). Carbon dioxide effects in the mushroom *Collybia velutipes*. *Mycologia*, **58**, 319–22.

McLaughlin, D. J. (1982). Ultrastructure and cytochemistry of basidial and basidiospore development. In: *Basidium and Basidiocarp. Evolution, Cytology, Function and Development*, ed. K. Wells & E. K. Wells, pp. 37–74. New York: Springer-Verlag.

Moss, E. H. (1923). Developmental studies in the genus *Collybia*. *Transactions of the Royal Canadian Institute*, **14**, 321–32.

Niederpruem, D. J. & Wessels, J. G. H. (1969). Cytodifferentiation and morphogenesis in *Schizophyllum commune*. *Bacteriological Reviews*, **33**, 505–35.

O'Brien, T. P. & McCully, M. E. (1981). *The Study of Plant Structure. Principles and Selected Methods*. Melbourne: Termacarphi Pty. Ltd.

Pihakaski, S. (1972). Cell morphogenetic studies on *Flammulina velutipes*, with special reference to ultrastructure, nucleic acids and ribonucleases. *Annales Universitatis Turkuensis*, Series A.II., **48**, 1–60.

Plunkett, B. E. (1951). *Some aspects of the physiology of fruiting in the Hymenomycetes with special reference to* Collybia velutipes *(Curt.) Fr.* Ph.D. thesis, University of London.

Plunkett, B. E. (1953). Nutritional and other aspects of fruitbody production in pure cultures of *Collybia velutipes* (Curt.) Fr. *Annals of Botany*, N.S., **17**, 193–217.

Plunkett, B. E. (1956). The influence of factors of the aeration complex and light upon fruitbody form in pure culture of an agaric and polypore. *Annals of Botany, N.S.*, **20**, 563–86.

Read, N. D. (1983). A scanning electron microscopic study of the external features of perithecium development in *Sordaria humana*. *Canadian Journal of Botany*, **61**, 3217–29.

Read, N. D. & Beckett, A. (1985). The anatomy of the mature perithecium in *Sordaria humana*; and its significance for fungal multicellular development. *Canadian Journal of Botany*, **63** (in press).

Read, N. D., Porter, R. & Beckett, A. (1983). A comparison of preparative techniques for the examination of the external morphology of fungal material with the scanning electron microscope. *Canadian Journal of Botany*, **61**, 2059–78.

Reijnders, A. F. M. *Les Problèmes du Développement des Carpophores des Agaricales et de quelques groupes voisins*. The Hague: W. Junk.

Reynolds, E. S. (1963). The use of lead citrate at high pH as an electron-opaque stain in electron microscopy. *Journal of Cell Biology*, **17**, 208–12.

Schnepf, E. (1974). Gland cells. In *Dynamic Aspects of Plant Ultrastructure*, ed. A. W. Robards, pp. 331–57. Maidenhead: McGraw-Hill.

Schütte, K. H. (1956). Translocation in the fungi. *New Phytologist*, **55**, 164–82.

Stewart, F. C. (1918). The velvet-stemmed *Collybia* – a wild winter mushroom. *Bulletin of the New York Agricultural Experiment Station*, no. 448.

Tonomura, H. (1978). *Flammulina velutipes*. In *The Biology and Cultivation of Edible Mushrooms*, ed. S. T. Chang & W. A. Hayes, pp. 409–21. New York: Academic Press.

Valk, P. van der & Marchant, R. (1978). Hyphal ultrastructure in fruit body primordia of the basidiomycetes *Schizophyllum commune* and *Coprinus cinereus*. *Protoplasma*, **95**, 57–72.

Wong, W. M. & Gruen, H. E. (1977). Changes in cell size and nuclear number during elongation of *Flammulina velutipes*. *Mycologia*, **69**, 898–913.

19

Developmental genetics – from spore to sporophore

T.J.ELLIOTT

Glasshouse Crops Research Institute, Worthing Road, Littlehampton, West Sussex BN17 6LP, UK

Introduction

The central dogma of developmental biologists is that development is a consequence of differential gene expression, and understanding just how gene expression is regulated during development remains one of the major challenges of modern biology. In the agarics this challenge has been taken up in respect of fruit body development. The differentiation of a relatively complex structure, the fruit body, from a relatively simple structure, the mycelium, has attracted interest as a model for developmental studies in eukaryotes as a whole. In the meantime the genetics of other aspects of agaric development have been little studied.

Sussman (1965) defined development as '...a programmed sequence of phenotypic changes, under temporal, spatial, and quantitative control which is irreversible or reversible only with difficulty at least under ordinary environmental conditions. The sum total of these modifications constitutes the life-cycle of an organism.'

The ultimate aim of developmental genetics is therefore to define the life cycle of an organism in genetic terms.

The agaric life cycle

Three developmental phases can be readily delimited in the life cycle of an agaric: the spore, the colony and the fruit body. Developmental studies of these phases have been for the most part descriptive, with development being described in morphological or biochemical terms. These studies have been concentrated on the genus *Coprinus*, especially *C. cinereus*, and *Schizophyllum commune* (e.g. Wessels, 1965; Niederpruem & Wessels, 1969; Niederpruem & Jersild, 1972; Moore, Elhiti & Butler, 1979).

Agarics produce sexual basidiospores on their fruit bodies and asexual spores, such as oidia and chlamydospores, on their vegetative mycelia. The vegetative colony is the predominant phase in the life cycle and, unlike the spore and the fruit body, is not ephemeral. Knowledge of the vegetative colony is based on *in vitro* studies, little is known of the nature of the colony in the wild. The fruit body is the culmination of the agaric life cycle and as the fruit body is the principal means of identifying agarics its structure in different species has been described in detail.

Developmental genetics

The isolation and characterisation of mutants which on the whole have some profound effect on phenotype and thus are easily detected has made an important contribution to our understanding of developmental processes in agarics. As an illustrative example, in the cultivated mushroom, *Agaricus bisporus*, homoallelism for a single recessive gene can drastically modify fruit body structure (Fig. 19.1; Miller, Robbins & Kananen, 1976). It is not known whether any of the many mutations characterised in a range of species are in genes which are actually concerned with the regulation or modulation of development.

Most published work on the developmental genetics of agarics has dealt

Fig. 19.1. The effect of homoallelism for the broad stipe (*bs*) gene on fruit body morphology in the cultivated mushroom, *Agaricus bisporus*.

with the fruit body and consideration of this will therefore comprise the major part of this chapter. However, to follow the developmental sequence in full, the spore is a convenient starting point.

The spore

In the ascomycetes an extensive range of mutants affecting ascospore shape and colour have been characterised (Srb *et al.*, 1963). No similar group of mutants has been isolated or selected in the basidiomycetes. Whether this reflects some fundamental difference in the developmental biology of ascospores and basidiospores or, as is perhaps more likely, is because insufficient effort has been directed at recovering such mutants is not known. The prospect of determining genetic ratios directly by looking at the gill surface is an appealing one, and spore colour mutants in agarics would be especially useful for teaching purposes.

Germination in fungi has been studied extensively and those involved tend to see germination as a consequence of differential gene expression. The first event that can be recognised in germination is a rapid increase in protein synthesis, and cytoplasmic protein synthesis is essential in all fungi for germ-tube growth (Van Etten, Dahlberg & Russo, 1983).

The early development of the spore can be separated into two phases: germination itself, that is the gearing up of the cell's metabolic machinery; and germ-tube formation. No genes which affect these processes have been characterised in agarics.

Germination may be the inevitable consequence of a spore finding itself in a suitable environment. Agaric spores will often germinate on water agar but in some species germination depends on a specific trigger. In *A. bisporus*, germination depends on the presence of a specific, volatile, long-chain fatty acid – isovaleric acid – which is produced by the actively growing mycelium (Losel, 1967). This volatile is produced by other fungi so *Agaricus* spores are not limited to germinating where their mycelium is already growing. Such spore germination thus provides a specific example of a well-defined environmental trigger which presumably switches on the 'germination' gene or genes. Nothing is known, however, about the biochemistry of the switching mechanism but the incorporation of radioactively labelled isovalerate into the spores has been demonstrated (Rast & Stauble, 1970).

The transition from germ-tube to normal hyphal growth might be considered a simple extension of the development programme controlling germ-tube growth. This may be true for some fungi: in others, however, the conditions most favouring germination are not the same as those favouring growth. The transition from germ-tube formation to vegetative growth may require the activation of additional genes.

The production of oidia is normally associated with the homokaryon but they may also be produced by dikaryons. The production of oidia varies with environment and from homokaryon to homokaryon. Nothing is known of the genetics of oidia production or how the developmental programme regulating their production is initiated. Chlamydospores are intercalary and characteristically have thickened walls. Their formation is very much a response to environment, and again varies from strain to strain. Chlamydospores are usually produced by aged mycelium. In the cultivated mushroom, for example, they are found in old compost-grown cultures. Again, nothing is known of the genetics of chlamydospore formation or how the developmental sequence which controls the production of cells with thickened walls is initiated.

Agaric mycelia may also produce sclerotial resting structures. Sclerotia and fruit body initials may be alternate outcomes of the same developmental pathway (Moore, 1981) so the genetics of sclerotial development will therefore be considered below.

The colony

The development of a mycelial colony from a germinated spore is clearly a regulated process. Patterns of hyphal growth have been studied and modelled. Despite variation in such things as hyphal diameter and cell length between species, the basic kinetics of hyphal growth are similar in those species that have been studied in detail. Colony morphology may be modified by mutation, so clearly what is regarded as 'normal' growth is genetically regulated. Mutations may affect hyphal growth or branching or the rhythmicity of vegetative growth. Colonial mutants have been studied in most detail in the ascomycete *Neurospora crassa* but similar mutants are available in *Coprinus cinereus* and *Schizophyllum commune*. Some of the *Neurospora* mutants have been analysed in an attempt to determine the precise nature of the branching mechanism. The cell wall, the cell membrane, various enzymes such as glucose 6-phosphate dehydrogenase and phosphoglucomutase, and cyclic AMP have all been implicated (Prosser, 1983). The genetic bases of hyphal growth patterns are still little understood.

The fruit body

Dikaryosis and dikaryon morphogenesis and the role of the mating type or incompatibility factors are described in detail elsewhere

(Casselton & Economou: Chapter 8; Ullrich & Novotny: Chapter 20). The morphogenetic sequence which follows the mating of compatible homokaryons is well defined and various aspects of dikaryosis can be ascribed to either the *A*- or *B*-factor in bifactorial heterothallic species such as *Schizophyllum commune*. The *A*-factor regulates clamp connection formation and the *B*-factor nuclear migration. The full expression of the fruiting sequence with a meiotic division in the basidium will in normal circumstances only occur in a dikaryon.

Homokaryotic fruiting

The expression of the genes which control fruit body morphogenesis is not, however, restricted to the dikaryon. Homokaryotic fruiting is a feature of many agarics and it was in fact first recognised and described in *S. commune* in 1909 (Wakefield, 1909). The fruiting structures produced by homokaryotic mycelia vary enormously. At one extreme they may simply be masses of little-differentiated tissue whilst at the other extreme fully differentiated fruit bodies may be formed which morphologically are indistinguishable from their dikaryotic counterparts. Homokaryotic fruits in a species of oyster mushroom, *Pleurotus eous*, are shown in Fig. 19.2.

Kniep (1930) postulated that genetic controls of fruit body formation

Fig. 19.2. Homokaryotic fruits on agar medium in an oyster mushroom, *Pleurotus eous*.

Table 19.1. *Agarics and polypores in which homokaryotic fruiting has been reported (taken from Stahl & Esser, 1976, except where indicated)*

Agarics	Polypores
Agrocybe aegerita	*Fomes cajanderi*
Armillaria mucida	*Lenzites trabea*
Collybia tuberosa	*Peniophora ludoviciana*
C. velutipes	*Polyporus brumalis*
Coprinus curtus	*P. ciliatus*
C. ephemerus	*Sistrotrema brinkmannii*
C. fimetarius	*Stereum bicolor*
C. lagopus	
C. cinereus (as *C. macrorhizus*)	
C. niveus	
Hygrocybe ceracea	
H. conica	
Hygrophorus constans	
H. virgineus	
Mycena spp.	
M. galericulata	
M. pseudopicta	
M. quercus-ilicis	
M. speirea	
Omphalina ericetorum	
O. luteolilacina	
O. luteovitellina	
Panaeolus campanulatus	
Pholiota nameko	
Pleurotus eous (Elliott & Challen, unpublished)	
P. flabellatus (Samsudin & Graham, 1984)	
Psilocybe panaeoliformis	
Schizophyllum commune	
Typhula erythropus	

Samsudin, S. & Graham, K. M. (1984). Monokaryotic fruiting in *Pleurotus flabellatus*. *Malaysian Applied Biology*, **13**, 61–5.

and of the sexual cycle were essentially independent. This hypothesis found support in the extensive analysis of fruiting competence in *S. commune* performed by Raper & Krongelb (1958). They concluded that homokaryotic fruiting was a 'quantitatively inherited character distinct from dikaryotic fruiting competence'.

Karl Esser and colleagues have published extensively on the genetic regulation of homokaryotic fruiting from initiation to development. They too have stressed that the action of the genes controlling fruit body development does not depend on the incompatibility factors. Stahl & Esser

(1976) reviewed the reported incidence of homokaryotic fruiting and were able to list 33 species in which it had been observed and this list could no doubt be extended (Table 19.1). Twenty four of the species are agarics and the remainder polypores.

The ability to produce homokaryotic fruits varies among populations of homokaryons. In the polypore *Sistrotrema brinkmannii*, 22 out of 80 homokaryons fruited (Ullrich, 1973) whereas in *Schizophyllum commune* the progeny of only four dikaryons from a worldwide sample of 58 produced homokaryotic fruits (Raper & Krongelb, 1958). In *Pleurotus eous*, five of 81 homokaryons derived from a single fruit body produced some kind of fruiting structure (Elliott & Challen, unpublished).

Major gene control

A detailed genetic analysis of the phenomenon of homokaryotic fruiting has been made by Esser and co-workers in three different species, *Polyporus ciliatus*, *Agrocybe aegerita* and *Schizophyllum commune*, and a basically similar system of control has been recognised in each. In *P. ciliatus* three unlinked genes have been identified (Stahl & Esser, 1976): the gene fi^+ responsible for the initiation of fruiting; the gene fb^+ which regulates the shape of the fruit body; and the gene mod^+ which determines its fertility. *P. ciliatus* is a bifactorial heterothallic species and of the two incompatibility factors, the *B*-factor seemed capable of influencing the expression of fi^+ and fb^+. In common-*B* heterokaryons fruiting was observed whilst in dikaryons and common-*A* heterokaryons, homokaryotic fruiting was suppressed. For the dikaryon, mod^+ inhibited fruiting but fi^+ and fb^+ had no effect, even when homoallelic.

In the bipolar heterothallic *A. aegerita*, up to 60% of homokaryotic mycelia derived from single basidiospores from a fruit body collected from the wild will produce fruiting initials or fully developed fruit bodies (Meinhardt & Esser, 1983) and the basidia of these homokaryotic fruit bodies are two-spored. These basidiospores can be germinated – though they germinate less well than their dikaryotic equivalents. The mycelia produced are again homokaryotic and are all of the same mating type as the parental homokaryon. Cytological studies show that nuclear fusion and meiosis do not occur in the basidium. This confirms that fruit body morphogenesis can take place totally independently of the action of the mating type factors.

Genetic analysis in this species has shown that the production of homokaryotic fruit bodies depends on the action of at least three genes which operate sequentially (Esser & Meinhardt, 1977; Meinhardt & Esser,

1981). There is a threshold gene – su^+/su – which in the form su^+ suppresses differentiation. If the su allele is present, differentiation may take place. The gene fi^+ is responsible for the production of fruit body initials and the gene fb^+ is necessary for fruit body development to be completed. Homokaryotic fruiting therefore requires the genotype su, fi^+, fb^+. Any other combination of alleles does not permit fruiting to occur. The genotype su, fi^+, fb results only in the production of initials.

Further genetic analysis using homokaryons of known genotype to form dikaryons shows that these genes are also instrumental in regulating fruiting in the dikaryon. The allele su^+ must be present and the active allele of at least one of the fruiting initiation or fruit body development genes.

In *Schizophyllum commune* Esser and colleagues found the genetic regulation of homokaryotic fruiting to be genetically more complex. Four genes affecting fruiting were characterised (Esser, Saleh & Meinhardt, 1977): two genes controlling initiation, designated fi-1^+ and fi-2^+; a further gene, fb^+, controlling fruit body development; and a fourth gene, st^+, responsible for the formation of lumps of undifferentiated stromatic tissue. Either of the initiation genes allows the formation of initials, and they act together to permit stipe development; the fruit body gene is necessary for the full expression of homokaryotic fruiting. The stromatic gene is epistatic and suppresses the action of the other three genes. In the dikaryon this gene also prevents dikaryotic fruiting whilst the other three genes only affect the timing of fruiting and not the development of dikaryotic fruit bodies. Homoallelism for these three genes in a dikaryon resulted in fruiting within 5–7 days. A single dose of st delayed fruiting for more than 25 days.

These studies have led Esser and his co-workers to propose a general model for the genetic control of fruiting in both homokaryons and dikaryons of basidiomycetes (Esser, 1983; Meinhardt & Esser, 1983). Fruit body development is claimed to depend on the action of three genes: an su gene which determines if differentiation can occur; an fi gene regulating initiation; and a third gene, fb, controlling further development (Fig. 19.3). This major gene model for the control of fruiting in basidiomycetes has not been widely accepted.

The existence of a fruiting initiation gene in *Polyporus ciliatus* has been questioned by Prillinger & Six (1983). They used isogenic mycelial lines which were non-fruiting, whether homokaryotic or dikaryotic, and should therefore lack the active allele of the fruiting initiation gene and have the genotype fi rather than fi^+. No fruiting occurred when compatible isolates within a line were crossed, but crosses between lines produced numerous fruit bodies although presumably of genotype *fi/fi*. These results were confirmed using some of the non-fruiting strains used by Stahl & Esser (1976) in their original study. Prillinger & Six (1983) conclude that fruit body formation is a 'typical polygenic character'.

Polygenic control

Homokaryotic fruiting in *S. commune* has been investigated by other workers, mainly in the USA, who would also argue for polygenic control. These studies were prompted by the discovery in 1968 of fruiting-inducing substances (FIS) which promote homokaryotic fruiting (Leonard & Dick, 1968). A single gene was identified as being responsible for fruiting in response to injury (Leonard & Raper, 1969). In 1973 it was also shown that homokaryotic fruiting could be induced by mechanical injury (Leonard & Dick, 1973). Homokaryotic fruiting therefore occurs spontaneously, or in response to biochemical substances, or following mechanical injury.

Leslie & Leonard (1979 *a*, *b*) later made a detailed study of the initiation step in homokaryotic fruiting and have shown that the responses in the three types of homokaryotic fruiting are genetically separable and under polygenic control. They have identified at least eight genes in four different pathways of initiation (Leslie & Leonard, 1979 *b*; Fig. 19.4). These genes are designated *hap*: *hap 5* and *hap 6* control the initiation of spontaneous

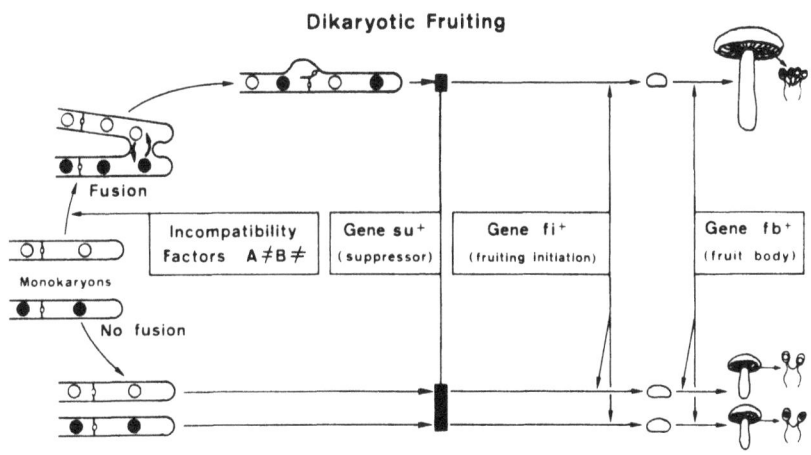

Fig. 19.3. Proposed model for the action of three major genes which control fruiting in both dikaryotic and homokaryotic mycelia of agarics. (From Meinhardt & Esser, 1983.)

fruits; *hap 1* alone or *hap 2*, *hap 3* and *hap 4* acting in concert control response to injury; *hap 7* and *hap 8* determine response to fruiting substances.

Leslie (1983) has extended this analysis and concludes that the initiation of homokaryotic fruits is controlled by *at least* five different sets of genes. One set of at least two genes is responsible for fruiting in response to an unknown biochemical substance or substances, termed FIS. One FIS is almost certainly a cerebroside (Kawai & Ikeda, 1982). Two additional sets of genes can initiate fruiting in response to mechanical injury. At least two, and possibly three, sets of genes can trigger spontaneous homokaryotic fruiting.

Other genes acting pleiotropically may prevent initiation. Leslie (1983) has reported that if mutations at the *mnd* or *thin* loci of *S. commune* are present, initiation does not occur. A further prerequisite for fruiting in a homokaryon is an active phenoloxidase.

A general model

The relationship between the *hap* genes of Leslie and Leonard and the two *fi* genes identified by Esser and co-workers in *Schizophyllum commune* is not known. Perhaps these genes are distinct and operate at different points in a developmental pathway. The *hap* genes exert their effect very early in the fruiting pathway, controlling the formation of cellular aggregates, whereas the *fi* genes control the differentiation of stipe-like structures. It would be encouraging if the two genetic systems recognised in *S. commune* could be reconciled.

Whatever the relationship between these two systems, the weight of evidence very much favours a polygenic control of homokaryotic fruiting. It seems premature to propose a major gene model for fruiting in dikaryons and homokaryons of agarics.

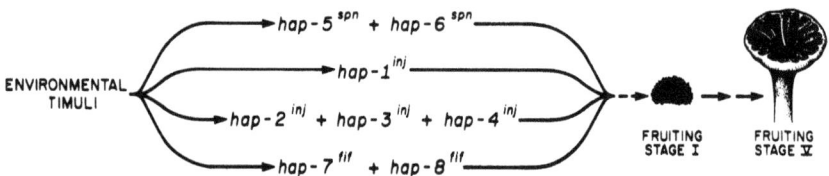

Fig. 19.4. Genes involved in homokaryotic fruiting in *Schizophyllum commune*. (From Leslie & Leonard, 1979*b*.) Additional genes regulating spontaneous homokaryotic fruiting have also been characterised (Leslie, 1983).

Dikaryotic fruiting

Dikaryotic fruiting proper has only been studied to any extent in *Coprinus cinereus*. In this species two different structures are produced by the dikaryotic mycelium: sclerotia and fruit bodies. The initial stages in the morphogenesis of both structures are very similar. Sclerotia are also produced by homokaryotic mycelia and Waters, Moore & Butler (1975) recovered strains unable to produce sclerotia. This character segregated in crosses as if controlled by a single major gene. Four such *scl* genes have now been characterised (Hereward & Moore, 1979), *scl-1* to *scl-4*. *Scl-4* behaves as a dominant gene, causing the abortion of primordia in heteroallelic dikaryons. The other three *scl* genes are recessive in heteroallelic dikaryons and have been mapped to existing linkage groups.

Dikaryons homoallelic for these genes have now been constructed (Moore, 1981). When homoallelic, the genes prevent both sclerotium and fruit body formation. When heteroallelic, the genes are not normally expressed but modifiers may permit their partial expression. On the basis of these experimental studies, Moore (1981) has proposed a developmental pathway for *C. cinereus* relating fruit body and sclerotium development (Fig. 19.5). Supporting evidence for the view that the sclerotium and the fruit body initial are alternative outcomes of the same developmental pathway has been provided by Henderson, Elliott & Ross (1983) in *Coprinus congregatus*. In this species, commitment to sclerotium production precludes the production of fruit body initials.

A series of genes specifically affecting fruit body development has been characterised in *C. cinereus* (under the name *C. macrorhizus*). Takemaru & Kamada (1969) macerated a dikaryon to produce hyphal fragments which were treated with ultraviolet radiation. Of the 5759 isolates recovered, 5490 were dikaryotic and 269 monokaryotic. Abnormal fruit bodies were

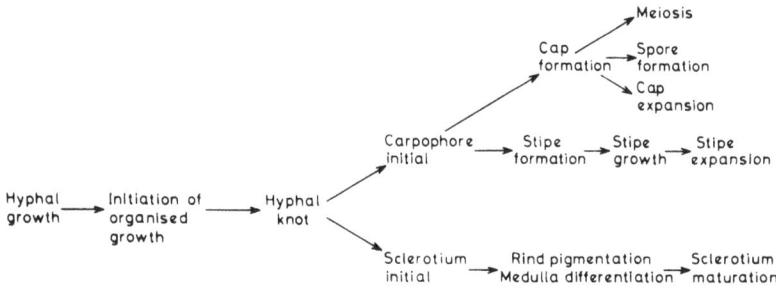

Fig. 19.5. Pathway for sclerotium and fruit body development in *Coprinus cinereus*. (From Moore, 1981.)

produced by 1045 of the dikaryons. They could be classified into seven basic types:

1. *Knotless*: hyphal aggregates not formed
2. *Primordiumless*: hyphal aggregates are formed but do not develop further
3. *Maturationless*: primordia are formed but do not develop further
4. *Elongationless*: cap development normal but stipe fails to elongate
5. *Expansionless*: stipe elongates normally but the cap fails to expand, although spores are formed
6. *Sporeless*: few or no spores are formed
7. *Autolysisless*: cap does not autolyse.

A genetic analysis of an elongationless variant and an expansionless variant showed that these characters were controlled by single dominant genes; these genes were designated *Eln* and *Exp*, respectively (Takemaru & Kamada, 1969).

The vast majority of the mutants had only one defect, although combinations of defects were also recovered. Expansionless–sporeless types were particularly common.

In a later paper (Takemaru & Kamada, 1972), two additional mutagens were used: nitrosoguanidine (NG) and bromouracil (BU). A similar range of mutants was produced by NG, but BU was ineffective. In total, 1582 mutants were recovered in 8574 survivors from the ultraviolet and NG treatments.

The production of mutants at such a high frequency is difficult to explain. The mutation rate approaches 20%, which is unprecedentedly high. Moore (1981) has suggested an explanation for this frequency of mutation. He suggests that the genes mutated in the original dikaryon are in the main modifiers of the expression of the developmental genes rather than the developmental genes themselves and in particular affect the dominance of pre-existing developmental variants. Whatever the genetic basis of these mutants they provide an excellent opportunity for studies on fruit body development. Elongationless mutants, for example, have been used in studies of the mechanism of stipe expansion (Kamada & Takemaru, 1977; Gooday: Chapter 12). Sporeless mutants have been used to demonstrate that the sequential production of enzymes found in fruit bodies is developmentally separable from the cellular events leading to basidiospore formation. The sporeless mutants had the same levels of various enzymes as the wild-type strains (Miyake, Takemaru & Ishikawa, 1980).

These mutants perhaps provide an indication of the genetic pathway leading to fruit body development. A sequential gene action can be proposed up to the formation of primordia. After primordium formation the developmental effect of the mutants operate independently. Cap expansion can occur without stipe elongation for example, and the converse is also true. These processes might be regarded as developmentally similar each involving a substantial amount of cell expansion but they are clearly under separate genetic control.

Conclusions

With the current state of our knowledge of developmental genetics in agarics, it is only possible to provide an outline sketch of genetic events during development. A number of genes with specific effects have been characterised and genes have been emphasised in this review at the expense of environmental and biochemical aspects of development. There is detailed information for some systems about the environmental stimuli that trigger development and of associated biochemical events. The role of light in fruit body morphogenesis is well documented in many agarics: in species such as *C. congregatus* it has been extensively studied (Robert & Durand, 1979; Ross: Chapter 14). Clearly, developmental systems which are regulated by well-defined environmental triggers are more amenable to analysis, both genetic and biochemical.

The central question of developmental genetics, that is how genes are turned on and off during development, can only be answered in the agarics, as elsewhere, by speculation. Picard-Bennoun (1982), for example, has suggested that the regulatory mechanism operating in the ascomycete *Podospora anserina* is an increase in the level of reading errors in DNA translation. She suggests that translational ambiguity increases at key points in cellular differentiation and that the increase in the misreading level allows sufficient readthrough or frameshifting for the synthesis of regulatory proteins.

Perhaps the answer to the problems of how the genes are turned on and off is most likely to be found through the techniques of recombinant DNA technology, and other contributors to this volume indicate the state of this art in agarics (Wessels, Dons & De Vries: Chapter 21; Pukkila *et al.*: Chapter 22).

References

Esser, K. (1983). Genetic control of fruiting in higher basidiomycetes. *Third International Mycological Congress (Abstracts)*, p. 59.

Esser, K. & Meinhardt, F. (1977). A common genetic control of dikaryotic and monokaryotic fruiting in the basidiomycete *Agrocybe aegerita*. *Molecular and General Genetics*, **155**, 113–15.

Esser, K., Saleh, F. & Meinhardt, F. (1979). Genetics of fruit-body production in higher basidiomycetes II. Monokaryotic and dikaryotic fruiting in *Schizophyllum commune*. *Current Genetics*, **1**, 85–8.

Henderson, L. E., Elliott, M. & Ross, I. K. (1983). Sclerotium formation in *Coprinus congregatus*. *Mycologia*, **75**, 738–41.

Hereward, F. V. & Moore, D. (1979). Polymorphic variation in the structure of aerial sclerotia of *Coprinus cinereus*. *Journal of General Microbiology*, **113**, 13–18.

Kamada, T. & Takemaru, T. (1977). Stipe elongation during basidiocarp maturation in *Coprinus macrorhizus*: mechanical properties of stipe cell wall. *Plant and Cell Physiology*, **18**, 831–40.

Kawai, G. & Ikeda, Y (1982). Fruiting-inducing activity of cerebrosides observed with *Schizophyllum commune*. *Biochimica et Biophysica Acta*, **719**, 612–18.

Kniep, H. (1930). Uber Selektionswirkungen in fertlaufenden Massenaussaaten von *Schizophyllum*. *Zeitschrift für Botanik*, **23**, 510–36.

Leonard, T. J. & Dick, S. (1968). Chemical induction of haploid fruiting in *Schizophyllum commune*. *Proceedings of the National Academy of Sciences, USA*, **59**, 745–51.

Leonard, T. J. & Dick, S. (1973). Induction of haploid fruiting by mechanical injury in *Schizophyllum commune*. *Mycologia*, **65**, 809–22.

Leonard, T. J. & Raper, J. R. (1969). *Schizophyllum commune*: gene controlling induced haploid fruiting. *Science*, **165**, 190.

Leslie, J. F. (1983). Initiation of monokaryotic fruiting in *Schizophyllum commune*: multiple stimuli, multiple genes. *Third International Mycological Congress (Abstracts)*, p. 163.

Leslie, J. F. & Leonard, T. J. (1979a). Three independent genetic systems that control initiation of a fungal fruiting body. *Molecular and General Genetics*, **171**, 257–60.

Leslie, J. F. & Leonard, T. J. (1979b). Monokaryotic fruiting in *Schizophyllum commune*: genetic control of the response to mechanical injury. *Molecular and General Genetics*, **175**, 5–12.

Losel, D. M. (1967). The stimulation of spore germination in *Agaricus bisporus* by organic acids. *Annals of Botany*, **31**, 417–25.

Meinhardt, F. & Esser, K. (1981). Genetic studies of the basidiomycete *Agrocybe aegerita*. 2. Genetic control of fruit body formation and its practical implications. *Theoretical and Applied Genetics*, **60**, 265–8.

Meinhardt, F. & Esser, K. (1983). Genetic aspects of sexual differentiation in fungi. In *Fungal Differentiation*, ed. J. E. Smith, pp. 537–57. New York: Marcel Dekker.

Miller, R. E., Robbins, W. A. & Kananen, D. L. (1976). Inheritance of sporophore colour and 'wild' morphology in *Agaricus bisporus*. *Mushroom Science*, **9**, 39–45.

Miyake, H., Takemaru, T. & Ishikawa, T. (1980). Sequential production of enzymes and basidiospore formation in fruiting bodies of *Coprinus macrorhizus*. *Archives of Microbiology*, **126**, 201–5.

Moore, D. (1981). Developmental genetics of *Coprinus cinereus*: genetic evidence that carpophores and sclerotia share a common pathway of initiation. *Current Genetics*, **3**, 145–50.

Moore, D., Elhiti, M. M. Y. & Butler, R. D. (1979). Morphogenesis of the carpophore of *Coprinus cinereus*. *New Phytologist*, **83**, 695–722.

Niederpruem, D. J. & Jersild, R. A. (1972). Cellular aspects of morphogenesis in the mushroom *Schizophyllum commune*. *CRC Critical Reviews in Microbiology*, **1**, 545–76.

Niederpruem, D. J. & Wessels, J. G. H. (1969). Cytodifferentiation and morphogenesis in *Schizophyllum commune*. *Bacterial Reviews*, **33**, 505–35.

Picard-Bennoun, M. (1982). Does translational ambiguity increase during cell differentiation? *FEBS Letters*, **149**, 167–70.

Prillinger, H. & Six, W. (1983). Genetic analysis of fruiting and speciation of basidiomycetes: genetic control of fruiting in *Polyporus ciliatus*. *Plant Systematics and Evolution*, **141**, 341–71.

Prosser, J. I. (1983). Hyphal growth patterns. In *Fungal Differentiation*, ed. J. E. Smith, pp. 357–96. New York: Marcel Dekker.

Raper, J. R. & Krongelb, G. S. (1958). Genetic and environmental aspects of fruiting in *Schizophyllum commune* Fr. *Mycologia*, **50**, 707–40.

Rast, D. & Stauble, E. J. (1970). On the mode of action of isovaleric acid in stimulating the germination of *Agaricus bisporus* spores. *New Phytologist*, **69**, 557–66.

Robert, J. C. & Durand, R. (1979). Light and temperature requirements during fruit-body development of a Basidiomycete mushroom, *Coprinus congregatus*. *Physiologia Plantarum*, **46**, 174–8.

Srb, A. M., Basl, M., Bobst, M. & Leary, J. V. (1973). Mutations in *Neurospora crassa* affecting ascus and ascospore development. *Journal of Heredity*, **64**, 242–6.

Stahl, U. & Esser, K. (1976). Genetics of fruit-body production in higher basidiomycetes I. Monokaryotic fruiting and its correlation with dikaryotic fruiting in *Polyporus ciliatus*. *Molecular and General Genetics*, **148**, 183–97.

Sussman, M. (1965). Developmental phenomena in microorganisms and in higher forms of life. *Annual Review of Microbiology*, **19**, 59–78.

Takemaru, T. & Kamada, T. (1969). The induction of morphogenetic variations in *Coprinus* basidiocarps by UV irradiation. *Report of the Tottori Mycological Institute*, **7**, 71–7.

Takemaru, T. & Kamada, T. (1970). Genetic analysis of UV-induced developmental variation in *Coprinus* basidiocarp: the elongationless and expansionless. *Report of the Tottori Mycological Institute*, **8**, 11–16.

Takemaru, T. & Kamada, T. (1972). Basidiocarp development in *Coprinus macrorhizus* I. Induction of developmental variations. *Botanical Magazine (Tokyo)*, **85**, 51–7.

Ullrich, R. C. (1973). Sexuality, incompatibility, and intersterility in the biology of the *Sistotrema brinkmannii* aggregate. *Mycologia*, **65**, 1234–49.

Van Etten, J. L., Dahlberg, K. R. & Russo, G. M. (1983). Fungal spore germination. In *Fungal Differentiation*, ed. J. E. Smith, pp. 235–66. New York: Marcel Dekker.

Wakefield, E. M. (1909). Uber die bedingungen der fruchtkörperbildung, sowie das auftreten fertiler und steriler stamme bei hymenomyceten. *Naturwissenschaftliche Zeitschrift fur Forst- und Landwirtschaft*, **7**, 521–51.

Waters, H., Butler, R. D. & Moore, D. (1975). Morphogenesis of aerial sclerotia of *Coprinus lagopus*. *New Phytologist*, **74**, 207–13.

Wessels, J. G. H. (1965). Morphogenesis and biochemical processes in *Schizophyllum commune*. *Wentia*, **13**, 1–113.

20

Nucleic acid studies in
Schizophyllum

ROBERT C. ULLRICH,
CHARLES P. NOVOTNY* and
CHARLES A. SPECHT*

*Department of Botany, Marsh Life Science Building, University of Vermont,
Burlington, Vermont 05405, USA*

**Department of Medical Microbiology, Given Medical Building, University of
Vermont, Burlington, Vermont 05405, USA*

Introduction

Studies on the nucleic acids of *Schizophyllum commune* derive from an interest in the regulation of gene expression in mating and development. A germinating spore develops into a mycelium of uninucleate cells with haploid nuclei, the *homokaryon*. A homokaryon may mate with another compatible homokaryon and differentiate into the fertile mycelium, the *dikaryon*. A dikaryon maintains two haploid nuclei in each cell, one derived from each mate (Raper, 1966). Two homokaryons are compatible if they have different mating types. Mating type is specified by two sets of genes called the *A* and *B* incompatibility factors (Fig. 20.1).

Each factor is composed of two closely linked loci, α and β. For each locus there are alternative alleles: 9 *Aα* alleles, 32 *Aβ*, 9 *Bα* and 9 *Bβ* have been recognised (Raper, Baxter & Ellingboe, 1960; Koltin, Raper & Simchen, 1967; Stamberg & Koltin, 1972). The mating type of a homokaryon is determined by the combination of alleles it possesses. To be fully compatible two mates must differ in at least one *A* factor allele (α or β) and one *B*-factor allele (α or β). The matings represented in Table 20.1 illustrate these features. Each incompatibility factor controls a specific developmental sequence. The events comprising each sequence were defined by the study of heterokaryons specifically active for one sequence or the other (Table 20.1). The two sequences together constitute the transition from homokaryon to dikaryon (see Casselton & Economou: Chapter 8). Transition to the dikaryon occurs only when mates are compatible in both the *A* and *B* factors. Analyses of developmental mutants suggest that dikaryosis in *S. commune* may occur via positive regulators that activate genes involved in development (Raper & Raper, 1973).

Table 20.1. *Pairings of homokaryons carrying mating type alleles in various combinations, their compatibility and developmental results*

Homokaryon	Homokaryon	Compatibility	Developmental result
$A\alpha_1\beta_1 B\alpha_1\beta_1 \times A\alpha_1\beta_2 B\alpha_2\beta_1$		Compatible	Dikaryon, A and B active
$A\alpha_1\beta_1 B\alpha_1\beta_2 \times A\alpha_2\beta_2 B\alpha_1\beta_2$		Compatible only for A	Heterokaryon, A active
$A\alpha_1\beta_1 B\alpha_1\beta_1 \times A\alpha_1\beta_1 B\alpha_1\beta_2$		Compatible only for B	Heterokaryon, B active
$A\alpha_3\beta_5 B\alpha_1\beta_1 \times A\alpha_3\beta_5 B\alpha_1\beta_1$		Incompatible	Homokaryon, no activity

Molecular models of regulation by products of the incompatibility genes

The mating type genes regulate more than 50 nuclear genes (Raper & Raper, 1966; Dubovoy, 1976). Two molecular events are essential for the determination of mating. The first event is recognition. By an unknown mechanism the specific $A\alpha$, $A\beta$, $B\alpha$, and $B\beta$ alleles of the two mates are recognised as identical or different. This determines the outcome of the second essential event, activation. If the alleles are identical, maintenance of the homokaryotic state continues. If the alleles are different, transition to heterokaryotic development is activated. In this respect the alleles and their products are molecular switches.

Two basic models have been proposed to explain the regulation of mating. The first proposes that switching occurs at the protein level. Each allele codes for a messenger RNA (mRNA) that is translated into a polypeptide. The various protein models postulate that the polypeptides interact in either of several ways (intragenic or intergenic complementation) that determine development based upon the self or non-self nature of the polypeptides that interact (Prévost, 1962; Raper, 1966; Kuhn & Parag, 1972).

The second type of model suggests that recognition and switching involve mobile, regulatory sequences of either DNA or RNA. These sequences are thought to interact in such a way that complementary or

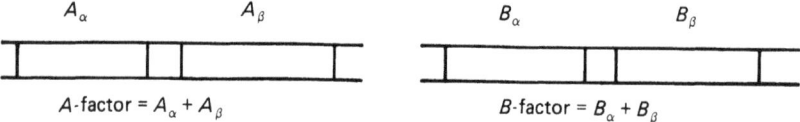

Fig. 20.1. Map of A and B incompatibility factors of *Schizophyllum commune*. Map distance between α and β is strain dependent and can vary from zero to a few percent recombination.

non-complementary base pairing yields molecules inactive or active in promoting heterokaryotic development, respectively (Ullrich, 1973, 1978). Active molecules would promote either transcription or translation, respectively, depending on whether they were DNA or RNA. In this model, the proteins derived from activated nucleic acid switches are thought to activate the expression of genes essential to heterokaryotic development. From the data available it is not clear if either model is correct. The study of nucleic acids may resolve the issue and lead to a better understanding of the regulation.

Mating type in ascogenous yeasts differs from that in basidiomycetes

Considerable progress has been made in the study of mating type in the ascogenous yeasts *Saccharomyces cerevisiae* and *Schizosaccharomyces pombe*. How much of this information applies to mating type in basidiomycetes?

In *Saccharomyces cerevisiae*, individual cells of homothallic strains are self-fertile by virtue of the gene *HO* which is unlinked to the mating type locus, *MAT*. *HO* confers the ability for high-frequency transfer of previously silent mating type information stored at the *HML* or *HMR* loci to the mating type expression site, *MAT*, which is linked to them (Hicks, Strathern & Herskowitz, 1977; Hicks, Strathern & Klar, 1979). Displacement of mating type *a* information by *α* information causes an *a* to *α* shift in mating type of the cell. The reciprocal displacement produces a reciprocal change in mating type (Klar *et al.*, 1981; Nasmyth *et al.*, 1981). Because *HO* effects informational displacement and mating type switching as frequently as each cell division, each clone of *HO* cells soon contains cells of each mating type. These may fuse to produce the vegetative diploid yeast cell which undergoes meiosis and ascospore formation under the proper environmental conditions. It is significant to note that even heterothallic (*ho*) *S. cerevisiae* cells undergo mating type switching at low frequency (10^{-6}). The mating system of *Schizosaccharomyces pombe* appears to be similar (Egel *et al.*, 1980; Beach, 1983).

At least five additional loci in homothallic and heterothallic strains of *S. cerevisiae* are of interest. These are known variously as *SIR*, *MAR* or *CMT* (Haber & George, 1979; Klar, Fogel & MacLeod, 1979; Rine *et al.*, 1979). Mutations in any of these loci permit expression of previously silent mating type information stored at *HML* and *HMR*. Products of the five *SIR*, *MAR* or *CMT* loci are thought to be trans-acting negative regulators.

Classical genetic studies of mating in basidiomycetes demonstrated clear differences from that in the ascogenous yeasts. First, mating type switching in basidiomycetes is unrecorded. Secondly, although homothallism does occur in close association with heterothallism within morphologically described species, its basis is genetically different. Heterothallic isolates do not routinely mate with, and become converted to dikaryons by, homothallic isolates. Furthermore, in *Sistotrema brinkmannii*, where hybridisation of heterothallic and primary homothallic isolates was achieved by selecting prototrophic hybrids, Ullrich (1973) and Ullrich & Raper (1975) observed several noteworthy features. Hybridisations of bipolar heterothallic strains with homothallic strains gave diminished (30–50%) frequencies of basidiospore germination compared to controls (90–100%). Monosporous, homothallic progeny from these hybrids exhibited abnormal sexual development as evidenced by lengthy delays in achieving the dikaryotic condition. Unlike the ascogenous yeasts, the homothallic basidiomycetes appear to be evolutionarily removed from their heterothallic cousins.

Ullrich (1973) and Ullrich & Raper (1975) characterised the pattern of sexuality in 613 monosporous progeny from four different bipolar heterothallic × homothallic hybrids; 332 progeny were bipolar heterothallic and 281 were homothallic. Each of the heterothallic progeny had a bipolar mating type identical to its heterothallic progenitor. The absence of new mating types suggests that basidiomycetes (homothallic or heterothallic) do not have silent mating type genes. Furthermore the 1:1 segregation of homothallism:heterothallism in these matings showed that there is no separable determinant of homothallism (such as *HO* in *Saccharomyces cerevisiae* or *h* in *Schizosaccharomyces pombe*) superimposed on heterothallism. The evidence suggests that homothallism in basidiomycetes, in contrast to that in ascomycetes, is determined by the mating type loci. Naturally occurring homothallic strains of *Sistotrema brinkmannii* are reminiscent of strains of *Schizophyllum commune* in which certain mutations of the mating type loci obliterate mating type and the mutants are constitutive for the sequence of developmental events controlled by those loci.

If silent mating type loci were present in basidiomycetes as they are in the ascogenous yeasts, these would have been revealed in the many mutational studies of mating (Parag, 1962; Raper, Boyd & Raper, 1965; Raper & Raper, 1966; Koltin, 1968; Dubovoy, 1976; Raudaskoski *et al.*, 1976; Simchen, personal communication). For example, selections for *Schizophyllum commune* constitutive mutants would have revealed trans-acting negative regulators like the *SIR* loci of *Saccharomyces cerevisiae* if they existed. In *Schizophyllum commune*, constitutive mating behaviour has been found only in mutants of the mating type loci themselves. In *Coprinus cinereus*, Day (1963) examined a suppressor mutation, *SuA*, that was unlinked to mating type loci and recessive in crosses when mated to homokaryotic testers with the same *A* factor. It is conceivable that this mutation was analogous to *SIR* and functioned by releasing expression of a silent mating type locus, but switching of mating type is also unrecorded in *C. cinereus*. Although *SuA* is a candidate for a *SIR*-like mutation, the absence of evidence for a previously silent mating type makes this unlikely. It is more likely that *SuA* is a recessive mutation that activates the *A* sequence by bypassing *A* factor control.

The above remarks document basic differences in the genetic basis of homothallism and the absence of silent mating type genes in basidiomycetes as opposed to ascomycetes. They do not address the more interesting questions of what the molecular mechanisms of self/non-self recognition are in the basidiomycetes and how the regulation of development is activated in these organisms when a non-self recognition occurs. It is likely that the molecular mechanisms of recognition and activation are quite different in the ascomycetes and basidiomycetes.

Nucleic acid studies

Studies of nucleic acids may lead to an understanding of mating and its regulation. Initially, DNA extraction procedures developed for plants (Britten, Graham & Neufeld, 1974) and *Coprinus* (Dutta *et al.*, 1971) were used. Kinetic analyses of DNA reassociations assayed in three different ways showed that the whole-cell DNA content of *Schizophyllum commune* was 3.6×10^7 base pairs (bp) with about 10% repetitive sequence (Ullrich *et al.*, 1980*a*). The guanine plus cytosine (GC) content of the DNA was estimated by thermal denaturation methods and isopycnic centrifugation to be 57% for the nuclear DNA component, and 22–27% for a lesser component now known to be mitochondrial (Ullrich *et al.*, 1980*a*). Other kinetic analyses established that repetitive sequences were present in one or a few clusters rather than interspersed in the unique-sequence DNA (Ullrich, Kohorn & Specht, 1980). These results were consistent with those reported by Wessels and his collaborators (Dons, De Vries & Wessels, 1979; Dons & Wessels, 1980; see also Wessels, Dons & De Vries: Chapter 21).

By 1975, reports of transformation in the ascomycete *Neurospora crassa* (Mishra & Tatum, 1973; Mishra, Szabo & Tatum, 1973; Mishra, 1976, 1977) suggested that transformation of *S. commune* might also be feasible. Transformation may make it possible to isolate the mating type genes from

S. commune. Unfortunately, the DNA employed for kinetic characterisations (Ullrich *et al.*, 1980*a*) was too short for transformation. The Gentle Extraction Method (GEM) which uses toluene and lengthy, but gentle, shaking to obtain milligram quantities of pure, high-molecular-weight DNA was developed (Specht *et al.*, 1982). This method provided DNA of length greater than 80 kb for use in transformation experiments and for cloning.

While constructing recombinant DNA vectors of various types it became clear that fractionation of the whole-cell DNA and characterisation of some of its components would be useful. Bisbenzimide was used in caesium chloride density gradient centrifugations after the methods of Hudspeth *et al.* (1980) to fractionate two satellite DNAs from the *S. commune* main band (Specht *et al.*, 1983). One of these was a light-density satellite, buoyant density 1.684 g/ml. DNA was also isolated from mitochondria prior to treatment with DNase I, and from mitochondria after treatment

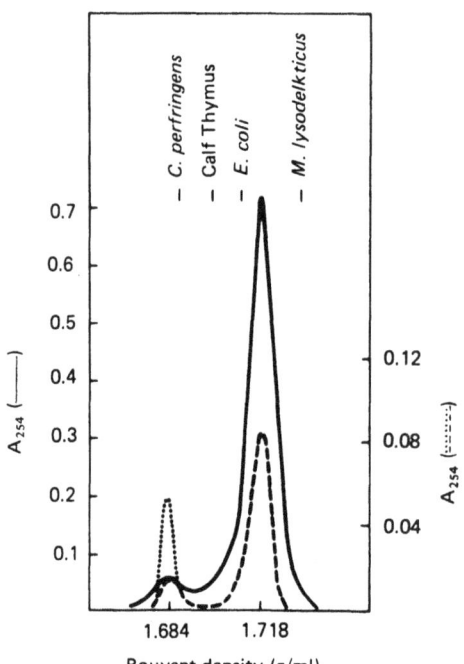

Fig. 20.2. Profiles of caesium chloride gradients containing total DNA extracted from homokaryotic strain 4–40 (solid line), DNA extracted from mitochondria without prior treatment with DNase I (dashed line) and DNA extracted from mitochondria after treatment with DNase I (dotted line).

with DNase I. These DNAs were subjected to isopycnic centrifugation in caesium chloride. The gradient profiles (Fig. 20.2) demonstrated that mitochondria contained DNA with buoyant density equal to that of the light-density satellite (Specht *et al.*, 1983). They also showed that mitochondrial preparations not treated with DNase I were contaminated with the main band DNA, but this could be removed by treatment with DNase prior to DNA extraction.

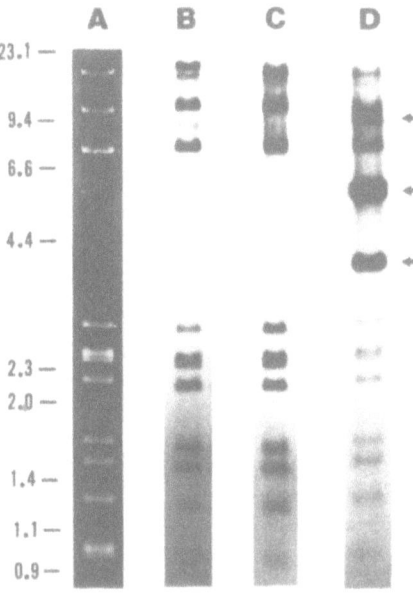

Fig. 20.3. Electrophoretogram and Southern hybridisation of restriction endonuclease-treated DNA of homokaryotic strain 4–40. Lane A: electrophoretogram of satellite DNA restricted with *Eco*R1 and stained with ethidium bromide. Lane B: Southern hybridisation of 4–40 total DNA (2 µg) restricted with *Eco*R1 and probed with ^{32}P nick-translated satellite DNA (10^6 cpm). Lane C: Southern hybridisation of 4–40 total DNA (2 µg) restricted with *Eco*R1 and probed with DNA from mitochondria pretreated with DNase I (10^6 cpm). Lane D: Southern hybridisation of 4–40 total DNA (2 µg) restricted with *Eco*R1 and probed with nick-translated 4–40 total DNA (6×10^6 cpm). Lanes B–D are composites of different-length exposures in order that each band may be visualised clearly. The bars to the left of lane A indicate the positions of *Hind*III restriction fragments of λ DNA and *Hae*III restriction fragments of φX174. Arrows to the right of lane D point to nuclear fragments that code for ribosomal RNA.

The light satellite and mitochondrial DNAs were characterised further as shown in Fig. 20.3. Light satellite DNA was restricted with *Eco*R1, electrophoresed and stained with ethidium bromide. The pattern of DNA fragments was identical to bands appearing in Southern blots of *Eco*R1-restricted whole-cell DNA probed with light-density satellite or DNA extracted from mitochondria pretreated with DNase I. This proved not only that the light-density satellite DNA was mitochondrial DNA, but that it contained all of the mitochondrial DNA sequences (Specht *et al.*, 1983).

Lane D of Fig. 20.3 shows a blot of whole-cell DNA restricted with *Eco*R1 and probed with ^{32}P nick-translated whole-cell DNA. Repetitive sequences are apparent. The genome size of *S. commune* is small, and restriction fragments containing repetitive sequences are visualised in Southern blots even when using whole-cell DNA as the probe (Specht *et al.*, 1983). The arrows in Fig. 20.3 lane D indicate three *Eco*R1 restriction fragments that contained repetitive sequences not present in the mitochondrial DNA. These fragments contained the ribosomal DNA (rDNA) unit repeat (see below).

Identification of the mitochondrial DNA restriction fragments was important when certain recombinant DNA plasmid vectors containing autonomously replicating sequences (*ars*) were constructed. It was theorised that filamentous fungi might be transformed with plasmids that replicate autonomously as is the case in the yeasts *Saccharomyces cerevisiae* (Hinnen, Hicks & Fink, 1978; Beggs, 1978) and *Schizosaccharomyces pombe* (Beach & Nurse, 1981). Plasmid YIp5 and *Saccharomyces cerevisiae* SHY2 cells were used to obtain plasmids containing *Schizophyllum commune ars* sequences that could replicate autonomously in yeast after the methods of Stinchcomb, Struhl & Davis (1979). YIp5 contains the *Saccharomyces cerevisiae URA3* gene; it complements the *ura3* mutation in SHY2, but only when YIp5 contains an *ars* sequence to allow autonomous replication. *Eco*R1 restriction fragments of *Schizophyllum commune* whole-cell DNA were cloned into the *Eco*R1 site of plasmid YIp5. After transformation of SHY2, prototrophic cells were selected that could grow because they had contained YIp5 plasmid containing an *S. commune ars*. The *Schizophyllum* inserts in replicating plasmids were then characterised. Despite the fact that mitochondrial DNA accounts for only 3–4% of whole-cell DNA, 95% of the *ars*-containing plasmids had mitochondrial, not nuclear, DNA sequences (Specht, unpublished data). At least six of 12 different *S. commune* mitochondrial *Eco*R1 fragments contained *ars* activity. Although nuclear DNA sequences that function as *ars* were isolated in this way, the majority of *ars* sequences were mitochondrial.

Bisbenzimide–caesium chloride gradients allowed the isolation of a second satellite DNA (Specht *et al.*, 1984; Fig. 20.4A, band R). This satellite was slightly lighter than main-band DNA (Fig. 20.4A, band N). Three fractions of DNA corresponding to bands M, R and N were recovered from the gradient, restricted with *Hind* III and electrophoresed. The restriction pattern (Fig. 20.4B, lane M) was that of mitochondrial DNA; as demonstrated above, the light-density satellite was mitochondrial DNA. The main band DNA restricted with *Hind* III (lane N) gave a smeared pattern expected for nuclear DNA. DNA from band R was restricted with *Hind* III and a Southern blot was probed with end-labelled

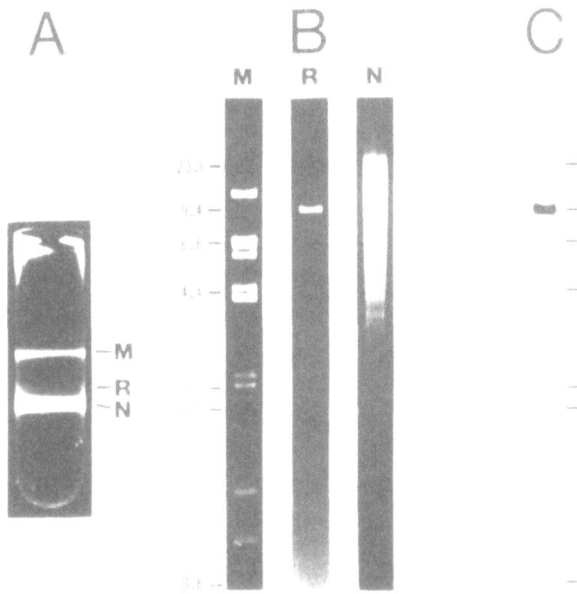

Fig. 20.4. Caesium chloride–bisbenzimide gradient of strain 4–40 whole-cell DNA and restriction endonuclease analysis of the resultant components. A: the gradient illuminated with 370 nm ultraviolet light. B: fractions removed from the gradient, restricted with *Hind*III and electrophoresced. Gels were stained with 1 μg/ml ethidium bromide. Lanes N and M are nuclear and mitochondrial DNA, respectively (Specht *et al.*, 1983). C: Southern blot of band R DNA restricted and electrophoresed as in lane R and probed with rRNA. Positions and sizes (kb) of *Hind*III fragments of λ DNA as indicated with dashes.

ribosomal RNA (rRNA). Positive hybridisation showed that the band R satellite DNA was rDNA (Fig. 20.4C). The length of the rDNA unit repeat was estimated at 9.2 kb. Lane C also confirmed that the rDNA was located in one, or at most a few, cluster(s) within the genome.

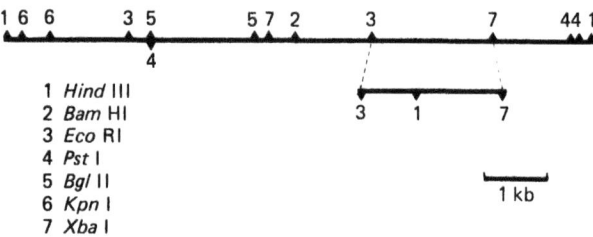

Fig. 20.5. Restriction map of the rDNA repeat of strains 4–40, 1–50, 1–54 and 1–106. The upper line represents the restriction map from strain 4–40. Strains 1–54 and 1–106 are identical except that an extra 200 bp and a *Hind*III restriction site is found in the region indicated by the lower line. Strain 1–50 rDNA contains 400 bp more than strain 4–40 and an additional *Hind*III site in this same region.

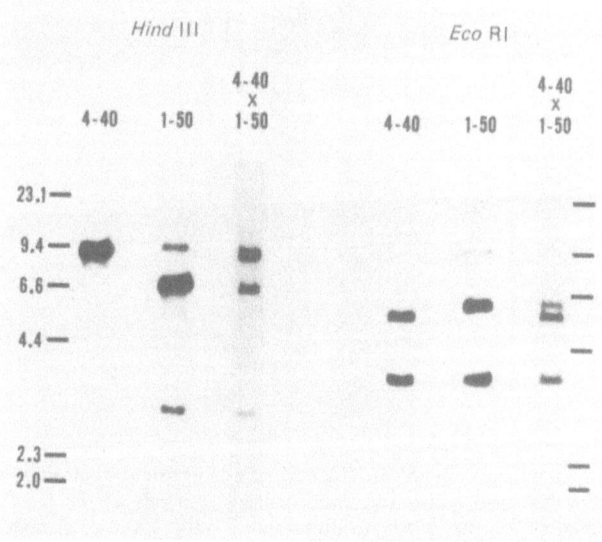

Fig. 20.6. Total DNA from homokaryotic strains 4–40 and 1–50 and dikaryotic strain 4–40 × 1–50 restricted with *Eco*R1 or *Hind*III and blotted to nitrocellulose for Southern hybridisations. ^{32}P-labelled rRNA was hybridised to the homokaryon lanes and ^{32}P-labelled rDNA cloned in plasmid YIp-5 was hybridised to the dikaryon lanes. *Hind*III fragments of λ DNA are shown as size standards.

Strains showed size and restriction polymorphisms in their rDNA unit repeats. A restriction map of four strains is shown in Fig. 20.5. Strains 1–54 and 1–106 differed from strain 4–40 by the insertion of an additional 200 bp of DNA within the *Eco*R1–*Xba*I fragment of strain 4–40. This addition carried a *Hind* III site not present in the rDNA of strain 4–40. Strain 1–50 contained a 400 bp addition to this same region; it also contained an additional *Hind* III site (see also Pukkila *et al.*: Chapter 22). The band intensities in Fig. 20.6 show that the two polymorphic rDNA unit repeats contributed by the two homokaryons in a mating were maintained in

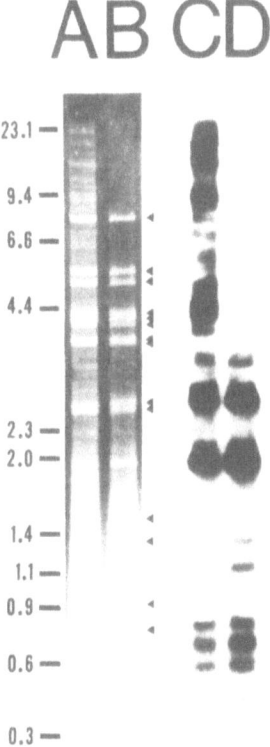

Fig. 20.7. Restriction analyses of strain 4–40 total DNA with *Msp*I and *Hpa*II. Lanes A and B: ethidium bromide-stained gels of total DNA restricted with *Hpa*II (A) and *Msp*I (B). Lanes C and D: Southern hybridisations of total DNA restricted with *Hpa*II (C) and *Msp*I (D) probed with ^{32}P-labelled rDNA cloned with plasmid YIp5. Arrow heads indicate mitochondrial DNA restriction fragments. Size markers are *Hind*III fragments of λ DNA and *Hae*III fragments of φX174 DNA.

approximately the same frequencies in the dikaryons as in the homokaryons. Fig. 20.6 also demonstrates the utility of haploid cells when trying to size and map a DNA sequence for which size and map polymorphisms exist. It is clear that interpretation of the restriction pattern of DNA from the haploid cell is simpler.

To determine if *Schizophyllum commune* rDNA contains methylated

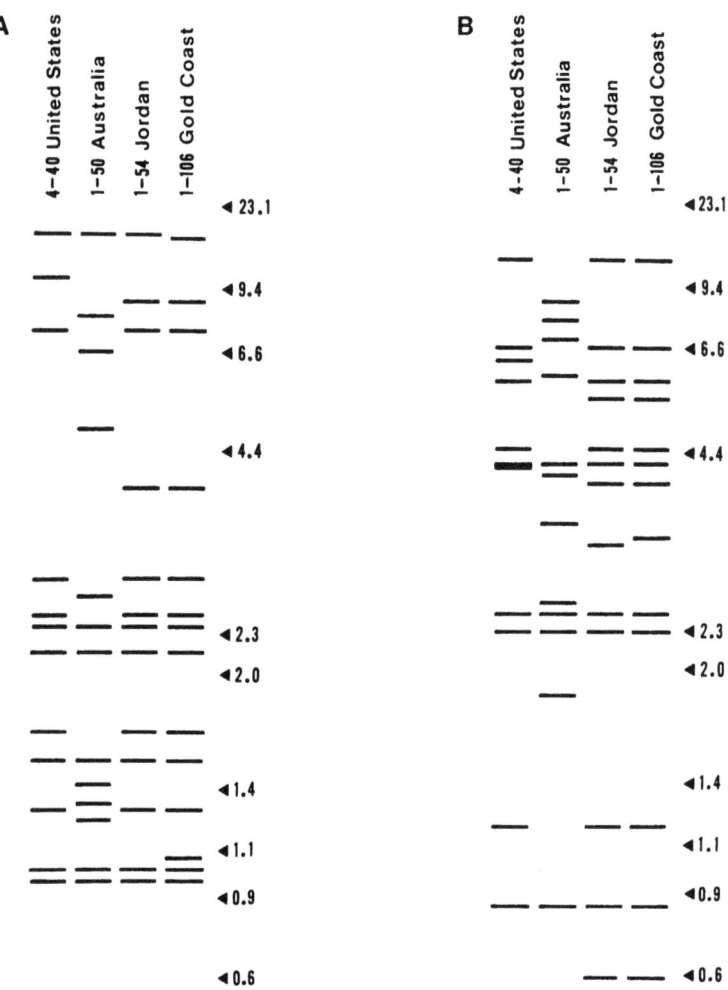

Fig. 20.8. *Eco*R1 (A) and *Hind*III (B) digests of mitochondrial DNA from strains 4–40, 1–50, 1–54 and 1–106. The origin of each strain is indicated. Standard fragment sizes from *Hind*III digests of λ DNA and *Hae*III digests of ϕX174 DNA are in kilobases.

Table 20.2. *Size of mitochondrial DNA from three strains of*
Schizophyllum commune *as calculated from restriction*
endonuclease digests

| Strain | Endonuclease | | Average size (kb) |
	*Hind*III	*Eco*R1	
1–50	50.2[a]	50.4	50.3
1–54	51.9	52.5	52.2
1–106	51.9	52.5	52.2

[a] Average of at least two determinations.

cytosine, it was cut with *Msp*I and *Hpa*II restriction endonucleases (Fig.
20.7). *Msp*I cleaves CCGG restriction sites regardless of methylation at
the internal C, whereas *Hpa*II cleaves only if the internal C is unmethylated.
Fig. 20.7 shows that the rDNA was methylated (compare patterns in lanes
C and D). Lane C also shows that the unit repeats were differentially
methylated; the methylation pattern within one unit repeat differed from
that in adjacent repeats (N.B. bands greater than 9.2 kb). To our knowledge
this is the first demonstration of differential methylation of the rDNA unit
repeat for any fungus (Specht *et al.*, 1984).

Size and restriction polymorphisms also exist in mitochondrial DNA.
Restriction digests of mitochondrial DNA from homokaryotic strains
4–40, 1–50, 1–54 and 1–106 are shown in Fig. 20.8. In the comparison of
any two strains, some mitochondrial fragments were found in common and
some were different. The total size of mitochondrial DNA for isogenic
strains 4–39 and 4–40 was estimated to be 49.85 kb (range 49.25–50.50 kb)
from restrictions using any of five endonucleases. Size estimates for three
other strains are shown in Table 20.2. These polymorphisms were used to
determine if mitochondria were exchanged and migrated during heterokaryosis. Restriction analyses did not reveal any DNA patterns indicative
of heteroplasmons; either they were not formed or they were lost by
segregation during the growth of the mycelium that was necessary to
extract a sufficient amount of DNA for analysis (Specht, unpublished
data).

Transformation experiments

Technology for manipulating the DNA of basidiomycetes is now
available, and the general properties of the DNA of *Schizophyllum
commune* are known. The goal is to develop transformation and to use it

to identify plasmids of an *S. commune* gene library that contain the mating type genes. The approach is to develop a transformation vector for a selectable trait such as prototrophy or resistance to antibiotic. *S. commune* DNA fragments could be introduced into the vector, and DNA from this plasmid gene library would be used to eliminate non-transformed cells, and changes in morphology associated with sexual development would be used to identify transformants for mating type.

The methods of De Vries & Wessels (1972) were modified to make protoplasts. Approximately 10^{10} basidiospores are germinated in a complete liquid glucose–salts–yeast extract medium overnight at 30 °C. Germlings from this treatment, or about 5.0 g mycelium grown in the same complete liquid medium, are resuspended in the magnesium osmoticum of De Vries & Wessels (1972) to which is added 10 mg ml^{-1} Cellulase T.v. concentrate (Miles Laboratories) and 10 mg ml^{-1} Novozym 234 (Novo Laboratories). The protoplasts are recovered by centrifugation and repeatedly washed in buffer. DNA is presented to protoplasts either naked or encapsulated in synthetic liposomes. The liposomes are synthesised according to the methods of Dellaporta & Fraley (1981). Use of liposomes is preferred because nuclease activity (about four cuts of double-stranded, linear DNA 20 kb in length per 30 min at 20 °C) is still present after repeated washing of protoplasts. DNA encapsulated in liposomes is protected from exogenous nuclease activity. Between 0.1 and 15.0% of the protoplasts regenerate after various treatments that may be employed in a transformation protocol.

One approach for developing a transformation vector is to include a gene that complements a metabolic deficiency in the host cell (a gene that converts an auxotrophic mutant to prototrophy). The identification and isolation of such a gene has proved difficult, and it is one reason that the development of transformation for basidiomycetes lags behind that in ascomycetes. The biochemical genetics of *S. commune* is undeveloped and the enzyme deficiencies of particular mutants are largely unknown. This creates uncertainty when trying to match a given auxotrophic mutant with a gene which will complement it. The efficiency of transformation anticipated in the development of a transformation system for eukaryotes is too low to expect the identification and retrieval of a complementing gene using DNA from an entire gene bank. Several non-reverting *S. commune* mutants deficient in the pyrimidine biosynthetic pathway were studied and *ura*1 mutants were shown to lack orotidylate decarboxylase activity. Attempts to hybridise the cloned orotidylate decarboxylase genes from other organisms with *S. commune* DNA (*URA*3 from yeast and *PYR*4 of *N. crassa*) were unsuccessful even at conditions of low stringency. The cloned *LEU2* and *ADE1* genes from *Saccharomyces cerevisiae* and *TRPC* from *Aspergillus nidulans* also failed to hybridise with *Schizophyllum commune* DNA. Attempts to cross-hybridise general housekeeping genes of different species of fungi have, in general, proved difficult.

Another means of identifying a *Schizophyllum commune* gene that may complement an *S. commune* mutant is to transform (using an *S. commune* gene bank) auxotrophic mutants of an organism for which high-efficiency transformation has been established. With this rationale, *Escherichia coli* (*hisB, leuB, pyrF, argH, aroD, lacZ, trpA* and *purC, D, E, F, G, H*) and *Saccharomyces cerevisiae* (*trp1, ura3, leu2* and *his3*) mutants were transformed with gene banks of *Schizophyllum commune* DNA cloned in plasmids. No complementing *S. commune* genes were identified.

Using an alternative rationale to identify a gene that would complement a *Schizophyllum commune* deficiency, plasmid vectors with identified genes from *Saccharomyces cerevisiae* (*URA3, LEU2* and *ADE1, 3, 4, 5–7, 8*) were used to attempt transformation of *S. commune* auxotrophs. These experiments were also unsuccessful, possibly because (1) the enzymatic deficiency of the *S. commune* recipient did not match the cloned gene, or (2) genes from a foreign species may not function in *S. commune*.

An alternative of transforming to prototrophy is to transform to antibiotic resistance. *S. commune* is sensitive to the deoxystreptamine antibiotic Geneticin (G418) at low levels of magnesium ions. The kanamycin resistance gene (*KmR*) from transposons Tn*5* and Tn*601* codes for a phosphoribosyl transferase that provides resistance to kanamycin, G418 and related deoxystreptamine antibiotics. *KmR* has been used to confer resistance to G418 in yeast (Jiminez & Davies, 1980), *Dictyostelium* (Barclay & Meller, 1983), mammalian cells (Colbere-Garapin *et al.*, 1981; Southern & Berg, 1982) and plant cells (Bevan, Flavel & Chilton, 1983).

Several vectors containing *KmR* were constructed for transformation of *Schizophyllum commune*. *S. commune ars* sequences were included in some of the vectors to provide replication function. By the same logic rDNA sequences were included in others. In other constructions, rDNA or random *S. commune* 'helper' sequences were inserted in vectors to increase the likelihood of integration of vector DNA into the *S. commune* genome by homologous recombination. In yet other constructions, random *S. commune* fragments were introduced into sites 5′ of genes in order to provide a functional promoter. Preliminary results of transformation with these constructions are not yet clear.

Attempts to transform *S. commune* have been laborious and intensive, but much has been learned about this fungus. Studies to characterise the mitochondrial and the ribosomal sequences, the variation of these repetitive sequences between different strains and the differential methylation of the rDNA are continuing. Much has also been learned about the formation, handling, processing and regeneration of protoplasts. Progress in all of these areas bodes well for the future of *Schizophyllum* research.

Acknowledgements. This paper is a contribution of the Agricultural Experiment Station, University of Vermont, Journal Article No. 561. The studies described were supported by NSF Grants PCM-7616959, PCM-8004589, PCM-8203496, Grant No. 7800090 from the Science and Education Administration of the United States Department of Agriculture, University of Vermont Biomedical Research Grant No. 43 and equipment purchase, and Institutional Grant BRSG-PHS-5429-18.

References

Barclay, S. & Meller, E. (1983). Efficient transformation of *Dictyostelium discoideum* amoebae. *Molecular and Cellular Biology*, **3**, 2117–30.

Beach, D. H. & Nurse, P. (1981). High-frequency transformation of the fission yeast *Schizosaccharomyces pombe. Nature*, **290**, 140–2.

Beach, D. H. (1983). Cell type switching by DNA transposition in fission yeast. *Nature*, **305**, 682–8.

Beggs, J. D. (1978). Transformation of yeast by replicating hybrid plasmid. *Nature*, **275**, 104–9.

Bevan, M. W., Flavel, R. B. & Chilton, M.-D. (1983). A chimaeric antibiotic resistance gene as a selectable marker for plant cell transformation. *Nature*, **304**, 184–7.

Britten, R. J., Graham, D. E. & Neufeld, B. R. (1974). Analysis of repeating DNA sequences by reassociation. In *Methods in Enzymology*, vol. 59, ed. L. Grossman & K. Moldave, pp. 363–417. New York: Academic Press.

Colbere-Garapin, F., Horodniceanu, F., Kourilsky, P. & Garapin, A. (1981). A new dominant hybrid selective marker for higher eukaryotic cells. *Journal of Molecular Biology*, **150**, 1–14.

Day, P. R. (1963). Mutations affecting the *A* mating-type locus in *Coprinus lagopus*: wild alleles. *Genetical Research*, **4**, 323–5.

De Vries, O. M. H. & Wessels, J. G. H. (1972). Release of protoplasts from *Schizophyllum commune* by a lytic enzyme preparation from *Trichoderma viride. Journal of General Microbiology*, **73**, 13–22.

Dellaporta, S. L. & Fraley, R. T. (1981). Delivery of liposome-encapsulated nucleic acids into plant protoplasts. *Plant Molecular Biology Newsletter*, **2**, 59–66.

Dons, J. J. M., De Vries, O. M. H. & Wessels, J. G. H. (1979). Characterization of the genome of the basidiomycete *Schizophyllum commune. Biochimica et Biophysica Acta*, **563**, 100–12.

Dons, J. J. M. & Wessels, J. G. H. (1980). Sequence organization of the nuclear DNA of *Schizophyllum commune. Biochimica et Biophysica Acta*, **607**, 385–96.

Dubovoy, C. A. (1976). A class of genes affecting *B* factor-regulated development in *Schizophyllum commune*. *Genetics*, **82**, 423–8.

Dutta, S. K., Penn, S. R., Knight, A. R. & Ojha, M. (1971). Characterization of DNAs from *Coprinus lagopus* and *Mucor azygospora*. *Experientia*, **28**, 582–4.

Egel, R., Kohli, J., Thuriaux, P. & Wolf, K. (1980). Genetics of the fission yeast *Schizosaccharomyces pombe*. *Annual Review of Genetics*, **14**, 77–108.

Haber, J. E. & George, J. P. (1979). A mutation that permits expression of normally silent copies of mating-type information in *Saccharomyces cerevisiae*. *Genetics*, **93**, 13–35.

Hicks, J., Strathern, J. N. & Herskowitz, I. (1977). The cassette model of mating-type interconversion. In *DNA Insertion Elements, Plasmids and Episomes*, ed. A. Bukhari, J. Shapiro & S. Adhaya, pp. 457–62. New York: Cold Spring Harbor.

Hicks, J., Strathern, J. N. & Klar, A. J. S. (1979). Transposable mating type genes in *Saccharomyces cerevisiae*. *Nature*, **289**, 478–83.

Hinnen, A., Hicks, J. & Fink, G. (1978). Transformation of yeast. *Proceedings of the National Academy of Sciences, USA*, **75**, 1929–33.

Hudspeth, M. E. S., Shumard, D. S., Tatti, K. M. & Grossman, L. I. (1980). Rapid purification of yeast mitochondrial DNA in high yield. *Biochimica et Biophysica Acta*, **610**, 221–8.

Jimenez, A. & Davies, J. (1980). Expression of a transposable antibiotic resistance element in *Saccharomyces*. *Nature*, **287**, 869–71.

Klar, A. J. S., Fogel, S. & MacLeod, K. (1979). *MAR1* – A regulator of the *HMa* and *HMα* loci in *Saccharomyces cerevisiae*. *Genetics*, **93**, 37–50.

Klar, A. J. S., Strathern, J. N., Broach, J. R. & Hicks, J. B. (1981). Regulation of transcription in expressed and unexpressed mating type cassettes of yeast. *Nature*, **289**, 239–44.

Koltin, Y. (1968). The genetic structure of the incompatibility factors of *Schizophyllum commune*: comparative studies of primary mutations in the *B* factor. *Molecular and General Genetics*, **102**, 196–203.

Koltin, Y., Raper, J. R. & Simchen, G. (1967). Genetic structure of the incompatibility factors of *Schizophyllum commune*: the *B* factor. *Proceedings of the National Academy of Sciences, USA*, **57**, 55–63.

Kuhn, J. & Parag, Y. (1972). Protein-subunit aggregation model for self-incompatibility in higher fungi. *Journal of Theoretical Biology*, **35**, 77–91.

Mishra, N. C. (1976). Episome-like behavior of donor DNA in transformed strains of *Neurospora crassa*. *Nature*, **264**, 251–3.

Mishra, N. C. (1977). Characterization of the new osmotic mutants (*os*) which originated during genetic transformation in *Neurospora crassa*. *Genetical Research*, **29**, 9–19.

Mishra, N. C., Szabo, G. & Tatum, E. L. (1973). Nucleic acid-induced genetic changes in *Neurospora*. In *The Role of RNA in Reproduction and Development*, ed. M. C. Niu & S. J. Segal, pp. 259–68. Amsterdam: North-Holland Publishing Co.

Mishra, N. C. & Tatum, E. L. (1973). Non-Mendelian inheritance of DNA-induced inositol independence in *Neurospora*. *Proceedings of the National Academy of Sciences, USA*, **70**, 3875–9.

Nasmyth, K. A., Tatchell, K., Hall, B. D., Astell, C. & Smith, M. (1981). A position effect in the control of transcription at yeast mating-type loci. *Nature*, **289**, 244–50.

Parag, Y. (1962). Mutations in the *B* incompatibility factor in *Schizophyllum commune*. *Proceedings of the National Academy of Sciences, USA*, **48**, 743–50.

Prévost, G. (1962). *Etude génétique d'un Basidiomycete:* Coprinus radiatus. Ph.D. thesis, Université de Paris.

Raper, C. A. & Raper, J. R. (1966). Mutations modifying sexual morphogenesis in *Schizophyllum. Genetics*, **54**, 1151–68.

Raper, C. A. & Raper, J. R. (1973). Mutational analysis of a regulatory gene for morphogenesis in *Schizophyllum. Proceedings of the National Academy of Sciences, USA*, **70**, 1427–31.

Raper, J. R. (1966). *Genetics of Sexuality in Higher Fungi.* New York: The Ronald Press.

Raper, J. R., Baxter, M. G. & Ellingboe, A. H. (1960). The genetic structure of the incompatibility factors of *Schizophyllum commune*: the *A* factor. *Proceedings of the National Academy of Sciences, USA*, **44**, 889–900.

Raper, J. R., Boyd, D. H. & Raper, C. A. (1965). Primary and secondary mutations at the incompatibility loci in *Schizophyllum. Proceedings of the National Academy of Sciences, USA*, **53**, 1324–32.

Raudaskoski, M., Stamberg, J., Bawnik, N. & Koltin, Y. (1976). Mutational analysis of natural alleles at the *B* incompatibility factor of *Schizophyllum commune* $\alpha 2$ and $\beta 6$. *Genetics*, **83**, 507–16.

Rine, J., Strathern, J. N., Hicks, J. B. & Herskowitz, I. (1979). A suppressor of mating-type locus mutations in *Saccharomyces cerevisiae*: evidence for and identification of cryptic mating-type loci. *Genetics*, **93**, 877–901.

Southern, P. J. & Berg, P. (1982). Transformation of mammalian cells to antibiotic resistance with a bacterial gene under control of the SV40 early promoter region. *Journal of Molecular and Applied Genetics*, **1**, 327–41.

Specht, C. A., DiRusso, C., Novotny, C. P. & Ullrich, R. C. (1982). A method for extracting high molecular weight deoxyribonucleic acid from fungi. *Analytical Biochemistry*, **119**, 158–63.

Specht, C. A., Novotny, C. P. & Ullrich, R. C. (1983). Isolation and characterization of mitochondrial DNA from the basidiomycete *Schizophyllum commune. Experimental Mycology*, **7**, 336–43.

Specht, C. A., Novotny, C. P. & Ullrich, R. C. (1984). Strain specific differences in ribosomal DNA from the fungus *Schizophyllum commune. Current Genetics*, **8**, 219–22.

Stamberg, J. & Koltin, Y. (1972). The organization of the incompatibility factors in higher fungi: the effects of structure and symmetry on breeding. *Heredity*, **30**, 15–26.

Stinchcomb, D. T., Struhl, K. & Davis, R. W. (1979). Isolation and characterization of a yeast chromosomal replicator. *Nature*, **282**, 39–43.

Ullrich, R. C. (1973). *Genetic determination of sexual diversity in the* Sistotrema brinkmannii *aggregate*. Ph.D. thesis, Harvard University.

Ullrich, R. C. (1978). On the regulation of gene expression: incompatibility in *Schizophyllum. Genetics*, **88**, 709–22.

Ullrich, R. C., Droms, K. A., Doyon, J. D. & Specht, C. A. (1980a). Characterization of DNA from the basidiomycete *Schizophyllum commune. Experimental Mycology*, **4**, 123–34.

Ullrich, R. C., Kohorn, B. D. & Specht, C. A. (1980b). Absence of short-period repetitive-sequence interspersion in the basidiomycete *Schizophyllum commune. Chromosoma*, **81**, 371–8.

Ullrich, R. C. & Raper, J. R. (1975). Primary homothallism – relation to heterothallism in the regulation of sexual morphogenesis in *Sistotrema. Genetics*, **80**, 311–21.

21

Molecular biology of fruit body formation in *Schizophyllum commune*

J. G. H. WESSELS, J. J. M. DONS and
O. M. H. DE VRIES
*Biologisch Centrum, Rijksuniversiteit Groningen, Vakgroep Plantenfysiologie,
Kerklaan 30, 9751 NN Haren, Nederland*

General aspects of fruit body development

The two species of basidiomycete that have been most intensively investigated with respect to fruit body formation are *Schizophyllum commune* (Aphyllophorales) and *Coprinus cinereus* (Agaricales). It is clear that the development of the fruit body of *S. commune* is not at all like that in agarics. In an agaric, like *Coprinus*, all the differentiated structures of the mature fruit body are formed at the button stage and subsequent development is mainly a process of hyphal expansion and differentiation (Moore, Elhiti & Butler, 1979; Watling: Chapter 11; Rosin, Horner & Moore: Chapter 13). In *Schizophyllum*, the fruit body is essentially a tiny cup, the inside of which is lined with hymenium producing basidiospores. The well-known fan-shaped fruit bodies develop, not by hyphal expansion, but by hyphal proliferation at the margin of the cup. At the same time, infoldings of the margin produce the typical split gills which often seem to radiate out from one point when growth is unilateral. The species has, therefore, been removed from the Agaricales (Donk, 1964).

Attempts have been made to divide fruit body development of *Schizophyllum commune* into five successive stages, namely the formation of a knot of intertwined hyphae (stage I), the formation of a small stalk with parallel growing hyphae (stage II), the formation on top of this stalk of an apical pit lined with hymenium (stage III), lateral expansion of the pit and formation of gills (stage IV), and negative tropism of the expanding hymenium (stage V) (Leonard & Dick, 1968; Schwalb, 1978*a*). However, a recent study of early development by scanning electron microscopy (Raudaskoski & Vauras, 1982) suggests that this description of development is not entirely correct. The rim of the fruit body primordium and the

485

centrally located hymenium are simultaneously and directly formed from branching and aggregating hyphae with short cells. The cup is then formed by parallel growing hyphae which at the tip bend towards the centrally located hymenium (van der Valk & Marchant, 1978; Raudaskoski & Vauras, 1982). Early development, ending at the cup stage, will, therefore, be called phase I morphogenesis (Fig. 21.1). The cup-shaped fruit bodies can sporulate and may possess a few infoldings of the rim (gills). During phase II morphogenesis a number of cups enlarge their sporulating surfaces and form many gills by hyphal proliferation.

Although under certain conditions starvation for nutrients may induce fruit body initiation, the whole developmental sequence as depicted in Fig. 21.1 can proceed on a nutritive medium. However, when the nitrogen source in the medium runs out, phase I fruit bodies can still increase in size at the expense of nitrogenous compounds in the vegetative mycelium but phase II morphogenesis is inhibited until the carbon source (glucose) in the medium is also exhausted (Wessels, 1965). When this occurs, the enlargement of the fruit bodies in phase II is accompanied by the formation of β-glucanases that degrade massive amounts of water-soluble and alkali-insoluble $(1 \rightarrow 3)$-β-$/(1 \rightarrow 6)$-β-glucans present in the medium and in the hyphal walls of vegetative mycelium and stunted phase I fruit bodies (Wessels & Sietsma, 1979).

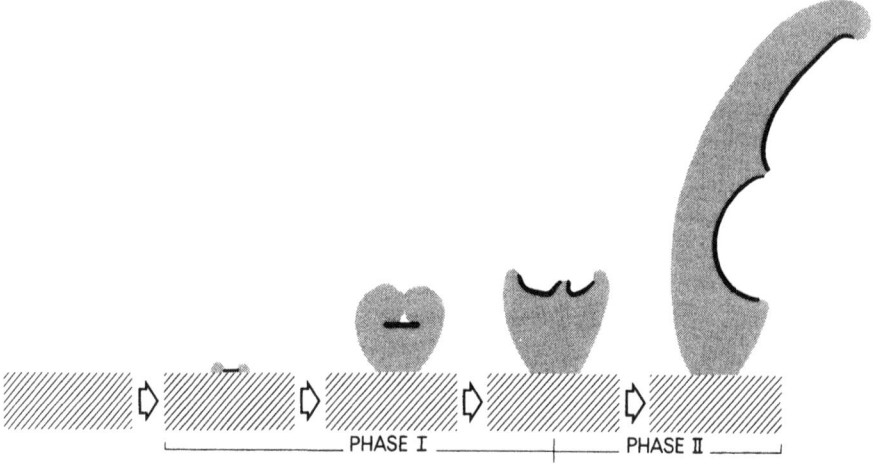

Fig. 21.1. Diagrammatic representation of fruit body development in *Schizophyllum commune*. The hatched area represents the vegetative mycelium, the shaded area the plectenchymatic tissue, and the thick line the hymenium. The morphogenetic sequence is simplified from observations made by Raudaskoski & Vauras (1982), van der Valk & Marchant (1978), Volz & Niederpruem (1969) and Wessels (1965).

As in other basidiomycetes (Manachère, 1980), light and carbon dioxide are the most important environmental factors controlling fruiting in *Schizophyllum commune* (Niederpruem & Wessels, 1969). A short exposure to light enhances shortening, branching and aggregation of hyphae in phase I morphogenesis (Perkins, 1969; Raudaskoski & Viitanen, 1982) while low carbon dioxide concentrations also appear necessary for these early processes (Raudaskoski & Viitanen, 1982). At the biochemical level, a high carbon dioxide concentration (5%) strongly promotes the formation of water-soluble β-glucans (mucilage) while decreasing the synthesis of wall glucans (Sietsma, Rast & Wessels, 1977). In particular, the synthesis of the alkali-insoluble β-glucan is diminished and its structure seems to be modified so that it is less susceptible to enzymic degradation.

Various other enzyme activities have been shown to vary during the development of fruit bodies in *S. commune* (Niederpruem & Wessels, 1969; Schwalb, 1978*a*) but their relevance to morphogenesis is not yet clear. A possible exception can be made for the assumed role of phenoloxidases. Although high phenoloxidase activities can be found in monokaryons and dikaryons not proceeding through phase I morphogenesis, this phase is always associated with an increase in phenoloxidase activity (Leonard & Philips, 1973; Philips & Leonard, 1976). Bu'lock (1967) suggested that phenoloxidases may play a role in oxidative cross-linking within the hyphal matrix of the fruit bodies, strengthening adhesion between hyphae. Since increased phenoloxidase activities are generally found associated with fruit body formation in both ascomycetes and basidiomycetes (Leatham & Stahmann, 1981), this is an intriguing hypothesis. Using a histochemical stain for laccases (Hermann, Kurtz & Champe, 1983), intense staining in young phase I fruit body primordia of *S. commune* has been noted.

With respect to genetic factors involved in fruiting in *Schizophyllum commune*, no extensive mutational analysis has been made (see Elliott: Chapter 19). From the analysis of naturally occurring alleles the importance of certain genes has been inferred. In a number of monokaryotic isolates fruiting occurs spontaneously, as a response to injury, or in response to treatment with an unidentified fruiting inducing substance. At least eight genes (*ha 1* to *8*) have been implicated in these processes (Leslie & Leonard, 1979). Another group (Esser, Saleh & Meinhardt, 1979) has stressed the importance of three genes (fi_1^+, fi_2^+ and fb^+) in haploid fruiting. However, the abundant formation of normal fruit bodies is a trait of the dikaryon and the haploid fruiting genes only seem to affect the time of fruiting in the dikaryon (Esser *et al.*, 1979). Apparently, the interaction between the *A* and *B* incompatibility factors ($A \neq B \neq$) is somehow conducive for the expression of genes involved in fruiting. Mutations in these factors cause the dikaryotic phenotype in homokaryons and these fruit abundantly (Raper, 1978), emphasising the importance of control by the incompatibility factors. In addition, a mutation (*coh*) has been described that interferes with hyphal aggregations in the dikaryon when homozygous (Perkins & Raper, 1970). With respect to the final shape of fruit bodies, various alleles have been described that cause a deviation from the normal pathway (Raper & Krongelb, 1958). Among these are *bse*, producing simple cups, and *med*, with extremely long stipes. The allele *bse* causes an increase in cyclic AMP in both monokaryons and dikaryons (Schwalb, 1978 *b*). A strain with a similar phenotype (*cup*) has been shown to be unable to degrade the alkali-insoluble wall glucan in vegetative mycelium and produces stunted fruit bodies (Wessels, 1965). In this case the absence of a steady flow of carbohydrates at low concentration, which would result from the breakdown of the wall glucan, was held responsible for the blockage of pileus growth.

Changes in the patterns of polypeptide synthesis during fruit body development

In shaking cultures, in which no fruit bodies can be formed, the monokaryons and the derived dikaryon of *S. commune* differ only slightly in the [^{35}S]-methionine-labelled polypeptides they synthesise (De Vries, Hoge & Wessels, 1980; Wessels *et al.*, 1981) and in the patterns of isozymes present in the mycelia (Ullrich, 1977). This is somewhat surprising in view of the conspicuous differences in hyphal morphology of the two types of mycelium. In contrast, when the monokaryon and the dikaryon are grown in surface cultures, allowing for the formation of fruit bodies in the dikaryon, the two types of mycelia develop much more distinct differences in their patterns of protein synthesis (De Vries & Wessels, 1984). In these experiments the dikaryon and monokaryon were grown in such a way that numerous fruit bodies developed in the dikaryon on a confluent mat of mycelium while the monokaryon only formed a mycelial mat. The fruit bodies proceeded through phase I but did not enter phase II under these conditions. Among 400 polypeptides pulse-labelled with [^{35}S]-sulphate and analysed on two-dimensional gels, only eight polypeptides appeared to be synthesised exclusively in the monokaryons but the fruiting dikaryon synthesised 37 polypeptides not present in the monokaryons (Fig. 21.2). In the same series of experiments very few differences in polypeptide synthesis were found between monokaryons and the dikaryon when they were grown in shaking cultures, indicating that the differences arose as a corollary of surface growth and/or fruit body formation in the dikaryon.

Fig. 21.2 shows that 15 of the novel polypeptides of the dikaryon were exclusively synthesised in the fruit bodies while three of the polypeptides which were synthesised in the vegetative mycelia of the monokaryons and the dikaryon were absent from the fruit bodies. The remaining 22 novel polypeptides of the dikaryon were synthesised in both the vegetative mycelium and the fruit bodies. It is also noteworthy that nine of these, being of low molecular weights (between 10000 and 26000), were abundantly excreted into the medium. Two of the excreted polypeptides (molecular weights 18000 and 26000) were also found firmly associated with the hyphal walls of the dikaryon. All these excreted polypeptides were absent in shaking cultures of the dikaryon which then produced the same extracellular polypeptides as the monokaryons.

In evaluating these results it should be noted that the numbers of polypeptides shown to be differentially synthesised necessarily represent minimum values. Detection of polypeptides was limited to those abundantly synthesised, containing sulphur, having isoelectric points between 4.5 and 6.5, and apparent molecular weights between 130000 and 10000. However, the number of differences between the mycelia with respect to abundant

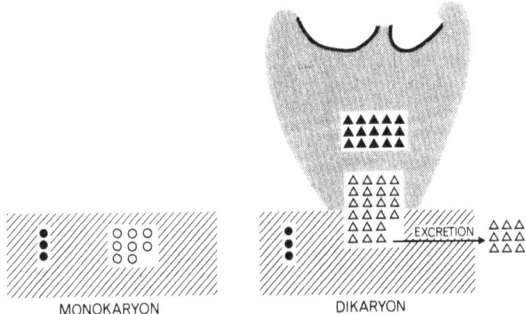

Fig. 21.2. Polypeptides specifically synthesised in the monokaryon and the fruiting dikaryon of *Schizophyllum commune* after 4 days in surface culture. Of 400 polypeptides examined most were the same. Each symbol refers to a single polypeptide differentially synthesised. ○, polypeptides exclusively synthesised in the monokaryon; ●, polypeptides synthesised in both the monokaryon and in the vegetative mycelium of the dikaryon, but not in the fruit bodies; △, polypeptides not synthesised in the monokaryon but synthesised in both the vegetative mycelium and in the fruit bodies of the dikaryon; ▲, polypeptides exclusively synthesised in fruit bodies. (Based on data from De Vries & Wessels, 1984.)

mRNAs, as measured by *in vitro* translations and RNA/cDNA hybridisations (see below), was similar to the number of differentially synthesised polypeptides detected *in vivo*. Therefore, the actual number of polypeptides differentially synthesised may be close to that detected.

Genome structure and mRNA production

The nuclear genome of *Schizophyllum commune*, probably contained in 11 chromosomes (Carmi *et al.*, 1978), has a size of $36\text{–}37 \times 10^6$ base pairs (Dons, De Vries & Wessels, 1979; Ullrich *et al.*, 1980*a*); that is, about eight times the size of the genome of *Escherichia coli*. Our studies (Dons *et al.*, 1979; Dons & Wessels, 1980; Wessels *et al.*, 1981) revealed the following characteristics. The nuclear DNA with a $G+C$ content of 57% contains little repetitive DNA (7%) and this is mainly taken up by rRNA cistrons. In agreement with this, no repetitive DNA sequences interspersed with unique sequences could be detected in the nuclear genome. The mitochondrial DNA has a size of about 5×10^4 base pairs and is characterised by interspersed $(A+T)$-rich sequences accounting for its low $G+C$ content (22%). The studies of Ullrich *et al.* (1980*a*, *b*) and Specht, Novotny & Ullrich (1983) have confirmed and extended these data. In addition, Specht *et al.* (1984) have shown that, due to restriction polymorphism, a previous estimate of the size of the rRNA cistrons (Dons & Wessels, 1980) was probably too high and should be in the range of 9.2 to 9.4 kilo base pairs (see Ullrich & Novotny: Chapter 20).

From the work on the formation of mRNAs in *Schizophyllum commune* (Zantinge, Dons & Wessels, 1979; Zantinge, Hoge & Wessels, 1981; Hoge *et al.*, 1982*b*) the following summary can be made. The complexities of total RNA, poly(A)-containing RNA and polysomal RNA are the same within the limits of accuracy of RNA : single-copy DNA and RNA : complementary DNA hybridisations. These measurements indicate the presence at all stages of the life cycle of about 10 000 to 13 000 different mRNAs of number-average lengths of 1100 nucleotides, while only 25% of the complex RNAs appear to carry a short poly(A)-tail of 33 nucleotides on average. These RNAs are complementary to 16.5% of the nuclear DNA indicating that a substantial part (33%) of the nuclear genome is transcribed into translatable RNAs. Although these findings indicate the absence of a sizable fraction of heterogeneous nuclear RNAs, at least some genes may still carry introns. It has recently been found that the 1G2 gene of *S. commune* (see below) which codes for a mRNA of 650 nucleotides carries three small introns of approximately 50 base pairs each (Dons *et al.*, 1984*a*).

Changes in mRNAs during fruit body development

Because fruit bodies normally develop only on dikaryotic mycelia which are derived from monokaryotic mycelia through interactions of their *A* and *B* incompatibility factors, RNA populations were first examined in these mycelia in the absence of fruit body formation by using shaking cultures. Using competition hybridisations of RNA with single-copy DNA and homologous and heterologous hybridisations with complementary DNA (cDNA), it was not possible to detect any qualitative differences in either the class of abundantly occurring RNA sequences or the class of rare RNA sequences (Zantinge *et al.*, 1979, 1981; Hoge *et al.*, 1982*b*).

For the abundantly occurring RNA sequences, the absence of qualitative differences in shaking cultures of the monokaryon and the dikaryon was clearly evident. This class of RNAs, comprising about 500 sequences each present in about 150 copies per cell (Zantinge *et al.*, 1981), can also be

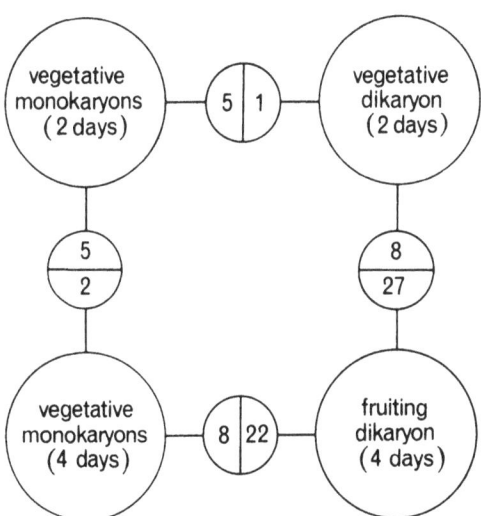

Fig. 21.3. Diagram of differences between *in vitro* translation patterns of total-RNA preparations isolated from monokaryons and dikaryon of *Schizophyllum commune* growing in surface cultures. The total number of polypeptides analysed on two-dimensional gels was 411. Numbers denote apparent differences in mRNAs in the paired cultures. For example, 27 mRNAs present in the 4-day-old fruiting dikaryon were absent in the 2-day-old vegetative dikaryon while eight mRNAs present in the vegetative dikaryon were absent in the fruiting dikaryon. Before translation, the RNAs from the two co-isogenic monokaryons were mixed. At an age of 2 days the two monokaryons showed a difference in four translatable RNAs. At an age of 4 days, six differences were noted. (From Hoge *et al.*, 1982*a*).

analysed by *in vitro* translation. No significant differences between mono-karyon and dikaryon were observed when 400 translation products of total-RNA preparations (Zantinge *et al.*, 1981) or polysomal-RNA prep-arations (Hoge *et al.*, 1982*b*) were analysed on two-dimensional gels. This suggests that slight differences observed in the patterns of polypeptides synthesised *in vivo* by shaking cultures of monokaryons and dikaryon (De Vries *et al.*, 1980) are not due to differences in the mRNAs and polypeptides actually synthesised but are brought about by differential modification of identical polypeptides. Of course this does not exclude possible differences in a few abundant mRNAs coding for products that cannot be visualised on two-dimensional gels.

With regard to the absence of differences in the class of rare DNA sequences, the number of sequences in this class is so large (about 13 000 sequences occurring in four to seven copies per cell (Zantinge *et al.*, 1981)) that a change in fewer than about 100 sequences cannot be detected by hybridisation studies. In fact, by using cloned cDNA sequences as probes, it has recently been found (Dons *et al.*, 1984*b* and unpublished) that at least two mRNAs abundantly present in the fruiting dikaryon are also present in low concentration in the dikaryon, but not in the monokaryon, when growing in shaking cultures.

In contrast to shaking cultures, *in vitro* translations of mRNAs isolated from surface cultures showed that the mRNAs from monokaryon and dikaryon progressively deviate when fruit bodies are formed in the dikaryon (Hoge, Springer & Wessels, 1982*a*). Whilst the dikaryon essentially maintains the complement of translatable RNAs present in the parent monokaryons, some 25 new mRNAs appear in the dikaryon at the time of fruit body formation (Fig. 21.3). This was confirmed by homologous and heterologous hybridisations of RNA with complementary DNA (cDNA). RNA prepared from 4-day-old monokaryons failed to hybridise 5% of the reactable cDNA made on RNA from 4-day-old fruiting dikaryon, while the reverse experiment indicated that the fruiting dikaryon essentially contained all the sequences present in the monokaryons. It was also shown that the difference resided only in the class of abundant RNAs, in this case totalling 250 sequences comprising 34% of the mass of complex RNA. Assuming equal frequency of all sequences in this class, this would mean that about 35 mRNAs are uniquely synthesised in the fruiting dikaryon. The apparent absence of differences in the class of approximately 13 000 rare RNA sequences was confirmed by RNA/single-copy DNA hybridisations. But, as discussed earlier, these experiments cannot exclude the possibility that some regulation occurred in this class.

To examine the developmentally regulated abundant mRNAs in more detail, specific clones complementary to these sequences were isolated (Dons *et al.*, 1984*b*). A preparation of cDNA made on poly(A)-RNA of the fruiting dikaryon was cloned into the *Pst*1 site of pBR327 and 2000 bacterial colonies were screened with radioactive cDNA made on the RNAs of the monokaryons and the fruiting dikaryon. As expected, most colonies hybridised similarly with the two cDNA preparations and hybridisation with dikaryotic cDNA was almost completely abolished by prehybridisation with excess monokaryotic RNA. This confirmed that the RNA populations of the monokaryon and the fruiting dikaryon were largely similar. Of those colonies hybridising specifically to cDNA of the fruiting dikaryon, ten were homologous to the same cDNA sequence. Assuming random cloning efficiency, it could be calculated that this sequence must occupy 0.6% of the mass of complex RNA (i.e. 1100 copies per cell). One of these clones (1G2 or 7D5) was used to probe the concentration of the corresponding mRNA. For comparison a clone (1D10) was used which was also abundantly represented in the cDNA bank but was homologous to both monokaryotic and dikaryotic cDNA. Fig. 21.4 shows the hybridisation patterns of the two cDNA clones to electrophoresed RNAs isolated at various times from surface cultures of the monokaryon and the dikaryon. The 1D10 mRNA (size 775 nucleotides, encoding a polypeptide of molecular weight 15000 as determined on SDS gels after hybrid-release translation) was invariably present in both the monokaryon and the dikaryon, although its concentration varied with age of the cultures. However, not a trace of the 1G2 mRNA (size 650 nucleotides, encoding a polypeptide of molecular weight 9842 as determined by a sequence study (Dons *et al.*, 1984*a*)) could be found in the monokaryons, while the concentration of this mRNA strongly increased during the establishment of fruit bodies in the dikaryon.

Exactly the same pattern of appearance in the fruiting dikaryon as that found for the 1G2 mRNA (Fig. 21.4A) was observed for another larger mRNA (1200 nucleotides) by using the specific cDNA clone 17B5. Significantly, both the 1G2 and 17B5 clones also gave weak hybridisation signals with RNAs isolated from the dikaryon grown in shaking culture. Apparently, transcription of the genes for these mRNAs is somehow conditioned by the presence of heteroallelic incompatibility factors but full expression only occurs at the time of fruit body formation.

The gene for 1G2 itself was located on a *Pst*1 generated chromosomal fragment of 9000 base pairs which was cloned. This genomic clone only hybridised to the 1G2 mRNA indicating that the gene for 1G2 is not

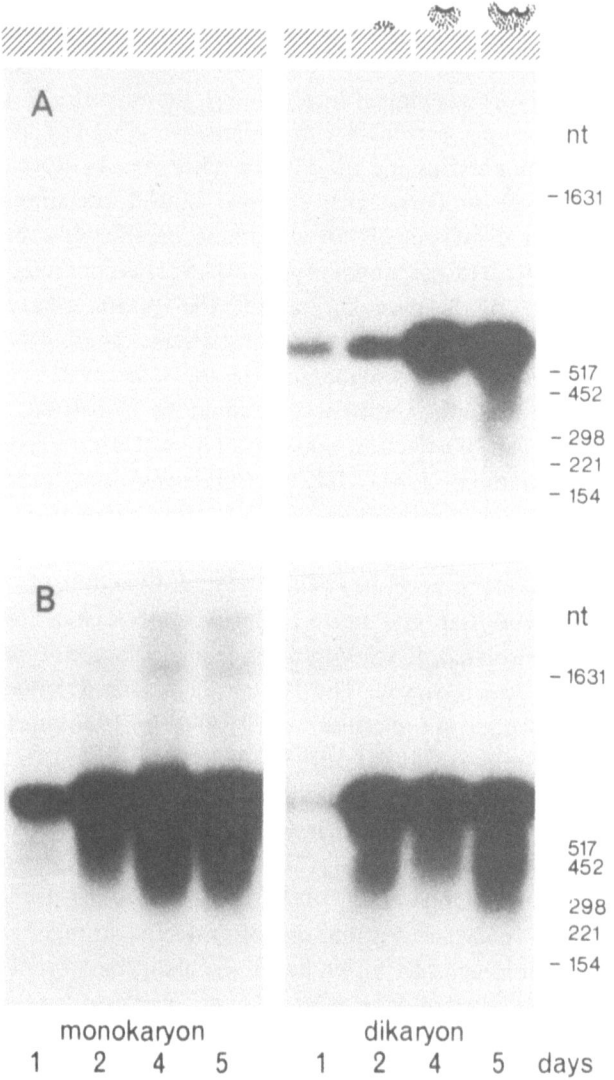

Fig. 21.4. Expression of cloned mRNA sequences in monokaryon and dikaryon of *Schizophyllum commune* growing in surface cultures. Poly(A)-RNAs were denatured, run on 1.5% agarose gels, blotted onto filters, and hybridised to radioactive cDNA clones 7D5 (homologous to clone 1G2, panel A) and 1D10 (panel B). Each lane was loaded with the same amount of poly(A)-RNA. Markers are *Hinf*1 fragments of plasmid pBR327 visualised by the clones. (Adapted from Dons *et al.*, 1984*b*.)

surrounded by other genes abundantly expressed under the culture conditions used (Dons *et al.*, 1984*b*). The 1G2 gene and its mRNA have been completely sequenced (Dons *et al.*, 1984*a*) and functional analysis of the gene product is anticipated.

Concluding remarks

Perhaps the most notable aspects of the work reviewed are that (1) the monokaryon–dikaryon transition, as governed by the incompatibility genes, is not accompanied by major qualitative changes in gene transcription; and (2) that the subsequent formation of fruit bodies in the dikaryon is accompanied by a change in the expression of a few genes that produce abundant mRNAs. There is also a good correspondence between the number of these regulated mRNAs and the number of new major polypeptides that appear during fruiting in the dikaryon.

Although the use of cloned sequences may reveal that some regulation also occurs in the class of rare mRNAs, it does appear that the number of genes regulated during fruit body formation is sufficiently small to permit a comprehensive analysis. Further studies should reveal the timing of regulation of these genes whilst their isolation and base sequence determination may provide clues as to the mechanism of their regulation. The fact that certain genes that are regulated produce abundant mRNAs should also make it possible to bridge the gap between gene regulation and the proteins that function to execute as yet unknown steps in fruit body morphogenesis.

References

Bu'lock, J. D. (1967). *Essays in Biosynthesis and Microbial Development. Rutgers University Institute of Microbiology. E. R. Squibb Lectures on Chemistry of Microbial Products.* New York: Wiley (see especially pp. 1–18).

Carmi, P., Holm, P. B., Koltin, Y., Rasmussen, S. W., Sage, J. & Zickler, D. (1978). The pachytene karyotype of *Schizophyllum commune* analyzed by three dimensional reconstruction of synaptonemal complexes. *Carlsberg Research Communications*, **43**, 117–32.

De Vries, O. M. H., Hoge, J. H. C. & Wessels, J. G. H. (1980). Regulation of the pattern of protein synthesis in *Schizophyllum commune*. *Developmental Biology*, **74**, 22–36.

De Vries, O. M. H. & Wessels, J. G. H. (1984). Pattern of polypeptide synthesis in non-fruiting monokaryons and a fruiting dikaryon of *Schizophyllum commune*. *Journal of General Microbiology*, **130**, 145–54.

Donk, M. A. (1964). A conspectus of the families of Aphyllophorales. *Persoonia*, **3**, 199–324.

Dons, J. J. M , De Vries, O. M. H. & Wessels, J. G. H. (1979). Characterization of the genome of the basidiomycete *Schizophyllum commune. Biochimica et Biophysica Acta*, **563**, 100–12.

Dons, J. J. M., Mulder, G. H., Rouwendal, G. J. A., Springer, J., Bremer, W. & Wessels, J. G. H. (1984*a*). Sequence analysis of a split gene involved in fruiting from the fungus *Schizophyllum commune. The EMBO Journal*, **3**, 2101–6.

Dons, J. J. M., Springer, J., De Vries, S. C. & Wessels, J. G. H. (1984*b*). Molecular cloning of a gene abundantly expressed during fruit-body initiation in *Schizophyllum commune. Journal of Bacteriology*, **157**, 802–8.

Dons, J. J. M. & Wessels, J. G. H. (1980). Sequence organization of the nuclear genome of *Schizophyllum commune. Biochimica et Biophysica Acta*, **607**, 385–96.

Esser, K., Saleh, F. & Meinhardt, F. (1979). Genetics of fruit-body production in higher basidiomycetes. II. Monokaryotic and dikaryotic fruiting in *Schizophyllum commune. Current Genetics*, **1**, 85–8.

Hermann, T. E., Kurtz, M. B. & Champe, S. P. (1983). Laccase localized in hülle cells and cleistothecial primordia of *Aspergillus nidulans. Journal of Bacteriology*, **154**, 955–64.

Hoge, J. H. C., Springer, J. & Wessels, J. G. H. (1982*a*). Changes in complex RNA during fruit-body initiation in the fungus *Schizophyllum commune. Experimental Mycology*, **6**, 233–43.

Hoge, J. H. C., Springer, J., Zantinge, B. & Wessels, J. G. H. (1982*b*). Absence of differences in polysomal RNAs from vegetative monokaryotic and dikaryotic cells of the fungus *Schizophyllum commune. Experimental Mycology*, **6**, 225–32.

Leatham, G. F. & Stahmann, M. A. (1981). Studies on the laccase of *Lentinus edodes*: specificity, localization and association with the development of fruiting bodies. *Journal of General Microbiology*, **125**, 147–57.

Leonard, T. J. & Dick, S. (1968). Chemical induction of haploid fruiting in *Schizophyllum commune. Proceedings of the National Academy of Sciences, USA*, **59**, 745–51.

Leonard, T. J. & Philips, L. E. (1973). Studies of phenoloxidase activity during the reproductive cycle in *Schizophyllum commune. Journal of Bacteriology*, **114**, 7–10.

Leslie, J. F. & Leonard, T. J. (1979). Monokaryotic fruiting in *Schizophyllum commune*: genetic control of the response to mechanical injury. *Molecular and General Genetics*, **175**, 5–12.

Manachère, G. (1980). Conditions essential for controlled fruiting of macromycetes. A review. *Transactions of the British Mycological Society*, **75**, 255–70.

Moore, D., Elhiti, M. M. Y. & Butler, R. D. (1979). Morphogenesis of the carpophore of *Coprinus cinereus. New Phytologist*, **83**, 695–722.

Niederpruem, D. J. & Wessels, J. G. H. (1969). Cytodifferentiation and morphogenesis in *Schizophyllum commune. Bacteriological Reviews*, **33**, 505–35.

Perkins, J. H. (1969). Morphogenesis in *Schizophyllum commune*. I. Effects of white light. *Plant Physiology*, **44**, 1706–11.

Perkins, J. H. & Raper, J. R. (1970). Morphogenesis in *Schizophyllum commune*. III. A mutation that blocks initiation of fruiting. *Molecular and General Genetics*, **196**, 151–4.

Philips, L. E. & Leonard, T. J. (1976). Extracellular and intracellular phenoloxidase activity during growth and development in *Schizophyllum commune. Mycologia*, **68**, 268–76.

Raper, C. A. (1978). Control of development by the incompatibility system in basidiomycetes. In *Genetics and Morphogenesis in the Basidiomycetes*, ed. M. N. Schwalb & P. G. Miles, pp. 3–29. New York: Academic Press.

Raper, J. R. & Krongelb, G. S. (1958). Genetic and environmental aspects of fruiting in *Schizophyllum commune* Fr. *Mycologia*, **50**, 707–40.

Raudaskoski, M. & Vauras, R. (1982). Scanning electron microscope study of fruit body differentiation in *Schizophyllum commune*. *Transactions of the British Mycological Society*, **78**, 475–81.

Raudaskoski, M. & Viitanen, H. (1982). Effects of aeration and light on fruit-body induction in *Schizophyllum commune*. *Transactions of the British Mycological Society*, **78**, 89–96.

Schwalb, M. N. (1978a). Regulation of fruiting. In *Genetics and Morphogenesis in the Basidiomycetes*, ed. M. N. Schwalb & P. G. Miles, pp. 135–65. New York: Academic Press.

Schwalb, M. N. (1978b). A developmental mutant affecting 3′:5′-cyclic AMP metabolism in the basidiomycete *Schizophyllum commune*. *FEMS Microbiology Letters*, **3**, 107–10.

Sietsma, J. H., Rast, D. & Wessels, J. G. H. (1977). The effect of carbon dioxide on fruiting and on the degradation of a cell wall glucan in *Schizophyllum commune*. *Journal of General Microbiology*, **192**, 385–9.

Specht, C. A., Novotny, C. P. & Ullrich, R. C. (1983). Isolation and characterization of mitochondrial DNA from the basidiomycete *Schizophyllum commune*. *Experimental Mycology*, **7**, 336–43.

Specht, C. A., Novotny, C. P. & Ullrich, R. C. (1984). Strain specific differences in ribosomal DNA from the fungus *Schizophyllum commune*. *Current Genetics*, **8**, 219–22.

Ullrich, R. C. (1977). Isozyme patterns and cellular differentiation in *Schizophyllum commune*. *Molecular and General Genetics*, **156**, 157–61.

Ullrich, R. C., Droms, K. A., Doyon, J. D. & Specht, C. A. (1980a). Characterization of DNA from the basidiomycete *Schizophyllum commune*. *Experimental Mycology*, **4**, 123–34.

Ullrich, R. C., Kohorn, B. D. & Specht, C. A. (1980b). Absence of short-period repetitive-sequence interspersion in the basidiomycete *Schizophyllum commune*. *Chromosoma*, **81**, 371–8.

Valk, P. van der & Marchant, R. (1978). Hyphal ultrastructure in fruit-body primordia of the Basidiomycetes *Schizophyllum commune* and *Coprinus cinereus*. *Protoplasma*, **95**, 57–72.

Volz, P. A. & Niederpruem, D. J. (1969). Dikaryotic fruiting in *Schizophyllum commune* Fr.: Morphology of developing basidiocarp. *Archiv für Mikrobiologie*, **68**, 246–58.

Wessels, J. G. H. (1965). Morphogenesis and biochemical processes in *Schizophyllum commune*. *Wentia*, **23**, 1–113.

Wessels, J. G. H. & Sietsma, J. H. (1979). Wall structure and growth in *Schizophyllum commune*. In *Fungal walls and Hyphal Growth* (British Mycological Society Symposium 2), ed. J. H. Burnett & A. P. J. Trinci, pp. 27–48. Cambridge University Press.

Wessels, J. G. H., Dons, J. J. M., Hoge, J. H. C., Springer, J., De Vries, O. M. H. & Zantinge, A. (1981). Genetic regulation of RNA and protein patterns in the monokaryon–dikaryon transition. In *The Fungal Nucleus* (British Mycological Society Symposium 5), ed. K. Gull & S. J. Oliver, pp. 295–314. Cambridge University Press.

Zantinge, A., Dons, J. J. M. & Wessels, J. G. H. (1979). Comparison of poly(A)-containing RNAs in different cell types of the lower eukaryote *Schizophyllum commune*. *European Journal of Biochemistry*, **101**, 251–60.

Zantinge, A., Hoge, J. H. C. & Wessels, J. G. H. (1981). Frequency and diversity of RNA sequences in different cell types of the fungus *Schizophyllum commune*. *European Journal of Biochemistry*, **113**, 381–9.

22

Meiosis and genetic recombination in *Coprinus cinereus*

PATRICIA J. PUKKILA, DAVID M. BINNINGER,
JEANE R. CASSIDY, BEVERLY M. YASHAR
and MIRIAM E. ZOLAN

Department of Biology and Curriculum in Genetics, University of North Carolina, Chapel Hill, North Carolina 27514, USA

Introduction

Homologous chromosome pairing and genetic recombination occur in a highly controlled manner in meiotic cells, yet virtually nothing is known at the molecular level concerning the regulation of these processes. Since the meiocyte is engaged in a variety of other tasks relating to its eventual production of haploid progeny, it is not surprising that functions relating specifically to meiotic chromosome behaviour have remained difficult to study. The meiotic process in *Coprinus cinereus* offers a unique set of advantages for molecular studies which are outlined below. We have attempted to develop cytological methods appropriate for sensitive monitoring of basidial differentiation in *Coprinus*, and we next sought to identify a class of polyadenylated RNAs which are differentially expressed during fruit body development. The third goal of these studies was to use DNA sequence polymorphisms to monitor crossing-over at the DNA level within the tandemly repeated ribosomal RNA genes.

Synchronous development in *Coprinus*

There are several features of the *Coprinus* life history which we have used to advantage. We routinely grow *Coprinus* on the simple synthetic medium of Rao & Niederpruem (1969) and subsequent fruit body development can be manipulated easily by altering temperature and light (Madelin, 1956; Tsusué, 1969; Matthews & Niederpruem, 1972; Morimoto & Oda, 1973, 1974; Lu, 1974; Kamada, Kurita & Takemaru, 1978). Vegetative growth is optimal at 37 °C, and fruit body development is inhibited unless a light stimulus is provided. If the cultures are kept in the dark at 37 °C until the vegetative mycelial growth is extensive and then shifted to 25 °C under a 16 h light/8 h dark regime, 10–20 fruit bodies

Table 22.1. *List of strains used*

Strain	ATCC number	Geographical origin	Source
Brum a	42721	England	Moore
C692		England	Casselton
FDG-1		England	North
H9	18064	England	Moore
Java a	42722	Java	Moore
Penn a	42726	USA	Moore
12890 #3	12890	USA (?)	Moore

FDG-1 is a dikaryon; the rest are monokaryons.

develop synchronously in each 50×90 mm dish within the next few days. Fertile strains also produce a second crop of synchronous fruit bodies a few days after the first complete their development. We can vary the time of day that fruit bodies at any desired stage appear by altering the time at which the light cycle begins. Two cycles are routinely used: in the A cycle the 16 h light period begins at 05.00, and in the B cycle it begins at 17.00. The strains used in these experiments are described in Table 22.1.

Buller (1909, 1924) was the first to emphasise the striking synchrony of meiosis and spore formation in *Coprinus*. He supplied an interesting rationale for this synchrony by suggesting that the bulk of the fruit body tissue was involved in spore production rather than in supplying the support that would be necessary if the gills were horizontally positioned. Spore discharge cannot proceed efficiently from vertical gills, however, and *Coprinus* solves this mechanical problem by a striking morphogenetic change associated with autodigestion of the gills (Rosin *et al.*: Chapter 13). Multiple generations of developing basidia are thus avoided, and the 10^7–10^8 meiocytes present in each fruiting body develop in a highly synchronised fashion. A major technical advantage which *Coprinus* offers is that meiotic chromosomes are visible using the light microscope so that cytological criteria can be used in addition to morphological criteria to define the stage of development of the material. To assess synchrony within and among a series of cultures, it is most convenient to examine the nuclear contents without extensively disrupting gill structure (Fig. 22.1). Watchmakers forceps were used to peel away single hymenial layers from carpophore tissue fixed according to Lu & Raju (1970). These 'half-gill' segments were kept intact through acid hydrolysis and ion haematoxylin staining, rinsed briefly in 70% ethanol to remove precipitated stain, and

mounted in glycerol. Fig. 22.1 A illustrates nuclei of the long basidia
near the end of meiotic prophase 14 h after the onset of karyogamy (the
nuclei of the interspersed short basidia are below the plane of focus here,
but exhibited an identical morphology). Fig. 22.1 B shows a particularly
useful cytological landmark: the entry into metaphase I, 15 h after the
onset of karyogamy. A few prophase nuclei still remain, and some nuclei
have continued into metaphase II, but sterigmata are not evident.
Typically, sterigmata did not appear until 19 h after the start of karyogamy,
and spores did not begin to form until 20 h. Spores were not shed until
35 h. Examination of these preparations at lower focal planes revealed that
the paraphyses expanded markedly around the time of metaphase I. This
expansion is reflected by the relative lack of crowding and overlap of the
basidia in Fig. 22.1 B when compared to those in 22.1 A.

As is apparent in Fig. 22.1 A, details of chromosome morphology were
difficult to discern when the nuclei remained confined within the basidia.
A further drawback of this method was that acid hydrolysis was required
to differentiate the chromosomes from the cytoplasm. These considerations
prompted us to try to adapt the surface microspreading and silver staining
techniques which have been used successfully to follow synapsis and
desynapsis in animals and plants (Moses, 1977; Gillies, 1981; Moses,
Dresser & Poorman, 1984). When gill segments were fixed in 4% formaldehyde (freshly prepared from paraformaldehyde and buffered to pH 8
with 0.02 M sodium borate) then transferred to 45% acetic acid, the
basidial walls were easily broken by tapping sharply on the slide. The
swollen nuclear contents adhered well to slides which had been coated with
polylysine (Mazia, Schatten & Sale, 1975). After freezing the slide in liquid
nitrogen and removing the coverslip, the squash was stained with silver
nitrate (Howell & Black, 1980), and individual chromosomes could be
distinguished (Fig. 22.1 C, D). The flattened basidial contents in Fig. 22.1 C
and D, squashed out of the cell walls, are shown at the same magnification
as the intact basidia in Fig. 22.1 A and B. It has recently been confirmed,
by examining such preparations in the electron microscope, that lateral
components of the synaptonemal complex are selectively stained by this
method (Pukkila & Lu, 1985). The chromosomes in Fig. 22.1 C and D
were somewhat stretched by the squashing procedure, as three-dimensional
reconstruction of synaptonemal complexes in *Coprinus* has indicated that
the total complex length is 48 μm in such zygotene nuclei (Holm *et al.*,
1981). We are currently attempting to achieve equivalent chromosome
separation without excessive stretching. This simple and rapid method can

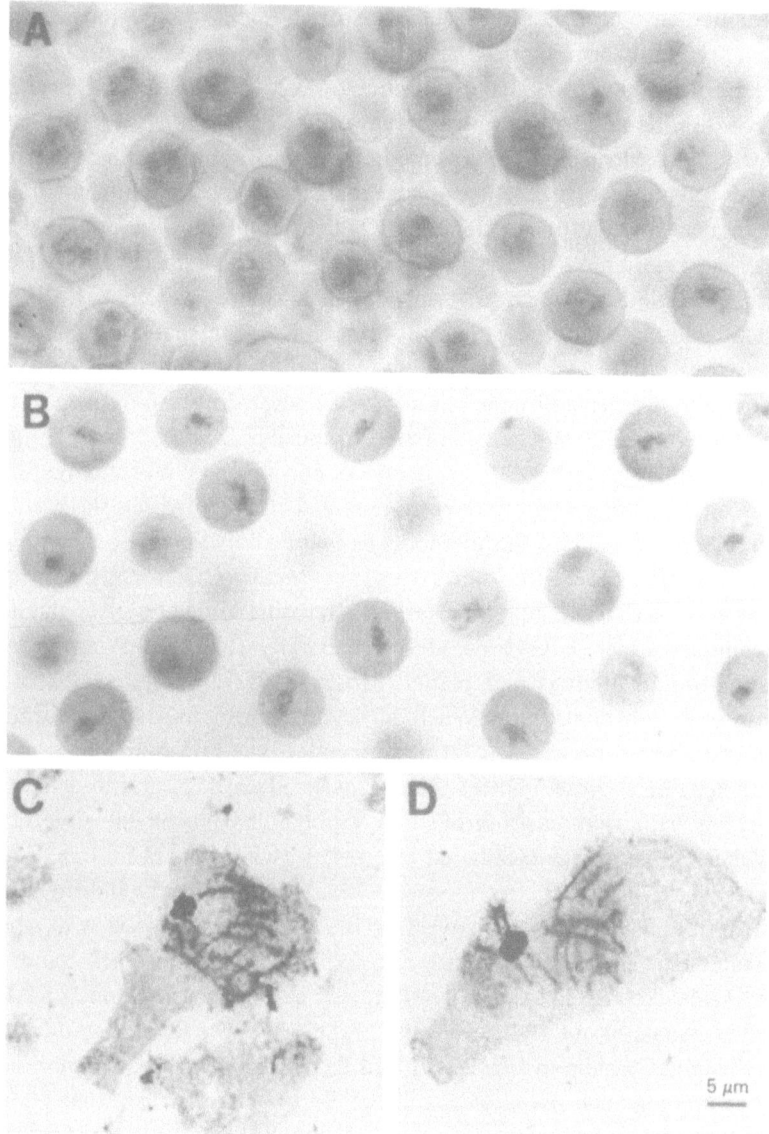

Fig. 22.1. Surface views (A and B) and squash preparations (C and D) of *Coprinus* basidia. Fixation, staining and culture conditions are as described in the text. A: late prophase, from strain FGD-1 in culture cycle A. B: metaphase I, from Brum a × 12890 #3 in B cycle. C: late zygotene, from FGD-1 in A cycle; D: mid zygotene, from Brum a × 12890 #3 in A cycle.

be used to monitor the process of chromosome synapsis in a single fruit body since removal of gill segments does not affect development in the remaining tissue.

We conclude that the natural meiotic synchrony and relatively long duration of these developmental processes allow sharply demarcated stages to be easily harvested. In particular, fruit bodies engaged in basidial differentiation, meiotic prophase and spore formation can be harvested separately quite readily. If controlling mechanisms for these processes are expressed in a similar temporal sequence, before the extensive degradative changes associated with spore release occur, then these should be amenable to molecular analysis.

Cloning sequences differentially expressed during fruit body morphogenesis

The small nuclear genome of *Coprinus* (37 500 kilobases (kb)) and the presence of easily separable tissues greatly facilitate the detection of changes in polyadenylated RNAs associated with morphogenesis. The technique of differential colony hybridisation has been used to identify cloned genomic segments which are expressed to differing extents in different developmental states. By examining the patterns of hybridisation of specific DNA fragments in selected cloned intervals, the genomic organisation of regulated transcripts was also studied.

When bacterial colonies containing different *Coprinus* genomic fragments inserted into a plasmid vector were screened by hybridisation to radioactive DNA copies of polyadenylated RNAs extracted from a particular tissue, some colonies hybridised to greater extents than others. Most of the differences in intensity of the hybridisation signal were a consequence of differences in prevalence of individual RNA species in the heterogeneous probe (Lasky *et al.*, 1980). Genomic sequences whose transcripts are abundant in the heterogeneous probe bind more radioactive molecules than genomic sequences which code for rare transcripts. By screening a duplicate array of bacterial colonies with heterogeneous cDNA probes from two different tissues, clones which react to the two probes to different extents could be detected easily (Fig. 22.2). This method was first used by St John & Davis (1979), and has subsequently been used to examine steady-state levels of individual RNA species and identify developmentally regulated genes in a variety of plants, animals and fungi.

In our experiments, a genomic clone library which contains a collection of 13 500 *Coprinus* DNA fragments averaging 10 kb in length has been used. These fragments were isolated from strain H9 by partial digestion

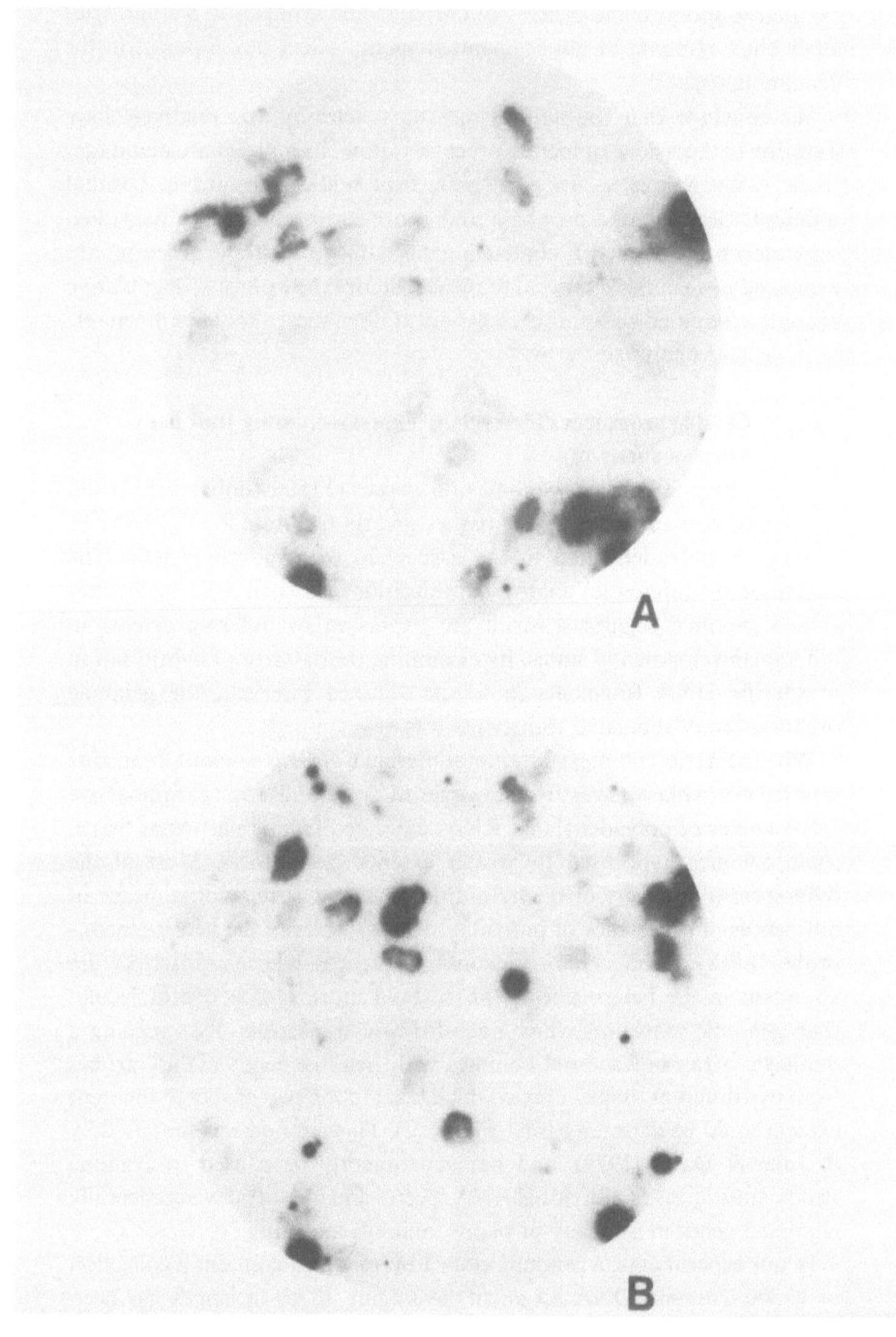

Fig. 22.2. Differential colony hybridisation. Filter A was probed with fruit body cDNA and filter B with dikaryon cDNA from Penn a × Java a. The duplicate clone arrays derive from H9. (From Pukkila, Yashar and Binninger, 1984.)

with restriction enzymes *Mbo*I and *Sau*3A, and sedimentation through sucrose gradients. The fragments were ligated into the yeast shuttle vector YRp12 (Scherer & Davis, 1979). There is a 97% probability that any particular *Coprinus* sequence will be represented in this library (Pukkila, Yashar & Binninger, 1984).

We first determined if the set of polyadenylated RNAs present in vegetative dikaryons was identical to that present in fruit bodies in meiotic prophase. RNAs were extracted from pulverised frozen tissue by homogenisation in guanidinium thiocyanate (Chirgwin *et al.*, 1979). Polyadenylated molecules were selected using oligo-dT cellulose (Aviv & Leder, 1972), and cDNA copies were synthesised using reverse transcriptase (Kacian *et al.*, 1971). Duplicate arrays of genomic clones were then hybridised with these two probes. As shown in Fig. 22.2, numerous examples of differential hybridisation were obvious. We analysed 654 individual genomic clones in this way (17% of the genome) and concluded that 7% of these clones exhibited at least a ten-fold difference in response to these two probes (Yashar & Pukkila, 1985).

These results were confirmed in two ways. First, 25 clones were chosen which exhibited a wide range of hybridisation intensity to the fruit body probe. DNA isolated from these clones was digested with restriction enzyme *Eco*R1 and specific DNA fragments hybridising to one or both probes were identified using Southern's (1975) method of transfer. All clones contained specific fragments which hybridised to the fruit body probe, and the wide range in signal intensities seen in the colony hybridisation experiments were again seen using this more sensitive technique (Fig. 22.3). That the amount of hybridisation to a given cloned sequence reflected the abundance of that particular transcript in the heterogeneous probe was confirmed by labelling individual cloned sequences and using these to probe unlabelled RNAs which had been extracted from various cell types, electrophoresed through denaturing gels, and transferred to nitrocellulose (results not shown).

Experiments involving these 25 genomic clones have revealed three types of transcripts as illustrated in Fig. 22.3. At least 17 transcripts were detected which appeared to be equally abundant in vegetative dikaryons and fruit bodies (Fig. 22.3, lanes b, h and m). At least 12 transcripts appeared to be more abundant in the dikaryon (Fig. 22.3, lane l and the 2.5-kb band in lane a), and at least 11 transcripts appeared to be more abundant in the fruit bodies (Fig. 22.3, lanes e, g and k). These results were confirmed by comparison with data obtained from 43 genomic clones selected at random. In addition to the three hybridisation patterns seen

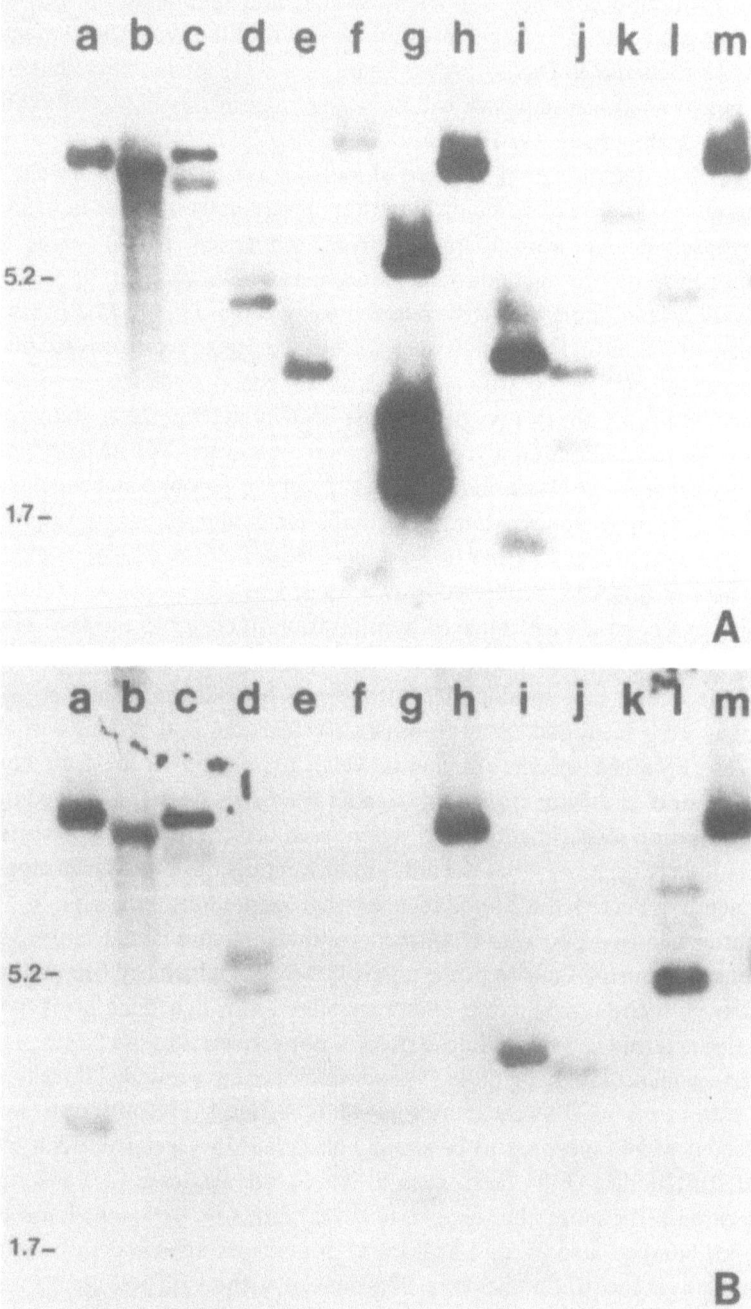

Fig. 22.3. Southern analysis of clones expressed in fruit bodies. DNA was extracted from individual clones and digested with restriction enzyme *Eco*R1. Filter A was probed with fruit body cDNA and filter B with dikaryon cDNA. Size markers (kb) are indicated. (From Pukkila, Yashar and Binninger, 1984.)

in the first set of clones, five clones failed to react to detectable levels with either probe, although the experiments would not have detected RNAs present at less than 0.01% of the total polyadenylated RNA.

These experiments have also indicated that transcripts arising from nearby genomic intervals were often differentially regulated (Fig. 22.3, lanes a, c and f). Such single cloned intervals evidently contained fragments which hybridised to the two probes to the same extent in addition to fragments which did not. The transcripts we have analysed to date reached their maximum concentrations in different tissues and at different times in fruit body development. Thus it appears that different RNA sets were selectively utilised during development in this basidiomycete.

Studies of recombination at the DNA level

The small genome size of *Coprinus* and the relatively high frequency of recombination in meiosis led us to search for DNA polymorphisms which would be useful to follow crossing-over *in vivo* (Wu, Cassidy & Pukkila, 1983; and see Ullrich & Novotny: Chapter 20). We have utilised two polymorphisms in the ribosomal RNA genes to determine if these repeated genes segregate as a single Mendelian locus and to monitor crossing-over between the genes during meiosis.

A map of restriction enzyme sites present in the cloned 9.2-kb fragment which codes for 26S, 18S, 5.8S and 5S ribosomal RNA is shown in Fig. 22.4A. This clone, pCc 1, was isolated from strain H9. We have determined that there are 60–90 copies of this repeat in each haploid genome (results not shown). Fig. 22.4A also shows the location of one of the polymorphisms studied, found in strain 12890 #3, in which an addition *Bam* H1 site is present in each repeat. When DNA was extracted from this strain, digested with *Bam* H1, and probed with the rDNA clone, two bands of hybridisation (at 5.6 and 3.9 kb) were seen (Fig. 22.4B, lane b), instead of the single band at 9.3 kb found in most strains, including strain C692 (Fig. 22.4B, lane a). When a dikaryon heterozygous for this polymorphism was constructed (C692 × 12890 #3) and the inheritance of rDNA digestion patterns examined in tetrads of basidiospores, it was found that all progeny contained a single parental pattern, never a mixture. The patterns from five such tetrads are shown in Fig. 22.4B, lanes c–v. It is concluded that virtually all rDNA copies were located in a single cluster (or more than one closely linked cluster) in the genome since they segregated as a single Mendelian locus. We have examined all members of 16 tetrads with distinguishable rDNAs from this and another cross, and we have found no examples of reciprocal exchange in these genes (Cassidy, Moore, Lu & Pukkila,

1984). If recombination were occurring at the expected frequency in these genes, the probability of our failing to detect it in this sample is less than 0.003. We conclude that crossing-over was inhibited in this tandem array.

Discussion

The *Coprinus* fruit body appears to have evolved to produce a maximum number of meiotic products at a specific time in development. We have developed a variety of methods to monitor chromosome behaviour in meiotic cells. Our recent success in visualising meiotic chromosomes released from the confines of the basidial cell wall without harsh treatments such as extensive fixation or acid hydrolysis should allow the application of more powerful analytical techniques in the future. These preparations are similar to those widely used for indirect immunofluorescence to identify antigenic determinants associated with particular nuclear and cytoplasmic components (Silver & Elgin, 1976; Moses *et al.*, 1984). The method could also be used in place of three-dimensional reconstruction to enable the recognition and mapping of chromosomal rearrangements

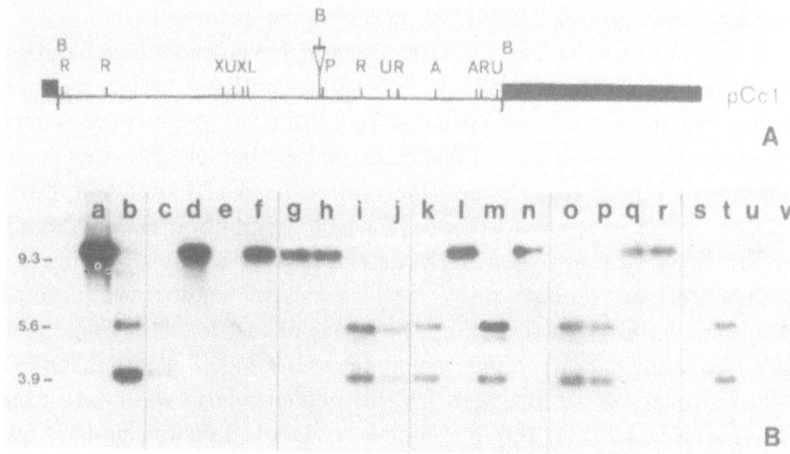

Fig. 22.4. Mendelian segregation of tandem rRNA genes in *Coprinus*. The restriction map is shown in A, and the Southern analysis in B. DNA was isolated from each strain, digested with *Bam* H1, and probed with the ribosomal DNA clone. Lane a: C692. Lane b: 12890 #3. Lanes c–f: haploid segregants from tetrad 1. Lanes g–j: haploid segregants from tetrad 2. Lanes k–n: haploid segregants from tetrad 3. Lanes o–r: haploid segregants from tetrad 4. Lanes s–v: haploid segregants from tetrad 5. Size markers (kb) are indicated.

such as large inversions and translocations, and would therefore be useful in correlating genetic and physical maps in *Coprinus* (Zickler *et al.*, 1984).

We have also demonstrated that differentiation in *Coprinus* is associated with extensive changes in concentration of particular polyadenylated RNA species. The primary cause for these changes is unknown. Differential polyadenylation and differential rates of turnover of particular RNAs cannot yet be distinguished from differential transcription of particular sequences. Meiosis in many organisms is associated with extensive turnover of RNAs, variously attributed to responses to the starvation necessary to induce meiosis (Esposito & Klapholz, 1981) or to 'cytoplasmic cleansing' appropriate to this developmental process (Porter, Parry & Dickinson, 1983). Such degradative changes in *Coprinus* appear to occur well after meiotic prophase, as judged from cytochemical analysis (McLaughlin, 1982); the large class of sequences that we have observed which remain at the same concentration in dikaryons and fruit bodies is consistent with this interpretation. Differential transcription has been extensively studied in another basidiomycete, *Schizophyllum commune* (Zantinge, Dons & Wessels, 1979; Zantinge, Hoge & Wessels, 1981; Hoge, Springer & Wessels, 1982; Wessels *et al.*: Chapter 21). These workers also found evidence of changes in polyadenylated RNAs associated with morphogenesis, although they reported fewer changes than we have detected. It will be of interest to determine to what extent changes in morphogenetic complexity in the two organisms are associated with changes in gene expression.

DNA sequence polymorphisms can be powerful aids to linkage analysis (Botstein, White, Skolnick & Davis, 1980; Rose, Baillie, Candido, Beckenbach & Nelson, 1982). We have used naturally occurring polymorphisms in the ribosomal RNA genes to show that recombination is suppressed in this genomic interval which comprises 1.5–2.2% of the genome. Similar findings have been reported in yeast for both tandem ribosomal RNA genes (Petes & Botstein, 1977) and tandem copies of copper-chelatin genes (Fogel, Welsh, Cathala & Karin, 1983). Suppression of crossing-over between genes in tandem arrays may be a widespread phenomenon since unequal exchanges, which would have deleterious consequences, can be avoided easily by such a mechanism and would be difficult to avoid otherwise. Thus, genes present as stable tandem arrays may lack sequences capable of serving as efficient substrates for the initiation or completion of meiotic recombination. The relative stability of tandem arrays in eukaryotes in contrast to their instability in prokaryotes may be a consequence of the precise regulation of recombination frequency and position in these large genomes (Thuriaux, 1977).

Acknowledgements. We are grateful for the productive collaborations with L. A. Casselton, D. Moore and B. C. Lu which were initiated during their visits to our laboratory, and to L. A. Casselton, D. Moore and J. North for providing strains used. We also thank M. Dresser and M. Moses for advice. This work was supported by the National Science Foundation and the National Institutes of Health.

References

Aviv, H. & Leder, P. (1972). Purification of biologically active globin messenger RNA by chromatography on oligothymidylic acid–cellulose. *Proceedings of the National Academy of Sciences, USA*, **69**, 1408–12.

Botstein, D., White, R., Skolnick, M. & Davis, R. (1980). Construction of a genetic linkage map in man using restriction fragment length polymorphisms. *American Journal of Human Genetics*, **32**, 314–31.

Buller, A. H. R. (1909). *Researches on Fungi*, vol. 1. London: Longmans, Green and Co.

Buller, A. H. R. (1924). *Researches on Fungi*, vol. 3. London: Longmans, Green and Co.

Cassidy, J. R., Moore, D., Lu, B. C. & Pukkila, P. J. (1984). Unusual organization and lack of recombination in the ribosomal RNA genes of *Coprinus cinereus*. *Current Genetics*, **8**, 607–13.

Chirgwin, J. M., Przybyla, A. E., MacDonald, R. J. & Rutter, W. J. (1979). Isolation of biologically active ribonucleic acid from sources enriched in ribonuclease. *Biochemistry*, **18**, 5294–9.

Esposito, R. E. & Klapholz, S. (1981). Meiosis and ascospore development. In *The Molecular Biology of the Yeast Saccharomyces. Life Cycle and Inheritance*, ed. J. N. Strathern, E. W. Jones & J. R. Broach, pp. 211–87. Cold Spring Harbor, USA: Cold Spring Harbor Laboratory.

Fogel, S., Welch, J. W., Cathala, G. & Karin, M. (1983). Gene amplification in yeast: CUP1 copy number regulates copper resistance. *Current Genetics*, **7**, 347–55.

Gillies, C. B. (1981). Electron microscopy of spread maize pachytene synaptonemal complexes. *Chromosoma*, **83**, 575–91.

Hoge, J. H. C., Springer, J. & Wessels, J. G. H. (1982). Changes in complex RNA during fruit-body initiation in the fungus *Schizophyllum commune*. *Experimental Mycology*, **6**, 233–43.

Holm, P. B., Rasmussen, S. W., Zickler, D., Lu, B. C. & Sage, J. (1981). Chromosome pairing, recombination nodules and chiasma formation in the basidiomycete *Coprinus cinereus*. *Carlsberg Research Communications*, **46**, 305–46.

Howell, W. M. & Black, D. A. (1980). Controlled silver-staining of nucleolus organizer regions with a protective colloidal developer: a 1-step method. *Experientia*, **36**, 1014–15.

Kacian, D. C., Watson, K. F., Burny, A. & Spiegelman, S. (1971). Purification of the DNA polymerase of avian myeloblastosis virus. *Biochemica et Biophysica Acta*, **246**, 365–83.

Kamada, T., Kurita, R. & Takemaru, T. (1978). Effects of light on basidiocarp maturation in *Coprinus macrorhizus*. *Plant and Cell Physiology*, **19**, 263–75.

Lasky, L. A., Lev, Z., Xin, J-H., Britten, R. J. & Davidson, E. H. (1980). Messenger RNA prevalence in sea urchin embryos measured with cloned cDNA. *Proceedings of the National Academy of Sciences, USA*, **77**, 5317–21.

Lu, B. C. (1974). Meiosis in *Coprinus*. V. The Role of light on basidiocarp initiation, mitosis, and hymenium differentiation in *Coprinus lagopus*. *Canadian Journal of Botany*, **52**, 299–305.

Lu, B. C. & Raju, N. B. (1970). Meiosis in *Coprinus*, II: Chromosome pairing and the lampbrush diplotene stage in meiotic prophase. *Chromosoma*, **29**, 305–16.

Madelin, M. F. (1956). The influence of light and temperature on fruiting of *Coprinus lagopus* in pure culture. *Annals of Botany*, *N.S.*, **20**, 467–80.

Matthews, T. R. & Niederpruem, D. J. (1972). Differentiation in *Coprinus lagopus*. I. Control of fruiting and cytology of initial events. *Archiv für Mikrobiologie*, **87**, 257–68.

Mazia, D., Schatten, G. & Sale, W. J. (1975). Adhesion of cells to surfaces coated with polylysine. *Journal of Cell Biology*, **66**, 198–200.

McLaughlin, D. J. (1982). Ultrastructure and cytochemistry of basidial and basidiospore development. In *Basidium and Basidiocarp: Evolution, Cytology, Function, and Development*, ed. E. Wells & E. K. Wells, pp. 37–74. New York: Springer-Verlag.

Morimoto, N. & Oda, Y. (1973). Effects of light of fruiting body formation in a basidiomycete, *Coprinus macrorhizus*. *Plant and Cell Physiology*, **14**, 217–25.

Morimoto, N. & Oda, Y. (1974). Photo-induced karyogamy in a basidiomycete, *Coprinus macrorhizus*. *Plant and Cell Physiology*, **15**, 183–6.

Moses, M. J. (1977). Synaptonemal complex karyotyping in spermatocytes of Chinese hamster (*Cricetulus griseus*). I. Morphology of the autosomal complement in spread preparations. *Chromosoma*, **60**, 99–125.

Moses, M. J., Dresser, M. E. & Poorman, P. A. (1984). Composition and role of the synaptonemal complex. In *Controlling Events in Meiosis* (38th Symposium of the Society for Experimental Biology, Sept. 1983), ed. C. W. Evans & H. G. Dickinson. Cambridge University Press. (in press)

Petes, T. D. & Botstein, D. (1977). Simple Mendelian inheritance of the reiterated ribosomal DNA of yeast. *Proceedings of the National Academy of Sciences, USA*, **74**, 5091–5.

Pukkila, P. J. & Lu, B. C. (1985). Silver staining of meiotic chromosomes in the fungus *Coprinus cinereus*. *Chromosoma*, **91**, 108–12.

Porter, E. K., Parry, D. & Dickinson, H. G. (1983). Changes in poly(A)$^+$ RNA during male meiosis in *Lilium*. *Journal of Cell Science*, **62**, 177–86.

Pukkila, P. J., Yashar, B. M. & Binninger, D. M. (1984). Analysis of meiotic development in *Coprinus cinereus*. In *Controlling Events in Meiosis* (38th Symposium of the Society for Experimental Biology, Sept. 1983), ed. C. W. Evans & H. G. Dickinson. Cambridge University Press. (in press)

Rao, P. S. & Niederpruem, D. J. (1969). Carbohydrate metabolism during morphogenesis of *Coprinus lagopus* (*sensu* Buller). *Journal of Bacteriology*, **100**, 1222–8.

Rose, A. M., Baillie, D. L., Candido, E. P. M., Beckenbach, K. A. & Nelson, D. (1982). The linkage mapping of cloned restriction fragment length differences in *Caenorabditis elegans*. *Molecular and General Genetics*, **188**, 286–91.

Scherer, S. & Davis, R. W. (1979). Replacement of chromosome segments with altered DNA sequences constructed *in vitro*. *Proceedings of the National Academy of Sciences, USA*, **76**, 4951–5.

Silver, L. M. & Elgin, S. C. R. (1976). A method for determination of the *in situ* distribution of chromosomal proteins. *Proceedings of the National Academy of Sciences*, **73**, 423–7.

Southern, E. M. (1975). Detection of specific sequences among DNA fragments separated by gel electrophoresis. *Journal of Molecular Biology*, **98**, 503–17.

St John, T. P. & Davis, R. W. (1979). Isolation of galactose-inducible DNA sequences from *Saccharamyces cerevisiae* by differential plaque filter hybridization. *Cell*, **16**, 443–52.

Thuriaux, P. (1977). Is recombination confined to structural genes on the eukaryotic genome? *Nature*, **268**, 460–2.

Tsusué, Y. M. (1969). Experimental control of fruit-body formation in *Coprinus macrorhizus*. *Development, Growth and Differentiation*, **11**, 164–78.

Wu, M. J., Cassidy, J. R. & Pukkila, P. J. (1983). Polymorphisms in DNA of *Coprinus cinereus*. *Current Genetics*, **7**, 385–92.

Yashar, B. M. & Pukkila, P. J. (1985). Changes in polyadenylated RNA sequences associated with fruiting body morphogenesis in *Coprinus cinereus*. *Transactions of the British Mycological Society*, **84**, 215–26.

Zantinge, B., Dons, H. & Wessels, J. G. H. (1979). Comparison of poly (A)-containing RNAs in different cell types of the lower eukaryote, *Schizophyllum commune*. *European Journal of Biochemistry*, **101**, 251–60.

Zantinge, B., Hoge, J. H. & Wessels, J. G. H. (1981). Frequency and diversity of RNA sequences in different cell types of the fungus *Schizophyllum commune*. *European Journal of Biochemistry*, **113**, 381–9.

Zickler, D., Leblon, G., Haedens, V., Collard, A. & Thuriaux, P. (1984). Linkage group–chromosome correlations in *Sordaria macrospora*: chromosome identification by three dimensional reconstruction of their synaptonemal complex. *Current Genetics*, **8**, 57–68.

23
Strategies for mushroom breeding

CARLENE A. RAPER

*Department of Medical Microbiology, University of Vermont,
School of Medicine, Burlington, Vermont 05405, USA*

Introduction

Perhaps 'breed is stronger than pasture', as George Eliot once claimed, but even if overstated, the relevance of ancestry to what one is cannot be denied. This applies to a fungus as well as any other organism, yet systematic programmes for breeding superior strains of edible mushrooms are not yet evident.

The benefits of such breeding to production of commercially valuable plants and animals have been amply demonstrated since the turn of the century. Acreage yield of maize, in the USA for example, increased from 30 bushels per acre in 1933 to 84 bushels in 1965, largely as a result of breeding hybrids to fit the changing conditions of improved crop management (Sprague, 1972). There are numerous other examples, but none is to be found among cultivated fungi.

Over the two centuries or so in which the commercial white mushroom, *Agaricus bisporus*, has been cultivated, advances in production have come about largely through improved conditions for cultivation and the occasional selection of chance variants. Although such methods have proved remarkably successful, they have required much time and effort and lack the finesse of a strategy in which desirable genetic traits are combined through controlled crossing and progeny selection. The application of breeding strategies to *A. bisporus* may yet hold promise for strain improvement in such traits as flavour, yield, vigour, stability, resistance to disease and efficient use of inexpensive substrate. Similarly, breeding may benefit other species such as *A. bitorquis*, *Pleurotus ostreatus*, *Lentinus edodes*, *Flammulina velutipes* and *Pholiota nameko*.

The application of known genetic concepts to breeding programmes is timely because the relevant background information and techniques of

513

manipulation are known for each species. This relevant information will be summarised and strategies will be suggested for the design of specific breeding programmes applicable to three cultivated basidiomycetes having different breeding systems: a heterothallic species with bipolar (unifactorial) control; a heterothallic species with tetrapolar (bifactorial) control; and a secondary homothallic species with bipolar control.

Requisites for a breeding programme

The objective is to improve the quality of extant strains in the most efficient way through controlled crossing, and progeny selection. The desired result should be clearly defined in terms of desired traits expressed within a chosen set of conditions employed consistently throughout the breeding programme.

The primary requisite for achieving this objective is a knowledge of the life cycle, including the sexual phase that can be exploited in controlled crosses and methods of manipulating the organism in the laboratory under sterile, controlled conditions. To assess feasibility, some knowledge of the pattern of inheritance of known genes such as mating type, nutritional requirements or enzyme variants should also be available. It is important that heterokaryons be distinguished from homokaryons, either morphologically for example by the presence or absence of clamp connections, genetically, through some evidence of complementation between genetic markers or enzymatically by detecting the presence of a pair of enzyme polymorphisms. To detect nutritional gene (auxotrophic) complementation, it must be possible to grow cultures in defined minimal medium, whereas detection of a gene for a protein polymorphism requires comparison of electrophoretically separated proteins extracted from the putative heterokaryon and the two component homokaryons as well. (DNA polymorphisms may also be useful in this respect; see Pukkila *et al.*: Chapter 22; Reijnders & Moore: Chapter 27.) In addition to identification of heterokaryons, a means of separating the nuclear types to regenerate the two component homokaryons is also important. Several methods could be used: (1) isolation of uninucleate vegetative spores of the heterokaryon, as is possible in *F. velutipes* (Takemaru, 1954) and in *P. nameko* (Arita, 1978); (2) isolation of hyphal tip cells, as in *P. nameko* (Arita, 1978); and (3) enzymatic digestion of heterokaryotic cell walls to generate protoplasts, and subsequent selection, isolation and regeneration of the uninucleate protoplast fraction to recover the component homokaryotic mycelia. The latter has been used with *Schizophyllum commune* (Wessels *et al.*, 1976;

Table 23.1. *Features of some cultivated basidomycetes*

Species	Breeding system	Homokaryon Nuclei per cell	Heterokaryon Nuclei per cell	Clamps	Distinct morphology
Agaricus bitorquis	Heterothallic, bipolar	Many (variable)	2 (usually)	No	Yes
Pholiota nameko	Heterothallic, bipolar	1	2	Yes	Yes
Pleurotus ostreatus	Heterothallic, tetrapolar	1	2	Yes	Yes
Flammulina velutipes	Heterothallic, tetrapolar	1	2	Yes	Yes
Lentinus edodes	Heterothallic, tetrapolar	1	2	Yes	Yes
Agaricus bisporus	Secondary homothallic, bipolar	Many (variable)	Many (variable)	No	No

Raper, 1983) and *A. bisporus* (Anderson, personal communication); theoretically it should work for any fungus.

All or most of these requisites have been established for each of the commercial mushrooms to be discussed here and for some others.

Background information

The relevant background information on the nature of each fungus to be considered is outlined in Table 23.1. In addition to this information it must be kept in mind that all fungi listed are typical of basidiomycetes in that they are haploid throughout the greatest portion of their life cycles, from sexual spore formation through spore germination, mycelial formation and heterokaryosis, to the point of nuclear fusion in the basidial cells of the fruit body which then give rise to the haploid basidiospores. Homokaryons have a single type of haploid nucleus; fertile heterokaryons have two types of haploid nuclei (at least with respect to mating type alleles). Both homokaryons and heterokaryons can be propagated vegetatively indefinitely.

Genetic markers have been identified and analysed through meiotic segregation in each of the species listed. All have been analysed for segregation of alleles at the mating type locus or loci and these show typical Mendelian inheritance. The breeding systems are typical of basidiomycetes in that there are multiple alleles at each mating type locus and compatibility requires different alleles for any given pair of mates. Compatible pairings in bipolar heterothallic species will be $A_1 \times A_2$, $A_3 \times A_4$, $A_1 \times A_3$, etc. Whereas incompatibile pairings will be $A_1 \times A_1$ or $A_2 \times A_2$, etc. Compatible pairings in tetrapolar heterothallic species require different alleles in the genes of the two unlinked factors A and B, e.g. $A_1 B_1 \times A_2 B_2$ or $A_{35} B_{42} \times A_{10} B_{12}$, etc. Pairings in which the alleles at either or both of these factors are identical are incompatible. In all tetrapolar heterothallic species carefully analysed so far, the A and B factors segregate independently and either one or both consist(s) of two linked, multiply allelic, sub-loci, $A\alpha$ and $A\beta$, $B\alpha$ and $B\beta$ (for details see Raper, 1966a,b; Raper, 1978; Casselton & Economou: Chapter 8; Ullrich & Novotny: Chapter 20).

Other genes relevant to mushroom breeding have been identified. Auxotrophic markers have been generated and analysed through meiotic segregation in *A. bisporus* and *A. bitorquis* (Raper *et al.*, 1972; Raper, 1976). Several genes for polymorphic enzymes were identified and traced through heterokaryosis and meiotic segregation in *A. bisporus* (Royse & May, 1982a; Spear *et al.*, 1983). Alleles for fruit body morphology, so-called wild-type (dominant) *versus* normal white (recessive), were shown to segregate as alleles of a single gene in *A. bisporus* (Miller & Kananen, 1972). A mutant for fruit body morphology (sporeless) was identified in *Pleurotus ostreatus*, although it is not clear that only a single gene was involved (Eger *et al.*, 1976). Genetic variation in growth rates was observed among monosporous cultures isolated from a single fruit body in *F. velutipes* (Simchen, 1965). Mutations for hyphal aberration and for fruit body abnormalities, e.g. sporelessness and white pilei, have been analysed in *L. edodes* (Komatsu & Kimura, 1964, 1968; Murakami & Takemaru, 1975). More comprehensive analyses of a variety of single gene and multigenic traits for nutritional competence, hyphal morphology, fruit body morphology, fruiting competence and growth rate have been carried out in *Coprinus cinereus* and *Schizophyllum commune*. Most auxotrophic mutations in both species are defined as single gene mutants and expressed as deficient alleles which segregate one to one with the wild-type allele. The mutants which alter hyphal morphology are determined either by one or two genes. Mutations affecting fruit body morphology may be in a single gene or two genes (see Raper, 1978, for review). Fruiting competence, on the other hand, as analysed extensively in 80 strains of *S. commune*, is a polygenic trait with 'good fruiting' alleles generally dominant over 'poor fruiting' alleles (Raper & Krongelb, 1958). Similar conclusions have been drawn from less-extensive data in *P. ostreatus* (Eger *et al.*, 1976). A single deleterious mutant allele in any one of the genes for fruiting competence, however, may block fruiting (Raper & Raper, 1966; Perkins & Raper, 1969; Esser, 1978).

Objectives and genetic components for breeding

There are many possible objectives and strategies for breeding mushrooms. For simplicity only one objective will be considered that might be accomplished using a strategy which can be adapted for the three different breeding systems found in cultivated species.

The chosen objective is to combine the trait of desirable mushroom morphology from a low-yielding strain with that of high yield from a strain with undesirable mushroom morphology. The assumptions are made that mushroom morphology is determined by alternate alleles of a single gene, that the allele for undesirable morphology (e.g. long thick stipes or stalk, small cap or pileus) is dominant to the allele for desirable morphology (shorter stipe, large, denser cap), and that mushroom yield is determined by alternate alleles of several genes scattered throughout the genome. These assumptions will have been verified in a preliminary study by genetic analyses which showed for example, that a sample of monosporous homokaryotic isolates from the small-cap, high-yielding strain, when mated in all possible pairs, varied continuously in yield from low to perhaps not as high relative to the parental heterokaryon, and that small-cap morphology was expressed in all of these pairings. Similarly a sample of monosporous homokaryotic isolates from the large-cap, low yielding strain, when mated in all possible pairs, expressed continuous variation in relatively low yield and large-cap morphology in all pairings. Dominance of the allele for small cap would have been verified in a subsequent cross between a monosporous homokaryotic isolate carrying the allele for large cap from the large-cap strain and a monosporous homokaryotic isolate carrying the allele for small cap from the small-cap strain. Such a cross would produce small-cap mushrooms and analysis of the F_1 generation would have revealed a 1:1 segregation for small cap *versus* large cap.

The original small-cap, high-yielding heterokaryon, therefore, can be designated as containing the genome S, for small cap, and $y1^+$, $y2^+$, $y3^-$, $y4^+$..., for yield, in one nucleus and S, $y1^-$, $y2^+$, $y3^+$, $y4^-$...in the other nucleus. The monosporous homokaryotic isolates would all carry S and various combinations of the parental y alleles. The original large-cap, low-yielding strain would contain s, for large cap, in the genomes of both nuclei, but different combinations of alleles of the various yield genes, e.g.

s, $y1^-$, $y2^-$, $y3^+$, $y4^-$...and s, $y1^-$, $y2^+$, $y3^-$, $y4^-$....The $+$ and $-$ superscripts do not necessarily imply superiority *versus* inferiority; for some genes homozygosity for the $+$ allele may be advantageous, for others, heterozygosity for the $+$ and $-$ alleles may be advantageous. In the latter case, the phenomenon known as hybrid vigour or heterosis would prevail (for discussion see Sprague, 1972; Mather & Jinks, 1982). The genetic component for yield would be indicated by the fact that replicates of each strain perform consistently as high *versus* low yielders under identical environmental conditions. Good evidence for polygenic inheritance would be the idealised distribution curves for yield as illustrated in Fig. 23.1 (see Mather & Jinks, 1982, for detailed explanation of polygenic inheritance).

The task is to incorporate the recessive allele for large-cap morphology (s) into each of the two genomes which together determine high yield ($y1^+$, $y2^+$, $y3^-$, $y4^+$...and $y1^-$, $y2^+$, $y3^+$, $y4^-$...). Both the preliminary analysis to determine the genetic basis for the traits to be considered and the application of a strategy for breeding would be far more difficult and complicated in the secondary homothallic *A. bisporus* than in any of the heterothallic species. This is discussed below.

The starting material for homothallic species as well as heterothallic species would have been identified in collections of stocks from nature and/or laboratories. The stocks would have been screened for yield under standard, controlled conditions in a pilot laboratory study. For

IDEALIZED EVIDENCE FOR POLYGENIC INHERITANCE OF YIELD

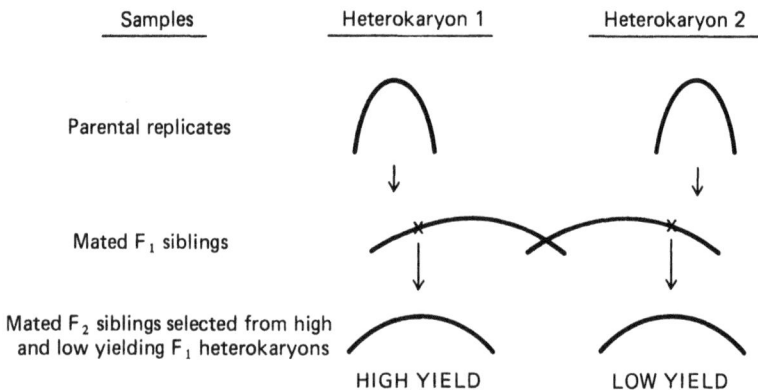

Fig. 23.1. Curves reflect distribution for yield under identical culturing conditions. Horizontal axis represents yield from high (left) to low (right). Vertical axis represents number in sample.

heterothallic fungi, interstock matings between progeny of fruiting heterokaryons from diverse locations might provide the best chance of identifying high yielding combinations of genomes. This proved to be the case in studies with *Schizophyllum commune* (Raper & Krongelb, 1958) and *Agaricus bitorquis* (Raper & Miller, unpublished). Because of the extensive series of multiple alleles at each mating type locus, mating type specificities would be expected to differ between stocks collected from different locations. The two homokaryotic components of each stock are likely, therefore, to be compatible with one another. For the secondary homothallic species, *A. bisporus*, fruiting heterokaryons from a variety of laboratories would be the primary source of stocks to be tested because this species is rarely found in nature.

Strategy for breeding heterothallic fungi

A logical strategy for achieving the proposed objective in heterothallic fungi would include the following steps. (1) Obtain the homokaryotic components of each of the two starting heterokaryons. (2) Select one homokaryon carrying the gene for large cap from the large-cap, low-yielding strain. (3) Backcross this homokaryon in two series to each of the two homokaryons which, when mated, constitute the small-cap, high-yielding heterokaryon. At each generation the progeny would be test mated with parental homokaryons of known mating type to identify mating type specificities compatible with the parental homokaryon to which the progeny were being backcrossed. They would also be tested for presence of the *s* allele by mating with a compatible homokaryon known to carry the *s* allele and observing fruit body morphology; an *s* plus *s* combination would produce large-cap mushrooms and an *S* plus *s* combination would produce small-cap mushrooms. (4) The last step is to mate the final products of each backcross series with each other. After several generations of backcrossing and progeny selection for mating type specificity and the *s* allele, these final products should be compatible and should carry the genotypes *s*, $y1^+$, $y2^+$, $y3^-$, $y4^+$...and *s*, $y1^-$, $y2^+$, $y3^+$, $y4^-$....This final step should establish a heterokaryon with as high a yield as the starting high-yielding stock, but which produces large-cap mushrooms of desirable shape and density (Fig. 23.2).

The size of the progeny sample tested at each generation should be large enough to assure the probability that a homokaryon carrying the *s* allele for desirable fruit body morphology and the appropriate allele(s) for mating type specificity is included in the sample. Since *s* and *S* segregate 1:1, the probability of progeny carrying *s* is 1/2. With respect to mating type genes, the probability of obtaining a progeny which is compatible with the backcross parent depends on the number of mating type loci involved. In bipolar species, with alternate alleles at a single locus, this probability is $1/2$; therefore the appropriate progeny combining *s* with a mating type allele different from the backcross parent is $1/2 \times 1/2 = 1/4$. A sample size of 8–12 would include the desired genotype with reasonable assurance. The larger the sample, the greater the chance of selecting progeny not only with the appropriate genotypes but with superior characteristics, such as vegetative vigour, short fruiting time. In a series of ten backcross generations performed in *S. commune*, for example, a sample size of ten was routinely chosen to assure the inclusion of a genotype of the desired type which by chance should have occurred once in a sample of four. In some generations, particularly the later ones, this sample size was inadequate because some progeny with the appropriate genotype for compatibility failed to fruit when mated with the backcross parent. This was attributed to the likelihood that heterozygosity for some genes is necessary for

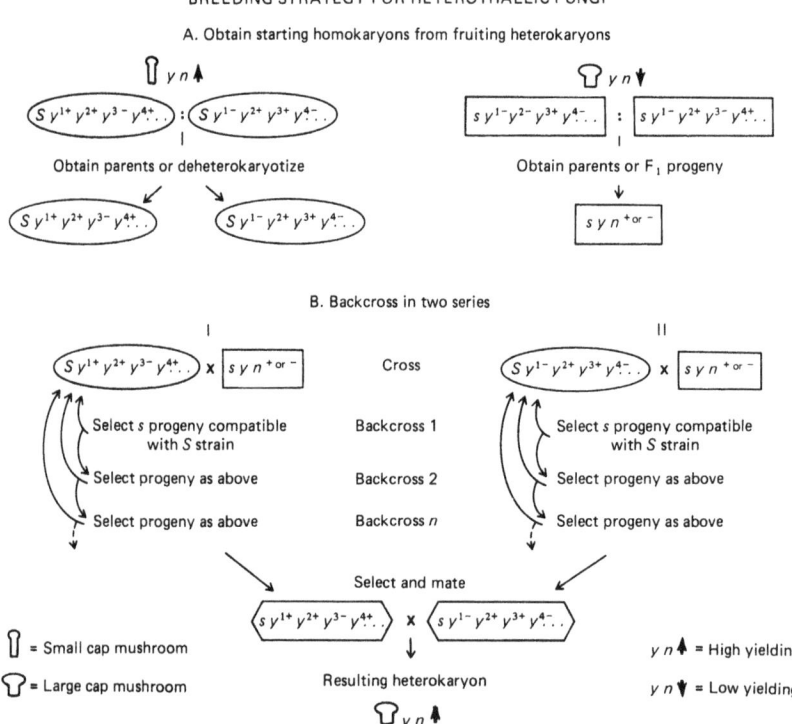

Fig. 23.2. Breeding strategy for heterothallic fungi.

fruiting and as two homokaryons become more isogenic with successive backcrossing they become homozygous for these particular fruiting genes. It appears that a minimum degree of heterozygosity must be preserved to permit fruiting (Raper, unpublished).

A somewhat larger sample size would be required to assure the inclusion of the desired genotype when breeding tetrapolar heterothallic fungi. The probability of obtaining a progeny with the *s* allele would still be 1/2 but the probability for the appropriate mating type would be reduced to slightly more than 1/4 because there are two mating type factors, *A* and *B*, and one or both is comprised of two linked genes α and β. *A* and *B* segregate independently and the α and β genes recombine in frequencies ranging from less than 0.1% to about 25% (Raper, 1978). A minimum sample size of progeny selected for testing at each backcross generation in the breeding of tetrapolar species would therefore be almost twice that for bipolar species, about 20.

The number of backcrosses required in each series to assure the likelihood of including all the desirable alleles for yield, along with the selected *s* allele for fruit body morphology, is calculated by a formula which takes into consideration the number of chromosomes, total map length and the number of chiasmata per nucleus. These data are not available for the edible basidiomycetes being considered here but are available in some higher organisms, such as the plant *Papaver dubium*. An application of the appropriate data in this species to the formula devised by Fisher and Bennet (see Gale, 1980) indicates that if the parents of the starting cross are 100% heterozygous at all loci, by the eighth generation backcross, 32% of the progeny will be homozygous for all loci relative to the backcross parent and to each other and there is a probability of only 39% that heterozygosity will remain at any given locus. This degree of homozygosity would be achieved earlier in fungi with their relatively smaller genomes (Leslie, 1981). The time of achieving homozygosity for the several theoretical genes for yield would depend on linkage. It would occur earlier for those loci which are not linked than for those which are. In a breeding programme one could begin testing for the desired end result, a high-yielding, large-cap heterokaryon, by mating selected progeny from the fourth or fifth generation onward.

Breeding a secondary homothallic fungus

Breeding in the secondary homothallic species *A. bisporus* can be accomplished but with greater difficulty than in heterothallic species and possible benefits should be carefully weighed against the cost in effort.

Studies on the repetitive incidence of specific auxotrophic markers (Miller, unpublished) and enzymatic markers (Royse & May, 1982a) indicate relatively little variation in the gene pool of extant stocks of *A. bisporus*. It has been suggested that most of the stocks available in laboratories today were derived from one or a few common ancestors and are consequently homozygous at most genetic loci. With little available genetic variation, the prospects for improvement by recombining alleles through a breeding programme are limited.

Nevertheless, a strategy for achieving the proposed objective through breeding *A. bisporus* might consist of the following steps. (1) Obtain the two starting heterokaryons, one with high yield, undesirable fruit body morphology and one with low yield, desirable fruit body morphology. (2) Separate the two homokaryotic components of each starting heterokaryon possibly by protoplast formation and regeneration as described previously. The method of distinguishing homokaryons from heterokaryons among the regenerated protoplasts might be an electrophoretic assay for known polymorphic enzymes (the two homokaryotic components of the heterokaryon carrying alternate alleles for a polymorphic enzyme, when the two alleles, expressed simultaneously in heterokaryotic cells, produce a dimer as well as the two monomeric enzymes (Royse & May, 1982a, b)) or by use of complementing auxotrophic markers as also described previously (Raper, Raper & Miller, 1972). (3) Mate, in two series, a single homokaryon carrying the *s* allele for large cap, and derived from the large-cap low-yielding strain, with each of the two homokaryons derived from the small-cap, high-yielding strain. In this case the method differs from that employed for heterothallic species. Instead of continually backcrossing isolated homokaryotic progeny to the high-yielding homokaryons (a task of considerable magnitude in a secondary homothallic species), it might be possible to achieve a relatively high yielding large-cap strain by continuous selfing and progeny selection throughout several generations in each of two series. The majority of basidiospores in most strains of *A. bisporus* contain two postmeiotic nuclei of compatible mating type which develop into fertile heterokaryons, thus selfing occurs automatically (Evans, 1959; Raper et al., 1972). The isolation of homokaryotic progeny at each generation, although possible through selection of spores from rare four-spored basidia (Miller, 1971; Elliott, 1972) or via protoplast manipulations, would be very labour-intensive. In order to assure selection of fertile progeny homozygous for the *s* allele, the sample size of isolated and germinated basidiospores would have to exceed six (2/3 chance for heterozygosity at mating type locus, according to calculations by Langton

& Elliott (1980), and 1/4 chance for homozygosity of the *s* allele). Beyond this, alleles for good yield, relative to those in the poor yielding parental strain, would have to be selected in fruiting tests. Since we are dealing with alleles of several genes of unknown number for yield, the probability of obtaining a given progeny with high yield cannot be calculated. Sample size, therefore, should be as large as possible and each monosporous progeny should be tested for fruiting competence, morphology and yield, always selecting the highest yielding progeny with large-cap fruit body morphology. (4) At the end of the two series of successive selfings, yield might be further enhanced by separating the homokaryotic components of each heterokaryotic progeny which had been finally selected, and crossing these homokaryons in all possible combinations between series to establish new heterokaryons, that is: homokaryon 1 from series I × homokaryon 1 from series II; 1–I × 2–II, 2–I × 1–II, and 2–I × 2–II. The success of such crosses would depend on compatibility for mating type.

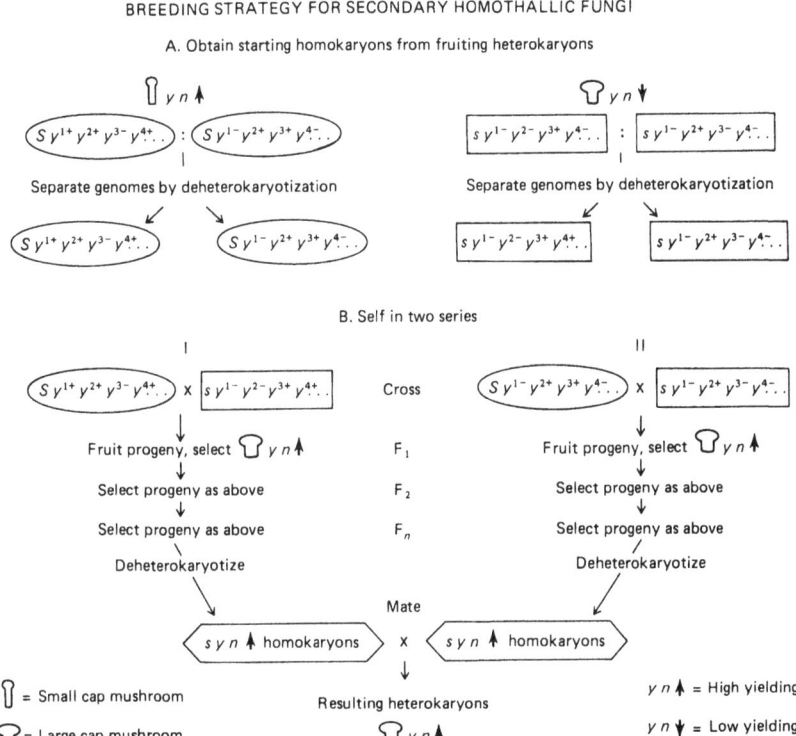

Fig. 23.3. Breeding strategy for secondary homothallic fungi.

The number of mating type alleles has not yet been determined for *A. bisporus*. Assuming the mating type alleles in both starting heterokaryons are different from one another, the establishment of all four heterokaryons would be possible at the end of this breeding programme. All should produce large-cap mushrooms and each should be tested for yield. With *luck*, a relatively high-yielding heterokaryon with large-cap mushrooms could be selected (Fig. 23.3).

The success of this strategy for breeding depends upon the fortuitous selection of alleles of several genes for high yield at each generation. Some could be lost along the way, also the phenomenon of heterosis might interfere with the final objective of combining the desired set of alleles for high yield in each homokaryotic end product. It is a programme with far less certainty than that suggested for the breeding of heterothallic fungi. In order to apply the more reliable technique of backcrossing to *A. bisporus* it would be necessary to start with strains which, atypically, produced a large proportion of tetrasporic basidia and consequently a large proportion of homokaryotic spores. This might be possible with exceptional strains (see Spear *et al.*, 1983) but not with the great majority of stocks in existence. Alternatively, the method of deheterokaryotisation through protoplast manipulations may become routinely simple and this could be used to separate homokaryotic components of large samples of heterokaryotic progeny at each generation. The homokaryons could then be used for backcrossing to the high-yielding parent homokaryon. Even so, the certain identification of homokaryons as contrasted to heterokaryons, and the detection of successful mating in *A. bisporus*, require cumbersome procedures in which the presence of genetic markers are monitored and fruiting trials are conducted. Breeding, while possible, would be a relatively costly procedure in this species.

Conclusions and discussion

A relatively simple example of an objective for breeding edible mushrooms to combine a single-gene trait with a polygenic trait has been presented. The strategies suggested for accomplishing this were tailored to the nature of the fungus so that species representing the three types of breeding systems in basidiomycetes have been considered. Although a variety of other objectives and strategies are conceivable, it is hoped that the examples discussed will provide some useful guidelines for procedures applicable to any breeding programme for each type of fungus.

The degree of difficulty with which breeding can be accomplished in edible basidiomycetes depends upon the nature of the fungus. Of the six species listed in Table 23.1, *Flammulina velutipes, Pleurotus ostreatus, Agaricus bitorquis* and *Pholiota nameko* can be bred most easily. All are heterothallic and hence can be backcrossed. All can be fruited under controlled conditions and will complete the life cycle within 3–8 weeks. *F. velutipes* and *P. nameko* have the advantage of being deheterokaryotised through isolation of vegetative spores; *F. velutipes* and *Pleurotus ostreatus* have relatively short life cycles; and *A. bitorquis* and *Pholiota nameko* have bipolar control of mating, thus requiring smaller samples of progeny throughout successive crosses in order to assure inclusion of the desired mating type specificities. *Lentinus edodes*, although heterothallic, has the disadvantage of a very long life cycle. A period of 2 years is normally required for fruiting of this species. Conditions for shortening this period for fruiting, however, have recently been reported (see Chapter 17). For reasons discussed in the previous section, *Agaricus bisporus* is the most difficult species to breed, though it is the most valuable cultivated crop (Flegg: Chapter 24; Wu: Chapter 25).

The haploid nature of all these fungi constitutes some advantage as compared to higher organisms. In contrast to the diploid condition in higher organisms in which genotype cannot often be ascertained from phenotype, progeny of heterothallic species express a phenotype strictly relevant to genotype. Although the phenotype of a fungal heterokaryon reflects expression of two genomes, each genome is contained separately in different nuclei and thus can be separated intact from one another. As is evident from the proposed strategies for breeding these fungi, deheterokaryotisation to separate parental genomes is a feasible and useful process which is not possible in diploid organisms.

Methods beyond those applicable to a standard breeding programme in which mechanisms natural to the species are employed have not been discussed. Processes which would not otherwise occur in nature, such as protoplast fusion, to achieve hybridisation between species, or transformation with cloned DNA, to incorporate foreign genes of known function, are currently possible and have been demonstrated in other fungi (Case *et al.*, 1979; Minuth & Esser, 1983). A consideration of these methods in a breeding programme, however, is somewhat premature in view of the fact that standard strategies for crossing and progeny selection have not yet been fully exploited in any of the edible basidiomycetes.

Finally, some cautionary notes: although prospects for strain improvement through inbreeding or backcrossing are reasonable, the consequent production of strains isogenic for most loci may be deleterious in heterokaryons derived from the mating of such co-isogenic homokaryons.

The possible advantage of heterosis might be reduced and homozygosity for deleterious alleles in polygenic systems could affect adversely such qualities as vegetative vigour, fruiting competence, or disease resistance. Such phenomena could limit the extent to which each series of backcrossings or selfings can be carried before establishment of fertile heterokaryons is no longer possible. As in the breeding of maize (Sprague, 1972), one might expect reduced vigour with continued backcrossing or selfing; but the mating of the final product selected at the end of two different series of crosses should re-establish hybrid vigour. Furthermore, assuming superior strains are developed through breeding, there could be a problem in the widespread use of one or a few such strains for commercial production. Again the example of maize is appropriate (Mangelsdorf, 1974). A single hybrid type bred to contain the cytoplasmic genetic factor for male sterility happened to carry genetic factors for increased susceptibility to the pathogenic fungus, *Helminthosporium maydis*, the causative agent of southern corn leaf blight. The hybrid was so successful that 70–90% of all hybrid corn grown in the USA by 1969 was of this type. An epidemic of the leaf blight occurred in 1970, however, and caused a widespread crop loss of 13% that year. Such a loss through disease could occur in selected strains of mushrooms as well. The use of a variety of strains, bred separately for yield, would reduce the risk of such large-scale losses. Finally, the experience of breeding in a variety of organisms has demonstrated the wisdom of preserving the starting strains used in all breeding programmes for the purpose of maintaining a pool of genetically diverse individuals. Such a practice would allow the possibility of carrying out additional breeding programmes to either improve or restore strains in current use.

References

Arita, I. (1978). *Pholiota nameko*. In *The Biology and Cultivation of Edible Mushrooms*, ed. S. T. Chang & W. A. Hayes, pp. 475–96. New York: Academic Press.

Case, M. E., Schweizer, M., Kushner, S. & Giles, N. H. (1979). Efficient transformation of *Neurospora crassa* by utilizing hybrid plasmid DNA. *Proceedings of the National Academy of Sciences of the USA*, **76**, 5359–63.

Eger, G., Eden, G. & Wissig, E. (1976). *Pleurotus ostreatus* – breeding potential of a new cultivated mushroom. *Theoretical and Applied Genetics*, **47**, 155–63.

Elliott, T. J. (1972). Sex and the single spore. *Mushroom Science*, **8**, 11–18.

Esser, K. (1978). Genetic control of fruiting body formation in higher basidiomycetes. *Mushroom Science*, **10**, 1–12.

Evans, H. J. (1959). Nuclear behavior in the cultivated mushroom. *Chromosoma*, **10**, 411–19.

Gale, J. S. (1980). *Population Genetics*. London: Wiley.

Komatsu, M. & Kimura, K. (1964). Studies of abnormal fruit bodies of the hymenomycetous fungi. III. Fruit bodies with brownish gills of *Lentinus edodes* (Berk.) Sing. *Reports of the Tottori Mycological Institute (Japan)*, **3**, 6–17.

Komatsu, M. & Kimura, K. (1968). Studies on abnormal fruit bodies of the hymenomycetous fungi. V. Fruit bodies with white pilei of *Lentinus edodes* (Berk.) Sing. *Reports of the Tottori Mycological Institute (Japan)*, **6**, 9–17.

Langton, F. A. & Elliott, T. J. (1980). Genetics of secondarily homothallic basidiomycetes. *Heredity*, **45**, 99–106.

Leslie, J. F. (1981). Inbreeding for isogenicity by backcrossing to a fixed parent in haploid and diploid eukaryotes. *Genetical Research*, **37**, 239–52.

Mangelsdorf, P. C. (1974). *Corn its Origin, Evolution and Improvement*. Cambridge, Mass., USA: The Belknap Press of Harvard University Press.

Mather, K. & Jinks, J. L. (1982). *Biometrical Genetics, The Study of Continuous Variation*, 3rd edn. London: Chapman & Hall.

May, B. J. & Royse, D. J. (1982). Confirmation of crosses between lines of *Agaricus brunnescens* by isozyme analysis. *Experimental Mycology*, **6**, 283–92.

Miller, R. E. (1971). Evidence of sexuality in the cultivated mushroom, *Agaricus bisporus*. *Mycologia*, **63**, 630–4.

Miller, R. E. & Kananen, D. L. (1972). Bipolar sexuality in the mushroom. *Mushroom Science*, **8**, 713–18.

Minuth, W. & Esser, K. (1983). Intraspecific, interspecific and intergenic recombination in β-lactam producing fungi via protoplast fusion. *European Journal of Applied Microbiology and Biotechnology*, **18**, 38–46.

Murakami, S. & Takemaru, T. (1975). 'Puff' mutation induced by uv irradiation in *Lentinus edodes* (Berk.) Sing. *Reports of the Tottori Mycological Institute*, **12**, 47–51.

Perkins, J. H. & Raper, J. R. (1969). Morphogenesis in *Schizophyllum commune*. III. A mutation that blocks initiation of fruiting. *Molecular and General Genetics*, **106**, 151–4.

Raper, C. A. (1976). Sexuality and life cycle of the edible, wild *Agaricus bitorquis*. *Journal of General Microbiology*, **95**, 54–66.

Raper, C. A. (1978). Sexuality and breeding. In *The Biology and Cultivation of Edible Mushrooms*, ed. S. T. Chang & W. A. Hayes, pp. 83–117. New York: Academic Press.

Raper, C. A. (1983). Controls for development and differentiation of the dikaryon in basidiomycetes. In *Secondary Metabolism and Differentiation in Fungi*, ed. J. W. Bennett & A. Ciegler, pp. 195–238. New York and Basel: Marcel Dekker.

Raper, C. A. & Raper, J. R. (1966). Mutations modifying sexual morphogenesis in *Schizophyllum*. *Genetics*, **54**, 1151–68.

Raper, C. A., Raper, J. R. & Miller, R. E. (1972). Genetic analysis of the life cycle of *Agaricus bisporus*. *Mycologia*, **64**, 1088–117.

Raper, J. R. (1966*a*). Life cycles, basic patterns of sexuality and sexual mechanisms. In *The Fungi*, ed. G. C. Ainsworth & A. S. Sussman, vol. 2, pp. 473–511. New York: Academic Press.

Raper, J. R. (1966*b*). *Genetics of Sexuality in Higher Fungi*. New York: The Ronald Press.

Raper, J. R. & Krongelb, G. S. (1958). Genetic and environmental aspects of fruiting in *Schizophyllum commune* Fr. *Mycologia*, **50**, 707–40.

Royse, D. J. & May, B. (1982*a*). Genetic relatedness and its application in selective breeding of *Agaricus brunnescens. Mycologia,* **74**, 569–75.

Royse, D. J. & May, B. (1982*b*). Use of isozyme variation to identify genotype classes of *Agaricus brunnescens. Mycologia,* **74**, 93–102.

Simchen, G. (1965). Variation in a dikaryotic population of *Collybia velutipes. Genetics,* **51**, 709–21.

Spear, M. C., Royse, D. J. & May, B. (1983). Atypical meiosis and joint segregation of biochemical loci in *Agaricus brunnescens. Journal of Heredity,* **74**, 417–20.

Sprague, G. F. (1972). The genetics of corn breeding. *Stadler Symposium, University of Missouri,* Columbia, Missouri, **4**, 69–81.

Takemaru, T. (1954). Genetics of *Collybia velutipes.* II. Dedikaryotization and its genetical implication. *Japanese Journal of Genetics,* **29**, 1–7.

Wessels, J. G. H., Hoeksema, H. L. & Stemerding, D. (1976). Reversion of protoplasts from dikaryotic mycelium of *Schizophyllum commune. Protoplasma,* **89**, 317–21.

24

Biological and technological aspects of commercial mushroom growing

P. B. FLEGG

formerly of Glasshouse Crops Research Institute, Littlehampton, West Sussex, UK*

Introduction

The cultivation of *Agaricus bisporus* seems to have originated early in the seventeenth century, perhaps even earlier, but the first known record of how mushrooms can be grown is by de Bonnefons (1650). Other early references include Culpeper (1652) and de Tournefort (1707).

A. bisporus is by no means the oldest cultivated mushroom. The shiitake (*Lentinus edodes*), which is grown mainly in Japan on cut tree logs, has a history dating back over 2000 years (Ainsworth, 1976; Leatham: Chapter 17). Probably the third most important cultivated mushroom, the paddy straw mushroom, *Volvariella volvacea*, is grown on rice straw mainly in South-East Asia. While the cultivation of *A. bisporus* is generally worldwide, the growing of other edible fungi is somewhat localised.

Total world production of cultivated edible mushrooms is currently around 1.4–1.5 million tonnes annually. Of this, *A. bisporus* accounts for about 1 million tonnes. Because of this supremacy in tonnage and because *A. bisporus* is the most important cultivated mushroom in Europe, the Americas and Australasia, I shall concentrate on that fungus.

The most important cultivated mushroom in the world has not suffered from lack of attention by taxonomists. Over the years it has been known variously as *Psalliota arvensis*, *P. hortensis*, *P. bispora*, *Agaricus campestris*, *A. hortensis* and *A. arvensis*; most recently, a claim for it to be called *A. brunnescens* has been put forward. Atkins, who did so much to found the modern UK mushroom industry, stated clearly in 1966 that the cultivated mushroom should be known as *Agaricus bisporus*. To minimise confusion (see also Elliott, 1983), this chapter retains that name.

* Present address: 50 St Flora's Road, Littlehampton, West Sussex, UK.

Table 24.1. *Mushroom crop production*

Operation	Process	Approximate duration (days)	
Composting phase 1	Wetting and mixing horse manure and wheat straw in long stacks on a concrete yard, temperatures up to 80 °C	7–14	
phase 2	Completion of composting process and pasteurisation in purpose-built rooms, temperatures 50–60 °C	4–8	
	Total compost preparation		11–22
Spawning	Inoculation of compost with mushroom mycelium	< 1	
Spawn-running	Colonisation of compost by mushroom mycelium, temperature 24 °C	10–14	
	Total compost colonisation		10–14
Casing	Covering of colonised compost with 3–5 cm layer of moist peat and chalk mixture and move to cropping room	< 1	
Pre-cropping	Colonisation of casing layer by mycelium and initiation and development of mushroom fruit bodies	18–21	
Cropping	Harvesting of the first and successive 'flushes' of fruit bodies at 7–10-day intervals	30–35	
Cook-out	End-of-crop heat (steam) treatment of cropping room and contents, temperature 60 °C	1–2	
Emptying	Compost and containers removed from room, clean-up ready for next crop	1–3	
	Total period in cropping room		51–72

Although the cultivation of *A. bisporus* was pioneered in France, mushroom growing has spread to about 80 countries. The major producer is now the USA at over 200000 tonnes annually; other important producers include France, the UK, the Netherlands, South Korea and China.

In the UK, mushrooms are one of the most important horticultural crops. About 250 mushroom farms produce some 70000 tonnes annually, valued at well over £90 million to the grower and nearly £200 million at retail prices. The UK mushroom industry has steadily increased in size since the end of the Second World War and is now greater in financial value than the tomato crop and the market for baked beans.

Commercial mushroom growing

The main operations in growing a crop of mushrooms are outlined in Table 24.1. A single crop occupies the cropping room for about 7–10 weeks so that it is possible to obtain successively about five to six crops from each house annually. Crop production goes on throughout the year and it is common practice to begin production of a new crop each week so attaining about 50 crops *per annum*. Thus, the average mushroom farmer produces more mushroom crops in a year than the wheat farmer can grow wheat crops in a lifetime. Plant pathologists will readily recognise that such a system of monoculture, with all stages of crop production being present on a farm at any one time, combined with a need to enter most cropping houses on most days of the year, is a near-perfect recipe for the maintenance of a continuous pest and disease problem.

Yields of mushrooms are customarily quoted per unit of bed area and per tonne of compost spawned. While there is considerable variation, a satisfactory yield for a single crop in 6 weeks of harvesting is about 15–25 kg m^{-2} and 150–250 kg tonne of prepared compost. The yield of mushrooms per tonne of compost is generally a measure of the quality of the compost, its suitability in nutritional value and physical characteristics for the mushroom, whereas yields per unit bed area usually reflect the amount of compost in the beds.

Mushrooms are generally grown in wooden trays or on shelves. The trays vary in size between about 0.5 and 2.5 m^2 and the shelves are usually about 1.5 m wide and run the length of the cropping house. Bed depth is usually 15–25 cm. Trays and shelves are stacked one above the other, some three to six high, with just enough space between tiers to allow the mushrooms, which develop on the upper surface of the beds, to be picked.

In broad outline, the main operations referred to in Table 24.1 have changed little recently. However, the materials and equipment used have changed. While mushroom growing may be considered reasonably profitable in most years, a high capital input is usually required and despite advances in our understanding of mushroom biology, the mushroom grower still has to rely on his accumulated skill and experience (see Wu: Chapter 25).

Composts and composting

Agaricus bisporus is a heterotrophic organism and is grown traditionally on a composted mixture of horse manure and wheat straw stable bedding. The purpose of composting is to prepare a medium which will favour the growth of the mushroom in preference to other organisms. During composting, microorganisms are encouraged to use the more readily available nutrients leaving the more resistant ones, such as hemicellulose, cellulose and lignin. The mushroom, having the requisite enzymes, can flourish on these while many of its competitors cannot.

The subject of mushroom composts and composting is large and cannot be dealt with other than briefly in this account. For further reading I commend the reviews by Lambert (1938), Stoller (1943), Flegg (1961), Hayes & Nair (1975) and Smith (1981).

Composting for mushroom growing probably remained little changed until about 50 years ago. As recently as 1950, composting usually consisted of wetting the straw and manure and stacking it, by hand, into heaps roughly 2 m wide and high and as long as necessary. The stack was forked over to mix the outer cool parts with the inner hot portion and, as was thought then, to allow air into the stack. These 'turns' were done at about weekly intervals until eventually, after 4 or 5 weeks, the compost was judged ready for inoculating with mushroom spawn when it was free of gaseous ammonia, which is toxic to the mushroom, and when it showed little or no tendency to heat up any further. Nowadays practically all the hard physical labour of turning compost is done by large, expensive self-propelled machines.

Lambert (1941) found that the mushroom grew best in compost taken from parts of the stack which had been at between 50 and 60 °C. This observation led to the development of a 'conditioning' phase towards the end of compost preparation, done in specially constructed buildings, in which the compost was maintained at between 50 and 60 °C until ready for spawning. The first phase, carried out on the concrete yard, is now known as 'phase 1' and is intended mainly to ensure adequate mixing of the ingredients and thorough wetting to about 70% moisture content, and to allow some softening of the straw. The second phase, carried out under controlled conditions, is known as 'phase 2', 'peak-heating' or 'sweat-out' and is aimed mainly at completing the composting process and enabling the compost to be pasteurised at 60 °C to eliminate mushroom pathogens.

Sinden & Hauser (1950, 1953) demonstrated that the duration of phase 1 could be reduced to about 7–8 days and phase 2 to about 5 days. This dramatic reduction in the total duration of the composting process reduced wastage of raw materials and led to a more consistent end product. Also at about this time several workers, for example, Sinden (1938), Lambert (1941), Stoller (1943) and Edwards (1949), showed that animal manures were not essential for successful mushroom compost production and a wide range of raw materials, such as wheat straw, lucerne hay, corn cobs, brewers grains and nitrogenous fertilizers, were incorporated into so-called synthetic composts. Deep-litter chicken manure is now one of the most popular compost additives either to manure or to synthetic composts (Gerrits, 1974).

An important recent development has been a method of carrying out phase 2 on the compost in bulk before it is transferred to trays or shelves (Derks, 1973). The compost is moved from the concrete yard into so-called 'tunnels' 2 m wide, 4 m high and 20 m or more long. The floor of the tunnel is slatted and below is an air plenum into which air is forced by a powerful fan and thence vertically through the compost into the air space above. Fresh air is mixed with the recirculating air to maintain aerobic conditions and the air is heated as required to maintain a temperature of 50–60 °C. Rapid circulation of air through the compost ensures more or less even temperatures throughout the depth and length of quantities of compost, often exceeding 25 tonnes. Control over this phase of composting using this bulk pasteurisation technique is so good that it readily lends itself to automation and computer control.

These more practical developments and improvements, backed up by an increasing understanding of the chemical and microbiological changes occurring in compost stacks, led to increased efficiency in compost production and a greatly improved and consistent end product. Waksman and his many co-workers (Waksman & Cordon, 1939; Waksman, Cordon & Hulpoi, 1939) showed that lignin tends to accumulate while cellulose and hemicellulose are decomposed more rapidly.

Microbial activity results in the synthesis of protein and a reduction in the amount of free ammonia. Changes in the carbon:nitrogen ratio during composting have been studied by Burrows (1951); and Hayes & Randle (1968) have shown the need for a balance between nitrogen and carbon compounds. Hayes (1972) claimed that the optimum carbon:nitrogen ratio of a prepared compost should be about 17:1.

Interest has concentrated more recently on the microbial changes occurring in mushroom composts. Staněk (1972) found that composts containing thermophilic actinomycetes and bacteria supported mushroom mycelial growth. A suggestion by Eddy & Jacobs (1976) that the microbial biomass accumulating during composting forms an important food source for the mushroom has been given greater weight by the work of Fermor & Wood (1981) who found that the mushroom thrives on media in which killed bacterial cells form the sole source of nutrient.

A closely related aspect of compost microbiology is the often-mentioned but little-studied phenomenon of compost selectivity. Well-prepared mushroom compost will allow vigorous and extensive growth of mushroom mycelium but will support the growth of few other microorganisms. The compost is said to be selective for the mushroom. Ross & Harris (1983 *a*, *b*) have shown that this selectivity can be destroyed by heat and by antimicrobial chemicals such as ethanol and chloroform. These workers proposed that selectivity depends on the presence of a viable but dormant biomass. There is also the possibility that the growth of competitive fungi in compost, e.g. *Chaetomium globosum*, may be inhibited by the production of antibiotics by certain thermophilic bacteria (Tautorus & Townsley, 1983).

The enrichment of prepared composts has recently been reviewed by Randle (1983). Much of the work has been done in the USA (Schisler & Sinden, 1962; Sinden & Schisler, 1962) and work on compost enrichment with delayed-release nutrients (Carroll & Schisler, 1974) has led to a considerable commercial application of this technique, especially in the USA. These materials are basically protein-rich meals which have been treated with formaldehyde to denature the protein. The general outlook for this development is promising; yield increases of up to 30% are possible. Work at the Glasshouse Crops Research Institute is in progress searching for novel and cheaper materials to use for compost enrichment.

The crop and cropping

Mushroom spawn, essentially a pure culture of mushroom on a cereal grain substrate, is mixed thoroughly into the prepared compost at the rate of about 0.5% by weight, and at about 25 °C the compost is completely colonised within about 10–14 days.

Producers of mushroom spawn offer a wide range of spawns, but often the differences between spawn strains are not always clear cut and much can depend on the conditions under which the crop is grown.

When the compost is colonised by mushroom mycelium it is covered by a layer about 3–5 cm deep of moist peat and chalk. The function of this layer, known as the casing layer, is to induce fruit body initiation. Although understanding of this process has improved in recent years, there is much more to be learnt. The physical and chemical properties and characteristics of the casing material are important. It must be capable of retaining water but allow oxygen and carbon dioxide to pass through it; it should be slightly alkaline but have a low salt content; and a low level of nutrients seems to be preferred. In 1961, Eger showed a relation between the presence of bacteria in the casing layer and mushroom fruit body initiation. It was later found that, although the mushroom does not generally fruit in the absence of microorganisms, their function can be replaced in sterile casing mixtures by mixing in activated charcoal. Hayes, Randle & Last (1969) demonstrated that some of the volatile metabolites of the mushroom, e.g. ethanol, encouraged the growth of a bacterium, *Pseudomonas putida*, which was capable of inducing fruit body initiation. Several workers have shown other organisms to be involved, but despite continuing studies little real progress has been made. Fortunately for the grower, the mushroom is able to produce fruit body initials with little difficulty when the correct conditions are provided.

In changing the mushroom from vegetative to reproductive growth the mushroom grower requires skill and experience. If fruit body initiation occurs while the mycelium has grown only a little way into the casing layer the mushrooms will develop below the surface of the peat and chalk and emerge dirty because they are covered in particles from the casing layer. Initiation is inhibited by carbon dioxide concentrations over 0.2%. The mushroom grower therefore allows carbon dioxide produced by the growth of mushroom in the compost to accumulate in his growing room and remain at a high level until the mycelium has reached the upper surface of the casing layer. At this point he reduces the carbon dioxide concentration by increasing the intake of fresh air. As a result the mushroom begins to produce fruit body initials on the surface of the casing layer where they grow completely free of contaminating peat and chalk particles.

Fruit body initiation is a particularly crucial stage of crop production, for the evidence is that nearly all the mushroom initials for the next 6 or more weeks of harvesting are produced at this time (Flegg, 1978). Overwatering the casing layer can result in the death of many potential mature mushrooms.

Not all the fruit body initials develop at once. Mushrooms appear on the beds at intervals of about 7–10 days, each fresh batch of mushrooms being known as a 'flush'. A flush of mushrooms is usually picked over a period of about 3–5 days, by hand. The interval between flushes depends largely on the stage of maturity of the mushrooms at harvest. When the mushrooms are cut before the veil has broken, cropping may appear to be almost continuous as successive flushes merge. Waiting until the veil is completely broken and the cap forms a 'T' with the stem before harvesting will cause the flushes to be well separated and there will be days on which no mushrooms are visible on the beds (Cooke & Flegg, 1962).

The timing of flushes in relation to the days on which markets are open and when labour is available can provide many problems for the grower. Weekends, particularly lengthened holiday weekends, are especially difficult times.

Harvesting can be made a much more efficient operation when control over flushing patterns in all cropping rooms is good. If it could be arranged that all the mushrooms of a flush were ready to harvest on the same day, harvesting would be much more efficient. Mechanical harvesting, already feasible and in use on a limited scale, would be made much easier.

A method of synchronising the development of mushrooms within a flush has recently been demonstrated (Flegg, 1980). It makes use of the requirement of most commercial white strains of mushroom for a lowered temperature, from 24 to 16 °C, while the fruit bodies develop from a diameter of 2 mm to about 6 mm. In passing through this stage, the mycelium within the fruit body changes from being an apparently tangled mass to being oriented more or less vertically in the stem and radially in the cap, as Bonner, Kane & Levy (1956) have shown. The crop is grown normally until most of the mushrooms comprising the first flush are greater than 6 mm in diameter and some will have reached 20 mm or more. At this point the temperature is raised from 16 to 24 °C. All mushrooms greater than 6 mm in diameter will continue to grow normally, but those not yet through the 2–6-mm phase of development, the stragglers of the flush, will either cease to grow or die. Thus uniformity of developmental stage within the flush will be increased and most mushrooms comprising the flush will reach a commercially harvestable stage at about the same time 1 or 2 days later. As soon as the flush is harvested, the temperature is returned to 16 °C for the development of the next flush of mushrooms, and again raised to 24 °C at the appropriate moment.

Systems of production

The earliest mushroom beds were laid on the floor. Sometimes the compost was heaped up in ridges perhaps 0.6 m or more high. Most mushroom farms now operate the tray or shelf system of growing. New developments in growing methods have usually arisen somewhat haphazardly and in response to economic pressures. However, realising that the productivity of a mushroom farm is in part related to the quantity of compost used for each crop, Flegg & Smith (1978) deliberately set about devising new methods of growing using increased quantities of compost in each cropping room. Several methods were proposed and one which is now used successfully, albeit so far on a limited scale, is the deep trough system. Troughs of compost about 1 m deep and 1.5–2 m wide are set up on a wooden slatted base with an air plenum beneath. A fan forces air through the compost via the plenum and so removes excess heat. Apart from being commercially viable this method of growing has led to interesting studies on the transport of nutrients through mushroom mycelium. It appears that the mushroom uses food supplies nearest the top of the bed first while nutrients furthest from the casing layer and developing mushrooms are utilised more slowly.

Conclusions and future prospects

In this brief account it has been possible to refer to only a few aspects of the biology and technology of mushroom growing. Barely mentioned are the problems of pests and diseases and omitted completely are references to the value of the mushroom as a source of human food or of pharmaceutical products. Further, the problems of post-harvest storage have not been considered.

The mushroom industry, while important, is quite different from other horticulture; generally speaking it is profitable, is expanding and has a good future. It is vigorous, being represented by an active Mushroom Growers Association, is continually calling for increased research effort, and is usually quick to seize on new developments. Opportunities for research and development work to have important effects on the mushroom industry exist in several fields of biological study.

Tschierpe (1983) has recently described differences between spawn strains in their responses to environmental factors and in characteristics contributing to quality of the harvested product. Backed up by an increased understanding of the nutritional requirements of the mushroom, is it possible to develop new strains which can utilise traditional composts more efficiently or even grow successfully on materials which do not require composting?

The introduction of mechanical harvesting would be facilitated if a better understanding of the response of the mushroom to environmental factors could be combined with a breeding programme to develop strains which would be induced to produce flushes of mushrooms all ready to harvest at the same time, evenly spaced and of the same height. Our experience so far suggests that this goal may not be entirely unattainable.

Evidence of reduced susceptibility or resistance to pests and disease in some mushroom strains exists (Gandy & Spencer, 1981), but these phenomena have yet to be fully exploited. A new development is the work of Elliott & Challen (1983) in producing mushroom strains able to withstand levels of fungicide application capable of killing pathogens yet producing a full crop of mushrooms.

The effects of breeding mushroom strains based on the results of research into mushroom genetics are only just beginning to be felt in the mushroom industry. The potential for such new developments is enormous.

References

Ainsworth, G. C. (1976). *Introduction to the History of Mycology*. Cambridge University Press.

Atkins, F. C. (1966). *Mushroom Growing Today*, 5th edn. London: Gaber & Faber.

Bonner, J. T., Kane, K. K. & Levy, R. H. (1956). Studies on the mechanics of growth in the common mushroom (*Agaricus campestris*). *Mycologia*, **48**, 13–19.

Burrows, S. (1951). The chemistry of mushroom composts. II. Nitrogen changes during the composting and cropping process. *Journal of the Science of Food and Agriculture*, **2**, 403–10.

Carroll, A. D. & Schisler, L. C. (1974). Delayed release nutrients for mushroom culture. *Mushroom News*, **22**, 10–11.

Cooke, D. & Flegg, P. B. (1962). The relation between yield of the cultivated mushroom and stage of maturity at picking. *Journal of Horticultural Science*, **37**, 167–74.

Culpeper, N. (1952). *Complete Herbal*. London: Foulsham 1952 reprint.

de Bonnefons, N. (1650). *Le Jardinière Français*. Paris.

de Tournefort, M. (1707). Observations sur la naissance et sur la culture des champignons. *Mémoires de l'Académie Royale des Sciences*, **1707**, 58–66.

Derks, G. (1973). 3-phase-1. *Mushroom Journal*, **9**, 396–403.

Eddy, B. P. & Jacobs, L. (1976). Mushroom compost as a source of food for *Agaricus bisporus*. *Mushroom Journal*, **38**, 56–67.

Edwards, R. L. (1949). M.R.A. Report on synthetic composts, *MGA Bulletin*, **15**, 84–8.

Eger, G. (1961). Untersuchungen über die Function der Deckschicht bei der Fruchtköperbildung des Kulturchampignons, *Psalliota bispora* Lge. *Archiv für Mikrobiologie*, **39**, 313–34.

Elliott, T. J. (1983). The cultivated mushroom: is it *Agaricus brunnescens*? *Mushroom Journal*, **122**, 69.

Elliott, T. J. & Challen, M. P. (1983). Mushroom genetics; strain improvement methods. *Glasshouse Crops Research Institute, Annual Report*, **1981**, 137–8.

Fermor, T. F. & Wood, D. A. (1981). Degradation of bacteria by *Agaricus bisporus* and other fungi. *Journal of General Microbiology*, **126**, 377–87.

Flegg, P. B. (1961). Mushroom composting: a review of the literature. *Glasshouse Crops Research Institute, Annual Report*, **1960**, 125–34.

Flegg, P. B. (1978). Effect of temperature on sporophore initiation and development in *Agaricus bisporus*. *Mushroom Science*, **10**, 595–602.

Flegg, P. B. (1980). Temperature induced synchronization of sporophore production in the mushroom *Agaricus bisporus*. *Scientia Horticulturae*, **13**, 307–14.

Flegg, P. B. & Smith, J. F. (1978). Future development in mushroom growing methods. *Mushroom Science*, **10**, 159–71.

Gandy, D. G. & Spencer, D. M. (1981). Fungicide evaluation for control of dry bubble, caused by *Verticillium fungicola*, on commercial mushroom strains. *Scientia Horticulturae*, **14**, 107–15.

Gerrits, J. P. G. (1974). Organic supplementation of mushroom compost. *Mushroom News*, **22**, 4–15.

Hayes, W. A. (1972). Nutritional factors in relation to mushroom production. *Mushroom Science*, **8**, 663–74.

Hayes, W. A. & Nair, N. G. (1975). The cultivation of *Agaricus bisporus* and other edible mushrooms. In *The Filamentous Fungi*, vol. 1, *Industrial Mycology*, ed. J. E. Smith & D. R. Berry, pp. 212–48. London: Edward Arnold.

Hayes, W. A. & Randle, P. E. (1968). The use of water soluble carbohydrates and methyl bromide in the preparation of mushroom composts. *MGA Bulletin*, **218**, 81–97.

Hayes, W. A., Randle, P. E. & Last, F. T. (1969). The nature of the microbial stimulus affecting sporophore formation in *Agaricus bisporus* (Lange) Sing. *Annals of Applied Biology*, **64**, 177–87.

Lambert, E. B. (1938). Principles and problems of mushroom culture. *Botanical Review*, **4**, 397–426.

Lambert, E. B. (1941). Studies on the preparation of mushroom compost. *Journal of Agricultural Research*, **62**, 415–22.

Randle, P. E. (1983). Supplementation of mushroom compost – a review. *Crop Research*, **23**, 51–69.

Ross, R. C. & Harris, P. J. (1983a). An investigation into the selective nature of mushroom compost. *Scientia Horticulturae*, **19**, 55–64.

Ross, R. C. & Harris, P. J. (1983b). The significance of thermophilic fungi in mushroom compost preparation. *Scientia Horticulturae*, **20**, 61–70.

Schisler, L. C. & Sinden, J. W. (1962). Nutrient supplementation of mushroom compost at spawning. *Mushroom Science*, **5**, 150–64.

Sinden, J. W. (1938). Synthetic Compost for mushroom growing. *Pennsylvania Agricultural Experiment Station Bulletin*, No. 365.

Sinden, J. W. & Hauser, E. (1950). The short method of composting. *Mushroom Science*, **1**, 52–9.

Sinden, J. W. & Hauser, E. (1953). The nature of the composting process and its relation to short composting. *Mushroom Science*, **2**, 123–31.

Sinden, J. W. & Schisler, L. C. (1962). Nutrient supplementation of mushroom compost at casing. *Mushroom Science*, **5**, 267–80.

Smith, J. F. (1981). The development of mushroom composting techniques – a review. *Glasshouse Crops Research Institute, Annual Report*, **1980**, 171–83.

Staněk, M. (1972). Micro organisms inhabiting mushroom compost during fermentation. *Mushroom Science*, **8**, 797–811.

Stoller, B. B. (1943). Preparation of synthetic composts for mushroom culture. *Plant Physiology*, **18**, 397–414.

Tautorus, T. E. & Townsley, P. M. (1983). Biological control of olive green mould in *Agaricus bisporus* cultivation. *Applied and Environmental Microbiology*, **45**, 511–15.

Tschierpe, H. J. (1983). Environmental factors and mushroom strains. *Mushroom Journal*, **132**, 417–29.

Waksman, S. A. & Cordon, T. C. (1939). Thermophilic decomposition of plant residues in composts by pure and mixed cultures of microorganisms. *Soil Science*, **47**, 217–25.

Waksman, S. A., Cordon, T. C. & Hulpoi, N. (1939). Influence of temperature upon the microbiological population and decomposition process in composts of stable manure. *Soil Science*, **47**, 83–113.

25

Composting technology

LUNG-CHI WU

Campbell Institute for Research & Technology, Napoleon, Ohio 43545, USA

Introduction

In the mushroom industry, compost is the major source of mushroom nutrition. Mushroom composts and composting methods have been reviewed previously by Lambert (1938), Stoller, (1954), Flegg (1961), Smith (1981) and Nair (1982). A seminar on this subject was held recently (Hayes, 1977) at the University of Aston to discuss improvements and future prospects.

Composting is an ancient art, probably practised by man since before the dawn of recorded history. It is a microbial process that depends on the enzymic reaction of microorganisms which are indigenous to various organic wastes. The historical aspects of composting have been discussed in reviews by Gotaas (1956), Golueke (1972) and Poincelot (1975). Recently, composting has come into widespread use as a sewage sludge treatment process (Finstein, 1983). Proper management of organic wastes can improve soil productivity (Parr & Willson, 1980). Recent advances in composting technology have been reviewed by Gray, Sherman & Biddlestone (1971 *a*, *b*), Gray, Biddlestone & Clark (1973), Poincelot (1975), Willson, Epstein & Parr (1976), Haug (1980), Crawford (1983) and Finstein *et al.* (1983).

The composting technology presented here is limited to substrate for mushroom production, especially for commercial operations in North America.

Mushroom production and economic structure

Mushrooms have been used as food for centuries. Of more than 2000 species of edible fungi known, only a few have been investigated thoroughly from the standpoint of their commercial potential. The oldest

541

Table 25.1. *Mushroom production in the USA*

Crop year	Area in production (m²)	Fresh market (kg)	Processing (kg)	Total production (kg)
1980–81	13 043 309	124 762 769	88 235 508	212 998 277
1981–82	13 134 387	144 757 326	89 818 561	234 575 887
1982–83	12 583 178	152 966 525	69 668 874	222 635 398

Data from Crop Reporting Board, Statistical Reporting Service, USDA (1983).

Table 25.2. *Mushroom production and value by regions in the USA*

Crop year	Region	Total production (kg)	Price (US$/kg)	Total value (US$)
1980–81	East	133 059 512	1.58	209 990 000
	Central	26 741 813	1.79	47 926 000
	West	53 196 952	2.19	116 583 000
	Total	212 998 277	—	374 499 000
1981–82	East	150 008 618	1.55	232 240 000
	Central	27 483 444	1.79	49 088 000
	West	57 083 825	2.40	137 769 000
	Total	234 575 887	—	419 097 000
1982–83	East	136 803 048	1.67	228 030 000
	Central	26 218 815	2.07	54 195 000
	West	59 613 535	2.50	149 177 000
	Total	222 635 398	—	431 402 000

Data from Crop Reporting Board, Statistical Reporting Service, USDA (1983).

crops of cultivated mushrooms are those of the Japanese shiitake, *Lentinus edodes*, and the Chinese straw mushroom, *Volvariella volvacea*, whereas the first European species to be cultivated, the white mushroom, *Agaricus bisporus*, has now a much larger area of cultivation (Wu & Stahmann, 1975).

The white mushroom has been the primary mushroom grown in North America since late in the nineteenth century (Wuest, 1983). Mushrooms are an important crop in the USA; of the 27 vegetables listed in Agricultural Statistics for 1981, only lettuce and tomatoes have a higher cash value (Farr, 1983). Commercial mushroom production is dispersed

Table 25.3. *Mushroom production in Canada*[a]

Crop year	Area in production (m²)	Fresh market (kg)	Processing (kg)	Total production (kg)
1975	1 533 129	10 775 000	8 523 000	19 298 000
1978	1 815 452	17 221 000	6 012 000	23 233 000
1979	2 000 137	18 647 000	6 587 000	25 234 000
1981[b]	2 280 297	23 253 000	9 457 000	32 710 000

[a] Ingratta & Blom (1980).
[b] Canadian Mushroom Growers Association (1983).

throughout the USA. In the 1982–83 marketing season, beginning July 1, 223 000 tonnes (t) of commercial mushrooms, valued at approximately US$431 million, were produced (Tables 25.1 and 25.2). Fresh sales accounted for 69% of US production. Yields averaged 17.72 kg m^{-2}. An upward trend in the proportion of fresh mushroom sales continued. This increase indicates a growing interest in mushrooms as food and suggests that there may be a market for additional mushroom species. Shiitake and *Pleurotus* (oyster mushroom) appear to be the most likely candidates for increased commercial production in the USA since oriental food stores and restaurants provide a ready market (Wu & Stahmann, 1975; Farr, 1983).

Eastern States accounted for 61% of the total US production, Central States, 12%, and Western States, 27% (Table 25.2). The first major mushroom-producing area of the USA was located in the vicinity of New York City. Close to the turn of the century, greenhouse operators in Kennett Square, Pennsylvania, started raising mushrooms. Due primarily to this early start in the use of improved techniques, Pennsylvania became the major mushroom-producing State in the country (Horowitz, 1979). This leading State grew 112 000 t of mushrooms, representing over 50% of the US 1982–83 crop. The typical mushroom farmer in Pennsylvania grows mushrooms in a house 20 m long × 12 m wide × 4 m high in which four tiers of six beds provide a growing area of 740 m² (Wuest, 1983). This is called a 'double' in contrast to a 'single' which is 370 m² square.

Mushrooms are produced commercially in almost all the agricultural areas of Canada (Blum, 1977). Mushroom production in Canada reached 33 000 t in 1981 (Table 25.3). The figures shown are on a calendar year basis and an upward trend in fresh market sales is apparent. Fresh mushroom sales accounted for 71% of the total production. Yields averaged 14.34 kg m^{-2}. Mushroom production in 1982 was over 35 000 t, rising 7%

Table 25.4. *Mushroom production and value by regions in Canada*

Crop year	Region	Total production (kg)	Price (Can.$/kg)	Total value (Can.$)
1981	East	21774614	2.49	54216000
	Central	2057530	3.11	6401000
	West	8877859	2.47	21947000
	Total	32710003	—	82564000
1982	East	21564144	2.57	55411000
	Central	2244413	3.32	7442000
	West	11341814	2.33	26380000
	Total	35150371	—	89233000

Data from Canadian Mushroom Growers Association (1983).

over 1981 levels. The total value of the 1982 crop was Can.$89 million. Data for 1982 were preliminary. Eastern Provinces accounted for 61% of the total production in Canada, Central Provinces, 7%, and the Western Provinces, 32% (Table 25.4). Ontario is the leading Province. The mushroom industry in Ontario is concentrated in York, Peel, Halton and Wentworth counties, which account for about 79% of the Province's mushroom production facilities. The crop ranked third in value of production after potatoes and tomatoes (Blum, 1977). Ontario's 14000 t in 1979 were valued at Can.$31 million which represented 56% of the Canadian total. Few doubles are found in Ontario. A typical house includes two tiers of six beds providing a growing area of 360 m² (Ingratta & Blom, 1980).

Compost costs are estimated to amount to 20–25% of total operating expenses (Royse & Schisler, 1980). As shown in Table 25.5, the cost of compost delivered to a farm amounted to 20% of total operating expenses at a standard Pennsylvania farm in the 1980–81 crop year. At a Canadian Farm, 33% of total operating expenses was for growing materials of which compost ingredients and its additives accounted for 57% (Tables 25.6 and 25.7). The breakdown of compost material cost is shown in Table 25.8. This is a typical compost formula in North America.

As shown in Table 25.9, labour costs for compost preparation ranked second in production expenditures in commercial operations. Filling, steaming, spawning, casing, spraying, emptying and cleaning operations are usually carried out by contract labour. Approximately 71% of the farms purchased compost from specialised compost suppliers (Owens *et*

Table 25.5. *Operation expenses on the US farm (1980–81)*

Item	Cost per farm (US$)	Percent of total cost
Compost	4140	20
Filling	400	2
Steam	150	1
Spawn	896	4
Spawning	240	1
Supplement	1000	5
Heating/cooling	2640	13
Peat moss	860	4
Casing	190	1
Picking	3038	14
Pest control	860	4
Emptying	500	2
Maintenance	525	2
Management	2000	10
Steam	480	2
Indirect costs	3100	15
Total cost	21019	100

Data from Wuest (1983).

Table 25.6. *Operation expenses on the Canadian farm (1976)*

Item	Cost per year (US$)	Percent of total cost
Growing materials	140010	33
Maintenance and repairs	72232	17
Energy	63434	15
Packing	95557	22
General business	56007	13
Total	427240	100

Data from Horowitz (1979).

al., 1982). Those farms preparing their own compost must spend more for labour. The decision of whether to prepare or purchase compost depends on a combination of the relative costs of compost from either source and subjective grower preferences.

To make compost, horse manure is purchased from a manure broker for US$25–US$30 t^{-1}, hay at US$55–US$65 t^{-1}, US$45 t^{-1} for corn cobs, and poultry manure for US$38 t^{-1}. The last three materials are purchased

Table 25.7. *Material costs on the Canadian farm (1976)*

Material	Cost per year (US$)	Percent of total cost
Compost ingredients	22767	16
Compost additives	57172	41
Spawn	39827	28
Casing material	16742	12
Chemicals	3502	3
Total	140010	100

Data from Horowitz (1979).

Table 25.8. *Compost material costs on the Canadian farm (1976)*

Material	Cost per year (US$)	Percent of total cost
Manure	7690	10
Hay	11377	14
Corn cobs	3700	5
Brewers' grain and soya	50700	63
Ammonium nitrate	1550	2
Muriate of potash	1120	1
Gypsum	3802	5
Total	79939	100

Data from Horowitz (1979).

from farmers, or at feed mills or auctions. Farmers who decide to prepare their own compost require a reinforced concrete wharf, a compost turner (US$18000–US$85000), an industrial-grade front-end loader (US$65000), a water supply and a waste water disposal system for run-off (Wuest, 1983). A common problem facing most small (less than 4645 m²) and medium-sized (4645–13935 m²) farms is the difficulty in accumulating enough capital to invest in new sophisticated equipment (Horowitz, 1979). Centralisation of a compost facility is a remedy, particularly on a cooperative basis. The feasibility of this has been demonstrated by the Fraser Valley Mushroom Growers' Cooperative Association in a western Province of Canada, where the most economical mushroom compost has been prepared and delivered to their mushroom growers. Its Phase One Division was formed in 1970–71 with deliveries of 765 m³ of compost per week. In the 1982–83 crop year, the production was 3060 m³ per week.

Table 25.9. *Production wages on the Canadian farm* (*1974–75*)

Item	Cost per farm (US$)	Percent of total cost
Compost preparation	40952	16
Filling beds	11231	5
Phase 2	6159	2
Spawning	9573	4
Casing	12926	5
Watering	13829	6
Spraying	9039	4
Picking and packing	116393	46
Emptying beds and cleaning	12149	5
Maintenance	16783	7
Total	249034	100

Data from Horowitz (1979).

They currently have 20 employees: Manager, Assistant Manager, Office Clerk, Laboratory Technician and 16 employees on the wharf (approximately 20235 m² of cement wharf which is covered by approximately 16188 m² of roof). The equipment consists of five Clark loaders, two Pannell turners, two Spiders (supplement hoppers), two Okie turners and one scale. Very little of this data is available in published literature.

Composting process

The first mushroom farmers used straw-bedded horse manure as the main compost ingredient. Today, the same material is still the basic ingredient for the preparation of mushroom compost in North America (Ingratta & Blom, 1980; Wuest & Bengtson, 1982). This is simply because of availability and price of compost ingredients and additives in the area where the mushroom composts are prepared (Kinrus, 1978). Some typical formulae used in North America are shown in Table 25.10 (Schisler, 1980; Wuest & Bengtson, 1982).

In the early days, mushroom compost was prepared for a minimum of 21 days and this has since become known as 'long compost'. Peak-heating, to maintain compost temperatures at 55–60 °C, took 2–3 days. Only one change of air in 24 h was considered necessary because long compost was comparatively inactive. This was often achieved by opening the door for a short period. Fans were used to recirculate the air to eliminate temperature gradients within the house (Randle, 1974). Sinden & Hauser

Table 25.10. *Typical compost formulae in North America*

Material	Percent (based on dry mass)			
	HMC	SYC-1	SYC-2	SYC-3
Horse manure	83.68	—	—	—
Straw	—	—	—	57.31
Hay	—	43.84	43.99	—
Corn cobs	—	43.84	21.99	—
Hardwood bark	—	—	21.99	—
Cottonseed hulls	—	—	—	17.91
Chicken manure	10.04	8.22	8.25	17.19
Cottonseed meal	—	—	—	5.73
Brewers' grain	4.19	—	—	—
Ammonium nitrate	—	1.03	0.69	—
Potash	—	1.03	1.03	—
Gypsum	2.09	2.05	2.06	1.86

Data from Schisler (1980).

(1950) introduced the 'short composting' method. Since then, the composting process has become a two-stage process (see Chapter 24).

The details of this process are as follows (Sinden & Hauser, 1953). Phase 1 is outdoors in a pile of limited cross-sectional dimensions and phase 2 is in an insulated room with beds or trays. Both are interdependent parts of a coordinated process necessary to the final development of the mushroom compost. The size, shape and compactness of the pile, as well as its water and nitrogen content, are important. Heat produced is mostly confined to the pile by the insulating, compacted sides. Attempts to overcome the core condition by vents, placed in the centre of the pile, result in an inner zone of biological decomposition, thus reducing the homogeneity of the pile. Only the adjustment of the width and compactness of the pile are advantageous to obtain optimal and uniform conditions. A short composting system of a 10- to 14-day phase 1 followed by a 12- to 15-day phase 2 has become the most popular method employed by mushroom growers. The sequential steps in the composting process for the preparation of mushroom compost are summarised in Table 25.11.

The unpredictable nature of compost quality imposes a major limitation on mushroom production. Variability in results of phase 1 composting arises from differences in the compost ingredients and environmental conditions during the composting process. An automatic phase 2 composting process for the Mushroom Test Demonstration Facility of the Pennsylvania State University, constructed in 1969, was designed to

Table 25.11. *Composting process for mushroom compost*

Sequence	Duration	Description
Phase 1[a,b]		
1. Preblending	1–2 days	Wet bulking ingredients
2. Blending	1–2 days	Wet and mix bulking ingredients
3. Ricking	8–11 days	Rick and turn rectangular pile
4. Filling	1 day	Fill compost in bed, tray or tunnel
Phase 2[b]		
1. Prepeak-heating	1–3 days	Equilibrate compost condition
2. Peak-heating	4–9 h	Pasteurise compost
3. Postpeak-heating	5–12 days	Condition compost for specificity
4. Cooling	1–2 days	Cool compost for spawning

[a] Horowitz (1979).
[b] Ingratta & Blom (1980).

overcome the effects of the variability of phase 1 composting in a minimum time. This system has been used successfully and has been widely adopted by the industry (Royse & Schisler, 1980).

Another major technological breakthrough for all mushroom growers is bulk pasteurisation, i.e. mass pasteurisation of mushroom compost in highly insulated rooms called 'tunnels'. Bulk pasteurisation enables the grower to perform the phase 2 composting process in a more controlled environment. Bulk spawn-run can also take place in similar tunnels. The compost layer, approximately 2 m high, is placed on a grid floor. Air is blown through the mass of compost to supply fresh air and to keep the optimum temperature, both during phase 2 composting and during spawn-run (Vedder, 1979). The Dutch exploited this technique on a commercial scale and a large composting cooperative unit was built at Ottersum to supply mushroom growers with a reliable ready-spawned compost (Gerrits, 1981; van Zaayen & Gerrits, 1982).

Factors affecting the composting process
Water

When horse manure, straw, hay, cobs, bark and hulls are ready to be used for compost preparation, water is added to bring the moisture content of these ingredients to the optimum range of 75–80% (Jeris & Regan, 1973 b). This must be accomplished without having water run from the pile, leaching out the soluble nutrients. A higher moisture content is permissible for a porous mixture of straw and manure. Composters of various types are used to chop, wet and assemble piles. Compost that is prepared from horse manure usually gets three turns, while synthetic compost may undergo two additional turns. The synthetic compost can be mixed with horse manure compost in a 30% : 70% ratio (Ingratta & Blom, 1980). This is called 'blend' compost, which is usually mixed the day before filling. The moisture content of compost at filling is 70–75% depending on compost conditions at that time (Wuest & Bengtson, 1982).

Of the many environmental factors, moisture content has been considered to be the most important criterion for optimum composting. Sufficient moisture in the compost is required for maximum efficiency of microbial activity (Jeris & Regan, 1973 b,c). When the moisture content is excessive, aeration in the compost pile is restricted. As the oxygen concentration is depleted, an anaerobic condition develops. However, microbial activity is greatly reduced at moisture contents below 30% (Gray *et al.*, 1971 b). Nonetheless, mushroom yield depends on the actual moisture content of compost at spawning (Gerrits, 1972). The optimum content at spawning is 65% (63–68%) compared with 71% (69–73%) at filling. When the moisture content of compost at spawning is higher or lower than the optimum, the yield decreases by about 0.5 kg m^{-2} for each percent moisture, and the efficiency of dry matter conversion also decreases.

Oxygen

A continuous supply of oxygen must be available to ensure aerobic composting. The rate of consumption of oxygen by aerobic microorganisms in compost depends on many factors. Finger, Hatch & Regan (1976) developed a conceptual model of aerobic composting and measured oxygen and temperature distribution experimentally at the Butler County Mushroom Farm, Pennsylvania, to test the validity of the assumptions based on steady-state microbial activity, and heat and mass transfer limitation. The compost formula consisted of ground corn husks, straw and horse manure. The data matched with the model excellently, with deviations between model predictions and actual temperature measurements never exceeding 3 °C. The predictions show that the maximum breakdown rate occurs for an optimum height of pile. Insulating the pile base increases the breakdown rate; increasing the external temperature increases the initial breakdown rate but decreases the uniformity of compost, and any increase in the external oxygen concentration increases the breakdown rate but decreases the uniformity of compost.

Based on this study, it is concluded that there is an optimum size of compost pile at about 2.44 m high for a height:width ratio of 0.93. The rate of substrate conversion is fastest for about this size of pile, while varying size has a minimal effect on compost uniformity. It is also concluded that density variations over the range of 384 to 449 kg m^{-3} do not significantly affect aerobic microbial activity in the compost pile. Although the steady-state rate is somewhat higher at the higher densities, the uniformity is slightly better at lower densities. Thus, density is not seen as a major factor.

Randle & Flegg (1978) also measured the changes in oxygen concentration and temperature in the compost pile during phase 1 composting. They identified distinct patterns of change in oxygen concentration in the compost pile. Oxygen concentrations were similar in the outer, middle and inner zones for each level, the only exception being the inner-bottom region where the oxygen concentration was low throughout composting. It was concluded that differences in oxygen concentration were greater horizontally than vertically. When the compost pile was turned, the physical environment was changed and existing temperature gradients were disturbed. Changes in oxygen concentrations in the compost between turns are the net result of oxygen being utilised by microorganisms and oxygen being replenished by mass air movement (convection) and gaseous exchange (diffusion) through the compost. The air used for ventilation and circulation during phase 2 composting has been reviewed by Tschierpe (1973) and Gerrits (1981).

Temperature

Temperature has frequently been used as a yardstick in composting to judge the efficiency and degree of stabilisation for waste treatment (Jeris & Regan, 1973 a). Aerobic composting generates a great deal of heat. Since composting materials are relatively good insulators, a temperature distribution occurs within the pile with highest temperatures being in the interior. Temperature distributions in actual compost piles have been measured by many workers (Lambert & Davis, 1934; Finger *et al.*, 1976; Randle & Flegg, 1978). Lambert (1941) found that compost prepared at temperatures between 50 and 60 °C was the most suitable for mushroom growing, providing conditions that favoured aerobic composting. It is interesting to note that the optimum temperature range for composting found by Lambert (1941) agrees with the optimum temperature conditions for waste treatment (Jeris & Regan, 1973 a), i.e. the composting rates increased with temperature, reaching maximum values near 60 °C. Recently, Finstein *et al.* (1983) reported that the threshold to limitation for composting process was approximately 60 °C, which was based on the interacting factors of microbial heat generation, temperature, ventilation and vaporisation of water.

After the house has been completely cleaned and washed following the previous crop, the beds or trays are loosely filled with the newly prepared phase 1 compost to a depth of about 30 cm. One tonne of compost fills approximately 9.29 m^2 of growing area. The compost is pasteurised using either a high- or a low-temperature cycle. In the former, all air openings in the mushroom houses are closed. In the enclosed house, temperatures rise to 66 °C due to the continuing microbial activity from phase 1. Steam may be injected into the room to stimulate this activity. This high temperature is maintained for only 6 h. The house is then ventilated to reduce the temperature by about 3 °C per day until the compost is completely cooled. This procedure takes between 12 and 15 days (Horowitz, 1979).

In a low-temperature cycle, steam is introduced into the house, raising the temperature to 52 °C. With ventilation, this temperature is reduced to 44 °C. Once again the air vents are closed and, using steam, the temperature is quickly increased to 60 °C. After about 4 h, the temperature is reduced with ventilation, and maintained at 43 °C. After one week, the compost is then completely cooled and prepared for spawning.

The aim of peak-heating is two-fold: pasteurisation and conditioning. The pasteurisation is to destroy a number of organisms noxious to mushroom growing. The conditioning is to shape the compost into a selective environment which favours the growth and development of mushrooms. The temperature level and the duration of this temperature level are both important (Overstijns, 1981). The pasteurisation is achieved by keeping the temperature of the air in the house as well as the temperature of the compost at 57–58 °C for 5–6 h. Theoretically, the most noxious organisms are destroyed, but there are some irregularities in practice. Conditioning transforms compost into a specific substrate for mushroom growing via microorganisms which develop optimally at temperatures of 45–55 °C (Overstijns, 1981; Ross & Harris, 1982).

One feature of mushroom compost is its 'selectivity' or 'specificity' which has received little attention until recently. Ross & Harris (1983a) destroyed the selectivity of mushroom compost by heat (temperatures from about 60 °C upwards) and antimicrobial chemicals such as chloroform and alcohol. Following heat and chemical treatment, they were able to restore the selectivity of mushroom compost and also to remove ammonia by giving a further period under phase 2 composting conditions. Reinoculation with untreated compost or with the thermophilic fungus *Torula thermophila*,

Table 25.12. *Approximate composition of mushroom compost at spawning*

Item	Percent
Nitrogen, Kjeldahl	2.20
Fat, ether extract	0.074
Fibre	8.70
Ash	9.87
Carbohydrate (difference)	6.90
Moisture	71.29

Data from Holtz, Smith & Barkate (1975).

greatly speeded up the process. Treatment of mushroom compost at temperatures of about 60 °C or upwards seems to destroy a significant population of microorganisms which result in the loss of compost specificity. Thus, phase 2 composting can have serious adverse effects on mushroom compost if temperature and its duration are not controlled accurately. In their later work, *T. thermophila* showed considerable potential as a rapid-composting organism (Ross & Harris, 1983b). The shorter composting technique is a definite general improvement in composting (Smith, 1981).

Nutrients

The approximate composition of mushroom compost at spawning, which is the average of 12 samples, is shown in Table 25.12. The largest component is ash, which forms the inorganic component of compost. The predominant organic component is fibre, accounting for 8.7%; it is the most abundant carbon source in compost. A nitrogen content of 2.2% is acceptable for mushroom growing.

Nitrogen is the major nutrient required by microorganisms in the assimilation of the carbon substrate in organic wastes. Phosphorus is next in importance, while potassium, magnesium, sulphur, calcium and trace quantities of several other elements all play a part in cell metabolism (Gray *et al.*, 1971b). To prepare mushroom compost, the nitrogen content of the compost pile is usually 1.5–1.7% at the beginning of phase 1 composting. It increases throughout composting and reaches 2.2–2.3% at the end of phase 2 (Wuest & Bengtson, 1982). The relation between the initial nitrogen content of mushroom compost and the duration of composting has been studied by Flegg & Randle (1981). The results indicate that with increasing duration of composting, a higher initial nitrogen content of the compost is acceptable, and probably desirable, for preparing a successful compost. In general, organic and inorganic additives are mixed into bulk compost ingredients to raise the nitrogen content of the compost to the desired level.

Calcium is one of the most important inorganic components in mushroom compost. Gerrits (1977) showed the influence of gypsum on mushroom yield. The optimum quantity of gypsum was proved to be 25 kg t^{-1} of horse manure. The time of application is not very important, but an early application favours a uniform distribution and a longer action.

Although fat content in mushroom compost only accounts for 0.074% of fresh compost, higher fat content in compost is beneficial to mushroom yield. Schisler & Patton (1970) showed the stimulation of mushroom yield by adding vegetable oils to mushroom compost prior to phase 2 composting. In practice, some growers add cottonseed oil at the rate of 38 l t^{-1} of compost at filling (Kinrus, 1978). Holtz, Smith & Barkate (1975) showed the changes in the lipid compositions during the composting process. The neutral lipid increased and polar lipids decreased during phase 1 composting. Composts at filling contained approximately 50% more ether-extractable fat than at spawning. They concluded that lipids in the compost remaining at the end of phase 2 are largely the intracellular lipids of the thermophilic microflora.

In the composting process, microorganisms rapidly utilise the available sugar and starches, and large losses of water-soluble materials follow. Lipids undergo major, and cellulose and hemicellulose intermediate, decomposition. Lignin is most resistant to degradation (Poincelot, 1975). Chemical changes in mushroom compost during composting have been studied (Waksman & Nissen, 1932; Sinden & Hauser, 1950; Gerrits, Bels-Koning & Muller, 1967; Hayes & Randle, 1968; Gerrits, 1969; Grabbe, 1972; Hsieh, Hu & Wu, 1972; Langar, Sehgal & Garcha, 1980). In the process of composting, there is a rapid development of microorganisms. Water-soluble substances, cellulose and hemicellulose are drastically decreased, whereas lignin, nitrogen and ash are increased both in manure compost (Waksman & Nissen, 1932) and in rice straw synthetic compost (Hsieh *et al.*, 1972).

Gerrits *et al.* (1967) showed that total lignin remained the same, while pentosan and cellulose decreased rapidly during the composting process in both phase 1 and phase 2. Hayes & Randle (1968) were able to conserve cellulose and hemicellulose by adding soluble carbohydrate in compost. Grabbe (1972) estimated that the humus and lignin made up 60–70% of the total organic matter in compost at the end of composting. About 80% of the insoluble nitrogen could be isolated in the lignin fraction and 10%

Table 25.13. *Chemical changes in straw compost during composting and cropping*

Item	Wheat straw compost		Rice straw compost	
	At filling (% dry weight)	End of cropping (% dry weight)	At filling (% dry weight)	End of cropping (% dry weight)
Neutral detergent fibre	56.7	52.2	85.3	75.9
Cell-soluble[a]	43.3	47.8	14.7	24.1
Acid detergent fibre	46.9	51.6	57.5	59.4
Cellulose	18.2	15.3	42.2	31.5
Hemicellulose[b]	9.8	0.6	27.8	16.5
Lignin	21.4	23.9	5.2	10.8

[a] 100 − neutral detergent fibre = cell-soluble.
[b] Neutral detergent fibre − acid detergent fibre = hemicellulose.
Data from Langar, Sehgal & Garcha (1980).

of the total nitrogen was lost as ammonia during composting. Gerrits *et al.* (1967) also indicated that lignin decreased rapidly after spawning. However, pentosan and cellulose decreased moderately during spawn-run through cropping. Further study showed that a major part of the lignin, estimated to be 63–92%, was utilised from spawning to the appearance of the first 'pinheads' (Gerrits, 1969). On the other hand, pentosan and cellulose decreased slowly during spawn-run but were more rapidly used when the mushroom fruit bodies were being formed. Langar *et al.* (1980) have cultivated *A. bisporus* on wheat straw and *V. volvacea* on rice straw, respectively. Hemicellulose in wheat straw was completely utilised by *A. bisporus* and approximately the same amount of rice straw hemicellulose was utilised by *V. volvacea* (Table 25.13).

Discussion

Mushrooms do not use the whole compost for nutrition and for energy needs. They seem to utilise certain organic complexes in preference to others. The complexes that the mushrooms use by preference are those which are accumulated in the composting process, e.g. cellulose, hemicellulose, lignin, and other organic complexes and some inorganic elements. A better understanding of the nutrition of mushrooms may lead to an improvement of composting technology in the future.

Wood & Fermor (1981) reviewed the nutrition of *A. bisporus* in compost where changes in degradation of compost nutrients were discussed in relation to the enzymic activity of laccase and cellulase. The significance of microbial biomass as a nutrient source for the mushroom mycelium was also examined. Microbial biomass does offer the mushroom mycelium a concentrated nutrient supply of both organic and inorganic nutrients. Gerrits *et al.* (1967) have suggested that the mushroom mycelium consumes the thermophilic biomass and the nitrogen-rich lignin–humus complex which are built up during composting processes. Fermor & Wood (1981) showed that mushroom mycelium was able to grow on 14 species from two groups of bacteria (Gram-negative and Gram-positive), including *Bacillus subtilis*. Degradation of bacteria by mushroom mycelium was determined with both light and electron microscopy. The mushroom produces a battery of enzymes which break down the carbohydrates, proteins, fats and various other compounds found in bacteria.

Eddy & Jacobs (1976) also showed the progressive build-up of an external layer of dark material on the straws taken at various stages in phase 1 composting. This layer persisted throughout phase 2 composting and could easily be scraped off from individual straws. A similar material was also found on the internal surface of straws. Microscopic examination of this deposit revealed the presence of bacterial cells, fungal spores and hyphal fragments embedded in a matrix of amorphous material. The staining characteristics of this substance resembled that of extracellular bacterial slime. Recently, an ecological succession of microorganisms on wheat straw from composting through cropping was revealed by Atkey & Wood (1983) with both transmission and scanning electron microscopy. These microscopic observations confirmed and extended previous microbiological and chemical studies of substrate preparation for mushroom production.

Once the nutritional requirement of mushrooms, including additional species of edible fungi (Farr, 1983), are determined, chemical criteria of compost quality can be established. A nutrient profile of the mushroom compost provides a guideline not only for analysing factors affecting the composting process, but also for determining applications of compost supplements at spawning and casing. In order to achieve these goals, rapid methods for compost analysis should be available for the mushroom industry.

The yield response of the mushroom to compost supplementation at spawning or thereafter varies occasionally. Differences in responses encountered were attributed to the compost formulae which contained different levels of lipids (Schisler, 1970). Interaction between ammonia and gypsum, resulting from supplementation with soybeans, was also suggested

(Gerrits, 1979). Differential response of mushroom species or even strains to compost supplement is well known. Growing environment can also alter their response (Tschierpe, 1983). Selection or breeding of strains with more efficient utilisation of compost nutrients and also strains less sensitive to environmental variables is another approach. Better understanding and effective management of composting processes are keys to overcoming the unpredictable nature of compost quality.

References

Atkey, P. T. & Wood, D. A. (1983). An electron microscope study of wheat straw composted as a substrate for the cultivation of the edible mushroom (*Agaricus bisporus*). *Journal of Applied Bacteriology*, **55**, 293–304.

Blum, H. (1977). *The Mushroom Industry in Ontario.* Canada: Economics Branch, Ontario Ministry of Agriculture and Food.

Canadian Mushroom Growers Association (1983). Mushroom growers survey 1982. *The Bulletin*, **3**, 18–19.

Crawford, J. H. (1983). Composting of agricultural wastes – a review. *Process Biochemistry*, **18** (1), 14–18.

Crop Reporting Board, Statistical Reporting Service, USDA, Washington (1983). *Mushrooms.* Vg 2-1-2 (8-83).

Eddy, B. P. & Jacobs, L. (1976). Mushroom compost as a source of food for *Agaricus bisporus. The Mushroom Journal*, **38**, 56–9, 67.

Farr, D. F. (1983). Mushroom industry: diversification with additional species in the United States. *Mycologia*, **75**, 351–60.

Fermor, T. R. & Wood, D. A. (1981). Degradation of bacteria by *Agaricus bisporus* and other fungi. *Journal of General Microbiology*, **126**, 377–87.

Finger, S. M., Hatch, R. T. & Regan, T. M. (1976). Aerobic microbial growth in semisolid matrices: heat and mass transfer limitation. *Biotechnology and Bioengineering*, **58**, 1193–218.

Finstein, M. S. (1983). Economic motives for managing the composting microbial ecosystem. *Bio/Technology*, **1**, 341–2.

Finstein, M. S., Miller, F. C., Strom, P. F., MacGregor, S. T. & Psarianos, K. M. (1983). Composting ecosystem management for waste treatment. *Bio/Technology*, **1**, 347–53.

Flegg, P. B. (1961). Mushroom composts and composting: a review of the literature. *Glasshouse Crops Research Institute, Annual Report*, **1960**, 125–34.

Flegg, P. B. & Randle, P. (1981). Relation between the initial nitrogen content of mushroom compost and the duration of composting. *Scientia Horticulturae*, **15**, 9–15.

Gerrits, J. P. G. (1969). Organic compost constituents and water utilized by the cultivated mushroom during spawn run and cropping. *Mushroom Science*, **7**, 111–26.

Gerrits, J. P. G. (1972). The influence of water in mushroom compost. *Mushroom Science*, **8**, 43–57.

Gerrits, J. P. G. (1977). The significance of gypsum applied to mushroom compost, in particular in relation to the ammonia content. *Netherland Journal of Agricultural Science*, **25**, 288–302.

Gerrits, J. P. G. (1979). Influence of pH and ammonia in mushroom compost. *Mushroom Science*, **10** (2), 15–29.

Gerrits, J. P. G. (1981). Factors in bulk pasteurization and spawn-running. *Mushroom Science*, **11** (1), 351–65.

Gerrits, J. P. G., Bels-Koning, H. C. & Muller, F. M. (1967). Changes in compost constituents during composting, pasteurization and cropping. *Mushroom Science*, **6**, 225–43.

Golueke, C. G. (1972). *Composting: a study of the process and its principles*. Emmaus, Pennsylvania: Rodale Press.

Gotaas, H. B. (1956). Composting: sanitary disposal and reclamation of organic wastes. *World Health Organization Monographs*, no. 31.

Grabbe, K. (1972). Vergleichende Untersuchungen zur Specifitaet von kompostiertem und nicht kompostiertem Champignonkultursubstrat. *Mushroom Science*, **8**, 533–52.

Gray, K. R., Biddlestone, A. J. & Clark, R. (1973). Review of composting – Part 3: Process and products. *Process Biochemistry*, **8** (10), 11–19, 30.

Gray, K. R., Sherman, K. & Biddlestone, A. J. (1971 a). A review of composting – Part 1. *Process Biochemistry*, **6** (6), 32–6.

Gray, K. R., Sherman, K. & Biddlestone, A. J. (1971 b). Review of composting Part 2 – The practical process. *Process Biochemistry*, **6** (10), 22–8.

Haug, R. T. (1980). *Compost engineering: principles and practice*. Ann Arbor, Michigan: Ann Arbor Science.

Hayes, W. A. (1977). *Composting: Improvement and Future Prospects*. Leeds: W. S. Maney & Son Ltd.

Hayes, W. A. & Randle, P. E. (1968). The use of water soluble carbohydrates and methyl bromide in the preparation of mushroom composts. *M.G.A. Bulletin*, **218**, 81–2, 87–92, 95–7.

Holtz, R. B., Smith, D. E. & Barkate, J. (1975). Lipid constituents of mushroom compost. *The Mushroom Journal*, **34**, 355–60.

Horowitz, M. A. (1979). Mushrooms: Production costs and feasibility. *Department of Agricultural Economics and Marketing, New Jersey Agricultural Experimental Station, Cook College*. A. E. 375.

Hsieh, Y.-L., Hu, K.-J. & Wu, L.-C. (1972). Chemical changes in rice straw compost during composting. *Memoirs of the College of Agriculture, National Taiwan University*, **13**, 122–31.

Ingratta, F. J. & Blom, T. J. (1980). Commercial mushroom growing. *Ontario Ministry of Agriculture and Food Publications*, no. 350.

Jeris, J. S. & Regan, R. W. (1973a). Controlling environmental parameters for optimum composting I. *Compost Science*, **14** (1), 10–15.

Jeris, J. S. & Regan, R. W. (1973b). Controlling environmental parameters for optimum composting II. *Compost Science*, **14** (2), 8–15.

Jeris, J. S. & Regan, R. W. (1973c). Controlling environmental parameters for optimum composting Part III. *Compost Science*, **14** (3), 16–22.

Kinrus, A. (1978). Different growing techniques used in mushroom growing throughout the United States. *Mushroom Science*, **10** (2), 149–58.

Lambert, E. B. (1938). Principles and problems of mushroom culture. *The Botanical Review*, **4**, 397–426.

Lambert, E. B. (1941). Studies on the preparation of mushroom compost. *Journal of Agricultural Research*, **62**, 415–22.

Lambert, E. B. & Davis, A. C. (1934). Distribution of oxygen and carbon dioxide in mushroom compost heaps as affecting microbial thermogenesis, acidity, and moisture therein. *Journal of Agricultural Research*, **48**, 587–601.

Langar, P. N., Sehgal, J. P. & Garcha, H. S. (1980). Chemical changes in wheat and paddy straws after fungal cultivation. *Indian Journal of Animal Science*, **50**, 942–6.

Nair, N. G. (1982). Substrates for mushroom production. In *Tropical Mushrooms*, ed. S. T. Chang & T. H. Quimio, pp. 47–61. Hong Kong: The Chinese University Press.

Overstijns, A. (1981). The conventional Phase 2 in trays or shelves. *The Mushroom Journal*, **97**, 5–17.

Owens, T. R., Garland, W. R., Kesecker, K. & Runyan, J. L. (1982). The U.S. mushroom industry: the import challenge. *Agricultural Cooperative Service, Agricultural Marketing Service, USDA, Marketing Research Report*, no. 1131.

Parr, J. F. & Willson, G. B. (1980). Recycling organic wastes to improve soil productivity. *HortScience*, **15**, 162–6.

Poincelot, R. P. (1975). The biochemistry and methodology of composting. *The Connecticut Agricultural Experiment Station Bulletin*, no. 754.

Randle, P. E. (1974). A review of peak-heating for mushroom composts. *The Mushroom Journal*, **22**, 388–93.

Randle, P. E. & Flegg, P. B. (1978). Oxygen measurements in a mushroom compost stack. *Scientia Horticulturae*, **8**, 315–32.

Ross, R. C. & Harris, P. J. (1982). Some factors involved in Phase II of mushroom compost preparation. *Scientia Horticulturae*, **17**, 223–9.

Ross, R. C. & Harris, P. J. (1983*a*). An investigation into the selective nature of mushroom compost. *Scientia Horticulturae*, **19**, 55–64.

Ross, R. C. & Harris, P. J. (1983*b*). The significance of thermophilic fungi in mushroom compost preparation. *Scientia Horticulturae*, **20**, 61–70.

Royse, D. J. & Schisler, L. C. (1980). Mushrooms. *Interdisciplinary Science Reviews*, **5**, 324–32.

Schisler, L. C. (1970). Supplementation of mushroom compost. *Mushroom News*, **18** (4), 13–18.

Schisler, L. C. (1980). Composting. *Mushroom News*, **28** (1), 5–13.

Schisler, L. C. & Patton, T. G. (1970). Stimulation of mushroom yield by supplementation with vegetable oils before Phase II of composting. *Journal of the American Society for Horticultural Science*, **95**, 595–7.

Sinden, J. W. & Hauser, E. (1950). The short method of composting. *Mushroom Science*, **1**, 52–9.

Sinden, J. W. & Hauser, E. (1953). The nature of the composting process and its relation to short composting. *Mushroom Science*, **2**, 123–31.

Smith, J. F. (1981). The development of mushroom composting techniques – a review. *Glasshouse Crops Research Institute, Annual Report*, **1980**, 171–83.

Stoller, B. B. (1954). Principles and practices of mushroom culture. *Economic Botany*, **8**, 48–95.

Tschierpe, H. J. (1973). Environmental factors and mushroom growing. *The Mushroom Journal*, **1**, 30–45; **2**, 77–94.

Tschierpe, H. J. (1983). Environmental factors and mushroom strains. *The Mushroom Journal*, **132**, 417–29.

Vedder, P. J. C. (1979). Shelf plants. *Mushroom News*, **27** (12), 16–29.

Waksman, S. A. & Nissen, W. (1932). On the nutrition of the cultivated mushroom *Agaricus campestris*, and the chemical changes brought about by this organism in the manure compost. *American Journal of Botany*, **19**, 514–37.

Willson, G. B., Epstein, E. & Parr, J. R. (1976). Recent advances in compost technology. In *Proceedings of the 3rd National Conference in Sludge Management and Disposal and Utilization*, pp. 167–72. Silver Springs, Maryland: Hazardous Materials Control Research Institute.

Wood, D. A. & Fermor, T. R. (1981). Nutrition of *Agaricus bisporus* in compost. *Mushroom Science*, **11**, 63–71.

Wu, L.-C. & Stahmann, M. A. (1975). Fungal protein. In *Papers from a workshop on unconventional sources of protein*, pp. 67–104. Madison, Wisconsin: College of Agricultural and Life Sciences, University of Wisconsin.

Wuest, P. J. (1983). Resources needed to form the 'Champignon'. *Mycologia*, **75**, 341–50.

Wuest, P. J. & Bengtson, G. D. (1982). *Penn State Handbook for Commercial Mushroom Growers:* University Park, Pennsylvania: College of Agriculture, Pennsylvania State University.

Zaayen, A. van & Gerrits, J. P. G. (1982). Hygiene in the preparation and the use of full-grown spawn-run compost in the Netherlands. *The Mushroom Journal*, **112**, 115–25.

26

Secondary metabolic products of selected agarics

NORMAN CLAYDON

Glasshouse Crops Research Institute, Worthing Road, Littlehampton, West Sussex, BN17 6LP, UK

Introduction

The phenomenon of secondary metabolism, widely encountered in fungi and higher plants, has long excited interest. Many secondary metabolites exhibit biological activity and some, such as antibiotics and mycotoxins, are of considerable economic importance. A vast number of fungal secondary metabolites have been recorded but only a relative few impinge upon economic activity; the rest remain largely as curiosities for the natural-product chemist.

Secondary metabolism is an enigmatic process since its occurrence may be strongly influenced by environmental factors and the metabolites, where produced, generally cannot be ascribed to any functional role in the biology of the fungus producing them. A particular metabolite or a series of analogues may be produced by a group of closely related fungi but species or even strain specificity is more commonly encountered. Secondary metabolism usually occurs after a period of hyphal growth has taken place and environmental constraints begin to apply. Morphogenetic changes may be coincident, especially in the formation of reproductive units and structures, and the natural speculation ensues that particular metabolites could play a crucial role in differentiation. This does, however, remain a largely speculative concept. Whatever the *raison d'être* for secondary metabolism it remains that a huge diversity of natural substances are excreted into the surroundings or may accumulate within the mycelium and other structures. The phenomenon of intracellular accumulation is of particular interest where the fungus concerned may be subject to human or domestic animal consumption.

The collection of fungi for food (Singer, 1961) and as a source of drugs (Wasson, 1962) has paralleled the history of mankind and has provided

561

a legacy of unreliable guidelines to edibility. Many orders of higher fungi, with their elaborate fruiting structures, have found widespread culinary or pharmacological favour, but the order Agaricales, containing so rich a diversity of mushrooms, is without doubt the most popular. The order contains, in addition to the mushrooms of commercial cultivation, a whole array of wild species, many of which are highly prized delicacies, a few notoriously hazardous. The capacity of fungi to produce biologically active secondary metabolites is well illustrated in the Agaricales and edibility does not imply the absence of such metabolites.

Agaricus bisporus

The white mushroom, *Agaricus bisporus*, is cultivated globally and, apart from baker's yeast, is the most commonly consumed fungus. For production on a commercial scale the medium used for mycelial growth is composted wheat straw. The composting process is a controlled microbial succession which produces a selective culture medium for the slow-growing mycelium of *A. bisporus* (Wood, 1979). This ability to colonise and dominate a non-sterile substrate has often prompted specu-lation into the possibility of antibiotic production and to some extent evidence has been presented to support this (Allendes, 1969; Zadrazil & Grabbe, 1983). Certain metabolites of *A. bisporus* have been isolated and antibiotic potential attributed to derivatives or analogues of them, but to date no specific antibiotics have been reported. Grove (1981) obtained from the steam distillate of *A. bisporus* mycelium terachloro-1, 4-dimethoxybenzene (drosophilin A methyl ether) (Fig. 26.1, R = CH$_3$), a known metabolite of several basidiomycete species (Singh & Rangaswami, 1966). The analogue tetrachloro-4-methoxyphenol (drosophilin A) (Fig. 26.1, R = H), also a known metabolite of basidiomycetes, shows antibiotic

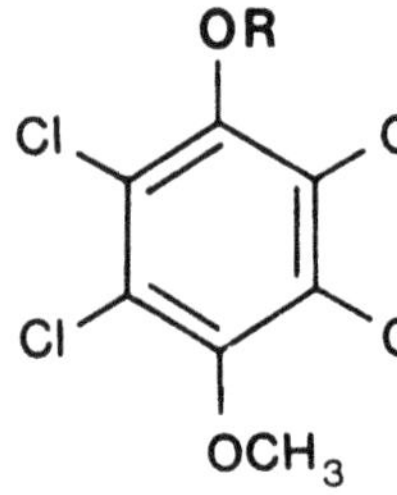

Fig. 26.1. Antibiotic metabolites of *Agaricus bisporus* and other basidio-mycetes; Drosophilin A methyl ether (R = CH$_3$), Drosophilin A (R = H).

activity against certain bacteria (Anon., 1968) but whether this activity is present in the methyl ether is not known. Bactericidal properties of an extract of freshly harvested fruit bodies were reported by Vogel *et al.* (1974).

Two substances associated with the pink-to-brown pigmentation change in the gill tissues were described as quinonoids with visible absorption maxima at 360 nm and 490 nm, respectively. The substances were active against a range of bacteria and one, termed the 490 quinone, was considered to be γ-glutaminyl-3,4-benzoquinone (GBQ) after earlier work by Weaver *et al.* (1971 *a, b*). Weaver and his associates had isolated a stable crystalline glutamate derivative from gill tissue, γ-L-glutaminyl-4-hydroxybenzene (GHB) (Fig. 26.2, R = glutamate, R_1 = H).

This compound had previously been described (Jadot *et al.*, 1960) as *p*-hydroxy(γ-L-glutamyl)-anilide, a metabolite of *Agaricus hortensis*. GHB was claimed to be a precursor of GBQ, the pinkish-red pigment found in copious quantities in young gills. Weaver contended that an enzymatic oxidation of GHB produced GBQ and proposed the 3,4-benzoquinone

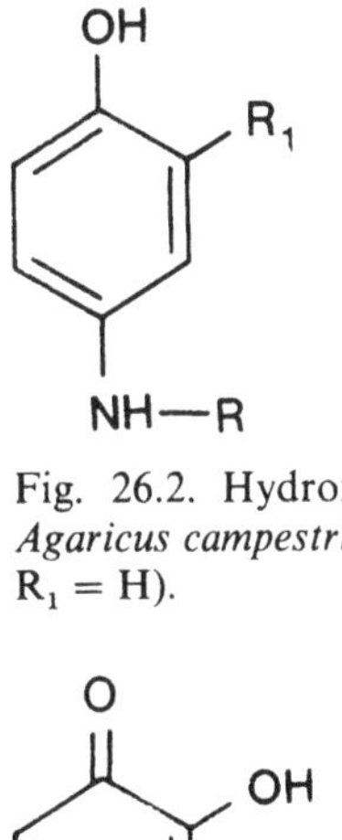

Fig. 26.2. Hydroxybenzenoid metabolites of *Agaricus bisporus* and *Agaricus campestris*. γ-L-glutaminyl-4-hydroxybenzene (R = glutamate, R_1 = H).

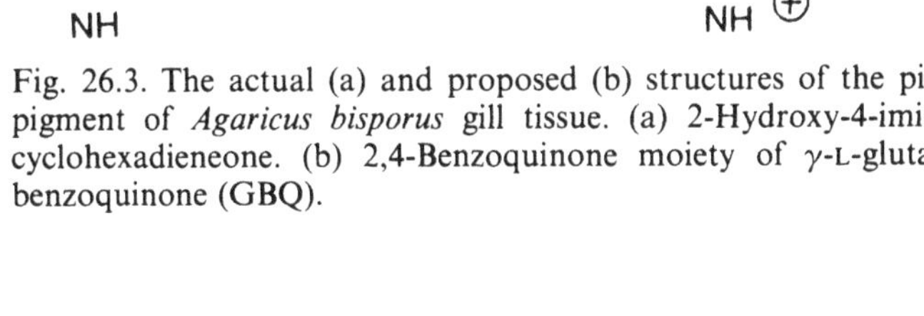

Fig. 26.3. The actual (a) and proposed (b) structures of the pink–red pigment of *Agaricus bisporus* gill tissue. (a) 2-Hydroxy-4-imino-2,5-cyclohexadieneone. (b) 2,4-Benzoquinone moiety of γ-L-glutaminyl-benzoquinone (GBQ).

structure (Fig. 26.3*b*). This was, however, an incorrect structural assignment as synthetically produced GBQ had an absorption maximum of 440 nm and not 490 nm (Tiffany *et al.*, 1978). The true structure was shown to be 2-hydroxy-4-imino-2,5-cyclohexadieneone (Fig. 26.3*a*) (Mize *et al.*, 1980).

Several papers reporting the biological properties of the '490 quinone' appeared before the true structural identity had been established and these all relied upon the use of GBQ as the identifying label. Observations that GBQ strongly inhibited the succinate dehydrogenase (SDH) of rat liver mitochondria (Weaver *et al.*, 1972) led to the hypothesis that GBQ and a related quinone were inducing agents of dormancy in *A. bisporus* spores (Weaver *et al.*, 1970; Vogel & Weaver, 1972; Vogel *et al.*, 1974). SDH activity in the mitochondria isolated from presporing gill tissue was also found to be inhibited by the GBQ-containing supernatant from homogenised lamellae. The degree of inhibition increased with the age of the gills and *in vivo* would gradually decrease the respiration rate of the gills to produce zero respiration in the spores. This hypothesis has been challenged (Rast *et al.*, 1979) in an elegant argument against GBQ involvement in spore dormancy. GBQ, it is claimed, is an artifact appearing as a consequence of the extraction procedure and Rast presented evidence that dormancy, due to SDS inhibition, was brought about by a carbon dioxide fixation product which reversibly interfered with the citric acid cycle. GBQ was never detected in the spores. The findings of the Rast group throw into doubt the conclusions reached by other workers concerning the possible roles of GHB-derived 'quinones'. One such role, related to plant-virus inhibition (Tavantzis & Smith, 1982) was based on the observation that the virus-like particles causing La France disease of mushrooms occurred at higher concentration in the spawn than in the fruit bodies derived from the same spawn. Extracted fruit bodies yielded GBQ which underwent further oxidation to form a substance identical in its properties to the '360 quinone' described by Vogel *et al.* (1974). The substance was found to have potent systemic inhibitory properties against tobacco mosaic and tobacco ringspot viruses in pinto bean and cowpea plants. In the light of Rast's work it is uncertain if such metabolites actually occur *in vivo* to act as mushroom virus inhibitors.

A catecholamine analogue of GHB has been obtained from *Agaricus campestris* and identified as 3,4-dihydroxy(γ-L-glutaminyl) anilide (Szent-Gyorgyi *et al.*, 1976) (Fig. 26.2, R = glutamate, R_1 = OH) and trivially named agaridoxine. In a recent report (Wheeler *et al.*, 1982) it is claimed that agaridoxine acts as an α-1 agonist of mammalian hypothalamic adenylate cyclase. Nothing else is known of the biological properties of this compound.

Several years prior to Weaver's isolation of GHB, another phenolic derivative of glutamic acid was isolated from the fruit bodies of *A. bisporus* (Levenberg, 1961, 1964). The metabolite was obtained in a pure state from the press juices of a commercial strain of the mushroom and tentatively identified as *β*-*N*-[*γ*-L(+)-glutamyl]-4-hydroxymethylphenylhydrazine. The trivial name agaritine was adopted and confirmation of the structure shortly followed (Daniels *et al.*, 1961) (Fig. 26.4, R = glutamate).

The age of the fruit body seemed to have a bearing upon the amount of agaritine obtained since young fruit bodies yielded relatively higher titres that subsequently diminished with age (Kelly *et al.*, 1962; Chiarlo & Acerbo, 1979). In addition to agaritine, Levenberg (1962) described the occurrence of 4-hydroxymethylbenzenediazonium ions (HMB) (Fig. 26.5a) in the basal stalk of the fruit body. The ion was not detected in the cap. The ability of *A. bisporus* to produce agaritine appears to be associated with the presence of an enzyme, *γ*-glutaminyl transferase (Gigliotti & Levenberg, 1964) which catalyses the irreversible hydrolysis of the *γ*-glutaminylhydrazide bond of agaritine *in vitro* to produce glutamate and 4-hydroxymethylphenylhydrazine(HMPH) (Fig. 26.4, R = H). The enzyme also possesses a *γ*-glutaminyl-transferring activity catalysing reactions involving arylhydrazines and their *γ*-glutaminyl derivatives as the acceptor and donor substrates, respectively. Agaritine will spontaneously undergo a similar cleavage to produce HMPH under mildly acidic conditions and is therefore only stable at or near neutrality.

The occurrence of hydrazines in nature is rare so it is of interest to note that compounds of this class are found in three species of edible fungi.

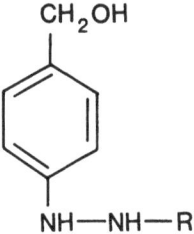

Fig. 26.4. Agaritine (R = glutamate): the hydroxymethylphenylhydrazine derivative of glutamic acid obtained from *Agaricus bisporus*. 4-hydroxymethylphenylhydrazine (HMPH) (R = H) and *N*-acetyl-HMPH (R = —oAc).

Agaritine has been obtained from both *A. bisporus* and the Japanese forest mushroom *Lentinus edodes* (*shiitake*) and a series of *N*-methyl,*N*-formyl hydrazones have been isolated from the false morel *Gyromitra esculenta* (Toth, 1979). The first of these to be isolated, gyromitrin (Fig. 26.6, R = CH$_3$CH) (List & Luft, 1967), serves as a useful model to demonstrate the release of highly toxic monomethylhydrazine from gyromitrin and its higher homologues.

During the latter years of the 1970s, Toth and various co-workers conducted studies on the carcinogenic potencies of hydrazine analogues known to be widespread in the environment as either natural or synthetic substances. Included in these studies were the agaric phenylhydrazines HMPH and HMB (Toth, 1979). HMPH is a relatively unstable compound *in vitro*, seemingly due to the facile 1,4 elimination of water across the phenyl ring (Toth *et al.*, 1978). Under reducing conditions HMPH undergoes dehydration and reduction to form 4-methylphenylhydrazine (MPH) (Fig. 26.7) but the significance of this *in vivo* is not obvious. Toth stabilised the HMPH by acetylation and examined the *N*-acetyl derivative (Fig. 26.4, R = OAc) and MPH hydrochloride for carcinogenesis in mice. Both substances were administered orally for the lifespan of albino mice and were related to the significant incidence of a variety of tumours found in different tissues. Lung and blood vessel tumours seemed to be particularly prevalent. Subcutaneous injections of MPH also resulted in a high

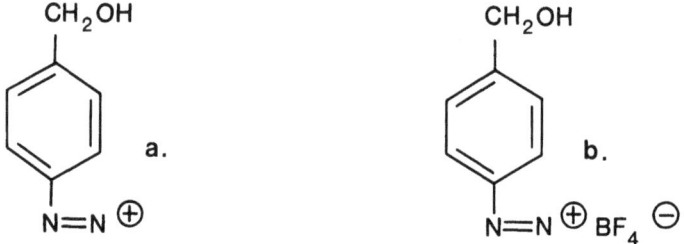

Fig. 26.5. Hydroxymethylbenzenediazonium ion (a): a potential carcinogen from the stipe of *Agaricus bisporus*. (b) The ion administered in toxicity tests in boron tetrafluoride solution.

Fig. 26.6. Gyromitrin (R = CH$_3$CH): a metabolite of *Gyromitra esculenta* which upon hydrolysis produces the toxic monomethylhydrazine.

incidence of tumours at the injection sites (Toth & Nagel, 1981) and it was postulated that a corresponding diazonium ion may have been formed locally and was ultimately responsible for cancer induction. Levenberg (1962) had suggested that the 4-hydroxymethylbenzenediazonium ion may be formed enzymatically from agaritine *in vivo* and this was later supported by Toth *et al.* (1981 *a*). Toth's group studied the carcinogenic potential of synthetic HMB, injected subcutaneously in boron tetrafluoride (Fig. 26.5 *b*). This, too, resulted in a significant incidence of local tumours. Despite the carcinogenicity of its derivatives, agaritine has never been shown to induce tumours. Experiments involving the lifespan oral feeding of agaritine to albino mice did not result in any carcinogenic activity but a substantial number of the mice developed convulsive seizures (Toth *et al.*, 1981 *b*).

Lentinus edodes

'Eat these mushrooms and you will live forever', was the rather extravagant claim in an American popular newspaper in response to an interview with Dr Henry Mee. Dr Mee had developed a relatively rapid process of cultivating the mushrooms of *L. edodes* but it is most doubtful that he would agree with the immortality factor (Mee, 1978). The fungus nevertheless has some remarkable properties.

L. edodes or shiitake has been grown for centuries in Asian forests and legend has it that Chinese emperors consumed the delicacy in large quantities to fend off old age. Other accounts tell us that ancient Japanese courts valued shiitake for its aphrodisiac qualities. The growing sites were well hidden and heavily guarded. Today, shiitake is widely cultivated in the Far East and the mushroom has excited a good deal of scientific interest. In keeping with the legendary elixir qualities of shiitake, antitumour, antiviral and hypolipidemic agents have all been isolated from the

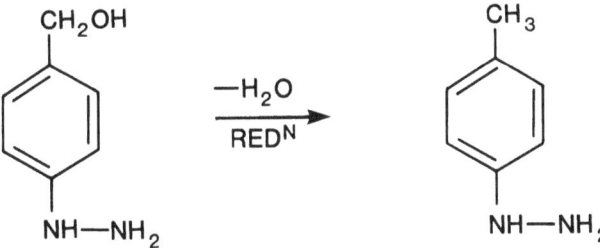

Fig. 26.7. Dehydration of hydroxymethylphenylhydrazine under reducing conditions to produce methylphenylhydrazine.

fruit bodies. The antitumour properties are attributed to polysaccharides: lentinans (Chihara *et al.*, 1970) and emitanin 1 (Yamamoto & Ikegawa, 1980), and the antiviral properties to an interferon-inducing double-stranded RNA (Suzuki *et al.*, 1974). Of more relevance to this review are the hypolipidemic properties, directly attributable to a secondary metabolic product. Several edible fungi have the ability to lower the levels of lipids in the blood but shiitake is particularly effective (Kaneda & Tokuda, 1966). Rats fed for 10 weeks on a diet supplemented with 5% ground dried fruit bodies showed a marked reduction of plasma cholesterol levels over the controls. Reductions in the order of 24% were recorded. Hypocholesterolemic effects have also been reported in man (Suzuki & Oshima, 1974) but the age of the individuals concerned seemed to influence efficacy. The metabolite in shiitake responsible for hypolipidemic activity has been identified as 2(R),3(R)-dihydroxy-4-(9-adenyl)-butyric acid and given the trivial name eritadenine (Fig. 26.8) (Chibata *et al.*, 1969; Kamiya *et al.*, 1969, 1972). The names lentysine and lentinacin have also been applied to this compound. Extensive bioassays revealed that eritadenine was effective in a range of mammals and birds and that plasma triglyceride and phospholipid levels were affected in addition to cholesterol levels (Takashima *et al.*, 1973). The action attributed to eritadenine in affecting cholesterol levels was the promotion of accelerated uptake by the tissues

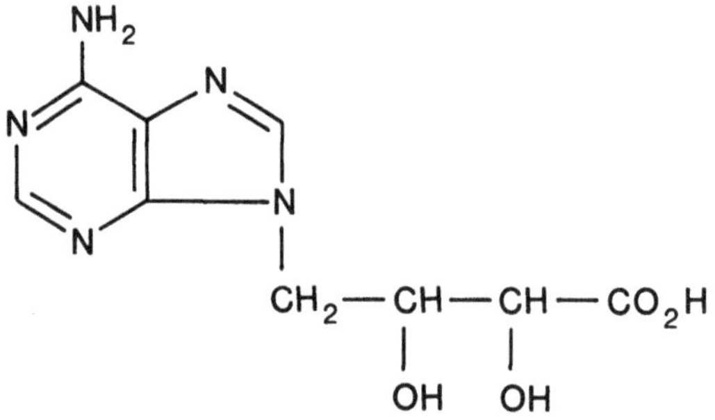

Fig. 26.8. D-Eritadenine: a hypocholesterolemic metabolite of *Lentinus edodes*.

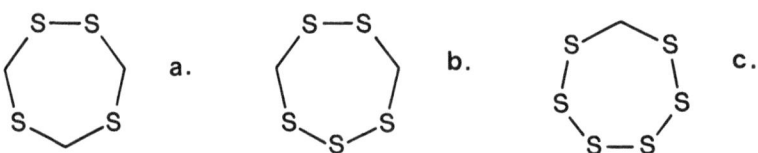

Fig. 26.9. Cyclic methylene polysulphides of *Lentinus edodes*.

(Takashima *et al.*, 1974) together with increased metabolism and excretion (Tokuda & Kaneda, 1976).

The two centres of asymmetry in the side chain of eritadenine allow for the possibility of three further stereoisomers and these have been synthesised (Hashimoto *et al.*, 1972). Hashimoto *et al.* commented that the stereochemistry of the side chain of this class of compound may be a factor of major importance in biological activity. Indeed, they found that the D-threo and L-threo isomers had less hypocholesterolemic activity than the natural D-eritadenine but more activity than the L isomer. A closely related compound, 2(R)-hydroxy-4-(9-adenyl)-butyric acid, was also isolated from shiitake but was found to possess only weak hypolipidemic activity (Tokita *et al.*, 1971).

Shiitake mushrooms have been said to produce a sulphury aftertaste and this observation is not without foundation. Three cyclic methylene polysulphides have been isolated from the fruit bodies (Fig. 26.9) (Morita & Kobayashi, 1966, 1967), and one, lenthionine, 1,2,3,5,6-pentathiepene (Fig. 26.9b), shows strong antibiotic activity against a range of bacteria and fungi.

Psilocybe mexicana

Several agaric species have been recorded as producers of hallucinogenic metabolites and this subject has already been extensively reviewed. The account presented here is but a brief synopsis, so for those whose interests require greater detail the works of Wasson (1962), Lincoff & Mitchel (1977), Rumack & Salzman (1978) and Shultes & Hofmann (1979) are recommended. *Psilocybe mexicana* and several other related species have in recent decades acquired a notoriety, particularly in Europe and North America, as convenient sources of psychoactive drugs: drugs which may produce a range of effects from alcoholic-type intoxications to hallucination and delirium. Whilst induction of hallucinations is largely a function of elevated doses of the active ingredient, lower doses may, for significant numbers of people, produce the desired effects of other social drugs. The source is convenient since consumption of the fresh, cooked or dried mushrooms is all that is required. The deliberate consumption of the mushrooms for this purpose is not merely a recent social problem but a phenomenon extending into the customs and divine practices of ancient civilisations.

The hallucinogenic metabolites of *P. mexicana* were isolated and structurally determined by Hofmann and his associates (Hofmann *et al.*, 1959). They were found to be *N*-methylated tryptamines and named

psilocybin and psilocin (Fig. 26.10). The pharmacological action of psilocin and its phosphate ester psilocybin are so similar, in both quantitative
and qualitative terms, that they may be described as one. Oral doses in
the order of 4–5 mg of psilocybin produce subjective changes in psychic
state after about 30 minutes and there is generally a pleasant sensation of
bodily relaxation and detachment. Higher doses of 6–20 mg, depending on
individual tolerance, produce profound psychic changes involving distor-
tions of space and time perception and changes in the awareness of the
self. These doses may be easily achieved by consumption of the fruit bodies.
The effects described, it must be assumed, are precisely those sought by
people who knowingly consume psilocibian mushrooms, since the biological activity of the indoles would preclude any normal culinary use. Consumption of the mushrooms does not, however, guarantee production of
the desired effects since individual susceptibility varies. Undesirable effects,
such as nausea, weakness, tremors, malaise and numerous others, are
frequently experienced (Lampe, 1978). Illusions and hallucinations may
also be experienced (Chilton, 1978), effects which closely resemble those
produced by mescaline, LSD and certain amphetamines (Hollister, 1972).
The conformational similarity between the hallucinogens and the biogenic
amines of the central nervous system has led to speculation on the
structure–activity relations of these substances but this, however, is still
uncertain.

The distribution of psilocybin is predominantly confined to *Panaeolus*
and *Psilocybe* species, but species of *Conocybe*, *Stropharia* and *Gymnopilus*
have also been associated with the indole. Chilton (1978) lists 20 *Psilocybe*
species reported to have produced psilocybin. The list includes *P. baeocystis*,
submerged cultures of which have been found to produce two analogues
of psilocybin: the unmethylated analogue, norbaeocystin, and the mono
N-methyl analogue, baeocystin (Fig. 26.10) (Leung & Paul, 1968). Baeo-

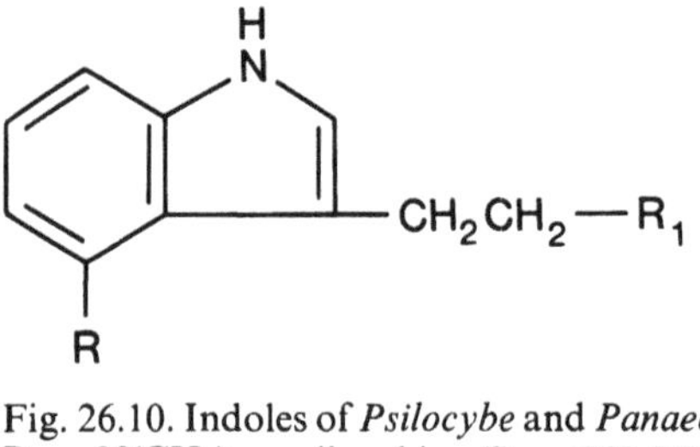

Fig. 26.10. Indoles of *Psilocybe* and *Panaeolus* species: psilocin (R = OH,
$R_1 = N(CH_3)_2$, psilocybin (R = OH_2PO_3, $R_1 = N(CH_3)_2$, baeocystin
(R = OH_2PO_3, $R_1 = NH_2CH_3$), norbaeocystin (R = OH_2PO_3,
$R_1 = NH_3$).

cystin has also been detected in *P. semilanciata* fruit bodies (Repke & Leslie, 1977).

The pharmacological action of baeocystin and norbaeocystin is not known, but a report of poisoning after consumption of an alleged *P. baeocystis* fruit body described many of the adverse effects otherwise attributed to psilocybin intoxication. Hallucination and delirium were not, however, reported (McCawley *et al.*, 1962). Some doubt does exist over the identity of the mushrooms consumed.

Amanita muscaria

Hallucinogenic properties have long been attributed to the fly agaric, *Amanita muscaria*, and the ceremonial rites of Siberian tribes, based upon the intoxicating properties of this mushroom, have been vividly described. Wasson's (1968) monograph, *Soma: Divine Mushroom of Immortality*, and Ramsbottom's (1953) *Mushrooms and Toadstools* are splendid accounts.

The hallucinogenic Amanitae are, like *Psilocybe*, ingested either fresh or dried, and diverse pharmacological action may ensue. The gross pharmacological properties, recorded in detail since the middle of the last century, are due to at least two distinct groups of compounds. The first to be isolated and structurally determined was muscarine (Fig. 26.11) (Kogl *et al.*, 1957) and three further isomers were subsequently synthesised. The L-(+)-isomer (Fig. 26.11) is by far the most active and produces the characteristic symptoms of muscarine poisoning: pupil contraction, blurred vision, perspiration, reduced heart rate and blood pressure, and asthmatic breathing. These symptoms are all due to the effect of muscarine on the cholinergic receptors of the nervous system.

Psychotropic activity has not been attributed to muscarine and the hallucinatory properties of *A. muscaria* rest with other metabolites which have the capability of acting upon the central nervous system. It should be added that although muscarine is found in a variety of *Amanita* species (Tyler, 1963), its concentration is usually so low that muscarinic effects are

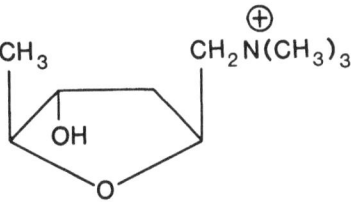

Fig. 26.11. Muscarine: a metabolite obtained from diverse agaric species.

rarely seen during poisoning by these species. Other agarics, particularly *Inocybe* and *Clitocybe*, accumulate far higher concentrations of the alkaloid (Brown *et al.*, 1962; Genest *et al.*, 1968; Catalfomo & Eugster, 1970) and it has been suggested that inocybin would be a more appropriate name than muscarine.

Investigations into the hallucinogenic metabolites of *A. muscaria* were coupled with similar research into insecticidal metabolites. *A. muscaria* has long been known in Europe for its fly-killing properties and similarly, in Japan, *A. pantherina* and *Tricholoma muscarium* have been used for the same purposes. This work led to simultaneous discoveries in Europe and Japan of a group of iso-oxazoles which accounted for the non-muscarinic activity demonstrated by these fungi (Takemoto & Nakajima, 1964; Takemoto *et al.*, 1964; Bowden *et al.*, 1965; Bowden & Drysdale, 1965; Muller & Eugster, 1965). A variety of synonyms were applied to these compounds but the names muscimol, ibotenic acid, muscazone and tricholomic acid were ultimately adopted (Eugster & Takemoto, 1966) (Fig. 26.12). Each of these compounds has been examined for its effects in man and other animals, and muscimol, ibotenic acid and muscazone appear to be qualitatively similar (Waser, 1967; Johnston *et al.*, 1968;

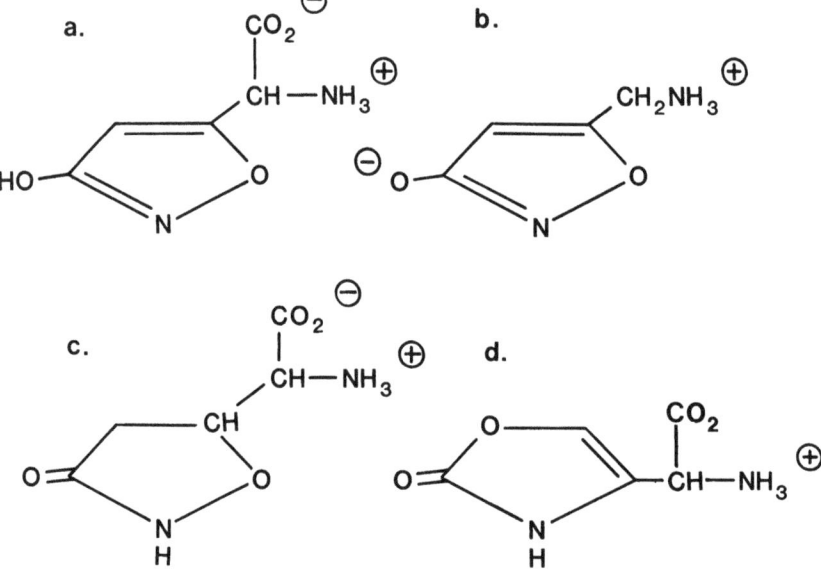

Fig. 26.12. Hallucinogenic and insecticidal metabolites obtained from *Amanita* and *Tricholoma* species: (a) ibotenic acid, (b) muscimol, (c) tricholomic acid, (sd) muscazone.

Koenig-Bersin *et al.*, 1970). Muscimol, however, shows the greatest potency and seems to occur in higher concentrations. Quantitative data on the relative amounts of muscimol and ibotenic acid in mushrooms are difficult to assess from the literature since there is considerable diversity in assay methods. Additionally, experiments to determine the effects of ibotenic acid are complicated due to the ease of decarboxylation of the compound to muscimol. It may be that a significant portion of the observed activity of administered ibotenic acid may be due to muscimol generated *in situ* (Chilton, 1978).

The insecticidal properties of *A. muscaria* and *A. pantherina* are due to the presence of ibotenic acid, whereas similar properties of *Tricholoma muscarium* derive from the analogue dihydroibotenic acid or tricholomic acid (Fig. 26.12c) (Takemoto *et al.*, 1964). Whilst experiments have shown that the insecticidal activity of tricholomic acid is undoubtedly superior to that of ibotenic acid, details of mammalian toxicity have not been forthcoming.

Coprinus atramentarius

Coprinus atramentarius has long been sought as a delicacy and appears to be a harmless mushroom producing no gastric or other discomforts. If, however, any alcoholic beverage is consumed with the mushroom or even up to 48 h later, one or more of several symptoms may be experienced. These include: flushing, metallic taste, paraesthesia, palpitations, hyperventilation, hypotension, nausea, and vomiting (Hatfield & Schaumberg, 1978). Similar effects result from a combination of alcohol with the drug disulfiram (used as an attempt to cure alcoholism) and this observation led to widespread belief that disulfiram was the active ingredient in the mushroom. Simandl & Franc (1956) claimed to have made the isolation but doubt was cast upon the claim when the findings could not be confirmed (Weir & Tyler, 1960).

A study of the effects of *C. atramentarius* on ethanol metabolism in mice

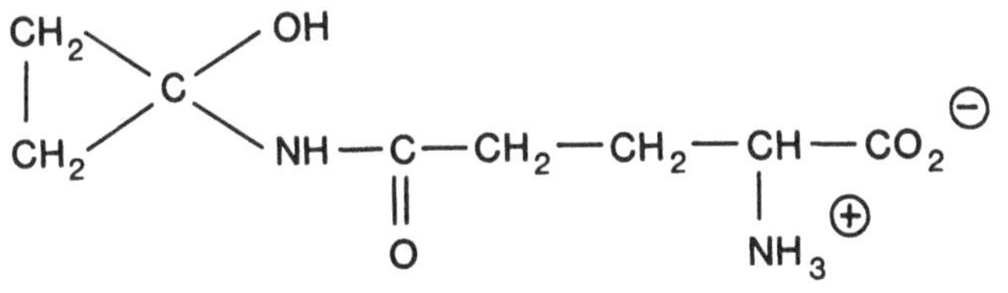

Fig. 26.13. Coprine: a cyclopropanol derivative of glutamine from *Coprinus atramentarius*.

(Coldwell *et al.*, 1969) revealed that after administration of both ethanol and an extract of the fungus, high levels of acetaldehyde were found in the blood. This implied that the mushroom in some way interfered with ethanol metabolism, possibly with aldehyde dehydrogenase. A constituent of the mushroom was isolated from a water-soluble fraction and purified by various chromatographic procedures. The purified fraction, named coprine, was the only component with significant disulfiram-like activity. The structure was determined and the substance was shown to be a cyclopropane derivative of glutamine: (*N*-[1-hydroxycyclopropyl])glutamine (Fig. 26.13) (Hatfield & Schaumberg, 1975; Lindberg *et al.*, 1975, 1977).

The hypersensitivity to ethanol produced by this compound appears to be due to hyperaldehydaemia and a biochemical investigation into the mode of action of coprine was made (Hatfield & Schaumberg, 1978). Disulfiram was known to be an active inhibitor of aldehyde dehydrogenase both *in vivo* and *in vitro* but coprine had no such action on the crude enzyme preparations from mouse liver. It was suggested that coprine may be metabolised *in vivo* to form an active inhibitor. This was confirmed by Tottmar & Lindberg (1977) who showed that metabolic activation of coprine to produce 1-aminocyclopropanol (Fig. 26.14) was indeed necessary to bring about the enzyme inhibition.

Discussion

The foregoing account highlights the capacity of a few selected agarics to produce and concentrate specific substances within their fruit bodies. The role of these substances, if indeed a role exists, remains obscure. All are produced in biosynthetic pathways distinct from, but closely linked to, the primary metabolic processes common to all fungi. The dividing line between primary and secondary metabolism is not however always clear. That the metabolites should show biological activity of one kind or another in organisms far removed from the organism of origin must be fortuitous, despite an extensive literature filled with speculation on evolutionary significance. One proposal suggests that

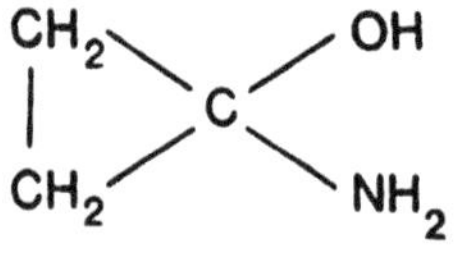

Fig. 26.14. 1-Aminocyclopropanol: an aldehyde dehydrogenase inhibitor derived from coprine.

secondary metabolism should be viewed as a form of chemical differentiation analogous to morphological differentiation and offers an attractive alternative to many of the more specific theories embracing the subject. It may be that further studies will reveal a closer link between chemical and morphological differentiation, particularly with secondary metabolites not too far removed in structure from their primary metabolic precursors.

References

Allendes, Y. P. (1968). Antibacterial activity of the higher fungi of Chile. *Anales de la Facultad de Quimica y Farmacia, Universidad de Chile*, **18**, 106–13.

Anon. (1968). Drosophilin A. *The Merck Index* (*Eighth edition*). Merck and Co.: New Jersey.

Bowden, K. & Drysdale, A. C. (1965). A novel constituent of *Amanita muscaria*. *Tetrahedron Letters*, **12**, 727–8.

Bowden, K., Drysdale, A. C. & Mogey, G. A. (1965). Constituents of *Amanita muscaria*. *Nature*, **206**, 1359–60.

Brown, J. K., Malone, M. H., Stuntz, D. E. & Tyler, V. E. (1962). Paper chromatographic determination of muscarine in *Inocybe* species. *Journal of Pharmaceutical Science*, **51**, 853–6.

Catalfomo, P. & Eugster, C. H. (1970). Muscarine and muscarine isomers in selected *Inocybe* species. *Helvetica Chimica Acta*, **53**, 848–51.

Chiarlo, C. E. & Acerbo, C. (1979). The presence of agaritine in a mushroom (*Agaricus bisporus*) commonly cultivated in Italy. *Fitoterapia*, **50**, 111–13.

Chibata, I., Okumura, K., Takeyama, S. & Kotera, K. (1969). Lentinacin: a new hypocholesterolemic substance in *Lentinus edodes*. *Experientia*, **25**, 1237–8.

Chihara, G., Hamuro, J., Maeda, Y., Arai, Y. & Fukuoka, F. (1970). Fractionation and purification of the polysaccharides with marked antitumour activity, especially lentinan from *Lentinus edodes* (an edible mushroom). *Cancer Research*, **30**, 2776–81.

Chilton, W. S. (1978). Chemistry and mode of action of mushroom toxins. In *Mushroom Poisoning: Diagnosis and Treatment*, ed. B. H. Rumack & E. Salzman, pp. 88–117. West Palm Beach, Florida: CRC Press.

Coldwell, B. B., Genest, K. & Hughes, D. W. (1969). Effect of *Coprinus atramentarius* on the metabolism of ethanol in mice. *Journal of Pharmacy and Pharmacology*, **21**, 176–9.

Daniels, E. G., Kelly, R. B. & Hinman, J. W. (1961). Agaritine: an improved isolation procedure and confirmation of structure by synthesis. *Journal of the American Chemical Society*, **83**, 3333–4.

Eugster, C. H. & Takemoto, T. (1966). Nomenclature of new compounds from *Amanita* species. *Helvetica Chimica Acta*, **50**, 126–7.

Genest, K., Hughes, D. W. & Rice, W. B. (1968). Muscarine in *Clitocybe* species. *Journal of Pharmaceutical Science*, **57**, 331–3.

Gigliotti, H. J. & Levenberg, B. (1964). γ-Glutaminyltransferase of *Agaricus bisporus*. *Journal of Biological Chemistry*, **239**, 2274–84.

Grove, J. F. (1981). Volatile compounds from the mycelium of the mushroom *Agaricus bisporus*. *Phytochemistry*, **20**, 2021–2.

Hashimoto, M., Saito, Y., Seki, H. & Kamiya, T. (1972). Hypocholesterolemic alkaloids of *Lentinus edodes* IV. Synthesis of three stereoisomers of eritadenine. *Chemical Pharmacology Bulletin*, **20**, 1374–9.

Hatfield, G. M. & Schaumberg, J. P. (1975). Isolation and structural studies of coprine, the disulfiram-like constituent of *Coprinus atramentarius*. *Lloydia*, **38**, 489–96.

Hatfield, G. M. & Schaumberg, J. P. (1978). The disulfiram-like effects of *Coprinus atramentarius* and related mushrooms. In *Mushroom Poisoning: Diagnosis and Treatment*, ed. B. H. Rumack & E. Saltzman, pp. 181–6. West Palm Beach, Florida: CRC Press.

Hofmann, A., Heim, R., Brack, A., Kobel, H., Frey, A., Ott, H., Petrzilka, T. & Troxler, F. (1959). Psilocybin and psilocin. *Helvetica Chimica Acta*, **42**, 1557–72.

Hollister, L. E. (1972). Clinical pharmacology of hallucinogens and marijuana in drug abuse. In *Proceedings of the International Conference on Drug Abuse, Ann Arbor, Michigan*, ed. C. J. Zaraphonetis, pp. 118–27. Philadelphia: Lea and Febiger.

Jadot, J., Casimir, J. & Renard, M. (1960). Séparation et characterisation du L-(+)-γ-(*p*-hydroxy) anilide de l'acide glutamique à partir de *Agaricus hortensis*. *Biochimica et Biophysica Acta*, **43**, 322–8.

Johnston, G. A., Curtis, D. R., de Groat, W. C. & Duggan, A. W. (1968). Central actions of ibotenic acid and muscimol. *Biochemical Pharmacology*, **17**, 2488–9.

Kamiya, T., Saito, Y., Hashimoto, M. & Seki, H. (1969). Structure and synthesis of lentysine, a new hypocholesterolemic substance. *Tetrahedron Letters*, **53**, 4729–32.

Kamiya, T., Saito, Y., Hashimoto, M. & Seki, H. (1972). Hypocholesterolemic alkaloids of *Lentinus edodes*. 1. Structure and synthesis of eritadenine. *Tetrahedron Letters*, **28**, 899–906.

Kaneda, T. & Tokuda, S. (1966). Effect of various mushroom preparations on cholesterol levels in rats. *Journal of Nutrition*, **90**, 371–6.

Kelly, R. B., Daniels, E. G. & Hinman, J. W. (1962). Agaritine: isolation degradation and synthesis. *Journal of Organic Chemistry*, **27**, 3229–31.

Koenig-Bersin, P., Waser, P. G., Langemann, H. & Lichtensteiger, W. (1970). Monoamines in the brain under the influence of muscimol and ibotenic acid, two psychoactive principles of *Amanita muscaria*. *Psychopharmacologia*, **18**, 1–10.

Kogl, F., Salemink, C. A., Schouten, H. & Jellinek, F. (1957). Muscarine 111. *Receuil des Travaux Chimiques des Pays-Bas*, **76**, 109–27.

Lampe, K. F. (1978). Pharmacology and therapy of mushroom intoxications. In *Mushroom Poisoning: Diagnosis and Treatment*, ed. B. H. Rumack & E. Salzman, pp. 125–69. West Palm Beach, Florida: CRC Press.

Leung, A. Y. & Paul, A. G. (1968). Baeocystin and norbaeocystin: new analogues of psilocybin from *Psilocybe baeocystis*. *Journal of Pharmaceutical Science*, **57**, 1667–71.

Levenberg, B. (1961). Structure and enzymic cleavage of agaritine, a phenylhydrazide of L-glutamic acid isolated from Agaricaceae. *Journal of the American Chemical Society*, **83**, 503–4.

Levenberg, B. (1962). An aromatic diazonium compound in the mushroom *Agaricus bisporus*. *Biochimica et Biophysica Acta*, **63**, 212–14.

Levenberg, B. (1964). Isolation and structure of agaritine, a γ-glutaminyl-substituted arylhydrazine derivative from Agaricaceae. *Journal of Biological Chemistry*, **239**, 2267–73.

Lincoff, G. & Mitchel, D. H. (1977). *Toxic and Hallucinogenic Mushroom Poisoning*. New York: Van Nostrand Reinhold & Co.

Lindberg, P., Bergman, R. & Wickberg, B. (1975). Isolation and structure of coprine, a novel physiologically active cyclopropane derivative from *Coprinus atramentarius* and its synthesis via 1-aminocyclopropanol *Journal of the Chemical Society, Chemical Communications*, 946–7.

Lindberg, P., Bergman, R. & Wickberg, B. (1977). Isolation and structure of coprine, the *in vivo* aldehyde dehydrogenase inhibitor of *Coprinus atramentarius*: synthesis of coprine and related cyclopropane derivatives. *Journal of the Chemical Society, Perkin Transactions, 1*, 684–91.

List, P. H. & Luft, P. (1967). Gyromitrin, the toxin of the spring lorchel *Gyromitra esculenta*. *Tetrahedron Letters*, **20**, 1893–4.

McCawley, E. L., Brummett, R. E. & Dana, G. W. (1962). Convulsions from *Psilocybe* mushroom poisoning. *Proceedings of the Western Pharmacological Society*, **5**, 27.

Mee, H. (1978). A mushroom that is magic. *National Examiner*, April 4.

Mize, P. D., Jeffs, P. W. & Boekelheide, K. (1980). Structure determination of the active sulfahydryl reagent in gill tissue of the mushroom *Agaricus bisporus*. *Journal of Organic Chemistry*, **45**, 3540–3.

Morita, K. & Kobayashi, S. (1966). Isolation and synthesis of lenthionine, an odourous substance of Shi-itake, an edible mushroom. *Tetrahedron Letters*, **6**, 573–7.

Morita, K. & Kobayashi, S. (1967). Isolation, structure and synthesis of lenthionine and its analogues. *Chemical and Pharmaceutical Bulletin (Tokyo)*, **15**, 988–93.

Muller, G. F. R. & Eugster, C. H. (1965). Muscimol, ein pharmakodynamisch wirksamer Stoff aus *Amanita muscaria*. *Helvetica Chimica Acta*, **48**, 910–26.

Ramsbottom, J. (1953). *Mushrooms and Toadstools*. London: Collins.

Rast, D., Stussi, H. & Zobrist, P. (1979). Self-inhibition of the *Agaricus bisporus* spore by CO_2 and/or γ-glutaminyl-4-hydroxybenzene and γ-glutaminyl-3, 4-benzoquinone: a biochemical analysis. *Physiologia Plantarum*, **46**, 227–34.

Repke, D. B. & Leslie, T. L. (1977). Baeocystin in *Psilocybe semilanceata*. *Journal of Pharmaceutical Science*, **66**, 113–14.

Rumack, B. H. & Salzman, E. (1978). *Mushroom Poisoning: Diagnosis and Treatment*. Florida: CRC Press.

Schultes, R. E. & Hofmann, A. (1979). Little flowers of the gods. In *Plants of the Gods: Origins of Hallucinogenic Use*, pp. 144–53. London: Hutchinson.

Simandl, J. & Franc, J. (1956). Isolace tetraethylthiuramdisulfide z hniku inkuostoveha (*Coprinus atramentarius*). *Chemica Listy*, **50**, 1862–3.

Singer, R. (1961). *Mushrooms and Truffles. Botany, Cultivation and Utilisation*. London: Leonard Hill.

Singh, P. & Rangaswami, S. (1966). Occurrence of *O*-methyl-drosophilin A in *Fomes fastuosus* (Lev.). *Tetrahedron Letters*, **11**, 1229–31.

Suzuki, F., Koide, T., Tsunoda, A. & Ishida, N. (1974). Mushroom extracts as an interferon inducer. Biological and physiochemical properties of spore extracts of *Lentinus edodes*. *Mushroom Science*, **9**, 509–20.

Suzuki, S. & Oshima, S. (1974). Influence of Shi-ta-ke (*Lentinus edodes*) on human serum cholesterol. *Mushroom Science*, **9**, 463–7.

Szent-Gyorgyi, A., Chung, R. H., Boyajian, M. J., Tischler, M., Arison, B. H., Schoenewaldt, E. F. & Wittick, J. J. (1976). Agaridoxin, a mushroom metabolite. Isolation, structure and synthesis. *Journal of Organic Chemistry*, **41**, 1603–6.

Takashima, K., Izami, K., Iwai, H. & Takeyama, S. (1973). The hypocholesterolemic action of eritadenine in the rat. *Atherosclerosis*, **17**, 491–502.

Takashima, K., Saito, C., Sasaki, Y., Morita, T. & Takeyama, S. (1974). Effect of eritadenine on cholesterol metabolism in the rat. *Biochemical Pharmacology*, **23**, 433–8.

Takemoto, T. & Nakajima, T. (1964). Isolation of the insecticidal constituent from *Tricholoma muscarium. Yakugaku Zasshi*, **84**, 1230–2.

Takemoto, T., Nakajima, T. & Sakuma, R. (1964). Isolation of an insecticidal constituent, ibotenic acid, from *Amanita muscaria* and *Amanita pantherina. Yakugaku Zasshi*, **84**, 1233–4.

Tavantzis, S. M. & Smith, S. H. (1982). Isolation and evaluation of a plant-virus inhibiting quinone from sporophores of *Agaricus bisporus. Phytopathology*, **72**, 619–21.

Tiffany, S. M., Graham, D. G., Vogel, S. F., Cass, M. W. & Jeffs, P. W. (1978). Investigation of the structure–function relationships of cytotoxic quinones of natural and synthetic origin. *Cancer Research*, **38**, 3230–5.

Tokita, F., Shibukawa, N., Yasumoto, T. & Kaneda, T. (1971). Effect of mushrooms on cholesterol metabolism in rats. IV. Separation and chemical structure of the plasma cholesterol reducing substances from mushrooms. *Eiyo To Shokuryo*, **24**, 92–5.

Tokuda, S. & Kaneda, T. (1976). Reducing mechanism of plasma cholesterol by Shi-itake. *Mushroom Science*, **9**, 445–62.

Toth, B. (1979). Mushroom hydrazines: occurrence, metabolism, carcinogenesis and environmental implications. In *Naturally occurring Carcinogens–mutagens and Modulators of Carcinogenesis*, ed. E. C. Miller, pp. 57–65. Baltimore: Japanese Scientific Society Press.

Toth, B. & Nagel, D. (1981). Studies on the Tumourigenic potential of 4-substituted phenylhydrazines by subcutaneous route. *Journal of Toxicology and Environmental Health*, **8**, 1–9.

Toth, B., Nagel, D., Patil, K., Erickson, J. & Antonson, K. (1978). Tumour induction with the *N'*-acetyl derivative of 4-hydroxymethyl phenylhydrazine, a metabolite of agaritine of *Agaricus bisporus. Cancer Research*, **38**, 177–80.

Toth, B., Patil, K. & Jae, H. (1981*a*). Carcinogenesis of 4-(hydroxymethyl) benzenediazonium ion (tetrafluroborate) of *Agaricus bisporus. Cancer Research*, **41**, 2444–9.

Toth, B., Raha, C. R., Wallcave, L. & Nagel, D. (1981*b*). Attempted tumour induction with agaritine in mice. *Anticancer Research*, **1**, 255–8.

Tottmar, O. & Lindberg, P. (1977). Effects on rat liver acetaldehyde dehydrogenase *in vitro* and *in vivo* by coprine, the disulfiram-like constituent of *Coprinus atramentarius. Acta Pharmacologica and Toxicologica*, **40**, 476–81.

Tyler, V. E. (1963). Poisonous mushrooms. *Progress in Chemical Toxicology*, **1**, 339–84.

Vogel, F. S., McGarry, S. J., Kemper, L. A. K. & Graham, D. G. (1974). Bacteriocidal properties of a class of quinoid compounds related to sporulation in the mushroom *Agaricus bisporus. American Journal of Pathology*, **76**, 165–74.

Vogel, F. S. & Weaver, R. F. (1972). Concerning the induction of dormancy in spores of *Agaricus bisporus. Experimental Cell Research*, **75**, 95–104.

Waser, P. G. (1967). The pharmacology of *Amanita muscaria. US Public Health Service Publications*, 1645, 419–39.

Wasson, R. G. (1962). The hallucinogenic mushrooms of Mexico and psilocybin: a biography. *Botanical Museum Leaflet, Harvard University*, 20, 25–73.

Wasson, R. G. (1968). *Soma: Divine Mushroom of Immortality*. New York: Harcourt, Brace & World.

Weaver, R. F., Rajagopalan, K. V. & Handler, P. (1972). Mechanism of action of a respiratory inhibitor from gill tissue of the sporulating common mushroom *Agaricus bisporus. Archives of Biochemistry and Physics*, **149**, 541–8.

Weaver, R. F., Rajagopalan, K. V., Handler, P. & Byrne, W. L. (1971*a*).

γ-L-Glutaminyl-3,4-benzoquinone: structural studies and enzymatic synthesis. *Journal of Biological Chemistry*, **246**, 2015–20.

Weaver, R. F., Rajagopalan, K. V., Handler, P., Jeffs, P., Byrne, W. L. & Rosenthal, D. (1970). Isolation of γ-L-glutaminyl-4-hydroxybenzene and γ-L-glutaminyl-3, 4-benzoquinone: a natural sulfahydryl reagent from sporulating gill tissue of the mushroom *Agaricus bisporus*. *Proceedings of the National Academy of Sciences, USA*, **67**, 1050–6.

Weaver, R. F., Rajagopalan, K. V., Handler, P., Rosenthal, D. & Jeffs, P. W. (1971 b). Isolation from the mushroom *Agaricus bisporus* and chemical synthesis of γ-L-glutaminyl-4-hydroxybenzene. *Journal of Biological Chemistry*, **246**, 2010–14.

Weir, J. K. & Tyler, V. E. (1960). An investigation of *Coprinus atramentarius* for the presence of disulfiram. *Journal of the American Pharmaceutical Association, Science Edition*, **49**, 426.

Wheeler, M., Tischler, M. & Bitensky, M. W. (1982). Agaridoxine: a fungal catecholamine which acts as an alpha-1 agonist of mammalian hypothalamic adenylate cyclase. *Brain Research*, **231**, 387–98.

Wood, D. A. (1979). Degradation of composted straw by the edible mushroom *Agaricus bisporus*. Enzyme activities associated with mycelial growth and fruit body formation. In *Straw Decay and its Effect on Disposal and Utilisation*, ed. E. Grossbard, pp. 95–104. New York: John Wiley & Sons.

Yamamoto, H. & Ikegawa, T. (1980). *Antitumour Agent Emitanin-1 Production*. Tokyo Koho, Japan, patent 80–15995.

Zadrazil, F. & Grabbe, K. (1983). Edible mushrooms. In *Biotechnology*, ed. H. J. Rehm & G. Reed, pp. 145–87. Weinheim: Verlag Chemie.

27

Developmental biology of agarics – an overview

A.F.M.REIJNDERS and DAVID MOORE*

De Schuilenburght B72, Schuilenburgerplein 1, 3816 TD Amersfoort, The Netherlands
**Department of Botany, The University, Manchester M13 9PL, UK*

Introduction

We are not attempting a comprehensive survey in this chapter, and the references we will quote are offered for illustration, not as part of an extensive review. Rather, we will present some personal views and observations about the subject.

The following aspects will be dealt with: (1) primordium initiation; (2) cell formation by cross wall formation, and nuclear numbers; (3) differentiation of cells corresponding with their location; (4) cell inflation; (5) the basal plectenchyma and bulb-tissue. We will not deal with wall formation, nor with the influence of environmental factors on primordium initiation, both of which have been considered in recent reviews (Burnett & Trinci, 1979; Manachère, 1978, 1980; Robert & Durand, 1979).

It is obvious that morphogenetic research is only at the start of its development: the possibilities to tackle problems of this kind have greatly increased in the last decade. Fruit bodies of hymenomycetes are particularly interesting in this respect because histogenesis is accomplished here by the cooperation of individual hyphal elements and, though this applies also to many ascomycetes and Rhodophyceae, the size and complexity of the structures formed by hymenomycetes are in general much greater. The development of any organised fungal structure requires that hyphae grow toward one another and cooperate in formation of the differentiating organ; this is the diametrically reversed character of the invasive, 'undifferentiated' mycelium. We are almost totally ignorant of the factors which control this peculiar reversal in behaviour. Autotropic agents and surface-active molecules must be involved. Such molecules are known from classic researches with yeasts (the glycoproteins which determine mating type-specific agglutination), water moulds (sex hormones) and *Mucor* (trisporic

581

acid synthesis) (for review see Moore, 1984), but apart from some ultrastructural indications of a surface architecture on basidiomycete cells (McLaughlin, 1982) we lack conclusive information about surface chemistry of basiodomycete hyphae and this will be necessary to explain how they can cooperate in the formation of an organ like the fruit body, sclerotium or rhizomorph. These fungi provide extremely good systems for the fundamental study of phenomena involved in cell homing and specific cell-to-cell adhesion, since they change from one state to the other as a normal part of their developmental pathway. In most other systems in which the biology of the cell surface is being studied, individual cells can only be investigated by culturing them in unnatural conditions.

Primordium initiation

Light microscopy does not reveal many differences between the generative hyphae (which in mass have been called protenchyma) and mycelial hyphae. They have mostly a minimum number of nuclei (i.e. increase in the number of nuclei is an aspect of cell differentiation in fruit bodies of Agaricales) and may or may not bear clamp connections. The protenchyma of the primordial fruit body displays from the very first two types of organisation: a bundle of nearly parallel hyphae, and interwoven hyphae which form a plectenchyma. But also in primordia, which are in the beginning plectenchymatous, bundles of strictly parallel hyphae (called meristemoids) show up later in the stipe and in the lateral parts of the pileus. This gives rise to the question whether the factors which control such a directed growth in primordia are comparable with those which cause the same behaviour in strands, coremia, and ramified fruit bodies (Clavariaceae, etc., see Watkinson (1979)). The hyphae of these aggregates are generally embedded in a slimy coating (see Rayner *et al.*: Chapter 10) and it may be supposed that this sheathing has an important morphogenetic function with regard to binding the hyphae together and the transmission of stimuli. We should note that this may be relevant both to the earliest stages of primordium formation (van der Valk & Marchant, 1978) and to the more mature structure. Newer methods of preparation for the scanning electron microscope show the fruit body to be covered, and the subhymenium to be filled, with material which could serve as an adhesive and through which soluble control factors could be transmitted (Williams, Beckett & Read: Chapter 18).

Little attention has been directly devoted to the factors responsible for organising the shift from a ramifying to an aggregating mode of hyphal growth, though the more indirect agents which promote the formation of primordia (such as nutritional requirements and light) have attracted much interest. Nutrient depletion of the medium has long been known to be involved (see Leatham: Chapter 17), but this does not equate with starvation of the mycelium because nutrients are stored in the mycelium and then redistributed to the developing fruit bodies. There are good data for this with regard to nitrogen metabolism (especially for the polypore *Favolus* (Kitamoto *et al.*, 1980)) and carbohydrate metabolism (particularly in *Coprinus* (Madelin, 1960)) but we are not aware of any serious attempt to look at lipids. This is worth remarking upon in view of the useful storage function which lipids can serve and the argument developed below for the importance of membrane-associated processes in primordium initiation. An extreme difficulty here is the problem of identifying *causal* events. For example, the data available on cyclic AMP metabolism (Uno & Ishikawa, 1982) and glycogen metabolism (Moore, Elhiti & Butler, 1979) in *Coprinus* are very promising, yet there is no necessary connection between these metabolites and fruiting, and no ready way of distinguishing causal from consequential events. Indeed, from knowledge of the nutritional control of fruit body initiation, coupled with the other environmental stimuli which are commonly involved, particularly the frequent requirement for light stimuli (Ross: Chapter 14) and the apparently universal need for a temperature shift-down, attention is focused on conformational shifts in important molecules and especially membrane-associated processes. Both temperature and light can be viewed as influencing metabolism via membrane architecture; the former through effects on membrane fluidity (which may affect the rate at which transmembrane phenomena occur) and the latter through direct interaction with membrane-localised receptors. Nutritional effects too, however, can also be seen as having a membrane relationship. We think it significant that many plasma membrane transport mechanisms are biphasic, having one phase which characterises the 'high external metabolite concentration' and another which characterises the 'low concentration' state (Scarborough, 1970 *a*,*b*; Neville, Suskind & Roseman, 1971; Schneider & Wiley, 1971; Beever & Burns, 1977; Moore & Devadatham, 1979). The shift from one to the other, which in some cases involves derepression of previously unexpressed genes and in other cases involves conformational changes in existing proteins, and which is frequently associated with altered energisation states of the transport process, could well be integrated with differentiation processes. The corollary, of course, is that we need to know more about transport mechanisms and the way they change during the different phases of the life cycle of the organism.

Primordium initiation evokes many morphogenetic problems which have only been partially studied despite the large quantity of work which has already been done. An equally serious deficiency is in the study of development with particular regard to the *comparison* of different systematic groups (Watling: Chapter 11). The number of species which have been subjected to morphogenetic research is still small: *Schizophyllum commune*, *Agaricus bisporus*, a few species of *Coprinus*, and *Flammulina velutipes*. Even study of aspects of somatic growth detectable with light microscopy is far from being adequate, let alone complete, and there is tremendous scope for biochemical and molecular investigation.

Cell formation

We distinguished above between meristemoids, with rows of cells in parallel hyphae where cell division is coordinated, and plectenchymatous tissue, where ramification and cross wall formation often occur in adjacent hyphal knots where short branches and cells are surrounded by coiled or straight hyphae with larger cells (Reijnders, 1977). A fact of fundamental importance is that in primordia of Agaricales cell division takes place exclusively by transverse cross wall formation. Corner (for example, in his monograph on *Clavaria*, 1950) called all cross walls which are not formed by the tip cell 'secondary walls', and this designation has been generally adopted. The formation of longitudinal walls seems to be almost impossible in agarics. This might correspond with the polarity of the tip cell and probably with the occurrence of dolipores in the cross walls. We have never seen longitudinal walls or walls in orientations other than transverse in primordia: neither in the stipe, nor in the pseudoparenchymatous subhymenium. When the stipe (or any other basidiomycete structure) increases in girth and inflation is insufficient to provide for this, more generative hyphae which in a young stage are always present in the axial portion of the stipe are utilised. Even the pseudoparenchymatous tissues of many sclerotia may arise only by ramification, as demonstrated for three different sclerotial types by Townsend & Willetts (1954). Although some exceptions seem to occur, it is nevertheless significant that cell division by longitudinal wall formation occurs only in very specialised cells such as the basidia of Tremellales or the multicellular spores of certain ascomycetes (Pleosporales), though longitudinal septa have also been reported in walls of pycnidia and perhaps some perithecia (Lohwag, 1941). Otherwise, longitudinal walls are not found.

There is only one apparently reliable description of a real meristem which is manifest in an organ of an agaric; Motta (1969) claimed that cell divisions in a narrow zone about 25 μm behind the extremity of a rhizomorph of *Armillaria* were effected by walls being formed in diverse directions. His comparisons with shoot- and root-tips of phanerogams led him to suggest the existence of histogens analogous to those proposed for higher plants. As we have seen (Rayner *et al.*: Chapter 10), there is reason to doubt the existence of a true meristem in *Armillaria* rhizomorphs, but Motta also described secondary meristems situated underneath the postulated apical meristem at the sides of the rhizomorph and, in these parts, cell formation proceeded by transverse cross wall formation.

Whatever may be realistic in these conceptions about the histology of the *Armillaria* rhizomorph there is certainly one striking conformity between this organ and cell division in the upper part of the *Coprinus* stipe. Motta states that one or two nuclei occur in 'meristem' cells, but the number of nuclei in cells which originate from these increases rapidly. This is a common characteristic of the cells of *Coprinus* stipes (Reijnders, 1979), but is observed also in many other agarics, though not in all as Kühner (1958) has pointed out. The number of nuclei in such 'coenocytes' can run up to 40–60 or even 100 or more (see also Wong & Gruen, 1977). The term nuclear bodies may be preferred to the description of these organelles as nuclei, in view of the disparity between their increase in number and the increase in the amount of DNA in the stipe (Gooday: Chapter 12). When first formed, at the onset of stipe formation in *Coprinus*, the cells have only one or two nuclei, and a narrow zone where this reduced number is retained is still evident in the very upper portion of the stipe for some time. Kühner (1977) supposes that the large number of nuclei is to a certain extent correlated with the volume of the cell, but Wong & Gruen (1977) found no correlation between the number of nuclei and cell size in *Flammulina*. It should be noted that such a multiplication of nuclei appears to be absent in other cells of the primordium, even those which enlarge considerably, e.g. cells of the veil. The universal veil in *Coprinus* often consists of rows of cells which arise at the circumference of the cap; we could never observe in this tissue a similar nuclear behaviour, the mother cells usually contained two nuclei which gradually disappeared in the huge cells of the veil (there is an exception in *Armillaria*: when hyphae of the surface of the stipe pass into the veil, cells of the latter have more nuclei in some cases). For the sake of comparison it will be important to consider the heteromerous trama of the Russulaceae, where the multinuclear elements arise in a very specific way; is this multinuclear condition caused by the same morphogenetic factors? The descriptions of Motta are accompanied by a series of illustrative electron micrographs but we doubt whether extensive documentation of this sort exists for stipe cells of Agaricales. It is certainly lacking for trama tissues of *Russula* and *Lactarius* and in these cases would be extremely welcome to enable comparison of the ultrastructural data of the induction-hyphae and the spherocytes with those of ordinary stipe cells. The ultrastructure of elements of the hymenium has often been studied, but the available details of somatic cells are in this respect far from being sufficient. Understanding of the way cells and tissues differentiate is dependent upon knowledge in precise detail of the structure and histological relationships between cells; in too many cases this knowledge is lacking.

Differentiation of cells corresponding with their location

The cells of phanerogams and fungi, however different may be their origins, differentiate in relation to their location in the plant. This is one of the most important principles of morphogenesis. We often see that the cells of one recognisable hypha become abruptly different because they are influenced by another morphogenetic factor (the nature of which is obscure). Thus, in *Leucocoprinus cepaestipes* the lower cells of hyphae at the surface of the cap form the pileodermium and the narrower upper part of the same hyphae merge into the universal veil (Reijnders, 1948); and in *Coprinus poliomallus* the narrower hyphae of the gill trama suddenly widen and form isodiametric cells when they pass into the lipsanenchyma (Reijnders, 1979). Such differences in adjacent cells of continuous hyphae can be detected everywhere in primordia, in mature tissues (notably the differentiation of hymenial elements from the tramal hyphae (Moore *et al.*, 1979)) and in vegetative structures (such as the distinction between thick-walled and thin-walled cells in sclerotia of *Coprinus cinereus* (Waters, Butler & Moore, 1975)). What signals are involved in directing such differentiation and how, in a basically hyphal structure, are they localised? Adjacent hyphal cells are separated by the dolipore septum (Moore: Chapter 7) which is obviously capable of extremely rapid response to experimental stress (Todd & Aylmore: Chapter 9) and must be involved in partitioning regulatory signals between cell compartments in the same hyphal strand. The structure of the dolipore/parenthesome septum is sufficiently complex to anticipate quite sophisticated involvement in localising differentiation signals in a longitudinal direction (relative to the long axis of the parent hypha). But even though the hypha is the very basis of the sorts of structures considered here, control of the longitudinal communication of organisational signals is probably insufficient to account for the regulation of the levels of differentiation which can be observed.

The scope of the differentiation seen in fruiting and vegetative fungal structures alike is every bit as complex as that of higher plant and animal systems. A consequence of the fungal dependence on hyphal organisation is that lateral communication between cells contributing to the same tissue must involve export and import of control signals. As discussed above, longitudinal secondary walls are not found; so lateral communication between adjacent cells must take place across two hyphal membranes and two mature hyphal walls. There can be no denying the probability (indeed, one is tempted to say the fact) that lateral communication plays an important role in defining the positional information on which tissue differentiation in fruiting and vegetative structures depends. Yet despite numerous ultrastructural studies of various fungal structures, there is no evidence for anything akin to plasmodesmata or gap junctions in the lateral walls of fungi (although channels through thickened cell walls in peridioles of *Nidularia* (Reijnders, 1976) and pit-like structures in sclerotia of *Penicillium* (Lohwag, 1941) have been reported). In general, a cytoplasmic route for signals capable of conveying positional information is excluded (see Rosin, Horner & Moore: Chapter 13).

Another very interesting aspect is the apparent fact that most of the changes in shape which characterise the later stages of fruit body maturation depend on cell expansion. It follows from this that the distinction between cell division and cell expansion (in terms of their contribution to development) is an important one; yet we have very little information about it. The *Coprinus* species are probably the best served in this respect but even these data are fragmentary. More modern techniques may help in this type of analysis; even embedding specimens in resin allowing thinner, and ideally serial, sections to be examined with the light microscope would be rewarding. A considerable research effort has been devoted to the metabolism which underlies the cell expansions important in maturation (whether or not cell division is still occurring). The account is particularly well developed for *Coprinus cinereus* but we have only recently recognised how paraphyses insert into the basidial layer of the young hymenium and that a proportion of the paraphyseal population insert later in development (see Chapter 13). Nevertheless, the available biochemical evidence quite clearly associates amplification of the tricarboxylic acid cycle in the maturing fruit body cap of *C. cinereus* with specific derepression of an ammonium-scavenging system and amplification of the urea cycle, the whole metabolic shift leading to accumulation of urea as an osmotic metabolite which serves to drive water into the expanding cells of the hymenium (Ewaze, Moore & Stewart, 1978; Moore *et al.*, 1979). It is interesting that urea has often been associated with basidiomycete fruits, so one wonders how general might be the specific metabolism identified so far in *Coprinus* (especially the potentially novel means of assimilating ammonium, the use of glutamate decarboxylation as part of the tricarboxylic acid cycle, and accumulation of urea). However, it must be recognised that cell inflation in different tissues (cap *versus* stipe in *Coprinus*) and in different organisms (*Coprinus versus Agaricus*) depends on different metabolic pathways for the provision of osmotic metabolites. These differences illustrate the versatility of intermediary metabolism; but, and we say this in the belief that the fundamental events underlying cell expansion are essentially similar wherever the process is encountered, such metabolic differences can direct attention away from the fundamental control processes. Recognising that those differences exist and that nevertheless there are similarities in the events observed can help in our search for those fundamental control mechanisms.

In these biochemical processes and in a few other cases (for example, involvement of laccase in fruiting of *Agaricus*, *Lentinus* and *Coprinus congregatus*, and of mannitol dehydrogenase and glucose 6-phosphate dehydrogenase in fruiting of *Agaricus* (Ross: Chapter 14; Wood: Chapter 15; Hammond: Chapter 16; Leatham: Chapter 17)) there are some excellent candidates for the application of recombinant DNA methods to study of the genes and gene transcripts involved in development-related processes and, indeed, it is encouraging to see that work is developing in this sort of direction (Ullrich & Novotny: Chapter 20; Wessels *et al.*: Chapter 21; Pukkila *et al.*: Chapter 22). However, this work can only fulfil its promise if it is firmly associated with particular metabolic steps which have particular morphogenetic consequences. We need a better and broader picture of the metabolism and biochemistry of developing structures and *comparative* details are essential. This means not only comparison between different species (so that, hopefully, causal events can be recognised among the complex of metabolic reactions) but also comparisons between tissues of the same structure. Far too often fruit bodies and other structures are dealt with as though they are homogeneous; they are not homogeneous and the value of data obtained is often considerably reduced if this fact is ignored.

There is another aspect of molecular analysis which deserves to be stressed. This is the use of restriction fragment length polymorphisms (which are naturally occurring polymorphisms in the DNA) as genetic markers. Their use as such was first suggested as an aid in construction of the human linkage map (Botstein *et al.*, 1980) and very recently some successful associations have been made between such molecular markers and particular human genetic disorders. The arguments which make this approach attractive as a means of studying human genetics also apply to the study of those basidiomycetes in which classic genetic approaches cannot be used very easily. This group includes the most important cultivated species, *Agaricus bisporus*. Casselton & Economou (Chapter 8) illustrate the use of mitochondrial DNA polymorphisms to monitor recombination between mitochondrial genomes, and Pukkila *et al.* (Chapter 22) demonstrate the segregation in spore tetrads of nuclear polymorphisms in ribosomal DNA sequences. Both of these studies used *Coprinus cinereus* which is an ideal organism for all types of experimental genetics. As pointed out by Raper (Chapter 23), *A. bisporus* has biological characteristics which make it an unsuitable candidate for even the simplest genetic exercise. However, DNA polymorphisms seem to be sufficiently common for one to expect to be able to recognise them fairly readily among the numerous strains of *A. bisporus* which are available. Such polymorphisms are attractive because they are natural deviations in genome structure that characterise specific genetic loci. Once the appropriate target sequence has been cloned, these molecular polymorphisms can be easily identified. They could be of immediate use in establishing the nuclear constitution of putative 'hybrid' heterokaryons, and in monitoring gene segregations in this two-spored, secondarily homothallic species. In the longer term, DNA molecular markers could also be associated with specific characters of importance to the value of this commercial crop; and since they can be scored in genomic DNA prepared from small samples of the mycelium they could offer a distinct advantage over larger scale fruiting trials which are otherwise necessary to score crop characters (Chapter 23). Thus as well as the intrinsic interest of this sort of study there could be genuine commercial advantage in moving in this direction.

Cell inflation

There are two types of cell inflation: a slow process which is often encountered in primordia, and a more rapid one involved especially in stipe and cap maturation. Strong inflation must represent a kind of differentiation of the cell and a mark of specialisation. The narrow generative hyphae, i.e. the basic tissue of the primordium, never have inflated cells. Large cells occur principally in fleshy fungi and are characteristic for Agaricales, Clavariaceae and some gasteromycetes; primordia of the most specialised genera of the Agaricales, e.g. *Mycena*, *Coprinus*, *Conocybe*, *Bolbitius* and *Pluteus*, have remarkably inflated cells even at very young stages. The elongation of the cells in primordia of Agaricales starts in general immediately after their formation. Motta (1969) makes the same observation for the rhizomorphs of *Armillaria*. Reijnders (1963) relates this early inflation to the general shape of the primordium; the different zones of the primordium enlarge proportionally; they do not impede the growth of other parts, and compressed tissues are seldom observed. Besides this continuous and slow inflation, whose morphogenetic control must be a complicated mechanism, the period of rapid expansion has attracted much more attention from as early as 1842 (Schmitz), through the work of Bonner, Kane & Levey (1956) and on to work such as that of Gooday (Chapter 12). This phenomenon has its parallels in higher plants. Modern research has examined the role of inflation in some selected species (Gooday, 1974, 1982, and Moore *et al.*, 1979, on *Coprinus cinereus*; Bret, 1977, on *C. congregatus*; Wong & Gruen, 1977, on *Flammulina*; Craig, Gull & Wood, 1977, on *Agaricus bisporus*). We now know a lot more about these processes in the fruit bodies of these particular species. Wong & Gruen (1977) produced data accounting for the distribution of inflation over the whole stipe and for the correlation between length of stipe cells and that of the whole stipe. Other studies have focused on cell wall changes (especially chitin deposition) and osmotic agents. Although the subject is complicated the analyses done so far are promising and should be continued and expanded. We stress the need for further comparative studies and we draw attention once more to the heteromerous trama of the Russulaceae, to the specific trama (lateral branches, etc.) of the Amanitaceae; and to the fact that comparative studies of inflation would probably be of taxonomic interest.

The basal plectenchyma

The lowest part of the youngest agaric fruit body primordium is already made up of an entangled mass of cells; in older specimens this tissue often has an almost pseudoparenchymatous character. It represents an autonomous organ as it does not, in fact, belong to the stipe proper. It is not present at the inception of the primordium, when the lattice of ramifying hyphae is formed by mycelial threads (Matthews & Niederpruem, 1972), but it is one of the first parts which is subjected to differentiation. It is characterised by adjacent coils which are almost always present; these have in the centre short elements: cells whose ramifications deliver the elements for an extending plectenchyma. Corner (1950) depicted this interwoven tissue several times at the foot of clavarioid fungi, so that it appears that this phenomenon is widespread. Interwoven tissue has been found several times in the basal portion of poroid Aphyllophorales, but here no inflation occurred. The large deposits of polysaccharide which have been demonstrated in the lower part of the fruit body of *Coprinus cinereus* by Matthews & Niederpruem (1972) and Moore *et al.* (1979) are probably stored in this basal plectenchyma and not in the stipe itself (which is generally composed of parallel hyphae). Closer examination of these deposits reveals the fact that differentiation in this region is quite extensive, the polysaccharide accumulations being localised in a cup-shaped structure at the stipe base.

During development of the *Coprinus* fruit body, glycogen is first accumulated in this structure in the stipe base. Subsequently, glycogen levels decline in the stipe and glycogen is accumulated in the hymenium or subhymenium (Moore *et al.*, 1979). The implication is that translocation occurs and the basal organ in which the glycogen is initially accumulated is involved. Metabolism to glucose or trehalose just for translocation appears pointless though it is not demanding energetically, but better experimental approaches remain to be examined. Nutrient translocation in general, both towards and within the fruit body (and discrimination of which initials will be allowed to develop), are further aspects of this same problem. One can ask how translocation is organised on a number of levels, but a fundamental question is the organisation at the metabolic level. Unless we do know the metabolism involved we have no hope of getting to grips with the molecular mechanisms.

When this basal plectenchyma acquires a larger extension we call it a bulb. Reijnders (1977) could show that the general features of the tissue in such bulbs are the same as in the basal plectenchyma. Still more conspicuous are protocarpic tubers which sometimes resemble sclerotia or pseudosclerotia and are sometimes described as such. They can rest for a long time but are able to produce normal fruit bodies (the stone fungus, *Pietra fungaja*). Besides these normal bulbous formations there exist what Singer (1975) has called 'carpophoroids', bodies arising from 'a primordium which, when maturing, fails to ever achieve the last agaricoid stage of its individual development after it has reached the endocarpous stage or before it forms an exposed hymenium'. They consist of sterile hyphal tissue similar to that of the normal fruit body but are either entirely sterile or with noticeably reduced fertility, and without visible function. Singer discusses at length these various modifications of probably homologous structures. Reijnders (1977) had the opportunity to examine similar formations of a mutant of *Agaricus bisporus* isolated by Fritsche & von Sengbusch (1963): their histological composition was equal to that of the primordial bulb, except for the specific cortex. The supposed homology of all these formations can probably be better established by physiological and genetic data than by histological observations. Vegetative and fertile structures have much in common in cellular terms, but we need to investigate their relationships at a much deeper level than this. In *Coprinus cinereus* there is genetic evidence to show that sclerotia and fruit bodies share the same initiation pathway (Moore, 1981). Information like this is of both phylogenetic and ontogenetic value.

Conclusions

There is a finite number of ways of modifying the structure of a hyphal cell during differentiation, but there may be an infinite number of ways in which those differentiated cells can be assembled into different structures. We are led to the conclusion that in the normal fruit body of an agaric, some genes are operative which account for a gasteromycetous bulb-like body and others regulate the formation and expansion of the normal 'mushroom' type of fruit body. It may be supposed that these genes are present in all species of Agaricales (they all have a basal plectenchyma) and that a normal development depends on the sequence of gene activity, which in some cases may be blocked by internal (mutational or regulatory) or external (environmental) factors. As every agaric can produce gasteromycete-like forms by the presence of the relevant genes (referring here only to the gross morphology) it cannot be reasonable to use these relations on behalf of phylogenetic speculations (derivation of Agaricales from gasteromycetes or *vice versa*), though we do not wish to pretend that every speculation on this problem is pointless. This conclusion that caution is required is corroborated by certain anomalies in Agaricales (Watling, 1971) which must be caused by disorder of genome activity. These anomalies are quite interesting, for the mycelium of a mutant of *Psilocybe* produces gasteromycete-like bulbs as well as specimens with a cyphelloid-habit, where growth and elongation of the hyphae is not inhibited and development is of the diffuse type (Reijnders, 1983).

We have to acknowledge that mushrooms are commercially very important, and we must accept that scientists must often do research for which they can get funding. The message is that there is much *interesting* research still to be done on the crop species and their relatives. Another message is that the mushroom industry invests far too little of its profits in scientific improvements of the crop. Mushroom producers are interested in controlling fruit body initiation (so that they can determine the synchrony of the crop production process) and in fruit body maturation (again to enable control of cropping, but also to control shelf life of the crop). In both of these processes much fundamental research remains to be done.

We do not want to imply that the work that needs to be done is all 'molecular', exciting and timely though this research may be. Knowledge of the details of histology and biochemistry in many species is essential before we can even make guesses about the ways tissue patterns are established and controlled. Such knowledge will reveal the potential routes for control signals and the identities of regulatory compounds, and can also suggest appropriate regulatory strategies. There is plenty of scope for research of this sort and we look forward to seeing its results.

Neither would we want to imply that only the fruit body is of interest. There are a number of vegetative structures which contribute to the biology of basidiomycetes and have intrinsic interest in themselves, and which are suited to particular research interests. Naturally, we would claim that a part of their description is an account of their relationship to the fruit body. But really, all structures, whether sexual or vegetative, impose similar developmental requirements on the hyphae of which they are composed; so knowledge of one is bound to improve knowledge of the others.

References

Beever, R. E. & Burns, D. J. W. (1977). Adaptive changes in phosphate uptake by the fungus *Neurospora crassa* in response to phosphate supply. *Journal of Bacteriology*, **132**, 520–5.

Bonner, J. F., Kane, K. K. & Levey, R. H. (1956). Studies on the mechanics of growth in the common mushroom, *Agaricus campestris. Mycologia*, **48**, 13–19.

Botstein, D., White, R. L., Skolnick, M. & Davis, R. W. (1980). Construction of a genetic linkage map in man using restriction fragment length polymorphisms. *American Journal of Human Genetics*, **32**, 314–31.

Bret, J. P. (1977). Respective role of cap and mycelium on stipe elongation of *Coprinus congregatus. Transactions of the British Mycological Society*, **68**, 363–9.

Burnett, J. H. & Trinci, A. P. J. (1979). *Fungal Walls and Hyphal Growth*. British Mycological Society Symposium 2. Cambridge University Press.

Corner, E. J. H. (1950). *A monograph on* Clavaria *and allied genera*. Annals of Botany Memoirs, no. 1. Oxford: Oxford University Press.

Craig, G. D., Gull, K. & Wood, D. A. (1977). Stipe elongation in *Agaricus bisporus. Journal of General Microbiology*, **102**, 337–47.

Ewaze, J. O., Moore, D. & Stewart, G. R. (1978). Co-ordinate regulation of enzymes involved in ornithine metabolism and its relation to sporophore morphogenesis in *Coprinus cinereus. Journal of General Microbiology*, **107**, 343–57.

Fritsche, G. & Sengbusch, R. von (1963). Beispiel der spontänen Entwicklung neuer Fruchtkörperformen beim Kulturchampignon. *Züchter*, **33**, 270–4.

Gooday, G. W. (1974). Control of development of excised fruit bodies and stipes of *Coprinus cinereus. Transactions of the British Mycological Society*, **62**, 391–9.

Gooday, G. W. (1982). Metabolic control of fruitbody morphogenesis in *Coprinus cinereus*. In *Basidium and Basidiocarp*, ed. K. Wells & E. K.Wells, pp. 157–73. New York: Springer-Verlag.

Kitamoto, Y., Matsumoto, T., Hosoi, N., Terashita, T., Kono, M. & Ichikawa, Y. (1980). Nitrogen metabolism of *Favolus arcularius*: changes in cellular nitrogen compounds during development of the mycelium and fruit bodies. *Transactions of the Mycological Society of Japan*, **21**, 237–44.

Kühner, R. (1958). Le comportement nucleaire daus les articles du stipe des Agarics et des Bolets. *Annales de l'Universite de Lyon*, **10**, 5–20.

Kühner, R. (1977). Variation of nuclear behaviour in the homobasidiomycetes. *Transactions of the British Mycological Society*, **68**, 1–16.

Lohwag, H. (1941). Anatomie der Asco- und Basidiomyceten. In *Handbuch der Pflanzenanatomie*, ed. K. Linsbauer, Band 4, Abteilung 2. Berlin: Borntraeger.

Madelin, M. F. (1960). Visible changes in the vegetative mycelium of *Coprinus lagopus* Fr. at the time of fruiting. *Transactions of the British Mycological Society*, **43**, 105–10.

Manachère, G. (1978). Morphogenèse des carpophores de Basidiomycètes supérieurs. Connaissances actuelles. *Revue de Mycologie, Paris*, **42**, 191–252.

Manachère, G. (1980). Conditions essential for controlled fruiting of macromycetes – a review. *Transactions of the British Mycological Society*, **75**, 255–70.

Matthews, T. R. & Niederpruem, D. J. (1972). Differentiation in *Coprinus lagopus*. I. Control of fruiting and cytology of initial events. *Archiv für Mikrobiologie*, **87**, 257–68.

McLaughlin, D. J. (1982). Ultastructure and cytochemistry of basidial and basidiospore development. In *Basidium and Basidiocarp*, ed. K. Wells & E. K. Wells, pp. 37–74. New York: Springer-Verlag.

Moore, D. (1981). Developmental genetics of *Coprinus cinereus*: genetic evidence that carpophores and sclerotia share a common pathway of initiation. *Current Genetics*, **3**, 145–50.

Moore, D. (1984). Positional control of development in fungi. In *Positional controls in Plant Development*, ed. P. W. Barlow & D. J. Carr, pp. 107–35. Cambridge University Press.

Moore, D. & Devadatham, M. S. (1979). Sugar transport in *Coprinus cinereus*. *Biochimica et Biophysica Acta*, **550**, 515–26.

Moore, D., Elhiti, M. M. Y. & Butler, R. D. (1979). Morphogenesis of the carpophore of *Coprinus cinereus*. *New Phytologist*, **83**, 695–722.

Motta, J. J. (1969). Cytology and morphogenesis in the rhizomorph of *Armillaria mellea*. *American Journal of Botany*, **56**, 610–19.

Neville, M. M., Suskind, S. R. & Roseman, S. (1971). A derepressible active transport system for glucose in *Neurospora crassa*. *Journal of Biological Chemistry*, **246**, 1294–301.

Reijnders, A. F. M. (1948). Etudes sur le développement et l'organisation histologique des carpophores dans les Agaricales. *Recueil des Travaux botaniques de la Néerlande*, **41**, 213–396.

Reijnders, A. F. M. (1963). *Les Problèmes du Développement des Carpophores des Agaricales et de quelques Groupes Voisins*. The Hague: Junk.

Reijnders, A. F. M. (1976). Sur le développement de trois espèces de gastéromycètes et l'origine coralloide ou lacunaire de la gleba. *Bulletin Trimestriel de la Société Mycologique de France*, **92**, 169–88.

Reijnders, A. F. M. (1977). The histogenesis of bulb- and trama tissue of the higher basidiomycetes and its phylogenetic implications. *Persoonia*, **9**, 329–61.

Reijnders, A. F. M. (1979). Developmental anatomy of *Coprinus*. *Persoonia*, **10**, 383–424.

Reijnders, A. F. M. (1983). Le développement de *Tectella patellaris* (Fr.) Murr. et la nature des basidiocarpes cupuliformes. *Bulletin de la Société Mycologique de France*, **99**, 109–26.

Robert, J. C. & Durand, R. (1979). Light and temperature requirements during fruit-body development of a basidiomycete mushroom, *Coprinus congregatus*. *Physiologia Plantarum*, **46**, 174–8.

Scarborough, G. A. (1970*a*). Sugar transport in *Neurospora crassa*. *Journal of Biological Chemistry*, **245**, 1694–8.

Scarborough, G. A. (1970*b*). Sugar transport in *Neurospora crassa*. II. A second glucose transport system. *Journal of Biological Chemistry*, **245**, 3985–7.

Schneider, R. P. & Wiley, W. R. (1971). Kinetic characteristics of the glucose transport system in *Neurospora crassa*. *Journal of Bacteriology*, **106**, 479–86.

Schmitz, J. (1842). Mycologische Beobachtungen als Beiträge zur Lebens-und Entwicklungsgeschichte einiger Schwämme aus der Klasse der Gasteromyceten un Hymenomyceten. *Linnaea*, **16**, 141–215.

Singer, R. (1975). *The Agaricales in Modern Taxonomy*, 3rd edn. Vaduz: Cramer.

Townsend, B. B. & Willetts, H. J. (1954). The development of sclerotia by certain fungi. *Transactions of the British Mycological Society*, **37**, 213–21.

Uno, I. & Ishikawa, T. (1982). Biochemical and genetic studies on the initial events of fruitbody formation. In *Basidium and Basidiocarp*, ed. K. Wells & E. K. Wells, pp. 113–23. New York: Springer-Verlag.

Valk, P. van der & Marchant, R. (1978). Hyphal ultrastructure in fruit-body primordia of the basidiomycetes *Schizophyllum commune* and *Coprinus cinereus*. *Protoplasma*, **95**, 57–72.

Waters, H., Butler, R. D. & Moore, D. (1975). Structure of aerial and submerged sclerotia of *Coprinus lagopus*. *New Phytologist*, **74**, 199–205.

Watkinson, S. C. (1979). Growth of rhizomorphs, mycelial strands, coremia and sclerotia. In *Fungal Walls and Hyphal Growth*, ed. J. H. Burnett & A. P. J. Trinci, pp. 93–113. British Mycological Society Symposium 2. Cambridge University Press.

Watling, R. (1971). Polymorphism in *Psilocybe merdaria*. *New Phytologist*, **70**, 307–26.

Wong, W. M. & Gruen, H. E. (1977). Changes in cell size and nuclear number during elongation of *Flammulina velutipes* fruitbodies. *Mycologia*, **69**, 899–913.

Species index

References are to chapter numbers. 'Checklist' refers to: Dennis, R. W. G., Orton, P. D. & Hora, F. B. (1960). New check list of British Agarics and Boleti. *Transactions of the British Mycological Society*, **43**, suppl., 1–225.

Acaulospora scrobiculata Trappe, 2
Agaricostilbum, 7
Agaricus, 11, 15, 16
Agaricus bisporus, 1, 2, 12, 14, 15, 16, 17, 18, 19, 23, 24, 25, 26, 27
Agaricus bitorquis, 23
Agaricus brunnescens Peck s. Malloch = *A. bisporus*, 7, 24, 25
Agaricus campestris, 1, 26
Agaricus sylvicola, 10, 18
Agaricus vaporarius, 11
Agrocybe, 11
Agrocybe acericola (Peck) Singer, 11
Agrocybe aegerita = *A. cylindrica*, 19
Agrocybe arvalis, 1
Agrocybe cylindrica, 1, 11
Agrocybe firma, 11
Agrocybe parasitica, 1
Agrocybe praecox, 1
Akenomyces costatus Hornby, 7
Amanita, 2, 5, 11, 26
Amanita gemmata, 6
Amanita muscaria, 1, 5, 26
Amanita pantherina, 26
Amauroderma, 2
Armillaria, 1, 2, 3, 10, 11, 27
Armillaria bulbosa (Barla) Kile & Watling, 1, 3, 10, 11
Armillaria elegans Heim, 10
Armillaria hiemii Pegler, 1
Armillaria mellea, 1, 2, 3, 6, 10
Armillaria ostoyae (Romagn.) Herink, 1, 3
Armillaria tabescens, 2, 3
Articularia quercina (Peck) von Höhnel, 7
Ascobolus stercorarius (Bull.) Schroet., 10
Aspergillus nidulans (Eidam) Wint., 20
Athelia, 7
Atractiella, 7
Atractogloea, 7

Aureobasidium pullulans (de Bary) Arnaud, 10
Auricularia, 7
Austroboletus, 1, 11

Baeospora myosura, 1
Biannularia, 11
Bjerkandera adusta, 7
Bolbitius, 1, 11, 27
Bolbitius titubans, 11
Boletellus, 11
Boletellus ananiceps (Berk.) Singer, 1
Boletinus, 11
Boletinus cavipes (Opat.) Kalch., 5
Boletochaete, 11
Boletus, 1, 5, 6, 11
Boletus armeniacus Quel., 11
Boletus parasiticus, 1
Boletus porosporus Imler ex Watling, 11
Boletus spadiceus Fries, 1
Botryobasidium, 7
Bullera, 7
Bullera alba (Hanna) Derx, 7

Calocybe, 1, 11
Calvatia sculpta (Hark.) Lloyd, 10
Calyptella, 11
Calyptella campanula (Fr.) Cke., 11
Camarophyllus, 11
Cantharellula, 11
Cantharellus, 2, 5, 11
Cenococcum graniforme (= *C. geophilum* Fr.), 6
Ceratobasidium, 7
Chalciporus, 1
Chionosphaera apobasidialis Cox, 7
Chlorophyllum molybdites (Meyer: Fr.) Mass., 2
Christiansenia, 7

Clavaria, 27
Clitocybe, 11, 26
Clitocybe flaccida, 1, 10
Clitocybe infundibuliformis, 1
Clitocybe langei, 1
Clitocybe nebularis, 1
Clitocybe odora, 1
Clitopilus, 11
Clitopilus prunulus, 1, 11
Collybia, 11
Collybia butyracea, 1
Collybia cirrhata, 11
Collybia confluens, 1
Collybia cookei, 11
Collybia dryophila, 10
Collybia fibrosipes (Berk. & Curt.) Dennis,
    2
Collybia maculata, 11
Collybia peronata, 1, 10, 11
Collybia tuberosa, 1
Collybia (= *Flammulina*) *velutipes,* 8
Coniodictyum chevalieri Har. & Pat., 7
Coniophora puteana (Schum.) Karst., 10
Conocybe, 1, 11, 26, 27
Conocybe coprophila, 11
Conocybe halophila Singer, 1
Conocybe pubescens, 11
Coprinus, 1, 11, 14, 15, 16, 18, 19, 20, 27
Coprinus atramentarius, 26
Coprinus bisporus, 1
Coprinus brassicae Peck (= *C. urticicola*),
    11
Coprinus cinereus, 1, 7, 8, 9, 10, 11, 12, 13,
    18, 19, 20, 21, 22, 23, 27
Coprinus clastophyllus Maniotis, 1
Coprinus comatus, 1
Coprinus congregatus, 11, 12, 14, 27
Coprinus cordisporus Gibbs, 11
Coprinus corthurnatus Godey, 11
Coprinus cubensis Berk. & Curt., 2, 11
Coprinus curtus, 11, 12
Coprinus delicatulus Apinis, 1
Coprinus disseminatus, 11
Coprinus domesticus, 11, 12
Coprinus ellisii, 11
Coprinus ephemeroides, 11
Coprinus ephemerus, 11
Coprinus filamentifer Kühner, 11
Coprinus gonophyllus, 11
Coprinus heptemerus, 11
Coprinus hexagonosporus Joss, 11
Coprinus lagopus, 11
Coprinus macrocephalus, 11
Coprinus macrorhizus (Pers.: Fr.) Rea
    (= *C. cinereus*), 11, 19
Coprinus miser, 11
Coprinus narcoticus, 11

Coprinus niveus, 11, 12
Coprinus patouillardii, 8, 11
Coprinus phaeosporus Karst., 11
Coprinus plicatilis, 11
Coprinus poliomallus Romagn., 27
Coprinus radians, 11
Coprinus radiatus, 11, 12
Coprinus sclerotiger Watling, 1, 11
Coprinus stellatus, 11
Coprinus stercorarius, 11
Coprinus sterquilinus, 1, 12
Coprinus stiriacus Knoll, 12
Coprinus tomentosus (Bull.: Fr.) Fr., 11
Coprinus tuberosus Quélet, 11
Coprinus xenobius P. D. Orton, 11
Coriolus versicolor, 7, 9
Corticium, 7
Cortinarius, 2, 5, 11
Cortinarius alboviolaceus, 11
Cortinarius anomalus, 11
Cortinarius callisteus, 11
Cortinarius lepidopus, 11
Cortinarius paleaceus, 11
Cortinarius zakii Ammirati & A. H. Smith,
    1, 6
Craterellus, 11
Crepidotus cuneiformis Pat., 2
Crinipellis, 2
Crinipellis eggersii Pat., 2
Crinipellis perniciosa (Stahel) Sing., 1, 2, 4,
    6, 7
Crinipellis trinitatis Dennis, 2
Cryptococcus, 7
Cystoderma, 11
Cystoderma amianthinum, 2
Cystoderma luteohemisphericum Dennis, 2
Cystofilobasidium capitatum Oberwinkler,
    Bandoni, Blanz & Kisimova-Horovitz, 7
Cyttarophyllum, 1

Dendrosporomyces prolifer Nawawi,
    Webster & Davey, 7
Dicellomyces, 7
Dictyonema, 7
Dictyostelium, 20
Dictyostelium discoideum Raper, 12, 16
Digitatispora marina Doguet, 7
Drosella, 11

Endogone flammicorona Trappe &
    Gerdemann, 6
Entoloma arbortivum (Berk. & Curt.)
    Donk, 1, 11
Entoloma vernum Lundell, 11
Eocronartium, 7
Erisyphe pisi (DC.) St. Amans, 6
Exobasidium, 7

Favolaschia cinnabarina (Berk. & Curt.)
  Pat., 2
Favolaschia dybowskyana (Sing.) Sing., 2
Favolus, 27
Filobasidiella, 7
Filobasidiella depauperata (T. Petch) R. A.
  Samson, J. A. Stalpers & A. C. M.
  Weijman, 7
Filobasidiella neoformans Kwon-Chang, 7
Filobasidium, 7
Filobasidium capsuligenum (Fell, Statzell,
  Hunter & Phaff) Rodrigues, 7
Filobasidium floriforme Olive, 7
Fistulinella, 1
Flammulaster carphophila (Fr.: Fr.) Watling
  (= *Flocculina*), 1
Flammulina, 11
Flammulina velutipes, 1, 2, 8, 18, 23, 27
Flocculina aff. *carphophila*, 1, 11
Fuscoboletinus, 1
Fuscoboletinus aeruginascens (Secr.)
  Pomerl. & Smith, 6

Galerina, 1, 11
Galerina ampullaceocystis, 11
Galerina mycenopsis, 11
Galeropsis, 1
Geopetalum (= *Hohenbuehelia*), 11
Gerronema, 2
Gerronema icterinum (Sing.) Sing., 2
Gliocladium roseum (Link) Bainier, 2
Gomphidius, 11
Gomphidius roseus, 1
Gymnopilus, 11, 26
Gymnopilus aculeatus (Bres.) Sing., 2
Gymnopilus chrysopellus (Berk. & Curt.)
  Murr., 2
Gymnopilus hybridus, 11
Gymnopilus junonius, 11
Gymnopilus penetrans, 11
Gyrodon, 11
Gyromitra esculenta (Pers.) Fr., 26
Gyroporus, 11

Hebeloma, 5, 6, 11
Hebeloma cavipes Huijsman, 11
Hebeloma crustuliniforme, 5
Hebeloma populinum Romagn., 11
Hebeloma radicosum, 1
Hebeloma sarcophyllum (Peck) Sacc., 1
Hebeloma vinosophyllum Hongo, 1
Heimiella, 11
Helicobasidium mompa Tanaka, 7
Helostroma album (Desm.) Pat.
  (= *Microstroma*), 7
Hemimycena, 2
Herpobasidium, 7

Hexagona, 2
Hirneola, 7
Hirschioporus, 7
Hirschioporus abietinus, 7
Hirschioporus fuscoviolaceus, 7
Hirschioporus laricinus (Karst.) Teramoto,
  7
Hirschioporus pargamenus (Fr.) Bond. &
  Sing., 7
Hirschioporus subchartaceus (Murr.) Bond.
  & Sing., 7
Hohenbuehelia, 1, 11
Hohenbuehelia geogina, 11
Hyalodendron, 7
Hyalodendron lignicola Diddens, 7
Hydnum rufescens Pers.: Fr. 11
Hydropus, 2
Hydropus paraensis Sing., 2
Hygrocybe (= *Hygrophorus* pro parte in
  Checklist), 11
Hygrocybe cf. *ceracea*, 11
Hygrocybe lilacina, 11
Hygrocybe nivea, 11
Hygrocybe psittacina, 11
Hygrophorus, 11
Hygrophoropsis, 11
Hygrotrama, 11
Hypholoma, 11
Hypholoma capnoides, 1
Hypholoma elongatum, 11
Hypholoma fasciculare, 1, 9
Hypholoma marginatum, 11
Hypholoma subericaeum, 11
Hypholoma sublateritium, 1

Inocybe, 5, 11, 26
Inocybe fastigiata, 11
Inocybe lanuginella, 5
Inocybe leptocystis Atk., 11
Inocybe longicystis, 5
Inonotus hispidus, 7

Jola, 7

Kriegeria, 7
Kuehneromyces mutabilis (= *Galerina*), 11

Laccaria, 2, 5, 11
Laccaria bicolor, 11
Laccaria laccata, 5
Laccaria tortilis, 5
Lacrymaria, 11
Lactarius, 2, 5, 6, 11, 27
Lactarius deterrimus Groger, 11
Lactarius glyciosmus, 11
Lactarius pubescens, 5, 11
Lactarius rufus, 5
Lactarius torminosus, 11

Leccinum, 1, 5, 11
Leccinum versipelle, 11
Lentinellus, 11
Lentinellus cochleatus, 11
Lentinula edodes (Berk.) Pegler, 11
Lentinus, 11
Lentinus edodes (= *Lentinula*), 15, 17, 23,
    24, 25, 26
Lentodium squamulosum Morgan, 11
Lepiota, 2, 11
Lepiota (*Macrolepiota*) *bohemica*
    Wichansky, 11
Lepiota clypeolaria, 1
Lepiota pseudoroseola Dennis, 2
Lepiota termitophila Heim, 2
Lepiota xanthophylla, 11
Lepista, 1, 11
Lepista nuda, 1
Leptonia howellii (Peck) Dennis, 2
Leucoagaricus, 1, 2, 11
Leucoagaricus bresadolae, 11
Leucoagaricus carnefolius (Gill.) Wasser, 11
Leucocoprinus, 1, 2, 11
Leucocoprinus birnbaumii (Corda) Sing., 11
Leucocoprinus cepaestipes, 2, 27
Leucocortinarius, 11
Leucopaxillus, 11
Leucosporidium scottii Fell, Statzell,
    Hunter & Phaff, 7
Limacella, 11
Lyophyllum, 1, 2, 11
Lyophyllum connatum, 11
Lyophyllum ulmarium (Bull.: Fr.) Kühn., 1

'*Macrolepiota*' *bohemica* = *Lepiota*
    bohemica Wichansky, 11
Marasmiellus, 2, 11
Marasmiellus cocophilus, 2
Marasmiellus nigripes (Schwein.) Sing., 2
Marasmiellus ramealis (= *Marasmius* in
    Checklist), 1
Marasmiellus semiustus (Berk. & Curt.)
    Sing., 2
Marasmius, 1, 2, 6, 10, 11
Marasmius androsaceus, 1, 2
Marasmius cladophyllus Berk., 2
Marasmius coniatus Berk. & Br., 6
Marasmius crinesqui Müller ex Kalchbr., 2
Marasmius cyphella Dennis & Reid, 2
Marasmius epiphyllus, 11
Marasmius equicrinus Müller apud Berk., 2
Marasmius ferrugineus (Berk.) Berk. &
    Curt., 2
Marasmius foliicola Sing., 2
Marasmius griseoviolaceus Petch, 2
Marasmius haematocephalus (Mont.)
    Sing., 2

Marasmius hudsonii, 11
Marasmius pallescens Murr., 2
Marasmius perniciosa = (*Crinipellis*), 4
Marasmius pulcher, 2
Marasmius rotula, 1
Megacollybia (included in *Tricholomopsis*
    in Checklist), 1
Melanoleuca, 1, 11
Melanophyllum, 11
Melanotus, 11
Melanotus vorax Horak, 11
Micromphale, 11
Micromphale perforans, 1
Microstroma juglandis (Béreng) Sacc., 7
Moniliella, 7
Moniliophthora roreri (Cif.) Evans,
    Stalpers, Samson & Benny, 2, 7
Montagnea, 1
Mycena, 2, 6, 11, 27
Mycena citricolor (Berk. & Curt.) Sacc., 1,
    2
Mycena crocata, 1
Mycena epipterygia, 1
Mycena fibula, 11
Mycena galericulata, 11
Mycena galopus, 1, 2, 11
Mycena haematopus, 1
Mycena osmundicola, 2
Mycena pelianthina, 1
Mycena polyadelpha, 2
Mycena pullata, 11
Mycena pura, 1
Mycena purpureofusca (Peck) Sacc., 11
Mycena veneta Stevenson, 11
Mycogloea, 7
Mylittopsis, 7
Myxomphalia, 11
Myxomphalia maura, 11

Naucoria, 1, 11
Nematoctonus, 1
Neosartorya fischeri (Wehmer) Malloch &
    Cain, 7
Neotyphula, 7
Neurospora, 9
Neurospora crassa Shear & Dodge, 7, 8,
    10, 12, 19, 20
Nia vibrissa R. T. Moore & Meyers, 7
Nyctalis (= *Asterophora* in Checklist), 1,
    11
Nyctalis asterophora, 1

'*Omphalia*' *flavella* (Berk. & Curt.) Sacc.,
    2
Omphalina, 1, 11
Omphalina cupulatoides P. D. Orton, 11
Omphalina ericetorum, 11

Omphalina hudsoniana Jennings, 11
Omphalina luteovitellina, 11
Onnia, 7
Onnia circinata (Fr.) Karst., 7
Onnia leporina (Fr.) H. Jahn, 7
Onnia tomentosa (Fr.) Karst., 7
Oudemansiella, 6, 11
Oudemansiella radicata, 10
Oudemansiella steffendii (Rick.) Sing., 2

Panaeolina, 11
Panaeolus, 11, 26
Panaeolus antillarum (Fr.) Dennis, 11
Panaeolus phaleanarum, 11
Panaeolus subbalteatus, 11
Panellus, 11
Paraphelaria, 7
Paxillus, 1, 5, 11
Paxillus atrotomentosus, 11
Paxillus involutus, 5, 6
Penicillium, 10
Peronospora trifoliorum De Bary, 6
Phaeolepiota, 11
Phallus impudicus L.: Pers., 10
Phanerochaete, 9
Phanerochaete velutina (Fr.) Karst., 9, 10
Phellinus torulosus, 7
Phleogena, 7
Pholiota, 1, 6, 11
Pholiota adiposa, 7
Pholiota alnicola, 11
Pholiota highlandensis (Peck) Smith &
  Hesler (= *P. carbonaria*), 11
Pholiota lubrica, 11
Pholiota nameko (T. Ito) S. Ito & Imai, 7,
  23
Pholiota pusilla (Peck) Smith & Hesler, 11
Pholiota scamba, 11
Pholiota squarrosa, 1
Pholiota terrestris Overholts, 7, 11
Pholiota tuberculosa, 11
Phycomyces, 12
Phylloporia, 7
Phyllotopsis, 11
Physalacria, 11
Pietra fungaja, 27
Pisolithus tinctorius (= *arrhizus* (Pers.)
  Rauschert), 5
Pleurotellus, 11
Pleurotus, 1, 11
Pleurotus cornucopiae, 1
Pleurotus cystidiosus O. K. Miller, 7
Pleurotus eryngii DC.: Fr., 1
Pleurotus eugrammus (Mont.) Dennis
  (= *Nothopanus*), 2
Pleurotus flabellatus (Berk. & Br.) Sacc., 11
Pleurotus fossulatus (Cooke) Sacc., 1

Pleurotus ostreatus, 12, 17, 23
Pluteus, 2, 11, 27
Pluteus laetifrons (Berk. & Curt.) Sacc., 2
Podaxis, 2
Pogonomyces (= *Hexagona* fide Pegler), 2
Polyporus, 11
Polyporus biennis (= *Heteroporus*), 7
Polyporus brumalis, 12
Polyporus ciliatus Fr., 19
Polyporus squamosus, 11
Porphyrellus, 1
Psathyra, 11
Psathyrella, 2, 11
Psathyrella ammophila, 1
Psathyrella coprophila, 11
Psathyrella epimyces (Peck) A. H. Smith, 1
Psathyrella polycystis Romagn., 11
Psathyrella spadiceogrisea, 11
Pseudotulasnella guatemalensis Lowy, 7
Psilocybe, 11, 26, 27
Psilocybe baeocystis Sing. & Smith, 26
Psilocybe cyanescens, 11
Psilocybe fimetaria (P. D. Orton) Watling,
  11
Psilocybe merdaria, 11
Psilocybe mexicana Heim, 26
Pterospora, 6
Pterula plumosa (Schw.) Fr., 2
Pulveroboletus, 1, 11

Resupinatus, 11
Resupinatus cyphelliformis, 11
Rhodocybe, 11
Rhodotus, 11
Rhodotus palmatus, 1
Rhizoctonia, 6, 7, 9
Rhizoctonia crocorum (Pers.) DC., 7
Rhizopogon, 5
Rhodosporidium capitatum Fell, Hunter &
  Tallman, 7
Rhodosporidium toruloides Banno, 7
Ripartites, 11
Rogersiomyces okefenokeensis Crane &
  Schoknecht, 7
Rozites, 11
Rozites caperatus, 1
Russula, 2, 5, 6, 11, 27
Russula adusta, 1
Russula annulata Heim, 1
Russula brevipes Peck, 5
Russula emetica, 5
Russula fellea, 11
Russula ochroleuca, 1
Russula xerampelina, 1

Saccharomyces cerevisiae Hansen, 8, 20
Schizophyllum, 1, 11

Schizophyllum commune, 1, 2, 7, 8, 9, 10, 12, 14, 15, 16, 17, 18, 19, 20, 21, 22, 23, 27
Schizosaccharomyces pombe Linder, 20
Scleroderma, 1, 5
Sclerotium hydrophilum, Sacc., 7
Septobasidium, 7
Septobasidium burtii Lloyd, 7
Serpula lacrimans (Wulf.: Fr.) Schroet., 10
Sistotrema brinkmanii (Bres.) J. Erikss. (= *Trechispora*), 19, 20
Sordaria humana (Fuckel) Winter, 7, 18
Sphaerobolus stellatus Tode: Pers, 7
Sphaerostilbe repens Berk. & Br., 10
Squamanita, 11
Squamanita odorata, 1
Stereum hirsutum (Willd.: Fr.) S. F. Gray, 10
Stilbotulasnella conidiophora Bandoni & Oberw., 7
Stilbum, 7
Strobilomyces, 1, 11
Strobilomyces polypyramis Hook f. apud Berk., 11
Strobilurus tenacellus (Pers.: Fr.) Sing. (= *Pseudohiatula*), 11
Stropharia, 11, 26
Stropharia cyanea (Bolt: Secr.) Tuomikoski, 11
Suillus (= *Boletus* pro parte of Checklist), 1, 5, 11
Suillus bovinus, 1, 6
Suillus granulatus, 5
Suillus grevillei, 1, 6
Suillus luteus, 5
Syzygospora, 7

Tectella, 11
Tephrocybe, 1, 11
Tephrocybe palustris (Peck) Donk, 1, 11
Termitomyces, 1, 2, 10, 11
Thelephora terrestris Ehrb.: Fr., 5
Tilletiaria, 7
Tremella, 7
Tremella mesenterica Retz: Hook, 7
Trichoderma viride Tode: Fr., 2

Tricholoma, 1, 5, 11
Tricholoma aurantium, 5
Tricholoma fulvum, 1
Tricholoma imbricatum, 11
Tricholoma muscarium Kawamura, 26
Tricholoma pachymeres (Berk. & Br.) Sacc., 2
Tricholoma sulphureum, 1
Tricholoma vaccinum, 11
Tricholoma virgatum, 11
Tricholomopsis, 11
Tricholomopsis decora, 11
Tricholomopsis platyphylla (= *Megacollybia*), 1, 10
Trichosporon, 7
Trogia, 2, 11
Tubaria, 11
Tubaria autochthona, 1
Tubaria conspersa, 11
Tulasnella, 7
Tylopilus, 1, 2, 11
Typhula trifolii Rostr., 9

Ustilago violacea (Pers.) Roussel, 7
Uthatobasidium, 7

Venturia inaequalis (Cke.) Winter, 6
Volvariella, 11
Volvariella bombycina, 2
Volvariella esculenta (Mass.) Sing., 11
Volvariella volvacea, 24, 25

Waitea, 7
Wallemia sebi (Fr.) v. Arx, 7

Xenogloea, 7
Xerocomus (= *Boletus* pro parte of Checklist), 1, 2, 11
Xerocomus astraeicola Imazeki, 1
Xerocomus parasiticus (Bull.: Fr.) Quél., 1
Xerocomus radicicola Singer & Araujo apud Singer, 1
Xerocomus subtomentosus (L.: Fr.) Quél., 6
Xylaria, 10
Xylaria longipes Nits., 10
Xylaria polymorpha (Pers.: Mérat) Grev., 10

Subject index

Numbers in **bold type** refer to pages on which relevant figures or tables appear.

A₁-layer agarics, 66–7
abiotic factors, 30, 162
  which affect morphogenesis, 256–8
N-acetyl-β-D-glucosaminidases, 405
N-acetyl-muramidase, 376
N-acetyl-N[1-³H]glucosamine, developing septa labelled with, 179
A-factor, 263
agaric communities,
  structure in tropical forest of, 77–82
agaric symbiont, 146
agarics,
  A₁-layer, 66–7
  biotrophy in, **5**, 42
  classification of, 31
  developmental biology of, 451, 499–503, 581–93
  developmental characters of, 281–307
  ecological strategies of, 30–5
  ecology of tropical, 55–67, **78**
  and ecosystems, 42–4
  and ectomycorrhizas, **48**
  F₁-layer, 65–6
  F₂-layer, 66
  facultative, 47
  folicolous, 54, 60–2, 76, 77
  geophiloid, 301
  heteromictic species of, 16
  and homokaryotic fruiting, **456**
  homomictic or heteromictic species of, 16
  L-layer, 62–5
  life cycle of, 451–2
  lignicolous, 54, 58–60, 76
  mycorrhizal associations of, 141–64
  necrotrophy in, **6**, 42, **50**
  non-component-restricted, 18
  r- and K-selected, 80, 81

replacement interactions between litter-inhabiting, **259**
resource utilisation, 67–70
resource-selected, 44
saprotrophy in, **8**, **9**, 12, 43, 44, 77
secondary metabolic products of, 561–75
taxon selectivity in, **11, 13**
terricolous, 54, 76, 77
tropical, 41–82
ubiquitous, 44
Agaricus, biochemistry of fructification, 389–99
Agaricus bisporus,
  axenic and non-axenic cultures of, 379
  commercial growing of, 531
  crop and cropping of, 534–6
  crop production in, 530, 541–7
  enzymes excreted by, 376
  flush of, 535
  initiation-inhibiting substance, 391
  modulation of endocellulase activity during fruiting of, 382–3
  mushroom breeding, 513–26
  mycelium and fruit body in, 391–6
  primordia, of, 394
  secondary metabolic products of, 562–67
  system of production of, 536–7
  and taxonomists, 529
agaridoxine, 564, **565**
agaritine, **565**
alleles, 467, 518
  for fruit body morphology, 516
  'good fruiting', 516
  non-self, 216
  'poor fruiting', 516
Amanitaceae, developmental characters of, 297–8

Amanita muscaria, secondary metabolites of, 571–3
angiocarpy, 303
  definition of, 287
aphyllophores, 192
apothecium, 177, 271
arbutoid hosts, ectendomycorrhizas on, 164
arbutoid mycorrhizas, 25, 141, 144, 154–7,
arginase, 335, 336
arginine, 335, 336
Armillaria,
  basidiospores of, 98
  and colonisation of natural substrates, 92–7
  and *Fraxinus*, 87
  genera attacked by *A. mellea*, 88
  genera attacked by *A. ostoyea*, 88
  growth in colonised material, 95–6
  and inoculation of broad-leaved species, **92**
  mycelial sheets of, 92
  as necrotrophic parasite, 141
  ordering of structure in rhizomorphs of, 265–74
  pathogenicity tests using, 90–2
  resources and hosts of, 87–100
  species of, 88
  species isolated from butt-rotted trees and stumps, **90**
  species isolated from conifers, **89**
Armillaria bulbosa, 99
  as effective saprotroph, 100
  rhizomorphs of, 100
Armillaria mellea, 97–8
  extensive host range of, 100
Armillaria ostoyae, 98–9
  and conifers, 100
Armillaria tabescens, 99–100
arthroconidium (oidium), 16, 34, 35
ascomycetes, 177, 178
  septal sealing in, 231–2
auricularialian complex, 198–200
autonomously replicating sequences, 474
auxotrophic markers, 516, 522

B-factor, 215, 263, 467–8
baeocystin, 570
basidia, 175, 312, 347, 349, 439
  apo-, **202–3**
  clavate, 195
  cruciate-septate, 196
  developmental stages of phragmo-, **202**
  dimorphic/trimorphic of *Coprinus cinereus*, 289
  holo-, 195, 199, **202**
  hypo-, 195
  NADP–GDH activity in, 338

parallel-septate, 196
  pro-, **202**
basidioma, 2, 281
basidiomycetes, 177, 178
  bipolar and tetrapolar breeding systems in, 215–16
  clamp-forming, 157
  extracellular enzymes during morphogenesis of, 375–99
  features of some cultivated, **515**
  multi-allelic incompatibility control in, **214**
  nucleotide sequence of 5S rRNAs of, **200**
Basidiomycota, 194
  diagram of dolipore/parenthesome septal types, **192**
basidiospores, 16, 124, 194, 311
  of *Armillaria mellea*, 98
  compressed fusiform, 32
  of *Crinipellis perniciosa*, 103, 105, 109
  floriform clusters of, **202–3**
Betula,
  agaricoid ectomycorrhizas of, 146
  and *Armillaria*, 87
  mycorrhizal formation on seedlings of, **124**
  and mycorrhizal fruit body species, **120**, 121
  number of root tips of different mycorrhizal types, **125**
biotic factors, which affect morphogenesis, 258–60
biotrophic parasite, 160
biotrophs, 4, 142
  ectomycorrhizal, 47–9, 77
  mutualistic, 141
  parasitic, 141
biotrophy, 4, 22, 117
  among agarics, **5**, 45
  definition of, 2
  and formation of ectomycorrhizas, 4
  and mutualistic and parasitic associations, 4
  mutualistic symbiotic, 117
  occurrence of, 2–4
  selective, 32
bipolar breeding system, 219
  in higher basidiomycetes, 215–16
bipolar (unifactorial) control, 514
bivelangiocarpy, 283–305
  definition of, 287
boletes, 293, 294
  classification of, 31–3
breeding programme,
  for heterothallic fungi, 519–21
  in mushrooms, 514–15, 517–19
  for secondary homothallic fungi, 519–24

breeding system, 213, 215–16
  bipolar and tetrapolar in higher
    basidiomycetes, 215–16
bulbangiocarpy, **283**, 296, 306
butt-rot,
  and *Armillaria* species, **90**
  of *Picea abies*, 87

C-selection, 30
cabecitas, see stilboid, pedicillate
calcium, in mushroom compost, 554
callose-like encasement, 154
callose-like material, **149**, 150
carbohydrate metabolism by fruit body,
  396–8
carbohydrates,
  flushing of, 394
  and mycorrhizal development, 129
carpophoroids, 591
cell,
  antepenultimate, 176
  apical, 219
  clamp, 219
  donor, **242–3**
  palisade, 340
  penultimate, 176, 234
  recipient, **242–3**
  sending, 350
  transfer, 150–62
  ultimate, 176
cell formation, 584–6
cell inflation, 589–90
cell wall metabolism, and stipe elongation
  in *Coprinus cinereus*, 317–20
cell wall microfibrils, disorganised, 142
cell wall precursors, 158, 159
cellulose, 376, 532
  in composts, 554
  oligo-dT, 505
Chanter model of fruit body initiation,
  395
chitin, 317, 318
  synthesis of, 179
chlamydospores, 16, 17, 452
chromatin, condensation of, 241–3, **244–5**
U-[^{14}C]-citrulline, 335
clamp connections, **155**, 186, 195, 219, 354,
  514, 582
  in *Crinipellis perniciosa* hyphae, 104
  false, **222**
  formation of, 235–8, **240–1**
  intercalary, 241
  section through in *Polyporus biennis*,
    180–1
clamp fusion, **237–8**, 239
clamp initial, **237–8**, **240–1**
cleistothecium, 176

cloning sequences, involved in fruit body
  morphogenesis, 503–7
cocoa,
  Amelonado seedlings and *Crinipellis
    perniciosa*, **105**, 111, **112**
  annual production and decomposition of
    litter, **57**
  and *Crinipellis perniciosa*, 103–14
  Crinipellis perniciosa as pathogen of,
    53–4
  EET 400 seedlings and *C. perniciosa*,
    106, 108
  hyperplasia and hypertrophy in, 107
  Moniliophthora roreri as pathogen of, 69
  Scavina 6 seedlings and *C. perniciosa*,
    105–9
cocoa litter, **61**, **63**, **64**, **68**, 79
  fruit bodies in, **71**, **72**, **75**
coenocytes, 585
colony hybridisation, differential, **504**
combative strategy, 30, 31
composting, 532–4
  nutrients and, 553–5
  phases in, 532, 548
composting process, 547–9, 549–55
composting technology, 541–7
compost selectivity, 534
compost tunnels, 533, 549
composts, 532–4
  long, 547
  short, 548
concanavalin-A (con-A), 360
conidium, 35, 177
  arthric, 196
  catenate, 196
Coprinaceae (inky caps), developmental
  characters of, 298–301
coprine, **573**, 574
Coprinus, development of species of, **300**,
  301
Coprinus cinereus,
  cystidia of, **346**
  development of fruit body structure in,
    339–48
  developmental stages in, **334**
  elongation of stipe of, 311–27
  elongationless strain of, 312, 316, 318
  enzyme activities in fruit bodies, of, **335**,
    336
  as experimental species for dikaryon
    formation, 213
  fruit body cap of, 333–50
  growth hormones in, 324
  hymenium structure, 345–8
  paraphyses in, 336, 347, 348, 349
  pathway for sclerotium and fruit body
    development in, **461**

Coprinus cinereus (*cont.*)
  phototropism and geotropism in, 320–4
  sporeless strain of, 312, 316, 327
  squash preparations of basidia of, **502**
  stipe elongation, **316**
  synchronous development in, 499–503
Coprinus congregatus,
  differentiation in, 353–71
  primordium development and
    translocation events in, **368**
cords, 18, 19, 20, 26, 35, 45
  mucilage in, 269
  of *Phanerochaete velutina*, 254
  as type of vegetative organ, 250
Coriolus versicolor, post-fusion events in,
  238–45
cortical cell, **143**, 144, **145**, **149**
  outer, **147–8**, **149**, **155**
Cortinariaceae, developmental
  characters of, 302–5
Crinipellis perniciosa,
  basidiospores of, 103, 105
  as biotrophic parasite, 141
  as cocoa pathogen, 53–4
  growth on living and dead cocoa tissue,
    103–14
  parasitic mycelium of, 104
  reactions between paired isolates of,
    113
  saprotrophic mycelium of, 104
  saprotrophic phase,
    environment–saprotroph relationships,
      10–14
crozier, in *Neosartorya fischeri*, **176**
cusps, meristematic, **202–3**
cyclic AMP, 263, 354, 369, 391
cyclic AMP metabolism, 583
cyclic AMP synthesising enzyme, adenylate
  cyclase as, 391
cystidia, **346**, 349, **433**, **434**
  of *Flammulina velutipes*, 433–43
cytochrome oxidase deficiency, 224
cytology, of hyphal interactions and
  reactions in *Schizophyllum commune*,
    231–46

DAPI, 325, 326
deheterokaryotisation, 524, 525
Diazonium Blue B, 177
dikaryon, 182, 241, 263, 318, 326, 353, 458
  definition of, 213
  formation of, 213–27
  reciprocal, **224**
  of *Schizophyllum commune*, 467
dikaryotic fruiting, 461–3
disease,
  horsehair blight, 52
  thread blight, 52
  witches broom, 53, 67, 103–14
disulfiram, 573
DNA, 325, 327, 397, 415
  cloned, 525
  light satellite, 474, 475
  mitochondrial, **227**, 474, 478, **479**, 589
DNA sequence polymorphisms, 509
DNA I, 472
cDNA, 491, 492, 503
  dikaryotic, 493
rDNA, 476, 478, 479, 481, 507, 589
DNase, 376, 473
dolipore formation, **181**
dolipore/parenthesome septum, 586
  ascomycete *versus* basidiomycete, 191–2
  challenge of, 175–204
  diagram of rust, **197**
  diagram of types in Basidiomycota, **192**
  of *Helicobasidium mompa*, **199**
  outer cap of, **190**
dolipore septum, **143**, **145**, **149**, 156,
  179–83, 219, **232–3**, **237–8**, 271, 584,
  586
  barrel-shaped, 179
  interpretation of homobasidiomycete,
    184
  in monokaryon, **220**
drosophin A, 562
drosophin A methyl ether, 562

economics, of mushroom production,
  543–7
ecosystems, and agarics, 42–4
ectendomycorrhizas, 154
  on arbutoid hosts, 164
ectomycorrhizas, 45, 119, 141, 142–51, 162,
  163
  associated with agarics, **48**
  carbohydrate requirement of, 126–30
  compatible interactions of, 142–6
  formation of, 25
  incompatible interactions of, 146–51
  obligate, 32, 33
  between *Pinus sylvestris* and *Suillus
    bovinus*, **149**
  and root exudates, 159
  and ultrastructure of host–argaric
    interface in, **143**
ectomycorrhizal biotroph, 47–9, 77
electrolucent material, 178, 179
electron-dense material, 142–53, 156, 177,
  179, 191, 233, **242–3**, 435
electron-dense plug, **232–3**, 234
electropaque layers, 178
electrophoretogram, **473**
Emerson's YpSs medium, 363

emitanin, 568
endocellulase, 376, 383
endoplasmic reticulum, 177, **180–1**, 185,
    186, 190, 442
    continuities of, 187
    elements of, 196
    parenthesome-generating, 191
environmental stress, 28
enzymatic markers, 522
enzymes,
    and basidiomycetes, 404–6
    diffusible hydrolytic, 144
    excreted by *Agaricus bisporus*, 376
enzyme activity, extracellular, 377–8
enzyme inactivation systems, 380–1
enzyme production, control and
    localisation of, 377–83
enzyme-mediated penetration, 144
epibasidium 196
ericoid mycorrhiza, 141–58
eritadenine, 568, 569
exit patterns by fungus, 27–30
    via mycelium, 29–30
    via reproduction, 27–9
exocellulase, 133, 376
extracellular material, 447
    of *Flammulina velutipes*, 439–44
    functions of, in *F. velutipes*, 444

FIS (fruiting inducing substances), 459, 460
fixation, 367
    definition of, 356
Flammulina velutipes,
    fruit body differentiation of, 429–48
    fruit body structure and development
        in **430**
    sequence of development of, 431–4
    structural integrity of young fruit body,
        444–7
    structural organisation of developing
        fruit body, 434–4
    structure of young fruit body, 447–8
flavonoids, 24
flushing, 395
    mathematical model of, 394
flushing cycle, 397
foliicolous agarics, 54, 60–2, 76, 77
forest succession, diagrammatic
    representation of, **118**
fruit body,
    abnormal types of, 462
    carbohydrate metabolism in, 391–6
    and developmental genetics, 454–63
    differentiation of *Flammulina velutipes*,
        429–47
    formation in *Schizophyllum commune*,
        485–95

initiation in *Coprinus congregatus*, 355
    and gene control, 457–9
    of higher fungi in cocoa litter, **71**
    under *Pinus contorta*, **122**
    and spore production, 500
    structure of *Flammulina velutipes*, 447–8
fruit body cap, of *Coprinus cinereus*,
    333–50
fruit body development, 485–8
    and polypeptide synthesis, 488–90
fruit body morphogenesis, cloning
    sequences during, 503–7
fruit body production, 70–3
fruiting, 41–82
    of *Agaricus bisporus*, 382–3
    dikaryotic, 461–3
    genes which control, **459**
fungal establishment and exploitation,
    interactions during, 20–6
    microenvironmental factors during,
        22–3
    spread and patterns of, 25–7
    stimulation and inhibition by chemical
        factors, 23–5
fungal peg, 144, **149**, 151, **152**, **155**, 161
    degenerating, 154
    membraneous sac in, 153
    in *Monotropa*, 153
fungal plasmalemma, **145**
fungal sheath, **147–8**
fungi,
    cup, 177
    early-stage, 124, 125, 127, 128, 130, 134
    growth of mycorrhizal, 127
    late-stage, 124–34
    point growth of, 261, **262**
    post-fusion event in, 231

G6PD (glucose 6-phosphate
    dehydrogenase), 393, 398
gap junctions, 186, 349, 587
    diagrams of thin sections of, **187**
    intercellular, **188**
    interpretation of structure of, **189**
GBQ (γ-glutaminyl-3,4-benzoquinone),
    563, 564
GC (guanine plus cytosine), 471
genes,
    copper-chelatin, 509
    which control fruiting, **459**
    fi, 458, 460
    hap, 460
    in haploid fruiting, 487
    incompatibility, 468–9
    mating type, 215–16, 468
    mitochondrial, **225**
    in mushroom breeding, 516

genes (*cont.*)
    segregation of mitochondrial, **226**
    tandem rRNA, **508**, 509
genetic components, for mushroom
    breeding, 517–19
genetic markers, 588
genetic recombination, 213
    in *Coprinus cinereus*, 499–510
Geneticin, 481
genetics, developmental, 451–63
genome, mitochondrial, 589
genome structure, in *Schizophyllum
    commune*, 490
geophiloid agarics, developmental
    characters of, 302
geotropic swinging, **322**, 323
geotropism, in *Coprinus cinereus*, 320–4
GHB (γ-L-glutaminyl-4-hydroxybenzene),
    563, 564, 565
gill,
    cytochemistry of, 338
    DNA content of, 397
    histological structure of, 345–8
gill cavity, **344**
glucan, 317, 319, 487
R-glucan–chitin, in walls of
    Schizophyllum commune, 183
S-glucan, 183
glucanase, 319
R-glucanase, 183
glucosaminidase, 376
^{14}C-glucose, 127, 128, 159
β-glucosidase, 376, 405
glutamate decarboxylation loop, 337
glutamine, 129, 574
glycogen, **143**, **147–8**, 394, 395, 396
    and stipe elongation in *Coprinus
    cinereus*, 315–17
glycoproteins, 163
Golgi body, **149**, 375
GS (glutamine synthetase), 334, 337,
    338
gymnangiocarpy, **284**, 299, **300**
gymnocarpy, 282–305
    definition of, 287
    Lentinellus as species, 296
gyromitrin, 566

habitat,
    considerations of, 44–5
    as criterion in description of fungal
        species, 12
    definition of, 1 (footnote)
    lignocellulose, 42
    of tropical agarics, 57–8
    type of, 1–15
half-gill segments, 500
haplo-dikaryotic life cycle, 175

haploid fruiting, 487
Hartig net, 25, 142–64
haustorial sheath, 158
haustorium, 2, 148, 150, 153, 161
hemiangiocarpy, definition of, 287
hemibiotroph, 161
hemicellulose, 376, 532, 554
heteroallelic dikaryon, 461
heterobasidia, 194
heterokaryon, 219, 353, 467, 515
heterothallic species, breeding system in,
    514, 519–21
HMB
    (4-hydroxymethylbenzenediazonium),
    565, 566, 567
HMPH
    (4-hydroxymethylphenylhydrazine),
    565, 566
holobasidia, 195
homoallelism, **452**, 458
homokaryon, 263, 264, 458
    diploid, 353
    pairing of, 468
    of *Schizophyllum commune*, 467
homokaryon interaction, **218**
homokaryotic fruiting, 455–7
homothallic species,
    breeding strategy for secondary, 521–4
    breeding system in, 514
host cell remains, **143**, **147–8**
host cell wall, **143**, **147–8**, 151, **152**
    invaginated, **149**, **152**, **155**
    thickened, **149**, 150
host–agaric interface, 153
    monotropoid, 151, **152**
host–fungus interface, **143**, 163
    role of, between symbionts of, 157–64
    ultrastructural analysis of, 141–64
host–fungus interface recognition, 24
host–fungus specificity, 119
host–pathogen interactions, between
    Crinipellis perniciosa and *Theobroma
    cacao*, 104–14
host plasma membrane, 153
host plasmalemma, **145**, **155**, 156, 157,
    163
host preferences, of *Armillaria*, 87–90
hosts,
    of *Armillaria*, 87–100
    of *Crinipellis perniciosa*, 103
hymenium, 176, 177, 191, 347, 439, 591
    outer, 190
hymenocarpy, **291**, 297, 301, 302, 306
    definition of, 289
hymenoderm, 298
hymenophore, 298, 340, 342, 345
hymenopileocarpy, 301
    definition of, 289

hyperplasia, in cocoa, 107
hypertrophy, in cocoa, 107
hyphae, **143**, **145**, **147–8**
  acute-angled branching marginal,
    261
  apical, 251
  ascogenous, 177, 271
  aspergilloid, 431
  bifurcation of, **236**
  collateral growth of, 260, 261
  compaction of, 260
  fusion bridge of, **236**
  generative, 582, 589
  induction, 586
  inflating, 295
  inner sheath, 142
  intracellular, **147–8**, **155**
  invading, 142
  leading, 233
  monokaryotic, 217
  mycelial, 582
  protenchyme, 344, 345, 347
  tertiary, 178
  tip-to-side contact of, **236**, **240–1**
  trama, 339, 347
hyphal aggregation, **432**
hyphal anastomosis, 224, 235, **236**, 446
hyphal cohesion, 446
hyphal extension, model of, 235
hyphal fusion, 235–8
hyphal interactions and reactions, cytology
  of, 231–46
hyphal interweaving, 446
Hypholoma fasciculare, mycelial and effuse
  growth phases of, 16
hypobasidia, 195
hypovelangiocarpy, 288

ibotenic acid, 572, 573
incompatibility, 219–21
  cytoplasmically determined, 227
  somatic, 231
incompatibility factors, 456, 488
  A and B, 467, **468**, 487
  effect of mutation in, **222**
incompatibility genes, molecular models of
  regulation by products of, 468–9
incompatibility mechanisms, 213
incompatible pairings, 516
inoculum potential, 18
inocybin, 572
inositol hexaphosphate-P, **132**, 133
interactions and modes of arrival of
  fungus, 15–20
interface,
  between Hartig net and host cortical
    cells of *Pinus*, 144–5
  core–jacket, 182

host–agaric, 146, 151
host–fungus, 148, 150
interfacial matrix, **155**, 158
interfacial zone, 158
intracellular hyphae, **147–8**, **155**
involving layer, **145**, **149**, 156
  in ectomycorrhizas, 158
isocarpy 290–306,
  definition of, 289
isocitrate dehydrogenase, 337
isogenic mycelial lines, 458
isovaleric acid, 453

Japanese Paper Pot culture, **129**

K-selection, 30, 80, 118, **126**, 130, 131
K-selected temperate agarics, 81

laccase, 376, 487
  in basidiomycete morphogenesis, 381–2
  extracellular and basidiomycete
    morphogenesis, 378–80
  high activity, 379
  and *Lentinus edodes*, 420–1
  low activity, 379
laminarinase, 376, 405
lectins, 163
lens effect, 321
lentinacin, 568
lentinans, 568
Lentinus edodes, 529
  growth and development of, 409–10
  laccase and, 420–1
  phosphatases and, 419
  protein production, localisation and
    resistance to proteases by, 413–15
  saccharidases and, 421–3
  secondary metabolic products of, 567–9
  stability of total protein and enzymes in,
    417
  technique for fractionating whole
    cultures of, **408**
  uptake and utilisation of carbon and
    nitrogen source, 410–12, 412–13
  see also shiitake mushroom
lentionine, 569
lentysine, 568
levhymenial, 300, 341, 342
  definition of, 288
Levine ridges, 344
light induction,
  of primordia, 356–9
  inhibition studies of, 359–60
light reception, and phenoloxidase, 360–6
lignicolous agarics, 54, 76
  communities of, 58–60
lignin, 118, 130, 376, 532
  in composts, 554

lignocellulose, 81
lipase, 376
lipsanenchyma, 298, 300, 586
litter, 57–8
  cocoa, **61, 63**
  F₁-layer, 63
  upper F₂-layer, 53
litter fungi, non-component-restricted,
    12–13
litter-inhabiting agarics, **259**
L-layer agarics, 62–5
L-layer material, 79
LRC (light receptor complex), 365, 367,
    369

mannitol, 391–4
mannitol dehydrogenase, 392, 396
MAR (mass action ratio), 396
mating function, 221, 222, 223
mating type, 467, 514
mating type gene mutations, 221–3
mating type genes, 215–6
  silent, 470
mating type loci, 467, 470
mating type switching, 470
medium,
  chemically defined for *Lentinus edodes*,
    403–24
  Emerson's YpSs, 363
  Hagem's, **132**
  malt extract agar, 429
  oakwood/oatmeal, **414**
  PDA, 110
  Trione's, 110
medullary region, of rhizomorph, **252, 266**
meiocyte, 175, 176, 200
meiosis, 213, 499–510
meiospore, 177
  production of ustilage-type, 201
membrane,
  centripetally invaginated cytokinetic, 175
  invaginating, 179
membrane cisternae, 177
membrane matrix, 198
Mendelian inheritance, 515
Mendelian locus, 507
Mendelian segregation, of tandem rRNA
    genes, **508**
meristem, apical, 142, 585
meristemoids, 582, 584
metavelangiocarpy, 282–99, 304
  definition of, 287
  of pleurotoid fungi, 295
method, of deheterokaryotisation, 524, 525
M-factor, and stimulation of mycelial
    growth and spore germination, 20
microfilaments, 184, 185, 271

microsclerotium, 17, 250
microscopy,
  light and preparative methods, 342, **431**
  scanning electron, **431**
  transmission electron, 445, 447, 481
microtubules, 184, 185
mitochondria, 237–8, 269, 473
  cross between two genes, **225**
  genes of, **225, 226**
  inheritance of, 223–7
  recombinant DNA of, **225**
  restriction of DNA, **227**
mixangiocarpy, **282, 292, 294**
  definition of, 287–8
  of pleurotoid fungi, 295
molecular switches, 468
Moniliophthora roreri, as cocoa pathogen,
    69
monokaryon, 178, 182, 318
  definition of, 213
monokaryon/dikaryon transition, 495
Monotropa,
  haustorial tip of, 161
  mycorrhizal root of, **155**
monotropoid mycorrhizas, 141, 151–4, 162
monovelangiocarpy, 283–304
  definition of, 287
morphogenesis,
  of basidiomycete fungi, 375–99
  in dikaryon formation, 216–21
  endogenous control of, 260–74
  exogenous control of, 256–60
  of the fruit body, 457
  genetic control of, 261–3
  metabolic control of, 333–8
  role of laccase in basidiomycete, 381–2
  of vegetative organs, 249–75
MPH (4-methylphenylhydrazine), 566
mucilaginous region, of rhizomorph, **252,
    267, 268,** 269
muscarine, **571**
muscazone, 572
muscimol, 572, 573
mushroom,
  breeding, 513–26
  commercial growing, 529–38
  crop production, **530,** 541–7
  flush of, 535
  objectives and genetic components for,
    517–19
  oyster, 543
  paddy straw, 529, 542
  shiitake, 529, 542, 543 (*see also Lentinus
    edodes*)
  single-gene and polygenic traits in, 524
mutagen,
  bromouracil as, 462

mutagen (*cont.*)
  nitrosoguanidine as, 462
  treatment of heterokaryon with, 221
mycelial colony, 454
mycelial cords, **253**, 255
mycelial domain, 27
mycelial fan, 255
mycelial front, slow dense/fast effuse
    transitions in, 260–3
mycelial modes, 26
mycelial outgrowth patterns
  spectrum of, **251**
mycelial strands, 255
mycelial syrrotia, 255
mycelium,
  arrival by, 18–19
  ascogeneous, 177
  carbohydrate metabolism, in, 391–6
  interstitial, 18
  parasitic, 104
  secondary, 353
  threads of, 255
Mycena citricolor, **51**
mycobiont,
  agaric or basidiomycete, 47
  dolipore/parenthesome septa in, 191
  non-taxon-selective, 49
  Termitomyces as, 44
  vesicular-arbuscular, 77
mycoparasitism, 7
mycorrhizal development, and
    carbohydrate, 129
mycorrhizal dynamics,
  changes in, 119–26
  during forest tree development, 117–35
mycorrhizal formation, recognition of, 24
mycorrhizal fungal symbiont, 118, 127
mycorrhizal fungi,
  growth at three glucose levels, **127**
  strain differences in, 12
  succession of, **126**
  vesicular-arbuscular, 47, 119
mycorrhizal succession, 199
  and resources, 126–33
mycorrhizas, 45, 154
  arbutoid, 141
  developing, 146
  development and fertilizer, **129**
  E-strain, 124
  ericoid, 141
  mature, 146
  monotropoid, 141
  number from *Betula pubescens*, **123**
  orchidaceous, 141
  vesicular-arbuscular, 47, 49, 141

NAD-linked enzyme, 337

NAD/NADP transhydrogenase system,
    392
NADP, 392, 393, 397
  production of, 396
NADP-GDH, 334, 337
  in basidia, 338
NADPH:NADP ratio, 398
necrotrophic parasite, 144
necrotrophy, 4–7, 49–53, 117
  in agarics, **6**, 45
  on animals, 7
  as combative mechanism, 4
  definition of, 2
  occurrence of, 2–4
Neurospora crassa,
  septa of, 177–8
  snowflake morphological mutant of, 271
nitrogen, and development of mycorrhizas,
    129
nitrosoguanidine, 462
non-component-restricted agarics, 18
norbaeocystin, 570
Nothofagus forest, 41, 47
nuclear migration, 223–7
nucleus, **145**, **149**, **155**
  degeneration of, 241–3
  donor cell, **244–5**
  recipient cell, **244–5**
nutrients, availability and uptake, 21–2
nutritional modes, 45–55
nutritional studies, and *Coprinus
    congregatus*, 362–4

OAT (ornithine acetyltransferase), 334
occlusion reaction, 234
occlusions,
  bipartite, 195
  within dolipore, 179, 183–5
OCT (ornithine carbamyltransferase), 334
oidia, 452
oligosaccharide determinants, 163
orchidaceous mycorrhizas, 156, 157
ornithine acetyltransferase, *see* OAT
ornithine carbamyltransferase, *see* OCT
orotidylate decarboxylase, 480
orthophosphate-P, **132**
osmiophiles, 23
osmiophilic material, vesicle-derived, 185
osmiophilic ring, **152**, 153, **155**
ostiole, 177
2-oxoglutarate amination, 337
oxygen, and composting, 550–1

palisade cells, 340, 341, 345, 348
palisade ridges, 344
palisododerm, 298
paracrystalline inclusions, 198

paramural body, **149**, 159
paraphyses, in *Coprinus cinereus*, 336, 347, 348, 349, 501, 587
parasite,
  biotrophic, 141
  facultative, 200
  necrotrophic, 141, 144
  obligate, 199
  weak, 3
parasitic mycelium, of *Crinipellis perniciosa*, 104
parasitism, as a *derived* condition, 200
parasol mushrooms, developmental characters of, 297–8
paravelangiocarpy, 283–304, 437
  definition of, 287
parenthesome membrane, **220**
parenthesomes, 179, 185–90, **232–3**, 234
  imperforate, 187, **193**
  perforate, 187, 194
  perforate versus imperforate, 192–5
  vesiculate, 195–6, **202–3**
pectin, 144
pectin–hemicellulose layer, 157
penetration point, 143
pentose phosphate pathway flux, 396
Percoll gradient, 360
perithecial cell walls, 271
perithecium, 176, 177, 584
phanerogam, 585, 586
phenolic material, **145**
phenolics, 81
Phenoloxidase, *see* PhO
PhO (phenoloxidase), 354, 360–6, 404, 405
  and phase I fruit body development, 487
phosphatase, 404
  acid, **132**
  and *Lentinus edodes*, 419
phosphoglucose isomerase, 396
phosphorus, and development of mycorrhizas, 129
phototropism, in *Coprinus cinereus*, 320–4
phycomycetes, septum of, 175
phytase, 132
phytoalexins, 22, 24
Pichilingue, **75**
  annual rainfall and temperature patterns at, **56**
  climate and plant community structure at, 55–6
  cocoa at, **57**
pilangiocarpy, **282**, **285**, **290**, 294, 296, 305
  definition of, 287–8
pileipelli, 296, 298, 302, 435, **443**
pileocarpy, **290**, **291**, 297, 306
  definition of, 288
pileocystidia, 437, 442

pileodermium, 586
pileostipitocarpy, 290–304
  definition of, 289
pileus, 298
  of *Flammulina velutipes*, 437–8, **440**
  Pinus sylvestris, and *Armillaria*, 90, **91**
plasma membrane invagination, 175
plasmalemma, 182, **232–3**, 269
  fungus, **145**
  host, **145**, 151–7, 163
plasmalemma-bound ATPase, 160
plasmid, 474
plasmodesmata, 349, 587
plectenchyma, 249, 582, 584
  basal, 590–2
pleurotoid fungi, developmental characters of, 295–6
plug,
  electron-dense, **232–3**
  electron-dense, pulley-shaped, 198
  zones of, in *Septobasidium*, 198
point-growth, of fungi, 261, **262**
polygenic control, and the fruit body, 459–60
polyoxin D, 318
polypeptide synthesis, during fruit body development, 488–90
polyphenol, 79, 118, 130
polyphenoloxidases, 133
polyphosphate granule, **149**
polypores, and homokaryotic fruiting, 456
polysaccharide mucigel, 142, **143**
polysaccharides, 158, 404, 405
pore, occlusions within, 179
pore caps, 177
post-Hartig net zone, 159, 160
post-mating function, 220, 222
pre-Hartig net zone, 142
primordium, 353, 354, 390
  of *Flammulina velutipes*, 429–42
  and light, 356, 357, 358, 359, 365
  stage 2, 346
primordium initiation, 582–4
propagules, 15–17
proteases, 376, 404
  and *Lentinus edodes*, 413–16
  production and localisation of, **416**
protein, 376
  and stipe elongation in *Coprinus cinereus*, 315–17
protenchyme hyphae, 344, 345, 347
protenchyme tissue, 340, 582
pseudoangiocarpy, 288
pseudorhiza, 250, 255
pseudosclerotial plates, 257, 258
pseudosclerotium, 17, 250, 591
psilocin, 570

psilocybin, 570
ptophagy, 161
ptysomes, 161
pycnidia, 584
pyrimidine biosynthetic pathway, 480
Pyrola rotundifolia ssp. *maritima*,
  black pseudopharenchymatous sheath
    in, 156
  two kinds of mycorrhizas on, 154–6

race tubes, 359
recombination, studies at DNA level,
    507–8
Reijnder's agaric types of development,
    286–9, 290–2
repetitive fusion (serial plasmogamy),
    176
resource capture,
  effective, 26
  secondary, 3
resource grouping, of higher fungi, **59**
resource pool, 27
resource quality, a determinant of nutrient
    availability, 130–3
resource relations, of tropical agarics,
    41–82
resource unit, definition of, 1 (footnote)
resource unit restriction, 15
  in saprotrophic agarics, 12
resources, 1–15
  of *Armillaria*, 87–100
  definition of, 1 (footnote)
  longevity of, 28
  role in mycorrhizal succession, 126–33
  selectivity of, 19, 23
restriction fragment length polymorphism
    588
restriction map, 477
rhizomorphic organs,
  induction in *Stereum hirsutum*, 263–5
  outgrowths of, **264**
rhizomorphs, 18, 19, 20, 26, 45, 52, 154
  of *Armillaria bulbosa*, 100
  extension of, 272
  in F₁ and F₂ litter layer, 63
  fringing mycelium in, **267**
  growth of *Armillaria*, **96**, 97
  longitudinally bisected apex of, **252**
  ordering of structure in *Armillaria*,
    265–72
  type, 45, 64
  as vegetative organ, 250
rhizosphere, 162
Rhodosporae
  developmental characters of, 296–7
RNA, 327
  poly(A)-containing, 490, 503, 505, 509

mRNA, 371, 375, 468, 492, 493
  changes in, during fruit body
    development, 491–5
  production of, in *Schizophyllum*
    commune, 490
rRNA, 476, 507
5SrRNA, **200**, 201, 204
rRNA genes, 509
  tandem, **508**, 509
  RNA/single copy DNA hybridisations,
    492
RNAse, 376
root-knot nematodes, 162
roots,
  cap cells of, 142
  growth of *Armillaria* in, 92–5
r-selection, 30, 118, **126**, 130, 131
R-selection, 30
ruderal selection, 30, 31, 80
rupthymenial, **300**, **301**, 341, 342, 345
  definition of, 288
Russula, and polyphenoloxidase system, 22
rusts,
  cowpea, 148
  diagram of dolipore/parenthesome
    septum of, **197**
  green islands produced by, 158
  haustorial neck of, 153

saccharidases, 404, 405
  and *Lentinus edodes*, 421–3
saprotroph, 141, 144, 162
  Armillaria bulbosa as, 100
  colonisation strategy of, 52
  obligate, 3
saprotrophic mycelium, of *Crinipellis*
    perniciosa, 104
saprotrophy, 7–9, 54–5, 117, 131
  in agarics, 8–9, 12, 45
  definition of, 2
  occurrence of, 2–4
Schiff's reagent, periodic acidic-, 439, **443**
schizohymenial, 297
  definition of, 288
Schizophyllum commune,
  auxotrophs of, 481
  behaviour of nuclei following fusion in,
    242–3, **244–5**
  clamp connection development and
    fusion of, **237–8**
  cytology of hyphal interactions and
    reactions in, 231–46
  dikaryotic hyphae of, **236**
  as experimental species for dikaryon
    formation, 213
  five stages of fruit body development,
    485–6

Schizophyllum commune (cont.)
  fusion between genetically identical
    dekaryotic hyphae of, **240**
  genes involved in fruiting in, **460**
  genome structure and mRNA
    production in, 490–5
  incompatibility factors in, **468**
  and laccase activity, 378
  molecular biology of fruit body
    formation in, 485–95
  nuclear replacement reaction in, **239**
  nucleic acid studies in, 467–82
  occurrence in tropics, 15
  polypeptides synthesised in, 489
sclerotium, 17, 250, 434, 586, 591
  mucilage in, 269
SDH (succinate dehydrogenase), 564
selectivity
  definition of, 10
septagenesis, 178
septal sealing, 231–4
septomycete sporothallus, 175, 178
septomycetes, 178
sheath, **143**, **147–8**, **149**, 150, **152**
  ectomycorrhizal, 250
shiitake mushroom (*Lentinus edodes*), 529,
    566, 567–9
signal peptide, 375
silver staining, 501
slow-dense/fast-effuse transitions,
  of mycelial front, 260–3, 264
somatic incompatibility reactions, 16
somatogamy, 353, 354
source–sink relations, 130
Southern's analysis of clones, **506**
Southern's hybridisation, **473**
Southern's transfer method, 505
Southwood's *r-K-A*-matrix, 133, 134
spherocytes, 586
  velar, 298
spore, and developmental genetics, 453–4
sporidiomycete, 195
sporothallus, 175
  dikaryotic, 176
S-selection, 30
stems, growth of *Armillaria* in, 92–5
Stereum hirsutum, induction of
  rhizomorphic organs in, 263–5
sterigma, 501
sterigma-like outgrowths, 194
Stichophragmobasidiomycetidae, 198, 203
stilbenes, 24
stilboid, 51
  pedicillate, 50
stipe, 298, 299, 304
  cortex cells from 324–7
  DNA, RNA and protein in, 327

elongation of *Coprinus cinereus*, 311–27
  of *Flammulina velutipes*, 434–7, **440**, **445**
  protein, glycogen and, 315–17
stipe cells, nuclear numbers in, 324–7
stipitangiocarpy, **285**, 303, 304, 305
  definition of, 287–8
stipitocarpy, 290–7, 303, 304
  of boletes, 294
  definition of, 288
stress-tolerant strategy, 30, 31
stretch receptors, 324
stromata, 250, 271
suberin, 142, **143**
subhymenium, 190, 191, 345, 347, 582, 591
  pseudoparenchymatous, 584
substrate, definition of, 1 (footnote)
substratum, definition of, 1 (footnote)
succession, in forest **118**
succinate dehydrogenase, 334, 337
Suillus bovinus, and ectomycorrhizas, **149**
Suillus grevillei, ectomycorrhizas aseptically
  synthesised with, **147**
suppressor mutation, 471
surface microspreading, 501
switch mechanisms, in control of
  morphogenesis, 260
symbiont,
  ectomycorrhizal, 119
  mycorrhizal fungal, 118, 127, 141
  phycomycetous, 141
  role of host–fungus interface between,
    157–64
symbiosis, unclassified, 9–10
synaptonemal complex, 501
synergistic resource capture, 16
syngamy, 175
synthesis, aseptic, 146, **147**

T_0 (minimum stress-relaxation time), 315
tannins, 24, 79, 81, 148
taxon selectivity, 10–14
  in biotrophic agarics, **11**
  in saprotrophic agarics, **13**
taxonomic relationships, 30–5
TCA (tricarboxylic acid), 334, 337, 338
telemorphotic behaviour, 235
temperature, and composting, 551–3
terpenoids, 24
terricolous (humicolous) agarics, 54, 76
tetrapolar breeding system
  in higher basidiomycetes, 215–16
tetrapolar (bifactorial) control, 514
tetrapolar incompatibility system, 223
tetrapolar species, homokaryon interaction
  in, **218**
Theobroma cacao, *see* cocoa
thermophile, 23

Thiery's reaction, 153
  for demonstration of periodate-sensitiv
    polysaccharides, 150
trama,
  bilateral hymenophoral, 297
  heteromerous, 585, 590
  hyphae of, 339
  of gills of *Agrocybe praecox*, 190
transcriptase, reverse, **505**
transfer cell, 150–62
transformation experiments, with
    Schizophyllum commune, 479–82
transformation vector, 480
translation patterns, **491**, 492
translocation,
  in *Coprinus congregatus*, **368**, 369
  and linear organs, 273–4
trehalose, 383, 394–7
  in *Coprinus cinereus* stipes, 314
Tremella complex, 196–8
tricarboxylic acid, *see* TCA
Tricholoma, 33–4
Tricholoma fulvum, and ligninocellulolytic
    capacity, 22
Tricholomataceae, developmental
    characters of, 302–5
tricholomic acid, 572
Trinci's peripheral growth zone, 245
Trione's medium, 110
tropical agarics, ecology of, 55–67
tropolones, 24
tuber, protocarpic, 34
turgor gradient, 273
turgor pressure, 393

ultrastructure, of *Flammulina velutipes*,
    429–48
universal veil, 297, 300, 301, 302, 585, 586
urase, 324
ustidium, definition of, 201
Ustomycota, 200–03
ustomycete, 199
ustospore, 175

VA mycorrhiza, *see* mycorrhizas,
    vesicular-arbuscular

vegetative organs,
  factors which initiate, 257
  linear, 250–6
  migratory, 250
  morphogenesis of, 249–75
  nonlinear, 256
  types of, 250–6
vesicles, at site of clamp connection
    initiation, **237–8**
vesicular bodies, 375
vesicular-arbuscular mycorrhizal plants,
    127, 128

wall apposition, **147–8**
wall architecture, in *Schizophyllum
    commune*, **182**, 183
wall ingrowth, **149**, 150, **152**, 154, 159
  degenerating, **155**
water, and composting, 549–50
witches broom (abnormal vegetative
    growth), 53, 67
  caused by *Crinipellis perniciosa*, 103–14
  fruit bodies of *C. perniciosa* on, **111**, **112**
  humidity in, 110–11
  hyperplasia and hypertrophy in cocoa,
    107
Woronin bodies, paracrystalline, 191

Xerocomus subtomentosus,
  differing abilities in, 144
  penetration by hyphae of, **145**
xerophiles 23, 43
xylanase, 376
xylomannans, 317

yeast, mating type in ascogenous, 469–71
yeast shuttle vector, **505**
yield, polygenic inheritance for, **518**

zone,
  definition of, 355–6
  induced, **368**, 379
  Step II-fixed, 367
zone of apposition, 158
zygotropic behaviour, 235

For EU product safety concerns, contact us at Calle de José Abascal, 56–1°, 28003 Madrid, Spain or eugpsr@cambridge.org.

www.ingramcontent.com/pod-product-compliance
Ingram Content Group UK Ltd.
Pitfield, Milton Keynes, MK11 3LW, UK
UKHW040618240426
470322UK00010B/191